Lecture Notes in Mathematics

2123

T0236462

More information about this series at
http://www.springer.com/series/304

Catherine Donati-Martin • Antoine Lejay •
Alain Rouault

Editors

Séminaire de Probabilités XLVI

 Springer

Editors
Catherine Donati-Martin
Laboratoire de Mathématiques
Université de Versailles-St-Quentin
Versailles, France

Antoine Lejay
INRIA
Nancy-Université
Vandoeuvre-lès-Nancy, France

Alain Rouault
Laboratoire de Mathématiques
Université de Versailles-St-Quentin
Versailles, France

ISBN 978-3-319-11969-4   ISBN 978-3-319-11970-0 (eBook)
DOI 10.1007/978-3-319-11970-0
Springer Cham Heidelberg New York Dordrecht London

Lecture Notes in Mathematics ISSN print edition: 0075-8434
ISSN electronic edition: 1617-9692

Library of Congress Control Number: 2014958308

Mathematics Subject Classification (2010): 60G, 60J, 60K

Printed on acid-free paper

Springer is part of Springer Science+Business Media (www.springer.com)

Preface

Marc Yor est décédé le 9 janvier 2014.

C'est avec une profonde tristesse et une immense reconnaissance que nous saluons la mémoire de celui qui anima le Séminaire pendant plus de 25 ans avec une énergie infatigable, et dont les qualités scientifiques et humaines faisaient l'admiration de tous.

Le Volume XLVII, actuellement en préparation, sera dédié à sa mémoire.

Le présent volume, incluant en particulier un article de Marc Yor, continue la tradition d'exposés divers en théorie des probabilités.

———————————

Marc Yor passed away on January 9, 2014.

With deep sorrow and immense thankfulness, we pay tribute to the memory of him who tirelessly animated the Séminaire for more than 25 years, and whose scientific and human qualities were admired by everyone.

Volume XLVII, now in preparation, will be dedicated to his memory.

The present volume, including in particular an article by Marc Yor, keeps the tradition of various lectures in probability theory.

Versailles, France   Catherine Donati-Martin
Vandoeuvre-lès-Nancy, France   Antoine Lejay
Versailles, France   Alain Rouault
May 2014

Contents

Branching Random Walk in an Inhomogeneous Breeding Potential 1
Sergey Bocharov and Simon C. Harris

**The Backbone Decomposition for Spatially Dependent
Supercritical Superprocesses** 33
A.E. Kyprianou, J-L. Pérez, and Y.-X. Ren

On Bochner-Kolmogorov Theorem 61
Lucian Beznea and Iulian Cîmpean

**Small Time Asymptotics for an Example of Strictly
Hypoelliptic Heat Kernel** 71
Jacques Franchi

**Onsager-Machlup Functional for Uniformly Elliptic
Time-Inhomogeneous Diffusion** 105
Koléhè A. Coulibaly-Pasquier

**G-Brownian Motion as Rough Paths and Differential
Equations Driven by G-Brownian Motion** 125
Xi Geng, Zhongmin Qian, and Danyu Yang

Flows Driven by Banach Space-Valued Rough Paths 195
Ismaël Bailleul

Some Properties of Path Measures 207
Christian Léonard

Semi Log-Concave Markov Diffusions 231
P. Cattiaux and A. Guillin

**On Maximal Inequalities for Purely Discontinuous Martingales
in Infinite Dimensions** 293
Carlo Marinelli and Michael Röckner

Admissible Trading Strategies Under Transaction Costs.................... 317
Walter Schachermayer

Potentials of Stable Processes 333
A.E. Kyprianou and A.R. Watson

Unimodality of Hitting Times for Stable Processes.......................... 345
Julien Letemplier and Thomas Simon

On the Law of a Triplet Associated with the Pseudo-Brownian Bridge.... 359
Mathieu Rosenbaum and Marc Yor

**Skew-Product Decomposition of Planar Brownian Motion
and Complementability**.. 377
Jean Brossard, Michel Émery, and Christophe Leuridan

On the Exactness of the Lévy-Transformation............................. 395
Vilmos Prokaj

Multi-Occupation Field Generates the Borel-Sigma-Field of Loops 401
Yinshan Chang

Ergodicity, Decisions, and Partial Information 411
Ramon van Handel

**Invariance Principle for the Random Walk Conditioned
to Have Few Zeros** 461
Laurent Serlet

A Short Proof of Stein's Universal Multiplier Theorem 473
Dario Trevisan

On a Flow of Operators Associated to Virtual Permutations 481
Joseph Najnudel and Ashkan Nikeghbali

Branching Random Walk in an Inhomogeneous Breeding Potential

Sergey Bocharov and Simon C. Harris

Abstract We consider a continuous-time branching random walk in the inhomogeneous breeding potential $\beta| \cdot |^p$, where $\beta > 0$, $p \geq 0$. We prove that the population almost surely explodes in finite time if $p > 1$ and doesn't explode if $p \leq 1$. In the non-explosive cases, we determine the asymptotic behaviour of the rightmost particle.

1 Introduction and Main Results

We consider a branching system with single particles moving independently according to a continuous-time random walk on $\mathbb{Z}$. The random walk makes jumps of size 1 up or down at constant rate $\lambda > 0$ in each direction. A particle currently at position $y \in \mathbb{Z}$ is independently replaced by two new particles at the parent's position at instantaneous rate $\beta|y|^p$, where $\beta > 0$ and $p \geq 0$ are some given constants.

We denote the set of particles present in the system at time t by N_t. If $u \in N_t$ then the position of a particle u at time t is X_t^u and its path up to time t is $(X_s^u)_{0 \leq s \leq t}$. The law of the branching process started with a single initial particle at $x \in \mathbb{Z}$ is denoted by P^x with the corresponding expectation E^x and the natural filtration of the process is denoted by $(\mathcal{F}_t)_{t \geq 0}$.

Let us define the explosion time of the population as

$$T_{explo} = \sup\{t : |N_t| < \infty\}.$$

We have the following dichotomy for T_{explo} in terms of p, the exponent of the breeding potential.

Theorem 1.1 (Explosion Criterion) *For the inhomogeneous BRW started at any* $x \in \mathbb{Z}$:

a) If $p \leq 1$ then $T_{explo} = \infty$ P^x-a.s.
b) If $p > 1$ then $T_{explo} < \infty$ P^x-a.s.

S. Bocharov • S.C. Harris (✉)
Department of Mathematical Sciences, University of Bath, Claverton Down, Bath, BA2 7AY, UK
e-mail: s.c.harris@bath.ac.uk

© Springer International Publishing Switzerland 2014   1
C. Donati-Martin et al. (eds.), *Séminaire de Probabilités XLVI*, Lecture Notes in Mathematics 2123, DOI 10.1007/978-3-319-11970-0_1

Let us also define the process of the rightmost particle as

$$R_t := \sup_{u \in N_t} X_t^u, \qquad t \geq 0.$$

For $p \in [0, 1]$, we prove the following result about the asymptotic behaviour of R_t.

Theorem 1.2 (Rightmost Particle Asymptotics) *For the inhomogeneous BRW and any $x \in \mathbb{Z}$:*

a) If $p = 0$ then

$$\lim_{t \to \infty} \frac{R_t}{t} = \lambda(\hat{\theta} - \frac{1}{\hat{\theta}}) \qquad P^x\text{-a.s.,} \tag{1}$$

where $\hat{\theta}$ is the unique solution of

$$\left(\theta - \frac{1}{\theta}\right) \log \theta - \left(\theta + \frac{1}{\theta}\right) + 2 = \frac{\beta}{\lambda} \quad \text{on } (1, \infty) \tag{2}$$

b) If $p \in (0, 1)$ then

$$\lim_{t \to \infty} \left(\frac{\log t}{t}\right)^{\hat{b}} R_t = \hat{c} \qquad P^x\text{-a.s.,} \tag{3}$$

where $\hat{b} = \frac{1}{1-p}$ and $\hat{c} = \left(\frac{\beta(1-p)^2}{p}\right)^{\hat{b}}$.

c) If $p = 1$ then

$$\lim_{t \to \infty} \frac{\log R_t}{\sqrt{t}} = \sqrt{2\beta} \qquad P^x\text{-a.s.} \tag{4}$$

Note that Part a) of Theorem 1.2 is a special case of a well known result of Biggins [3,4].

Theorems 1.1 and 1.2 for this inhomogeneous branching random walk should be compared with some analogous known results for Branching Brownian Motion. Consider a model for branching Brownian motion in an inhomogeneous potential where single particles move as standard Brownian motions, each branching into two new particles at instantaneous rate $\beta|x|^p$ when at position x, where $\beta > 0$, $p \geq 0$. This inhomogeneous BBM has been considered in Itô and McKean [9], Harris and Harris [8] and Berestycki et al. [1, 2] where, in particular, the following results can be found:

Theorem 1.3 (Itô and McKean [9], Section 5.14) *Consider a BBM in the potential $\beta| \cdot |^p$, $\beta > 0$, $p \geq 0$ started from $x \in \mathbb{R}$:*

a) If $p \leq 2$ then $T_{explo} = \infty$ P^x-a.s.
b) If $p > 2$ then $T_{explo} < \infty$ P^x-a.s.

Theorem 1.4 (Harris and Harris [8]) *Consider the BBM model with $\beta > 0$, $p \in [0, 2]$, $x \in \mathbb{R}$.*

a) If $p \in [0, 2)$ then

$$\lim_{t \to \infty} \frac{R_t}{t^{\hat{b}}} = \hat{a} \qquad P^x\text{-}a.s. \tag{5}$$

where $\hat{b} = \frac{2}{2-p}$ and $\hat{a} = \left(\frac{\beta}{2}(2 - p)^2 \right)^{\frac{1}{2-p}}$.
b) If $p = 2$ then

$$\lim_{t \to \infty} \frac{\log R_t}{t} = \sqrt{2\beta} \qquad P^x\text{-}a.s. \tag{6}$$

Despite the general similarities, it can be seen that the inhomogeneous Branching Random Walk shows different behaviour from the inhomogeneous Branching Brownian Motion in terms of both the precise explosion criteria and the asymptotic growth of the rightmost particle position.

We shall give a heuristic argument to help explain Theorems 1.1–1.4 in Sect. 2. The rest of the paper will then contain the detailed proofs of Theorems 1.1 and 1.2. In Sect. 3 we introduce a family of one-particle martingales. We also present some other relevant one-particle results, which will be used in later sections. Section 3 is self-contained and can be read out of the context of branching processes. In Sect. 4 we recall some standard techniques used in the analysis of branching systems, which include spines, additive martingales and martingale changes of measure. In Sect. 5 we prove Theorem 1.1 about the explosion time using standard spine methods. Section 6 is devoted to the proof of Theorem 1.2 about the rightmost particle using the spine methods again.

Remark 1.1 The methods of proof that we use in this article could be applied to a large number of other symmetric breeding potentials. In particular, the proof could be adapted to show that the results of Theorems 1.1 and 1.2 hold as long as the breeding potential is continuous and satisfies $\beta(x) \sim \beta|x|^p$ as $x \to \pm\infty$.

2 Heuristics

Theorems 1.1–1.4 are concerned with *almost sure* explosion and *almost sure* rightmost particle asymptotics. We can informally recover analogous *expectation* results with careful use of the well known Many-to-One Lemma (for example, see [7]), which reduces the expectation of the sum of functionals of particles alive at time t to the expectation of a single particle.

In particular, the expected number of particles alive at time t in the branching system is

$$E^x|N_t| = \mathbb{E}^x e^{\int_0^t \beta(X_s)\mathrm{d}s} = \mathbb{E}^x e^{\int_0^t \beta|X_s|^p \mathrm{d}s} \tag{7}$$

where $(X_t)_{t\geq 0}$ is the single-particle process under $\mathbb{P}^x$. It is then relatively straightforward to check that if $(X_t)_{t\geq 0}$ is a Brownian motion, the expected number of particles at time t is: finite for all $t > 0$ if $p < 2$; finite for $t < \hat{t}$ and infinite for $t \geq \hat{t}$ for some constant $\hat{t}$ when $p = 2$; and, infinite for all $t > 0$ if $p > 2$. Whereas, if $(X_t)_{t\geq 0}$ is a continuous-time random walk then the expected number of particles at time t is: finite for all $t > 0$ if $p \leq 1$; and, infinite for all $t > 0$ if $p > 1$. These computations give the critical value of p for explosion of the expected numbers of particles, and suggest the almost sure explosion criteria found in Theorems 1.1 and 1.3

The Many-to-One Lemma also reveals that the expected number of particles following *close* to a given trajectory f up to time t is, roughly speaking, given by

$$E^x\left(\sum_{u\in N_t} \mathbf{1}_{\{X_s^u \approx f(s)\ \forall s\in[0,t]\}}\right) = \mathbb{E}^x\left(\mathbf{1}_{\{X_s \approx f(s)\ \forall s\in[0,t]\}} e^{\int_0^t \beta|X_s|^p \mathrm{d}s}\right). \tag{8}$$

If $(X_t)_{t\geq 0}$ is a continuous-time random walk then using heuristic methods which involve large deviations theory for Lévy processes (for example, see [6]), we find

$$\log \mathbb{E}^x\left(\mathbf{1}_{\{X_s \approx f(s)\ \forall s\in[0,t]\}} e^{\int_0^t \beta|X_s|^p \mathrm{d}s}\right) \sim I_t(f) := \int_0^t \beta f(s)^p - \Lambda\big(f'(s)\big)\mathrm{d}s,$$

where $\Lambda : [0,\infty) \to [0,\infty)$ is the rate function of the random walk given by

$$\Lambda(x) = 2\lambda + x\log\left(\frac{\sqrt{x^2 + 4\lambda^2} + x}{2\lambda}\right) - \sqrt{x^2 + 4\lambda^2} \sim x\log x \text{ as } x \to \infty.$$

(See Schilder's theorem for large deviations of paths in Brownian motion, where $\Lambda(x) = \frac{1}{2}x^2$.) Hence the expected number of particles following the curve f either grows exponentially or decays exponentially in t depending on the growth rate of f.

Further, we anticipate that the almost sure number of particles that have stayed close to path f over large time period $[0,t]$ will be roughly of order $\exp\{I_t(f)\}$ *as long as there have not been any extinction events along the path*, corresponding to the growth rate always remaining positive with $I_s(f) > 0$ for all $s \in (0,t]$. See Berestycki et al. [2] where such almost sure growth rates along paths are made rigorous for inhomogeneous BBM.

Thus, in order to find the almost sure asymptotic rightmost particle position, for t large we would like to find $\sup f(t)$ where the supremum is taken over all paths such that no extinction occurs, that is, over paths f with $I_s(f) > 0$ for all $s \in (0,t]$. In fact, it turns out that the optimal path f^* for the rightmost position then satisfies

$I_s(f^*) = 0$ for all $s \in (0, t]$, that is, f^* solves the equation

$$\Lambda\big(f^{*\prime}(s)\big) = \beta f^*(s)^p.$$

Solving this equation for the inhomogenous BRW leads exactly to the asymptotics of the rightmost particle as given in Theorem 1.2. Although we will not make the above heuristics rigorous for the BRW in this article, our more direct proof of Theorem 1.2, which we give in Sect. 6, will involve showing that there almost surely exists a particle staying close to the critical curve f^*.

3   Single-Particle Results

In this section we introduce a family of martingales for continuous-time random walks. Throughout this section the time set for all the processes is assumed to be $[0, T)$, where $T \in (0, \infty]$ is deterministic.

Suppose we are given a Poisson process $(Y_t)_{t \in [0,T)} \overset{d}{=} PP(\lambda)$ under a probability measure $\mathbb{P}$. Let us denote by J_i the time of the ith jump of $(Y_t)_{t \in [0,T)}$. Then we have the following result.

Lemma 3.1 *Let* $\theta : [0, T) \rightarrow [0, \infty)$ *be a locally-integrable function. That is,* $\int_0^t \theta(s)\mathrm{d}s < \infty \ \forall t \in [0, T)$. *Then the following process is a* $\mathbb{P}$-*martingale:*

$$M_t := e^{\int_0^t \log \theta(s)\mathrm{d}Y_s + \lambda \int_0^t (1-\theta(s))\mathrm{d}s} = \Big(\prod_{i : J_i \leq t} \theta(J_i) \Big) e^{\lambda \int_0^t (1-\theta(s))\mathrm{d}s} \, , \, t \in [0, T),$$

where for any function f, $\int_0^t f(s)\mathrm{d}Y_s := \sum_{i : J_i \leq t} f(J_i)$.

The next result tells what effect the martingale $(M_t)_{t \in [0,T)}$ has on the process $(Y_t)_{t \in [0,T)}$ when used as a Radon-Nikodym derivative.

Lemma 3.2 *Let* $(\hat{\mathscr{F}}_t)_{t \in [0,T)}$ *be the natural filtration of* $(Y_t)_{t \in [0,T)}$. *Define the new measure* $\mathbb{Q}$ *via*

$$\frac{\mathrm{d}\mathbb{Q}}{\mathrm{d}\mathbb{P}}\bigg|_{\hat{\mathscr{F}}_t} = M_t \quad , t \in [0, T).$$

Then under the new measure $\mathbb{Q}$

$$(Y_t)_{t \in [0,T)} \overset{d}{=} IPP\big(\lambda\theta(t)\big),$$

where $IPP\big(\lambda\theta(t)\big)$ *stands for time-inhomogeneous Poisson process of instantaneous jump rate* $\lambda\theta(t)$.

Proof (Outline of the Proof of Lemmas 3.1 and 3.2) As an intermediate step one can check by standard calculations that the following identity holds:

$$\mathbb{E}\left(e^{\int_0^t \log\theta(s)dY_s}\mathbf{1}_{\{Y_t=k\}}\right) = e^{-\lambda t}\frac{\lambda^k}{k!}\left(\int_0^t \theta(s)ds\right)^k \quad \forall k \in \mathbb{N}, \tag{9}$$

where $\mathbb{E}$ is the expectation associated with $\mathbb{P}$.

The martingale property of $(M_t)_{t \in [0,T)}$ then follows immediately.

To verify that under $\mathbb{Q}$, $(Y_t)_{t \in [0,T)}$ is a time-inhomogeneous Poisson process one can check the finite-dimensional distributions. $\square$

For the next few results suppose that $(Y_t)_{t \in [0,T)} \overset{d}{=} IPP(r(t))$, where $r : [0, T) \to [0, \infty)$ is a locally-integrable function. That is, $(Y_t)_{t \in [0,T)}$ is a time-inhomogeneous Poisson process with instantaneous jump rate $r(t)$.

The following identity is a standard integration by-parts-formula which is trivial to prove.

Proposition 3.1 (Integration by Parts for Time-Inhomogeneous Poisson Processes) *Let $f \in C^1([0, T))$. Then almost surely*

$$\int_0^t f(s)dY_s = f(t)Y_t - \int_0^t f'(s)Y_s ds.$$

Since $(Y_t)_{t \in [0,T)} \overset{d}{=} (Z_{R(t)})_{t \in [0,T)}$, where $R(t) := \int_0^t r(s)ds$ and $(Z_t)_{t \geq 0} \overset{d}{=} PP(1)$ we also have the following useful result.

Proposition 3.2 (SLLN for Time-Inhomogeneous Poisson Processes) *If* $\lim_{t \to T} \int_0^t r(s)ds = \infty$ *then*

$$\frac{Y_t}{\int_0^t r(s)ds} \to 1 \; a.s. \; as \; t \to T.$$

The next result combines Propositions 3.1 and 3.2.

Proposition 3.3 *Let $f : [0, T) \to [0, \infty)$ be differentiable such that $f'(t) \geq 0$ for t large enough. Suppose r and f satisfy the following two conditions:*

1. $\int_0^t r(s)ds \to \infty$ *as* $t \to T$
2. $\limsup_{t \to T} \frac{f(t)\int_0^t r(s)ds}{\int_0^t f(s)r(s)ds} < \infty$

Then

$$\frac{\int_0^t f(s)dY_s}{\int_0^t f(s)r(s)ds} \to 1 \; a.s. \; as \; t \to T.$$

Note that the second condition is generally rather restrictive, but it will be satisfied by the functions that we consider in this article.

Proof Observe that by Proposition 3.1 we have

$$\frac{\int_0^t f(s)\mathrm{d}Y_s}{\int_0^t f(s)r(s)\mathrm{d}s} = \frac{f(t)Y_t - \int_0^t f'(s)Y_s\mathrm{d}s}{\int_0^t f(s)r(s)\mathrm{d}s}.$$

Then apply Proposition 3.2 and use the deterministic integration-by-parts formula.

□

Let us now consider a continuous-time random walk $(X_t)_{t\in[0,T)}$ defined under some probability measure $\mathbb{P}$ as it was described in the introduction. It can be written as a difference of two independent Poisson processes of rate λ:

$$X_t = X_t^+ - X_t^- , t \in [0, T),$$

where $(X_t^+)_{t\to[0,T)}$ is the process of positive jumps and $(X_t^-)_{t\in[0,T)} \overset{d}{=} PP(\lambda)$ is the process of negative jumps. From Lemmas 3.1 and 3.2 we get the following result.

Proposition 3.4 *Let θ^+, $\theta^- : [0, T) \to [0, \infty)$ be two locally-integrable functions. Then the following process is a $\mathbb{P}$-martingale:*

$$M_t := e^{\int_0^t \log\theta^+(s)\mathrm{d}X_s^+ + \lambda\int_0^t(1-\theta^+(s))\mathrm{d}s + \int_0^t \log\theta^-(s)\mathrm{d}X_s^- + \lambda\int_0^t(1-\theta^-(s))\mathrm{d}s} , t \in [0, T).$$

$$(10)$$

Moreover, if we define the new measure $\mathbb{Q}$ as

$$\frac{\mathrm{d}\mathbb{Q}}{\mathrm{d}\mathbb{P}}\bigg|_{\hat{\mathscr{F}}_t} = M_t \quad , t \in [0, T),$$

where $(\hat{\mathscr{F}}_t)_{t\in[0,T)}$ is the natural filtration of $(X_t)_{t\in[0,T)}$, then under $\mathbb{Q}$

$$(X_t^+)_{t\in[0,T)} \overset{d}{=} IPP(\lambda\theta^+(t)), \quad (X_t^-)_{t\in[0,T)} \overset{d}{=} IPP(\lambda\theta^-(t)).$$

In other words the martingale M used as the Radon-Nikodym derivative has the effect of scaling the upward jumps by the factor of $\theta^+(t)$ and the rate of downward jumps by the factor $\theta^-(t)$ at time t.

Furthermore from Propositions 3.2 and 3.3 we know that $\mathbb{Q}$-a.s.

$$\lim_{t\to T} \frac{X_t^+}{\int_0^t \lambda\theta^+(s)\mathrm{d}s} = 1, \quad \lim_{t\to T} \frac{X_t^-}{\int_0^t \lambda\theta^-(s)\mathrm{d}s} = 1,$$

$$\lim_{t\to T} \frac{\int_0^t f(s)\mathrm{d}X_s^+}{\int_0^t \lambda\theta^+(s)f(s)\mathrm{d}s} = 1, \quad \lim_{t\to T} \frac{\int_0^t f(s)\mathrm{d}X_s^-}{\int_0^t \lambda\theta^-(s)f(s)\mathrm{d}s} = 1$$

provided that θ^+, θ^- and f satisfy the conditions of Propositions 3.2 and 3.3.

4   Spines and Additive Martingales

In this section we give a brief overview of the main spine tools. The major reference for this section is the work of Hardy and Harris [7] where all the proofs and further references can be found.

Firstly, let us take the time set of our model to be $[0, T)$ for some deterministic $T \in (0, \infty]$. We assume in this section that the branching process starts from 0.

We let $(\mathscr{F}_t)_{t \in [0,T)}$ denote the natural filtration of our branching process as described in the introduction. We define $\mathscr{F}_T := \sigma(\cup_{t \in [0,T)} \mathscr{F}_t)$.

Let us now extend our branching random walk by identifying an infinite line of descent, which we refer to as the spine, in the following way. The initial particle of the branching process begins the spine. When it splits into two new particle, one of them is chosen with probability $\frac{1}{2}$ to continue the spine. This goes on in the obvious way: whenever the particle currently in the spine splits, one of its children is chosen uniformly at random to continue the spine.

The spine is denoted by $\xi = \{\varnothing, \xi_1, \xi_2, \cdots\}$, where $\varnothing$ is the initial particle (both in the spine and in the entire branching process) and ξ_n is the particle in the $(n+1)$st generation of the spine. Furthermore, at time $t \in [0, T)$ we define:

- $node_t(\xi) := u \in N_t \cap \xi$ (such u is unique). That is, $node_t(\xi)$ is the particle in the spine alive at time t.
- $n_t := |node_t(\xi)|$. Thus n_t is the number of fissions that have occurred along the spine by time t.
- $\xi_t := X_t^u$ for $u \in N_t \cap \xi$. So $(\xi_t)_{t \in [0,T)}$ is the path of the spine.

The next important step is to define a number of filtrations of our sample space, which contain different information about the process.

Definition 4.1 (Filtrations)

- $\mathscr{F}_t$ was defined earlier. It is the filtration which knows everything about the particles' motion and their genealogy, but it knows nothing about the spine.
- We also define $\tilde{\mathscr{F}}_t := \sigma(\mathscr{F}_t, node_t(\xi))$. Thus $\tilde{\mathscr{F}}$ has all the information about the process together with all the information about the spine. This will be the largest filtration.
- $\mathscr{G}_t := \sigma(\xi_s : 0 \le s \le t)$. This filtration only has information about the path of the spine process, but it can't tell which particle $u \in N_t$ is the spine particle at time t.
- $\tilde{\mathscr{G}}_t := \sigma(\mathscr{G}_t, (node_s(\xi) : 0 \le s \le t))$. This filtration knows everything about the spine including which particles make up the spine, but it doesn't know what is happening off the spine.

Note that $\mathscr{G}_t \subset \tilde{\mathscr{G}}_t \subset \tilde{\mathscr{F}}_t$ and $\mathscr{F}_t \subset \tilde{\mathscr{F}}_t$. We shall be using these filtrations throughout the whole article for taking various conditional expectations.

We let $\tilde{P}$ be the probability measure under which the branching random walk is defined together with the spine. Hence $P = \tilde{P}|_{\mathscr{F}_T}$. We shall write $\tilde{E}$ for the expectation with respect to $\tilde{P}$.

Under $\tilde{P}$ the entire branching process (with the spine) can be described in the following way.

- The initial particle (the spine) moves like a random walk.
- At instantaneous rate $\beta| \cdot |^p$ it splits into two new particles.
- One of these particles (chosen uniformly at random) continues the spine. That is, it continues moving as a random walk and branching at rate $\beta| \cdot |^p$.
- The other particle initiates a new independent P-branching processes from the position of the split.

It is not hard to see that under $\tilde{P}$ the spine's path $(\xi_t)_{t\in[0,T)}$ is itself a continuous-time random walk.

Also, conditional on the path of the spine, $(n_t)_{t\in[0,T)}$ is a time-inhomogeneous Poisson process (or a Cox process) with instantaneous jump rate $\beta|\xi_t|^p$. That is, conditional on $\mathscr{G}_t$, k splits take place along the spine by time t with probability

$$\tilde{P}(n_t = k|\mathscr{G}_t) = \frac{(\int_0^t \beta|\xi_s|^p \mathrm{d}s)^k}{k!} e^{-\int_0^t \beta|\xi_s|^p \mathrm{d}s}.$$

The next result (see e.g. [7]) has already been mentioned in the introduction.

Theorem 4.1 (Many-to-One Theorem) *Let $f(t) \in m\mathscr{G}_t$, that is, $f(t)$ is $\mathscr{G}_t$-measurable. Suppose it has the representation*

$$f(t) = \sum_{u\in N_t} f_u(t)\mathbf{1}_{\{node_t(\xi)=u\}},$$

where $f_u(t) \in m\mathscr{F}_t$, then

$$E\left(\sum_{u\in N_t} f_u(t)\right) = \tilde{E}\left(f(t)e^{\int_0^t \beta|\xi_s|^p \mathrm{d}s}\right).$$

Now let $\theta = (\theta^+, \theta^-)$, where $\theta^+, \theta^- : [0, T) \to [0, \infty)$ are two locally-integrable functions. In view of Proposition 3.4 we define the following $\tilde{P}$-martingale w.r.t filtration $(\hat{\mathscr{G}}_t)_{t\in[0,T)}$:

$$\tilde{M}_\theta(t) := e^{-\beta\int_0^t |\xi_s|^p \mathrm{d}s} 2^{n_t} \times \exp\left(\int_0^t \log\theta^+(s)\mathrm{d}\xi_s^+ + \int_0^t \lambda(1-\theta^+(s))\mathrm{d}s\right.$$

$$\left. + \int_0^t \log\theta^-(s)\mathrm{d}\xi_s^- + \int_0^t \lambda(1-\theta^-(s))\mathrm{d}s\right), \tag{11}$$

where $(\xi_t^+)_{t\in[0,T)}$ is the process of positive jumps of the spine process and $(\xi_t^-)_{t\in[0,T)}$ is the process of its negative jumps.

Note that $\tilde{M}_\theta$ is the product of two $\tilde{P}$-martingales, the first of which doubles the branching rate along the spine, and the second biases the rates of upward and downward jumps of the spine process. If we define the probability measure $\tilde{Q}_\theta$ as

$$\frac{\mathrm{d}\tilde{Q}_\theta}{\mathrm{d}\tilde{P}}\Bigg|_{\tilde{\mathscr{F}}_t} = \tilde{M}_\theta(t), \qquad t \in [0, T) \tag{12}$$

then under $\tilde{Q}_\theta$ the branching process has the following description:

Proposition 4.1 (Branching Process Under $\tilde{Q}_\theta$)

- *The initial particle (the spine) moves like a biased random walk. That is, at time t it jumps up at instantaneous rate $\lambda\theta^+(t)$ and jumps down at instantaneous rate $\lambda\theta^-(t)$.*
- *When it is at position x it splits into two new particles at instantaneous rate $2\beta|x|^p$.*
- *One of these particles (chosen uniformly at random) continues the spine. I.e. it continues moving as a biased random walk and branching at rate $2\beta| \cdot |^p$.*
- *The other particle initiates an unbiased branching process (as under P) from the position of the split.*

Note that although (12) only defines $\tilde{Q}_\theta$ on events in $\cup_{t\in[0,T)}\tilde{\mathscr{F}}_t$, Carathéodory's extension theorem tells that $\tilde{Q}_\theta$ has a unique extension on $\tilde{\mathscr{F}}_T := \sigma(\cup_{t\in[0,T)}\tilde{\mathscr{F}}_t)$ and thus (12) implicitly defines $\tilde{Q}_\theta$ on $\tilde{\mathscr{F}}_T$.

Proposition 4.2 (Additive Martingale) *Let $Q_\theta := \tilde{Q}_\theta|_{\mathscr{F}_T}$. Then*

$$\frac{\mathrm{d}Q_\theta}{\mathrm{d}P}\Bigg|_{\mathscr{F}_t} = M_\theta(t), \qquad t \in [0, T), \tag{13}$$

where $M_\theta(t)$ is the additive martingale

$$M_\theta(t) = \sum_{u \in N_t} \exp\left(\int_0^t \log\theta^+(s)\mathrm{d}X_u^+(s) + \int_0^t \log\theta^-(s)\mathrm{d}X_u^-(s) \right.$$
$$\left. + \int_0^t \lambda\big(2 - \theta^+(s) - \theta^-(s)\big)\mathrm{d}s - \beta \int_0^t |X_u(s)|^p\mathrm{d}s \right) \tag{14}$$

and $(X_u^+(s))_{0\leq s\leq t}$ is the process of positive jumps of particle u, $(X_u^-(s))_{0\leq s\leq t}$ is the process of its negative jumps.

Let us recall the following measure-theoretic result, which gives Lebesgue's decomposition of Q_θ into absolutely-continuous and singular parts. It can for example be found in the book of R. Durrett [5] (Section 4.3).

Lemma 4.1 *For events $A \in \mathscr{F}_T$*

$$Q_\theta(A) = \int_A \limsup_{t\to T} M_\theta(t)\mathrm{d}P + Q_\theta\big(A \cap \{\limsup_{t\to T} M_\theta(t) = \infty\}\big). \tag{15}$$

In view of this lemma one will be interested in identifying the set of values of θ for which $\limsup_{t \to T} M_\theta(t) < \infty$ Q_θ-a.s., in which case $Q_\theta \ll P$ on $\mathscr{F}_T$. An important tool for doing this is the so-called spine decomposition.

Lemma 4.2 (Spine Decomposition)

$$E^{\tilde{Q}_\theta}\left(M_\theta(t)|\mathscr{G}_T\right) = \exp\left(\int_0^t \log \theta^+(s)\mathrm{d}\xi_s^+ + \int_0^t \log \theta^-(s)\mathrm{d}\xi_s^-\right.$$

$$+ \lambda \int_0^t (2 - \theta^+(s) - \theta^-(s))\mathrm{d}s - \beta \int_0^t |\xi_s|^p \mathrm{d}s\Big)$$

$$+ \sum_{u < node_t(\xi)} \exp\left(\int_0^{S_u} \log \theta^+(s)\mathrm{d}\xi_s^+ + \int_0^{S_u} \log \theta^-(s)\mathrm{d}\xi_s^-\right.$$

$$+ \lambda \int_0^{S_u} (2 - \theta^+(s) - \theta^-(s))\mathrm{d}s - \beta \int_0^{S_u} |\xi_s|^p \mathrm{d}s\Big),$$

$$\tag{16}$$

where $\{S_u : u \in \xi\}$ is the set of fission times along the spine.

The first term is called the spine term or spine(t) and the second one is called the sum term or sum(t).

5 Explosion: Proof of Theorem 1.1

5.1 Case $p \leq 1$

Firstly, we shall prove that $P^x(T_{explo} = \infty) = 1$ if the exponent of the branching rate p is ≤ 1. As in the proof of Theorem 1.3 a) from [9] for the BBM model it will be sufficient to show that $E|N_t| < \infty$ for some $t > 0$ as it is explained below.

Let us begin by proving a few useful facts about the explosion time that we are going to use in the proof of both parts of Theorem 1.1.

Proposition 5.1 (Properties of T_{explo})

a) We either have $P^x(T_{explo} = \infty) = 0$ $\forall x \in \mathbb{Z}$ or $P^x(T_{explo} = \infty) = 1$ $\forall x \in \mathbb{Z}$.

b) If $\exists y \in \mathbb{Z}$, $t > 0$, such that $P^y(T_{explo} < t) = 0$ then $P^x(T_{explo} = \infty) = 1$ $\forall x \in \mathbb{Z}$.

Proof a) Take any x and $y \in \mathbb{Z}$ and start the branching random walk from x. Let T_y be the first passage time of the process to level y. That is,

$$T_y := \inf\{t : \exists u \in N_t \text{ s.t. } X_t^u = y\}.$$

Note, $T_y < \infty$ P^x a.s., since a random walk started from any level x will hit any level y. Then, by the strong Markov property of the branching process, the subtree initiated from y at time T_y has the same law as a branching random walk started from y. Consequently, if the explosion does not happen in the tree started from x, it cannot happen in its subtree started from y. Thus

$$P^x \left(T_{explo} = \infty \right) \leq P^y \left(T_{explo} = \infty \right).$$

Since x and y were arbitrary it follows that

$$P^x \left(T_{explo} = \infty \right) = P^y \left(T_{explo} = \infty \right) \qquad \forall x, y \in \mathbb{Z}.$$

Now let X_1 be the position of the first split of the branching process started from x. Then from the branching property and the previous equation we have

$$P^x \left(T_{explo} = \infty \right) = E^x \left(\left(P^{X_1} \left(T_{explo} = \infty \right) \right)^2 \right) = \left(P^x \left(T_{explo} = \infty \right) \right)^2.$$

Thus $P^x(T_{explo} = \infty) \in \{0, 1\}$.

b) Consider the branching process started from y. Take any $\epsilon \in (0, t)$ and $x \in \mathbb{Z}$. Let T_x be the hitting time of level x as in part a). Then there is a positive probability that the process will hit level x before time ϵ. Then

$$0 = P^y \left(T_{explo} < t \right) \geq P^y \left(T_{explo} < t, T_x < \epsilon \right) \geq P^y \left(T^x_{explo} < t - \epsilon, T_x < \epsilon \right)$$

$$= E^y \left(P^y \left(T^x_{explo} < t - \epsilon, T_x < \epsilon | T_x \right) \right) = P^y \left(T_x < \epsilon \right) P^x \left(T_{explo} < t - \epsilon \right),$$

where T^x_{explo} is the explosion time of the subtree started from x. Thus, since $P^y (T_x < \epsilon) > 0$, we find that $P^x \left(T_{explo} < t - \epsilon \right) = 0$. Since ϵ was arbitrary, letting $\epsilon \downarrow 0$ gives that

$$P^x \left(T_{explo} < t \right) = 0.$$

Thus we have shown that if $\exists y \in \mathbb{Z}$ and $t > 0$ such that $P^y \left(T_{explo} < t \right) = 0$ then $P^x \left(T_{explo} < t \right) = 0 \ \forall x \in \mathbb{Z}$.

The result of part b) now follows by induction and the previous statement since if the original tree started from y almost surely does not explode by time t then none of its subtrees initiated at time t will explode by time $2t$ and one can repeat this argument any number of times.   □

In particular, Proposition 5.1 a) says that the starting position of the branching process is not important in Theorem 1.1 and we shall assume it to be 0 for the rest of this section writing P for P^0.

Proof (Proof of Theorem 1.1 a)) We wish to show that if $p \leq 1$ then $P(T_{explo} = \infty) = 1$. From Proposition 5.1 b), it is sufficient to show that $E(|N_t|) < \infty$ for some $t > 0$.

By the Many-to-One Theorem (Theorem 4.1)

$$E\left(|N_t|\right) = E\left(\sum_{u \in N_t} 1\right) = \tilde{E}\left(e^{\int_0^t \beta |\xi_s|^p \, ds}\right),$$

where $(\xi_t)_{t \geq 0}$ is a continuous-time random walk under $\tilde{P}$. Recall, $\xi_t = \xi_t^+ - \xi_t^-$, where $(\xi_t^+)_{t \geq 0}$ and $(\xi_t^-)_{t \geq 0}$ are two independent Poisson processes with jump rate λ. Then

$$\tilde{E}\left(e^{\int_0^t \beta |\xi_s|^p \, ds}\right) \leq \tilde{E}\left(e^{t\beta \sup_{0 \leq s \leq t} |\xi_s|^p}\right)$$

$$= \tilde{E}\left(e^{t\beta \sup_{0 \leq s \leq t} |\xi_s^+ - \xi_s^-|^p}\right) \leq \tilde{E}\left(e^{t\beta \sup_{0 \leq s \leq t} \left((\xi_s^+)^p \vee (\xi_s^-)^p\right)}\right)$$

$$= \tilde{E}\left(e^{t\beta \left((\xi_t^+)^p \vee (\xi_t^-)^p\right)}\right) \leq \tilde{E}\left(e^{t\beta \left((\xi_t^+)^p + (\xi_t^-)^p\right)}\right)$$

$$= \left[\tilde{E}\left(e^{t\beta(\xi_t^+)^p}\right)\right]^2 \leq \left[\tilde{E}\left(e^{t\beta \xi_t^+}\right)\right]^2$$

since ξ^+ is supported on $\{0, 1, 2, \ldots\}$ whence $(\xi_t^+)^p \leq \xi_t^+$ for $p \in [0, 1]$. Then

$$\tilde{E}\left(e^{t\beta \xi_t^+}\right) = \sum_{n=0}^{\infty} e^{\beta t n} \frac{(\lambda t)^n}{n!} e^{-\lambda t} = \exp\left\{e^{\beta t} \lambda t - \lambda t\right\} < \infty \quad \forall t \geq 0.$$

Thus $E(|N_t|) < \infty$ for all $t > 0$. □

5.2 Case $p > 1$

Proof (Proof of Theorem 1.1 b)) We wish to show that if $p > 1$ then $P(T_{explo} < \infty) = 1$. By Proposition 5.1 a) this is equivalent to $P(T_{explo} < \infty) > 0$. It would be sufficient to prove that $P(T_{explo} \leq T) > 0$ for some $T > 0$ (since $\{T_{explo} \leq T\} \subseteq \{T_{explo} < \infty\}$). For a contradiction let us suppose that $\forall T > 0$

$$P(T_{explo} \leq T) = 0. \tag{17}$$

We choose an arbitrary $T > 0$ and fix it for the rest of this subsection. Under the assumption (17) that there is no explosion before time T we can perform the usual spine construction on $[0, T)$. The key steps of the proof can then be summarised as follows:

1. We choose $\theta^+, \theta^- : [0, T) \to [0, \infty)$ such that at time T

    (A)   the spine process ξ_t goes to ∞ under $\tilde{Q}_\theta$
    (B)   the martingale M_θ from (14) satisfies $\lim \sup_{t \to T} M_\theta(t) < \infty$ Q_θ-a.s.

2. We deduce that $Q_\theta \ll P$ on $\mathscr{F}_T$, whence with positive P-probability one particle goes to ∞ at time T giving infinitely many births along its path.
3. We get a contradiction to (17).

We take $\theta^-(\cdot) \equiv 1$. That is, we leave the negative jumps of the spine process unaltered under $\tilde{Q}_\theta$. $\theta^+(\cdot)$ needs to be chosen carefully such that both (A) and (B) above are satisfied. One such choice is

$$\theta^+(s) = (T - s)^{-c} , s \in [0, T),\tag{18}$$

where $c > \frac{p}{p-1}$ (e.g. take $c = \frac{p}{p-1} + 1$).

The additive martingale (14) in this case takes the following form (with $\theta^+(\cdot)$ defined above)

$$M_\theta(t) = \sum_{u \in N_t} \exp\left(\int_0^t \log \theta^+(s)dX_u^+(s) + \int_0^t \lambda\big(1 - \theta^+(s)\big)ds\right.$$
$$\left. - \beta \int_0^t |X_u(s)|^p ds\right), \ t \in [0, T),\tag{19}$$

If we can now show that

$$\lim_{t \to T} \sup M_\theta(t) < \infty \qquad Q_\theta\text{-a.s.}\tag{20}$$

it would follow from Lemma 4.1 that $Q_\theta \ll P$ on $\mathscr{F}_T$.

To prove (20) it is sufficient to show that

$$\lim_{t \to T} \sup E^{\tilde{Q}_\theta}\left(M_\theta(t)\big|\tilde{\mathscr{G}}_T\right) < \infty \qquad \tilde{Q}_\theta\text{-a.s.},\tag{21}$$

since if (21) holds then by Fatou's lemma

$$E^{\tilde{Q}_\theta}\left(\lim_{t \to T} \inf M_\theta(t)\big|\tilde{\mathscr{G}}_T\right) \le \lim_{t \to T} \inf E^{\tilde{Q}_\theta}\left(M_\theta(t)\big|\tilde{\mathscr{G}}_T\right)$$
$$\le \lim_{t \to T} \sup E^{\tilde{Q}_\theta}\left(M_\theta(t)\big|\tilde{\mathscr{G}}_T\right) < \infty \quad \tilde{Q}_\theta\text{-a.s.},$$

therefore $\liminf_{t \to T} M_\theta(t) < \infty$ $\tilde{Q}_\theta$-a.s. and hence also Q_θ-a.s. Then since $\frac{1}{M_\theta(t)}$ is a positive Q_θ-supermartingale on $[0, T)$, it must converge Q_θ-a.s., hence

$$\limsup_{t \to T} M_\theta(t) = \liminf_{t \to T} M_\theta(t) < \infty \qquad Q_\theta\text{-a.s.}$$

So let us now prove (21). Recall the spine decomposition (16):

$$E^{\tilde{Q}_\theta}\left(M_\theta(t)|\tilde{\mathscr{G}}_T\right) = spine(t) + sum(t),$$

where

$$spine(t) = \exp\left(\int_0^t \log\theta^+(s)\mathrm{d}\xi_s^+ + \int_0^t \lambda\left(1 - \theta^+(s)\right)\mathrm{d}s - \int_0^t \beta|\xi_s|^p\mathrm{d}s\right)$$

and

$$sum(t) = \sum_{u < node_t(\xi)} spine(S_u).$$

We start by proving the following assertion about the spine term.

Proposition 5.2 *There exist some $\tilde{Q}_\theta$-a.s. finite positive random variables C', C'' and a random time $T' \in [0, T)$ such that $\forall t > T'$*

$$spine(t) \leq C'\exp\left(-C''(T - t)^{-p(c-1)+1}\right).$$

Proof (Proof of Proposition 5.2) From Proposition 3.4 under $\tilde{Q}_\theta$ the process $(\xi_t^+)_{t \in [0,T)}$ is a time-inhomogeneous Poisson process of rate $\lambda\theta^+(t)$ and $(\xi_t^-)_{t \in [0,T)}$ is a Poisson process of rate λ.

Using the standard integration-by-parts formula one can check that

$$\int_0^t (T - s)^{-c}\log(T - s)\mathrm{d}s \sim \frac{1}{c - 1}(T - t)^{-c+1}\log(T - t) \text{ as } t \to T.$$

Hence for θ^+ defined as in (18)

$$\limsup_{t \to T} \frac{\log\theta^+(t)\int_0^t \lambda\theta^+(s)\mathrm{d}s}{\int_0^t \log\theta^+(s)\,\lambda\theta^+(s)\mathrm{d}s} = 1.$$

Also $\int_0^t \lambda\theta^+(s)\mathrm{d}s = \lambda(c - 1)^{-1}(T - t)^{-c+1} \to \infty$ as $t \to T$ and $\log\theta^+(\cdot)$ is increasing. Thus from Propositions 3.2 and 3.3 we have that $\tilde{Q}_\theta$-a.s.

$$\frac{\xi_t^+}{\int_0^t \lambda\theta^+(s)\mathrm{d}s} \to 1, \qquad \frac{\int_0^t \log\theta^+(s)\mathrm{d}\xi_s^+}{\int_0^t \log\theta^+(s)\,\lambda\theta^+(s)\mathrm{d}s} \to 1. \tag{22}$$

Also, $\xi_t \sim \xi_t^+$ as $t \to T$ since ξ_t^- is bounded on $[0, T)$.

Combining these observations we get that $\forall \epsilon > 0 \; \exists \; \tilde{Q}_\theta$-a.s. finite time T_ϵ such that $\forall t > T_\epsilon$ the following inequalities are true:

$$\int_0^t \log \theta^+(s) d\xi_s^+ < (1+\epsilon) \int_0^t \log \theta^+(s) \, \lambda \theta^+(s) ds$$

$$= (1+\epsilon) \int_0^t -c \log(T-s) \lambda (T-s)^{-c} ds;$$

$$|\xi_t| > (1-\epsilon) \int_0^t \lambda \theta^+(s) ds = (1-\epsilon) \frac{\lambda}{c-1} (T-t)^{-c+1};$$

$$\lambda\big(1 - \theta^+(t)\big) < 0;$$

$$-\log(T-t)(T-t)^{-c} \leq \frac{1}{2} \frac{\beta(\frac{\lambda}{c-1}(1-\epsilon))^p}{\lambda c (1+\epsilon)} (T-t)^{-(c-1)p}.$$

The last inequality follows from the fact that $c > \frac{p}{p-1}$. Thus, for $t > T_\epsilon$ we have

$$spine(t) = \exp \left(\int_0^t \log \theta^+(s) d\xi_s^+ + \int_0^t \lambda\big(1 - \theta^+(s)\big) ds - \int_0^t \beta |\xi_s|^p ds \right)$$

$$\leq C_\epsilon \exp \left\{ (1+\epsilon) \int_0^t -c\lambda \log(T-s)(T-s)^{-c} ds \right.$$

$$\left. - \beta \int_0^t \left(\frac{\lambda(1-\epsilon)}{c-1}(T-s)^{-c+1} \right)^p ds \right\}$$

$$\leq C_\epsilon' \exp \left\{ -\frac{1}{2}\beta \left(\frac{\lambda(1-\epsilon)}{c-1} \right)^p \frac{1}{p(c-1)-1} (T-t)^{-(c-1)p+1} \right\},$$

where C_ϵ and C_ϵ' are some $\tilde{Q}_\theta$-a.s. finite random variables, which don't depend on t. Letting $T' = T_\epsilon$, $C' = C_\epsilon'$ and $C'' = \frac{1}{2}\beta \left(\frac{\lambda(1-\epsilon)}{c-1} \right)^p \frac{1}{p(c-1)-1}$ we finish the proof of Proposition 5.2. $\square$

We now look at the *sum* term:

$$sum(t) = \sum_{u < node_t(\xi)} spine(S_u)$$

$$= \left(\sum_{u < node_t(\xi), \, S_u \leq T'} spine(S_u) \right) + \left(\sum_{u < node_t(\xi), \, S_u > T'} spine(S_u) \right)$$

$$\leq \sum_{u<node_t(\xi),\ S_u\leq T'} spine(S_u)$$

$$+ \sum_{u<node_t(\xi),\ S_u>T'} C'\exp\left(-C''(T-S_u)^{-p(c-1)+1}\right)$$

using Proposition 5.2. The first sum is $\tilde{Q}_\theta$-a.s. bounded since it only counts births up to time T'. Call an upper bound on the first sum C_1. Then we have

$$sum(t) \leq C_1 + C'\sum_{n=1}^{\infty}\exp\left(-C''(T-S_n)^{-p(c-1)+1}\right), \qquad (23)$$

where S_n is the time of the nth birth on the spine.

Under $\tilde{Q}_\theta$ the birth process along the spine $(n_t)_{t\in[0,T)}$ conditional on the path of the spine is time-inhomogeneous Poisson process (or Cox process) with birth rate $2\beta|\xi_t|^p$ at time t. Also, from (22), $\xi_t \sim \xi_t^+ \sim \int_0^t \lambda\theta^+(s)ds = \frac{\lambda}{c-1}(T-t)^{-c+1}$ as $t \to T$. Thus, recalling Proposition 3.2, we have that as $t \to T$, almost surely under $\tilde{Q}_\theta$

$$n_t \sim \int_0^t 2\beta|\xi_s|^p ds \sim 2\beta\left(\frac{\lambda}{c-1}\right)^p \frac{1}{p(c-1)-1}(T-t)^{-p(c-1)+1}, \qquad (24)$$

hence, letting $t = S_n$,

$$n \sim 2\beta\left(\frac{\lambda}{c-1}\right)^p \frac{1}{p(c-1)-1}(T-S_n)^{-p(c-1)+1}.$$

So for some $\tilde{Q}_\theta$-a.s. finite positive random variable C_2 we have

$$(T-S_n)^{-p(c-1)+1} \geq C_2 n \quad \forall n.$$

Then substituting this into (23) we get

$$sum(t) \leq C_1 + C'\sum_{n=1}^{\infty}e^{-C''C_2 n},$$

which is bounded $\tilde{Q}_\theta$-a.s. We have thus shown that

$$\limsup_{t\to T} E^{\tilde{Q}_\theta}\left(M_\theta(t)|\tilde{\mathscr{G}}_T\right) = \limsup_{t\to T}\left(spine(t) + sum(t)\right) < \infty \qquad \tilde{Q}_\theta\text{-a.s.}$$

proving (21) and consequently (20).

From Lemma 4.1 it now follows that for events $A \in \mathscr{F}_T$

$$Q_\theta(A) = \int_A \limsup_{t \to T} M_\theta(t) \mathrm{d}P.$$

Thus $Q_\theta(A) > 0 \Rightarrow P(A) > 0$. Let us consider the event $\{|N_t| \to \infty \text{ as } t \to T\}$. From (24) we have $\tilde{Q}_\theta(n_t \to \infty \text{ as } t \to T) = 1$, so $Q_\theta(|N_t| \to \infty \text{ as } t \to T) = 1$ and then $P(|N_t| \to \infty \text{ as } t \to T) > 0$. Thus $P(T_{explo} \leq T) > 0$, which contradicts the initial assumption (17). Therefore, $P(T_{explo} \leq T) > 0$, $\forall T > 0$ and hence by Proposition 5.1 a)

$$T_{explo} < \infty \ P\text{-a.s.}$$

This completes the proof of Theorem 1.1 $\square$

6   The Rightmost Particle: Proof of Theorem 1.2

In this section we consider a branching random walk in the potential $\beta| \cdot |^p$, $\beta > 0$, $p \in [0, 1]$. By Theorem 1.1 there is no explosion of the population and so we take the time set of the branching process to be $[0, \infty)$. That is, in the set-up presented in Sect. 4 we let $T = \infty$.

Just like with the explosion probability in Sect. 5, the starting position of the branching process does not affect the behaviour of the rightmost particle in Theorem 1.2. For example in part a) suppose we know that $P^x(\lim_{t \to \infty} t^{-1} R_t = \lambda(\hat{\theta} - \hat{\theta}^{-1})) = 1$ for some $x \in \mathbb{Z}$. Take some $y \in \mathbb{Z}$. Then a branching process started from x will contain a subtree started from y. Hence $P^y(\limsup_{t \to \infty} t^{-1} R_t \leq \lambda(\hat{\theta} - \hat{\theta}^{-1})) = 1$. Also a branching process started from y will contain a subtree started from x. Hence $P^y(\liminf_{t \to \infty} t^{-1} R_t \geq \lambda(\hat{\theta} - \hat{\theta}^{-1})) = 1$ and so $P^y(\lim_{t \to \infty} t^{-1} R_t = \lambda(\hat{\theta} - \hat{\theta}^{-1})) = 1$. We shall thus take the starting position of the branching process to be 0 in the forthcoming proof presented in Sects. 6.1–6.3.

Our proof follows a similar approach as was used for the BBM model in J. Harris and S. Harris in [8].

6.1   *Convergence Properties of M_θ Under Q_θ*

We let M_θ be the additive martingale as defined in (14) for a given parameter θ. Note that since each M_θ is a positive P-martingale it must converge P-almost surely to a finite limit $M_\theta(\infty)$. We are interested in those values of θ for which $M_\theta(\infty)$ is strictly positive. The following result deals with this question.

Theorem 6.1

Case A ($p = 0$), *homogeneous branching:*
Recall $\hat{\theta}$ from (2) which solves (uniquely)

$$\left(\theta - \frac{1}{\theta}\right) \log \theta - \left(\theta + \frac{1}{\theta}\right) + 2 = \frac{\beta}{\lambda} \quad on \ (1, \infty)$$

Consider $\theta = (\theta^+, \theta^-)$, where $\theta^+(\cdot) \equiv \theta_0$ and $\theta^-(\cdot) \equiv \frac{1}{\theta_0}$ for some constant $\theta_0 > 1$. Then

i) $\theta_0 < \hat{\theta} \Rightarrow M_\theta$ *is UI and* $M_\theta(\infty) > 0$ *a.s. (under P).*
ii) $\theta_0 > \hat{\theta} \Rightarrow M_\theta(\infty) = 0$ *P-a.s.*

Case B ($p \in (0, 1)$), *inhomogeneous subcritical branching:*

$$Let \ \hat{b} = \frac{1}{1-p}, \quad \hat{c} = \left(\frac{\beta(1-p)^2}{p}\right)^{\hat{b}} \ as \ in \ (3).$$

Consider $\theta = (\theta^+, \theta^-)$, where $\theta^-(\cdot) \equiv 1$, and for a given $c > 0$,

$$\theta^+(s) := \frac{c}{\lambda(1-p)} \frac{s^{\hat{b}-1}}{(\log(s+2))^{\hat{b}}}, \ s \geq 0.$$

Then

i) $c < \hat{c} \Rightarrow M_\theta$ *is UI and* $M_\theta(\infty) > 0$ *P-a.s.*
ii) $c > \hat{c} \Rightarrow M_\theta(\infty) = 0$ *P-a.s.*

Case C ($p = 1$), *inhomogeneous near-critical branching:*
Consider $\theta = (\theta^+, \theta^-)$, where $\theta^-(\cdot) \equiv 1$, and for a given $\alpha > 0$,

$$\theta^+(s) := e^{\alpha\sqrt{s}}, \ s \geq 0.$$

Then

i) $\alpha < \sqrt{2\beta} \Rightarrow M_\theta$ *is UI and* $M_\theta(\infty) > 0$ *P-a.s.*
ii) $\alpha > \sqrt{2\beta} \Rightarrow M_\theta(\infty) = 0$ *P-a.s.*

The importance of this Theorem comes from the fact that if M_θ is P-uniformly integrable and $M_\theta(\infty) > 0$ P-a.s. then, as it follows from Lemma 4.1, the measures P and Q_θ are equivalent on $\mathscr{F}_\infty$. Since under $\tilde{Q}_\theta$ the spine process satisfies

$$\frac{\xi_t}{\int_0^t \lambda(\theta^+(s) - \theta^-(s))ds} \to 1 \text{ a.s. as } t \to \infty$$

it would then follow that under P there is a particle with such asymptotic behaviour too. That would give the lower bound on the rightmost particle:

$$\liminf_{t \to \infty} \frac{R_t}{\int_0^t \lambda(\theta^+(s) - \theta^-(s))ds} \geq 1,$$

which we can then optimise over suitable θ^+ and θ^-.

The upper bound on the rightmost particle needs a slightly different approach, which we present in the last subsection.

Remark 6.1 Let us note that the only important feature of $\theta^+(\cdot)$ in cases **B** and **C** is its asymptotic growth. By this we mean that we have freedom in defining $\theta(\cdot)$ as long as we keep

$$\theta^+(t) \sim \frac{c}{\lambda(1-p)} \frac{t^{b-1}}{(\log t)^b}, \text{ as } t \to \infty \text{ in Case } \mathbf{A}$$

and

$$\log \theta^+(t) \sim \alpha \sqrt{t} \text{ as } t \to \infty \text{ in Case } \mathbf{B}.$$

Remark 6.2 Parts A ii), B ii) and C ii) of Theorem 6.1 will not be used in the proof of our main result, Theorem 1.2. We included them to better illustrate the behaviour of martingales M_θ.

Recall Lemma 4.1, which says that for events $A \in \mathscr{F}_\infty$

$$Q_\theta(A) = \int_A \limsup_{t \to \infty} M_\theta(t) dP + Q_\theta\left(A \cap \{\limsup_{t \to \infty} M_\theta(t) = \infty\}\right) \qquad (25)$$

Immediate consequences of this (after taking $A = \Omega$) are:

1) $Q_\theta(\limsup_{t \to \infty} M_\theta(t) = \infty) = 1 \Leftrightarrow \limsup_{t \to \infty} M_\theta(t) = 0$ P-a.s. So to prove parts A ii), B ii) and C ii) of Theorem 6.1 we need to show that $\limsup_{t \to \infty} M_\theta(t) = \infty$ Q_θ-a.s.

2) $Q_\theta(\limsup_{t \to \infty} M_\theta(t) < \infty) = 1 \Leftrightarrow EM_\theta(\infty) = 1$ in which case $P(M_\theta(\infty) > 0) > 0$ and M_θ is L^1-convergent w.r.t P as it follows from Scheffe's Lemma. Thus M_θ is P-uniformly integrable. So to prove the uniform integrability in parts A i), B i) and C i) of Theorem 6.1 we need to show that $\limsup_{t \to \infty} M_\theta(t) < \infty$ Q_θ-a.s.

The fact that $P(M_\theta(\infty) > 0) = 1$ (in parts A i), B i) and C i)) requires additionally a certain zero-one law, which we shall give at the end of this subsection.

Proof (Proof of Theorem 6.1: Uniform Integrability in A i), B i), C i)) We start with proving that for the given values of θ in A i), B i) and C i) M_θ is UI. As we just said above, it is sufficient to prove that

$$\limsup_{t\to\infty} M_\theta(t) < \infty \; Q_\theta\text{-a.s.} \tag{26}$$

for the given paths θ. We have already seen how to do this using the spine decomposition in Sect. 5. Just as before it is sufficient for us to check that

$$\limsup_{t\to\infty} E^{\tilde{Q}_\theta}(M_\theta(t)|\tilde{\mathscr{G}}_\infty) = \limsup_{t\to\infty} \big(spine(t) + sum(t)\big) < \infty \; \tilde{Q}_\theta\text{-a.s.} \tag{27}$$

Let us outline the main steps of proving (27) in cases A, B and C.

Case A ($p = 0$), homogeneous branching:

We note that under $\tilde{Q}_\theta$, $(\xi_t^+)_{t\geq 0} \overset{d}{=} PP(\lambda\theta_0)$ and $(\xi_t^-)_{t\geq 0} \overset{d}{=} PP(\frac{\lambda}{\theta_0})$. Hence

$$\frac{\xi_t^+}{t} \to \lambda\theta_0 \text{ and } \frac{\xi_t^-}{t} \to \frac{\lambda}{\theta_0} \quad \tilde{Q}_\theta\text{-a.s.}$$

Then using the above convergence results the reader can check that there exist some positive constant C'' and a $\tilde{Q}_\theta$-a.s. finite time T' such that $\forall t > T'$

$$spine(t) = \exp\left(\log\theta_0(\xi_t^+ - \xi_t^-) + \big(\lambda(2 - \theta_0 - \frac{1}{\theta_0}) - \beta\big)t\right) \leq e^{-C''t}. \tag{28}$$

Then we have

$$sum(t) = \sum_{u<node_t(\xi)} spine(S_u)$$

$$\leq \left(\sum_{u<node_t(\xi),\, S_u\leq T'} spine(S_u)\right) + \left(\sum_{u<node_t(\xi),\, S_u>T'} e^{-C''S_u}\right),$$

where the first sum, call it C_1, is $\tilde{Q}_\theta$-a.s. bounded since it only counts births up to time T'. Thus

$$sum(t) \leq C_1 + \sum_{n=1}^{\infty} e^{-C''S_n}, \tag{29}$$

where S_n is the time of the nth birth on the spine.

The birth process along the spine $(n_t)_{t\in[0,\infty)}$ is a Poisson process with rate 2β. Therefore $t^{-1}n_t \to 2\beta \; \tilde{Q}_\theta$-a.s. as $t \to \infty$ and hence $n^{-1}S_n \to (2\beta)^{-1} \; \tilde{Q}_\theta$-a.s.

as $n \to \infty$. So for some $\tilde{Q}_\theta$-a.s. finite positive random variable C_2 we have $S_n \geq C_2 n$ $\forall n$. Then substituting this into (29) we get

$$sum(t) \leq C_1 + \sum_{n=1}^{\infty} e^{-C'' C_2 n} < \infty \quad \tilde{Q}_\theta\text{-a.s.},$$

which gives (27).

Case B ($p \in (0, 1)$), inhomogeneous subcritical branching:

From Proposition 3.4 under $\tilde{Q}_\theta$ the process $(\xi_t^+)_{t \in [0,\infty)}$ is a time-inhomogeneous Poisson process with jump rate $\lambda \theta^+(t)$ and $(\xi_t^-)_{t \in [0,\infty)}$ is a Poisson process of rate λ. Then from Propositions 3.2 and 3.3 we find that, $\tilde{Q}_\theta$-a.s.,

$$\frac{\xi_t^+}{\int_0^t \lambda \theta^+(s)ds} \to 1, \quad \frac{\xi_t^-}{\lambda t} \to 1, \quad \frac{\int_0^t \log \theta^+(s)d\xi_s^+}{\int_0^t \log \theta^+(s)\, \lambda \theta^+(s)ds} \to 1. \tag{30}$$

It can then be checked in a similar way as before that there exist some $\tilde{Q}_\theta$-a.s. finite positive random variables C', C'' and T' such that, $\forall t > T'$,

$$spine(t) \leq C' \exp\left(-C'' \int_0^t \frac{s^{\hat{b}p}}{(\log(s+2))^{\hat{b}p}} ds\right).$$

For the sum term of the spine decomposition we have when $t > T'$

$$sum(t) \leq \sum_{\substack{u < node_t(\xi), \\ S_u \leq T'}} spine(S_u) + \sum_{\substack{u < node_t(\xi), \\ S_u > T'}} C' \exp\left(-C'' \int_0^{S_u} \frac{s^{\hat{b}p}}{(\log(s+2))^{\hat{b}p}} ds\right)$$

The first sum is a $\tilde{Q}_\theta$-a.s. finite random variable which doesn't depend on t, and which we call C_1. Then

$$sum(t) \leq C_1 + C' \sum_{n=1}^{\infty} \exp\left(-C'' \int_0^{S_n} \frac{s^{\hat{b}p}}{(\log(s+2))^{\hat{b}p}} ds\right), \tag{31}$$

where S_n is the time of the nth birth on the spine.

Under $\tilde{Q}_\theta$ the birth process along the spine $(n_t)_{t \in [0,\infty)}$ conditional on the path of the spine is time-inhomogeneous Poisson process (or Cox process) with jump rate $2\beta|\xi_t|^p$ at time t.

Also, from (30), $\xi_t = \xi_t^+ - \xi_t^- \sim \xi_t^+ \sim \int_0^t \lambda\theta^+(s)ds \sim c(t/\log(t+2))^{\hat{b}}$ as $t \to \infty$. Thus, recalling Proposition 3.2, we find that

$$n_t \sim 2\beta \int_0^t |\xi_s|^p ds \sim 2\beta c^p \int_0^t \frac{s^{\hat{b}p}}{(\log(s+2))^{\hat{b}p}} ds \quad \tilde{Q}_\theta\text{-a.s. as } t \to \infty.$$

So for some $\tilde{Q}_\theta$-a.s. finite positive random variable C_2 we have

$$\int_0^{S_n} \frac{s^{\hat{b}p}}{(\log(s+2))^{\hat{b}p}} ds \geq C_2 n \quad \forall n.$$

Then substituting this into (31) we verify that (27) again holds.

Case C ($p = 1$), inhomogeneous near-critical branching:

As in the previous case, under $\tilde{Q}_\theta$ the process $(\xi_t^+)_{t\in[0,\infty)}$ is a time-inhomogeneous Poisson process with jump rate $\lambda\theta^+(t)$ and $(\xi_t^-)_{t\in[0,\infty)}$ is a Poisson process of rate λ. Then $\tilde{Q}_\theta$-a.s. we have

$$\frac{\xi_t^+}{\int_0^t \lambda\theta^+(s)ds} \to 1 \ , \ \frac{\xi_t^-}{\lambda t} \to 1 \ , \ \frac{\int_0^t \log\theta^+(s)d\xi_s^+}{\int_0^t \log\theta^+(s)\,\lambda\theta^+(s)ds} \to 1.$$

One can check that there exist some $\tilde{Q}_\theta$-a.s. finite positive random variables C', C'' and T' such that, $\forall t > T'$,

$$spine(t) \leq C' \exp\left(-C'' \int_0^t \sqrt{s}e^{\alpha\sqrt{s}}ds\right).$$

Then for $t > T'$

$$sum(t) \leq C_1 + C' \sum_{n=1}^\infty \exp\left(-C'' \int_0^{S_n} \sqrt{s}e^{\alpha\sqrt{s}}ds\right), \tag{32}$$

where $C_1 < \infty$ and S_n is the time of the nth birth on the spine. Since $\xi_t = \xi_t^+ - \xi_t^- \sim \xi_t^+ \sim \int_0^t \lambda\theta(s)ds \sim (2\lambda/\alpha)\sqrt{t}e^{\alpha\sqrt{t}}$ as $t \to \infty$, the birth process along the spine $(n_t)_{t\in[0,\infty)}$ satisfies

$$n_t \sim \int_0^t 2\beta|\xi_s|ds \sim \frac{4\beta\lambda}{\alpha}\int_0^t \sqrt{s}e^{\alpha\sqrt{s}}ds \quad \tilde{Q}_\theta\text{-a.s. as } t \to \infty.$$

So for some $\tilde{Q}_\theta$-a.s. finite positive random variable C_2 we have

$$\int_0^{S_n} \sqrt{s}e^{\alpha\sqrt{s}}ds \geq C_2 n \quad \forall n.$$

Then substituting this into (32) we again find that (27) holds.

Thus we have completed the proof of uniform integrability and the fact that $P(M_\theta(\infty) > 0) > 0$ in Theorem 6.1. $\qquad\square$

Proof (Proof of Theorem 6.1: Parts A ii), B ii), C ii)) Since one of the particles at time t is the spine, we have

$$M_\theta(t) \geq \exp\left(\int_0^t \log(\theta^+(s))\mathrm{d}\xi_s^+ + \int_0^t \log(\theta^-(s))\mathrm{d}\xi_s^- \right.$$
$$\left. + \lambda \int_0^t (2 - \theta^+(s) - \theta^-(s))\mathrm{d}s - \beta \int_0^t |\xi_s|^p \mathrm{d}s \right) = spine(t).$$

For the paths θ in parts ii) of Theorem 6.1 one can check (following the same analysis as in the proof of parts i) of the Theorem) that $spine(t) \to \infty$ $\tilde{Q}_\theta$-a.s. Thus

$$\limsup_{t\to\infty} M_\theta(t) = \infty \quad \tilde{Q}_\theta\text{-a.s.}$$

and so also Q_θ-a.s. Recalling (25) we see that $M_\theta(\infty) = 0$ P-a.s. for the proposed choices of θ. $\square$

It remains to show that in Theorem 6.1 $P(M_\theta(\infty) > 0) = 1$ when M_θ is UI. The following 0–1 law will do the job. A similar result for Branching Brownian Motion can for example be found in [8].

Lemma 6.1 *Let* $q : \mathbb{Z} \to [0,1]$ *be such that* $M_t := \prod_{u\in N_t} q(X_u(t))$ *is a* P-*martingale. Then* $q(x) \equiv q \in \{0,1\}$.

Proof (Proof of Lemma 6.1) Since M_t is a martingale and one of the particles alive at time t is the spine we have

$$q(x) = E^x M_t = \tilde{E}^x M_t \leq \tilde{E}^x q(\xi_t).$$

So $q(\xi_t)$ is a positive $\tilde{P}$-submartingale. Since it is bounded it converges $\tilde{P}$-a.s. to some limit q_∞. We also know that under $\tilde{P}$, $(\xi_t)_{t\geq 0}$ is a continuous-time random walk. Recurrence of $(\xi_t)_{t\geq 0}$ implies that $q_\infty \equiv q(0)$ and that $q(x)$ is constant in x.

Now suppose for contradiction that $q(0) \in (0,1)$. Then

$$M_t = \prod_{u\in N_t} q(X_u(t)) = q(0)^{|N_t|} \to 0$$

because $|N_t| \to \infty$. Since M is bounded it is uniformly integrable, so $q(0) = EM_\infty = 0$, which is a contradiction. So $q(0) \notin (0,1)$ and thus $q(0) \in \{0,1\}$. $\square$

Proof (Proof of Theorem 6.1: Positivity of Limits in A i), B i), C i)) We apply Lemma 6.1 to $q(x) = P^x(M_\theta(\infty) = 0)$. By the tower property of conditional expectations and the branching Markov property we have

$$q(x) = E^x\left(P^x\big(M_\theta(\infty) = 0\big|\mathscr{F}_t\big)\right) = E^x\left(\prod_{u\in N_t} q\big(X_u(t)\big)\right)$$

whence $\prod_{u \in N_t} q(X_u(t))$ is a P-martingale. Also $E(M_\theta(\infty)) = M_\theta(0) = 1 > 0$. Therefore $P(M_\theta(\infty) = 0) \neq 1$. So by Lemma 6.1 $P(M_\theta(\infty) = 0) = 0$. □

6.2 Lower Bound on the Rightmost Particle

Proposition 6.1 *Let $\hat{\theta}$, $\hat{b}$ and $\hat{c}$ be as defined in Theorem 1.2. Then*

Case A ($p = 0$):

$$\liminf_{t \to \infty} \frac{R_t}{t} \geq \lambda(\hat{\theta} - \frac{1}{\hat{\theta}}) \; P\text{-a.s.}$$

Case B ($p \in (0, 1)$):

$$\liminf_{t \to \infty} \left(\frac{\log t}{t}\right)^{\hat{b}} R_t \geq \hat{c} \; P\text{-a.s.}$$

Case C ($p = 1$):

$$\liminf_{t \to \infty} \frac{\log R_t}{\sqrt{t}} \geq \sqrt{2\beta} \; P\text{-a.s.}$$

Proof **Case A ($p = 0$):**
We consider $\theta = (\theta^+, \theta^-)$, where $\theta^+(\cdot) \equiv \theta_0$, $\theta^-(\cdot) \equiv \frac{1}{\theta_0}$ and $\theta_0 < \hat{\theta}$. Take the event

$$B_{\theta_0} := \left\{\exists \text{ infinite line of descent } u : \liminf_{t \to \infty} \frac{X_u(t)}{t} = \lambda(\theta_0 - \frac{1}{\theta_0})\right\} \in \mathscr{F}_\infty.$$

We know that $\tilde{Q}_\theta(\lim_{t \to \infty} \frac{\xi_t}{t} = \lambda(\theta_0 - \frac{1}{\theta_0})) = 1$. Hence $Q_\theta(B_{\theta_0}) = \tilde{Q}_\theta(B_{\theta_0}) = 1$. Since Q_θ and P are equivalent it follows that $P(B_{\theta_0}) = 1$. Thus $P\left(\liminf_{t \to \infty} t^{-1} R_t \geq \lambda(\theta_0 - \theta_0^{-1})\right) = 1$. Taking the limit $\theta_0 \nearrow \hat{\theta}$ we get

$$P\left(\liminf_{t \to \infty} \frac{R_t}{t} \geq \lambda(\hat{\theta} - \frac{1}{\hat{\theta}})\right) = 1.$$

Case B ($p \in (0, 1)$):

Consider $\theta = (\theta^+, \theta^-)$, where $\theta^-(\cdot) \equiv 1$, $\theta^+(s) = \dfrac{c}{\lambda(1 - p)} \dfrac{s^{\hat{b}-1}}{(\log(s + 2))^{\hat{b}}}$
and $c < \hat{c}$. Take the event

$$B_c := \left\{\exists \text{ infinite line of descent } u : \liminf_{t \to \infty} \left(\frac{\log t}{t}\right)^{\hat{b}} X_u(t) = c\right\}.$$

Same argument as above gives that $P(B_c) = 1$ and hence $P\Big(\liminf_{t\to\infty}$ $\big(t^{-1}\log t\big)^{\hat{b}} R_t \geq c \Big) = 1$ for all $c < \hat{c}$. Letting $c \nearrow \hat{c}$ proves the result.

Case C ($p = 1$):

Consider $\theta = (\theta^+, \theta^-)$, where $\theta^-(\cdot) \equiv 1$, $\theta^+(s) = e^{\alpha\sqrt{s}}$ and $\alpha < \sqrt{2\beta}$. Take the event

$$B_\alpha := \Big\{ \exists \text{ infinite line of descent } u : \liminf_{t\to\infty} \frac{\log X_u(t)}{\sqrt{t}} = \sqrt{2\beta} \Big\}.$$

Again, the same argument as above gives $P(B_\alpha) = 1$ and hence for all $\alpha < \sqrt{2\beta}$ we find that $P\Big(\liminf_{t\to\infty} t^{-1/2} \log R_t \geq \alpha \Big) = 1$. Letting $\alpha \nearrow \sqrt{2\beta}$ proves the result.   □

6.3   Upper Bound on the Rightmost Particle

To complete the proof of Theorem 1.2 and hence the whole section we need to prove the following proposition.

Proposition 6.2 *Let $\hat{\theta}$, $\hat{b}$ and $\hat{c}$ be as defined in Theorem 1.2. Then for different values of p we have the following.*

Case A ($p = 0$):

$$\limsup_{t\to\infty} \frac{R_t}{t} \leq \lambda(\hat{\theta} - \frac{1}{\hat{\theta}}) \ P\text{-}a.s.$$

Case B ($p \in (0, 1)$):

$$\limsup_{t\to\infty} \Big(\frac{\log t}{t} \Big)^{\hat{b}} R_t \leq \hat{c} \ P\text{-}a.s.$$

Case C ($p = 1$):

$$\limsup_{t\to\infty} \frac{\log R_t}{\sqrt{t}} \leq \sqrt{2\beta} \ P\text{-}a.s.$$

To prove Proposition 6.2 we shall assume for contradiction that it is false. Then we shall show that under such assumption certain additive P-martingales will diverge to ∞ contradicting the Martingale Convergence Theorem.

We start by proving the following 0–1 law.

Lemma 6.2 *Let $g : [0, \infty) \to \mathbb{R}$ be increasing, $f : [0, \infty) \to [0, \infty)$ be such that $\forall s \geq 0,\ \frac{f(t)}{f(s+t)} \to 1$ as $t \to \infty$ and $a > 0$. Then*

$$P\left(\limsup_{t \to \infty} \frac{g(R_t)}{f(t)} \leq a \right) \in \{0, 1\}.$$

Proof We consider

$$q(x) = P^x\left(\limsup_{t \to \infty} \frac{g(R_t)}{f(t)} \leq a \right).$$

Then, it is easy to see that

$$q(x) = E^x\left(P^x\left(\limsup_{t \to \infty} \frac{g(R_{t+s})}{f(t+s)} \leq a \Big| \mathscr{F}_s \right) \right)$$

$$= E^x\left(P^x\left(\limsup_{t \to \infty} \frac{g(\max_{u \in N_s} R_t^u)}{f(t+s)} \leq a \Big| \mathscr{F}_s \right) \right)$$

$$= E^x\left(P^x\left(\max_{u \in N_s} \{ \limsup_{t \to \infty} \frac{g(R_t^u)}{f(t+s)} \} \leq a \Big| \mathscr{F}_s \right) \right)$$

$$= E^x\left(\prod_{u \in N_s} P^{X_u(s)}\left(\limsup_{t \to \infty} \frac{g(R_t)}{f(t+s)} \leq a \right) \right)$$

$$= E^x\left(\prod_{u \in N_s} P^{X_u(s)}\left(\limsup_{t \to \infty} \frac{g(R_t)}{f(t)} \leq a \right) \right)$$

$$= E^x\left(\prod_{u \in N_s} q(X_u(s)) \right),$$

where $(R_t^u)_{t \geq 0}$ is the position of the rightmost particle of a subtree started from $X_u(s)$.

Thus $\prod_{u \in N_t} q(X_u(t))$ is a martingale. Applying Lemma 6.1 to $q(\cdot)$ we obtain the required result. □

Proof (Proof of Proposition 6.2) The first step of the proof is slightly different for cases A, B and C, so we do it for the three cases separately.

Case A $(p = 0)$

Let us suppose for contradiction that $\exists \theta_0 > \hat{\theta}$ such that

$$P\left(\limsup_{t \to \infty} \frac{R_t}{t} > \lambda(\theta_0 - \frac{1}{\theta_0}) \right) = 1. \tag{33}$$

Choose any $\theta_A \in (\hat{\theta}, \theta_0)$ and take $\theta = (\theta^+, \theta^-)$, where $\theta^+(\cdot) \equiv \theta_A$, $\theta^-(\cdot) = 1/\theta_A$. Let

$$f_A(s) := \lambda(\theta_A - \frac{1}{\theta_A})s, \quad s \geq 0.$$

Case B $(p \in (0, 1))$

Let us suppose for contradiction that $\exists c_0 > \hat{c}$ such that

$$P\left(\limsup_{t \to \infty} \left(\frac{\log t}{t}\right)^{\hat{b}} R_t > c_0 \right) = 1. \tag{34}$$

Choose any $c_1 \in (\hat{c}, c_0)$ and take $\theta = (\theta^+, \theta^-)$, where $\theta^+(s) = \theta_B(s)$, $\theta^-(s) = 1/\theta_B(s)$ and

$$\theta_B(s) = \frac{c_1}{\lambda(1-p)} \frac{s^{\hat{b}-1}}{(\log(s+2))^{\hat{b}}}, \quad s \geq 0.$$

Let

$$f_B(s) := c_1 \left(\frac{s}{\log(s+2)}\right)^{\hat{b}}, \quad s \geq 0.$$

Case C $(p = 1)$

Let us suppose for contradiction that $\exists \alpha_0 > \sqrt{2\beta}$ such that

$$P\left(\limsup_{t \to \infty} \frac{\log R_t}{\sqrt{t}} > \alpha_0 \right) = 1. \tag{35}$$

Choose any $\alpha_1 \in (\sqrt{2\beta}, \alpha_0)$ and take $\theta = (\theta^+, \theta^-)$, where $\theta^+(s) = \theta_C(s)$, $\theta^-(s) = 1/\theta_C(s)$ and

$$\theta_C(s) = \frac{1}{\sqrt{s+1}} e^{\alpha_1 \sqrt{s}}, \quad s \geq 0.$$

Let

$$f_C(s) := e^{\alpha_1 \sqrt{s}}, \quad s \geq 0.$$

The next step in the proof is the same in all cases.

Let us write f to denote f_A, f_B and f_C. Note that f grows linearly in case A and faster than linearly in cases B and C (since $\hat{b} > 1$ and $\alpha_1 > 0$). We define $D(f)$ to be the space-time region bounded above by the curve $y = f(t)$ and below by the curve $y = -f(t)$.

Under P the spine process $(\xi_t)_{t\geq 0}$ is a continuous-time random walk and so $t^{-1}|\xi_t| \to 0$ P-a.s. as $t \to \infty$. Hence there exists an a.s. finite random time $T' < \infty$ such that $\xi_t \in D(f)$ for all $t > T'$.

Since $(\xi_t)_{t\geq 0}$ is recurrent it will spend an infinite amount of time at position $y = 1$. During this time it will be giving birth to offspring at rate β. This assures us of the existence of an infinite sequence $\{T_n\}_{n\in\mathbb{N}}$ of birth times along the path of the spine when it stays at $y = 1$ with $0 \leq T' \leq T_1 < T_2 < \dots$ and $T_n \nearrow \infty$.

Denote by u_n the label of the particle born at time T_n, which does not continue the spine. Then each particle u_n gives rise to an independent copy of the Branching random walk under P started from ξ_{T_n} at time T_n. Almost surely, by assumptions (33), (34) and (35), each u_n has some descendant that leaves the space-time region $D(f)$.

Let $\{v_n\}_{n\in\mathbb{N}}$ be the subsequence of $\{u_n\}_{n\in\mathbb{N}}$ of those particles whose first descendent leaving $D(f)$ does this by crossing the upper boundary $y = f(t)$. Since the breeding potential is symmetric and the particles u_n are born in the upper half-plane, there is at least probability $\frac{1}{2}$ that the first descendant of u_n to leave $D(f)$ does this by crossing the positive boundary curve. Therefore P-a.s. the sequence $\{v_n\}_{n\in\mathbb{N}}$ is infinite.

Let w_n be the descendent of v_n, which exits $D(f)$ first and let J_n be the time when this occurs. That is,

$$J_n = \inf\{t : X_{w_n}(t) \geq f(t)\}.$$

Note that the path of particle w_n satisfies

$$|X_{w_n}(s)| < f(s) \quad \forall s \in [T', J_n).$$

Clearly $J_n \to \infty$ as $n \to \infty$. To obtain a contradiction we shall show that the additive martingale M_θ fails to converge along the sequence of times $\{J_n\}_{n\geq 1}$, where θ was defined above differently for cases A, B and C. Thus for the last bit of the proof we have to look at cases A, B and C separately again.

Case A $(p = 0)$

$M_\theta(J_n)$

$$= \sum_{u\in N_{J_n}} \exp\left\{\log\theta_A X_u^+(J_n) + \log(\frac{1}{\theta_A})X_u^-(J_n) + \lambda(2 - \theta_A - \frac{1}{\theta_A})J_n - \beta J_n\right\}$$

$$\geq \exp\left\{\log\theta_A X_{w_n}(J_n) + \lambda(2 - \theta_A - \frac{1}{\theta_A})J_n - \beta J_n\right\}$$

$$\geq \exp\left\{\left(\lambda(\theta_A - \frac{1}{\theta_A})\log\theta_A + \lambda(2 - \theta_A - \frac{1}{\theta_A}) - \beta\right)J_n\right\}$$

$$= \exp\left\{\left(\lambda g(\theta_A) - \beta\right)J_n\right\},$$

where $g(\theta) = (\theta - \theta^{-1}) \log \theta - (\theta + \theta^{-1}) + 2$, $\theta \in [1, \infty)$. Then since $g(\cdot)$ is increasing, $\theta_A > \hat{\theta}$ and $g(\hat{\theta}) = \frac{\beta}{\lambda}$ it follows that

$$\lambda g(\theta_A) - \beta > 0$$

and thus $M_\theta(J_n) \to \infty$ as $n \to \infty$, which is a contradiction.
Therefore assumption (33) is wrong and we must have that $\forall \theta_0 > \hat{\theta}$

$$P\left(\limsup_{t \to \infty} \frac{R_t}{t} > \lambda(\theta_0 - \frac{1}{\theta_0}) \right) \neq 1.$$

It follows from Lemma 6.2 that $\forall \theta_0 > \hat{\theta}$ $P\left(\limsup_{t \to \infty} t^{-1} R_t > \lambda(\theta_0 - \theta_0^{-1}) \right) = 0$.
Hence $P\left(\limsup_{t \to \infty} t^{-1} R_t \leq \lambda(\theta_0 - \theta_0^{-1}) \right) = 1$ and after letting $\theta_0 \searrow \hat{\theta}$ we get

$$P\left(\limsup_{t \to \infty} \frac{R_t}{t} \leq \lambda(\hat{\theta} - \frac{1}{\hat{\theta}}) \right) = 1.$$

Case B ($p \in (0, 1)$)

$$M_\theta(J_n) = \sum_{u \in N_{J_n}} \exp\left\{ \int_0^{J_n} \log \theta_B(s) dX_u^+(s) + \int_0^{J_n} \log\left(\frac{1}{\theta_B(s)}\right) dX_u^-(s) \right.$$

$$\left. + \lambda \int_0^{J_n} \left(2 - \theta_B(s) - \frac{1}{\theta_B(s)}\right) ds - \beta \int_0^{J_n} |X_u(s)|^p ds \right\}$$

$$\geq \exp\left\{ \int_0^{J_n} \log \theta_B(s) dX_{w_n}^+(s) + \int_0^{J_n} \log\left(\frac{1}{\theta_B(s)}\right) dX_{w_n}^-(s) \right.$$

$$\left. + \lambda \int_0^{J_n} \left(2 - \theta_B(s) - \frac{1}{\theta_B(s)}\right) ds - \beta \int_0^{J_n} |X_{w_n}(s)|^p ds \right\}.$$

Applying the integration by parts formula from Proposition 3.1 we get

$$\exp\left\{ \log \theta_B(J_n) X_{w_n}^+(J_n) - \int_0^{J_n} \frac{\theta_B'(s)}{\theta_B(s)} X_{w_n}^+(s) ds \right.$$

$$- \log \theta_B(J_n) X_{w_n}^-(J_n) + \int_0^{J_n} \frac{\theta_B'(s)}{\theta_B(s)} X_{w_n}^-(s) ds$$

$$\left. + \lambda \int_0^{J_n} \left(2 - \theta_B(s) - \frac{1}{\theta_B(s)}\right) ds - \beta \int_0^{J_n} |X_{w_n}(s)|^p ds \right\}$$

$$= \exp\left\{ \log \theta_B(J_n) X_{w_n}(J_n) - \int_0^{J_n} \frac{\theta_B'(s)}{\theta_B(s)} X_{w_n}(s) ds \right.$$

$$+ \lambda \int_0^{J_n} \left(2 - \theta_B(s) - \frac{1}{\theta_B(s)}\right) ds - \beta \int_0^{J_n} |X_{w_n}(s)|^p ds \right\}$$

$$\geq C \exp \left\{ \log \theta_B(J_n) f_B(J_n) - \int_0^{J_n} \frac{\theta_B'(s)}{\theta_B(s)} f_B(s) ds \right.$$

$$\left. + \lambda \int_0^{J_n} \left(2 - \theta_B(s) - \frac{1}{\theta_B(s)}\right) ds - \beta \int_0^{J_n} f_B(s)^p ds \right\}$$

using the facts that $X_{w_n}(J_n) \geq f_B(J_n)$ and $|X_{w_n}(s)| < f_B(s)$ for $s \in [T', J_n)$ and where C is some P-a.s positive random variable. Now asymptotic properties of $\theta_B(\cdot)$ and $f_B(\cdot)$ give us that for any $\epsilon > 0$ and n large enough the above expression is

$$\geq C_\epsilon \exp \left\{ (\hat{b} - 1) c_1 \frac{(J_n)^{\hat{b}}}{(\log J_n)^{\hat{b}-1}} (1 - \epsilon) - \beta c_1^p \frac{1}{\hat{b}} \frac{(J_n)^{\hat{b}}}{(\log J_n)^{\hat{b}-1}} (1 + \epsilon) \right\}$$

for some P-a.s. positive random variable C_ϵ. Then since $c_1 > \hat{c} = \left(\frac{\beta(1-p)^2}{p} \right)^{(1-p)^{-1}}$

$$(\hat{b} - 1) c_1 (1 - \epsilon) - \beta c_1^p \frac{1}{\hat{b}} (1 + \epsilon) = c_1^p (\hat{b} - 1)(1 - \epsilon) \left(c_1^{1-p} - \hat{c}^{1-p} \frac{1 + \epsilon}{1 - \epsilon} \right) > 0$$

for ϵ small enough. Thus $M_\theta(J_n) \to \infty$ as $n \to \infty$, which is a contradiction. Therefore assumption (34) is wrong and we must have that $\forall c_0 > \hat{c}$

$$P\left(\limsup_{t \to \infty} \left(\frac{\log t}{t}\right)^{\hat{b}} R_t > c_0 \right) \neq 1.$$

It follows from Lemma 6.2 that $\forall c_0 > \hat{c}$

$$P\left(\limsup_{t \to \infty} \left(\frac{\log t}{t}\right)^{\hat{b}} R_t \leq c_0 \right) = 1.$$

Hence taking the limit $c_0 \searrow \hat{c}$ proves Proposition 6.2 in Case B.

Case C ($p = 1$)

Essentially the same argument as in Case B gives that for any $\epsilon > 0$ and n large enough

$$M_\theta(J_n) \geq C_\epsilon \exp \left\{ (1 - \epsilon) \alpha_1 \sqrt{J_n} e^{\alpha_1 \sqrt{J_n}} - (1 + \epsilon) \frac{2\beta}{\alpha_1} \sqrt{J_n} e^{\alpha_1 \sqrt{J_n}} \right\}$$

for some $C_\epsilon > 0$ P-a.s. Then since $\alpha_1 > \sqrt{2\beta}$

$$(1 - \epsilon)\alpha_1 - (1 + \epsilon)\frac{2\beta}{\alpha_1} > 0$$

for ϵ chosen sufficiently small. Therefore $M_\theta(J_n) \to \infty$, which is a contradiction. Hence $\forall \alpha_0 > \sqrt{2\beta}$

$$P\left(\limsup_{t \to \infty} \frac{\log R_t}{\sqrt{t}} \leq \alpha_0 \right) = 1$$

and therefore

$$P\left(\limsup_{t \to \infty} \frac{\log R_t}{\sqrt{t}} \leq \sqrt{2\beta} \right) = 1.$$

This finishes the proof of Proposition 6.2 and also Theorem 1.2.   □

References

1. J. Berestycki, É. Brunet, J. Harris, S. Harris, The almost-sure population growth rate in branching Brownian motion with a quadratic breeding potential. Stat. Probab. Lett. **80**, 1442–1446 (2010)
2. J. Berestycki, É. Brunet, J. Harris, S. Harris, M. Roberts, Growth rates of the population in a branching Brownian motion with an inhomogeneous breeding potential. arXiv:1203.0513
3. J.D. Biggins, The growth and spread of the general branching random walk. Ann. Appl. Probab. **5**(4), 1008–1024 (1995)
4. J.D. Biggins, *How Fast Does a General Branching Random Walk Spread?* Classical and Modern Branching Processes. IMA Vol. Math. Appl., vol. 84 (Springer, New York, 1996), pp. 19–39
5. R. Durrett, *Probability: Theory and Examples*, 2nd edn. (Duxbury Press, Belmont, 1996)
6. J. Feng, T.G. Kurtz, *Large Deviations for Stochastic Processes* (American Mathematical Society, Providence, 2006)
7. R. Hardy, S.C. Harris, *A Spine Approach to Branching Diffusions with Applications to L^p-Convergence of Martingales*. Séminaire de Probabilités, XLII, Lecture Notes in Math., vol. 1979 (Springer, Berlin, 2009)
8. J.W. Harris, S.C. Harris, Branching Brownian motion with an inhomogeneous breeding potential. Ann. de l'Institut Henri Poincaré (B) Probab. Stat. **45**(3), 793–801 (2009)
9. K. Itô, H.P. McKean, *Diffusion Processes and Their Sample Paths*, 2nd edn. (Springer, Berlin, 1974)

The Backbone Decomposition for Spatially Dependent Supercritical Superprocesses

A.E. Kyprianou, J-L. Pérez, and Y.-X. Ren

Abstract Consider any supercritical Galton-Watson process which may become extinct with positive probability. It is a well-understood and intuitively obvious phenomenon that, on the survival set, the process may be pathwise decomposed into a stochastically 'thinner' Galton-Watson process, which almost surely survives and which is decorated with immigrants, at every time step, initiating independent copies of the original Galton-Watson process conditioned to become extinct. The thinner process is known as the *backbone* and characterizes the genealogical lines of descent of prolific individuals in the original process. Here, prolific means individuals who have at least one descendant in every subsequent generation to their own.

Starting with Evans and O'Connell (Can Math Bull 37:187–196, 1994), there exists a cluster of literature, (Engländer and Pinsky, Ann Probab 27:684–730, 1999; Salisbury and Verzani, Probab Theory Relat Fields 115:237–285, 1999; Duquesne and Winkel, Probab Theory Relat Fields 139:313–371, 2007; Berestycki, Kyprianou and Murillo-Salas, Stoch Proc Appl 121:1315–1331, 2011; Kyprianou and Ren, Stat Probab Lett 82:139–144, 2012), describing the analogue of this decomposition (the so-called *backbone decomposition*) for a variety of different classes of superprocesses and continuous-state branching processes. Note that the latter family of stochastic processes may be seen as the total mass process of superprocesses with non-spatially dependent branching mechanism.

In this article we consolidate the aforementioned collection of results concerning backbone decompositions and describe a result for a general class of supercritical superprocesses with spatially dependent branching mechanisms. Our approach

A.E. Kyprianou (✉)
Department of Mathematical Sciences, University of Bath, Claverton Down, Bath, BA2 7AY, UK
e-mail: a.kyprianou@bath.ac.uk

J-L. Pérez
Department of Probability and Statistics, IIMAS, UNAM, C.P. 04510 Mexico, D.F., Mexico
e-mail: garmendia@sigma.iimas.unam.mx

Y.-X. Ren
LMAM School of Mathematical Sciences & Center for Statistical Science, Peking University, Beijing 100871, P. R. China
e-mail: yxren@math.pku.edu.cn

© Springer International Publishing Switzerland 2014
C. Donati-Martin et al. (eds.), *Séminaire de Probabilités XLVI*, Lecture Notes in Mathematics 2123, DOI 10.1007/978-3-319-11970-0_2

exposes the commonality and robustness of many of the existing arguments in the literature.

1 Superprocesses and Markov Branching Processes

This paper concerns a fundamental decomposition which can be found amongst a general family of superprocesses and has, to date, been identified for a number of specific sub-families thereof by a variety of different authors. We therefore start by briefly describing the general family of superprocesses that we shall concern ourselves with. The reader is referred to the many, and now classical, works of Dynkin for further details of what we present below; see for example [5–9]. The books of Le Gall [21], Etheridge [12] and Li [22] also serve as an excellent point of reference.

Let E be a domain of $\mathbb{R}^d$. Following the setting of Fitzsimmons [16], we are interested in strong Markov processes, $X = \{X_t : t \geq 0\}$ which are valued in $\mathcal{M}_F(E)$, the space of finite measures with support in E. The evolution of X depends on two quantities $\mathscr{P}$ and ψ. Here, $\mathscr{P} = \{\mathscr{P}_t : t \geq 0\}$ is the semi-group of an $\mathbb{R}^d$-valued diffusion killed on exiting E, and ψ is a so-called branching mechanism which, by assumption, takes the form

$$\psi(x, \lambda) = -\alpha(x)\lambda + \beta(x)\lambda^2 + \int_{(0,\infty)} (e^{-\lambda z} - 1 + \lambda z)\pi(x, dz), \tag{1}$$

where α and $\beta \geq 0$ are bounded measurable mappings from E to $\mathbb{R}$ and $[0, \infty)$ respectively and for each $x \in E$, $\pi(x, dz)$ is a measure concentrated on $(0, \infty)$ such that $x \to \int_{(0,\infty)} (z \wedge z^2)\pi(x, dz)$ is bounded and measurable. The latter ensure that the total mass of X is finite in expectation at each time. For technical reasons, we shall additionally assume that the diffusion associated to $\mathscr{P}$ satisfies certain conditions. These conditions are lifted from Section II.1.1 (Assumptions 1.1A and 1.1B) on pages 1218–1219 of [6].[1] They state that $\mathscr{P}$ has associated infinitesimal generator

$$L = \sum_{i,j} a_{i,j} \frac{\partial^2}{\partial x_i \partial x_j} + \sum_i b_i \frac{\partial}{\partial x_i},$$

where the coefficients $a_{i,j}$ and b_j are space dependent coefficients satisfying:

(Uniform Elliptically) There exists a constant $\gamma > 0$ such that

$$\sum_{i,l} a_{i,j} u_i u_j \geq \gamma \sum_i u_i^2$$

for all $x \in E$ and $u_1, \cdots, u_d \in \mathbb{R}$.

---

[1]The assumptions on $\mathscr{P}$ may in principle be relaxed. The main reason for this imposition here comes in the proof of Lemma 5 where a comparison principle is used for diffusions.

(**Hölder Continuity**) The coefficients $a_{i,j}$ and b_i are uniformly bounded and Hölder continuous in such way that there exist a constants $C > 0$ and $\alpha \in (0, 1]$ with

$$|a_{i,j}(x) - a_{i,j}(y)|, \quad |b_i(x) - b_i(y)| \leq C|x - y|^\alpha$$

for all $x, y \in E$. Throughout, we shall refer to X as the $(\mathscr{P}, \psi)$-superprocess.

For each $\mu \in \mathscr{M}_F(E)$ we denote by $\mathbb{P}_\mu$ the law of X when issued from initial state $X_0 = \mu$. The semi-group of X, which in particular characterizes the laws $\{\mathbb{P}_\mu : \mu \in \mathscr{M}_F(E)\}$, can be described as follows. For each $\mu \in \mathscr{M}_F(E)$ and all $f \in bp(E)$, the space of non-negative, bounded measurable functions on E,

$$\mathbb{E}_\mu(e^{-\langle f, X_t\rangle}) = \exp\left\{-\int_E u_f(x, t)\mu(dx)\right\} \qquad t \geq 0, \tag{2}$$

where $u_f(x, t)$ is the unique non-negative solution to the equation

$$u_f(x, t) = \mathscr{P}_t[f](x) - \int_0^t ds \cdot \mathscr{P}_s[\psi(\cdot, u_f(\cdot, t - s))](x) \qquad x \in E, t \geq 0. \tag{3}$$

See for example Theorem 1.1 on pages 1208–1209 of [6] or Proposition 2.3 of [16]. Here we have used the standard inner product notation,

$$\langle f, \mu\rangle = \int_E f(x)\mu(dx),$$

for $\mu \in \mathscr{M}_F(E)$ and any f such that the integral makes sense.

Suppose that we define $\mathscr{E} = \{\langle 1, X_t\rangle = 0 \text{ for some } t > 0\}$, the event of *extinction*. For each $x \in E$ write

$$w(x) = -\log \mathbb{P}_{\delta_x}(\mathscr{E}). \tag{4}$$

It follows from (2) that

$$\mathbb{E}_\mu(e^{-\theta\langle 1, X_t\rangle}) = \exp\left\{-\int_E u_\theta(x, t)\mu(dx)\right\} \qquad t \geq 0, \tag{5}$$

Note that $u_\theta(t, x)$ is increasing in θ and that $\mathbb{P}_\mu(\langle 1, X_t\rangle = 0)$ is monotone increasing. Using these facts and letting $\theta \to \infty$, then $t \to \infty$, we get that

$$\mathbb{P}_\mu(\mathscr{E}) = \lim_{t\to\infty} \mathbb{P}_\mu(\langle 1, X_t\rangle = 0) = \exp\left\{-\int_E \lim_{t\to\infty}\lim_{\theta\to\infty} u_\theta(x, t)\mu(dx)\right\}. \tag{6}$$

By choosing $\mu = \delta_x$, with $x \in E$, we see that

$$\mathbb{P}_\mu(\mathscr{E}) = \exp\left\{-\int_E w(x)\mu(\mathrm{d}x)\right\}. \tag{7}$$

For the special case that ψ does not depend on x and $\mathscr{P}$ is conservative, $\langle 1, X_t \rangle$ is a continuous state branching process. If $\psi(\lambda)$ satisfy the following condition:

$$\int^\infty \frac{1}{\psi(\lambda)}\mathrm{d}\lambda < \infty,$$

then $\mathbb{P}_\mu$ almost surely we have $\mathscr{E} = \{\lim_{t\to\infty}\langle 1, X_t\rangle = 0\}$, that is to say the event of *extinction* is equivalent to the event of *extinguishing*, see [1] and [20] for examples.

By first conditioning the event $\mathscr{E}$ on $\mathscr{F}_t := \sigma\{X_s : s \leq t\}$, we find that for all $t \geq 0$,

$$\mathbb{E}_\mu(e^{-\langle w, X_t\rangle}) = e^{-\langle w, \mu\rangle}.$$

The function w will play an important role in the forthcoming analysis and henceforth we shall assume that it respects the following property.

(A): w is locally bounded away from 0 and ∞.

Note that the notion of supercriticality is implicitly hidden in the assumption above, specifically in that w is locally bounded away from 0. This ensures that the extinction probability in (7) is not unity. We point out that we do not need local compact support property of X. The reason we consider diffusion as our spatial motion is that we will use the comparison principle of some integral equation, see (47) below. We expect that our results hold for more general superprocesses, for example, super-Lévy processes.

The pathwise evolution of superprocesses is somewhat difficult to visualise on account of their realisations at each fixed time being sampled from the space of finite measures. However a related class of stochastic processes which exhibit similar mathematical properties to superprocesses and whose paths are much easier to visualise is that of Markov branching processes. A Markov branching process $Z = \{Z_t : t \geq 0\}$ takes values in the space $\mathscr{M}_a(E)$ of finite atomic measures in E taking the form $\sum_{i=1}^n \delta_{x_i}$, where $n \in \mathbb{N} \cup \{0\}$ and $x_1, \cdots, x_n \in E$. To describe its evolution we need to specify two quantities, $(\mathscr{P}, F)$, where, as before, $\mathscr{P}$ is the semi-group of a diffusion on E and F is the so-called branching generator which takes the form

$$F(x, s) = q(x)\sum_{n\geq 0} p_n(x)(s^n - s), \qquad x \in E, s \in [0, 1], \tag{8}$$

where q is a bounded measurable mapping from E to $[0, \infty)$ and, the measurable sequences $\{p_n(x) : n \geq 0\}$, $x \in E$, are probability distributions. For each $\nu \in$

$\mathcal{M}_a(E)$, we denote by P_ν the law of Z when issued from initial state $Z_0 = \nu$. The probability P_ν can be constructed in a pathwise sense as follows. From each point in the support of ν we issue an independent copy of the diffusion with semi-group $\mathcal{P}$. Independently of one another, for $(x, t) \in E \times [0, \infty)$, each of these particles will be killed at rate $q(x)\mathrm{d}t$ to be replaced at their space-time point of death by $n \geq 0$ particles with probability $p_n(x)$. Relative to their point of creation, new particles behave independently to one another, as well as to existing particles, and undergo the same life cycle in law as their parents.

By conditioning on the first split time in the above description of a $(\mathcal{P}, F)$-Markov branching process, it is also possible to show that for any $f \in \mathrm{bp}(E)$,

$$\mathrm{E}_\nu(e^{-\langle f, Z_t \rangle}) = \exp\left\{-\int_E v_f(x, t)\nu(\mathrm{d}x)\right\} \qquad t \geq 0,$$

where $v_f(x, t)$ solves

$$e^{-v_f(x,t)} = \mathcal{P}_t[e^{-f}](x) + \int_0^t \mathrm{d}s\, \mathcal{P}_s[F(\cdot, e^{-v_f(\cdot, t-s)})](x) \qquad x \in E, t \geq 0. \quad (9)$$

Moreover, it is known, cf. Theorem 1.1 on pages 1208–1209 of [6], that the solution to this equation is unique. This shows a similar characterisation of the semi-groups of Markov branching processes to those of superprocesses.

The close similarities between the two processes become clearer when one takes account of the fact that the existence of superprocesses can be justified through a high density scaling procedure of Markov branching processes. Roughly speaking, for a fixed triplet, $(\mu, \mathcal{P}, \psi)$, one may construct a sequence of Markov branching processes, say $\{Z^{(n)} : n \geq 1\}$, such that the n-th element of the sequence is issued with an initial configuration of points which is taken to be an independent Poisson random measure with intensity $n\mu$ and branching generator F_n satisfying

$$F_n(x, s) = \frac{1}{n}[\psi(x, n(1 - s)) + \alpha(x)n(1 - s)], \qquad x \in E, s \in [0, 1].$$

It is not immediately obvious that the right-hand side above conforms to the required structure of branching generators as stipulated in (8), however this can be shown to be the case; see for example the discussion on p.93 of [22]. It is now a straightforward exercise to show that for all $f \in \mathrm{bp}(E)$ and $t \geq 0$ the law of $\langle f, n^{-1}Z_t^{(n)} \rangle$ converges weakly to the law of $\langle f, X_t \rangle$, where the measure X_t satisfies (2). A little more work shows the convergence of the sequence of processes $\{n^{-1}Z^{(n)} : n \geq 1\}$ in an appropriate sense to a $(\mathcal{P}, \psi)$-superprocess issued from an initial state μ.

Rather than going into the details of this scaling limit, we focus instead in this paper on another connection between superprocesses and branching processes which explains their many similarities without the need to refer to a scaling limit. The basic idea is that, under suitable assumptions, for a given $(\mathcal{P}, \psi)$-superprocess, there exists a related Markov branching process, Z, with computable characteristics such that at each fixed $t \geq 0$, the law of Z_t may be coupled to the law of X_t in such a way that, given X_t, Z_t has the law of a Poisson random measure with intensity $w(x)X_t(\mathrm{d}x)$, where w is given by (4). The study of so-called *backbone decompositions* pertains to how the aforementioned Poisson embedding may be implemented in a pathwise sense at the level of processes.

The remainder of this paper is structured as follows. In the next section we briefly review the sense and settings in which backbone decompositions have been previously studied. Section 3 looks at some preliminary results needed to address the general backbone decomposition that we deal with in Sects. 4, 5 and 6.

2   A Brief History of Backbones

The basic idea of a backbone decomposition can be traced back to the setting of Galton-Watson trees with ideas coming from Harris and Sevast'yanov; cf Harris [18]. Within any supercritical Galton-Watson process with a single initial ancestor for which the probability of survival is not equal to 0 or 1, one may identify prolific genealogical lines of descent on the event of survival. That is to say, infinite sequences of descendants which have the property that every individual has at least one descendant in every subsequent generation beyond its own. Together, these prolific genealogical lines of descent make a Galton-Watson tree which is thinner than the original tree. One may describe the original Galton-Watson process in terms of this thinner Galton-Watson process, which we now refer to as a *backbone*, as follows. Let $0 < p < 1$ be the probability of survival. Consider a branching process which, with probability $1 - p$, is an independent copy of the original Galton-Watson process conditioned to become extinct and, with probability p, is a copy of the backbone process, having the additional feature that every individual in the backbone process immigrates an additional random number of offspring, each of which initiate independent copies of the original Galton-Watson process conditioned to become extinct. With an appropriate choice of immigration numbers, the resulting object has the same law as the original Galton-Watson process.

In Evans and O'Connell [15], and later in Engländer and Pinsky [11], a new decomposition of a supercritical superprocess with quadratic branching mechanism was introduced in which one may write the distribution of the superprocess at time $t \geq 0$ as the result of summing two independent processes together. The first is a copy of the original process conditioned on extinction. The second process is understood as the superposition of mass from independent copies of the original process conditioned on extinction which have immigrated 'continuously' along the path of an auxiliary dyadic branching particle diffusion which starts with a random number of initial ancestors whose cardinality and spatial position is governed by an independent Poisson point process. The embedded branching particle system is known as the *backbone* (as opposed to the *spine* or *immortal particle* which appears in another related decomposition, introduced in Roelly-Coppoletta and Rouault [24] and Evans [14]). In both [15] and [11] the decomposition is seen through the semi-group evolution equations which drive the process semi-group. However no pathwise construction is offered.

A pathwise backbone decomposition appears in Salisbury and Verzani [23], who consider the case of conditioning a super-Brownian motion as it exits a given domain such that the exit measure contains at least n pre-specified points in its support. There it was found that the conditioned process has the same law as the superposition of mass that immigrates in a Poissonian way along the spatial path of a branching particle motion which exits the domain with precisely n particles at the pre-specified points. Another pathwise backbone decomposition for branching particle systems is given in Etheridge and Williams [13], which is used in combination with a limiting procedure to prove another version of Evan's immortal particle picture.

In Duquesne and Winkel [3] a version of the Evans-O'Connell backbone decomposition was established for more general branching mechanisms, albeit without taking account of spatial motion. In their case, quadratic branching is replaced by a general branching mechanism ψ which is the Laplace exponent of a spectrally positive Lévy process and which satisfies the conditions $0 < -\psi'(0+) < \infty$ and $\int^{\infty} 1/\psi(\xi)d\xi < \infty$. Moreover, the decomposition is offered in the pathwise sense and is described through the growth of genealogical trees embedded within the underling continuous state branching process. The backbone is a continuous-time Galton Watson process and the general nature of the branching mechanism induces three different kinds of immigration. Firstly there is continuous immigration which is described by a Poisson point process of independent processes along the trajectory of the backbone where the rate of immigration is given by a so-called excursion measure which assigns zero initial mass, and finite life length of the immigrating processes. A second Poisson point process along the backbone describes the immigration of independent processes where the rate of immigration is given by the law of the original process conditioned on extinguishing and with a positive initial volume of mass randomised by an infinite measure. This accounts for so-called discontinuous immigration. Finally, at the times of branching of the backbone, independent copies of the original process conditioned on extinguishing are immigrated with randomly distributed initial mass which depends on the number of offspring at the branch point. The last two forms of immigration do not occur when the branching mechanism is purely quadratic.

Concurrently to the work of [3] and within the class of branching mechanisms corresponding to spectrally positive Lévy processes with paths of unbounded variation (also allowing for the case that $-\psi'(0+) = \infty$), Bertoin et al. [2] identify the aforementioned backbone as characterizing prolific genealogies within the underling continuous state branching process.

Berestycki et al. [1] extend the results of [15] and [3], showing that for superprocesses with relatively general motion and non-spatial branching mechanism corresponding to spectrally positive Lévy processes with finite mean, a pathwise backbone decomposition arises. The role of the backbone is played by a branching particle diffusion with the same motion operator as the superprocesses and, like

Salisbury and Verzani [23], additional mass immigrates along the trajectory of the backbone in a Poissonian way. Finally Kyprianou and Ren [20] look at the case of a continuous-state branching process with immigration for which a similar backbone decomposition to [1] can be shown.

As alluded to in the abstract, our objective in this article is to provide a general backbone decomposition which overlaps with many of the cases listed above and, in particular, exposes the general effect on the backbone of spatially dependent branching. It is also our intention to demonstrate the robustness of some of the arguments that have been used in earlier work on backbone decompositions. Specifically, we are referring to the original manipulations associated with the semi-group equations given in Evans and O'Connell [15] and Engländer and Pinsky [11], as well as the use the Dynkin-Kuznetsov excursion measure, as found in Salisbury and Verzani [23], Berestyki et al. [1] and Kyprianou and Ren [20], to describe the rate of immigration along the backbone.

3 Preliminaries

Before stating and proving the backbone decomposition, it will first be necessary to describe a number of mathematical structures which will play an important role.

3.1 Localisation

Suppose that the stochastic process $\xi = \{\xi_t : t \geq 0\}$ on $E \cup \{\dagger\}$, where $\dagger$ is its cemetery state, is the diffusion in E corresponding to the semi-group $\mathscr{P}$. We shall denote its probabilities by $\{\Pi_x : x \in E\}$. Throughout this paper, we shall take $\mathrm{bp}(E \times [0,t])$ to be the space of non-negative, bounded measurable functions on $E \times [0,t]$, and it is implicitly understood that for all functions $f \in \mathrm{bp}(E \times [0,t])$, we extend their spatial domain to include $\{\dagger\}$ and set $f(\{\dagger, s\}) = 0$.

Definition 1 For any open, bounded set D compactly embedded in E (written $D \subset\subset E$), and $t \geq 0$, there exists a random measure $\tilde{X}_t^D$ supported on the boundary of $D \times [0,t)$ such that, for all $f \in \mathrm{bp}(E \times [0,t])$ with the additional property that the value of $f(x,s)$ on $E \times [0,t]$ is independent of s and $\mu \in \mathscr{M}_F(D)$, the space of finite measures on D,

$$-\log \mathbb{E}_\mu \left(e^{-\langle f, \tilde{X}_t^D \rangle} \right) = \int_E \tilde{u}_f^D(x,t)\mu(\mathrm{d}x), \qquad t \geq 0, \tag{10}$$

where $\tilde{u}_f^D(x,t)$ is the unique non-negative solution to the integral equation

$$\tilde{u}_f^D(x,t) = \Pi_x[f(\xi_{t \wedge \tau^D}, t \wedge \tau^D)] - \Pi_x \left[\int_0^{t \wedge \tau^D} \psi(\xi_s, \tilde{u}_f^D(\xi_s, t-s))\mathrm{d}s \right], \tag{11}$$

and $\tau^D = \inf\{t \geq 0, \xi_t \in D^c\}$. Note that, here, we use the obvious notation that $\langle f, \tilde{X}_t^D \rangle = \int_{\partial(D \times [0,t))} f(x,s) \tilde{X}_t^D(dx, ds)$. Moreover, with a slight abuse of notation, since their effective spatial domain is restricted to $D \cup \{\dagger\}$ in the above equation, we treat ψ and $\tilde{u}_f^D$ as functions in bp$(E \times [0,t])$ and accordingly it is clear how to handle a spatial argument equal to $\dagger$, as before. In the language of Dynkin [8], $\tilde{X}_t^D$ is called an exit measure.

Now we define a random measure X_t^D on D such that $\langle f, X_t^D \rangle = \langle f, \tilde{X}_t^D \rangle$ for any $f \in$ bp(D), the space of non-negative, bounded measurable functions on D, where, henceforth, as is appropriate, we regard f as a function defined on $E \times [0, \infty)$ in the sense that

$$f(x,t) = \begin{cases} f(x), & x \in D \\ 0, & x \in E \setminus D. \end{cases} \tag{12}$$

Then for any $f \in$ bp(D) and $\mu \in \mathcal{M}_F(D)$,

$$-\log \mathbb{E}_\mu \left(e^{-\langle f, X_t^D \rangle} \right) = \int_E u_f^D(x,t) \mu(dx), \qquad t \geq 0, \tag{13}$$

where $u_f^D(x,t)$ is the unique non-negative solution to the integral equation

$$u_f^D(x,t) = \Pi_x[f(\xi_t); t < \tau^D] - \Pi_x \left[\int_0^{t \wedge \tau^D} \psi(\xi_s, u_f^D(\xi_s, t - s)) ds \right], \quad x \in D. \tag{14}$$

As a process in time, $\tilde{X}^D = \{\tilde{X}_t^D : t \geq 0\}$ is a superprocess with branching mechanism $\psi(x, \lambda) \mathbf{1}_D(x)$, but whose associated semi-group is replaced by that of the process ξ absorbed on ∂D. Similarly, as a process in time, $X^D = \{X_t^D : t \geq 0\}$ is a superprocess with branching mechanism $\psi(x, \lambda) \mathbf{1}_D(x)$, but whose associated semi-group is replaced by that of the process ξ killed upon leaving D. One may think of X_t^D as describing the mass at time t in X which *historically* avoids exiting the domain D. Note moreover that for any two open bounded domains, $D_1 \subset\subset D_2 \subset\subset E$, the processes $\tilde{X}^{D_1}$ and $\tilde{X}^{D_2}$ (and hence X^{D_1} and X^{D_2}) are consistent in the sense that

$$\tilde{X}_t^{D_1} = (\widetilde{\tilde{X}_t^{D_2}})^{D_1}, \tag{15}$$

for all $t \geq 0$ (and similarly $X_t^{D_1} = (X_t^{D_2})^{D_1}$ for all $t \geq 0$).

3.2   Conditioning on Extinction

In the spirit of the relationship between (10) and (11), we have that w is the unique solution to

$$w(x) = \Pi_x[w(\xi_{t \wedge \tau^D})] - \Pi_x\left[\int_0^{t \wedge \tau^D} \psi(\xi_s, w(\xi_s))ds\right], \qquad x \in D, \qquad (16)$$

for all open domains $D \subset\subset E$. Again, with a slight abuse of notation, we treat w with its spatial domain $E \cup \{\dagger\}$ as a function on $E \times [0, t]$ and $w(\dagger) := 0$. From Lemma 1.5 in [6] we may transform (16) to the equation

$$w(x) = \Pi_x\left[w(\xi_{t \wedge \tau_D}) \exp\left\{-\int_0^{t \wedge \tau_D} \frac{\psi(\xi_s, w(\xi_s))}{w(\xi_s)} ds\right\}\right], \qquad x \in D,$$

which shows that for all open bounded domains D,

$$w(\xi_{t \wedge \tau^D}) \exp\left\{-\int_0^{t \wedge \tau^D} \frac{\psi(\xi_s, w(\xi_s))}{w(\xi_s)} ds\right\}, \qquad t \geq 0, \qquad (17)$$

is a martingale.

The function w can be used to locally describe the law of the superprocess when conditioned on *global extinction* (as opposed to extinction on the sub-domain D). The following lemma outlines standard theory.

Lemma 1 *Suppose that $\mu \in \mathcal{M}_F(E)$ satisfies $\langle w, \mu \rangle < \infty$ (so, for example, it suffices that μ is compactly supported). Define*

$$\mathbb{P}_\mu^*(\cdot) = \mathbb{P}_{\gtreqless}(\cdot | \mathscr{E}).$$

Then for any $f \in \mathrm{bp}(E \times [0, t])$ with the additional property that the value of $f(x, s)$ on $E \times [0, t]$ is independent of s and $\mu \in \mathcal{M}_F(D)$,

$$-\log \mathbb{E}_\mu^*\left(e^{-\langle f, \tilde{X}_t^D \rangle}\right) = \int_D \tilde{u}_f^{D,*}(x, t)\mu(dx),$$

where $\tilde{u}_f^{D,*}(x, t) = \tilde{u}_{f+w}^D(x, t) - w(x)$ and it is the unique solution of

$$\tilde{u}_f^{D,*}(x, t) = \Pi_x[f(\xi_{t \wedge \tau^D})] - \Pi_x\left[\int_0^{t \wedge \tau_D} \psi^*(\xi_s, \tilde{u}_f^{D,*}(\xi_s, t - s))ds\right], \qquad x \in D,$$
$$(18)$$

where $\psi^*(x, \lambda) = \psi(x, \lambda + w(x)) - \psi(x, w(x))$, restricted to D, is a branching mechanism of the kind described in the introduction and for each $\mu \in \mathcal{M}_F(E)$,

$(\tilde{X}, \mathbb{P}_\mu^*)$ *is a superprocess. Specifically, on* E,

$$\psi^*(x, \lambda) = -\alpha^*(x)\lambda + \beta(x)\lambda^2 + \int_{(0,\infty)} (e^{-\lambda z} - 1 + \lambda z)\pi^*(x, dz), \qquad (19)$$

where

$$\alpha^*(x) = \alpha(x) - 2\beta(x)w(x) - \int_{(0,\infty)} (1 - e^{-w(x)z})z\pi(x, dz)$$

and

$$\pi^*(x, dz) = e^{-w(x)z}\pi(x, dz) \quad on \ E \times (0, \infty).$$

Proof For all $f \in \mathrm{bp}(\partial(D \times [0, t)))$ with the additional property that the value of $f(x, s)$ on $E \times [0, t]$ is independent of s, we have

$$\begin{aligned}
\mathbb{E}_\mu^*(e^{-\langle f, \tilde{X}_t^D \rangle}) &= \mathbb{E}_\mu(e^{-\langle f, \tilde{X}_t^D \rangle} | \mathscr{E}) \\
&= e^{\langle w, \mu \rangle}\mathbb{E}_\mu(e^{-\langle f, \tilde{X}_t^D \rangle}1_{\mathscr{E}}) \\
&= e^{\langle w, \mu \rangle}\mathbb{E}_\mu(e^{-\langle f, \tilde{X}_t^D \rangle}\mathbb{E}_{\tilde{X}_t^D}(1_{\mathscr{E}})) \\
&= e^{\langle w, \mu \rangle}\mathbb{E}_\mu(e^{-\langle f + w, \tilde{X}_t^D \rangle}) \\
&= e^{-\langle \tilde{u}_{f+w}^D(\cdot, t) - w, \mu \rangle}.
\end{aligned}$$

Using (11) and (16) then it is straightforward to check that $\tilde{u}_f^{D,*}(x, t) = \tilde{u}_{f+w}^D(x, t) - w(x)$ is a non-negative solution to (18), which is necessarily unique. The proof is complete as soon as we can show that $\psi^*(x, \lambda)$, restricted to D, is a branching mechanism which falls into the appropriate class. One easily verifies the formula (19) and that the new parameters α^* and π^*, restricted to D, respect the properties stipulated in the definition of a branching mechanism in the introduction. □

Corollary 1 *For any bounded open domain* $D \subset\subset E$, *any function* $f \in \mathrm{bp}(D)$ *and any* $\mu \in \mathcal{M}_F(D)$ *satisfying* $\langle w, \mu \rangle < \infty$,

$$-\log \mathbb{E}_\mu^* \left(e^{-\langle f, X_t^D \rangle}\right) = \int_D u_f^{D,*}(x, t)\mu(dx),$$

where $u_f^{D,*}(x,t) = \tilde{u}_{f+w}^D(x,t) - w(x)$ and it is the unique solution of

$$u_f^{D,*}(x,t) = \Pi_x[f(\xi_t); t < \tau^D] - \Pi_x\left[\int_0^{t \wedge \tau_D} \psi^*(\xi_s, u_f^{D,*}(\xi_s, t-s))\mathrm{d}s\right], \quad x \in D, \tag{20}$$

where ψ^* is defined by (19).

3.3 Excursion Measure

Associated to the law of the processes X, are the measures $\{\mathbb{N}_x^* : x \in E\}$, defined on the same measurable space as the probabilities $\{\mathbb{P}_{\delta_x}^* : x \in E\}$ are defined on, and which satisfy

$$\mathbb{N}_x^*(1 - e^{-\langle f, X_t \rangle}) = -\log \mathbb{E}_{\delta_x}^*(e^{-\langle f, X_t \rangle}) = u_f^*(x,t), \tag{21}$$

for all $f \in \mathrm{bp}(E)$ and $t \geq 0$. Intuitively speaking, the branching property implies that $\mathbb{P}_{\delta_x}^*$ is an infinitely divisible measure on the path space of X, that is to say the space of measure-valued cadlag functions, $\mathbb{D}([0, \infty) \times \mathcal{M}(E))$, and (21) is a 'Lévy-Khinchine' formula in which $\mathbb{N}_x^*$ plays the role of its 'Lévy measure'. Such measures are formally defined and explored in detail in [10].

Note that, by the monotonicity property, for any two open bounded domains, $D_1 \subset\subset D_2 \subset\subset E$,

$$\langle f, X_t^{D_1} \rangle \leq \langle f, X_t^{D_2} \rangle \qquad \mathbb{N}_x^*\text{-a.e.,}$$

for all $f \in \mathrm{bp}(D_1)$ understood in the sense of (12), $x \in D_1$ and $t \geq 0$. Moreover, for an open bounded domain D and f as before, it is also clear that $\mathbb{N}^*(1 - e^{-\langle f, X_t^D \rangle}) = u_f^{D,*}(x,t)$.

The measures $\{\mathbb{N}_x^* : x \in E\}$ will play a crucial role in the forthcoming analysis in order to describe the 'rate' of a Poisson point process of immigration.

3.4 A Markov Branching Process

In this section we introduce a particular Markov branching process which is built from the components of the $(\mathscr{P}, \psi)$-superprocess and which plays a central role in the backbone decomposition.

Recall that we abuse our notation and extend the domain of w with the implicit understanding that $w(\dagger) = 0$. Note, moreover, that thanks to (17), we have that, for $x \in E$, $w(x)^{-1}w(\xi_t) \exp\left\{-\int_0^t \psi(\xi_s, w(\xi_s))/w(\xi_s)\mathrm{d}s\right\}$ is in general a positive local martingale (and hence a supermartingale) under Π_x. For each $t \geq 0$, let $\mathscr{F}_t^\xi = \sigma(\xi_s : s \leq t)$. Let $\zeta = \inf\{t > 0 : \xi_t \in \{\dagger\}\}$ be the life time of ξ. The formula

$$\left.\frac{d\Pi_x^w}{d\Pi_x}\right|_{\mathscr{F}_t^\xi} = \frac{w(\xi_t)}{w(x)} \exp\left\{-\int_0^t \frac{\psi(\xi_s, w(\xi_s))}{w(\xi_s)}ds\right\} \qquad \text{on } \{t < \zeta\}, \quad t \geq 0, x \in E,$$

(22)

uniquely determines a family of (sub-)probability measures $\{\Pi_x^w : x \in E\}$. It is known that under these new probabilities, ξ is a right Markov process on E; see [4, Section 10.4], [19] or [25, Section 62]. We will denote by $\mathscr{P}^w$ the semi-group of the $E \cup \{\dagger\}$-valued process ξ whose probabilities are $\{\Pi_x^w : x \in E\}$.

Remark 1 Equation (16) may formally be associated with the equation $Lw(x) - \psi(x, w(x)) = 0$ on E, and the semi-group $\mathscr{P}^w$ corresponds to the diffusion with generator

$$L_0^w := L^w - w^{-1}Lw = L^w - w^{-1}\psi(\cdot, w),$$

where $L^w u = w^{-1}L(wu)$ for any u in the domain of L. Intuitively speaking, this means that the dynamics associated to $\mathscr{P}^w$, encourages the motion of ξ to visit domains where the global survival rate is high and discourages it from visiting domains where the global survival rate is low. (Recall from (7) that larger values of $w(x)$ make extinction of the $(\mathscr{P}, \psi)$-superprocess less likely under $\mathbb{P}_{\delta_x}$.)

Henceforth the process $Z = \{Z_t : t \geq 0\}$ will denote the Markov branching process whose particles move with associated semi-group $\mathscr{P}^w$. Moreover, the branching generator is given by

$$F(x, s) = q(x) \sum_{n \geq 0} p_n(x)(s^n - s),$$

(23)

where

$$q(x) = \psi'(x, w(x)) - \frac{\psi(x, w(x))}{w(x)},$$

(24)

$p_0(x) = p_1(x) = 0$ and for $n \geq 2$,

$$p_n(x) = \frac{1}{w(x)q(x)} \left\{\beta(x)w^2(x)1_{\{n=2\}} + w^n(x)\int_{(0,\infty)} \frac{y^n}{n!}e^{-w(x)y}\pi(x, dy)\right\}.$$

Here we use the notation

$$\psi'(x, w(x)) := \left.\frac{\partial}{\partial \lambda}\psi(x, \lambda)\right|_{\lambda=w(x)}, \qquad x \in E.$$

Note that the choice of $q(x)$ ensures that $\{p_n(x) : n \geq 0\}$ is a probability mass function. In order to see that $q(x) \geq 0$ for all $x \in E$ (but $q \neq 0$), write

$$q(x) = \beta(x)w(x) + \frac{1}{w(x)} \int_{(0,\infty)} (1 - e^{-w(x)z}(1 + w(x)z))\pi(x, \mathrm{d}z) \qquad (25)$$

and note that $\beta \geq 0$, $w > 0$ and $1 - e^{-\lambda z}(1 + \lambda z)$, $\lambda \geq 0$, are all non-negative.

Definition 2 In the sequel we shall refer to Z as the $(\mathscr{P}^w, F)$-backbone. Moreover, in the spirit of Definition 1, for all bounded domains D and $t \geq 0$, we shall also define $\tilde{Z}_t^D$ to be the atomic measure, supported on $\partial(D \times [0, t))$, describing particles in Z which are first in their genealogical line of descent to exit the domain $D \times [0, t)$.

Just as with the case of exit measures for superprocesses, we define the random measure, $Z^D = \{Z_t^D : t \geq 0\}$, on D such that $\langle f, Z_t^D \rangle = \langle f, \tilde{Z}_t^D \rangle$ for any $f \in \mathrm{bp}(D)$, where we remind the reader that we regard f as a function defined on $E \times [0, \infty)$ as in (12). As a process in time, Z^D is a Markov branching process, with branching generator which is the same as in (23) except that the branching rate $q(x)$ is replaced by $q^D(x) := q(x)\mathbf{1}_D(x)$, and associated motion semi-group given by that of the process ξ killed upon leaving D. Similarly to the case of superprocesses, for any two open bounded domains, $D_1 \subset\subset D_2 \subset\subset E$, the processes $\tilde{Z}^{D_1}$ and $\tilde{Z}^{D_2}$ (and hence Z^{D_1} and Z^{D_2}) are consistent in the sense that

$$\tilde{Z}_t^{D_1} = (\widetilde{\tilde{Z}_t^{D_2}})^{D_1}$$

for all $t \geq 0$ (and similarly $Z_t^{D_1} = (Z_t^{D_2})^{D_1}$ for all $t \geq 0$).

4   Local Backbone Decomposition

We are interested in immigrating $(\mathscr{P}, \psi^*)$-superprocesses onto the path of an $(\mathscr{P}^w, F)$-backbone within the confines of an open, bounded domain $D \subset\subset E$ and initial configuration $\nu \in \mathscr{M}_a(D)$, the space of finite atomic measures in D of the form $\sum_{i=1}^n \delta_{x_i}$, where $n \in \mathbb{N} \cup \{0\}$ and $x_1, \cdots, x_n \in D$. There will be three types of immigration: continuous, discontinuous and branch-point immigration which we now describe in detail. In doing so, we shall need to refer to individuals in the process Z for which we shall use classical Ulam-Harris notation, see for example p290 of Harris and Hardy [17]. Although the Ulam-Harris labelling of individuals is rich enough to encode genealogical order, the only feature we really need of the Ulam-Harris notation is that individuals are uniquely identifiable amongst $\mathscr{T}$, the set labels of individuals realised in Z. For each individual $u \in \mathscr{T}$ we shall write b_u and d_u for its birth and death times respectively, $\{z_u(r) : r \in [b_u, d_u]\}$ for its spatial trajectory and N_u for the number of offspring it has at time d_u. We shall also write $\mathscr{T}^D$ for the set of labels of individuals realised in Z^D. For each $u \in \mathscr{T}^D$ we shall also define

$$\tau_u^D = \inf\{s \in [b_u, d_u], z_u(s) \in D^c\},$$

with the usual convention that $\inf \emptyset := \infty$. Note that if $u \in \mathscr{T}^D$, we denote by ω its historical path on $[0, d_u]$ (the spatial motion of its ancestors, including itself). Then we have $\inf\{t \geq 0 : \omega(t) \in D^c\} \geq b_u$.

Definition 3 For $\nu \in \mathscr{M}_a(D)$ and $\mu \in \mathscr{M}_F(D)$, let Z^D be a Markov branching process with initial configuration ν, branching generator which is the same as in (23), except that the branching rate $q(x)$ is replaced by $q^D(x) := q(x)\mathbf{1}_D(x)$, and associated motion semi-group given by that of $\mathscr{P}^w$ killed upon leaving D. Let $X^{D,*}$ be an independent copy of X^D under $\mathbb{P}_\mu^*$. Then we define the measure valued stochastic process $\Delta^D = \{\Delta_t^D : t \geq 0\}$ such that, for $t \geq 0$,

$$\Delta_t^D = X_t^{D,*} + I_t^{D,\mathbb{N}^*} + I_t^{D,\mathbb{P}^*} + I_t^{D,\eta}, \tag{26}$$

where $I^{D,\mathbb{N}^*} = \{I_t^{D,\mathbb{N}^*} : t \geq 0\}$, $I^{D,\mathbb{P}^*} = \{I_t^{D,\mathbb{P}^*} : t \geq 0\}$ and $I^{D,\eta} = \{I_t^{D,\eta} : t \geq 0\}$ are defined as follows.

i) (**Continuous immigration:**) The process $I^{D,\mathbb{N}^*}$ is measure-valued on D such that

$$I_t^{D,\mathbb{N}^*} = \sum_{u \in \mathscr{T}^D} \sum_{b_u < r \leq t \wedge d_u \wedge \tau_u^D} X_{t-r}^{(D,1,u,r)},$$

where, given Z^D, independently for each $u \in \mathscr{T}^D$ such that $b_u < t$,

$$\sum_{b_u < r \leq t \wedge d_u \wedge \tau_u^D} \delta_{(r, X^{(D,1,u,r)})}$$

is a Poisson point process on $[b_u, t \wedge d_u \wedge \tau_u^D] \times \mathbb{D}([0, \infty) \times \mathscr{M}(E))$ with intensity

$$dr \times 2\beta(z_u(r))d\mathbb{N}_{z_u(r)}^*.$$

ii) (**Discontinuous immigration:**) The process $I^{D,\mathbb{P}^*}$ is measure-valued on D such that

$$I_t^{D,\mathbb{P}^*} = \sum_{u \in \mathscr{T}^D} \sum_{b_u < r \leq t \wedge d_u \wedge \tau_u^D} X_{t-r}^{(D,2,u,r)},$$

where, given Z^D, independently for each $u \in \mathscr{T}^D$ such that $b_u < t$,

$$\sum_{b_u < r \leq t \wedge d_u \wedge \tau_u^D} \delta_{(r, X^{(D,2,u,r)})}$$

is a Poisson point process on $[b_u, t \wedge d_u \wedge \tau_u^D] \times \mathbb{D}([0, \infty) \times \mathscr{M}(E))$ with intensity

$$\mathrm{d}r \times \int_{y \in (0,\infty)} y e^{-w(z_u(r))y} \pi(z_u(r), \mathrm{d}y) \times \mathrm{d}\mathbb{P}^*_{y\delta_{z_u(r)}}.$$

iii) (**Branch point biased immigration:**) The process $I^{D,\eta}$ is measure-valued on D such that

$$I_t^{D,\eta} = \sum_{u \in \mathscr{T}^D} \mathbf{1}_{\{d_u \leq t \wedge \tau_u^D\}} X_{t-d_u}^{(D,3,u)},$$

where, given Z^D, independently for each $u \in \mathscr{T}^D$ such that $d_u < t \wedge \tau_u^D$, the processes $X^{(D,3,u)}$ are independent copies of the canonical process X^D issued at time d_u with law $\mathbb{P}^*_{Y_u \delta_{z_u(d_u)}}$ such that, given u has $n \geq 2$ offspring, the independent random variable Y_u has distribution $\eta_n(z_u(d_u), \mathrm{d}y)$, where

$$\eta_n(x, \mathrm{d}y)$$
$$= \frac{1}{q(x)w(x)p_n(x)} \left\{ \beta(x)w^2(x)\delta_0(\mathrm{d}y)\mathbf{1}_{\{n=2\}} + w(x)^n \frac{y^n}{n!} e^{-w(x)y} \pi(x, \mathrm{d}y) \right\}.$$
$$(27)$$

It is not difficult to see that Δ^D is consistent in the domain D in the sense of (15). Accordingly, we denote by $\mathbf{P}_{(\mu,\nu)}$ the law induced by $\{\Delta_t^D, D \in \mathscr{O}(E), t \geq 0\}$, where $\mathscr{O}(E)$ is the collection of bounded open sets in E.

The so-called backbone decomposition of $(X^D, \mathbb{P}_\mu)$ for $\mu \in \mathscr{M}_F(D)$ entails looking at the process Δ^D in the special case that we randomise the law $\mathbf{P}_{(\mu,\nu)}$ by replacing the deterministic choice of ν with a Poisson random measure having intensity measure $w(x)\mu(\mathrm{d}x)$. We denote the resulting law by $\mathbf{P}_\mu$.

Theorem 1 *For any $\mu \in \mathscr{M}_F(D)$, the process $(\Delta^D, \mathbf{P}_\mu)$ is Markovian and has the same law as $(X^D, \mathbb{P}_\mu)$.*

5 Proof of Theorem 1

The proof involves several intermediary results in the spirit of the non-spatially dependent case of Berestycki et al. [1]. Localisation will be an important part of the process, allowing us to make use of Assumption A and uniqueness properties for certain integral equations. Accordingly, throughout we take D as an open, bounded domain such that $D \subset\subset E$. Any function f defined on D will be extended to E by defining $f = 0$ on $E \setminus D$.

Lemma 2 *Suppose that* $\mu \in \mathscr{M}_F(D)$, $\nu \in \mathscr{M}_a(D)$, $t \geq 0$ *and* $f \in \mathrm{bp}(D)$. *We have*

$$\mathbf{E}_{(\mu,\nu)}\left(e^{-\langle f, I_t^{D,\mathbb{N}^*} + I_t^{D,\mathbb{P}^*}\rangle}|\{Z_s^D : s \leq t\}\right) = \exp\left\{-\int_0^t \langle \phi(\cdot, u_f^{D,*}(\cdot, t-s)), Z_s^D\rangle ds\right\},$$

where

$$\phi(x, \lambda) = 2\beta(x)\lambda + \int_{(0,\infty)} (1 - e^{-\lambda y})z e^{-w(x)y} \pi(x, dy), \qquad x \in D, \lambda \geq 0. \tag{28}$$

Proof We write

$$\langle f, I_t^{D,\mathbb{N}^*} + I_t^{D,\mathbb{P}^*}\rangle = \sum_{u \in \mathscr{T}^D} \sum_{b_u < r \leq t \wedge d_u \wedge \tau_u^D} \langle f, X_{t-r}^{(D,1,u,r)}\rangle$$

$$+ \sum_{u \in \mathscr{T}^D} \sum_{b_u < r \leq t \wedge d_u \wedge \tau_u^D} \langle f, X_{t-r}^{(D,2,u,r)}\rangle.$$

Hence conditioning on Z^D, appealing to the independence of the immigration processes together with Campbell's formula and that $\mathbb{N}_x^*(1 - e^{-\langle f, X_s^D\rangle}) = u_f^{D,*}(x, s)$, we have

$$\mathbf{E}_{(\mu,\nu)}(e^{-\langle f, I_t^{D,\mathbb{N}^*}\rangle}|\{Z_s^D : s \leq t\})$$

$$= \exp\left\{-\sum_{u \in \mathscr{T}^D} 2\int_{b_u}^{t \wedge d_u \wedge \tau_u^D} \beta(z_u(r)) \cdot \mathbb{N}_{z_u(r)}^*(1 - e^{-\langle f, X_{t-r}^D\rangle})dr\right\}$$

$$= \exp\left\{-\sum_{u \in \mathscr{T}^D} 2\int_{b_u}^{t \wedge d_u \wedge \tau_u^D} \beta(z_u(r))u_f^{D,*}(z_u(r), t-r)dr\right\}. \tag{29}$$

On the other hand

$$\mathbf{E}_{(\mu,\nu)}(e^{-\langle f, I_t^{D,\mathbb{P}^*}\rangle}|\{Z_s^D : s \leq t\})$$

$$= \exp\left\{-\sum_{u \in \mathscr{T}^D} \int_{b_u}^{t \wedge d_u \wedge \tau_u^D} \int_0^\infty y e^{-w(z_u(r))y} \pi(z_u(r), dy)\mathbb{E}_{y\delta_{z_u(r)}}^*(1 - e^{-\langle f, X_{t-r}^D\rangle})dr\right\}$$

$$= \exp \left\{ - \sum_{u \in \mathscr{F}^D} \int_{b_u}^{t \wedge d_u \wedge \tau_u^D} \int_0^\infty (1 - e^{-u_f^{D,*}(z_u(r),t-r)y}) y e^{-w(z_u(r))y} \pi(z_u(r), \mathrm{d}y) \mathrm{d}r \right\}.$$
(30)

Combining (29) and (30) the desired result follows.   □

Lemma 3 *Suppose that the real-valued function $J(s, x, \lambda)$ defined on $[0, T) \times D \times \mathbb{R}$ satisfies that for any $c > 0$ there is a constant $A(c)$ such that*

$$|J(s, x, \lambda_1) - J(s, x, \lambda_2)| \le A(c)|\lambda_1 - \lambda_2|,$$

for all $s \in [0, T)$, $x \in D$ and $\lambda_1, \lambda_2 \in [-c, c]$. Then for any bounded measurable function $g(s, x)$ on $[0, T) \times D$, the integral equation

$$v(t, x) = g(t, x) + \int_0^t \Pi_x \left[J(t - s, \xi_s, v(t - s, \xi_s)); s < \tau^D \right] \mathrm{d}x, \quad t \in [0, T),$$

has at most one bounded solution.

Proof Suppose that v_1 and v_2 are two solutions, then there is a constant $c > 0$ such that $-c \le v_1, v_2 \le c$ and

$$\|v_1 - v_2\|(t) \le A(c) \int_0^t \|v_1 - v_2\|(s)\mathrm{d}s,$$

where $\|v_1 - v_2\|(t) = \sup_{x \in D} |v_1(t, x) - v_2(t, x)|$, $t \in (0, T)$. It follows from Gronwall's lemma (see, for example, Lemma 1.1 on page 1208 of [6]) that $\|v_1 - v_2\|(t) = 0, t \in [0, T)$.

Lemma 4 *Fix $t > 0$. Suppose that $f, h \in \mathrm{bp}(D)$ and $g_s(x)$ is jointly measurable in $(x, s) \in D \times [0, t]$ and bounded on finite time horizons of s such that $g_s(x) = 0$ for $x \in D^c$. Then for any $\mu \in \mathscr{M}_F(D)$, $x \in D$ and $t \ge 0$,*

$$e^{-W(x,t)} := \mathbf{E}_{(\mu, \delta_x)} \left[\exp \left(- \int_0^t \langle g_{t-s}, Z_s^D \rangle \mathrm{d}s - \langle f, I_t^{D, \eta} \rangle - \langle h, Z_t^D \rangle \right) \right]$$

is the unique $[0, 1]$-valued solution to the integral equation

$$w(x)e^{-W(x,t)} = \Pi_x \left[w(\xi_{t \wedge \tau_D})e^{-h(\xi_{t \wedge \tau_D})} \right]$$

$$+ \Pi_x \left[\int_0^{t \wedge \tau_D} [H_{t-s}(\xi_s, -w(\xi_s)e^{-W(\xi_s, t-s)}) - w(\xi_s)e^{-W(\xi_s, t-s)}g_{t-s}(\xi_s) \right.$$

$$\left. - \psi(\xi_s, w(\xi_s))e^{-W(\xi_s, t-s)}]\mathrm{d}s \right].$$
(31)

for $x \in D$, where

$$H_{t-s}(x, \lambda) = q(x)\lambda + \beta(x)\lambda^2$$

$$+ \int_0^\infty (e^{-\lambda y} - 1 + \lambda y)e^{-(w(x)+u_f^{D,*}(x,t-s))y}\pi(x, dy), \quad x \in D,$$

and $q(x)$ was defined in (24).

Proof Following Evans and O'Connell [15] it suffices to prove the result in the case when g is time invariant. To this end, let us start by defining the semi-group $\mathscr{P}^{h,D}$ by

$$\mathscr{P}_t^{h,D}[k](x) = \Pi_x \left(e^{-\int_0^{t\wedge\tau^D} h(\xi_s)ds} k(\xi_{t\wedge\tau^D}) \right) \quad \text{for } h, k \in bp(\overline{D}), \quad (32)$$

where, for convenience, we shall write

$$\mathscr{P}_t^D[k] = \mathscr{P}_t^{0,D}[k]. \tag{33}$$

Recall that for $h, k \in bp(D)$, $h(x) = k(x) = 0$ for $x \notin D$. Then we have

$$\mathscr{P}_t^{h,D}[k](x) = \Pi_x \left(e^{-\int_0^t h(\xi_s)ds} k(\xi_t); t < \tau^D \right) \quad \text{for } h, k \in bp(D). \tag{34}$$

Define the function $\chi(x) = \psi(x, w(x))/w(x)$. Conditioning on the first splitting time in the process Z^D and recalling that the branching occurs at the spatial rate $q^D(x) = 1_D(x)(\psi'(x, w(x)) - \chi(x))$ we get that for any $x \in D$,

$$e^{-W(x,t)} = \frac{1}{w(x)} \mathscr{P}_t^{g+q+\chi,D}[we^{-h}](x)$$

$$+ \Pi_x^w \left[\int_0^{t\wedge\tau^D} \exp\left(-\int_0^s (g+q)(\xi_r)dr \right) \right.$$

$$\left. \left\{ q(\xi_s) \sum_{n\geq 2} p_n(\xi_s)e^{-nW(\xi_s,t-s)} \int_{(0,\infty)} \eta_n(\xi_s, dy)e^{-yu_f^{D,*}(\xi_s,t-s)} \right\} ds \right]. \tag{35}$$

From (27) we quickly find that for $x \in D$,

$$\sum_{n\geq 2} p_n(x)e^{-nW(x,t-s)} \int_{(0,\infty)} \eta_n(x, dy)e^{-yu_f^{D,*}(x,t-s)}$$

$$= \frac{1}{q(x)w(x)} \left\{ H_{t-s}(x, -w(x)e^{-W(x,t-s)}) + w(x)q(x)e^{-W(x,t-s)} \right\}.$$

Using the above expression in (35) we have that

$$
w(x)e^{-W(x,t)} = \mathscr{P}_t^{g+q+\chi,D}[we^{-h}](x)
$$

$$
+\Pi_x\left[\int_0^{t\wedge\tau^D} \exp\left(-\int_0^s (g+q+\chi)(\xi_r)dr\right)\right.
$$

$$
\left.[(H_{t-s}(\xi_s,-w(\xi_s)e^{-W(\xi_s,t-s)}) + w(\xi_s)q(\xi_s)e^{-W(\xi_s,t-s)})]\,ds\right].
$$

Now appealing to Lemma 1.2 in Dynkin [9] and recalling that $\chi(\cdot) = \psi(\cdot,w(\cdot))/w(\cdot)$ on D, we may deduce that for any $x \in D$,

$$
w(x)e^{-W(x,t)} = \mathscr{P}_t^D[we^{-h}](x)
$$

$$
+\Pi_x\left[\int_0^{t\wedge\tau^D} [H_{t-s}(\xi_s,-w(\xi_s)e^{-W(\xi_s,t-s)}) - w(\xi_s)g(\xi_s)e^{-W(\xi_s,t-s)}\right.
$$

$$
\left. -\psi(\xi_s,w(\xi_s))e^{-W(\xi_s,t-s)}]ds\right] \tag{36}
$$

as required. Note that in the above computations we have implicitly used that w is uniformly bounded away from ∞ on D.

To complete the proof we need to show uniqueness of solutions to (36). Lemma 3 offers sufficient conditions for uniqueness of solutions to a general family of integral equations which includes (36). In order to check these sufficient conditions, let us define $\overline{w}^D = \sup_{y\in D} w(y)$. Thanks to Assumption (A) we have that $0 < \overline{w}^D < \infty$. For $s \geq 0$, $x \in D$ and $\lambda \in [0,\overline{w}^D]$, define the function $J(s,x,\lambda) := [H_s(x,-\lambda) - (g(x)+\chi(x))\lambda]$. We rewrite (36) as

$$
w(x)e^{-W(x,t)} = \mathscr{P}_t^D[we^{-h}](x)+\int_0^t \Pi_x\left[J(t-s,\xi_s,w(\xi_s)e^{-W(\xi_s,t-s)});s<\tau^D\right]ds.
$$

Lemma 3 tells us that (36) has a unique solution as soon as we can show that J is continuous in s and that for each fixed $T > 0$, there exists a $K > 0$ (which may depend on D and T) such that

$$
\sup_{s\leq T}\sup_{y\in D} |J(s,y,\lambda_1) - J(s,y,\lambda_2)| \leq K|\lambda_1 - \lambda_2|, \quad \lambda_1,\lambda_2 \in (0,\overline{w}^D].
$$

Recall that $g(y)$ is assumed to be bounded, moreover, Assumption (A) together with the fact that

$$
\sup_{y\in D}\left\{|\alpha(y)| + \beta(y) + \int_{(0,\infty)} (z\wedge z^2)\pi(y,dz)\right\} < \infty \tag{37}
$$

also implies that χ is bounded on D. Appealing to the triangle inequality, it now suffices to check that for each fixed $T > 0$, there exists a $K > 0$ such that

$$\sup_{s \leq T} \sup_{y \in D} |H_s(y, -\lambda_1) - H_s(y, -\lambda_2)| \leq K|\lambda_1 - \lambda_2|, \quad \lambda_1, \lambda_2 \in (0, \overline{w}^D]. \tag{38}$$

First note from Proposition 2.3 of Fitzsimmons [16] that

$$\sup_{s \leq T} \sup_{x \in D} u_f^{D,*}(x, s) < \infty. \tag{39}$$

Straightforward differentiation of the function $H_s(x, -\lambda)$ in the variable λ yields

$$-\frac{\partial}{\partial \lambda} H_s(x, -\lambda) = q(x) - 2\beta(x)\lambda + \int_{(0,\infty)} (1 - e^{\lambda z}) e^{-(w(x) + u_f^{D,*}(x,s))z} z\pi(x, dz).$$

Appealing to (37) and (39) it is not difficult to show that the derivative above is uniformly bounded in absolute value for $s \leq T$, $x \in D$ and $\lambda \in [0, \overline{w}^D]$, from which (38) follows by straightforward linearisation. The proof is now complete.

$\square$

Theorem 2 *For every $\mu \in \mathcal{M}_F(D)$, $v \in \mathcal{M}_a(D)$ and $f, h \in bp(D)$*

$$\mathbf{E}_{(\mu,v)} \left(e^{-\langle f, \Delta_t^D \rangle - \langle h, Z_t^D \rangle} \right) = e^{-\langle u_f^{D,*}(\cdot,t),\mu \rangle - \langle v_{f,h}^D(\cdot,t),v \rangle}, \tag{40}$$

where $e^{-v_{f,h}^D(x,t)}$ is the unique $[0, 1]$-solution to the integral equation

$$w(x)e^{-v_{f,h}^D(x,t)} = \Pi_x \left[w(\xi_{t \wedge \tau_D}) e^{-h(\xi_{t \wedge \tau_D})} \right]$$
$$+ \Pi_x \left[\int_0^{t \wedge \tau_D} [\psi^*(\xi_s, -w(\xi_s)e^{-v_{f,h}^D(\xi_s, t-s)} + u_f^{D,*}(\xi_s, t-s)) \right.$$
$$\left. - \psi^*(\xi_s, u_f^{D,*}(\xi_s, t-s))]ds \right]. \tag{41}$$

Proof Thanks to Corollary 1 it suffices to prove that

$$\mathbf{E}_{(\mu,v)}(e^{-\langle f, I_t^D \rangle - \langle h, Z_t^D \rangle}) = e^{-\langle v_{f,h}^D(\cdot,t),v \rangle},$$

where $I^D := I^{D,\mathbb{N}^*} + I^{D,\mathbb{P}^*} + I^{D,\eta}$, and $v_{f,h}^D$ solves (41). Putting Lemma 2 and Lemma 4 together we only need to show that, when $g_{t-s}(\cdot) = \phi(\cdot, u_f^{D,*}(\cdot, t - s))$ (where ϕ is given by (28)), we have that $\exp\{-W(x,t)\}$ is the unique $[0, 1]$-valued solution to (41). Again following the lead of [1], in particular referring to Lemma 5 there, it is easy to see that on D

$$H_{t-s}(\cdot, -w(\cdot)e^{-W(\cdot,t-s)}) - \phi(\cdot, u_f^{D,*}(\cdot, t-s))w(\cdot)e^{-W(\cdot,t-s)} - \frac{\psi(\cdot, w(\cdot))}{w(\cdot)}w(\cdot)e^{-W(\cdot,t-s)}$$

$$= \psi^*(\cdot, w(\cdot)e^{-W(\cdot,t-s)} + u_f^{D,*}(\cdot, t-s)) - \psi^*(\cdot, u_f^{D,*}(\cdot, t-s)),$$

which implies that $\exp\{-W(x,t)\}$ is the unique $[0,1]$-valued solution to (41). $\square$

Proof of Theorem 1 The proof is guided by the calculation in the proof of Theorem 2 of [1]. We start by addressing the claim that $(\varDelta^D, \mathbf{P}_\mu)$ is a Markov process. Given the Markov property of the pair $(\varDelta^D, Z^D)$, it suffices to show that, given $\varDelta_t^D$, the atomic measure Z_t^D is equal in law to a Poisson random measure with intensity $w(x)\varDelta_t^D$. Thanks to Campbell's formula for Poisson random measures, this is equivalent to showing that for all $h \in \mathrm{bp}(D)$,

$$\mathbf{E}_\mu(e^{-\langle h, Z_t^D \rangle}|\varDelta_t^D) = e^{-\langle w\cdot(1-e^{-h}), \varDelta_t^D \rangle},$$

which in turn is equivalent to showing that for all $f, h \in \mathrm{bp}(D)$,

$$\mathbf{E}_\mu(e^{-\langle f, \varDelta_t^D \rangle - \langle h, Z_t^D \rangle}) = \mathbf{E}_\mu(e^{-\langle w\cdot(1-e^{-h})+f, \varDelta_t^D \rangle}). \tag{42}$$

Note from (40) however that when we randomize ν so that it has the law of a Poisson random measure with intensity $w(x)\mu(dx)$, we find the identity

$$\mathbf{E}_\mu(e^{-\langle f, \varDelta_t^D \rangle - \langle h, Z_t^D \rangle}) = \exp\left\{-u_f^{D,*}(\cdot, t) - w \cdot (1 - e^{-v_{f,h}^D(\cdot,t)}), \mu\right\}.$$

Moreover, if we replace f by $w \cdot (1-e^{-h}) + f$ and h by 0 in (40) and again randomize ν so that it has the law of a Poisson random measure with intensity $w(x)\mu(dx)$ then we get

$$\mathbf{E}_\mu\left(e^{-\langle w\cdot(1-e^{-h})+f, \varDelta_t^D \rangle}\right)$$
$$= \exp\left\{-u_{w\cdot(1-e^{-h})+f}^{D,*}(\cdot, t) - w \cdot \left(1 - \exp\left\{-v_{w\cdot(1-e^{-h})+f,0}^D\right\}\right), \mu\right\}.$$

These last two observations indicate that (42) is equivalent to showing that, for all f, h as stipulated above and $t \geq 0$,

$$u_f^{D,*}(x,t) + w(x)(1 - e^{-v_{f,h}^D(x,t)}) = u_{w\cdot(1-e^{-h})+f}^{D,*}(x,t) + w(x)(1 - e^{-v_{w\cdot(1-e^{-h})+f,0}^D(x,t)}). \tag{43}$$

Note that both left and right-hand sides of the equality above are necessarily non-negative given that they are Laplace exponents of the left and right-hand sides of (42). Making use of (20), (16), and (41), it is computationally very straightforward to show that both left and right-hand sides of (43) solve (14) with initial condition $f + w(1 - e^{-h})$, which is bounded in $\overline{D}$. Since (14) has a unique solution with this initial condition, namely $u^D_{f+w\cdot(1-e^{-h})}(x, t)$, we conclude that (43) holds true. The proof of the claimed Markov property is thus complete.

Having now established the Markov property, the proof is complete as soon as we can show that $(\Delta^D, \mathbf{P}_\mu)$ has the same semi-group as $(X^D, \mathbb{P}_\mu)$. However, from the previous part of the proof we have already established that when $f, h \in bp(D)$,

$$\mathbf{E}_\mu\left(e^{-\langle h, Z^D_t\rangle - \langle f, \Delta^D_t\rangle}\right) = e^{-\langle u^D_{w(1-e^{-h})+f}(\cdot,t),\mu\rangle} = \mathbf{E}_\mu\left(e^{-\langle f+w(1-e^{-h}), X^D_t\rangle}\right). \quad (44)$$

In particular, choosing $h = 0$ we find

$$\mathbf{E}_\mu\left(e^{-\langle f, \Delta^D_t\rangle}\right) = \mathbb{E}_\mu\left(e^{-\langle f, X^D_t\rangle}\right), \qquad t \geq 0,$$

which is equivalent to saying that the semi-groups of $(\Delta^D, \mathbf{P}_\mu)$ and $(X^D, \mathbb{P}_\mu)$ agree.

$\square$

6 Global Backbone Decomposition

So far we have localized our computations to an open bounded domain D. Our ultimate objective is to provide a backbone decomposition on the whole domain E. To this end, let D_n be a sequence of open bounded domains in E such that $D_1 \subseteq D_2 \subseteq \cdots \subseteq D_n \subseteq \cdots \subseteq E$ and $E = \cup_{n\geq1} D_n$. Let X^{D_n}, Δ^{D_n} and Z^{D_n} be defined as in previous sections with D being replaced by D_n.

Lemma 5 *For any $h, f \in bp(E)$ with compact support and any $\mu \in \mathcal{M}_F(E)$, we have that for any $t \geq 0$, each element of the pair $\{(\langle h, Z^{D_n}_s\rangle, \langle f, \Delta^{D_n}_s\rangle) : s \leq t\}$ pathwise increases $\mathbf{P}_\mu$-almost surely as $n \to \infty$. The limiting pair of processes, here denoted by $\{((\langle h, Z^{\min}_s\rangle, \langle f, \Delta^{\min}_s\rangle) : s \leq t\}$, are such that $\langle f, \Delta^{\min}_t\rangle$ is equal in law to $\langle f, X_t\rangle$ and, given $\Delta^{\min}$, the law of $Z^{\min}_t$ is a Poisson random field with intensity $w(x)\Delta^{\min}_t(dx)$. Moreover, $Z^{\min}$ is a $(\mathscr{P}^w, F)$ branching process with branching generator as in (23) and associated motion semi-group given by (22).*

Proof Appealing to the stochastic consistency of Z^D and Δ^D in the domain D, it is clear that both $\langle h, Z^{D_n}_t, \rangle$ and $\langle f, \Delta^{D_n}_t\rangle$ are almost surely increasing in n. It therefore follows that the limit as $n \to \infty$ exists for both $\langle h, Z^{D_n}_t\rangle$ and $\langle f, \Delta^{D_n}_t\rangle$, $\mathbf{P}_\mu$-almost surely. In light of the discussion at the end of the proof of Theorem 1, the distributional properties of the limiting pair are established as soon as we show that

$$-\log \mathbf{E}_\mu\left(e^{-\langle h, Z^{\min}_t\rangle - \langle f, \Delta^{\min}_t\rangle}\right) = \int_E u_{w(1-e^{-h})+f}(x, t)\mu(dx), \quad t \geq 0. \quad (45)$$

Assume temporarily that suppμ, the support of μ, is compactly embedded in E so that there exists an $n_0 \in \mathbb{N}$ such that for $n \geq n_0$ we have that supp$\mu \subset D_n$ and $h = f = 0$ on D_n^c. Thanks to (44) and monotone convergence (45) holds as soon as we can show that $u_g^{D_n} \uparrow u_g$ for all $g \in \mathrm{bp}(E)$ satisfying $g = 0$ on D_n^c for $n \geq n_0$. By (13) and (14), we know that $u_g^{D_n}(x, t)$ is the unique non-negative solution to the integral equation

$$u_g^{D_n}(x, t) = \Pi_x[g(\xi_{t \wedge \tau_{D_n}})] - \Pi_x \left[\int_0^{t \wedge \tau_{D_n}} \psi(\xi_s, u_g^{D_n}(\xi_s, t - s)) ds \right]. \qquad (46)$$

Using Lemma 1.5 in [6] we can rewrite the above integral equation in the form

$$u_g^{D_n}(x, t) = \Pi_x \left[g(\xi_{t \wedge \tau_{D_n}}) \exp \left(\int_0^{t \wedge \tau_{D_n}} \alpha(\xi_s) ds \right) \right]$$
$$- \Pi_x \left[\int_0^{t \wedge \tau_{D_n}} \exp \left(\int_0^s \alpha(\xi_r) dr \right) \right.$$
$$\left. \left[\psi(\xi_s, u_g^{D_n}(\xi_s, t - s)) + \alpha(\xi_s u_g^{D_n}(\xi_s, t - s)) \right] ds \right]. \qquad (47)$$

Since $g = 0$ on D_n^c for $n \geq n_0$, we have

$$\Pi_x \left[g(\xi_{t \wedge \tau_{D_n}}) \exp \left(\int_0^{t \wedge \tau_{D_n}} \alpha(\xi_s) ds \right) \right] = \Pi_x \left[g(\xi_t) \exp \left(\int_0^t \alpha(\xi_s) ds \right) ; t < \tau_{D_n} \right],$$

which is increasing in n. By the comparison principle, $u_g^{D_n}$ is increasing in n (see Theorem 3.2 in part II of [6]). Put $\tilde{u}_g = \lim_{n \to \infty} u_g^{D_n}$. Note that $\psi(x, \lambda) + \alpha(x)\lambda$ is increasing in λ. Letting $n \to \infty$ in (47), by the monotone convergence theorem,

$$\tilde{u}_g(x, t) = \mathscr{P}_t^\alpha g(x) - \Pi_x \int_0^t \mathscr{P}_s^\alpha \left[\psi(\cdot, \tilde{u}_g(\cdot, t - s)) + \alpha(\cdot)\tilde{u}_g(\cdot, t - s)) \right] ds,$$

where

$$\mathscr{P}_t^\alpha g = \Pi_x \left[g(\xi_t) \exp \left(\int_0^t \alpha(\xi_s) ds \right) \right], \quad g \in \mathrm{bp}(E),$$

which in turn is equivalent to

$$\tilde{u}_g(x, t) = \mathscr{P}_s g(x) - \Pi_x \int_0^t \mathscr{P}_s \psi(\cdot, \tilde{u}_g(\cdot, t - s)) ds.$$

Therefore, $\tilde{u}_g$ and u_g are two solutions of (3) and hence by uniqueness they are the same, as required.

To remove the initial assumption that the support of μ is compactly embedded in E, suppose that μ_n is a sequence of compactly supported measures with mutually disjoint support such that $\mu = \sum_{k \geq 1} \mu_k$. By considering (45) for $\sum_{k=1}^{n} \mu_k$ and taking limits as $n \uparrow \infty$ we see that (45) holds for μ. Note in particular that the limit on the left hand side of (45) holds as a result of the additive property of the backbone decomposition in the initial state μ. $\qquad \square$

Note that, in the style of the proof given above (appealing to monotonicity and the maximality principle) we can easily show that the processes $X^{D_n, *}, n \geq 1$, converge distributionally at fixed times, and hence in law, to the process $(X, \mathbb{P}_\mu^*)$; that is, a $(\mathscr{P}, \psi^*)$-superprocess. With this in mind, again appealing to the consistency and monotonicity of the local backbone decomposition, our main result follows as a simple corollary of Lemma 5.

Corollary 2 *Suppose that $\mu \in \mathscr{M}_F(E)$. Let Z be a $(\mathscr{P}^w, F)$-Markov branching process with initial configuration consisting of a Poisson random field of particles in E with intensity $w(x)\mu(dx)$. Let X^* be an independent copy of $(X, \mathbb{P}_\mu^*)$. Then define the measure valued stochastic process $\Delta = \{\Delta_t : t \geq 0\}$ such that, for $t \geq 0$,*

$$\Delta_t = X_t^* + I_t^{\mathbb{N}^*} + I_t^{\mathbb{P}^*} + I_t^\eta, \tag{48}$$

where $I^{\mathbb{N}^*} = \{I_t^{\mathbb{N}^*} : t \geq 0\}$, $I^{\mathbb{P}^*} = \{I_t^{\mathbb{P}^*} : t \geq 0\}$ and $I^\eta = \{I_t^\eta : t \geq 0\}$ are defined as follows.

i) (**Continuum immigration:**) *The process $I^{\mathbb{N}^*}$ is measure-valued on E such that*

$$I_t^{\mathbb{N}^*} = \sum_{u \in \mathscr{T}} \sum_{b_u < r \leq t \wedge d_u} X_{t-r}^{(1,u,r)},$$

where, given Z, independently for each $u \in \mathscr{T}$ such that $b_u < t$,

$$\sum_{b_u < r \leq t \wedge d_u} \delta_{(r, X^{(1,u,r)})}$$

is a Poisson point process on $[b_u, t \wedge d_u] \times \mathbb{D}([0, \infty) \times \mathscr{M}(E))$ with intensity

$$dr \times 2\beta(z_u(r)) d\mathbb{N}_{z_u(r)}^*.$$

ii) (**Discontinuous immigration:**) *The process $I^{\mathbb{P}^*}$ is measure-valued on E such that*

$$I_t^{\mathbb{P}^*} = \sum_{u \in \mathscr{T}} \sum_{b_u < r \leq t \wedge d_u} X_{t-r}^{(2,u,r)},$$

where, given Z, independently for each $u \in \mathcal{T}$ such that $b_u < t$,

$$\sum_{b_u < r \leq t \wedge d_u} \delta_{(r, X^{(2,u,r)})}$$

is a Poisson point process on $[b_u, t \wedge d_u] \times \mathbb{D}([0, \infty) \times \mathcal{M}(E))$ with intensity

$$\mathrm{d}r \times \int_{y \in (0,\infty)} y e^{-w(z_u(r))y} \pi(z_u(r), \mathrm{d}y) \times \mathrm{d}\mathbb{P}^*_{y \delta_{z_u(r)}}.$$

iii) *(**Branch point biased immigration:**) The process I^η is measure-valued on E such that*

$$I_t^\eta = \sum_{u \in \mathcal{T}^D} \mathbf{1}_{\{d_u \leq t\}} X_{t-d_u}^{(3,u)},$$

where, given Z, independently for each $u \in \mathcal{T}$ such that $d_u < t$, the processes $X^{(3,u)}$ are independent copies of the canonical process X issued at time d_u with law $\mathbb{P}^*_{Y_u \delta_{z_u(d_u)}}$ such that, given u has $n \geq 2$ offspring, the independent random variable Y_u has distribution $\eta_n(z_u(d_u), \mathrm{d}y)$, where $\eta_n(x, \mathrm{d}y)$ is defined by (27).

Then $(\Delta, \mathbf{P}_\mu)$ is Markovian and has the same law as $(X, \mathbb{P}_\mu)$.

Acknowledgements We would like to thank Maren Eckhoff for a number of helpful comments on earlier versions of this paper. Part of this research was carried out whilst AEK was on sabbatical at ETH Zürich, hosted by the Forschungsinstitut für Mathematik, for whose hospitality he is grateful. The research of YXR is supported in part by the NNSF of China (Grant Nos. 11271030 and 11128101).

References

1. J. Berestycki, A.E. Kyprianou, A. Murillo-Salas, The prolific backbone decomposition for supercritical superdiffusions. Stoch. Proc. Appl. **121**, 1315–1331 (2011)
2. J. Bertoin, J. Fontbona, S. Martínez, On prolific individuals in a continuous-state branching process. J. Appl. Probab. **45**, 714–726 (2008)
3. T. Duquesne, M. Winkel, Growth of Lévy trees. Probab. Theory Relat. Fields **139**, 313–371 (2007)
4. E.B. Dynkin, *Markov Processes I (English translation)* (Springer, Berlin, 1965)
5. E.B. Dynkin, A probabilistic approach to one class of non-linear differential equations. Probab. Theory Relat. Fields **89**, 89–115 (1991)
6. E.B. Dynkin, Superprocesses and partial differential equations. Ann. Probab. **21**, 1185–1262 (1993)
7. E.B. Dynkin, *An Introduction to Branching Measure-Valued Processes.* CRM Monograph Series, vol. 6 (American Mathematical Society, Providence, 1994), 134 pp.
8. E.B. Dynkin, Branching exit Markov systems and superprocesses. Ann. Probab. **29**, 1833–1858 (2001)

9. E.B. Dynkin, *Diffusions, Superprocesses and Partial Differential Equations*. American Mathematical Society, Colloquium Publications, vol. 50 (Providence, Rhode Island, 2002)
10. E.B. Dynkin, S.E. Kuznetsov, N-measures for branching exit Markov systems and their applications to differential equations. Probab. Theory Relat. Fields **130**, 135–150 (2004)
11. J. Engländer, R.G. Pinsky, On the construction and support properties of measure-valued diffusions on $D \subseteq R^d$ with spatially dependent branching. Ann. Probab. **27**, 684–730 (1999)
12. A.M. Etheridge, *An Introduction to Superprocesses*. University Lecture Series, vol. 20 (American Mathematical Society, Providence, 2000), 187 pp.
13. A. Etheridge, D.R.E. Williams, A decomposition of the $(1 + \beta)$-superprocess conditioned on survival. Proc. Roy. Soc. Edin. **133A**, 829–847 (2003)
14. S.N. Evans, Two representations of a superprocess. Proc. Roy. Soc. Edin. **123A**, 959–971 (1993)
15. S.N. Evans, N. O'Connell, Weighted occupation time for branching particle systems and a representation for the supercritical superprocess. Can. Math. Bull. **37**, 187–196 (1994)
16. P.J. Fitzsimmons, Construction and regularity of measure-valued Markov branching processes. Isr. J. Math. **63**, 337–361 (1988)
17. R. Hardy, S.C. Harris, A spine approach to branching diffusions with applications to L^p-convergence of martingales. Séminaire de Probab. **XLII**, 281–330 (2009)
18. T. Harris, *The Theory of Branching Processes* (Springer, Berlin; Prentice-Hall, Englewood Cliffs, 1964)
19. K. Ito, S. Watanabe, Transformation of Markov processes by multiplicative functionals. Ann. Inst. Fourier Grenobl **15**, 13–30 (1965)
20. A.E. Kyprianou, Y-X. Ren, Backbone decomposition for continuous-state branching processes with immigration. Stat. Probab. Lett. **82**, 139–144 (2012)
21. J-F. Le Gall, *Spatial Branching Processes, Random Snakes and Partial Differential Equations*. Lectures in Mathematics (ETH Zürich, Birkhäuser, 1999)
22. Z. Li, *Measure-Valued Branching Markov Processes*. Probability and Its Applications (New York) (Springer, Heidelberg, 2011), 350 pp.
23. T. Salisbury, J. Verzani, On the conditioned exit measures of super Brownian motion. Probab. Theory Relat. Fields **115**, 237–285 (1999)
24. S. Roelly-Coppoletta, A. Rouault, Processus de Dawson-Watanabe conditioné par le futur lointain. C.R. Acad. Sci. Paris Série I **309**, 867–872 (1989)
25. M.J. Sharpe, *General Theory of Markov Processes* (Academic Press, San Diego, 1988)

On Bochner-Kolmogorov Theorem

Lucian Beznea and Iulian Cîmpean

Abstract We prove the Bochner-Kolmogorov theorem on the existence of the limit of projective systems of second countable Hausdorff (non-metrizable) spaces with tight probabilities, such that the projection mappings are merely measurable functions. Our direct and transparent approach (using Lusin's theorem) should be compared with the previous work where the spaces are assumed metrizable and the main idea was to reduce the general context to a regular one via some isomorphisms. The motivation of the revisit of this classical result is an application to the construction of the continuous time fragmentation processes and related branching processes, based on a measurable identification between the space of all fragmentation sizes considered by J. Bertoin and the limit of a projective system of spaces of finite configurations.

Keywords Bochner-Kolmogorov theorem • Space of finite configurations • Space of fragmentation sizes

AMS classification (2010): 60B05, 28A33, 60A10

1 Introduction

The main result of the paper is the existence of the limit of a projective system that consists of second countable Hausdorff topological spaces, not necessarily metrizable, with tight probabilities. Clearly, this is the case if the spaces are Lusin

L. Beznea (✉)
Simion Stoilow Institute of Mathematics of the Romanian Academy, Research unit No. 2, P.O. Box 1–764, RO-014700 Bucharest, Romania

Faculty of Mathematics and Computer Science, University of Bucharest, Bucharest, Romania
e-mail: lucian.beznea@imar.ro

I. Cîmpean
Simion Stoilow Institute of Mathematics of the Romanian Academy, Research unit No. 2, P.O. Box 1–764, RO-014700 Bucharest, Romania
e-mail: cimpean_iulian2005@yahoo.com

© Springer International Publishing Switzerland 2014
C. Donati-Martin et al. (eds.), *Séminaire de Probabilités XLVI*, Lecture Notes in Mathematics 2123, DOI 10.1007/978-3-319-11970-0_3

(or more general, Radon) topological spaces and it is straightforward to extend the result to Lusin measurable spaces.

Our motivation is an application in [2] of such a result to the construction of a branching process associated to a fragmentation one. A main step of this approach is the measurable identification between the canonical state space of the fragmentation processes considered by J. Bertoin in [1] and the limit of a projective system of spaces of finite configurations, the state spaces of the induced branching processes (see, e.g., [3, 4], and the classical articles [9] and [15]).

Recall that the classical Kolmogorov extension theorem guarantees that a compatible collection of finite-dimensional distributions will define a stochastic process. Bochner (see [5]) considered the abstract situation of projective systems of Hausdorff topological measure spaces and proved the existence of the limit, provided that the measures can be approximated by compacts and replacing the canonical projections by continuous mappings. In general, the limit space will not have a decent topology. However, Hausdorff compactness is preserved by taking projective limits and, in addition, if the measures are Baire then so is the projective measure. These results are due to Choksi (see [7]).

Recent applications of Bochner-Kolmogorov theorem in proving non-trivial results about spatial systems, such as the existence of Gibbs states (see for example [10] and [13]), require rather measure theoretical assumptions than topological ones. A significant improvement which is convenient to these applications was obtained by Parthasarathy (see [12], Chapter V, Theorem 3.2). He proved the existence of the limit of a projective system of measurable spaces indexed by the set of natural numbers, where the spaces are Lusin measurable and the projection mappings are merely measurable. The main idea to prove it is to reduce the context, via measurable isomorphisms, to compact spaces such that the projections are continuous (hence admitting a projective limit). Another proof of this result, in the case of separable metrizable spaces with tight probabilities may be found in [8], Chapter III, page 70, where the strategy is to reduce the projective system to a canonical one (so that the spaces are product spaces and the projection mappings are the canonical projections) and then to apply Kolmogorov extension theorem. We emphasize that, in contrast with our result, the metrizability of the underlying spaces is crucial for the proofs of the previous mentioned versions, since one of the ingredients is Souslin-Lusin theorem on direct images of measurable, respectively Souslin sets.

We also refer to [14] (Theorems 4.3, 4.5, and 4.7) for several characterizations for the existence of projective limits in terms of μ-pure fields, uniform σ-additivity, uniform integrable martingales, and Orlicz spaces.

The organization of the paper is as follows: in Sect. 2, after some preliminaries, we state the main results, namely Theorems 1 and 2. In the second part of this section we present the announced application (Corollary 1; see also Remark 2 for a more concrete connection to branching and fragmentation processes). All of the proofs are collected in Sect. 3.

2 The Main Results and the Application to Branching and Fragmentation Processes

Recall that (cf. e.g. [12]) if $\{(\Omega_\alpha, \mathscr{S}_\alpha), \alpha \in A\}$ is a family of measurable spaces indexed by an upper directed set A (i.e. A is ordered by an order relation "$\leq$" and for any $\alpha, \beta \in A$ there exists $\gamma \in A$ such that $\alpha \leq \gamma$ and $\beta \leq \gamma$) and for all $\alpha \leq \beta$ in A, $f_{\alpha\beta}$ is a mapping from Ω_β into Ω_α with the following properties:

(i) $f_{\alpha\beta}^{-1}(\mathscr{S}_\alpha) \subset \mathscr{S}_\beta$,
(ii) $f_{\alpha\beta} \circ f_{\beta\gamma} = f_{\alpha\gamma}$, $f_{\alpha\alpha} = $ identity for all $\alpha \leq \beta \leq \gamma$,
   then the family $\{(\Omega_\alpha, \mathscr{S}_\alpha, f_{\alpha\beta})\}$ is called a *projective system of measurable spaces*.
   We set

$$\Omega_\infty := \{\omega = (\omega_\alpha)_{\alpha \in A} \in \prod_{\alpha \in A} \Omega_\alpha \; \omega_\alpha \in \Omega_\alpha, \; \omega_\alpha = f_{\alpha\beta}(\omega_\beta) \text{ for all } \alpha \leq \beta \text{ in } A\}$$

and for each $\alpha \in A$ define the projection $f_\alpha : \Omega_\infty \to \Omega_\alpha$ by $f_\alpha(\omega) = \omega_\alpha$. Let $\mathscr{S}$ be the σ-algebra generated by the algebra $\mathscr{S}^* := \bigcup_{\alpha \in A} f_\alpha^{-1}(\mathscr{S}_\alpha)$. The measurable space $(\Omega_\infty, \mathscr{S})$ is called the *projective limit* of $\{(\Omega_\alpha, \mathscr{S}_\alpha, f_{\alpha\beta})\}$.

If for each $\alpha \in A$, P_α is a probability measure on $(\Omega_\alpha, \mathscr{S}_\alpha)$ such that the following additional compatibility condition

(iii) $P_\alpha = P_\beta \circ f_{\alpha\beta}^{-1}$ for all $\alpha \leq \beta$ in A

is satisfied, then the family $\{(\Omega_\alpha, \mathscr{S}_\alpha, P_\alpha, f_{\alpha\beta})\}$ is called a *projective system of probability spaces*.

Let $P^* : \mathscr{S}^* \to [0, 1]$ be defined as

$$P^*(B) = P_\alpha(f_\alpha(B)) \quad \text{for any } B \in f_\alpha^{-1}(\mathscr{S}_\alpha) \text{ and } \alpha \in A.$$

We claim that P^* is a well defined additive set function. Indeed, if $B = f_\alpha^{-1}(B_\alpha) = f_\beta^{-1}(B_\beta)$, for some α and β in A, there exists $\gamma \in A$ with $\alpha \leq \gamma$ and $\beta \leq \gamma$ such that if $B_\gamma = f_{\alpha\gamma}^{-1}(B_\alpha) \in \mathscr{S}_\gamma$ then $A = f_\gamma^{-1}(B_\gamma)$ and $P_\alpha(B_\alpha) = P_\gamma(B_\gamma) = P_\beta(B_\beta)$. In particular $f_\alpha^{-1}(\mathscr{S}_\alpha) \subset f_\gamma^{-1}(\mathscr{S}_\gamma)$ for every $\alpha \leq \gamma$ in A.

Whenever P^* may be extended to a σ-additive set function P on $\mathscr{S}$ then $(\Omega_\infty, \mathscr{S}, P)$ is called the *projective limit* of $\{(\Omega_\alpha, \mathscr{S}_\alpha, P_\alpha, f_{\alpha\beta})\}$.

We remark that even if each Ω_α is non-empty and all the mappings $f_{\alpha\beta}$ are surjective, Ω_∞ may be empty (see [6], chapter III, no.4, Exercise 4). Sufficient conditions for the non-triviality of the projective limit are given in [6] (ch. III, subsection 4, Proposition 5 and Theorem 1). The following further condition which assures that $\Omega_\infty \neq \emptyset$ is found necessary in almost all theorems concerning the existence of projective limit of probability measures:

A projective system $\{(\Omega_\alpha, \mathscr{S}_\alpha, P_\alpha, f_{\alpha\beta})\}$ of probability spaces is said to satisfy the *sequential maximality* condition if, given $\alpha_1 \leq \alpha_2 \leq \ldots \alpha_n \leq \ldots$ in A and $\omega_n \in \Omega_{\alpha_n}$ such that $f_{\alpha_n \alpha_{n+1}}(\omega_{n+1}) = \omega_n$ for all n, there exists $\omega \in \Omega_\infty$ with

$f_{\alpha_n}(\omega) = \omega_n$ for all n. In particular, sequential maximality implies that f_α and $f_{\alpha\beta}$ are surjective for all $\alpha \leq \beta$ in A, and if $A = \mathbb{N}$, by an inductive argument one can easily check that the converse is also true.

We can now state the main results of this paper.

Theorem 1 *Let $\{(\Omega_\alpha, \mathscr{S}_\alpha, P_\alpha, f_{\alpha\beta}), \ \alpha \leq \beta, \ \alpha, \beta \in A\}$ be a projective system of probability spaces satisfying the sequential maximality condition, where $\Omega_\alpha, \ \alpha \in A$ are second countable Hausdorff topological spaces. If $\mathscr{S}_\alpha$ contains the Borel sets of Ω_α and P_α is inner regular relative to the class of compact sets in Ω_α for every $\alpha \in A$, then the (unique) projective limit $(\Omega_\infty, \mathscr{S}, P)$ exists.*

The topological assumptions in Theorem 1 are justified by Lusin's theorem, which is the main ingredient of our proof and for the reader convenience we restate it here (see [11], Theorem 1 for a very short proof which avoids the Egorov's theorem): Let $(\Omega, \mathscr{S}, P)$ be a probability space such that Ω is a Hausdorff topological space, $\mathscr{S}$ is a σ-algebra containing all Borel subsets of Ω and P is inner regular (with respect to the class of compact sets in Ω). If f is a measurable function from Ω into a second countable topological space Y then for every $\varepsilon > 0$ there exists a compact set $K \subset \Omega$ such that f restricted to K is continuous and $P(\Omega \setminus K) < \varepsilon$. Clearly, if $(f_n)_{n \geq 1}$ is a sequence of measurable functions from Ω into Y and if $K_n \subset \Omega$ is a compact set such that f_n is continuous on K_n with $P(\Omega \setminus K_n) < \frac{\varepsilon}{2^n}$ for all $n \geq 1$ then $K := \bigcap_{n \geq 1} K_n$ is compact, $P(\Omega \setminus K) < \varepsilon$ and f_n is continuous on K for all $n \geq 1$.

Recall that a topological space is called *Lusin* (resp. *Radon*) if it is homeomorphic to a Borel (resp. universally measurable) subset of a compact metric space (or equivalently, of a Polish space). A measurable space is named Lusin (resp. Radon) if it is measurable isomorphic to a Lusin (resp. Radon) topological space, endowed with the Borel σ-algebra.

If we deal with Lusin (or Radon) measurable spaces then Theorem 1 leads to the following result (see [12], Chapter V, Theorem 3.2 and also [8], Chapter III, page 70).

Theorem 2 *If $\{(\Omega_\alpha, \mathscr{S}_\alpha, P_\alpha, f_{\alpha\beta}), \ \alpha \leq \beta, \ \alpha, \beta \in A\}$ is a projective system of probability spaces satisfying the sequential maximality condition, such that the underlying spaces are Lusin (resp. Radon) measurable spaces then the (unique) projective limit exists. If $A = \mathbb{N}$, it is a Lusin (resp. Radon) measurable space.*

2.1 Application to Branching and Fragmentation Processes

Let

$$S^\downarrow := \{\mathbf{x} = (x_k)_{k \geq 1} \subset [0, 1] : (x_k)_{k \geq 1} \text{ decreasing, } \lim_k x_k = 0\}.$$

Endowed with the uniform distance $d(\mathbf{x}, \mathbf{x}') := \max_{i \in \mathbb{N}} |x_i - x_i'|$, $S^{\downarrow}$ becomes a Polish space. The set $S^{\downarrow}$ is called the *space of fragmentation sizes* and it appears as the state space of the fragmentation processes (see [1], Section 1.1.3).

Let further $(d_n)_{n \geq 1} \subset (0, 1)$ be a fixed sequence strictly decreasing to zero and for each $n \geq 1$ consider the set S_n of all *finite configurations* of the segment $[d_n, 1]$,

$$S_n := \{ \sum_{k \leq k_0} \delta_{x_k} : k_0 \in \mathbb{N}^*, \ 1 \geq x_k \geq d_n \text{ for all } 1 \leq k \leq k_0 \}.$$

To each set S_n we add the zero measure $\mathbf{0}$,

$$S_n^0 := S_n \cup \{\mathbf{0}\}, \ n \geq 1.$$

The set S_n^0 is identified with the union of all symmetric k-th powers $[d_n, 1]^{(k)}, k \geq 0$, where $[d_n, 1]^{(0)} = \{\mathbf{0}\}$; see [2–4, 9], and [15]. The set S_n^0 is equipped with the canonical disjoint union topology so that it becomes a Lusin topological space (cf. [4], Lemma 2.3). We denote by $\mathscr{B}(S_n^0)$ the corresponding Borel σ-algebra on S_n^0.

It is useful to identify a sequence $\mathbf{x} = (x_k)_{k \geq 1} \in S^{\downarrow}$ with the σ-finite measure $\mu_{\mathbf{x}}$ on $[0, 1]$,

$$\mu_{\mathbf{x}} := \begin{cases} \sum_{x_k > 0} \delta_{x_k}, & \text{if } \mathbf{x} \neq \mathbf{0} \\ \mathbf{0} & , \text{ if } \mathbf{x} = \mathbf{0}. \end{cases}$$

In this way, S_n^0 becomes a subset of $S^{\downarrow}$ and we denote by $\mathscr{B}_u(S_n^0)$ the Borel σ-algebra associated to the metric on S_n^0 inherited from $S^{\downarrow}$.

Remark 1 i) Summarizing, we endowed S_n^0 with two σ-algebras $\mathscr{B}(S_n^0)$ and $\mathscr{B}_u(S_n^0)$, corresponding to the disjoint union topology and respectively, to the uniform distance topology. It turns out easily that these two σ-algebras are actually the same. Indeed, since S_n^0 with the uniform distance is a Polish space, by a well known result of Lusin (see for example [12], Corollary 3.9 or [8], page 49), the two σ-algebras will coincide once we show that the disjoint union topology is included in the topology of the uniform distance on S_n^0. To this end, let $(\mathbf{x}^m)_{m \geq 0} \subset S_n^0$ such that $\lim_m d(\mathbf{x}^m, \mathbf{x}^0) = 0$. If $\mathbf{x}^0 = \mathbf{0}$ then $\mathbf{x}^m$ equals $\mathbf{0}$ for sufficiently large m, hence $(\mathbf{x}^m)_{m \geq 1}$ converges to $\mathbf{x}^0$ also in the disjoint union topology. If $\mathbf{x}^m = (x_1^m, \ldots, x_{k(m)}^m, 0, \ldots)$ are in S_n so that $x_1^m \geq x_2^m \geq \ldots \geq x_{k(m)}^m \geq d_n$ for all $m \geq 0$, then for m large enough we have that $\mathbf{x}^m$ belongs to $[d_n, 1]^{(k(0))}$. So, we need to show that $(\mathbf{x}^m)_{m \geq 1}$ converges to $\mathbf{x}^0$ in the quotient topology on $[d_n, 1]^{(k(0))}$, but this is immediate since $\lim_m d(\mathbf{x}^m, \mathbf{x}^0) = 0$.

ii) Let $M([d_n, 1])$ be the space of all finite measures on $[d_n, 1]$ endowed with the weak*-topology and denote by $\mathscr{M}([d_n, 1])$ its Borel σ-algebra. One can easily see that the inclusion $S_n^0 \subset M([d_n, 1])$ is continuous, hence by the above mentioned result of Lusin we conclude that $\mathscr{B}(S_n^0)$ $(= \mathscr{B}_u(S_n^0)$, by i)) and $\mathscr{M}([d_n, 1])$ are the same.

For each $n \geq 1$ define the mapping $\alpha_n : S^\downarrow \to S_n^0$ by

$$\alpha_n(\mu_\mathbf{x}) := \mu_\mathbf{x}|_{[d_n,1]}, \quad \mathbf{x} \in S^\downarrow.$$

We use the same notation α_n also for the restriction $\alpha_n|_{S_{n+1}^0}$ and note that α_n is not continuous since it is discontinuous in δ_{d_n}. Define

$$S_\infty := \{(\mathbf{x}^n)_{n \geq 1} \in \prod_{n \geq 1} S_n^0 : \mathbf{x}^n = \alpha_n(\mathbf{x}^m) \text{ for all } m > n \geq 1\}$$

and the mapping $i : S^\downarrow \to S_\infty$ by

$$i(\mathbf{x}) := (\alpha_n(\mathbf{x}))_{n \geq 1}, \quad \mathbf{x} \in S^\downarrow.$$

According to [2], Proposition 4.5, i is a bijection. Let $\alpha_{n\infty} : S_\infty \to S_n^0$ be given by

$$\alpha_{n\infty}((\mathbf{x}^n)_{n \geq 1}) = \mathbf{x}^n$$

and $\mathscr{S}$ be the σ-algebra on S_∞ generated by $(\alpha_{n\infty})_{n \geq 1}$. Also, we denote by $\mathscr{B}(S^\downarrow)$ the Borel σ-algebra of the metric space $S^\downarrow$.

Corollary 1 $(S_n^0, \mathscr{B}(S_n^0), \alpha_n)_{n \geq 1}$ *is a projective system of Lusin spaces and its projective limit* $(S_\infty, \mathscr{S})$ *is measurably isomorphic to* $(S^\downarrow, \mathscr{B}(S^\downarrow))$ *via the mapping i.*

Remark 2 A sequence from $S^\downarrow$ should be seen as the sequence of sizes of the fragments of a unit mass block after it splits. The goal is to construct a right (Markov) process with state space $S^\downarrow$ which describes the dynamics of a non-interacting infinite particle system, whose particles behave independently for some life time until they split. The process goes on under the same law. Despite the obvious branching property of such a process, the difficulty is that $S^\downarrow$ is not a space of finite measures but σ-finite. The idea developed in [2] is to construct some branching transition functions $(\widehat{P_t^n})_{t \geq 0}$ on S_n^0 associated to a fragmentation kernel. Then, if $\mathbf{x} \in S^\downarrow$ and $\mathbf{x}^n := \alpha_n(\mathbf{x}) \in S_n^0$, it follows from Corollary 1 and [2], Proposition 4.7 that $(S_n^0, \mathscr{B}(S_n^0), \alpha_n, \delta_{\mathbf{x}^n} \circ \widehat{P_t^n})_{n \geq 1}$ is a projective system of probability spaces and its limit $(\widehat{P_t})_{t \geq 0}$ is a Markovian transition function on $(S^\downarrow, \mathscr{B}(S^\downarrow))$. The existence of a right (Markov) process on $S^\downarrow$ with transition function $(\widehat{P_t})_{t \geq 0}$ is further obtained based on some recent potential theoretical techniques (see [2], Theorem 5.2).

3 Proofs of the Main Results

As we noted in Sect. 2, P^* is additive, hence it only remains to show the σ-additivity. Suppose P^* is not σ-additive on $\mathscr{S}^*$ and let $\alpha > 0$ and $(S_n^*)_{n \geq 1} \subset \mathscr{S}^*$ such that $S_n^* \searrow \emptyset$ and $P^*(S_n^*) \geq \alpha$ for every $n \geq 1$. Let α_n and $S_n \in \mathscr{S}_{\alpha_n}$ satisfying $S_n^* = f_{\alpha_n}^{-1}(S_n)$, $n \geq 1$. Since A is upper directed we may assume that $(\alpha_n)_{n \geq 1}$ is increasing.

Proof of Theorem 1 Lusin's theorem ensures the existence of a compact set $\overline{C}_n \subset S_n$ such that

$$P_{\alpha_n}(\overline{C}_n) \geq P_{\alpha_n}(S_n) - \frac{\alpha}{2^n}$$

and $f_{\alpha_m \alpha_n}$ is continuous on $\overline{C}_n$ for every $m < n$. Let $\overline{C}_n^* := f_{\alpha_n}^{-1}(\overline{C}_n)$. Then $\overline{C}_n^* \in \mathscr{S}_{\alpha_n}^*$, $\overline{C}_n^* \subset S_n^*$ and if $C_n^* := \bigcap_{k=1}^{n} \overline{C}_n^* = \overline{C}_n^* \setminus [\bigcup_{k=1}^{n-1} (S_k^* \setminus \overline{C}_k^*)]$ then

$$C_n^* \subset S_n^*, \quad C_n^* \in \mathscr{S}_{\alpha_n}^*, \quad C_n^* \searrow \emptyset$$

and $P^*(C_n^*) \geq P^*(\overline{C}_n^*) - \sum_{k=1}^{n=1}[P^*(S_k^*) - P^*(\overline{C}_k^*)] \geq P_{\alpha_n}(S_n) - \frac{\alpha}{2} \geq \frac{\alpha}{2} > 0$.
Thus C_n^* is non-empty for every $n \geq 1$. If $C_n := f_{\alpha_n}(C_n^*)$, then

$$C_n = f_{\alpha_n}(\bigcap_{k=1}^{n}(f_{\alpha_k \alpha_n} \circ f_{\alpha_n})^{-1}(\overline{C}_k) = \bigcap_{k=1}^{n} f_{\alpha_k \alpha_n}^{-1}(\overline{C}_k)$$

and

$$C_n \subset f_{\alpha_n}(\overline{C}_n^*) = \overline{C}_n.$$

Consequently, by the continuity of $f_{\alpha_k \alpha_n}$ on $\overline{C}_n$ for every $k < n$, we obtain that C_n is compact (as a closed subset). Note that $f_{\alpha_n \alpha_m}(C_m) \subset C_n$ for every $n < m$. Choose $\omega_n \in C_n$ and let $\omega_n^m = f_{\alpha_n \alpha_m}(\omega_m)$ for every $n < m$. Then $(\omega_n^m)_m$ has a limit point ω_n^0 in C_n and passing to a subsequence we assume that $\lim_m \omega_n^m = \omega_n^0$. Hence

$$\omega_n^0 = \lim_m f_{\alpha_n \alpha_m}(\omega_m^m) = \lim_m f_{\alpha_n \alpha_k} \circ f_{\alpha_k \alpha_m}(\omega_m^m) = f_{\alpha_n \alpha_k}(\omega_k^0) \quad \text{for every } n < k.$$

By the sequential maximality condition there exists an $\omega^0 \in \Omega_\infty$, such that $\omega_n^0 = f_{\alpha_n}(\omega^0)$ for all n and $\omega^0 \in \bigcap_{n=1}^{\infty} C_n^*$, which leads to the contradiction $\bigcap_{n=1}^{\infty} C_n^* \neq \emptyset$.

$\square$

Proof of Theorem 2 We only consider the Radon case. As in [12] we transfer the problem to the topological case as follows: let $\sigma_\alpha : (\Omega_\alpha, \mathscr{S}_\alpha) \to (\overline{\Omega}_\alpha, \overline{\mathscr{S}}_\alpha)$ be a measurable isomorphism, where $\overline{\Omega}_\alpha$ is a Radon topological space and $\overline{\mathscr{S}}_\alpha$ is its Borel σ-algebra. For each $\alpha \in A$ define $\overline{P}_\alpha$ on $(\overline{\Omega}_\alpha, \overline{\mathscr{S}}_\alpha)$ by

$$\overline{P}_\alpha(B) = P_\alpha(\sigma_\alpha^{-1}(B)) \quad \text{for all } B \in \overline{\mathscr{S}}_\alpha$$

and the mappings $\overline{f}_{\alpha\beta} : \overline{\Omega}_\beta \to \overline{\Omega}_\alpha$ by

$$\overline{f}_{\alpha\beta} = \sigma_\alpha \circ f_{\alpha\beta} \circ \sigma_\beta^{-1}.$$

Let $\overline{S}_n^* = \overline{f}_{\alpha_n}^{-1}(\sigma_{\alpha_n}(S_n))$. Since $\overline{S}_n^* \searrow \emptyset$ and

$$P_{\alpha_n}(S_n) = P(S_n^*) = \overline{P}_{\alpha_n}(\sigma_{\alpha_n}(S_n)) = \overline{P}^*(\overline{S}_n^*)$$

the proof will be complete as soon as we show that the system $\{(\overline{\Omega}_\alpha, \overline{\mathscr{S}}_\alpha, \overline{P}_\alpha, \overline{f}_{\alpha\beta})\}$ satisfies the conditions in Theorem 1 and hence its projective limit exists. First, note that $\overline{f}_{\alpha\beta} \circ \overline{f}_{\beta\gamma} = \overline{f}_{\alpha\gamma}$, $\overline{f}_{\alpha\alpha} = $ identity, $\overline{f}_{\alpha\beta}^{-1}(\overline{\mathscr{S}}_\alpha) = \overline{\mathscr{S}}_\beta$ and $\overline{P}_\alpha = \overline{P}_\beta \circ \overline{f}_{\alpha\beta}^{-1}$, so the system is a projective system of probability spaces. Since any finite measure on a Radon topological space is inner regular it only remains to verify the sequential maximality condition. To this end, let $(\overline{\omega}_n)_{n\in\mathbb{N}} \in \prod_{n\geq 1} \overline{\Omega}_{\alpha_n}$ such that $\overline{f}_{\alpha_n\alpha_m}(\overline{\omega}_m) = \overline{\omega}_n$ for every $n \leq m$. Then $\omega_n := \sigma_{\alpha_n}^{-1}(\overline{\omega}_n)$ satisfies $f_{\alpha_n\alpha_m}(\omega_m) = \omega_n$ and by hypothesis there exists $\omega = (\omega_\alpha)_{\alpha\in A} \in \Omega_\infty$ such that $\omega_n = f_{\alpha_n}(\omega)$ for all $n \geq 1$. If $\overline{\omega} = (\sigma_\alpha(\omega_\alpha))_{\alpha\in A}$ then $\overline{\omega} \in \overline{\Omega}_\infty$ and $\overline{f}_{\alpha_n}(\overline{\omega}) = \overline{\omega}_n$ so the desired condition is fulfilled.

If $A = \mathbb{N}$ then $\prod_{n=1}^\infty \overline{\Omega}_n$ is a Radon topological space and the same is $(\overline{\Omega}_\infty, \overline{\mathscr{S}})$ since it is a Borel subset of $\prod_{n=1}^\infty \overline{\Omega}_n$ (cf. [12], Chapter V, Theorem 2.5). One can easily check that $(\Omega_\infty, \mathscr{S})$ is measurable isomorphic with $(\overline{\Omega}_\infty, \overline{\mathscr{S}})$ hence the projective limit is a Radon measurable space. $\square$

Note that a similar transfer approach was used by Rao (see [14], Theorem 3.7) to obtain a weaker version which asserts that any abstract projective system of probability spaces (without any kind of regularity) is the inverse image, under an isomorphic measure preserving mapping, of some regular projective system of compact Hausdorff probability spaces (which we know that admits a projective limit). The disadvantage is that the projective limit cannot be retransferred to the initial system since the isomorphic identification is not pointwise. In fact, even the isomorphic image is quite hard to control since the isomorphism is given by the Stone representation theorem.

Proof of Corollary 1 First, we show that α_n is $\mathscr{B}(S^\downarrow)/\mathscr{B}_u(S_n^0)$-measurable. Let $\mathbf{x} \equiv \mu_{\mathbf{x}} \in S_n^0$ and $r > 0$. If we denote by $B(\mu_{\mathbf{x}}, r)$ the set $\{\mu \in S_n^0 : d(\mu_{\mathbf{x}}, \mu) < r\}$, then

$$\alpha_n^{-1}(B(\mu_\mathbf{x}, r)) = \{(y_1, \ldots, y_l, \ldots) \in S^\downarrow : y_1 < d_n, d(\mathbf{0}, \mathbf{x}) < r\} \cup$$

$$\bigcup_{i \geq 1} \Big(\{(y_1, \ldots, y_i, \ldots) \in S^\downarrow : y_i \geq d_n\} \cap \{(y_1, \ldots, y_i, \ldots) \in S^\downarrow : y_{i+1} < d_n\} \cap$$

$$\{(y_1, \ldots, y_i, \ldots) \in S^\downarrow : d((y_1, \ldots, y_i, 0, \ldots), \mathbf{x}) < r\} \Big)$$

But each set in the right hand of the above equality is either open or closed in $S^\downarrow$, therefore $\alpha_n^{-1}(B(\nu, r)) \in \mathscr{B}(S^\downarrow)$. Since S_n with the uniform distance is separable, α_n is $\mathscr{B}(S^\downarrow) \backslash \mathscr{B}_u(S_n^0)$-measurable.

By Remark 1 it follows that α_n and hence the restrictions of α_n to S_m^0 are $\mathscr{B}(S^\downarrow) \backslash \mathscr{B}(S_n^0)$-measurable, and it is straightforward to check that they satisfy the compatibility condition ii) from the beginning of Sect. 2, so that $(S_n^0, \mathscr{B}(S_n^0), \alpha_n)_{n \geq 1}$ becomes a projective system of Lusin spaces.

To show that i is a measurable isomorphism we recall that by [12] (Chapter V, Theorem 2.5), $\mathscr{S}$ is the trace of the product σ-algebra $\bigotimes_{n=1}^{\infty} \mathscr{B}(S_n^0)$ on S_∞ hence i is $\mathscr{B}(S^\downarrow) \backslash \mathscr{S}$-measurable if and only if each α_n is measurable and this is true by the first part of the proof. The measurability of i^{-1} follows by the same result of Lusin mentioned in Remark 1. $\square$

Acknowledgements This work was supported by a grant of the Romanian National Authority for Scientific Research, CNCS–UEFISCDI, project number PN-II-RU-TE-2011-3-0259.

References

1. J. Bertoin, *Random Fragmentation and Coagulation Processes* (Cambridge University Press, New York 2006)
2. L. Beznea, M. Deaconu, O. Lupaşcu, Branching processes for the fragmentation equation, Stochastic Process. Appl., 2014, to appear
3. L. Beznea, O. Lupaşcu, Measure-valued discrete branching Markov processes, 2013 (preprint)
4. L. Beznea, A. Oprina, Nonlinear PDEs and measure-valued branching type processes. J. Math. Anal. Appl. **384**, 16–32 (2011)
5. S. Bochner, *Harmonic Analysis and the Theory of Probability* (University of California Press, Los Angeles, 1955)
6. N. Bourbaki, *Theory of Sets* (Hermann, Paris, 1968)
7. J.R. Choksi, Inverse limits of measure spaces. Proc. Lond. Math. Soc. **8**, 321–342 (1958)
8. C. Dellacherie, P.A. Meyer, *Probabilities and Potential A* (Hermann, Paris 1978)
9. N. Ikeda, M. Nagasawa, S. Watanabe, Branching Markov processes I. J. Math. Kyoto Univ. **8**, 232–278 (1968)
10. Y. Kondratiev, T. Pasurek, M. Röckner, Gibbs measures of continuous systems: An analytic approach. Rev. Math. Phys. **24**, 1250026 (2012)

11. P.A. Loeb, E. Talvila, Lusin's theorem and Bochner integration. Sci. Math. Jpn. **10**, 55–62 (2004)
12. K.R. Parthasarathy, *Probability Measures on Metric Spaces* (Academic Press, New York 1967)
13. C. Preston, *Specifications and Their Gibbs States*. Lecture Notes (Bielefeld University, Bielefeld, 2005)
14. M.M. Rao, Projective limits of probability spaces. J. Multivar. Anal. **1**, 28–57 (1971)
15. M.L. Silverstein, Markov processes with creation of particles. Z. Warsch verw. Geb. **9**, 235–257 (1968)

Small Time Asymptotics for an Example of Strictly Hypoelliptic Heat Kernel

Jacques Franchi

Abstract A small time asymptotics of the density is established for a simplified (non-Gaussian, strictly hypoelliptic) second chaos process tangent to the Dudley relativistic diffusion.

1 Introduction

The problem of estimating the heat kernel, or the density of a diffusion, particularly as time goes to zero, has been extensively studied for a long time. Let us mention only the articles [1–3, 11, 15, 18], and the existence of other works on that subject by Azencott, Molchanov and Bismut, quoted in [2].

To summary roughly, a very classical question addresses the asymptotic behavior (as $s \searrow 0$) of the density $p_s(x, y)$ of the diffusion (x_s) solving a Stratonovich stochastic differential equation

$$x_s = x + \sum_{j=1}^{k} \int_0^s V_j(x_\tau) \circ dW_\tau^j + \int_0^s V_0(x_\tau) \, d\tau \, ,$$

where the smooth vector fields V_j are supposed to satisfy a Hörmander condition ; the underlying space being $\mathbb{R}^d$ or some d-dimensional smooth manifold $\mathcal{M}$.

The elliptic case being very well understood for a long time [1, 18], the studies focussed then on the sub-elliptic case, that is to say, when the strong Hörmander condition (that the Lie algebra generated by the fields $V_1, \ldots, V_k$ has maximal rank everywhere) is fulfilled. In that case these fields generate a sub-Riemannian distance $d(x, y)$, defined as in control theory, by considering only C^1 paths whose tangent vectors are spanned by them. Then the wanted asymptotic expansion tends to have

J. Franchi (✉)
IRMA, Université de Strasbourg, Strasbourg, France
e-mail: jacques.franchi@math.unistra.fr

© Springer International Publishing Switzerland 2014   71
C. Donati-Martin et al. (eds.), *Séminaire de Probabilités XLVI*, Lecture Notes
in Mathematics 2123, DOI 10.1007/978-3-319-11970-0_4

the following Gaussian-like form:

$$p_s(x, y) = s^{-d/2} \exp\left(- d(x, y)^2/(2s)\right)\left(\sum_{\ell=0}^{n} \gamma_\ell(x, y)\, s^\ell + \mathscr{O}(s^{n+1})\right) \qquad (1)$$

for any $n \in \mathbb{N}^*$, with smooth γ_ℓ's and $\gamma_0 > 0$, provided (x, y) does not belong to the cut-locus (and uniformly within any compact set which does not intersect the cut-locus). See in particular [2, théorème 3.1]. Note that the condition of remaining outside the cut-locus is here necessary, as showed in particular by [3].

The methods used to get this or a similar result have been of different nature. In [2], G. Ben Arous proceeds by expanding the flow associated to the diffusion (in this direction, see also [6]) and using a Laplace method applied to the Fourier transform of x_s, then inverted by means of Malliavin's calculus.

The strictly hypoelliptic case, i.e., when only the weak Hörmander condition (requiring the use of the drift vector field V_0 to recover the full tangent space) is fulfilled, remains much more problematic, and then rarely addressed. There is a priori no longer any reason that in such case the asymptotic behavior of $p_s(x, y)$ remains of the Gaussian-like type (1), all the less as a natural candidate for replacing the sub-Riemannian distance $d(x, y)$ is missing. Indeed this already fails for the mere (however Gaussian) Langevin process $\left(\beta_s, \int_0^s \beta_\tau d\tau\right)$, where β stands for a standard scalar Brownian motion: the missing distance is replaced by a time-dependent distance which presents some degeneracy in one direction, namely $\frac{6}{s^3}\left|(x - y) - \frac{s}{2}(\dot{x} - \dot{y})\right|^2 + \frac{1}{2s}|\dot{x} - \dot{y}|^2$, see (11), (13) below. See also [9] for a more involved (non-curved, strictly hypoelliptic, perturbed) case where Langevin-like estimates hold (without precise asymptotics), roughly having the following Li-Yau-like form:

$$C^{-1}\, s^{-N}\, e^{-C\, d_s(x_s, y)^2} \le p_s(x, y) \le C\, s^{-N}\, e^{-C^{-1}\, d_s(x_s, y)^2}, \qquad \text{for } 0 < s < s_0. \tag{2}$$

Note that in such expression x_s stands classically enough for the transport of the initial condition x by the deterministic differential system associated to the first order vector field, and that the notation d_s emphasizes the multi-scale normalization of the components (as in the previous Gaussian example).

In this article, an interesting case of rather natural hypoelliptic diffusion is considered first: that of a relativistic diffusion, first constructed in Minkowski's space (see [10]), which makes sense on a generic smooth Lorentzian manifold, see [12–14]. In the simplest case of Minkowski's space, it consists in the pair $(\xi_s, \dot{\xi}_s) \in \mathbb{R}^{1,d} \times \mathbb{H}^d$ (parametrized by its proper time s, and analogous to a Langevin process), where the velocity $(\dot{\xi}_s)$ is a hyperbolic Brownian motion. In the general case this Dudley diffusion can be rolled without slipping from a reference tangent space to the Lorentzian manifold, see [12]. Note that even in the Minkowski space, there is a curvature constraint to be taken into account, namely that of the mass shell $\mathbb{H}^d$, at the heart of this framework. Moreover the relativistic diffusion is never sub-elliptic, but only hypoelliptic, and a priori a Gaussian-like asymptotic expansion as (1) does not even make sense, since there is no longer any natural candidate to replace the sub-Riemannian distance $d(x, y)$. See however [4], where some non-trivial information is extracted about the relativistic diffusion, by considering the sub-Riemannian distance generated by all fields $V_0, V_1, \ldots, V_k$ (i.e., not only $V_1, \ldots, V_k$). Talking of this, an important feature of the strictly hypoelliptic case, which is fulfilled in the relativistic framework, is when the graded geometry generated by the successive brackets of a given weight is (at least locally) constant (see [16], and also [17]), yielding homogeneity in the afore-mentioned time-dependent distance.

To proceed, we shall compute a Fourier-Laplace transform, which seems to be the only way of getting any quantitative access to the density kernel of the relativistic diffusion [2, relies already on the Fourier transform, but then the method followed by G. Ben Arous is based on a stochastic variation about a minimal geodesic, which does not exist here, and on the local strict convexity of the energy functional (due to the sub-ellipticity), which does not hold here]. Because of the singularity in the most natural polar coordinates, we first choose alternative, less intuitive but smooth coordinates, and using them, partially expand the relativistic diffusion to project it on the second Wiener chaos, thereby exhibiting a simplified "tangent process". For this simplified process the Fourier-Laplace transform is exactly computable. Then analyzing its inverse Fourier transform very carefully, using a saddle-point method and restricting to the case of a dominant normalized Gaussian contribution, allows to derive an off-diagonal asymptotic equivalent for the density q_s of this tangent process, as time s goes to zero (see Theorem 6.4 below). As in the Langevin or in the more sophisticated case of [9], the exponential term is given by a time-dependent distance, namely the same as the afore-mentioned Langevin one, the strictly second chaos coordinate appearing only in the off-exponent term, as a perturbative contribution. The initial analogous question about the relativistic diffusion remains open, as the degree of contact between both considered processes (the effective computation of the Fourier-Laplace transform being bounded to the second Wiener chaos) seems so far too weak, to allow to deduce a former asymptotic behaviour from the second one. The non-appearance of the non-Gaussian coordinates in the found asymptotic exponent lets however think that this could remain so for all higher order chaos terms of the Taylor expansion, at least as long as a perturbative regime is considered. Thence a tempting guess, resulting from both the sub-elliptic case (1) as solved by [2], the modified Li-Yau-like estimates (2) obtained in [9] and the present work, whose main result is Theorem 6.4 below, is that an expansion having the following form could (maybe generally, under consistency of the Lie graded geometry) hold:

$$p_s(x, y) = s^{-N} \exp\left(-d_s(x, y)^2\right)\left(\sum_{\ell=0}^{n} \gamma_\ell(x, y)\, s^\ell + \mathcal{O}(s^{n+1})\right).$$

Note however that such expansion and the methods used till now (in particular [2,9] and the present one), are of a perturbative nature (with respect to a minimizing geodesic or to a Gaussian contribution); this could be the limit of its validity.

In this first attempt we restrict to the simplest case of the five-dimensional Minkowski space $\mathbb{R}^{1,2} \times \mathbb{H}^2$ and then to its five-dimensional second chaos tangent process. The case of the generic Minkowski space $\mathbb{R}^{1,d} \times \mathbb{H}^d$ is actually very analogous, but would mainly bring notational difficulties without modifying the method. We hope that this particular toy example will allow to understand better what can happen and could be undertaken, concerning small time asymptotics of the relativistic diffusion itself, and then maybe in some more generic strictly hypoelliptic framework. However the method used here could be limited to results of a perturbative type, as explained above and in Sect. 6.3.

The content is organized as follows.

In Sect. 2 are mainly described the setting and the smooth parametrization used then.

In Sect. 3 the simplified "tangent process" (Y_s) to the relativistic diffusion (X_s) is exhibited. It is the second chaos projection of (X_s), which appears in the second order Taylor expansion of (X_s).

In Sect. 4 the Fourier-Laplace transform of the tangent process (Y_s) is computed.

In Sect. 5 a closed integral expression for the density q_s of (Y_s) is presented.

Section 6 contains the main result: Theorem 6.4, which yields a precise asymptotics for the density q_s as $s \to 0$, in some specific off-diagonal regimes, where the non-Gaussian (second chaos) contribution is small with respect to the main (Gaussian) one. It is indeed shown that the small time contribution to the density of such a nearly negligible (normalized) non-Gaussian component can be analyzed in a perturbative way, and modifies the basic Gaussian density by a off-exponent correction term.

Section 7 contains three rather technical proofs, which have been postponed till there to lighten the reading.

I wish to thank the anonymous referee for his careful reading and a judicious comment.

2   A Smooth Parametrization of $\mathbb{R}^{1,2} \times \mathbb{H}^2 \equiv T_+^1 \mathbb{R}^{1,2}$

The Dudley relativistic diffusion $X_s = (\dot{\xi}_s, \xi_s)$ (see [10]) lives in the future-directed unit tangent bundle $T_+^1 \mathbb{R}^{1,2} \equiv \mathbb{R}^{1,2} \times \mathbb{H}^2$ to the Minkowski space $\left(\mathbb{R}^{1,2}, \langle \cdot, \cdot \rangle\right)$ (or alternatively, in its frame bundle, isomorphic to the Poincaré isometry group $\mathscr{P}^3 = \mathrm{PSO}(1,2) \propto \mathbb{R}^{1,2}$). We classically identify the hyperbolic plane $\mathbb{H}^2 \subset \mathbb{R}^{1,2}$ with the upper sheet of the hyperboloid having equation $|\dot{\xi}^0|^2 - |\dot{\xi}^1|^2 - |\dot{\xi}^2|^2 \equiv \langle \dot{\xi}, \dot{\xi} \rangle = 1$ within $\mathbb{R}^{1,2}$ (endowed with its canonical basis (e_0, e_1, e_2)).

The velocity sub-diffusion $(\dot{\xi}_s)$ is a hyperbolic Brownian motion, and we merely have $d\xi_s = \dot{\xi}_s \, ds$. The parameter s is precisely the physical proper time.

We shall use the following smooth coordinates $(\lambda, \mu, x, y, z) \in \mathbb{R}^5$ on $T_+^1 \mathbb{R}^{1,2}$:

$$\dot{\xi}^0 = \operatorname{ch}\lambda \operatorname{ch}\mu \; ; \quad \dot{\xi}^1 = \operatorname{ch}\lambda \operatorname{sh}\mu \; ; \quad \dot{\xi}^2 = \operatorname{sh}\lambda \; ; \quad \xi^0 = x \; ; \quad \xi^1 = y \; ; \quad \xi^2 = z. \tag{3}$$

In these coordinates the Dudley diffusion $X_s \equiv (\lambda_s, \mu_s, x_s, y_s, z_s) \in \mathbb{R}^5$ satisfies the following system of stochastic differential equations (for independent real Brownian motions w, β):

$$d\lambda_s = \sigma \, dw_s + \tfrac{\sigma^2}{2} \operatorname{th}\lambda_s \, ds \; ; \quad d\mu_s = \sigma \, \frac{d\beta_s}{\operatorname{ch}\lambda_s} \; ; \tag{4}$$

$$dx_s = \operatorname{ch}\lambda_s \operatorname{ch}\mu_s \, ds \; ; \quad dy_s = \operatorname{ch}\lambda_s \operatorname{sh}\mu_s \, ds \; ; \quad dz_s = \operatorname{sh}\lambda_s \, ds. \tag{5}$$

The infinitesimal generator $\mathscr{L}$ of (X_s) reads in these coordinates:

$$\mathscr{L} = \frac{\sigma^2}{2}\left[\frac{\partial^2}{\partial\lambda^2} + \operatorname{th}\lambda\,\frac{\partial}{\partial\lambda} + (\operatorname{ch}\lambda)^{-2}\frac{\partial^2}{\partial\mu^2}\right] + \operatorname{ch}\lambda \operatorname{ch}\mu\,\frac{\partial}{\partial x} + \operatorname{ch}\lambda \operatorname{sh}\mu\,\frac{\partial}{\partial y} + \operatorname{sh}\lambda\,\frac{\partial}{\partial z}. \tag{6}$$

Consider the following smooth vector fields on $\mathbb{R}^5$:

$$V_0' := \frac{\partial}{\partial s} + \frac{\sigma^2}{2}\operatorname{th}\lambda\,\frac{\partial}{\partial\lambda} + \operatorname{ch}\lambda \operatorname{ch}\mu\,\frac{\partial}{\partial x} + \operatorname{ch}\lambda \operatorname{sh}\mu\,\frac{\partial}{\partial y} + \operatorname{sh}\lambda\,\frac{\partial}{\partial z},$$

$$V_1 := \sigma\,\frac{\partial}{\partial\lambda}, \; V_2 := \frac{\sigma}{\operatorname{ch}\lambda}\,\frac{\partial}{\partial\mu}, \quad \text{and also} \quad V_0 := V_0' - \frac{\partial}{\partial s}.$$

We have then

$$\frac{\partial}{\partial s} + \mathscr{L} = \tfrac{1}{2}\left(V_1^2 + V_2^2\right) + V_0',$$

$$[V_1, V_0'] = \frac{\sigma^3}{2\operatorname{ch}^2\lambda}\,\frac{\partial}{\partial\lambda} + \sigma \operatorname{sh}\lambda \operatorname{ch}\mu\,\frac{\partial}{\partial x} + \sigma \operatorname{sh}\lambda \operatorname{sh}\mu\,\frac{\partial}{\partial y} + \sigma \operatorname{ch}\lambda\,\frac{\partial}{\partial z},$$

$$[V_2, V_0'] = \frac{\sigma^3 \operatorname{th}^2\lambda}{2\operatorname{ch}\lambda}\,\frac{\partial}{\partial\lambda} + \sigma \operatorname{sh}\mu\,\frac{\partial}{\partial x} + \sigma \operatorname{ch}\mu\,\frac{\partial}{\partial y},$$

$$\big[V_2, [V_2, V_0']\big] = \frac{\sigma^2}{\operatorname{ch}\lambda}\Big(\operatorname{ch}\mu\,\frac{\partial}{\partial x} + \operatorname{sh}\mu\,\frac{\partial}{\partial y}\Big),$$

and then $\Big(V_0', V_1, V_2, [V_1, V_0'], [V_2, V_0'], \big[V_2, [V_2, V_0']\big]\Big)$ has full rank 6 at any point, so that the weak Hörmander condition holds.

Hence by hypoellipticity, $\mathscr{L}$ admits a smooth heat kernel $p_s(X_0; X)$ ($s \in \mathbb{R}_+$, $X_0, X \in \mathbb{R}^5$), with respect to the Liouville measure L, which reads $L(dX) = \mathrm{ch}\,\lambda\,d\lambda\,d\mu\,dx\,dy\,dz$:

$$\mathbb{E}_{X_0}\big[F(X_s)\big] = \int_{T^1_+ \mathbb{R}^{1,2}} p_s(X_0; X)\,F(X)\,L(dX)$$

$$= \int_{\mathbb{R}^5} p_s(X_0; \lambda, \mu, x, y, z)\,F(\lambda, \mu, x, y, z)\,\mathrm{ch}\,\lambda\,d\lambda\,d\mu\,dx\,dy\,dz.$$

An open question is to estimate $p_s(X_0; X)$, for small proper times s.

Up to apply some element of the Poincaré group $\mathscr{P}^3$, we can restrict to $X_0 = (e_0, 0) \equiv (0, 0, 0, 0, 0)$. Thus we have to deal with $p_s(0, X) \equiv p_s(X)$, for $X \equiv (\lambda, \mu, x, y, z) \in \mathbb{R}^5$.

The underlying unperturbed (deterministic) process $X^0 \equiv (\lambda^0, \mu^0, x^0, y^0, z^0) \in \mathbb{R}^5$ solves:

$$d\lambda^0_s = \tfrac{\sigma^2}{2}\,\mathrm{th}\,\lambda^0_s\,ds; \quad \mu^0_s = 0; \quad dx^0_s = \mathrm{ch}\,\lambda^0_s\,ds; \quad y^0_s = 0; \quad dz^0_s = \mathrm{sh}\,\lambda^0_s\,ds,$$

and then is merely given by the geodesic $X^0_s \equiv (0, 0, s, 0, 0)$ for any proper time s.

Up to change the speed of the canonical Brownian motion (w, β), by considering $(w_{\sigma^2 s}, \beta_{\sigma^2 s})$ instead of (w_s, β_s), we can absorb the speed parameter σ, and then suppose that $\sigma = 1$.

3   A Process Tangent to the Relativistic Diffusion (X_s)

The main Theorem 2.1 in [6] could apply here (beware however that V_0 is unbounded), yielding a full general Taylor expansion for the diffusion (X_s), in terms of the above vector fields V_1, V_2, V_0, their successive brackets, and of the iterated Stratonovich integrals with respect to (w, β). Indeed, Eqs. (4), (5) read equivalently:

$$dX_s = V_1(X_s)\,dw_s + V_2(X_s)\,d\beta_s + V_0(X_s)\,ds.$$

In [6], the successive remainders corresponding to the truncated Taylor expansion are controlled in probability. In this spirit and also almost surely, the process X_s is approached as follows.

Lemma 3.1   *(i)   For any $\varepsilon > 0$, almost surely as proper time $s \searrow 0$ we have:*

$$X_s = \left(w_s + \tfrac{1}{2}\int_0^s w_\tau\,d\tau,\ \beta_s + o(s^{3/2-\varepsilon}),\ s + \tfrac{1}{2}\int_0^s \big[\beta_\tau^2 + w_\tau^2\big]d\tau,\ \int_0^s \beta_\tau\,d\tau,\ \int_0^s w_\tau\,d\tau\right)$$

$$+ o(s^{5/2-\varepsilon}).$$

(ii) Setting $\quad R'_s := (x_s, y_s, z_s) - \left(s + \frac{1}{2}\int_0^s [\beta_\tau^2 + w_\tau^2]d\tau, \int_0^s \beta_\tau \, d\tau, \int_0^s w_\tau \, d\tau\right)$

and $R_s := (\lambda_s, \mu_s) - (w_s, \beta_s)$, *there exist* $c, \kappa > 0$ *such that for any* $R > c$
we have:

$$\lim_{s \searrow 0} \mathbb{P}\left[\sup_{0 \le t \le s} \|R_t\| \ge R\, s^{3/2}\right] \le e^{-R^\kappa/c} \text{ and } \lim_{s \searrow 0} \mathbb{P}\left[\sup_{0 \le t \le s} \|R'_t\| \ge R\, s^{5/2}\right] \le e^{-R^\kappa/c}.$$

The proof is postponed to Sect. 7.

Remark 3.2 More precisely, concerning the martingale (μ_s) we have

$$\mu_s = \int_0^s \frac{d\beta_\tau}{\mathrm{ch}\,\lambda_\tau} = \int_0^s \left(1 - \tfrac{1}{2}w_\tau^2 + o(\tau^{2-\varepsilon})\right)d\beta_\tau = \beta_s - \tfrac{1}{2}\int_0^s w_\tau^2 \, d\beta_\tau + o(s^{5/2-\varepsilon}).$$

But the method used then does not work with the non-quadratic martingale

$$\int_0^s w_\tau^2 \, d\beta_\tau = w_s^2 \beta_s - 2\int_0^s \beta_\tau w_\tau \, dw_\tau - \int_0^s \beta_\tau \, d\tau = o(s^{3/2-\varepsilon}).$$

As a consequence, we shall use a perturbation method, approaching (for small proper time s) the relativistic diffusion X_s by means of the $\mathbb{R}^5$-valued "tangent process":

$$Y_s := \left(w_s, \beta_s, \tfrac{1}{2}\int_0^s [\beta_\tau^2 + w_\tau^2]d\tau, \int_0^s \beta_\tau \, d\tau, \int_0^s w_\tau \, d\tau\right) =: \left(w_s, \beta_s, A_s, \zeta_s, \bar{z}_s\right) \tag{7}$$

which is not Gaussian, but has its third coordinate A_s in the second Wiener chaos. This actually yields the orthogonal projection of the process (X_s) onto the second Wiener chaos.

Remark 3.3 The fact that the second chaos term is needed in the approximation (without it, the tangent process would clearly not admit any density) makes a significant difference with the situation exhaustively investigated in [9], where the approaching process is Gaussian. This can no longer be the case in the present setting, though both settings share the feature of being strictly hypoelliptic. A difference between both is the curvature, at the heart of the relativistic realm (even in the present Minkowski-Dudley flat case), due to the mass shell constraint on velocities.

Note that for any fixed proper time $s > 0$ we have:

$$Y_s \overset{law}{\equiv} \left(\sqrt{s}\, w_1, \sqrt{s}\, \beta_1, \tfrac{s^2}{2}\int_0^1 [\beta_\tau^2 + w_\tau^2]d\tau, \sqrt{s^3}\int_0^1 \beta_\tau \, d\tau, \sqrt{s^3}\int_0^1 w_\tau \, d\tau\right). \tag{8}$$

Denote by $q_s = q_s(w, \beta, x, \zeta, z)$ the density of Y_s with respect to the Lebesgue measure $\Lambda(dY) = dw\, d\beta\, dx\, d\zeta\, dz$ (i.e., *not* the Liouville measure L) on $\mathbb{R}^2 \times \mathbb{R}_+^* \times \mathbb{R}^2$.

By the scaling property (8), it must satisfy:

$$q_s(w, \beta, x, \zeta, z) = \frac{1}{s^6}\, q_1\left(\frac{w}{\sqrt{s}}, \frac{\beta}{\sqrt{s}}, \frac{x}{s^2}, \frac{\zeta}{\sqrt{s^3}}, \frac{z}{\sqrt{s^3}}\right), \tag{9}$$

and otherwise : $\quad q_s(w, \beta, x, \zeta, z) = q_s(-w, \beta, x, \zeta, -z) = q_s(w, -\beta, x, -\zeta, z).$

4 Fourier-Laplace Transform of the Tangent Process

4.1 Fourier-Laplace Transform of the Simplified Process Z_s

We need information on the density at time s of the tangent process (Y_s) given in (7). By the independence of w, β, it will be enough to consider the density of

$$Z_s := \left(w_s,\, \int_0^s w_\tau\, d\tau,\, \int_0^s w_\tau^2\, d\tau\right) \overset{law}{\equiv} \left(\sqrt{s}\, w_1,\, \sqrt{s^3}\int_0^1 w_\tau\, d\tau,\, s^2 \int_0^1 w_\tau^2\, d\tau\right). \tag{10}$$

Note that this simplified tangent process Z_s does not have any component beyond the second chaos. Because of the scaling property (10) of Z_s, its density $q_s^0(w, z, x)$ satisfies

$$q_s^0(w, z, x) = \frac{1}{s^4}\, q_1^0\left(\frac{w}{\sqrt{s}}, \frac{z}{\sqrt{s^3}}, \frac{x}{s^2}\right) = q_s^0(-w, -z, x).$$

Of course, the Langevin process $\left(w_s,\, \int_0^s w_\tau\, d\tau\right)$ is Gaussian with covariance $K_s^0 = \begin{pmatrix} s & s^2/2 \\ s^2/2 & s^3/3 \end{pmatrix}$, so that it has the well-known density

$$(w, z) \longmapsto \frac{\sqrt{3}}{\pi\, s^2}\, e^{-(6z^2 - 6s\, zw + 2s^2 w^2)/s^3} = \frac{\sqrt{3}}{\pi\, s^2}\, \exp\left[-\frac{6}{s^3}\left(z - \frac{s}{2}w\right)^2 - \frac{w^2}{2s}\right]. \tag{11}$$

(In particular, the expected value of $\int_0^s w_\tau\, d\tau$, conditionally on $w_s = w$, is $sw/2$.)

The law of the variable Z_s is not at all that simple, but it is known (see [8, 19]) that its Fourier-Laplace transform is computable. The following lemma is proved in Sect. 7.

Lemma 4.1.1 *The law of the variable Z_s of (10) is given by: for any $s \geq 0$ and real r, c, b,*

$$
\mathbb{E}_0\left[\exp\left(\sqrt{-1}\left[r\,w_s + c\int_0^s w_\tau\, d\tau \right] - \frac{b^2}{2}\int_0^s w_\tau^2\, d\tau \right) \right]
$$
$$
= \frac{1}{\sqrt{\mathrm{ch}(bs)}}\, \exp\left[-\frac{\mathrm{th}(bs)}{2\,b}\, r^2 - 2\,\frac{\mathrm{sh}^2(bs/2)}{b^2\,\mathrm{ch}(bs)}\, rc - \frac{bs - \mathrm{th}(bs)}{2\,b^3}\, c^2 \right].
$$

For $b = 0$ we recover the Fourier transform of (11), namely $e^{-\left(r^2 + s\,rc + \frac{s^2}{3}c^2\right)s/2}$.

Proposition 4.1.2 *The x-Laplace transform of the variable Z_1 of (10) is given by: for any real w, z, b,*

$$
\int_0^\infty e^{-\frac{b^2}{2}x}\, q_1^0(w, z, x)\, dx = \frac{b^2 \exp\left[-\dfrac{\left[b\,z - \mathrm{th}(b/2)\,w\right]^2 + \coth b \left[b - 2\,\mathrm{th}(b/2)\right]w^2}{2\left[1 - (2/b)\,\mathrm{th}(b/2)\right]} \right]}{2\pi\sqrt{[b - 2\,\mathrm{th}(b/2)]\,\mathrm{sh}\,b}}
$$

$$
= \frac{b^2}{2\pi\sqrt{[b - 2\,\mathrm{th}(b/2)]\,\mathrm{sh}\,b}} \times \exp\left[\frac{b^2}{8} \times \frac{(w - 2z)^2}{1 - \frac{b}{2}\coth(\frac{b}{2})} - \frac{b^2}{2}z^2 - \frac{b}{4}\coth(\tfrac{b}{2})\,w^2 \right].
$$

This is of course consistent with (11), via $b \to 0$; and integrating with respect to $dw\,dz$, we recover $\displaystyle\int_0^\infty e^{-\frac{b^2}{2}x}\left[\iint q_1^0(w, z, x)\, dw\,dz \right] dx = (\mathrm{ch}\,b)^{-1/2}$, as it must be.

Proof We invert the Fourier transform in Lemma 4.1.1 by Plancherel's Formula:

$$
\int_0^\infty e^{-\frac{b^2}{2}x}\, q_1^0(w, z, x)\, dx
$$
$$
= \frac{1}{4\pi^2\sqrt{\mathrm{ch}\,b}} \int_{\mathbb{R}^2} e^{-\sqrt{-1}[wr+zc]} \exp\left[-\frac{\mathrm{th}\,b}{2\,b}\, r^2 - 2\,\frac{\mathrm{sh}^2(b/2)}{b^2\,\mathrm{ch}\,b}\, rc - \frac{b - \mathrm{th}\,b}{2\,b^3}\, c^2 \right] dr\,dc
$$
$$
= \frac{1}{4\pi^2\sqrt{\mathrm{ch}\,b}} \int_{\mathbb{R}^2} e^{-\sqrt{-1}[wr+zc]} \exp\left[-\frac{\mathrm{th}\,b}{2\,b}\left(r + \frac{\mathrm{th}(b/2)}{b}c \right)^2 - \frac{b - 2\,\mathrm{th}(b/2)}{2\,b^3}\, c^2 \right] dr\,dc
$$
$$
= \frac{1}{4\pi^2\sqrt{\mathrm{ch}\,b}} \int_{\mathbb{R}^2} e^{-\sqrt{-1}\left[wr + \left(z - \frac{\mathrm{th}(b/2)}{b}w\right)c\right]} \exp\left[-\frac{\mathrm{th}\,b}{2\,b}\, r^2 - \frac{b - 2\,\mathrm{th}(b/2)}{2\,b^3}\, c^2 \right] dr\,dc
$$
$$
= \sqrt{\frac{b}{\mathrm{sh}\,b}} \times \frac{e^{-\frac{b\,w^2}{2\,\mathrm{th}\,b}}}{2\pi} \int_{\mathbb{R}} e^{-\sqrt{-1}\left(z - \frac{\mathrm{th}(b/2)}{b}w\right)c} \exp\left[-\frac{b - 2\,\mathrm{th}(b/2)}{2\,b^3}\, c^2 \right] \frac{dc}{\sqrt{2\pi}}
$$

$$= \sqrt{\frac{b}{\operatorname{sh} b}} \times \frac{e^{-\frac{bw^2}{2\operatorname{th} b}}}{2\pi} \times \sqrt{\frac{b^3}{b - 2\operatorname{th}(b/2)}} \times \exp\left[-\frac{b^3}{2[b - 2\operatorname{th}(b/2)]}\left(z - \frac{\operatorname{th}(b/2)}{b}w\right)^2\right]$$

$$= \frac{b^2}{2\pi\sqrt{[b - 2\operatorname{th}(b/2)]\operatorname{sh} b}} \exp\left[-\frac{[bz - \operatorname{th}(b/2)w]^2 + \coth b[b - 2\operatorname{th}(b/2)]w^2}{2[1 - (2/b)\operatorname{th}(b/2)]}\right]. \quad \diamond$$

4.2 Fourier-Laplace Transform of the Tangent Process Y_s

We use Lemma 4.1.1 to express this Fourier-Laplace transform.

Lemma 4.2.1 *The law of the variable Y_s of (7) is given by : for any $s \geq 0$ and real r, ϱ, b, γ, c,*

$$\mathbb{E}_0\left[\exp\left(\sqrt{-1}\left[r\,w_s + \varrho\,\beta_s + \gamma\,\zeta_s + c\,\bar{z}_s\right] - b^2 A_s\right)\right]$$

$$= \frac{1}{\operatorname{ch}(bs)} \exp\left[-\frac{\operatorname{th}(bs)}{2b}(r^2 + \varrho^2) - 2\frac{\operatorname{sh}^2(bs/2)}{b^2\operatorname{ch}(bs)}(r\,c + \varrho\,\gamma)\right.$$

$$\left. -\frac{bs - \operatorname{th}(bs)}{2b^3}(c^2 + \gamma^2)\right].$$

In particular, the law of A_s is given by $\mathbb{E}_0\left[\exp(-b^2 A_s)\right] = 1/\operatorname{ch}(bs)$.

Proof This follows directly from Lemma 4.1.1, by independence of w and β. $\diamond$

Proposition 4.2.2 *The x-Laplace transform of the variable Y_1 of (7) is given by : for any real w, β, z, ζ, b,*

$$\int_0^\infty e^{-b^2 x}\,q_1(w, \beta, x, \zeta, z)\,dx =$$

$$\frac{b^4\,\exp\left[\frac{b^2}{8} \times \frac{(w - 2z)^2 + (\beta - 2\zeta)^2}{1 - \frac{b}{2}\coth(\frac{b}{2})} - \frac{b^2}{2}(z^2 + \zeta^2) - \frac{b}{4}\coth(\frac{b}{2})(w^2 + \beta^2)\right]}{8\pi^2\left[b\operatorname{ch}(\frac{b}{2}) - 2\operatorname{sh}(\frac{b}{2})\right]\operatorname{sh}(\frac{b}{2})} =: \Psi_{w,\beta,\zeta,z}(b).$$

$$(12)$$

In particular, $\Psi_{w,\beta,\zeta,z}(0) = \frac{3}{\pi^2}\exp\left[-\frac{w^2 + \beta^2}{2} - 6(z - w/2)^2 - 6(\zeta - \beta/2)^2\right]$ is the marginal density of $(w_1, \beta_1, \zeta_1, \bar{z}_1)$.

Proof As Lemma 4.2.1 follows from Lemma 4.1.1, this follows merely from Proposition 4.1.2 by independence of w and β. Indeed, for any test functions f, g

on $\mathbb{R}^2$ we have:

$$\int_{\mathbb{R}^4} f(w,z)\, g(\beta,\zeta) \left[\int_0^\infty e^{-b^2 x}\, q_1(w,\beta,x,\zeta,z)\, dx \right] dw\, dz\, d\beta\, d\zeta$$

$$= \mathbb{E}\left[f(w_1,\bar{z}_1)\, g(\beta_1,\zeta_1)\, e^{-\frac{b^2}{2}\left[\int_0^1 w_\tau^2\, d\tau + \int_0^1 \beta_\tau^2\, d\tau \right]} \right]$$

$$= \mathbb{E}\left[f(w_1,\bar{z}_1)\, e^{-\frac{b^2}{2} \int_0^1 w_\tau^2\, d\tau} \right] \times \mathbb{E}\left[g(\beta_1,\zeta_1)\, e^{-\frac{b^2}{2} \int_0^1 \beta_\tau^2\, d\tau} \right]$$

$$= \int_{\mathbb{R}^2} f(w,z) \int_0^\infty e^{-\frac{b^2}{2} x}\, q_1^0(w,z,x)\, dx\, dw\, dz \int_{\mathbb{R}^2} g(\beta,\zeta) \int_0^\infty e^{-\frac{b^2}{2} x}\, q_1^0(\beta,\zeta,x)\, dx\, d\beta\, d\zeta$$

$$= \int_{\mathbb{R}^4} f(w,z)\, g(\beta,\zeta) \left[\int_0^\infty e^{-\frac{b^2}{2} x}\, q_1^0(w,z,x)\, dx \int_0^\infty e^{-\frac{b^2}{2} x}\, q_1^0(\beta,\zeta,x)\, dx \right] dw\, dz\, d\beta\, d\zeta,$$

and the claim follows directly from Proposition 4.1.2. $\diamond$

Lemma 4.2.3 *All solutions* $z \in \mathbb{C}^*$ *of the equation* $z = \mathrm{th}\, z$ *belong to the imaginary axis, and form a sequence* $\mathscr{R} = \{\pm\sqrt{-1}\, y_n \mid n \in \mathbb{N}\}$, *with* $\frac{17\pi}{12} < y_0 < y_1 < \dots < y_n \nearrow \infty$.

Proof Writing $z = x + \sqrt{-1}\, y \in \mathbb{C}^*$, we have $z = \mathrm{th}\, z \Leftrightarrow e^{2z} = \frac{1+z}{1-z}$ and then equivalently

$$\cos(2y) = \frac{1-x^2-y^2}{(1-x)^2+y^2}\, e^{-2x} \quad \text{and} \quad \sin(2y) = \frac{2y}{(1-x)^2+y^2}\, e^{-2x},$$

whence

$$e^{4x}\left((1-x)^2 + y^2\right)^2 = 4y^2 + (1-x^2-y^2)^2.$$

The latter is equivalent either to $x = 0$, or to $y^2 = -(1-x)^2$ (which is excluded), or to $y^2 = \frac{(1+x)^2 - e^{4x}(1-x)^2}{e^{4x}-1}$. Then using this last value of y^2, by the above we must also have $\sin(2y) = \frac{y}{x}\,\mathrm{sh}\,(2x)$, which is impossible since for any $x, y \in \mathbb{R}^*$ we have $\frac{\sin(2y)}{y} < 2 < \frac{\mathrm{sh}(2x)}{x}$, and clearly z cannot be real. Hence we are left with $z = \pm\sqrt{-1}\, y$, with $y > 0$ and then $y = \mathrm{tg}\, y$. The claim follows, with moreover $\left((n+3/2)\pi - y_n\right) \searrow 0$, and $\frac{3\pi}{2} > y_0 > \frac{17\pi}{12}$ since $\mathrm{tg}\,\frac{17\pi}{12} = \mathrm{cotg}\,\frac{\pi}{12} = 2 + \sqrt{3} < \frac{17\pi}{12}$. $\diamond$

The Laplace transform $\Psi_{w,\beta,\zeta,z}(b)$ in Proposition 4.2.2 is a meromorphic function of $b \in \mathbb{C}$, with singularities at the points of $\sqrt{-1}\, 2\pi\, \mathbb{Z}^*$ and at the non-null zeros of $\left[b - 2\,\mathrm{th}(b/2)\right]$, that is to say at the points of $2\mathscr{R}$ (according to Lemma 4.2.3).

4.3   Complement : Density $\alpha_s(x)$ of the Variable A_s

Denote by $\alpha_s = \alpha_s(x)$ the density of the variable A_s, so that for any $s > 0$ we have

$$\alpha_s(x) = \int q_s(w, \beta, x, \zeta, z) \, dw \, d\beta \, d\zeta \, dz = \frac{1}{s^2} \alpha_1\left(\frac{x}{s^2}\right), \text{ by (9)}.$$

Lemma 4.3.1 *The density α_1 is smooth and bounded (with bounded derivatives), and we have*

$$\alpha_1(x) = \frac{4}{\pi} \int_0^\infty \frac{\cos(2x \, y^2 - y)}{\mathrm{sh}^2 y + \cos^2 y} \, y \, \mathrm{sh} \, y \, dy + \frac{4}{\pi} \int_0^\infty \frac{\cos y \, \cos(2x \, y^2)}{\mathrm{sh}^2 y + \cos^2 y} \, e^{-y} \, y \, dy.$$

Proof According to Lemma 4.2.1, for any $\eta > 0$ we have : $\quad \alpha_1(x) =$

$$\frac{e^{\eta x}}{2\pi} \int_{-\infty}^\infty \frac{e^{\sqrt{-1} \, x \, y}}{\mathrm{ch}\left(\sqrt{\eta + \sqrt{-1} \, y}\right)} \, dy$$

$$= \frac{e^{\eta x}}{2\pi} \int_{-\infty}^\infty \frac{e^{\sqrt{-1} \, x \, y}}{\mathrm{ch}\left(\sqrt{\frac{\sqrt{y^2+\eta^2}+\eta}{2}} + \sqrt{-1} \, \mathrm{sgn}(y)\sqrt{\frac{\sqrt{y^2+\eta^2}-\eta}{2}}\right)} \, dy.$$

Expanding and letting $\eta \searrow 0$ we obtain :

$$\alpha_1(x) = \frac{4}{\pi} \int_0^\infty \frac{\mathrm{ch} \, y \, \cos y \, \cos(2x \, y^2) + \mathrm{sh} \, y \, \sin y \, \sin(2x \, y^2)}{\mathrm{sh}^2 y + \cos^2 y} \, y \, dy$$

$$= \frac{4}{\pi} \int_0^\infty \frac{\cos(2x \, y^2 - y)}{\mathrm{sh}^2 y + \cos^2 y} \, y \, \mathrm{sh} \, y \, dy + \frac{4}{\pi} \int_0^\infty \frac{\cos y \, \cos(2x \, y^2)}{\mathrm{sh}^2 y + \cos^2 y} \, e^{-y} \, y \, dy.$$

As a consequence, α_1 is smooth, and bounded by $\dfrac{4}{\pi} \displaystyle\int_0^\infty \dfrac{y \, \mathrm{ch} \, y \, dy}{\mathrm{sh}^2 y + \cos^2 y}$. $\quad \diamond$

Remark 4.3.2 We have $\mathrm{sh}^2 z + \cos^2 z = 0 \Leftrightarrow \mathrm{ch}^2 z = \sin^2 z \Leftrightarrow z \in (1 \pm \sqrt{-1})\dfrac{\pi}{4}(1 + 2\mathbb{Z})$,

$$\text{i.e., } |z| \in \frac{\pi(1 + 2\mathbb{N})}{2\sqrt{2}}, \quad \mathrm{Arg} \, z \equiv \frac{\pi}{4} \ modulo \ \frac{\pi}{2}.$$

Proof There is clearly no real solution, since $\mathrm{ch} \, x > 1 \geq \sin x$ for any $x \in \mathbb{R}^*$. Consider then $z = (\sqrt{-1} + \alpha) x$, with $x, \alpha \in \mathbb{R}$. Up to change z into $-z$, we only have to consider the equation $\mathrm{ch} \, z = \sin z$. Then

$$\mathrm{ch} \, z = \sin z \Longleftrightarrow \mathrm{ch}(\alpha x) \cos x = \mathrm{ch} \, x \, \sin(\alpha x) \text{ and } \mathrm{sh}(\alpha x) \sin x = \mathrm{sh} \, x \, \cos(\alpha x)$$

$$\Rightarrow \text{ch}^2(\alpha x)\cos^2 x + \text{sh}^2(\alpha x)\sin^2 x = \text{ch}^2 x \, \text{sh}^2 x$$

$$\Longleftrightarrow \left(\text{ch}^2 x - \cos^2 x\right)\left(\text{sh}^2(\alpha x) - \text{sh}^2 x\right) = 0.$$

As $x = 0$ is not a solution, the only possibility is $\text{sh}(\alpha x) = \pm \text{sh}\, x$, whence $\alpha = \pm 1$, and then $\cos x = \alpha \sin x$. This yields the claim. ◇

Proposition 4.3.3 *We have* $\quad \alpha_1(x) = 2\pi \sum_{n \in \mathbb{N}} (-1)^n \left(n + \tfrac{1}{2}\right) e^{-\left(n+\frac{1}{2}\right)^2 \pi^2 x}, \quad$ *so that*

$$\pi \, e^{-\pi^2 x/4}\left(1 - 3\,e^{-2\pi^2 x}\right) \le \alpha_1(x) \le \pi \, e^{-\pi^2 x/4} \text{ for any } x \ge \pi^{-2}, \text{ and then}$$

$$\alpha_1(x) = \pi \, e^{-\pi^2 x/4}\left(1 - \mathcal{O}(e^{-2\pi^2 x})\right) \text{ as } x \to \infty.$$

Moreover, α_1 *decreases on* $[3\pi^{-2}, \infty[$.

Proof This results from Lemma 4.2.1 and [5, Table 1 continued and Section 3.3, in particular Formula (3.11), with $C_1 \equiv 2A_1$, got merely by expanding $1/(\text{ch}\,t)$]. For any $\pi^2 x > 1$, this alternate series has decreasing generic term $\left(n + \tfrac{1}{2}\right) e^{-\left(n+\frac{1}{2}\right)^2 \pi^2 x}$, whence the estimate. Finally, the same holds for the series $\left(n + \tfrac{1}{2}\right)^3 e^{-\left(n+\frac{1}{2}\right)^2 \pi^2 x}$ yielding $\alpha_1'(x)$, thereby garanteing $\alpha_1'(x) < 0$, as soon as $\pi^2 x > 3$. ◇

Lemma 4.2.1 and Proposition 4.3.3 at once entail the following.

Corollary 4.3.4 $\lambda \longmapsto \int_0^\infty e^{\lambda x} \alpha_1(x)\, dx$ *defines an analytic function on* $\Re(\lambda) < \pi^2/4$, *equal to* $\frac{1}{\cos \sqrt{\lambda}}$ *for* $0 \le \lambda < \frac{\pi^2}{4}$. *Moreover, we have* $\alpha_s(x) = \frac{\pi}{s^2} e^{-\pi^2 x/(4s^2)}\left(1 - \mathcal{O}(e^{-2\pi^2 x/s^2})\right)$ *(as* $\frac{x}{s^2} \to \infty$*).*

Proof By Lemma 4.2.1 the set of $b \in \mathbb{C}$ such that $\int_0^\infty e^{-b^2 x} \alpha_1(x)\, dx = \frac{1}{\text{ch}\, b}$ contains $\mathbb{R}$. By Proposition 4.3.3 $b \mapsto \int_0^\infty e^{-b^2 x} \alpha_1(x)\, dx$ is analytic on $\{|\Im(b)| < \pi/2\}$, and so is $b \mapsto \frac{1}{\text{ch}\, b}$ too. Hence we have $\int_0^\infty e^{b^2 x} \alpha_1(x)\, dx = \frac{1}{\cos b}$ for $|\Re(b)| < \pi/2$. ◇

Remark 4.3.5 1) The law of $2\,A_1$ is that of $\inf\{\tau > 0 \,|\, |\beta_\tau| = 1\}$ and also that of $\left(\max\{|\beta_\tau| \,|\, 0 \le \tau \le 1\}\right)^{-2}$ (See [5], Table 2, which also exhibits two other random variables having the same law as A_1). 2) Note that for $\Re(\lambda) < \pi^2/4$

(letting $\sqrt{\cdot} > 0$ on $\mathbb{R}_+^*$):

$$\frac{1}{\cos\sqrt{\lambda}} = \int_0^\infty \int e^{\lambda x} q_1(w, \beta, x, \zeta, z) \, dw \, d\beta \, d\zeta \, dz \, dx$$
$$= \int \Psi_{w,\beta,\zeta,z}(\sqrt{-\lambda}) \, dw \, d\beta \, d\zeta \, dz.$$

5 Integral Expression of the Density $q_s(w, \beta, x, \zeta, z)$

Let

$$\tilde{q}_s \equiv \tilde{q}_s(w, \beta, \zeta, z) := \frac{3}{\pi^2 s^4} \exp\left[-\frac{3\left[(sw - 2z)^2 + (s\beta - 2\zeta)^2\right]}{2 s^3} - \frac{w^2 + \beta^2}{2 s} \right]$$
(13)

denote the marginal Gaussian density of $\tilde{Y}_s := \left(w_s, \beta_s, \int_0^s \beta_\tau \, d\tau, \int_0^s w_\tau \, d\tau \right)$ (according to (11) or Proposition 4.2.2)).

Notation Consider the function $\Phi(\lambda) \equiv \Phi_{w,\beta,\zeta,z}(\lambda) := \Psi_{w,\beta,\zeta,z}(\sqrt{-\lambda})$, derived from Proposition 4.2.2. We systematically use the usual determination of the complex square root, cutting $\mathbb{C}$ along the negative real semi-axis and letting $\sqrt{\cdot} > 0$ on $\mathbb{R}_+^*$. By expression (12), we have:

$$\Phi_{w,\beta,\zeta,z}(\lambda) = \frac{\lambda^2 \exp\left[\frac{\lambda}{8} \times \frac{(w-2z)^2+(\beta-2\zeta)^2}{\frac{\sqrt{\lambda}}{2}\cot g(\frac{\sqrt{\lambda}}{2})-1} + \frac{\lambda}{2}(z^2 + \zeta^2) - (w^2 + \beta^2)\frac{\sqrt{\lambda}}{4}\cot g(\frac{\sqrt{\lambda}}{2}) \right]}{8\pi^2 \left[2\sin(\frac{\sqrt{\lambda}}{2}) - \sqrt{\lambda}\cos(\frac{\sqrt{\lambda}}{2}) \right] \sin(\frac{\sqrt{\lambda}}{2})}.$$
(14)

Lemma 5.1 *The function* $\lambda \mapsto \Phi_{w,\beta,\zeta,z}(\lambda)$ *is analytic for* $\Re(\lambda) < 4\pi^2$.

Proof Note that the functions $\sin(\frac{\sqrt{\lambda}}{2})\big/\frac{\sqrt{\lambda}}{2}$ and $\cos(\frac{\sqrt{\lambda}}{2})$ are plainly analytically continued for any $\lambda \in \mathbb{C}$, and that by the expression (14), $\Phi(\lambda)$ is analytically continued at $\lambda = 0$, and analytic at any $\lambda \in \mathbb{C}^*$ such that $\sin(\frac{\sqrt{\lambda}}{2}) \neq 0$ and $\operatorname{tg}(\frac{\sqrt{\lambda}}{2}) \neq \frac{\sqrt{\lambda}}{2}$, hence, according to Lemma 4.2.3, at those $\lambda \in \mathbb{C}$ not belonging to the sequence $\{4\pi^2(n + 1)^2, 4 y_n^2 \mid n \in \mathbb{N}\} \subset [4\pi^2, \infty[$. This shows the analyticity for $\Re(\lambda) < 4\pi^2$. $\diamond$

The following is proved in Sect. 7.

Proposition 5.2 *For any* (w, β, ζ, z) *we have*

$$\Phi_{w,\beta,\zeta,z}(\lambda) = \int_0^\infty e^{\lambda x} q_1(w, \beta, x, \zeta, z) \, dx, \quad for \ \Re(\lambda) < 4\pi^2, \ and$$

$$e^{r x} q_1(w, \beta, x, \zeta, z) = \int_{-\infty}^{\infty} e^{-\sqrt{-1} t x} \Phi(r + \sqrt{-1} t) \frac{dt}{2\pi}, \quad for \ x > 0 \ and \ r < 4\pi^2.$$

(15)

By scaling and using (15), for any s, $x > 0$, $r < 4\pi^2$ and $(w, \beta, \zeta, z) \in \mathbb{R}^4$ we have:

$$q_s(w, \beta, x, \zeta, z) = \frac{1}{s^6} q_1\left(\frac{w}{\sqrt{s}}, \frac{\beta}{\sqrt{s}}, \frac{x}{s^2}, \frac{\zeta}{\sqrt{s^3}}, \frac{z}{\sqrt{s^3}}\right)$$

$$= \frac{e^{-r x/s^2}}{2\pi s^6} \int_{-\infty}^{\infty} e^{-\sqrt{-1} x y/s^2} \Phi_{\left(\frac{w}{\sqrt{s}}, \frac{\beta}{\sqrt{s}}, \frac{\zeta}{\sqrt{s^3}}, \frac{z}{\sqrt{s^3}}\right)}(r + \sqrt{-1} y) \, dy.$$

Taking merely $r = 0$, and setting $\quad \tilde{\Psi}_s := e^{-\sqrt{-1} x y/s^2} \Phi_{\left(\frac{w}{\sqrt{s}}, \frac{\beta}{\sqrt{s}}, \frac{\zeta}{\sqrt{s^3}}, \frac{z}{\sqrt{s^3}}\right)}(\sqrt{-1} y)$

for convenience, for any $s, x > 0$ and $(w, \beta, \zeta, z) \in \mathbb{R}^4$ we have:

$$q_s(w, \beta, x, \zeta, z) = \frac{1}{2\pi s^6} \int_{-\infty}^{\infty} \tilde{\Psi}_s(y) \, dy = \frac{1}{2\pi s^6} \int_{0}^{\infty} \left(\tilde{\Psi}_s(y) + \tilde{\Psi}_s(-y)\right) dy.$$

(16)

Now according to (14) we have:

$$\Phi_{\left(\frac{w}{\sqrt{s}}, \frac{\beta}{\sqrt{s}}, \frac{\zeta}{\sqrt{s^3}}, \frac{z}{\sqrt{s^3}}\right)}(\sqrt{-1} y) = \exp\left[\frac{B_s^2 \sqrt{-1} y s^{-3}}{1 - \frac{2}{\sqrt{\sqrt{-1} y}} \mathrm{tg}\left(\frac{\sqrt{\sqrt{-1} y}}{2}\right)}\right] \times$$

$$\times \frac{-y^2 e^{B_s' \sqrt{-1} y s^{-2} - \frac{\sqrt{\sqrt{-1} y}}{4s} \cot\left(\frac{\sqrt{\sqrt{-1} y}}{2}\right)}(w^2 + \beta^2)}{8\pi^2\left[1 - \cos(\sqrt{\sqrt{-1} y}) - (\sqrt{\sqrt{-1} y}/2) \sin(\sqrt{\sqrt{-1} y})\right]},$$

in which we have set

$$B_s^2 := \frac{(sw - 2z)^2 + (s\beta - 2\zeta)^2}{8} \quad and \quad B_s' := \frac{4wz + 4\beta\zeta - s(w^2 + \beta^2)}{8}.$$

(17)

Let us now write out a more tractable expression of $\tilde{\Psi}_s$ introduced above.
First, for any real y, θ, setting $\xi := \sqrt{\frac{|y|}{2}}$, we successively have:

$$\sqrt{\sqrt{-1} y} = (1 + \mathrm{sgn}(y)\sqrt{-1})\xi; \quad \frac{2}{\sqrt{\sqrt{-1} y}} = (1 - \mathrm{sgn}(y)\sqrt{-1})/\xi;$$

$$\mathrm{tg}((1 + \sqrt{-1})\theta) = \frac{\sin(2\theta) + \sqrt{-1} \, \mathrm{sh}(2\theta)}{\mathrm{ch}(2\theta) + \cos(2\theta)}$$

$$\cotg\big((1+\sqrt{-1}\,)\theta\big) = \frac{\sin(2\theta) - \sqrt{-1}\,\sh(2\theta)}{\ch(2\theta) - \cos(2\theta)}\ ;$$

$$\tg\bigg(\frac{\sqrt{\sqrt{-1}\,y}}{2}\bigg) = \frac{\sin\xi + \sgn(y)\sqrt{-1}\,\sh\xi}{\ch\xi + \cos\xi}\ ;$$

$$\cotg\bigg(\frac{\sqrt{\sqrt{-1}\,y}}{2}\bigg) = \frac{\sin\xi - \sgn(y)\sqrt{-1}\,\sh\xi}{\ch\xi - \cos\xi}\ ;$$

$$\frac{2}{\sqrt{\sqrt{-1}\,y}}\,\tg\bigg(\frac{\sqrt{\sqrt{-1}\,y}}{2}\bigg) = \frac{(\sh\xi + \sin\xi) + \sgn(y)\sqrt{-1}\,(\sh\xi - \sin\xi)}{\xi\,(\ch\xi + \cos\xi)}\ ;$$

$$\frac{1}{1 - \dfrac{2}{\sqrt{\sqrt{-1}\,y}}\,\tg\Big(\frac{\sqrt{\sqrt{-1}\,y}}{2}\Big)} = \frac{\xi\,(\ch\xi + \cos\xi) - (\sh\xi + \sin\xi) + \sgn(y)\sqrt{-1}\,(\sh\xi - \sin\xi)}{\xi\,(\ch\xi + \cos\xi) - 2(\sh\xi + \sin\xi) + 2\xi^{-1}(\ch\xi - \cos\xi)}\ ;$$

$$\frac{\sqrt{-1}\,y}{1 - \dfrac{2}{\sqrt{\sqrt{-1}\,y}}\,\tg\Big(\frac{\sqrt{\sqrt{-1}\,y}}{2}\Big)} = \frac{y\,\xi\bigg[\sqrt{-1}\,\Big(\xi - \frac{\sh\xi+\sin\xi}{\ch\xi+\cos\xi}\Big) - \sgn(y)\Big(\frac{\sh\xi-\sin\xi}{\ch\xi+\cos\xi}\Big)\bigg]}{\Big(\xi - \frac{\sh\xi+\sin\xi}{\ch\xi+\cos\xi}\Big)^2 + \Big(\frac{\sh\xi-\sin\xi}{\ch\xi+\cos\xi}\Big)^2} =: U(y);$$

$$\frac{\sqrt{\sqrt{-1}\,y}}{2}\,\cotg\bigg(\frac{\sqrt{\sqrt{-1}\,y}}{2}\bigg) = \frac{(\sh\xi + \sin\xi) - \sgn(y)\sqrt{-1}\,(\sh\xi - \sin\xi)}{2\,\xi^{-1}\,(\ch\xi - \cos\xi)} =: V(y);$$

$$\cos\big(\sqrt{\sqrt{-1}\,y}\,\big) = \ch\xi\,\cos\xi - \sgn(y)\sqrt{-1}\,\sh\xi\,\sin\xi\ ;$$

$$\sin\big(\sqrt{\sqrt{-1}\,y}\,\big) = \ch\xi\,\sin\xi + \sgn(y)\sqrt{-1}\,\sh\xi\,\cos\xi\ ;$$

$$1 - \cos\big(\sqrt{\sqrt{-1}\,y}\,\big) - \big(\sqrt{\sqrt{-1}\,y}/2\big)\sin\big(\sqrt{\sqrt{-1}\,y}\,\big) =$$

$$1 - \ch\xi\,\cos\xi - \tfrac{\xi}{2}(\ch\xi\,\sin\xi - \sh\xi\,\cos\xi) + \sgn(y)\sqrt{-1}\bigg[\sh\xi\,\sin\xi - \tfrac{\xi}{2}(\ch\xi\,\sin\xi + \sh\xi\,\cos\xi)\bigg].$$

Set

$$F(y) := \frac{-1}{1 - \cos\big(\sqrt{\sqrt{-1}\,y}\,\big) - \big(\sqrt{\sqrt{-1}\,y}/2\big)\sin\big(\sqrt{\sqrt{-1}\,y}\,\big)}$$

$$= F_r(\xi) + \sgn(y)\sqrt{-1}\,F_i(\xi)\,,$$

with

$$F_r(\xi) := \frac{\ch\xi\,\cos\xi + \tfrac{\xi}{2}(\ch\xi\,\sin\xi - \sh\xi\,\cos\xi) - 1}{(\ch\xi - \cos\xi)\big[(\ch\xi - \cos\xi) - \xi\,(\sh\xi + \sin\xi) + \tfrac{\xi^2}{2}(\ch\xi + \cos\xi)\big]}\,,$$

and

$$F_i(\xi) := \frac{\operatorname{sh}\xi \sin\xi - \frac{\xi}{2}(\operatorname{ch}\xi \sin\xi + \operatorname{sh}\xi \cos\xi)}{(\operatorname{ch}\xi - \cos\xi)\left[(\operatorname{ch}\xi - \cos\xi) - \xi(\operatorname{sh}\xi + \sin\xi) + \frac{\xi^2}{2}(\operatorname{ch}\xi + \cos\xi)\right]} .$$

Thus

$$\tilde{\Psi}_s \equiv \tilde{\Psi}_s(y) = \frac{y^2 F(y)}{8\pi^2} \exp\left[\sqrt{-1}\,\frac{B'_s - x}{s^2}\,y - \frac{w^2 + \beta^2}{2s}\,V(y) + \frac{B_s^2}{s^3}\,U(y)\right] = \tag{18}$$

$$\frac{y^2 F(y)}{8\pi^2} \exp\left[\sqrt{-1}\left(\frac{B_s^2}{s^3}U_i(\xi) + \frac{B'_s - x}{s^2} + \frac{w^2+\beta^2}{2s}V_i(\xi)\right)y - \frac{B_s^2}{s^3}U_r(\xi) - \frac{w^2+\beta^2}{2s}V_r(\xi)\right]$$

with

$$U_i(\xi) := \frac{\xi\left(\xi - \frac{\operatorname{sh}\xi + \sin\xi}{\operatorname{ch}\xi + \cos\xi}\right)}{\left(\xi - \frac{\operatorname{sh}\xi + \sin\xi}{\operatorname{ch}\xi + \cos\xi}\right)^2 + \left(\frac{\operatorname{sh}\xi - \sin\xi}{\operatorname{ch}\xi + \cos\xi}\right)^2} ; \quad V_i(\xi) := \frac{\operatorname{sh}\xi - \sin\xi}{4\xi(\operatorname{ch}\xi - \cos\xi)} ;$$

$$U_r(\xi) := \frac{2\xi^3\left(\frac{\operatorname{sh}\xi - \sin\xi}{\operatorname{ch}\xi + \cos\xi}\right)}{\left(\xi - \frac{\operatorname{sh}\xi + \sin\xi}{\operatorname{ch}\xi + \cos\xi}\right)^2 + \left(\frac{\operatorname{sh}\xi - \sin\xi}{\operatorname{ch}\xi + \cos\xi}\right)^2} ; \quad V_r(\xi) := \frac{\xi(\operatorname{sh}\xi + \sin\xi)}{2(\operatorname{ch}\xi - \cos\xi)} .$$

Using these auxiliary functions, (16) reads:

$$q_s(w, \beta, x, \zeta, z)$$

$$= \frac{1}{8\pi^3 s^6} \int_0^\infty \Re\left[F(y)\,e^{\sqrt{-1}\left[\frac{B_s^2}{s^3}U_i(\xi) + \frac{B'_s - x}{s^2} + \frac{w^2+\beta^2}{2s}V_i(\xi)\right]y}\right] e^{-\frac{B_s^2}{s^3}U_r(\xi) - \frac{w^2+\beta^2}{2s}V_r(\xi)} y^2\,dy$$

$$= \frac{\tilde{q}_s(w, \beta, \zeta, z)}{(3\pi/2)\,s^2} \int_0^\infty \Re\left[F(y)\,e^{\sqrt{-1}\left[\frac{B_s^2}{s^3}U_i(\xi) + \frac{B'_s - x}{s^2} + \frac{w^2+\beta^2}{2s}V_i(\xi)\right]2\xi^2}\right]$$

$$\times e^{-\frac{B_s^2}{s^3}[U_r(\xi) - 12] - \frac{w^2+\beta^2}{2s}[V_r(\xi) - 1]} \xi^5\,d\xi$$

$$= \frac{\tilde{q}_s(w, \beta, \zeta, z)}{3\pi\,s^2} \int_0^\infty G_s(\xi)\,e^{-\frac{B_s^2}{s^3}[U_r(\xi) - 12] - \frac{w^2+\beta^2}{2s}[V_r(\xi) - 1]} \xi^5\,d\xi , \tag{19}$$

with

$$G_s(\xi) := \left(F_r(\xi) + \sqrt{-1}\,F_i(\xi)\right)e^{\sqrt{-1}\,\Lambda_s(\xi)} + \left(F_r(\xi) - \sqrt{-1}\,F_i(\xi)\right)e^{-\sqrt{-1}\,\Lambda_s(\xi)}$$

$$= 2F_r(\xi)\cos\left(\Lambda_s(\xi)\right) - 2F_i(\xi)\sin\left(\Lambda_s(\xi)\right),$$

where we set $\quad \Lambda_s(\xi) := 2\,\xi^2\big(\frac{B_s^2}{s^3}\,U_i(\xi) + \frac{B_s'-x}{s^2} + \frac{w^2+\beta^2}{2s}\,V_i(\xi)\big).$

Note that the functions $F_r, F_i, U_r, U_i, V_r, V_i, G_s$ are all even.

6   Small Time Asymptotics for the Density $q_s(w,\beta,x,\zeta,z)$

We shall here use the expression (19) for $q_s(w,\beta,x,\zeta,z)$, to derive its asymptotics as $s \searrow 0$, proceeding by adapting the saddle-point method (see [7] for example). This will require the following lemma and asymptotics regarding the auxiliary functions entering that expression.

Lemma 6.1 *We have* $V_r(\xi) > 1$ *and* $U_r(\xi) > 12$, *for any real* $\xi \neq 0$.

Proof Indeed for $\xi > 0$ we have:

$$2\,(\mathrm{ch}\,\xi - \cos\xi)\,[V_r(\xi) - 1] = \xi\,(\mathrm{sh}\,\xi + \sin\xi) - 2\,(\mathrm{ch}\,\xi - \cos\xi) > 0$$

(since the first derivatives are $\xi(\mathrm{ch}\,\xi + \cos\xi) - (\mathrm{sh}\,\xi + \sin\xi)$ and $\xi\,(\mathrm{sh}\,\xi - \sin\xi) > 0$) and

$$\tfrac{1}{2}(\mathrm{ch}\,\xi + \cos\xi)\left[\left(\xi - \tfrac{\mathrm{sh}\,\xi+\sin\xi}{\mathrm{ch}\,\xi+\cos\xi}\right)^2 + \left(\tfrac{\mathrm{sh}\,\xi-\sin\xi}{\mathrm{ch}\,\xi+\cos\xi}\right)^2\right][U_r(\xi) - 12]$$

$$= \xi^3\,(\mathrm{sh}\,\xi - \sin\xi) - 6\,(\mathrm{ch}\,\xi + \cos\xi)\left[\left(\xi - \tfrac{\mathrm{sh}\,\xi+\sin\xi}{\mathrm{ch}\,\xi+\cos\xi}\right)^2 + \left(\tfrac{\mathrm{sh}\,\xi-\sin\xi}{\mathrm{ch}\,\xi+\cos\xi}\right)^2\right]$$

$$= (\mathrm{sh}\,\xi - \sin\xi)\xi^3 - 6(\mathrm{ch}\,\xi + \cos\xi)\xi^2 + 12(\mathrm{sh}\,\xi + \sin\xi)\xi - 12(\mathrm{ch}\,\xi - \cos\xi) > 0,$$

since its derivative is:

$$\big[(\mathrm{ch}\,\xi - \cos\xi)\,\xi - 3\,(\mathrm{sh}\,\xi - \sin\xi)\big]\xi^2 > 0,$$

because (by the preceding)

$$\frac{d}{d\xi}\big[(\mathrm{ch}\,\xi - \cos\xi)\,\xi - 3\,(\mathrm{sh}\,\xi - \sin\xi)\big] = \xi\,(\mathrm{sh}\,\xi + \sin\xi) - 2\,(\mathrm{ch}\,\xi - \cos\xi) > 0. \quad \diamond$$

6.1   Auxiliary Asymptotics

As $\xi \to \infty$ we have $\quad U_r(\xi) = 2\xi + 4 + \tfrac{4}{\xi} + \mathscr{O}(\tfrac{1}{\xi^3}); \quad U_i(\xi) = 1 + \mathscr{O}(\tfrac{1}{\xi});$

$V_r(\xi) = \tfrac{\xi}{2}[1 + \mathscr{O}(e^{-\xi})]; \quad V_i(\xi) = \tfrac{1+\mathscr{O}(e^{-\xi})}{4\xi}; \quad F(2\xi^2) = \mathscr{O}\big(\tfrac{e^{-\xi}}{\xi}\big).$

Then near 0 we successively have:

$$U_r(\xi) = \frac{2\,\xi^3 \times \frac{\xi^3}{6}\left(1 - \frac{17}{420}\xi^4 + \mathcal{O}(\xi^8)\right)}{\left(\frac{\xi^5}{30}\right)^2 + \left(\frac{\xi^3}{6}\left(1 - \frac{17}{420}\xi^4\right)\right)^2 + \mathcal{O}(\xi^{10})} = \frac{\frac{\xi^6}{3}\left(1 - \frac{17}{420}\xi^4 + \mathcal{O}(\xi^8)\right)}{\frac{\xi^6}{36}\left(1 - \frac{17}{210}\xi^4\right) + \mathcal{O}(\xi^{10})}$$

$$= 12 + \tfrac{17}{35}\,\xi^4 + \mathcal{O}(\xi^8);$$

$$U_i(\xi) = \frac{\frac{\xi^6}{30}\left(1 - \frac{113}{648}\xi^4 + \mathcal{O}(\xi^8)\right)}{\frac{\xi^6}{36}\left(1 - \frac{17}{210}\xi^4 + \mathcal{O}(\xi^8)\right)} = \frac{6}{5} - \frac{2119}{18900}\,\xi^4 + \mathcal{O}(\xi^8);$$

$$V_i(\xi) = \tfrac{1}{12} - \tfrac{\xi^4}{7560} + \mathcal{O}(\xi^8); \quad V_r(\xi) = 1 + \tfrac{\xi^4}{180} + \mathcal{O}(\xi^8);$$

$$F_r(\xi) = \frac{6}{\xi^4} - \frac{620659}{135600} + \mathcal{O}(\xi^4); \quad F_i(\xi) = \frac{1}{5} + \mathcal{O}(\xi^4);$$

$$\Lambda_s(\xi) = 2\left[\frac{B_s^2}{s^3}\left[\tfrac{6}{5} - \tfrac{2119}{18900}\,\xi^4 + \mathcal{O}(\xi^8)\right] + \frac{B_s' - x}{s^2} + \frac{w^2 + \beta^2}{2s}\left[\tfrac{1}{12} - \tfrac{1}{756}\xi^4 + \mathcal{O}(\xi^8)\right]\right]\xi^2$$

$$= 2\,\Theta_s\,\xi^2 - \left[\tfrac{2119}{9450}\tfrac{B_s^2}{s^3} + \tfrac{1}{63}\tfrac{w^2+\beta^2}{12\,s}\right]\left[\xi^6 + \mathcal{O}(\xi^{10})\right] = 2\,\Theta_s\,\xi^2 + \mathcal{O}(v_s)\,\xi^4,$$

where (recall that B_s^2, B_s' were given by (17)) we have set

$$\Theta_s := \frac{6B_s^2}{5\,s^3} + \frac{B_s' - x}{s^2} + \frac{w^2 + \beta^2}{24\,s}, \quad \text{and} \quad v_s := \frac{17\,B_s^2}{35\,s^3} + \frac{w^2 + \beta^2}{360\,s}. \quad (20)$$

Moreover

$$e^{-\frac{B_s^2}{s^3}[U_r(\xi)-12] - \frac{w^2+\beta^2}{2s}[V_r(\xi)-1]} = e^{-\frac{B_s^2}{s^3}\left[\frac{17}{35}\xi^4 + \mathcal{O}(\xi^8)\right] - \frac{w^2+\beta^2}{2s}\left[\frac{1}{180}\xi^4 + \mathcal{O}(\xi^8)\right]}$$

$$= e^{-v_s\,[\xi^4 + \mathcal{O}(\xi^8)]}.$$

Furthermore,

$$U_r(\xi) - 2\xi - 4 =$$

$$\frac{4\xi\left(1 - \frac{2(e^{-\xi}-\sin\xi)}{\mathrm{ch}\,\xi+\cos\xi}\right) - 8\left(\frac{\mathrm{ch}\,\xi-\cos\xi}{\mathrm{ch}\,\xi+\cos\xi}\right) - 2\,\xi^3\left(\frac{e^{-\xi}+\cos\xi+\sin\xi}{\mathrm{ch}\,\xi+\cos\xi}\right) - 4\,\xi^2\left(\frac{e^{-\xi}+\cos\xi-\sin\xi}{\mathrm{ch}\,\xi+\cos\xi}\right)}{(\xi-1)^2 + 1 + 2\xi\left(\frac{e^{-\xi}+\cos\xi-\sin\xi}{\mathrm{ch}\,\xi+\cos\xi}\right) - \frac{4\cos\xi}{\mathrm{ch}\,\xi+\cos\xi}}$$

$$\left(= \frac{4\xi - 8 + \mathcal{O}(\xi^3 e^{-\xi})}{\xi^2 - 2\xi + 2 + \mathcal{O}(\xi\,e^{-\xi})}\right)$$

$$> \frac{4\xi - 8 - \frac{4(\xi+1)^3}{\mathrm{ch}\,\xi+\cos\xi}}{(\xi-1)^2 + 1 + \frac{4(\xi+1)}{\mathrm{ch}\,\xi+\cos\xi}} > \frac{4\xi - 8 - 9(\xi+1)^3 e^{-\xi}}{(\xi-1)^2 + 1 + 9(\xi+1)e^{-\xi}} > \frac{4\xi - 9}{\xi^2 - 2\xi + 2.2} > 0$$

for $\xi \geq 2\pi$, and similarly

$$V_r(\xi) - \frac{\xi}{2} = \frac{\xi \, (\cos \xi + \sin \xi - e^{-\xi})}{2 \, (\mathrm{ch}\,\xi - \cos \xi)} > \frac{-\xi}{\mathrm{ch}\,\xi - \cos \xi} > -3\xi \, e^{-\xi} > \frac{-1}{20},$$

so that $V_r(\xi) > \pi - \frac{1}{20} > 3$ for $\xi \geq 2\pi$.

Therefore, in a small neighbourhood $[2\pi - \varepsilon, \infty[+ \sqrt{-1}\ [-\varepsilon, \varepsilon]$ of $[2\pi, \infty[$ we shall have

$$\Re\big[U_r(\xi) - 2\xi\big] \geq 4 \quad \text{and} \quad \Re\big[V_r(\xi)\big] \geq 3. \tag{21}$$

6.2   Changes of Contour and Saddle-Point Method

Note that by Lemmas 4.2.3, 5.1, (14) and the changes of variable: $\lambda = \sqrt{-1}\ y = \pm 2\sqrt{-1}\ \xi^2$, the poles of the integrand in (19) are located at $e^{\sqrt{-1}\,k\pi/2}(1+\sqrt{-1})\, y_n$ and $e^{\sqrt{-1}\,k\pi/2}(1 + \sqrt{-1})(n + 1)\, \pi$, with $n, k \in \mathbb{N}$ and $\frac{17\pi}{12} < y_0 < y_1 < \ldots$, so that this integrand is analytic at 0 with convergence radius $\sqrt{2}\,\pi$.

In particular, we may perform the following changes of contour in (19):

$$3\pi \, s^2 \, \frac{q_s(w, \beta, x, \zeta, z)}{\bar{q}_s(w, \beta, \zeta, z)} = \int_0^\infty G_s(\xi)\, e^{-\frac{B_\xi^2}{s^3}[U_r(\xi)-12]-\frac{w^2+\beta^2}{2s}[V_r(\xi)-1]} \xi^5 \, d\xi$$

$$= \int_0^{(1+\sqrt{-1})\Theta_s^{-\eta}} \big(F_r(\xi) + \sqrt{-1}\ F_i(\xi)\big)\, e^{\sqrt{-1}\,\Lambda_s(\xi) - \frac{B_\xi^2}{s^3}[U_r(\xi)-12]-\frac{w^2+\beta^2}{2s}[V_r(\xi)-1]} \xi^5 \, d\xi$$

$$+ \int_0^{(1-\sqrt{-1})\Theta_s^{-\eta}} \big(F_r(\xi) - \sqrt{-1}\ F_i(\xi)\big)\, e^{-\sqrt{-1}\,\Lambda_s(\xi) - \frac{B_\xi^2}{s^3}[U_r(\xi)-12]-\frac{w^2+\beta^2}{2s}[V_r(\xi)-1]} \xi^5 d\xi$$

$$+ \int_{(1+\sqrt{-1})\Theta_s^{-\eta}}^{(1+\sqrt{-1})\Theta_s^{-\eta}+2\pi} \big(F_r(\xi) + \sqrt{-1}\ F_i(\xi)\big) e^{\sqrt{-1}\,\Lambda_s(\xi) - \frac{B_\xi^2}{s^3}[U_r(\xi)-12]-\frac{w^2+\beta^2}{2s}[V_r(\xi)-1]} \xi^5 d\xi$$

$$+ \int_{(1-\sqrt{-1})\Theta_s^{-\eta}}^{(1-\sqrt{-1})\Theta_s^{-\eta}+2\pi} \big(F_r(\xi) - \sqrt{-1}\ F_i(\xi)\big) e^{-\sqrt{-1}\,\Lambda_s(\xi) - \frac{B_\xi^2}{s^3}[U_r(\xi)-12]-\frac{w^2+\beta^2}{2s}[V_r(\xi)-1]} \xi^5 d\xi$$

$$+ \int_{(1+\sqrt{-1})\Theta_s^{-\eta}+2\pi}^{(1+\sqrt{-1})\Theta_s^{-\eta}+\infty} \big(F_r(\xi) + \sqrt{-1}\ F_i(\xi)\big) e^{\sqrt{-1}\,\Lambda_s(\xi) - \frac{B_\xi^2}{s^3}[U_r(\xi)-12]-\frac{w^2+\beta^2}{2s}[V_r(\xi)-1]} \xi^5 d\xi$$

$$+ \int_{(1-\sqrt{-1})\Theta_s^{-\eta}+2\pi}^{(1-\sqrt{-1})\Theta_s^{-\eta}+\infty} \big(F_r(\xi) - \sqrt{-1}\ F_i(\xi)\big) e^{-\sqrt{-1}\,\Lambda_s(\xi) - \frac{B_\xi^2}{s^3}[U_r(\xi)-12]-\frac{w^2+\beta^2}{2s}[V_r(\xi)-1]} \xi^5 d\xi$$

$$=: J_s^0 + \bar{J}_s^0 + J_s^\pi + \bar{J}_s^\pi + J_s^\infty + \bar{J}_s^\infty,$$

where $\eta > \frac{1}{4}$ will be specified further.

Note that the estimate (21) and the control $F(2\xi^2) = \mathscr{O}\left(\frac{e^{-\xi}}{\xi}\right)$ ensure the vanishing of the unmentioned limiting contribution (for any small enough fixed s, provided $\lim\limits_{s\to 0}\Theta_s = \infty$) in the above changes of contour:

$$\int_R^{R\pm(1+\sqrt{-1})\Theta_s^{-\eta}} \left(F_r(\xi) + \sqrt{-1}\,F_i(\xi)\right) e^{\sqrt{-1}\,\Lambda_s(\xi)-\frac{B_s^2}{s^3}[U_r(\xi)-12]-\frac{w^2+\beta^2}{2s}[V_r(\xi)-1]}\,\xi^5\,d\xi$$

goes to 0 as $R \to \infty$.

Now on the one hand, setting $\xi = (1+\sqrt{-1})\Theta_s^{-\eta}\,t$ by the above we have:

$$J_s^0 = \frac{8}{\Theta_s^{6\eta}}\int_0^1 \left[F_i(\xi) - \sqrt{-1}\,F_r(\xi)\right] e^{\left[\sqrt{-1}\,\Lambda_s(\xi)-\frac{B_s^2}{s^3}[U_r(\xi)-12]-\frac{w^2+\beta^2}{2s}[V_r(\xi)-1]\right]}t^5 dt$$

$$= \frac{8}{5\Theta_s^{6\eta}}\int_0^1 \left[1 + \frac{15\sqrt{-1}}{2\Theta_s^{-4\eta}t^4} + \frac{620659\sqrt{-1}}{27120} + \mathscr{O}(\Theta_s^{-4\eta})\right] e^{-4(\Theta_s^{1-2\eta}t^2-v_s\Theta_s^{-4\eta})(1+\mathscr{O}(\Theta_s^{-4\eta}))}t^5 dt$$

$$= \frac{8}{5\,\Theta_s^{6\eta}}\left(1 + (\Theta_s^{1-2\eta} + v_s)\mathscr{O}(\Theta_s^{-4\eta})\right)$$

$$\times \int_0^1 \left[1 + \frac{15\sqrt{-1}}{2\Theta_s^{-4\eta}t^4} + \frac{620659\sqrt{-1}}{27120} + \mathscr{O}(\Theta_s^{-4\eta})\right] e^{-4\Theta_s^{1-2\eta}t^2}\,t^5\,dt\,,$$

so that for $\frac{1}{4} < \eta < \frac{1}{2}$:

$$J_s^0 + \bar{J}_s^0 = \frac{16}{5\Theta_s^{6\eta}}\left(1 + \mathscr{O}(v_s\Theta_s^{-4\eta})\right)\int_0^1 e^{-4\Theta_s^{1-2\eta}t^2}\,t^5\,dt$$

$$= \frac{1 + \mathscr{O}(v_s\,\Theta_s^{-4\eta})}{40\,\Theta_s^3}\int_0^{4\Theta_s^{1-2\eta}} e^{-u}\,u^2 du = \frac{1 + \mathscr{O}(\Theta_s^{1-4\eta})}{20\,\Theta_s^3}\,,$$

provided $\lim\limits_{s\to 0}\Theta_s = \infty$ and $v_s = \mathscr{O}(\Theta_s)$.

Then we deal with the intermediate part $(J_s^\pi + \bar{J}_s^\pi)$. By the above we know that we have

$$U_i(\xi) = \tfrac{6}{5} + \xi^4\,\tilde{U}_i(\xi), \quad V_i(\xi) = \tfrac{1}{12} + \xi^4\,\tilde{V}_i(\xi), \quad U_r(\xi) - 12 = \xi^4\,\tilde{U}_r(\xi),$$

$$V_r(\xi) - 1 = \xi^4\,\tilde{V}_r(\xi),$$

with even functions $\tilde{U}_i, \tilde{V}_i, \tilde{U}_r, \tilde{V}_r$ (of ξ alone) that are analytic outside the sequence $\{e^{\sqrt{-1}\,k\pi/2}(1 + \sqrt{-1})\,y_n \mid n,k \in \mathbb{N}, \frac{17\pi}{12} < y_0 < y_1 < \dots\}$ and then in the compact disc (centred at 0) of radius $\frac{17\pi}{6\sqrt{2}} > 2\pi$ (note that the proof of Lemma 5.1 shows that the points $e^{\sqrt{-1}\,k\pi/2}(1 + \sqrt{-1})(n + 1)\,\pi$ are poles only for the functions F_r, F_i). Therefore we can write the phase as follows:

$$\varphi_s(\xi) := \sqrt{-1}\,\Lambda_s(\xi) - \frac{B_s^2}{s^3}[U_r(\xi) - 12] - \frac{w^2 + \beta^2}{2s}[V_r(\xi) - 1] = 2\sqrt{-1}\,\Theta_s\xi^2 + \tilde{\varphi}_s(\xi)\xi^4$$

with

$$\tilde{\varphi}_s(\xi) := \frac{B_s^2}{s^3}\left[2\sqrt{-1}\,\tilde{U}_i(\xi) - \tilde{U}_r(\xi)\right] + \frac{w^2 + \beta^2}{2s}\left[2\sqrt{-1}\,\tilde{V}_i(\xi) - \tilde{V}_r(\xi)\right].$$

Moreover as before near 0 we have $\left(F_r(\xi) + \sqrt{-1}\,F_i(\xi)\right)\xi^4 = 6 + \mathscr{O}(\xi^4)$, whence for small s: $\quad J_s^\pi =$

$$\int_{(1+\sqrt{-1})\Theta_s^{-\eta}}^{(1+\sqrt{-1})\Theta_s^{-\eta}+2\pi} \left(F_r(\xi) + \sqrt{-1}\,F_i(\xi)\right) e^{\sqrt{-1}\,\Lambda_s(\xi) - \frac{B_s^2}{s^3}[U_r(\xi) - 12] - \frac{w^2 + \beta^2}{2s}[V_r(\xi) - 1]} \xi^5 d\xi$$

$$= \int_0^{2\pi} \mathscr{O}(1)\, e^{2\sqrt{-1}\,\Theta_s\left((1+\sqrt{-1})\Theta_s^{-\eta}+t\right)^2 + \left((1+\sqrt{-1})\Theta_s^{-\eta}+t\right)^4 \tilde{\varphi}_s\left((1+\sqrt{-1})\Theta_s^{-\eta}+t\right)} dt.$$

Now we have $\left((1+\sqrt{-1})\Theta_s^{-\eta} + t\right)^4 = \Theta_s^{-4\eta}\left[T^4 - 6T^2 + 1 + 4\sqrt{-1}\,(T^2 - 1)T\right]$, with $T := 1 + \Theta_s^\eta t \geq 1$, and on the other hand, by Lemma 6.1 and Sect. 6.1 we know that

$$\max_{[0,2\pi]} \Re\left[2\sqrt{-1}\,\tilde{U}_i - \tilde{U}_r\right] < 0 \quad \text{and} \quad \max_{[0,2\pi]} \Re\left[2\sqrt{-1}\,\tilde{V}_i - \tilde{V}_r\right] < 0,$$

so that by continuity, for some negative constant $-h$ and for small s:

$$\max_{t \in [0,2\pi]} \Re\left[\tilde{\varphi}_s\left((1+\sqrt{-1})\Theta_s^{-\eta} + t\right)\right] \leq -h\,\nu_s;$$

and of course $\max_{t \in [0,2\pi]} \left|\tilde{\varphi}_s\left((1+\sqrt{-1})\Theta_s^{-\eta} + t\right)\right| = \mathscr{O}(\nu_s)$. As a consequence, we have

$$\max_{t \in [0,2\pi]} \Re\left[\left((1+\sqrt{-1})\Theta_s^{-\eta} + t\right)^4 \tilde{\varphi}_s\left((1+\sqrt{-1})\Theta_s^{-\eta} + t\right)\right]$$

$$\leq \Theta_s^{-4\eta}\nu_s \max_{T \geq 1}\left[-h\,T^4 + \mathscr{O}(T^3)\right] = \mathscr{O}(\Theta_s^{1-4\eta}) \qquad \left(\text{provided } \nu_s = \mathscr{O}(\Theta_s)\right).$$

Therefore, provided $\lim_{s \to 0} \Theta_s = \infty$ and $\nu_s = \mathscr{O}(\Theta_s)$, we have

$$J_s^\pi = \mathscr{O}(1)\int_0^{2\pi} e^{-4\Theta_s^{1-2\eta} - 4\Theta_s^{1-\eta}t + \mathscr{O}(\Theta_s^{1-4\eta})} dt = \mathscr{O}\left(e^{-3\Theta_s^{1-2\eta}}\right).$$

The same of course holds for $\bar{J}_s^\pi$.

To deal with the remaining contribution $(J_s^\infty + \bar{J}_s^\infty)$, we use the lower estimates near infinity : (21) computed above, as follows : $J_s^\infty =$

$$\int_{(1+\sqrt{-1})\Theta_s^{-\eta}+2\pi}^{(1+\sqrt{-1})\Theta_s^{-\eta}+\infty} \left(F_r(\xi) + \sqrt{-1}\,F_i(\xi)\right) e^{\sqrt{-1}\,\Lambda_s(\xi) - \frac{B_\xi^2}{s^3}[U_r(\xi)-12] - \frac{w^2+\beta^2}{2s}[V_r(\xi)-1]} \xi^5 \, d\xi$$

$$= \int_{2\pi}^{\infty} \mathcal{O}\left[\frac{e^{-t}}{t}\right] e^{-\frac{B_\xi^2}{s^3}[2(\Theta_s^{-\eta}+t)-8] - \frac{w^2+\beta^2}{s}} (1+t)^5 dt$$

$$= \mathcal{O}(1)\, e^{\frac{8B_\xi^2}{s^3} - \frac{w^2+\beta^2}{s}} \int_{2\pi}^{\infty} e^{-\left[\frac{2B_\xi^2}{s^3}+1\right]t}\, t^4 \, dt$$

$$= \mathcal{O}(1)\, e^{\frac{8B_\xi^2}{s^3} - \frac{w^2+\beta^2}{s}} \left(\frac{2B_\xi^2}{s^3}+1\right)^{-5} \int_{2\pi\left[\frac{2B_\xi^2}{s^3}+1\right]}^{\infty} e^{-t}\, t^4 \, dt$$

$$= \mathcal{O}\left[\left(\frac{2B_\xi^2}{s^3}+1\right)^{-1}\right] e^{-4(\pi-2)\frac{B_\xi^2}{s^3} - \frac{w^2+\beta^2}{s}},$$

and the same of course holds for $\bar{J}_s^\infty$.

So far, we have obtained :

$$3\pi\, s^2\, \frac{q_s(w,\beta,x,\zeta,z)}{\tilde{q}_s(w,\beta,\zeta,z)} = J_s^0 + \bar{J}_s^0 + J_s^\pi + \bar{J}_s^\pi + J_s^\infty + \bar{J}_s^\infty$$

$$= \frac{1+\mathcal{O}(v_s\,\Theta_s^{-4\eta})}{20\,\Theta_s^3} + \mathcal{O}\left(e^{-3\Theta_s^{1-2\eta}}\right) + \mathcal{O}\left[\left(\frac{2B_\xi^2}{s^3}+1\right)^{-1}\right]e^{-4(\pi-2)\frac{B_\xi^2}{s^3} - \frac{w^2+\beta^2}{s}}$$

$$= \frac{1+\mathcal{O}(\Theta_s^{1-4\eta})}{20\,\Theta_s^3} \qquad \text{for } \tfrac{1}{4} < \eta < \tfrac{1}{2},$$

provided both $\lim_{s\to 0} \Theta_s = \infty$ and $v_s = \mathcal{O}(\Theta_s)$.

By (17) and (20), this condition holds as soon as both $\lim_{s\to 0} \Theta_s = \infty$ and $\frac{x}{s^2} \le \frac{z^2+\zeta^2}{s^3} + \frac{\Theta_s}{\varepsilon}$ (for some $\varepsilon > 0$), and then also as soon as both $\lim_{s\to 0} v_s = \infty$ and $2s\,x \le (z^2+\zeta^2) + \varepsilon s^2(w^2+\beta^2)$.

Finally, under this condition, as $s \searrow 0$ and for any positive ε, we have obtained :

$$q_s(w,\beta,x,\zeta,z) = \frac{1+\mathcal{O}(\Theta_s^{\varepsilon-1})}{60\,\pi\,s^2\,\Theta_s^3} \times \tilde{q}_s(w,\beta,\zeta,z).$$

The result of this section (and main result) is thus the following specific off-diagonal equivalent.

Theorem 6.4 *As* $s \searrow 0$, *for any positive* ε, *uniformly for* $x \geq 0$ *and* $(w, \beta, \zeta, z) \in \mathbb{R}^4$ *such that*:

$$\Theta_s \equiv \frac{3[(z - \frac{sw}{12})^2 + (\zeta - \frac{s\beta}{12})^2]}{5\,s^3} + \frac{w^2 + \beta^2}{16\,s} - \frac{x}{s^2} \longrightarrow \infty \; and \; \frac{x}{s^2} \leq \frac{z^2 + \zeta^2}{s^3} + \frac{\Theta_s}{\varepsilon},$$

$$(22)$$

we have

$$q_s(w, \beta, x, \zeta, z) = \frac{1 + \mathcal{O}(\Theta_s^{\varepsilon-1})}{20\pi^3\,s^6\,\Theta_s^3} \times \exp\left[-\frac{3[(sw - 2z)^2 + (s\beta - 2\zeta)^2]}{2\,s^3} - \frac{w^2 + \beta^2}{2\,s} \right].$$

An alternative condition (to (22) above) guaranteeing this asymptotic equivalent is

$$\frac{(sw - 2z)^2 + (s\beta - 2\zeta)^2}{s^3} + \frac{w^2 + \beta^2}{s} \longrightarrow \infty \; and \; 2s\,x \leq (z^2 + \zeta^2) + \varepsilon s^2(w^2 + \beta^2).$$

$$(23)$$

6.3 Final Remarks

Theorem 6.4 rather precisely yields the small time asymptotic behaviour of the heat kernel of the second chaos approximation (Y_s) to the Dudley relativistic diffusion (X_s), in so far as the normalized non-Gaussian component x/s^2 remains nearly negligible with respect to the normalized squared Gaussian one $\left(\frac{(w^2, \beta^2)}{s}, \frac{(z^2, \zeta^2)}{s^3} \right)$.

But this is not enough to derive any small time asymptotics for the density $p_s(\lambda, \mu, x, y, z)$ of the original process $X_s \equiv (\lambda_s, \mu_s, x_s, y_s, z_s)$, even for fixed (λ, μ, x, y, z).

On the one hand, the computations performed in Sect. 4 above cannot work beyond the second chaos, so that Sect. 3 cannot yield a sufficiently precise control on the gap between both tangent processes (X_s) and (Y_s). To be more specific, Sect. 3 and Theorem 6.4 heuristically yield:

$$p_s(w, \beta, x, \zeta, z) \approx \mathbb{P}[\lambda_s = w, \, \mu_s = \beta, \, x_s = x, \, y_s = \zeta, \, z_s = z] \approx$$

$$\mathbb{P}\left[(w_s, \beta_s, A_s, \zeta_s, \bar{z}_s) = (w, \beta, x, \zeta, z) + \mathcal{O}(R)(s^{3/2}, s^{3/2}, s^{5/2}, s^{5/2}, s^{5/2}) \right] + \mathcal{O}\left(e^{-R^\kappa/c}\right)$$

$$\approx \frac{\exp\left[-\frac{3\left[(sw - 2z + \mathcal{O}(s^{5/2}))^2 + (s\beta - 2\zeta + \mathcal{O}(s^{5/2}))^2 \right]}{2\,s^3} - \frac{w^2 + \beta^2 + \mathcal{O}(s^{3/2})}{2\,s} \right]}{20\,\pi^3\,s^6\,\Theta_s\left(w + \mathcal{O}(s^{3/2}), \beta + \mathcal{O}(s^{3/2}), x + \mathcal{O}(s^{5/2}), \zeta + \mathcal{O}(s^{5/2}), z + \mathcal{O}(s^{5/2}) \right)^3}$$

$$+ \mathcal{O}\left(e^{-R^\kappa/c}\right)$$

$$= \exp\left[-\frac{3\left[(sw - 2z)^2 + (s\beta - 2\zeta)^2 \right]}{2\,s^3} - \frac{w^2 + \beta^2}{2\,s} + \mathcal{O}(s^{-1/2}) \right] + \mathcal{O}\left(e^{-R^\kappa/c}\right),$$

which were not too bad only if the additive term $\mathcal{O}(e^{-R^\kappa/c})$ were not there to ruin such estimate. Indeed, even taking $R \asymp s^{-\gamma}$ would only control this correction term by (at best) $e^{-s^{-2\gamma/3}}$, which would be significant only for $\gamma \geq 9/2$, so that the remaining information would then reduce to nothing.

On the other hand, it is unclear whether even the knowledge of small time asymptotics for a further Taylor expansion of (X_s) would allow to derive a similar knowledge about (X_s) itself.

Finally, the method used here seems suitable to get perturbative results such as Theorem 6.4, but precise asymptotics under other regimes could demand the use of another method.

7 Proofs of Some Technical Results

We gather here the rather technical proofs of Lemmas 3.1 and 4.1.1 and Proposition 5.2.

Proof of Lemma 3.1 (i) Equation (4) entails that for small proper time s we have:

$$\lambda_s = w_s + \frac{1}{2}\int_0^s \mathrm{th}\big(w_\tau + o(\tau)\big)d\tau = w_s + \frac{1}{2}\int_0^s w_\tau\, d\tau + o(s^2) = w_s + o(s^{3/2-\varepsilon})$$

$$= w_s + \frac{1}{2}\int_0^s \mathrm{th}\big(w_\tau + o(\tau^{3/2-\varepsilon})\big)d\tau = w_s + \frac{1}{2}\int_0^s \big(w_\tau + o(\tau^{3/2-\varepsilon})\big)d\tau$$

$$= w_s + \frac{1}{2}\int_0^s w_\tau\, d\tau + o(s^{5/2-\varepsilon}).$$

Then

$$\mathrm{ch}\,\lambda_s = \mathrm{ch}\big[w_s + o(s^{3/2-\varepsilon})\big] = 1 + \tfrac{1}{2}w_s^2 + o(s^{2-\varepsilon}) = 1 + o(s^{1-\varepsilon}),$$

and by Eq. (5) we have:

$$\dot{z}_s = \mathrm{sh}\,\lambda_s = \mathrm{sh}\big[w_s + o(s^{3/2-\varepsilon})\big] = w_s + o(s^{3/2-\varepsilon}).$$

$$\left(\text{Also,}\quad \dot{z}_s = \int_0^s \sqrt{1+\dot{z}_\tau^2}\, dw_\tau + \int_0^s \ddot{z}_\tau\, d\tau.\right)$$

Similarly,

$$\mu_s = \int_0^s \frac{d\beta_\tau}{\mathrm{ch}\,[w_\tau + o(\tau^{3/2-\varepsilon})]} = \int_0^s \big(1 - o(\tau^{1-\varepsilon})\big)d\beta_\tau = \beta_s + o(s^{3/2-\varepsilon}),$$

$$\mathrm{ch}\,\mu_s = 1 + o(s^{1-\varepsilon}), \quad \mathrm{sh}\,\mu_s = \beta_s + o(s^{3/2-\varepsilon}).$$

$$\left(\text{Also,}\quad \mathrm{sh}\,\mu_s = \int_0^s \sqrt{\frac{1+\mathrm{sh}^2\mu_\tau}{1+\dot{z}_\tau^2}}\, d\beta_\tau + \int_0^s \frac{\mathrm{sh}\,\mu_\tau}{2(1+\dot{z}_\tau^2)}\, d\tau.\right)$$

Finally, the result follows at once from:

$$x_s = \int_0^s \left[1 + \tfrac{1}{2} w_\tau^2 + o(\tau^{2-\varepsilon})\right]\left[1 + \tfrac{1}{2} \beta_\tau^2 + o(\tau^{2-\varepsilon})\right] d\tau$$

$$= s + \tfrac{1}{2} \int_0^s \left[\beta_\tau^2 + w_\tau^2\right] d\tau + o(s^{3-\varepsilon}) \, ;$$

$$y_s = \int_0^s \left[1 + o(\tau^{1-\varepsilon})\right]\left[\beta_\tau + o(\tau^{3/2-\varepsilon})\right] d\tau = \int_0^s \beta_\tau \, d\tau + o(s^{5/2-\varepsilon}) \, ;$$

$$z_s = \int_0^s w_\tau \, d\tau + o(s^{5/2-\varepsilon}) \, .$$

$\left(\text{More precisely, } z_s = \int_0^s w_\tau \, d\tau + \int_0^s \int_0^\tau w_u \, du \, d\tau + \tfrac{1}{2} \int_0^s \int_0^\tau w_u^2 \, dw_u \, d\tau + o(s^{7/2-\varepsilon}).\right)$

(*ii*) For any $s \in [0,1]$ we almost surely have : $|\lambda_s| \leq \sup_{[0,s]} |w| + \tfrac{1}{2} \int_0^s |\lambda_\tau| \, d\tau$,

whence by Gronwall's Lemma : $\sup_{[0,s]} |\lambda| \leq \sup_{[0,s]} |w| \times e^{s/2}$. Then for $0 \leq s \leq$ $2 \log 2$:

$$\Lambda_s := \sup_{[0,s]} |\lambda - w| \leq \tfrac{1}{2} \int_0^s \sup_{[0,\tau]} |w| \times e^{\tau/2} d\tau \leq \int_0^s \sup_{[0,\tau]} |w| \, d\tau \, .$$

Hence for $R \geq 1$ and $0 \leq s \leq 2 \log 2$:

$$\mathbb{P}\left[\sup_{0 \leq t \leq s} |\Lambda_t| \geq R s^{\frac{3}{2}}\right] \leq \mathbb{P}\left[\int_0^1 \sup_{[0,\tau]} |w| \, d\tau \geq R\right] \leq \mathbb{P}\left[\sup_{[0,1]} |w| \geq R\right]$$

$$\leq 4 \, \mathbb{P}[w_1 \geq R] \leq 2 \, e^{-\frac{R^2}{2}} \, .$$

Then

$$\beta_s - \mu_s = 2 \int_0^s \frac{\mathrm{sh}^2(\lambda_\tau/2)}{\mathrm{ch}\,\lambda_\tau} \, d\beta_\tau \equiv 2 \, B\left[\int_0^s \frac{\mathrm{sh}^4(\lambda_\tau/2)}{\mathrm{ch}^2\lambda_\tau} \, d\tau\right],$$

so that

$$\mathbb{P}\left[\sup_{0 \leq t \leq s} |\beta_t - \mu_t| \geq R s^{\frac{3}{2}}\right] \leq 2 \, \mathbb{E}\left[\exp\left(-\frac{R^2 s^3}{8} \middle/ \int_0^s \frac{\mathrm{sh}^4(\lambda_\tau/2)}{\mathrm{ch}^2\lambda_\tau} \, d\tau\right)\right]$$

$$\leq 2 \, \mathbb{E}\left[\exp\left(-2R^2 s^3 \middle/ \int_0^s \lambda_\tau^4 \, d\tau\right)\right] \leq 2 \, \mathbb{E}\left[\exp\left(-\frac{R^2}{8 \sup_{[0,1]} |w|^4}\right)\right]$$

$$= 2 \int_0^1 \mathbb{P}\left[-\frac{R^2}{8 \sup_{[0,1]} |w|^4} > \log y \right] dy = 2 \int_0^1 \mathbb{P}\left[\sup_{[0,1]} |w|^4 > \frac{-R^2}{8 \log y} \right] dy$$

$$\leq 4 \int_0^1 e^{-\frac{1}{2}\sqrt{\frac{-R^2}{8 \log y}}} \, dy = 4 \int_0^1 e^{\frac{-R}{4\sqrt{2 \log(1/y)}}} \, dy = 4 \int_0^\infty e^{\frac{-R}{4\sqrt{2t}} - t} \, dt \leq 132 \, e^{-R^{2/3}/(32)^{1/3}}.$$

Then

$$\left| z_s - \int_0^s w_\tau \, d\tau \right| \leq \int_0^s \left| \mathrm{sh}\big(w_\tau + (\lambda_\tau - w_\tau)\big) - w_\tau \right| d\tau$$

$$= \int_0^s \left| \mathrm{sh}\, w_\tau - w_\tau + 2 \, \mathrm{sh}\big[\tfrac{\lambda_\tau - w_\tau}{2}\big] \mathrm{ch}\big[\tfrac{\lambda_\tau + w_\tau}{2}\big] \right| d\tau$$

$$\leq \int_0^s \left(R^3 \tau^{3/2} + 2 \, \mathrm{sh}\big[\tfrac{R\tau^{3/2}}{2}\big] \mathrm{ch}\big[2R\sqrt{\tau}\big] \right) d\tau$$

$$\leq \int_0^s (R^3 + 2R) \, \tau^{3/2} d\tau \leq R^3 s^{5/2} \quad \text{for } 0 \leq s \leq s_R > 0,$$

with probability $1 - \mathscr{O}(e^{-R^2/2})$.   Similarly

$$\left| y_s - \int_0^s \beta_\tau \, d\tau \right| \leq \int_0^s \left| \mathrm{ch}\big(w_\tau + (\lambda_\tau - w_\tau)\big) \, \mathrm{sh}\big(\beta_\tau + (\mu_\tau - \beta_\tau)\big) - \beta_\tau \right| d\tau =$$

$$\int_0^s \left| \big(1 + 2\mathrm{sh}^2\big[\tfrac{w_\tau}{2}\big] + 2\mathrm{sh}\big[\tfrac{\lambda_\tau - w_\tau}{2}\big]\mathrm{sh}\big[\tfrac{\lambda_\tau + w_\tau}{2}\big]\big)\big(\mathrm{sh}\, \beta_\tau + 2\mathrm{sh}\big[\tfrac{\mu_\tau - \beta_\tau}{2}\big]\mathrm{ch}\big[\tfrac{\mu_\tau + \beta_\tau}{2}\big]\big) - \beta_\tau \right| d\tau$$

$$\leq \int_0^s \left(R^3 \tau^{3/2} + R \, \tau^{3/2} + \mathscr{O}(R \, \tau^{5/2}) \right) d\tau \leq R^3 s^{5/2} \quad \text{for } 0 \leq s \leq s'_R > 0,$$

with probability $1 - \mathscr{O}(e^{-R^{2/3}/4})$.   Finally, in the same way we obtain:

$$\left| x_s - \Big(s + \tfrac{1}{2}\int_0^s [\beta_\tau^2 + w_\tau^2] d\tau\Big) \right|$$

$$\leq \int_0^s \left| \mathrm{ch}\big(w_\tau + (\lambda_\tau - w_\tau)\big) \, \mathrm{ch}\big(\beta_\tau + (\mu_\tau - \beta_\tau)\big) - 1 - \tfrac{1}{2}[\beta_\tau^2 + w_\tau^2] \right| d\tau =$$

$$\int_0^s \left| \big(\mathrm{ch}\, w_\tau + 2\mathrm{sh}\big[\tfrac{\lambda_\tau - w_\tau}{2}\big]\mathrm{sh}\big[\tfrac{\lambda_\tau + w_\tau}{2}\big]\big)\big(\mathrm{ch}\, \beta_\tau + 2\mathrm{sh}\big[\tfrac{\mu_\tau - \beta_\tau}{2}\big]\mathrm{sh}\big[\tfrac{\mu_\tau + \beta_\tau}{2}\big]\big) - 1 - \tfrac{\beta_\tau^2 + w_\tau^2}{2} \right| d\tau$$

$$\leq \int_0^s \left(\big|2\mathrm{sh}^2\big[\tfrac{w_\tau}{2}\big] - \tfrac{w_\tau^2}{2}\big| + \big|2\mathrm{sh}^2\big[\tfrac{\beta_\tau}{2}\big] - \tfrac{\beta_\tau^2}{2}\big| + \mathscr{O}(R^2 \tau^2) \right) d\tau$$

$$\leq \int_0^s \left(R^4 \tau^2 + \mathscr{O}(R^2 \tau^2) \right) d\tau \leq R^4 s^3 \leq s^{5/2} \quad \text{for } 0 \leq s \leq s''_R > 0,$$

with probability $1 - \mathscr{O}(e^{-R^{2/3}/4})$. In particular we can take $\kappa = 2/9$ in the statement. $\diamond$

Proof of Lemma 4.1.1 Let us use [19, Chapter (2) "The laws of some quadratic functionals of Brownian motion"], considering for any $b > 0$ the exponential martingale

$$M_s^b := \exp\left(-\tfrac{b}{2}\left(w_s^2 - w_0^2 - s\right) - \tfrac{b^2}{2}\int_0^s w_\tau^2\, d\tau\right) = \exp\left(-b\int_0^s w_\tau\, dw_\tau - \tfrac{b^2}{2}\int_0^s w_\tau^2\, d\tau\right),$$

and the new probability $\mathbb{P}^b$ having on $\mathscr{F}_s$ density M_s^b with respect to $\mathbb{P}$. As noticed in [19, (2.1.1)], Girsanov's Theorem yields a real $(\mathbb{P}^b, \mathscr{F}_s)$-Brownian motion $(B_u, 0 \le u \le s)$ such that $w_u = w_0 + B_u - b\int_0^u w_\tau\, d\tau$, which means that under $\mathbb{P}^b$, w has become an Ornstein-Uhlenbeck process, alternatively expressed by $w_u = e^{-bu}\left(w_0 + \int_0^u e^{b\tau}\, dB_\tau\right)$. Therefore

$$\mathbb{E}_0\left[\exp\left(\sqrt{-1}\int_0^s (a + c\,\tau)\, dw_\tau - \tfrac{b^2}{2}\int_0^s w_\tau^2\, d\tau\right)\right]$$

$$= e^{-bs/2} \times \mathbb{E}_0^b\left[\exp\left(\sqrt{-1}\int_0^s (a + c\,\tau)\, dw_\tau + b\, w_s^2/2\right)\right]$$

on the one hand, and on the other hand for any test-function f on $\mathbb{R}$:

$$\int_0^s f(u)\, dw_u = \int_0^s f(u)\left[dB_u - b\, e^{-bu}\left(\int_0^u e^{b\tau}\, dB_\tau\right) du\right]$$

$$= \int_0^s \left(f(\tau) - b\, e^{b\tau}\int_\tau^s f(u)\, e^{-bu}\, du\right) dB_\tau,$$

so that

$$\mathbb{E}_0^b\left[\left(\int_0^s f(\tau)\, dw_\tau\right)^2\right] = \int_0^s \left[f(\tau) - b\, e^{b\tau}\int_\tau^s f(u)\, e^{-bu}\, du\right]^2 d\tau.$$

Taking $f(\tau) = a + c\,\tau$, we obtain

$$\mathbb{E}_0^b\left[\left(a\, w_s + c\int_0^s \tau\, dw_\tau\right)^2\right]$$

$$= \int_0^s \left[a + c\,\tau + (a + c/b)(e^{b(\tau-s)} - 1) + c\,(s\, e^{b(\tau-s)} - \tau)\right]^2 d\tau$$

$$= b^{-2}\int_0^s \left[(ab + c + bc\, s)\, e^{b(\tau-s)} - c\right]^2 d\tau$$

$$= b^{-2} \int_0^s \left[(ab + c + bc\, s)^2 \, e^{2b(\tau - s)} - 2c\,(ab + c + bc\, s)\, e^{b(\tau - s)} + c^2 \right] d\tau$$

$$= b^{-3} \left[\tfrac{1}{2} (b\, a + (bs + 1)c)^2 \, (1 - e^{-2bs}) - 2c\,(b\, a + (bs + 1)c)\,(1 - e^{-bs}) + bs\, c^2 \right]$$

$$= \tfrac{1 - e^{-2bs}}{2\,b}\, a^2 + \tfrac{bs - 1 + 2e^{-b\, s} - (bs + 1)e^{-2bs}}{b^2}\, ac + \tfrac{b^2 s^2 - 3 + 4\,(bs + 1)\,e^{-bs} - (bs + 1)^2\, e^{-2bs}}{2\, b^3}\, c^2 \, .$$

This yields the covariance matrix of the $\mathbb{P}_0^b$-Gaussian variable $\left(w_s, \int_0^s \tau \, dw_\tau \right)$, hence its joint law. Namely the covariance matrix under $\mathbb{P}_0^b$ of $\left(\sqrt{2b}\, w_s, \right.$ $\left. \sqrt{2b^3} \int_0^s \tau \, dw_\tau \right)$ is $\quad K_{bs} =$

$$= \begin{pmatrix} 1 - e^{-2bs} & bs - 1 + 2e^{-bs} - (bs + 1)e^{-2bs} \\ bs - 1 + 2e^{-bs} - (bs + 1)e^{-2bs} & b^2 s^2 - 3 + 4\,(bs + 1)\, e^{-bs} - (bs + 1)^2\, e^{-2bs} \end{pmatrix},$$

and its determinant is $\delta_{bs} := 2(bs - 2) + 8\, e^{-b\, s} - 2(bs + 2)\, e^{-2b\, s}$, which increases with $bs > 0$ and then does not vanish. Therefore the density of $\left(\sqrt{2b}\, w_s \, , \, \sqrt{2b^3} \int_0^s \tau \, dw_\tau \right)$ is

$$(u, v) \longmapsto \frac{1}{2\pi \sqrt{\delta_{bs}}} \exp\left[\frac{-1}{2\,\delta_{bs}} \left(\alpha_{bs}\, u^2 + 2\,\gamma_{bs}\, uv + \left(1 - e^{-2bs}\right) v^2 \right) \right],$$

with

$$\alpha_x := x^2 - 3 + 4\,(x + 1)\, e^{-x} - (x + 1)^2\, e^{-2x} \quad \text{and} \quad \gamma_x := 1 - x - 2e^{-x} + (x + 1)e^{-2x} \, ,$$

so that $\delta_x = (1 - e^{-2x})\,\alpha_x - \gamma_x^2 \, .$ Hence

$$\mathbb{E}_0^b \left[\exp\left(\sqrt{-1} \int_0^s (a + c\,\tau)\, dw_\tau + b\, w_s^2 / 2 \right) \right] =$$

$$\frac{1}{2\pi \sqrt{\delta_{bs}}} \int_{\mathbb{R}^2} e^{\frac{a\sqrt{-1}}{\sqrt{2b}} u + \frac{c\sqrt{-1}}{\sqrt{2b^3}} v + u^2/4} \exp\left[\frac{-\alpha_{bs}\, u^2 - 2\gamma_{bs}\, uv - (1 - e^{-2bs}) v^2}{2\,\delta_{bs}} \right] du\, dv$$

$$= \frac{1}{\sqrt{2\pi(1 - e^{-2bs})}} \int_{\mathbb{R}} e^{\frac{a\sqrt{-1}}{\sqrt{2b}} u + \frac{\delta_{bs}}{2(1 - e^{-2bs})} \left(\frac{c\sqrt{-1}}{\sqrt{2b^3}} - \frac{\gamma_{bs}}{\delta_{bs}} u \right)^2 + \left(\frac{1}{4} - \frac{\alpha_{bs}}{2\,\delta_{bs}} \right) u^2} du$$

$$= \frac{e^{\frac{-\delta_{bs}\, c^2}{4(1 - e^{-2bs})b^3}}}{\sqrt{2\pi(1 - e^{-2bs})}} \int_{\mathbb{R}} e^{\left(\frac{a\sqrt{-1}}{\sqrt{2b}} - \frac{c\sqrt{-1}\,\gamma_{bs}}{(1 - e^{-2bs})\sqrt{2b^3}} \right) u - \left(\frac{\alpha_{bs}}{\delta_{bs}} - \frac{1}{2} - \frac{\gamma_{bs}^2}{(1 - e^{-2bs})\delta_{bs}} \right) u^2/2} du$$

$$= \frac{e^{\frac{-\delta_{bs}\, c^2}{4(1 - e^{-2bs})b^3}}}{\sqrt{2\pi(1 - e^{-2bs})}} \int_{\mathbb{R}} e^{\left(a - \frac{c\,\gamma_{bs}}{b\,(1 - e^{-2bs})} \right) \frac{\sqrt{-1}}{\sqrt{2b}} u - \left(\frac{1 + e^{-2bs}}{1 - e^{-2bs}} \right) u^2/4} du$$

$$= \frac{2\, e^{\frac{-\delta_{bs}\, c^2}{4(1-e^{-2bs})b^3}}}{\sqrt{2(1+e^{-2bs})}} \times e^{\frac{-1}{2b^3} \times \frac{1-e^{-2bs}}{1+e^{-2bs}} \left(ab - \frac{c\,\gamma_{bs}}{1-e^{-2bs}}\right)^2}$$

$$= \frac{e^{bs/2}}{\sqrt{\mathrm{ch}(bs)}}\, \exp\left[\frac{-1}{4b^3}\left(\frac{\delta_{bs}\, c^2}{1-e^{-2bs}} + 2\,\frac{1-e^{-2bs}}{1+e^{-2bs}}\left(a\,b - \frac{\gamma_{bs}\, c}{1-e^{-2bs}}\right)^2\right)\right]$$

$$= \frac{e^{bs/2}}{\sqrt{\mathrm{ch}(bs)}}\, \exp\left[\frac{-1}{4b^3(1+e^{-2bs})}\left(2\,(1-e^{-2bs})\,b^2 a^2 + (2\alpha_{bs} - \delta_{bs})\,c^2 - 4\,b\,\gamma_{bs}\, ac\right)\right].$$

Therefore for any $b, s > 0$ and real a, c we obtain:

$$\mathbb{E}_0\left[\exp\left(\sqrt{-1}\int_0^s (a + c\,\tau)\, dw_\tau - \frac{b^2}{2}\int_0^s w_\tau^2\, d\tau\right)\right]$$

$$= \frac{1}{\sqrt{\mathrm{ch}(bs)}}\, \exp\left[\frac{-1}{4b^3(1+e^{-2bs})}\left(2\,(1-e^{-2bs})\,b^2 a^2 + (2\alpha_{bs} - \delta_{bs})\,c^2 - 4\,\gamma_{bs}\,b\, ac\right)\right].$$

Then taking $a = r + c\,s$, we get:

$$\mathbb{E}_0\left[\exp\left(\sqrt{-1}\left[r\,w_s + c\int_0^s w_\tau\, d\tau\right] - \frac{b^2}{2}\int_0^s w_\tau^2\, d\tau\right)\right]$$

$$= \mathbb{E}_0\left[\exp\left(\sqrt{-1}\int_0^s (r + c\,s - c\,\tau)\, dw_\tau - \frac{b^2}{2}\int_0^s w_\tau^2\, d\tau\right)\right]$$

$$= \frac{1}{\sqrt{\mathrm{ch}(bs)}}\, \exp\left[-\frac{2(1-e^{-2bs})\,b^2(r + c\,s)^2 + (2\alpha_{bs} - \delta_{bs})\,c^2 + 4\gamma_{bs}\,b\,(r + c\,s)\,c}{4b^3(1+e^{-2bs})}\right]$$

$$= \frac{1}{\sqrt{\mathrm{ch}(bs)}}\, \exp\left[-\frac{(1-e^{-2bs})\,b^2(r^2 + 2s\,rc) + \left(bs\,(1+e^{-2bs}) - (1-e^{-2bs})\right)c^2 + 2b\,\gamma_{bs}\,rc}{2b^3(1+e^{-2bs})}\right]$$

$$= \frac{1}{\sqrt{\mathrm{ch}(bs)}}\, \exp\left[-\frac{\mathrm{th}(bs)}{2b}\, r^2 - \frac{bs - \mathrm{th}(bs)}{2b^3}\, c^2 - 2\,\frac{\mathrm{sh}^2(bs/2)}{b^2\,\mathrm{ch}(bs)}\, rc\right]. \qquad \diamond$$

Proof of Proposition 5.2 Denote by $\lambda_0 \in \mathbb{R}_+$ the abscissa of convergence of the integral, so that the map $\lambda \mapsto \displaystyle\int_0^\infty e^{\lambda x}\, q_1(w, \beta, x, \zeta, z)\, dx$ is analytic on $\{\Re(\lambda) < \lambda_0\}$. By Proposition 4.2.2 and Lemma 5.1 it is equal to $\Phi(\lambda)$ for $\Re(\lambda) < \min\{4\pi^2, \lambda_0\}$. Hence, for any real $\lambda < \min\{4\pi^2, \lambda_0\}$ and $t \in \mathbb{R}$ we have

$$\Phi(\lambda + \sqrt{-1}\, t) = \int_0^\infty e^{\sqrt{-1}\, t\, x}\, e^{\lambda x}\, q_1(w, \beta, x, \zeta, z)\, dx.$$

Let us show now that $t \mapsto \Phi(\lambda + \sqrt{-1}\, t)$ belongs to $L^1 \cap L^2(\mathbb{R})$, in order to inverse the above Fourier transform. Of course we have to deal here with the large values of

$|t|$, i.e., of $|\lambda + \sqrt{-1}\, t|$ in the expression (14):

$$\Phi_{w,\beta,\zeta,z}(\lambda) = \frac{\lambda^2\, e^{B'\lambda - \frac{\sqrt{\lambda}}{4}\cot g(\frac{\sqrt{\lambda}}{2})(w^2+\beta^2)}}{8\pi^2\big[1 - \cos(\sqrt{\lambda}) - (\sqrt{\lambda}/2)\sin(\sqrt{\lambda})\big]}\, \exp\left[\frac{B^2\lambda}{1 - \frac{2}{\sqrt{\lambda}}\,\mathrm{tg}(\frac{\sqrt{\lambda}}{2})}\right],$$

in which λ is to be replaced by $\lambda + \sqrt{-1}\, t$, and we have set $B^2 := \frac{(w-2z)^2+(\beta-2\zeta)^2}{8}$ and $B' := \frac{4wz+4\beta\zeta-w^2-\beta^2}{8} = \frac{z^2+\zeta^2}{2} - B^2$. Then for any $t \in \mathbb{R}$ we have:

$$\frac{\sqrt{\lambda+\sqrt{-1}\,t}}{2} =: \alpha + \sqrt{-1}\, b = \sqrt{\frac{|\lambda+\sqrt{-1}\,t|+\lambda}{8}} + \sqrt{-1}\,\mathrm{sign}(t)\sqrt{\frac{|\lambda+\sqrt{-1}\,t|-\lambda}{8}}$$

and

$$\mathrm{tg}\left[\frac{\sqrt{\lambda+\sqrt{-1}\,t}}{2}\right] = \frac{\mathrm{tg}\,\alpha + \sqrt{-1}\,\mathrm{th}\,b}{1 - \sqrt{-1}\,\mathrm{tg}\,\alpha\,\mathrm{th}\,b} = \frac{\sin(2\alpha) + \sqrt{-1}\,\mathrm{sh}(2b)}{\mathrm{ch}(2b) + \cos(2\alpha)}$$

$$= \sqrt{-1}\,\mathrm{sign}(t) + \mathcal{O}\left(e^{-\sqrt{|t|/2}}\right).$$

Therefore, for large $|t|$ we have:

$$\exp\left[B'(\lambda+\sqrt{-1}\,t)-(w^2+\beta^2)\frac{\sqrt{\lambda+\sqrt{-1}\,t}}{4}\cot g\left(\frac{\sqrt{\lambda+\sqrt{-1}\,t}}{2}\right) + \frac{B(\lambda+\sqrt{-1}\,t)}{1-\frac{2}{\sqrt{\lambda+\sqrt{-1}\,t}}\mathrm{tg}\left(\frac{\sqrt{\lambda+\sqrt{-1}\,t}}{2}\right)}\right]$$

$$= e^{\left(\frac{z^2+\zeta^2}{2}\right)\lambda}\exp\left[\left(\frac{z^2+\zeta^2}{2}\right)\sqrt{-1}\,t + \sqrt{-1}\,\mathrm{sign}(t)(\frac{w^2+\beta^2}{4}+2B)\sqrt{\lambda+\sqrt{-1}\,t}\left[1+\mathcal{O}\left(|t|^{-1/2}\right)\right]\right],$$

the modulus of which is

$$e^{(z^2+\zeta^2)\lambda/2}\exp\left[-(\tfrac{w^2+\beta^2}{4} + 2B^2)\sqrt{\frac{|\lambda+\sqrt{-1}\,t|-\lambda}{2}} + \mathcal{O}(1)\right].$$

Moreover

$$\cos(\sqrt{\lambda + \sqrt{-1}\, t}) = \mathrm{ch}(2b)\cos(2\alpha) - \sqrt{-1}\,\mathrm{sh}(2b)\sin(2\alpha)$$

and

$$\sin(\sqrt{\lambda + \sqrt{-1}\, t}) = \mathrm{ch}(2b)\sin(2\alpha) + \sqrt{-1}\,\mathrm{sh}(2b)\cos(2\alpha)$$

entail

$$\left|1 - \cos(\sqrt{\lambda + \sqrt{-1}\, t}) - (\sqrt{\lambda + \sqrt{-1}\, t}/2)\sin(\sqrt{\lambda + \sqrt{-1}\, t})\right|^2$$

$$= [\text{ch}(2b)-\cos(2\alpha)]^2 + (\alpha^2+b^2)[\text{ch}^2(2b)-\cos^2(2\alpha)]-2[\text{ch}(2b)-\cos(2\alpha)][b\,\text{sh}(2b)+\alpha\sin(2\alpha)]$$

$$= [\text{ch}(2b) - \cos(2\alpha) - b\,\text{sh}(2b) - \alpha\sin(2\alpha)]^2 + [\alpha\,\text{sh}(2b) - b\sin(2\alpha)]^2$$

$$= (\alpha^2 + b^2)\text{sh}^2(2b) + \mathscr{O}\big(\alpha\,\text{sh}^2(2b)\big) = \left[\frac{|\lambda+\sqrt{-1}\,t|}{4} + \mathscr{O}\big(\sqrt{|t|}\big)\right]\text{sh}^2\sqrt{\frac{|\lambda+\sqrt{-1}\,t|-\lambda}{2}}$$

for large $|t|$. So far, for $\lambda < 4\pi^2$ and for large $|t|$ we have:

$$\left|\Phi_{w,\beta,\zeta,z}(\lambda + \sqrt{-1}\,t)\right|$$

$$= (\lambda^2+t^2)^{3/4}\,e^{(z^2+\zeta^2)\lambda/2}\,\exp\left[-\left(\frac{w^2+\beta^2}{4}+2B^2+1\right)\sqrt{\frac{\sqrt{\lambda^2+t^2}-\lambda}{2}}+\mathscr{O}(1)\right]\left[1+\mathscr{O}\big(|t|^{-1/2}\big)\right]$$

$$= \mathscr{O}\big(e^{2\pi^2(z^2+\zeta^2)}\big)|t|^{3/2}\exp\left[-\left(\frac{w^2+\beta^2}{4}+2B^2+1+\mathscr{O}(|t|^{-1})\right)\sqrt{|t|/2}\right]$$

$$= \mathscr{O}\big(e^{2\pi^2(z^2+\zeta^2)}\big)\exp\left[-\left(\frac{w^2+\beta^2}{4}+2B^2+1/2\right)\sqrt{|t|/2}\right].$$

Hence we can inverse the above Fourier transform for $\lambda < \min\{4\pi^2, \lambda_0\}$, and thus we obtain the wanted (15), which holds a posteriori for $\lambda < 4\pi^2$:

$$e^{\lambda x}q_1(w,\beta,x,\zeta,z) = \int_{-\infty}^{\infty} e^{-\sqrt{-1}\,t\,x}\,\Phi(\lambda + \sqrt{-1}\,t)\,\frac{dt}{2\pi}, \quad \text{for } x > 0 \text{ and for } \lambda < 4\pi^2.$$

Thence, for any real $\lambda < 4\pi^2$, taking $\varepsilon < 4\pi^2 - \lambda$ for positive x we have

$$e^{\lambda x}\,q_1(w,\beta,x,\zeta,z) = \mathscr{O}\big(e^{-\varepsilon x}\big) \times \int_{-\infty}^{\infty} \left|\Phi(\lambda + \varepsilon + \sqrt{-1}\,t)\right| dt = \mathscr{O}\big(e^{-\varepsilon x}\big),$$

which entails the integrability of $x \mapsto e^{\lambda x}\,q_1(w,\beta,x,\zeta,z)$, so that finally $\lambda_0 \geq 4\pi^2$. $\diamond$

References

1. R. Azencott, *Densité des Diffusions en temps petit : Développements Asymptotiques*. Sém. Proba. XVIII, Lecture Notes, vol. 1059 (Springer, New York, 1984), pp. 402–498
2. G. Ben Arous, Développement asymptotique du noyau de la chaleur hypoelliptique hors du cut-locus. Ann. sci. É.N.S. Sér. 4, **21**(3), 307–331 (1988)
3. G. Ben Arous, Développement asymptotique du noyau de la chaleur hypoelliptique sur la diagonale. Ann. Inst. Fourier **39**(1), 73–99 (1989)
4. I. Bailleul, J. Franchi, Non-explosion criteria for relativistic diffusions. Ann. Probab. **40**(5), 2168–2196 (2012)
5. P. Biane, J. Pitman, M. Yor, Probability laws related to the Jacobi theta and Riemann zeta functions, and Brownian excursions. Bull. A.M.S. **38**(4), 435–465 (2001)
6. F. Castell, Asymptotic expansion of stochastic flows. Probab. Theor. Relat. Field **96**, 225–239 (1993)
7. E.T. Copson, *Asymptotic Expansions* (Cambridge University Press, Cambridge, 1965)

8. T. Chan, D.S. Dean, K.M. Jansons, L.C.G. Rogers, On polymer conformations in elongational flows. Commun. Math. Phys. **160**, 239–257 (1994)
9. F. Delarue, S. Menozzi, Density estimates for a random noise propagating through a chain of differential equations. J. Funct. Anal. **259**, 1577–1630 (2010)
10. R.M. Dudley, Lorentz-invariant Markov processes in relativistic phase space. Arkiv för Matematik **6**(14), 241–268 (1965)
11. A.F.M. ter Elst, D.W. Robinson, A. Sikora, Small time asymptotics of diffusion processes. J. Evol. Equ. **7**(1), 79–112 (2007)
12. J. Franchi, Y. Le Jan, Relativistic diffusions and schwarzschild geometry. Commun. Pure Appl. Math. **LX**(2), 187–251 (2007)
13. J. Franchi, Y. Le Jan, Curvature diffusions in general relativity. Commun. Math. Phys. **307**(2), 351–382 (2011)
14. J. Franchi, Y. Le Jan, *Hyperbolic Dynamics and Brownian Motion*. Oxford Mathematical Monographs (Oxford Science, Oxford, 2012)
15. R. Léandre, Intégration dans la fibre associée à une diffusion dégénérée. Probab. Theor. Relat. Field **76**(3), 341–358 (1987)
16. A. Nagel, E.M. Stein, S. Wainger, Balls and metrics defined by vector fields I : Basic properties. Acta Math. **155**, 103–147 (1985)
17. C. Tardif, A Poincaré cone condition in the Poincaré group. Pot. Anal. **38**(3), 1001–1030 (2013)
18. S.R.S. Varadhan, Diffusion processes in a small time interval. Commun. Pure Appl. Math. **20**, 659–685 (1967)
19. M. Yor, *Some Aspects of Brownian Motion. Part I. Some Special Functionals*. Lectures in Mathematics, E.T.H. Zürich (Birkhäuser Verlag, Basel, 1992)

Onsager-Machlup Functional for Uniformly Elliptic Time-Inhomogeneous Diffusion

Koléhè A. Coulibaly-Pasquier

Abstract In this paper, we will compute the Onsager-Machlup functional of an inhomogeneous uniformly elliptic diffusion process. This functional is very similar to the corresponding functional for homogeneous diffusions; indeed, the only difference come from the infinitesimal variation of the volume. We will also use the Onsager-Machlup functional to study small ball probability for weighted sup-norm of some inhomogeneous diffusion.

1 Introduction

Let M be a n-dimensional Riemannian manifold, and L_t be an inhomogeneous uniformly elliptic second order operator over M, without constant term. It is always possible to endow M with a time-dependent family of metrics $g(t)$ such that

$$L_t = \frac{1}{2}\Delta_t + Z(t), \tag{1}$$

where Δ_t is a Laplace Beltrami operator for the metric $g(t)$ and $Z(t,.)$ is a time-dependent vector field on M. Let $X_t(x_0)$ be a L_t-diffusion process on M, starting at the point x_0. An example of such a diffusion is the $g(t)$-Brownian motion, introduced in [2], where the family of metrics $(g(t))_{t\in[0,T]}$ comes from the Ricci flow on M.

Let $d(t, x, y)$ be the Riemannian distance on M according to the metric $g(t)$. Consider a smooth curve $\varphi : [0, T] \to M$, such that $\varphi(0) = x_0$. We are now interested in the asymptotic equivalent as ϵ goes to zero of the following probability

$$\mathbb{P}_{x_0}[\forall t \in [0, T] \quad d(t, X_t, \varphi(t)) \leq \epsilon].$$

K.A. Coulibaly-Pasquier (✉)
Institut Élie Cartan de Lorraine, BP 78239, 54506 Vandœuvre-lès-Nancy, France
e-mail: kolehe.coulibaly@univ-lorraine.fr

© Springer International Publishing Switzerland 2014
C. Donati-Martin et al. (eds.), *Séminaire de Probabilités XLVI*, Lecture Notes in Mathematics 2123, DOI 10.1007/978-3-319-11970-0_5

This asymptotic will be expressed as the product of two terms. The first one is a decreasing function of ϵ that does not depend on the curve nor geometries (except the dimension), while the second term depends on the geometries along the curve φ. This second term is expressed as a Lagrangian. So maximizing this term reduces to finding the most probable path of the diffusion. This term is usually called the Onsager-Machlup functional of the diffusion X_t.

To compute the O.M. functional, we will use both the technics introduced by Takahashi and Watanabe in [7], and the non-singular drift introduced by Hara. Using this drift, Hara and Takahashi made in [4] a substantial simplification of the latter proof (of O.M. functional) of Takahashi and Watanabe.

We propose here to introduce a time-dependent parallel transport along a curve, according to a family of metrics. It will allow us to compute the Onsager-Machlup functional in the time-inhomogeneous case.

Let $\mathrm{div}_{g(t)}$ and $R_{g(t)}$ be respectively the divergence operator and the scalar curvature with respect to the metric $g(t)$. Let H be a time-dependent function on the tangent bundle defined for $v \in T_x M$ as:

$$H(t, x, v) = \frac{1}{2} \|Z(t, x) - v\|_{g(t)}^2 + \frac{1}{2} \, \mathrm{div}_{g(t)}(Z)(t, x) - \frac{1}{12} R_{g(t)}(x)$$
$$+ \frac{1}{4} \, \mathrm{trace}_{g(t)}(\dot{g}(t)).$$

The main result of this paper is the following:

Theorem 1 *Let $X_t(x_0)$ be a L_t diffusion process starting at point x_0, where $L_t = \frac{1}{2}\Delta_t + Z(t, .)$. Then we have the following asymptotic:*

$$\mathbb{P}_{x_0}[\forall t \in [0, T], d(t, X_t, \varphi(t)) \leq \epsilon]$$
$$\sim_{\epsilon \downarrow 0} C \exp\{-\frac{\lambda_1 T}{\epsilon^2}\} \exp\{-\int_0^T H(t, \varphi(t), \dot{\varphi}(t)) \, dt\}.$$

Here C and λ_1 are explicit constants.

A similar result was obtained by [1, 4, 7] in the homogeneous case. Our contribution comes from the time-inhomogeneity of the diffusion.

The paper is organized as follows: first, we will define in Sect. 2 a parallel transport along a curve according to a family of metrics $(g(t)_t)$. This parallel transport will enable us to obtain a Fermi coordinates in a neighborhood of a smooth curve φ. We will also give a (local) development of a tensor that will be used in the following.

Then, we will introduce some useful tools in Sect. 3. They are not new, and clearly exposed by Capitaine in [1]. So we will keep the same notation as in [1] in this paper. In [1], the author has investigated the case of different norms, in the homogeneous case. In the literature, non smooth functions φ are also considered, but this will not be discussed here. In the second part of Sect. 3, we will establish the proof of Theorem 1. Finally, Sect. 4 is devoted to some applications. First, we will describe the most probable path of an inhomogeneous diffusion. Then, we will obtain a small ball estimate (for the weighted sup-norm) for inhomogeneous diffusions.

2 Parallel Transport Along a Curve, and Fermi Coordinate

Let $\varphi : [0, T] \longrightarrow M$ be a smooth curve. Suppose that the manifold M is endowed with a family of metrics $g(t)_{t \in [0,T]}$ which is C^1 in time and C^2 in space. This family of metrics induces a time dependent family of Levi-Civita connexions, denoted by ∇^t.

Suppose that A is a bilinear form on a given vector space E. Let v, w be in E. Suppose that there exists a scalar product $\langle ., . \rangle_{g(t)}$ on E. Then define $A^{\#g(t)}(v) \in E$ as the element of E such that $\langle A^{\#g(t)}v, w \rangle_{g(t)} = A(v, w)$.

Let v be a vector on $T_{\varphi(0)}M$ and define $\tau_t v$ as the solution of the following ODE:

$$\begin{cases} \nabla^t_{\dot{\varphi}(t)}(\tau_t v) = -\frac{1}{2}\dot{g}(t)^{\#g(t)}(\tau_t v) \\ \tau_0 v = v. \end{cases}$$

The map $\tau_t v$ is called a parallel transport of v along the curve φ according to the family of metrics $g(t)$.

Proposition 1 *The parallel transport τ_t is an isometry between the tangent space $(T_{\varphi(0)}M, g(0))$ and the tangent space $(T_{\varphi(t)}M, g(t))$. In particular, if $(e_1, e_2, \ldots, e_n)$ is an orthonormal basis of $T_{\varphi(0)}M$ for the metric $g(0)$, then $(\tau_t e_1, \tau_t e_2, \ldots, \tau_t e_n)$ is an orthonormal basis of $T_{\varphi(t)}M$ for the metric $g(t)$.*

Proof Let v, w be in $T_{\varphi(0)}M$, we have:

$$\frac{d}{dt}\langle \tau_t v, \tau_t w \rangle_{g(t)} = \nabla^t g(t)(\tau_t v, \tau_t w) + \langle \nabla^t \tau_t v, \tau_t w \rangle_{g(t)} + \langle \tau_t v, \nabla^t \tau_t w \rangle_{g(t)}$$

$$+ \dot{g}(t)(\tau_t v, \tau_t w)$$

$$= -\frac{1}{2}\langle \dot{g}(t)^{\#g(t)}(\tau_t v), \tau_t w \rangle_{g(t)} - \frac{1}{2}\langle \tau_t v, \dot{g}(t)^{\#g(t)}(\tau_t w) \rangle_{g(t)}$$

$$+ \dot{g}(t)(\tau_t v, \tau_t w) = 0.$$

We are now able to write the Fermi coordinates in a neighborhood of a curve. Let $\varphi : [0, T] \longrightarrow M$ be a smooth curve and let τ be the parallel transport above φ in the sense of Proposition 1, where we have fixed a $g(0)$-orthonormal basis $(e_1, \ldots, e_n)$

of $T_{\varphi(0)}M$. Consider the map

$$\Psi : [0, T] \times \mathbb{R}^n \longrightarrow [0, T] \times M$$

$$(t, v_1, \ldots, v_n) \longmapsto (t, \exp_{\varphi(t)}^{g(t)}(\tau_t \sum_{1}^{n} v_i e_i)),$$

where $\exp_x^{g(t)}$ means the exponential map for the metric $g(t)$. The map Ψ is clearly a diffeomorphism on some neighborhood U of $[0, T] \times 0$. Define now $V = \Psi(U)$. Remark that, for each fixed t, the map $\Psi(t, .)$ is the normal coordinates for the metric $g(t)$ in a neighborhood of the point $\varphi(t)$.

Let $X_t(x_0)$ be an L_t-diffusion starting at the point x_0, where L_t is a time-dependent operator as in (1). Using these Fermi coordinates, the time-dependent distance of Theorem 1 can be translated in terms of Euclidean norm, while the generator will be the pull back operator of L_t by Ψ. By assumption, the generator of (t, X_t) is $\partial_t + \frac{1}{2}\Delta_t + Z(t, .)$. We will now compute the generator of $\Psi^{-1}(t, X_t)$, or more precisely its local development:

$$\Psi^*(\partial_t + \frac{1}{2}\Delta_t + Z(t, .)) = \frac{\tilde{\partial}}{\partial_t} + \frac{1}{2}\tilde{\Delta}_t + \tilde{Z}(t, .). \tag{2}$$

The second term in the right hand side is computed in [2] as:

$$\tilde{\Delta}_t = g^{ij}(\Psi(t, .))\frac{\partial}{\partial x_i}\frac{\partial}{\partial x_j} - g^{kl}(\Psi(t, .))\Gamma_{kl}^i(\Psi(t, .))\frac{\partial}{\partial x_i},$$

where $(x_1^t, \ldots, x_n^t)$, $g_{ij}(\Psi(t, .))$, $g^{ij}(\Psi(t, .))$, and $\Gamma_{kl}^i(\Psi(t, .))$ are respectively the normal coordinates at the point $\varphi(t)$ for the metric $g(t)$ with respect to the vector basis $(\tau_t e_1, \ldots, \tau_t e_1)$, the coefficient of metric $g(t)$ in this basis, its inverse, and the Christoffel symbols of the Levi-Civita connexion of the metric $g(t)$ in this basis.

Clearly we have

$$\tilde{Z}(t, .) = \sum_{i=1}^{n} Z^i(t, .)\frac{\partial}{\partial x_i}$$

where $Z^i(t, .) = \langle Z(\Psi(t, .)), \frac{\partial}{\partial x_i^t}|_{\Psi(t,.)}\rangle_{g(t)}$.

Recall that V is a neighborhood of $\{(t, \varphi(t)), t \in [0, T]\}$ and $V = \Psi(U)$. For a point $(t, x) \in V$ such that Ψ induces a diffeomorphism in (t, x), we will write $\Psi^{-1}(t, x) = (t, x_1^t, \ldots, x_n^t) \in [0, T] \times \mathbb{R}^n$.

To study (2), we have to compute $\frac{\tilde{\partial}}{\partial_t}|_{(t,x)} = \sum_{i=1}^{n} a_i(t, x)\frac{\partial}{\partial x_i} + a_0(t, x)\frac{\partial}{\partial t}$. For any $1 \leq i \leq n$, we easily see that $a_i(t_0, x) = \frac{\partial}{\partial t}|_{t_0}(x_i^t)|_{\Psi(t_0, x)}$. For any fixed $x \in M$, we have the equality:

$$\frac{\partial}{\partial t}\left(\exp^{g(t)}(\varphi(t), \sum_{i=1}^{n} \tau_t e_i x_i^t)\right) = 0,$$

where for any $v \in T_x M$, $\exp^{g(t)}(x, v)$ is the exponential map for the metric $g(t)$ at the point x. The next technical result will be useful to compute the term $a_i(t, x) = \frac{\partial}{\partial t}(x_i^t)|_{\psi(t,x)}$.

Lemma 1 *Let $v \in T_x M$. Then*

$$\frac{\partial}{\partial t}\Big|_{t_0} \exp^{g(t)}(x, v) = O(\|v\|_{g(t_0)}^2).$$

Proof Let $x_i(t, s)$ be the i-th coordinate of the geodesic $\exp^{g(t)}(x, s.v)$ in the normal coordinates system centered at $\varphi(t_0)$ with respect to the metric $g(t_0)$. In the following, we will shorten the notation an write $\dot{x}(t, s)$ for $\frac{\partial}{\partial s} x(t, s)$. The usual equation of geodesics shows that:

$$\frac{\partial}{\partial t}\Big|_{t_0} x_i(t, s) = -\sum_{jk} \frac{\partial}{\partial t}\Big|_{t_0}\left[\int_0^s du \int_0^u dl\, \Gamma_{jk}^i(t, x(t, l))\dot{x}_j(t, l)\dot{x}_k(t, l)\right]$$

$$= \sum_{jk} -\int_0^s du \int_0^u dl\left(\frac{\partial}{\partial t}\Big|_{t_0} \Gamma_{jk}^i(t, x(t_0, l))\dot{x}_j(t_0, l)\dot{x}_k(t_0, l)\right.$$

$$+\langle d\Gamma_{jk}^i(t_0, .), \frac{\partial}{\partial t}\Big|_{t_0} x(t, l)\rangle \dot{x}_j(t_0, l)\dot{x}_k(t_0, l)$$

$$\left.+2\Gamma_{jk}^i(t_0, x(t_0, l))\frac{\partial}{\partial t}\Big|_{t_0}(\dot{x}_j(t, l))\dot{x}_k(t_0, l)\right).$$

Note that we have $\| \dot{x}(t_0, s) \|_{g(t_0)}^2 = \| v \|_{g(t_0)}^2$ and $\Gamma_{jk}^i(t_0, x) = O(\| x \|_{g(t_0)})$. In a neighborhood V of $\{(t, \varphi(t)), t \in [0, T]\}$, the quantities $| \frac{\partial}{\partial t}\Gamma_{jk}^i(t, .) |$ and $\| d\Gamma_{jk}^i(t, .) \|$ are bounded by some constant C, hence we get the equality:

$$\frac{\partial}{\partial t}\Big|_{t_0} x(t, s) := \frac{\partial}{\partial t}\Big|_{t_0}(x_1(t, s), \ldots, x_n(t, s))$$

$$= O(\|v\|_{g(t_0)}^2) + \int_0^s dl\, O(\|v\|_{g(t_0)}^2)\frac{\partial}{\partial t}\Big|_{t_0} x(t, l)$$

$$+ \int_0^s du \int_0^u dl\, O(\|v\|_{g(t_0)}^2)\frac{\partial}{\partial t}\Big|_{t_0} x(t, l).$$

By Gronwall's lemma we deduce that:

$$\| \frac{\partial}{\partial t}_{|_{t_0}} x(t,1) \| = O(\|v\|^2_{g(t_0)}).$$

Lemma 2 *Let* $(x_1(t), \ldots, x_n(t))$ *be the coordinates of*

$$\exp^{g(t_0)}(\varphi(t), \sum_{i=1}^{n} \tau_t e_i x_i^t)$$

in the normal coordinates system at the point $\varphi(t_0)$ *for the metric* $g(t_0)$ *with reference basis* $(\tau_{t_0} e_i)_{i=1,\ldots,n}$ *and* $\partial_i = \frac{\partial}{\partial_{x_i^{t_0}}}$ *be the associated vector field. Then we have:*

$$\frac{\partial}{\partial t}_{|_{t_0}} x_i(t) = \frac{\partial}{\partial t}_{|_{t_0}} x_i^t - \frac{1}{2}\frac{\partial}{\partial t}_{|_{t_0}} (g(t))_{\varphi(t_0)}(\partial_i, \sum_{j=1}^{n} x_j^{t_0}\partial_j) + \langle \frac{\partial}{\partial t}_{|_{t_0}} \varphi(t), \partial_i \rangle_{g(t_0)}$$

$$+ O(\| x^{t_0} \|^2).$$

Proof As in the previous proof, we write the geodesic

$$\exp^{g(t_0)}(\varphi(t), s. \sum_{i=1}^{n} \tau_t e_i x_i^t)$$

in normal coordinate. It satisfies the system:

$$\begin{cases} \ddot{x}_i(t,s) = -\sum_{jk} \Gamma^i_{jk}(t_0, x(t,s))\dot{x}_j(t,s)\dot{x}_k(t,s), \\[2mm] \dot{x}_i(t,0) = \langle \sum_{l=1}^{n} \tau_t e_l x_l^t, \partial_{i|_{\varphi(t)}} \rangle_{g(t_0)}, \\[2mm] x_i(t,0) = \varphi(t)^i. \end{cases}$$

Moreover, we have:

$$\frac{\partial}{\partial t}_{|_{t_0}} x_i(t,s) = -\int_0^s du \int_0^u dl \sum_{jk} \frac{\partial}{\partial t}_{|_{t_0}} \left[\Gamma^i_{jk}(t_0, x(t,l))\dot{x}_j(t,l)\dot{x}_k(t,l) \right]$$

$$+ s\frac{\partial}{\partial t}_{|_{t_0}} \dot{x}_i(t,0) + \frac{\partial}{\partial t}_{|_{t_0}} x_i(t,0). \tag{3}$$

Similarly to the latter proof, the equality (3) can be rewritten in a matrix. Then using again Gronwall's lemma, we see that $\frac{\partial}{\partial t}_{|_{t_0}} x_i(t,s)$ is bounded for any $s \in [0,1]$.

So the integral term of (3) is an $O(\|x^{t_0}\|_{g(t_0)}^2)$. Hence, we deduce that

$$\frac{\partial}{\partial t}_{|t_0} x_i(t,1) = O(\|x^{t_0}\|_{g(t_0)}^2) + \frac{\partial}{\partial t}_{|t_0} \langle \sum_{l=1}^{n} \tau_t e_l x_l^t, \partial_{i|_{\varphi(t)}} \rangle_{g(t_0)}$$

$$+ \langle \frac{\partial}{\partial t}_{|t_0} \varphi(t), \partial_{i|_{\varphi(t_0)}} \rangle_{g(t_0)}.$$

Remark that for $t = t_0$, we have $\partial_{i|_{\varphi(t_0)}} = \tau_{t_0} e_i$. So, we obtain the equality:

$$\frac{\partial}{\partial t}_{|t_0} x_i(t,1) = O(\|x^{t_0}\|_{g(t_0)}^2) + \frac{\partial}{\partial t}_{|t_0} x_i^t \delta_i^l$$

$$+ \sum_{l=1}^{n} x_l^{t_0} \frac{\partial}{\partial t}_{|t_0} \langle \tau_t e_l, \partial_{i|_{\varphi(t)}} \rangle_{g(t_0)} + \langle \frac{\partial}{\partial t}_{|t_0} \varphi(t), \partial_{i|_{\varphi(t_0)}} \rangle_{g(t_0)}.$$

By construction of the parallel transport τ, we have:

$$\frac{\partial}{\partial t}_{|t_0} \langle \tau_t e_l, \partial_{i|_{\varphi(t)}} \rangle_{g(t_0)} = \langle \nabla^{t_0} \tau_t e_l, \partial_{i|_{\varphi(t_0)}} \rangle_{g(t_0)} + \langle \tau_{t_0} e_l, \nabla^{t_0} \partial_i \rangle_{g(t_0)}$$

$$= -\frac{1}{2} \dot{g}(t_0) (\tau_{t_0} e_l, \partial_{i|_{\varphi(t_0)}})$$

$$= -\frac{1}{2} \dot{g}(t_0) (\partial_{l|_{\varphi(t_0)}}, \partial_{i|_{\varphi(t_0)}}).$$

The last term of the right hand side has vanished because ∂_i comes from normal coordinates for the metric $g(t_0)$. Putting all pieces together leads to the result.

Proposition 2 *We have*

$$\frac{\partial}{\partial t}_{|t_0} x_i^t = \frac{1}{2} \frac{\partial}{\partial t}_{|t_0} (g(t))_{\varphi(t_0)} (\partial_i, \sum_{j=1}^{n} x_j^{t_0} \partial_j) - \langle \frac{\partial}{\partial t}_{|t_0} \varphi(t), \partial_i \rangle_{g(t_0)} + O(\| x^{t_0} \|_{g(t_0)}^2).$$

Proof Recall that:

$$\frac{\partial}{\partial t} (\exp^{g(t)} (\varphi(t), \sum_{i=1}^{n} \tau_t e_i x_i^t)) = 0.$$

The expected equation follows from the previous two lemmas.

We can now conclude that the Taylor series of the generator is:

$$\frac{\tilde{\partial}}{\partial_t} + \tilde{L}_t := \Psi^* (\partial_t + \frac{1}{2} \Delta_t + Z(t,.))_{|(t,x)}$$

$$= \frac{\tilde{\partial}}{\partial_t} + \frac{1}{2} \sum_{i,j=1}^{n} g^{ij}(t,x) \frac{\partial}{\partial x_i} \frac{\partial}{\partial x_j} + \sum_{i=1}^{n} \tilde{b}^i(t,x) \frac{\partial}{\partial x_i}$$

$$= \frac{\partial}{\partial_t} + \sum_{i,j=1}^{n} \left(\frac{1}{2} \dot{g}(t) \left(\frac{\partial}{\partial x_i^t}, \frac{\partial}{\partial x_j^t} \right) x_j - \dot{\varphi}(t)^i \right) \frac{\partial}{\partial x_i}$$

$$- \frac{1}{2} \sum_{k,l,i=1}^{n} g^{kl}(t,x) \Gamma_{kl}^i(t,x) \frac{\partial}{\partial x_i} + \frac{1}{2} \sum_{i,j=1}^{n} g^{ij}(t,x) \frac{\partial}{\partial x_i} \frac{\partial}{\partial x_j}$$

$$+ \sum_{i=1}^{n} Z^i(t,x) \frac{\partial}{\partial x_i} + O(\|x\|^2),$$

where $g^{ij}(t,x)$ corresponds to the inverse of the metric $g(t)$ in the normal coordinates $(x_1^t, \ldots, x_n^t)$ evaluated at the point $\Psi(t,x)$, $\Gamma_{ij}^k(t,x)$ are the Christoffel symbols in these coordinates at the point $\Psi(t,x)$, $\dot{\varphi}^i(t)$ and $Z^i(t,x)$ are the coordinates of the corresponding vector in these normal coordinates.

Remark 1 We have no time-dependence term such as $O(\|.\|_{g(t)})$ because all metrics are equivalent on U.

3   Proof of the Main Result

3.1   A Useful Tool: Besselizing Drift

Let $X(t)$ be a L_t-diffusion, and $\tilde{T} = inf\{t \in [0,T], s.t. (t, X(t)) \notin V\}$. Let us define $\tilde{X}(t)$ a $\mathbb{R}^n$-valued process such that $(t \wedge \tilde{T}, \tilde{X}(t)) = \Psi^{-1}(t \wedge \tilde{T}, X(t \wedge \tilde{T}))$. Then for any small enough ϵ, we have:

$$\mathbb{P}_{x_0}[\sup_{t \in [0,T]} d(t, X(t), \varphi(t)) \le \epsilon] = \mathbb{P}_0[\sup_{t \in [0,T]} \|\tilde{X}(t)\| \le \epsilon].$$

It is obvious that $(t, \tilde{X}(t))$ is a $\frac{\tilde{\partial}}{\partial_t} + \tilde{L}_t$ diffusion. So there exists a $\mathbb{R}^n$-valued Brownian motion $\tilde{B}$, such that $\tilde{X}(t)$ is a solution of the following Itô stochastic differential equation:

$$\begin{cases} d\tilde{X}^i(t) = \sum_{j=1}^{n} \sqrt{\tilde{g}}^{ij}(t, \tilde{X}(t)) \, d\tilde{B}_t^j + \tilde{b}^i(t, \tilde{X}(t)) \, dt, \\ \tilde{X}(0) = 0. \end{cases}$$

Here $\sqrt{g^{ij}}(t,x)$ is the square root of the inverse metric $g(t)$ in the coordinates $(x_1^t, \ldots, x_n^t)$ at the point $\Psi(t,x)$ and the drift term is defined by:

$$\tilde{b}^i(t,x) = -\dot{\varphi}^i(t) - \frac{1}{2}\sum_{kl} g^{kl}(t,x)\Gamma_{kl}^i(t,x)$$

$$+\frac{1}{2}\dot{g}(t)|_{\varphi(t)}\left(\frac{\partial}{\partial x_i^t}, \sum_{j=1}^n x^j \frac{\partial}{\partial x_j^t}\right) + Z^i(t,x) + O(\|x\|^2).$$

The Onsager Machlup functional for the sup-norm was studied by Takahashi and Watanabe. They introduced a drift, which is singular at the origin. The smooth Besselizing drift we will use here has been found by Hara. Let us describe shortly the Hara Besselizing drift. Since the coordinates are normal, we use Gauss' Lemma to find that for any $i \in [1..n]$:

$$\sum_{j=1}^n g^{ij}(t,x)x_j = x_i, \quad \text{and} \quad \sum_{j=1}^n \sqrt{g^{ij}}(t,x)x_j = x_i.$$

The Hara drift γ is then defined by:

$$\gamma^i(t,x) = \frac{1}{2}\sum_{j=1}^n \frac{\partial g^{ij}}{\partial x_j}(t,x).$$

It satisfies the following equation:

$$\sum_{i=1}^n (1 - g^{ii}(t,x)) = 2\sum_j^n \gamma^j(t,x)x_j.$$

Let us denote $\tilde{\sigma}_{ij}(t,.) = \sqrt{g^{ij}}(t,x)$. We remind the reader that the process $\tilde{X}(t)$ satisfies the equation:

$$\begin{cases} d\tilde{X}(t) = \tilde{\sigma}(t, \tilde{X}(t))\, d\tilde{B}_t + \tilde{b}(t, \tilde{X}(t))\, dt \\ \tilde{X}(0) = 0. \end{cases} \tag{4}$$

Define the $\mathbb{R}^n$-valued process $Y(t)$ as the solution of the following Itô equation:

$$\begin{cases} dY(t) = \tilde{\sigma}(t, Y(t))\, d\tilde{B}_t + \gamma(t, Y(t))\, dt \\ Y(0) = 0. \end{cases} \tag{5}$$

By definition of the vector field γ, we get by Itô's formula:

$$d \parallel Y(t) \parallel^2 = 2 \sum_{k=1}^{n} Y^k(t) d\tilde{B}_t^k + n\, dt.$$

By Lévy's Theorem, we see that $B(t) = \sum_{k=1}^{n} \int_0^t \frac{Y^k(s)}{\parallel Y(s) \parallel} d\tilde{B}_s^k$ is a one dimensional Brownian motion in the filtration generated by $\tilde{B}$ and

$$d \parallel Y(t) \parallel^2 = 2 \parallel Y(t) \parallel dB_t + n\, dt.$$

It proves that $\parallel Y(t) \parallel$ is a n-dimensional Bessel process.

Let us define $\hat{Y}_t := (t, Y(t))$. The next step consists in finding a well-suited probability measure such that, under this measure, the process $Y(t)$ has the same distribution as the process $\tilde{X}(t)$. Let us define:

$$N_t = \int_0^t \langle \tilde{\sigma}^{-1}(\hat{Y}_t)(\tilde{b}(\hat{Y}_t) - \gamma(\hat{Y}_t)), \, d\tilde{B}_t \rangle,$$

$$M_t = \exp(N_t - \frac{1}{2} \langle N \rangle_t)$$

$$\mathbb{Q} = M_T.\mathbb{P}.$$

Girsanov's Theorem ensures that $(Y, \mathbb{Q})$ is a solution of (4). The uniqueness in law of such a solution then implies that:

$$\mathbb{P}_0[\sup_{t \in [0,T]} \parallel \tilde{X}(t) \parallel \le \epsilon] = \mathbb{Q}[\sup_{t \in [0,T]} \parallel Y(t) \parallel \le \epsilon]$$

$$= \mathbb{E}_{\mathbb{P}}[M_T; \sup_{t \in [0,T]} \parallel Y(t) \parallel \le \epsilon] \qquad (6)$$

$$= \mathbb{E}_{\mathbb{P}}[M_T \mid \sup_{t \in [0,T]} \parallel Y(t) \parallel \le \epsilon]\mathbb{P}[\sup_{t \in [0,T]} \parallel Y(t) \parallel \le \epsilon].$$

The term $\mathbb{P}[\sup_{t \in [0,T]} \parallel Y(t) \parallel \le \epsilon]$ is easily controlled by a stopping time argument. So finding the Onsager Machlup functional reduces to the study of the behavior of a conditioned exponential martingale, as in the paper [7]. We will study the behavior of:

$$\mathbb{E}_{\mathbb{P}}\left[\exp\left(\sum_{i,j=1}^{n} \int_0^T \sqrt{g}_{ij}(\hat{Y}_t)\delta^j(\hat{Y}_t) d\tilde{B}_t^i \right. \right.$$

$$\left. \left. - \frac{1}{2} \sum_{i,j=1}^{n} \int_0^T g_{ij}(\hat{Y}_t)\delta^i(\hat{Y}_t)\delta^j(\hat{Y}_t)dt \right) \right| \sup_{t \in [0,T]} \parallel Y(t) \parallel \le \epsilon \right]. \qquad (7)$$

Where $\delta^i(t,x) = \tilde{b}^i(t,x) - \gamma^i(t,x)$.

Remark 2 From Lemma 1 in [1] it is sufficient to control the exponential moments one by one in the following sense.

Let us recall briefly this lemma:

Lemma 3 ([1,5]) *Let $I_1, \ldots, I_n$ be n random variables, $\{A_\epsilon\}_{0<\epsilon}$ a family of events, and $a_1, \ldots, a_n$ some real numbers. If, for every real number c and every $1 \leq i \leq n$, we have*

$$\limsup_{\epsilon \to 0} \mathbb{E}[\exp(c I_i) \mid A_\epsilon] \leq \exp(c a_i),$$

then,

$$\lim_{\epsilon \to 0} \mathbb{E}(\exp(\sum_{i=1}^n I_i) \mid A_\epsilon) = \exp(\sum_{i=1}^n a_i).$$

Note that, in the case studied here, all the metrics $g(t)$ are equivalent. Recall Cartan's Theorem dealing with Taylor series of metric and curvature in normal coordinates. We have:

$$g_{ij}(t,x) = \delta_i^j - \frac{1}{3}\sum_{kl} R_{iklj}(t,0)x_k x_l + O(\|x\|^3),$$

where $R_{iklj}(t,0)$ are the components of the Riemannian curvature tensor, for the metric $g(t)$ in normal coordinates centered at the point $\varphi(t)$. We thus deduce the following equalities:

$$g^{ij}(t,x) = \delta_i^j + O(\|x\|^2),$$

$$\gamma^i(t,x) = -\frac{1}{6}\sum_{j=1}^n R_{ij}(t,0)x_j + O(\|x\|^2),$$

where $R_{ij}(t,0)$ are the component of the Ricci curvature tensor for the metric $g(t)$ in normal coordinates, at the point $\varphi(t)$. By definition of the Christoffel symbol, we have,

$$\begin{aligned}
\Gamma_{ij}^k(t,x) &= \frac{1}{2}(\frac{\partial}{\partial x_i}g_{jk}(t,x) + \frac{\partial}{\partial x_j}g_{ik}(t,x) - \frac{\partial}{\partial x_k}g_{ij}(t,x)) \\
&= -\frac{1}{3}\sum_{l=1}^n (R_{jlik}(t,0) + R_{iljk}(t,0))x_l + O(\|x\|^2).
\end{aligned} \tag{8}$$

So we obtain,

$$-\frac{1}{2}\sum_{i,j=1}^{n} g^{ij}(t,x)\Gamma_{ij}^{k}(t,x) = -\frac{1}{3}\sum_{l=1}^{n} R_{lk}(t,0)x_{l} + O(\|x\|^2),$$

and thus,

$$\delta^i(t,x) = -\dot{\varphi}^i(t) + \sum_{j=1}^{n}\left(\frac{1}{2}\dot{g}_{ij}(t,0) - \frac{1}{6}R_{ij}(t,0)\right)x_j + Z^i(t,x) + O(\|x\|^2),$$

$$= -\dot{\varphi}^i(t) + Z^i(t,0) + \sum_{j=1}^{n}\left(\frac{1}{2}\dot{g}_{ij}(t,0) - \frac{1}{6}R_{ij}(t,0) + \frac{\partial}{\partial x_j}Z^i(t,0)\right)x_j$$

$$+ O(\|x\|^2),$$

(9)

where $\dot{g}_{ij}(t,0) = \dot{g}(t)\left(\frac{\partial}{\partial x_i^t}|_{\varphi(t)}, \frac{\partial}{\partial x_j^t}|_{\varphi(t)}\right)$.

3.2 Proof of the Theorem 1

According to Lemma 3 we will separately estimate the terms of (7). The easiest one is the drift term. Namely, we have:

$$\limsup_{\epsilon\to 0}\mathbb{E}\left[\exp\{-\frac{c}{2}\int_0^T g_{ij}(\hat{Y}_t)\delta^i(\hat{Y}_t)\delta^j(\hat{Y}_t)dt\}\Big|\sup_{t\in[0,T]}\|Y(t)\|\le\epsilon\right]$$

$$\le \lim_{\epsilon\to 0}\exp\left[-\frac{c}{2}\int_0^T \delta_i^j(-\dot{\varphi}^i(t) + Z^i(t,0))^2 + O(\epsilon)\,dt\right] \qquad (10)$$

$$\le \exp\left(-\frac{c}{2}\int_0^T \delta_i^j(-\dot{\varphi}^i(t) + Z^i(t,0))^2\,dt\right),$$

where we have used in the second inequality the fact that $O(\epsilon)$ is uniform in t according to the uniform equivalence of the family of metrics $\{g(t)\}_{t\in[0,T]}$. So, it remains to control the first term in (7). To this aim, we will use the following Theorem established in [4].

Let us denote by $*d$ the Stratonovich differential.

Theorem 2 ([4])
  Let α be a one form on $[0,T]\times\mathbb{R}^n$, which does not depend on dt and $Y(t)$ be a diffusion process in $\mathbb{R}^n$ whose radial part is a Bessel process, and such that for any

$1 \leq i, j \leq n$,

$$\left\langle \int_0^{\cdot} Y^i dY^j - Y^j dY^i, \|Y\|_{\cdot} \right\rangle = 0.$$

Then the following estimate holds for the stochastic line integral $\int_{*d(t,Y_t)} \alpha$ (in the sense of Stratonovich integration of a one form along a process):

$$\mathbb{E}\left[\exp\left(\int_{*d(t,Y_t)} \alpha \right) \Big|_{\substack{\sup \\ t \in [0,T]}} \|Y(t)\| \leq \epsilon \right] = \exp(O(\epsilon)).$$

The proof of this Theorem is based on the stochastic Stokes theorem which is deduced from Stokes' theorem by using Stratonovich integrals and the Kunita-Watanabe theorem for orthogonal martingales.

To use the above Theorem we first have to write the first term of (7) in terms of Stratonovich integral of a one form along a Bessel radial part process. Using the definition of Y, (see (5)):

$$d\tilde{B}_t^i = \sum_{j=1}^n \tilde{\sigma}_{ij}^{-1}(\hat{Y}_t)\, dY_t^j - \sum_{j=1}^n \tilde{\sigma}_{ij}^{-1}(\hat{Y}_t)\gamma^j(\hat{Y}_t)\, dt,$$

so

$$\sum_{i,j=1}^n \int_0^T \sqrt{g}_{ij}(\hat{Y}_t)\delta^j(\hat{Y}_t)d\tilde{B}_t^i$$

$$= \sum_{i,j=1}^n \int_0^T g_{ij}(\hat{Y}_t)\delta^j(\hat{Y}_t)dY_t^i - \int_0^T g_{ij}(\hat{Y}_t)\delta^j(\hat{Y}_t)\gamma^i(\hat{Y}_t)\, dt$$

$$\tag{11}$$

$$= \sum_{i,j=1}^n \int_0^T g_{ij}(\hat{Y}_t)\delta^j(\hat{Y}_t) * dY_t^i$$

$$- \frac{1}{2}\sum_{i,j=1}^n \int_0^T \langle d(g_{ij}(\hat{Y}_t)\delta^j(\hat{Y}_t)), dY_t^i \rangle - \int_0^T g_{ij}(\hat{Y}_t)\delta^j(\hat{Y}_t)\gamma^i(\hat{Y}_t)\, dt.$$

Proposition 3 *Denote by A_ϵ the event* $\{\sup_{t \in [0,T]} \|Y(t)\| \leq \epsilon\}$. *Then, the following equalities hold for any $1 \leq i, j \leq n$ and $c \in \mathbb{R}$:*

(i) $\mathbb{E}[\exp(c \int_0^T \sum_{i,j=1}^n g_{ij}(\hat{Y}_t)\delta^j(\hat{Y}_t) * dY_t^i) \mid A_\epsilon] = \exp(O(\epsilon))$.

(ii)
$$\limsup_{\epsilon \to 0} \mathbb{E}[\exp(-\frac{c}{2}\int_0^T \langle d(g_{ij}(\hat{Y}_t)\delta^j(\hat{Y}_t)), dY_t^i \rangle) \mid A_\epsilon]$$

$$\leq \exp(-\frac{c}{2}\int_0^T \delta_i^j \{\frac{1}{2}\dot{g}_{ij}(t,0) - \frac{1}{6}R_{ij}(t,0) + \frac{\partial}{\partial x_j}Z^i(t,0)\}\, dt).$$

(iii) $\limsup_{\epsilon \to 0} \mathbb{E}[\exp(-c \int_0^T g_{ij}(\hat{Y}_t)\delta^j(\hat{Y}_t)\gamma^i(\hat{Y}_t)\,dt) \mid A_\epsilon] = 1.$

Proof (i) Let $\alpha = c \sum_{i,j=1}^n g_{ij}(t,x)\delta^j(t,x)dx^i$ be defined in the neighborhood $U \subset [0,T] \times \mathbb{R}^n$, and extend it to the whole space. The expected asymptotic expansion is straightforward corollary of Theorem 2.

(ii) Using Itô's formula, the definition of Y leads to, for any $1 \le i \le n$:

$$\limsup_{\epsilon \to 0} \mathbb{E}[\exp(-\frac{c}{2} \int_0^T \langle d(g_{ij}(\hat{Y}_t)\delta^j(\hat{Y}_t)), dY_t^i \rangle) \mid A_\epsilon]$$

$$= \limsup_{\epsilon \to 0} \mathbb{E}[\exp(-\frac{c}{2} \int_0^T \sum_{l=1}^n \frac{\partial}{\partial x_l}(g_{ij}(t,\cdot)\delta^j(t,\cdot))(Y_t)dY_t^l dY_t^i) \mid A_\epsilon]$$

$$= \limsup_{\epsilon \to 0} \mathbb{E}[\exp(-\frac{c}{2} \int_0^T \sum_{l=1}^n \frac{\partial}{\partial x_l}(g_{ij}(t,.)\delta^j(t,.))(Y_t)g_{il}(\hat{Y}_t)\,dt) \mid A_\epsilon]$$

$$\le \exp(-\frac{c}{2} \int_0^T (\frac{1}{2}\dot{g}_{ii}(t,0) - \frac{1}{6}R_{ii}(t,0) + \frac{\partial}{\partial x_i}Z^i(t,0))\,dt).$$

For the latter inequality, we have used the Taylor expansion computed in the last section.

(iii) We have:

$$\limsup_{\epsilon \to 0} \mathbb{E}[\exp(-c \int_0^T g_{ij}(t,Y(t)\delta^j(\hat{Y}_t))\gamma^i(t,Y(t))\,dt) \mid A_\epsilon] \tag{12}$$

$$= \limsup_{\epsilon \to 0} \mathbb{E}[\exp(-c \int_0^T O(\|Y(t)\|)\,dt) \mid A_\epsilon] = 1.$$

We are now ready to prove Theorem 1

Proof Theorem 1
Using Lemma 3, formula (6), (7), (10) and Proposition 3, we obtain:

$$\lim_{\epsilon \to 0} \mathbb{E}_\mathbb{P}[M_T \mid \sup_{t \in [0,T]} \|Y(t)\| \le \epsilon]$$

$$= \exp \Big(\int_0^T \{-\frac{1}{2}\|Z(t,\varphi(t)) - \dot{\varphi}(t)\|_{g(t)}^2 - \frac{1}{4}(\operatorname{Tr}_{g(t)}(\dot{g}(t)))_{\varphi(t)}$$

$$+ \frac{1}{12}R(t,\varphi(t)) - \frac{1}{2}\operatorname{div}_{g(t)} Z(t,\varphi(t))\}dt \Big) \tag{13}$$

$$= \exp \Big(-\int_0^T H(t,\varphi(t),\dot{\varphi}(t))\,dt \Big).$$

Since the second term of (6) is given by the scaling property of the Brownian motion, we see that

$$\mathbb{P}_0[\sup_{t \in [0,T]} \|Y(t)\| \le \epsilon] = \mathbb{P}_0[\tau_1^n(B) > \frac{T}{\epsilon^2}],$$

where $\tau_1^n(B)$ is the first hitting time of the ball of radius 1 by the n-dimensional Brownian motion. Thus, using arguments of stopping time, Dirichlet problem and spectral Theorem, we get the following:

$$\mathbb{P}_0[\sup_{t \in [0,T]} \|Y(t)\| \le \epsilon] \sim_{\epsilon \to 0} C \exp(-\lambda_1 \frac{T}{\epsilon^2}),$$

where λ_1 is the first eigenvalue of the Laplace operator $(-\frac{1}{2}\Delta_{\mathbb{R}^n})$ in the unit ball in $\mathbb{R}^n$ with Dirichlet's boundary conditions, and C is an explicit constant that only depends on the dimension, (see Lemma 8.1 [5]).

4   Applications

4.1   The Most Probable Path

In this section, we will deduce from Theorem 1 the equation of the "most likely"curve. Namely, we will find a second order differential equation for the critical curve of the Onsager Machlup functional $E[\varphi] = \int_0^T H(t, \varphi(t), \dot{\varphi}(t)) \, dt$. This will be done below, in Proposition 4. For the Ricci flow, for instance, the functional is clos to the $\mathscr{L}_0$ distance used by Lott in [6].

Let φ and ψ be two smooth curves in M such that $\varphi(0) = \psi(0)$ and $\varphi(T) = \psi(T)$. Using the notation of Theorem 1, it implies that, for any given initial x_0, we have:

$$\lim_{\epsilon \to 0} \frac{\mathbb{P}_{x_0}[\sup_{t \in [0,T]} \quad d(t, X(t), \varphi(t)) \le \epsilon]}{\mathbb{P}_{x_0}[\sup_{t \in [0,T]} \quad d(t, X(t), \psi(t)) \le \epsilon]}$$

$$= \frac{\exp\left(-\int_0^T H(t, \varphi(t), \dot{\varphi}(t)) \, dt \right)}{\exp\left(-\int_0^T H(t, \psi(t), \dot{\psi}(t)) \, dt \right)}.$$

Our goal consists in computing the critical curve of the functional:

$$E[\varphi] = \int_0^T H(t, \varphi(t), \dot{\varphi}(t)) \, dt,$$

when both the initial and ending points are fixed. In the next result, we determine the equation of this curve in the case of $g(t)$-Brownian motion (see [2]). The general case could be deduced by the same computation.

Proposition 4 *Let X_t be a $L_t := \frac{1}{2}\Delta_t$ diffusion, where Δ_t is the Laplace operator with respect to a family of metrics $g(t)$ coming from the Ricci flow $\partial_t g(t) = \alpha \, Ric_{g(t)}$, (as in [2]). Then the critical curve φ for the functional E satisfies the following second order differential equation:*

$$\nabla^t_{\partial_t}\dot{\varphi}(t) + \alpha \, Ric^{\#g(t)}(\dot{\varphi}(t)) + \frac{1-3\alpha}{12}\nabla^t R_t(\varphi(t)) = 0.$$

Proof Let φ be a critical curve for E and let exp be the exponential map according to some fixed metric. Then for all vector fields V over φ such that $V(0) = V(1) = 0$, we have:

$$\frac{\partial}{\partial s}_{|s=0} E[t \mapsto \exp_{\varphi(t)}(sV(t))] = 0.$$

Let us recall that the generator of (t, X_t) is given by $\partial_t + \frac{1}{2}\Delta_t + Z(t, \cdot)$. So when $L_t = \frac{1}{2}\Delta_t$, we have $Z(t, .) = 0$, and hence

$$H(t, x, v) = \frac{1}{2}\|v\|^2_{g(t)} - \frac{1-3\alpha}{12}R_{g(t)}(x).$$

Let us now denote the variation of the curve φ by $\varphi_V(t, s) := \exp_{\varphi(t)}(sV(t))$, and $\dot{\varphi}_V(t, s) := \frac{\partial}{\partial t}\varphi_V(t, s)$. The preceding equation of E becomes:

$$\begin{aligned}
0 &= \frac{\partial}{\partial s}_{|s=0} \int_0^T \frac{1}{2}\|\dot{\varphi}_V(t, s)\|^2_{g(t)} - \frac{1-3\alpha}{12}R_{g(t)}(\varphi_V(t, s))dt \\
&= \int_0^T \langle \dot{\varphi}_V(t, 0), \nabla^t_{\partial_s}\dot{\varphi}_V(t, 0)\rangle_{g(t)} \\
&\quad - \frac{1-3\alpha}{12}\langle \nabla^t R_{g(t)}(\varphi_V(t, 0)), \frac{\partial}{\partial s}_{|s=0}\varphi_V(t, s)\rangle_{g(t)} dt \\
&= \int_0^T \langle \dot{\varphi}(t), \nabla^t_{\partial_s}\dot{\varphi}_V(t, s)\rangle_{g(t)} - \frac{1-3\alpha}{12}\langle \nabla^t R_{g(t)}(\varphi(t)), V(t)\rangle_{g(t)} dt.
\end{aligned}$$

Since ∂_t and ∂_s commute, and since the connection ∇^t is torsion free, we have $\nabla^t_{\partial_s}\dot{\varphi}_V(t, s) = \nabla^t_{\partial_t}\frac{\partial}{\partial s}\varphi_V(t, s)$. So, for any vector field V such that $V(0) = V(T) = 0$, the critical curve satisfies:

$$\int_0^T \langle \dot{\varphi}(t), \nabla^t_{\partial_t}V(t)\rangle_{g(t)} - \frac{1-3\alpha}{12}\langle \nabla^t R_{g(t)}(\varphi(t)), V(t)\rangle_{g(t)} dt = 0. \qquad (14)$$

Moreover, a straightforward computation shows that,

$$\partial_t \langle \dot{\varphi}(t), V(t) \rangle_{g(t)} = \langle \nabla_{\partial_t}^t \dot{\varphi}(t), V(t) \rangle_{g(t)} + \langle \dot{\varphi}(t), \nabla_{\partial_t}^t V(t) \rangle_{g(t)}$$
$$+ \dot{g}(t) \Big(\dot{\varphi}(t), V(t) \Big).$$

The final condition of the vector field $V(0) = V(T) = 0$, gives:

$$\int_0^T \partial_t \Big(\langle \dot{\varphi}(t), V(t) \rangle_{g(t)} \Big) dt = 0.$$

Hence, for any vector field V such that $V(0) = V(T) = 0$, the preceding Eq. (14) becomes

$$\int_0^T \Big(\langle \nabla_{\partial_t}^t \dot{\varphi}(t), V(t) \rangle_{g(t)} + \langle \alpha \, \mathrm{Ric}^{\#g(t)}(\dot{\varphi}(t)), V(t) \rangle$$
$$+ \frac{1 - 3\alpha}{12} \langle \nabla^t R_{g(t)}(\varphi(t)), V(t) \rangle_{g(t)} \Big) dt = 0.$$

Thus we conclude that φ is a critical value of E if and only if it satisfies:

$$\nabla_{\partial_t}^t \dot{\varphi}(t) + \alpha \, \mathrm{Ric}^{\#g(t)}(\dot{\varphi}(t)) + \frac{1 - 3\alpha}{12} \nabla^t R_{g(t)}(\varphi(t)) = 0.$$

Remark 3 The choice $\alpha = \frac{1}{3}$ for the speed of the backward Ricci flow produces a simplification in the above expressions and makes the functional E positive for all time, for any fixed metric $g(0)$ (when the backward Ricci flow exists).

Remark 4 The more general case of $g(t)$-BM can be easily deduced by the same proof. Let X_t be a $L_t := \frac{1}{2}\Delta_t$ diffusion, where Δ_t is the Laplace operator with respect to a family of metric $g(t)$, then the E-critical curve φ satisfies:

$$\nabla_{\partial_t}^t \dot{\varphi}(t) + \dot{g}(t)^{\#g(t)}(\dot{\varphi}(t)) + \frac{1}{12} \nabla^t R_{g(t)}(\varphi(t)) - \frac{1}{4} \nabla^t (\mathrm{Tr}_{g(t)} \, \dot{g}(t))(\varphi(t)) = 0.$$

We can also use this formula for the Brownian motion induced by the mean curvature flow as in [3], and compute the most probable path for this inhomogeneous diffusion. This result can be used to compute the most probable path for the degenerated diffusion $Z(t)$ (see Remark 2.9 of [3]).

4.2   Small Ball Properties of Inhomogeneous Diffusion
for Weighted Sup Norm

Let $X_t(x)$ be a $L_t = \frac{1}{2}\Delta_t + Z(t)$ diffusion. Let $f \in C^1([0, T])$ be a positive function on $[0, T]$. In this paragraph, we wish to estimate the following probability

$$\mathbb{P}_{x_0}[\forall t \in [0, T] \quad d(t, X_t, \varphi(t)) \le \epsilon f(t)],$$

when ϵ is positive and close to 0. We deduce the following small ball estimate:

Proposition 5 *There exists an explicit positive constant $C > 0$ such that:*

$$\mathbb{P}_{x_0}[\forall t \in [0, T] \quad d(t, X_t, \varphi(t)) \le \epsilon f(t)]$$

$$\sim_{\epsilon \downarrow 0} C \exp\left\{-\frac{\lambda_1 \int_0^T \frac{1}{f^2(s)}\,ds}{\epsilon^2}\right\} \exp\left\{-\int_0^T \tilde{H}(t, \varphi(t), \dot{\varphi}(t))\,dt\right\}$$

where

$$\tilde{H}(t, x, v) = \|Z(t, x) - v\|_{g(t)}^2 + \frac{1}{2}\operatorname{div}_{g(t)}(Z)(t, x) - \frac{1}{12}R_{g(t)}(x)$$

$$+ \frac{1}{4}f^{-2}(t)\operatorname{trace}_{g(t)}(\dot{g}(t)) - \frac{1}{2}n(f'(t)f^{-3}(t)).$$

Proof Let $\tilde{g}(t) = \frac{1}{f^2(t)}g(t)$, and let $\tilde{d}(t, ., .)$ be the associated distance. Then the probability we wish to estimate is

$$\mathbb{P}_{x_0}[\forall t \in [0, T] \quad \tilde{d}(t, X_t, \varphi(t)) \le \epsilon].$$

Now after a change of time we will turn the L_t diffusion X into a $\tilde{L}_t$ diffusion, in order to apply Theorem 1. Let us define

$$\delta(t) = \left(\int_0^{\cdot} \frac{1}{f^2(s)}\,ds\right)^{-1}(t),$$

and let $\tilde{X}(t) := X_{\delta(t)}$. Then $\tilde{X}$ is a $\tilde{L}_t$ diffusion, where

$$\tilde{L}_t := \frac{1}{2}\Delta_{\tilde{g}(\delta(t))} + f^2(\delta(t))Z(\delta(t), .).$$

We deduce that:

$$\mathbb{P}_{x_0}[\forall t \in [0, T] \quad d(t, X_t, \varphi(t)) \le \epsilon f(t)]$$

$$= \mathbb{P}_{x_0}[\forall t \in [0, T] \quad \tilde{d}(t, X_t, \varphi(t)) \le \epsilon]$$

$$= \mathbb{P}_{x_0}[\forall t \in [0, \delta^{-1}(T)] \quad \tilde{d}(\delta(t), \tilde{X}_t, \varphi(\delta(t))) \le \epsilon]$$

$$\sim_{\epsilon \downarrow 0} C \exp\left\{-\frac{\lambda_1 \delta^{-1}(T)}{\epsilon^2}\right\} \exp\left\{-\int_0^{\delta^{-1}(T)} H(\delta(t), \varphi(\delta(t)), \dot{\delta}(t)\dot{\varphi}(\delta(t))) \, dt\right\}.$$

In the last line, we have used Theorem 1, and the Lagrangian H related to the diffusion $\tilde{X}$. After a change of variables we get the result.

Corollary 1 *It holds that*

$$\epsilon^2 \log\{\mathbb{P}_{x_0}[\forall t \in [0, T], \quad d(t, X_t, \varphi(t)) \le \epsilon f(t)]\} \xrightarrow[\epsilon \to 0]{} -\lambda_1 \int_0^T \frac{1}{f^2(s)} \, ds.$$

References

1. M. Capitaine, On the Onsager-Machlup functional for elliptic diffusion processes, in *Séminaire de Probabilités, XXXIV*. Lecture Notes in Mathematics, vol. 1729 (Springer, Berlin, 2000), pp. 313–328
2. A.K. Coulibaly-Pasquier, Brownian motion with respect to time-changing riemannian metrics, applications to Ricci flow. Ann. Inst. Henri Poincaré Probab. Stat. **47**(2), 515–538 (2011)
3. A.K. Coulibaly-Pasquier, Some stochastic process without birth, linked to the mean curvature flow. Ann. Probab. **39**(4), 1305–1331 (2011)
4. K. Hara, Y. Takahashi, Lagrangian for pinned diffusion process, in *Itô's Stochastic Calculus and Probability Theory* (Springer, Tokyo, 1996), pp. 117–128
5. N. Ikeda, S. Watanabe, *Stochastic Differential Equations and Diffusion Processes*. North-Holland Mathematical Library, vol. 24, 2nd edn. (North-Holland, Amsterdam, 1989)
6. J. Lott, Optimal transport and Perelman's reduced volume. arXiv:0804.0343v2
7. Y. Takahashi, S. Watanabe, The probability functionals (Onsager-Machlup functions) of diffusion processes, in *Stochastic Integrals, (Proc. Sympos., Univ. Durham, Durham, 1980)*. Lecture Notes in Mathematics, vol. 851 (Springer, Berlin, 1981), pp. 433–463

G-Brownian Motion as Rough Paths and Differential Equations Driven by G-Brownian Motion

Xi Geng, Zhongmin Qian, and Danyu Yang

Abstract The present article is devoted to the study of sample paths of G-Brownian motion and stochastic differential equations (SDEs) driven by G-Brownian motion from the viewpoint of rough path theory. As the starting point, by using techniques in rough path theory, we show that quasi-surely, sample paths of G-Brownian motion can be enhanced to the second level in a canonical way so that they become geometric rough paths of roughness $2 < p < 3$. This result enables us to introduce the notion of rough differential equations (RDEs) driven by G-Brownian motion in the pathwise sense under the general framework of rough paths. Next we establish the fundamental relation between SDEs and RDEs driven by G-Brownian motion. As an application, we introduce the notion of SDEs on a differentiable manifold driven by G-Brownian motion and construct solutions from the RDE point of view by using pathwise localization technique. This is the starting point of developing G-Brownian motion on a Riemannian manifold, based on the idea of Eells-Elworthy-Malliavin. The last part of this article is devoted to such construction for a wide and interesting class of G-functions whose invariant group is the orthogonal group. In particular, we establish the generating nonlinear heat equation for such G-Brownian motion on a Riemannian manifold. We also develop the Euler-Maruyama approximation for SDEs driven by G-Brownian motion of independent interest.

Keywords Euler-Maruyama approximation • G-Brownian motion • G-expectation • Geometric rough paths • Nonlinear diffusion processes • Nonlinear heat flow • Rough differential equations

AMS classification (2010): 60H10, 34A12, 58J65

X. Geng • D. Yang
Mathematical Institute, University of Oxford, Oxford OX2 6GG, UK

Oxford-Man Institute, University of Oxford, Oxford OX2 6ED, UK
e-mail: xi.geng@maths.ox.ac.uk; danyu.yang@maths.ox.ac.uk

Z. Qian (✉)
Exeter College, University of Oxford, Oxford OX1 3DP, UK
e-mail: qianz@maths.ox.ac.uk

© Springer International Publishing Switzerland 2014   125
C. Donati-Martin et al. (eds.), *Séminaire de Probabilités XLVI*, Lecture Notes
in Mathematics 2123, DOI 10.1007/978-3-319-11970-0_6

1 Introduction

The classical Feynman-Kac formula (see [14, 15]) provides us with a way to represent the solution of a linear parabolic PDE in terms of the conditional expectation of certain functional of a diffusion process (solution of an SDE). However, it works only for the linear case, mainly due to the linear nature of diffusion processes. To understand nonlinear parabolic PDEs from the probabilistic point of view, Peng and Pardoux (see [21–23]) initiated the study of backward stochastic differential equations (BSDEs) and showed that the solution of a certain type of quasilinear parabolic PDEs can be expressed in terms of the solution of BSDE. This result suggests that BSDE reveals a certain type of nonlinear dynamics, and was made explicit by Peng [24]. More precisely, Peng introduced a notion of nonlinear expectation called the g-expectation in terms of the solution of BSDE which is filtration consistent. However, it was developed under the framework of classical Itô calculus and did not capture the fully nonlinear situation.

Motivated from the study of fully nonlinear dynamics, Peng [25] introduced the notion of G-expectation in an intrinsic way which does not rely on any particular probability space. It reveals the probability distribution uncertainty in a fundamental way which is crucial in many situations such as modeling risk uncertainty in mathematical finance. The underlying mechanism corresponding to such kind of uncertainty is a fully nonlinear parabolic PDE. In [25, 26], he also introduced the concept of G-Brownian motion which is generated by the so-called nonlinear G-heat equation and related stochastic calculus such as G-Itô integral, G-Itô formula, SDEs driven by G-Brownian motion, etc. One of the major significance of such theory is the corresponding nonlinear Feynman-Kac formula proved by Peng [27], which gives us a way to represent the solution of a fully nonlinear parabolic PDE via the solution of a forward-backward SDE under the framework of G-expectation.

On the other hand, motivated from the study of integration against irregular paths and differential equations driven by rough signals, Lyons [17] proposed a theory of rough paths which reveals the fundamental way of understanding the roughness of a continuous path. He pointed out that to understand the evolution of a system whose input signal (driven path) is rough, a finite sequence of "iterated integrals" (higher levels) of the driving path which satisfy a certain type of algebraic relation (Chen identity) should be specified in advance. Such point of view is fundamental, if we look at the Taylor expansion for the solution of an ODE whose driving path is of bounded variation (see (6) and a more detailed introduction in the next section). In other words, it is essential to regard a path as an object valued in some tensor algebra which records the information of higher levels if we wish to understand the "differential" of the path. Moreover, Lyons [17] proved the so-called universal limit theorem (see Theorem 6 in the next section), which allows us to introduce the notion of differential equations driven by rough paths (simply called RDEs) in a rigorous way. The theory of rough paths has significant applications in classical stochastic analysis, as we can prove that the sample paths of many stochastic processes we've encountered are essentially rough paths with certain roughness. According to Lyons' universal limit theorem, we are able to establish RDEs driven by the sample paths of those stochastic processes in a pathwise manner. It provides us with a new way to understand SDEs, especially when the driving process is not the classical Brownian motion in which case a well-developed Itô SDE theory is still not available.

The case of classical Brownian motion is quite special, since we have a complete SDE theory in the L^2-sense, as well as the notion of Stratonovich type integrals and differential equations. The fundamental relation between the two types of stochastic differentials (one-dimensional case) can be expressed by

$$X \circ dY = XdY + \frac{1}{2}dX \cdot dY.$$

It is proved in the rough path theory (see [9, 18], and also [13, 28] from the viewpoint of Wong-Zakai type approximation) that the Stratonovich type integrals and differential equations are equivalent to the pathwise integrals and RDEs in the sense of rough paths. In other words, the following to types of differential equations driven by Brownian motion

$$dX_t = \sum_{\alpha=1}^{d} V_\alpha(X_t)dW_t^\alpha + b(X_t)dt, \quad \text{(Itô type SDE)}$$

$$dY_t = \sum_{\alpha=1}^{d} V_\alpha(Y_t)dW_t^\alpha + (b(Y_t) - \sum_{\alpha=1}^{d} \frac{1}{2}DV_\alpha(Y_t) \cdot V_\alpha(Y_t))dt, \quad \text{(RDE)}$$

which are both well-defined under some regularity assumptions on the generating vector fields, are equivalent in the sense that if their solutions X_t and Y_t satisfy $X_0 = Y_0$, then $X = Y$ almost surely.

Under the framework of G-expectation, SDEs driven by G-Brownian motion introduced by Peng, can be regarded as nonlinear diffusion processes in Euclidean spaces. The idea of constructing G-Itô integrals and SDEs driven by G-Brownian motion is similar to the classical Itô calculus, which is also an L^2-theory but under the G-expectation instead of probability measures. What is missing is the notion of Stratonovich type integrals, mainly due to the reason that the theory of G-martingales is still not well understood. In particular, we don't have the corresponding nonlinear Doob-Meyer type decomposition theorem and the notion of quadratic variation processes for G-martingales. However, by the key observation in the classical case that the Stratonovich type integrals and the pathwise integrals are essentially equivalent in the sense of rough paths, we can study the sample paths of G-Brownian motion and SDEs driven by G-Brownian motion from the viewpoint of rough path theory, once we prove that the sample paths of G-Brownian motion can be regarded as objects in some rough path space with certain roughness. This is in fact what the present article is mainly focused on. The basic language to describe path structure under the G-expectation is quasi-sure analysis and capacity theory, which was developed by Denis et al. [7]. They generalized the Kolmogorov continuity theorem and studied sample path properties of G-Brownian motion. In particular, they also studied the relation between G-expectation and upper expectation associated to a family of probability measures which defines a Choquet capacity and the relation between the corresponding two types of L^p-spaces. The pathwise properties and homeomorphic flows for SDEs driven by G-Brownian motion in the quasi-sure setting was studied by Gao [10].

There are two main goals of the present article. This first one is to study the geometric rough path nature of sample paths of G-Brownian motion so that we can define RDEs driven by G-Brownian motion (the Stratonovich counterpart in the classical case) in the pathwise sense, and establish the fundamental relation between two types of differential equations driven by G-Brownian motion. The second one is to understand nonlinear diffusion processes in a (Riemannian) geometric setting, from the viewpoint of paths and distributions (the generating nonlinear PDE).

The present article is organized in the following way. Section 2 is a basic review of the theory of G-expectation and rough paths, which provides us with the general framework and basic tools for our study. In Sect. 3 we study the Euler-Maruyama approximation scheme for SDEs driven by G-Brownian motion. In Sect. 4 we show that for quasi-surely, the sample paths of G-Brownian motion can be enhanced to the second level in a canonical way so that they become geometric rough paths of roughness $2 < p < 3$ by using techniques in rough path theory. In Sect. 5 we establish the fundamental relation between SDEs and RDEs driven by G-Brownian motion by using rough Taylor expansions. In Sect. 6 we introduce the notion of SDEs on a differentiable manifold driven by G-Brownian motion from the RDE point of view by using pathwise localization technique. In the last section, we study the infinitesimal diffusive nature and the generating PDE for nonlinear diffusion processes in a (Riemannian) geometric setting, which leads to the construction of G-Brownian motion on a Riemannian manifold. We restrict ourselves to compact manifolds only, although the general case can be treated in a similar way with more technical complexity.

Throughout the rest of this article, we will use standard geometric notation for differential equations. Moreover, we will use the Einstein convention of summation, that is, when an index α appears as both subscript and superscript in the same expression, summation over α is taken automatically.

2 Preliminaries on G-Expectation and Rough Path Theory

2.1 G-Expectation and Related Stochastic Calculus

We first introduce some fundamentals on G-expectation and related stochastic calculus. For a systematic introduction, see [25–27].

Let Ω be a nonempty set, and $\mathscr{H}$ be a vector space of functionals on Ω such that $\mathscr{H}$ contains all constant functionals and for any $X_1, \cdots, X_n \in \mathscr{H}$ and any

$\varphi \in C_{l,Lip}(\mathbb{R}^n)$,

$$\varphi(X_1, \cdots, X_n) \in \mathscr{H},$$

where $C_{l,Lip}(\mathbb{R}^n)$ denotes the space of functions φ on $\mathbb{R}^n$ satisfying

$$|\varphi(x) - \varphi(y)| \leqslant C(1 + |x|^m + |y|^m)(|x - y|), \ \forall x, y \in \mathbb{R}^n,$$

for some constant $C > 0$ and $m \in \mathbb{N}$ depending on φ. $\mathscr{H}$ can be regarded as the space of random variables.

Definition 1 A sublinear expectation $\mathbb{E}$ on $(\Omega, \mathscr{H})$ is a functional $\mathbb{E} : \mathscr{H} \to \mathbb{R}$ such that

1. if $X \leqslant Y$, then $\mathbb{E}[X] \leqslant \mathbb{E}[Y]$;
2. for any constant c, $\mathbb{E}[c] = c$;
3. for any $X, Y \in \mathscr{H}$, $\mathbb{E}[X + Y] \leqslant \mathbb{E}[X] + \mathbb{E}[Y]$;
4. for any $\lambda \geqslant 0$ and $X \in \mathscr{H}$, $\mathbb{E}[\lambda X] = \lambda \mathbb{E}[X]$.

The triple $(\Omega, \mathscr{H}, \mathbb{E})$ is called a sublinear expectation space.

The relation between sublinear expectations and linear expectations, which was proved by Peng [27], is contained in the following representation theorem.

Theorem 1 *Let* $(\Omega, \mathscr{H}, \mathbb{E})$ *be a sublinear expectation space. Then there exists a family of linear expectations (linear functionals)* $\{\mathbb{E}_\theta : \theta \in \Theta\}$ *on* $\mathscr{H}$, *such that*

$$\mathbb{E}[X] = \sup_{\theta \in \Theta} \mathbb{E}_\theta[X], \ \forall X \in \mathscr{H}.$$

Under the frame work of sublinear expectation space, we also have the notion of independence and distribution (law).

Definition 2 1. A random vector $Y \in \mathscr{H}^n$ is said to be independent from another random vector $X \in \mathscr{H}^m$ under the sublinear expectation $\mathbb{E}$, if for any $\varphi \in C_{l,Lip}(\mathbb{R}^m \times \mathbb{R}^n)$,

$$\mathbb{E}[\varphi(X, Y)] = \mathbb{E}[\mathbb{E}[\varphi(x, Y)]_{x=X}].$$

2. Given a random vector $X \in \mathscr{H}^n$, the distribution (or the law) of X is defined as the sublinear expectation

$$\mathbb{F}_X[\varphi] := \mathbb{E}[\varphi(X)], \ \varphi \in C_{l,Lip}(\mathbb{R}^n),$$

on $(\mathbb{R}^n, C_{l,Lip}(\mathbb{R}^n))$. By saying that two random vectors X, Y (possibly defined on different sublinear expectation spaces) are identically distributed, we mean that their distributions are the same.

Now we introduce the notion of G-distribution, which is the generalization of degenerate distributions and normal distributions. It captures the uncertainty of probability distributions and plays a fundamental role in the theory of sublinear expectation.

Let $S(d)$ be the space of $d \times d$ symmetric matrices, and let $G : \mathbb{R}^d \times S(d) \to \mathbb{R}$ be a continuous and sublinear function monotonic in $S(d)$ in the sense that:

1. $G(p + \bar{p}, A + \bar{A}) \leqslant G(p, A) + G(\bar{p}, \bar{A})$, $\forall p, \bar{p} \in \mathbb{R}^d$, $A, \bar{A} \in S(d)$;
2. $G(\lambda p, \lambda A) = \lambda G(p, A)$, $\forall \lambda \geqslant 0$;
3. $G(p, A) \leqslant G(p, \bar{A})$, $\forall A \leqslant \bar{A}$.

Definition 3 Let $X, \eta \in \mathcal{H}^d$ be two random vectors. (X, η) is called G-distributed if for any $\varphi \in C_{l,Lip}(\mathbb{R}^d \times \mathbb{R}^d)$, the function

$$u(t, x, y) := \mathbb{E}[\varphi(x + \sqrt{t}X, y + t\eta)], \ (t, x, y) \in [0, \infty) \times \mathbb{R}^d \times \mathbb{R}^d,$$

is a viscosity solution of the following parabolic PDE (called a G-heat equation):

$$\partial_t u - G(D_y u, D_x^2 u) = 0, \tag{1}$$

with Cauchy condition $u|_{t=0} = \varphi$.

Remark 1 From the general theory of viscosity solutions (see [4, 27]), the G-heat Eq. (1) has a unique viscosity solution. By solving the G-heat Eq. (1) (in some special cases, it is explicitly solvable), we can compute the sublinear expectation of some functionals of a G-distributed random vector. The case of convex functionals, for instance, the power function $|x|^k$, is quite interesting.

It can be proved that for such a function G, there exists a bounded, closed and convex subset $\Gamma \subset \mathbb{R}^d \times \mathbb{R}^{d \times d}$, such that G has the following representation:

$$G(p, A) = \sup_{(q,Q) \in \Gamma} \{\frac{1}{2}\mathrm{tr}(AQQ^T) + \langle p, q \rangle\}, \ \forall (p, A) \in \mathbb{R}^d \times S(d).$$

The set Γ captures the uncertainty of probability distribution (mean uncertainty and variance uncertainty) of a G-distributed random vector.

In particular, if G only depends on $p \in \mathbb{R}^d$, then there exists some bounded, closed and convex subset $\Lambda \subset \mathbb{R}^d$, such that

$$G(p) = \sup_{q \in \Lambda} \langle p, q \rangle.$$

In this case a G-distributed random vector η is called maximal distributed and is denoted by $\eta \sim N(\Lambda, \{0\})$. Similarly, if G only depends on $A \in S(d)$, then there exists some bounded, closed and convex subset $\Sigma \subset S_+(d)$ (the space of symmetric and nonnegative definite matrices) such that

$$G(A) = \frac{1}{2} \sup_{B \in \Sigma} \text{tr}(AB), \ \forall A \in S(d). \tag{2}$$

A G-distributed random vector X for such G is called G-normal distributed and is denoted by $X \sim N(\{0\}, \Sigma)$.

Now we introduce the concept of G-Brownian motion and related stochastic calculus.

From now on, let $G : S(d) \to \mathbb{R}$ be a function given by (2).

Definition 4 A d-dimensional process B_t is called a G-Brownian motion if

1. $B_0(\omega) = 0, \ \forall \omega \in \Omega$;
2. for each $s, t \geq 0$, $B_{t+s} - B_t \sim N(\{0\}, s\Sigma)$ and is independent from

$$(B_{t_1}, \cdots, B_{t_n})$$

for any $n \geq 1$ and $0 \leq t_1 < \cdots < t_n \leq t$.

Similar to the classical situation, a G-Brownian motion can be constructed explicitly on the canonical path space by using independent G-normal random vectors. We refer the readers to [27] for a detailed construction.

In summary, let $\Omega = C_0([0, \infty); \mathbb{R}^d)$ be the space of $\mathbb{R}^d$-valued continuous paths starting at the origin, and let $B_t(\omega) := \omega_t$ be the coordinate process. For any $T \geq 0$, define

$$L_{ip}(\Omega_T) := \{\varphi(B_{t_1}, \cdots, B_{t_n}) : n \geq 1, t_1, \cdots, t_n \in [0, T], \varphi \in C_{l,Lip}(\mathbb{R}^{d \times n})\},$$

and

$$L_{ip}(\Omega) := \bigcup_{n=1}^{\infty} L_{ip}(\Omega_n).$$

Then on $(\Omega, L_{ip}(\Omega))$ we can define the canonical sublinear expectation $\mathbb{E}$ such that the coordinate process B_t becomes a G-Brownian motion, which is usually called the G-expectation and denoted by $\mathbb{E}^G$. $(\Omega, L_{ip}(\Omega), \mathbb{E}^G)$ is also called the canonical G-expectation space. Throughout the rest of this article, we will restrict ourselves on the canonical G-expectation space and its completion (to be defined later on).

On $(\Omega, L_{ip}(\Omega), \mathbb{E}^G)$ we can introduce the notion of conditional G-expectation. More precisely, for

$$X = \varphi(B_{t_1}, B_{t_2} - B_{t_1}, \cdots, B_{t_n} - B_{t_{n-1}}) \in L_{ip}(\Omega),$$

where $0 \leqslant t_1 < t_2 < \cdots < t_n$, the G-conditional expectation of X under Ω_{t_j} is defined by

$$\mathbb{E}^G[X|\Omega_{t_j}] := \psi(B_{t_1}, B_{t_2} - B_{t_1}, \cdots, B_{t_j} - B_{t_{j-1}}),$$

where

$$\psi(x_1, \cdots, x_j) := \mathbb{E}^G[\varphi(x_1, \cdots, x_j, B_{t_{j+1}} - B_{t_j}, \cdots, B_{t_n} - B_{t_{n-1}})], \ x_1, \cdots, x_j \in \mathbb{R}^d.$$

The conditional G-expectation $\mathbb{E}^G[\cdot|\Omega_t]$ has the following properties: for any $X, Y \in L_{ip}(\Omega)$,

1. if $X \leqslant Y$, then $\mathbb{E}^G[X|\Omega_t] \leqslant \mathbb{E}^G[Y|\Omega_t]$;
2. $\mathbb{E}^G[X + Y|\Omega_t] \leqslant \mathbb{E}^G[X|\Omega_t] + \mathbb{E}^G[Y|\Omega_t]$;
3. for any $\eta \in L_{ip}(\Omega_t)$,

$$\mathbb{E}^G[\eta|\Omega_t] = \eta,$$
$$\mathbb{E}^G[\eta X|\Omega_t] = \eta^+ \mathbb{E}^G[X|\Omega_t] + \eta^- \mathbb{E}[-X|\Omega_t];$$

4. $\mathbb{E}^G[\mathbb{E}^G[X|\Omega_t]|\Omega_s] = \mathbb{E}^G[X|\Omega_{t \wedge s}]$. In particular, $\mathbb{E}^G[\mathbb{E}^G[X|\Omega_t]] = \mathbb{E}^G[X]$.

For any $p \geqslant 1$, let L_G^p (respectively, $L_G^p(\Omega_t)$)) be the completion of $L_{ip}(\Omega)$ (respectively, $L_{ip}(\Omega_t)$)) under the semi-norm $\|X\|_p := (\mathbb{E}^G[|X|^p])^{\frac{1}{p}}$. Then $\mathbb{E}^G$ can be continuously extended to a sublinear expectation on $L_G^p(\Omega)$ (respectively, $L_G^p(\Omega_t)$), still denoted by $\mathbb{E}^G$.

For $t < T \leqslant \infty$, the conditional G-expectation $\mathbb{E}^G[\cdot|\Omega_t] : L_{ip}(\Omega_T) \to L_{ip}(\Omega_t)$ is a continuous mapping under $\|\cdot\|_1$ and can be continuously extended to a mapping

$$\mathbb{E}^G[\cdot|\Omega_t] : L_G^1(\Omega_T) \to L_G^1(\Omega_t),$$

which can still be interpreted as the conditional G-expectation. It is easy to show that the properties (1)–(4) for the conditional G-expectation still hold true on $L_G^1(\Omega_T)$ as long as it is well-defined.

Now we introduce the related stochastic calculus for G-Brownian motion and (Itô type) stochastic differential equations (SDEs) driven by G-Brownian motion.

First of all, similar to the idea in the classical case, we still have the notion of Itô integral with respect to a one-dimensional G-Brownian motion. More precisely, consider $d = 1$, we can first define Itô integral of simple processes and then pass limit under the G-expectation $\mathbb{E}^G$ in some suitable functional spaces. Let $M_G^{p,0}(0, T)$ be the space of simple processes $\eta_t(\omega)$ on $[0, T]$ of the form

$$\eta_t(\omega) = \sum_{k=1}^{N} \xi_{k-1}(\omega) \mathbf{1}_{[t_{k-1}, t_k)}(t),$$

where $\pi_T^N := \{t_0, t_1, \cdots, t_N\}$ is a partition of $[0, T]$ and $\xi_k \in L_G^p(\Omega_{t_k})$, and introduce the semi-norm

$$\|\eta\|_{M_G^p(0,T)} := (\mathbb{E}^G[\int_0^T |\eta_t|^p dt])^{\frac{1}{p}}$$

on $M_G^{p,0}(0, T)$. Let $M_G^p(0, T)$ be the completion of $M_G^{p,0}(0, T)$ under $\|\cdot\|_{M_G^p(0,T)}$. It is straight forward to define Itô integral $\int_0^T \eta_t dB_t$ of simple processes. Moreover, such an integral operator is linear and continuous under $\|\cdot\|_{M_G^p(0,T)}$ and hence can be extended to a bounded linear operator

$$I : M_G^2(0, T) \to L_G^2(0, T).$$

The operator I is defined as the Itô integral operator with respect to a G-Brownian motion. For $0 \leqslant s < t \leqslant T$, define

$$\int_s^t \eta_u dB_u := \int_0^T \mathbf{1}_{[s,t]}(u)\eta_u dB_u.$$

We list some important properties of G-Itô integral in the following.

Proposition 1 *Let* $\eta, \theta \in M_G^2(0, T)$ *and let* $0 \leqslant s \leqslant r \leqslant t \leqslant T$. *Then*

1.

$$\int_s^t \eta_u dB_u = \int_s^r \eta_u dB_u + \int_r^t \eta_u dB_u;$$

2. if α *is bounded in* $L_G^1(\Omega_s)$, *then*

$$\int_s^t (\alpha\eta_u + \theta_u)dB_u = \alpha \int_s^t \eta_u dB_u + \int_s^t \theta_u dB_u;$$

3. for any $X \in L_G^1(\Omega)$,

$$\mathbb{E}^G[X + \int_r^T \eta_u dB_u | \Omega_s] = \mathbb{E}^G[X | \Omega_s];$$

4.

$$\underline{\sigma}^2 \mathbb{E}^G[\int_0^T \eta_t^2 dt] \leqslant \mathbb{E}^G[(\int_0^T \eta_t dB_t)^2] \leqslant \overline{\sigma}^2 \mathbb{E}^G[\int_0^T \eta_t^2 dt],$$

where $\overline{\sigma}^2 := \mathbb{E}^G[B_1^2]$ *and* $\underline{\sigma}^2 := -\mathbb{E}^G[-B_1^2].$

Secondly, we have the notion of quadratic variation process of G-Brownian motion. In the case of one-dimensional G-Brownian motion, the quadratic variation process $\langle B \rangle_t$ is defined as

$$\langle B \rangle_t := B_t^2 - 2 \int_0^t B_s dB_s,$$

which can be regarded as the L_G^2-limit of the sum $\sum_{j=1}^{k_N} (B_{t_j^N} - B_{t_{j-1}^N})^2$ as $\mu(\pi_t^N) \to 0$, where $\pi_t^N := \{t_j^N\}_{j=0}^{k_N}$ is a sequence of partitions of $[0, t]$ and

$$\mu(\pi_t^N) := \max\{t_j^N - t_{j-1}^N : \ j = 1, 2, \cdots, k_N\}.$$

It follows that $\langle B \rangle_t$ is an increasing process with $\langle B \rangle_0 = 0$.

Similar to the definition of G-Itô integral, we can define the integration with respect to $\langle B \rangle_t$ where B_t is a one-dimensional G-Brownian motion. We refer the readers to [27] for a detailed construction but we remark that the integral operator with respect to $\langle B \rangle_t$ is a continuous linear mapping

$$Q_{0,T} : \ M_G^1(0, T) \to L_G^1(\Omega_T).$$

The following identity can be regarded as the G-Itô isometry.

Proposition 2 Let $\eta \in M_G^2(0, T)$, then

$$\mathbb{E}^G[(\int_0^T \eta_t dB_t)^2] = \mathbb{E}^G[\int_0^T \eta_t^2 d \langle B \rangle_t].$$

Now consider the multi-dimensional case. Let B_t is a d-dimensional G-Brownian motion, and for any $v \in \mathbb{R}^d$, denote

$$B_t^v := \langle v, B_t \rangle,$$

where $\langle \cdot, \cdot \rangle$ is the Euclidean inner product. Then for $a, \bar{a} \in \mathbb{R}^d$, the cross variation process $\langle B^a, B^{\bar{a}} \rangle_t$ is defined as

$$\langle B^a, B^{\bar{a}} \rangle_t = \frac{1}{4}(\langle B^{a+\bar{a}}, B^{a+\bar{a}} \rangle_t - \langle B^{a-\bar{a}}, B^{a-\bar{a}} \rangle_t).$$

Similar to the case of quadratic variation process, we have

$$\langle B^a, B^{\bar{a}} \rangle_t = (L_G^2 -) \lim_{\mu(\pi_t^N) \to 0} \sum_{j=1}^{k_N} (B_{t_j^N}^a - B_{t_{j-1}^N}^a)(B_{t_j^N}^{\bar{a}} - B_{t_{j-1}^N}^{\bar{a}})$$

$$= B_t^a B_t^{\bar{a}} - \int_0^t B_s^a dB_s^{\bar{a}} - \int_0^t B_s^{\bar{a}} dB_s^a.$$

Note that unlike the classical case, the cross variation process is not deterministic. The following results characterizes the distribution of $\langle B \rangle_t := (\langle B^\alpha, B^\beta \rangle_t)_{\alpha,\beta=1}^d$, where B_t is a d-dimensional G-Brownian motion and B_t^α is the α-th component of B_t.

Proposition 3 *Recall that the function G has the representation (2). Then $\langle B \rangle_t \sim N(t\Sigma, \{0\})$.*

As in the classical case, we also have the important G-Itô formula under G-expectation, which takes a similar form to the classical one. The main difference is that $dB_t^\alpha \cdot dB_t^\beta$ should be $d\langle B^\alpha, B^\beta \rangle_t$ instead of $\delta_{\alpha\beta}dt$. We are not going to state the full result of G-Itô formula here. See [27] for a detailed discussion.

Now we introduce the notion of SDEs driven by G-Brownian motion.

For $p \geq 1$, let $\overline{M}_G^p(0, T; \mathbb{R}^n)$ be the completion of $M_G^{p,0}(0, T; \mathbb{R}^n)$ under the norm

$$\|\eta\|_{\overline{M}_G^p(0,T;\mathbb{R}^n)} := (\int_0^T \mathbb{E}^G[|\eta_t|^p]dt)^{\frac{1}{p}}.$$

It is easy to see that $\overline{M}_G^p(0, T; \mathbb{R}^n) \subset M_G^p(0, T; \mathbb{R}^n)$.

Consider the following N-dimensional SDE driven by G-Brownian motion over $[0, T]$:

$$dX_t = b(t, X_t)dt + \sum_{\alpha,\beta=1}^d h_{\alpha\beta}(t, X_t)d\langle B^\alpha, B^\beta \rangle_t + \sum_{\alpha=1}^d V_\alpha(t, X_t)dB_t^\alpha \tag{3}$$

with initial condition $\xi \in \mathbb{R}^N$. Here we assume that the coefficients $b^i, h_{\alpha\beta}^i, V_\alpha^i$ are Lipschitz functions in the space variable, uniformly in time. A solution of (3) is a process in $\overline{M}_G^2(0, T; \mathbb{R}^N)$ satisfying the Eq. (3) in its integral form.

The existence and uniqueness of (3) was studied by Peng [27].

Theorem 2 *There exists a unique solution $X \in \overline{M}_G^2(0, T; \mathbb{R}^N)$ to the SDE (3).*

Finally, we introduce the notion of quasi-sure analysis for G-expectation. It plays an important role in studying pathwise properties of stochastic processes under the framework of G-expectation.

First of all, on the canonical sublinear expectation space $(\Omega, L_{ip}(\Omega), \mathbb{E}^G)$, we can prove a refinement of Theorem 1: there exists a weakly compact family $\mathscr{P}$ of probability measures on $(\Omega, \mathscr{B}(\Omega))$, such that for any $X \in L_{ip}(\Omega)$ and $P \in \mathscr{P}$, $\mathbb{E}_P[X]$ is well-defined and

$$\mathbb{E}^G[X] = \max_{P \in \mathscr{P}} \mathbb{E}_P[X], \quad \forall X \in L_{ip}(\Omega),$$

where "max" means that the supremum is attainable (for each X). Moreover, there is an explicit characterization of the family $\mathscr{P}$. Let G be represented in the following way:

$$G(A) = \frac{1}{2} \sup_{Q \in \Gamma} \operatorname{tr}(AQQ^T),$$

for some bounded, closed and convex subset $\Gamma \subset \mathbb{R}^{d \times d}$, and let $\mathscr{A}_\Gamma$ be the collection of all Γ-valued $\{\mathscr{F}_t^W : t \geqslant 0\}$-adapted processes on $[0, \infty)$, where $\{\mathscr{F}_t^W : t \geqslant 0\}$ is the natural filtration of the coordinate process on Ω. Let $\mathscr{P}_0$ be the collection of probability laws of the following classical Itô integral processes with respect to the standard Wiener measure:

$$B_t^\gamma := \int_0^t \gamma_s dW_s, \ t \geqslant 0, \ \gamma \in A_\Gamma.$$

Then $\mathscr{P} = \overline{\mathscr{P}_0}$. For the proof of this result, please refer to [7].

For this particular family $\mathscr{P}$, define the set function c by

$$c(A) := \sup_{P \in \mathscr{P}} P(A), \ A \in \mathscr{B}(\Omega).$$

Then we have the following result.

Theorem 3 *The set function c is a Choquet capacity (for an introduction of capacity theory, see [3, 6]). In other words,*

1. *for any $A \in \mathscr{B}(\Omega)$, $0 \leqslant c(A) \leqslant 1$;*
2. *if $A \subset B$, then $c(A) \leqslant c(B)$;*
3. *if A_n is a sequence in $\mathscr{B}(\Omega)$, then $c(\cup_n A_n) \leqslant \sum_n c(A_n)$;*
4. *if A_n is increasing in $\mathscr{B}(\Omega)$, then $c(\cup A_n) = \lim_{n \to \infty} c(A_n)$.*

For any $\mathscr{B}(\Omega)$-measurable random variable X such that $\mathbb{E}_P[X]$ is well-defined for all $P \in \mathscr{P}$, define the upper expectation

$$\hat{\mathbb{E}}[X] := \sup_{P \in \mathscr{P}} \mathbb{E}_P[X].$$

Then we can prove that for any $0 \leqslant T \leqslant \infty$ and $X \in L_G^1(\Omega_T)$,

$$\mathbb{E}^G[X] = \hat{\mathbb{E}}[X].$$

For a detailed discussion and other related properties, please refer to [7].

The following Markov inequality and Borel-Cantelli lemma under the capacity c are important for us.

Theorem 4 *1. For any $X \in L_G^p(\Omega)$ and $\lambda > 0$, we have*

$$c(|X| > \lambda) \leqslant \frac{\mathbb{E}^G[|X|^p]}{\lambda^p}.$$

2. Let A_n be a sequence in $\mathscr{B}(\Omega)$ such that

$$\sum_{n=1}^{\infty} c(A_n) < \infty.$$

Then

$$c(\limsup A_n) = 0.$$

Definition 5 A property depending on $\omega \in \Omega$ is said to hold quasi-surely, if it holds outside a $\mathscr{B}(\Omega)$-measurable subset of zero capacity.

2.2 Rough Path Theory and Rough Differential Equations

Now we introduce some fundamentals in the theory of rough paths and rough differential equations. For a systematic introduction, please refer to [9, 18, 19].

For $n \geqslant 1$, define

$$T^{(\infty)}(\mathbb{R}^d) := \oplus_{k=0}^{\infty} (\mathbb{R}^d)^{\otimes k}$$

to be the infinite tensor algebra and

$$T^{(n)}(\mathbb{R}^d) := \oplus_{k=0}^{n} (\mathbb{R}^d)^{\otimes k}$$

to be the truncated tensor algebra of order n, equipped with the Euclidean norm. Let Δ be the triangle region $\{(s, t) : 0 \leqslant s < t \leqslant 1\}$. A functional $\mathbf{X} : \Delta \to T^{(n)}(\mathbb{R}^d)$ of order n is called multiplicative if for any $s < u < t$,

$$\mathbf{X}_{s,t} = \mathbf{X}_{s,u} \otimes \mathbf{X}_{u,t}.$$

Such a multiplicative structure is called the Chen identity. It describes the (nonlinear) additive structure of integrals over different intervals.

A control function ω is a nonnegative continuous function on Δ such that for any $s < u < t$,

$$\omega(s, u) + \omega(u, t) \leqslant \omega(s, t),$$

and for any $t \in [0, 1]$, $\omega(t, t) = 0$. An example of control function $\omega(s, t)$ is the 1-variation norm over $[s, t]$ of a path with bounded variation.

Let $p \geqslant 1$ be a fixed constant. A continuous and multiplicative functional

$$\mathbf{X}_{s,t} = (1, X_{s,t}^1, \cdots, X_{s,t}^n)$$

of order n has finite p-variation if for some control function ω,

$$|X_{s,t}^i| \leqslant \omega(s,t)^{\frac{i}{p}}, \; \forall i = 1, 2, \cdots, n, \; (s,t) \in \Delta. \tag{4}$$

$\mathbf{X}$ has finite p-variation if and only if for any $i = 1, 2, \cdots, n$,

$$\sup_D \sum_l |X_{t_{l-1},t_l}^i|^{\frac{p}{i}} < \infty,$$

where $\sup_D$ runs over all finite partitions of $[0, 1]$. We can also introduce the notion of finite p-variation for multiplicative functionals in $T^{(\infty)}(\mathbb{R}^d)$ by allowing $1 \leqslant i < \infty$ in (4). A continuous and multiplicative functional $\mathbf{X}$ of order $[p]$ with finite p-variation is called a rough path with roughness p. The space of rough paths with roughness p is denoted by $\Omega_p(\mathbb{R}^d)$.

The following Lyons lifting theorem (see [17]) shows that the higher levels of a rough path $\mathbf{X}$ with roughness p are uniquely determined by $\mathbf{X}$ itself.

Theorem 5 *Let $\mathbf{X}$ be a rough path with roughness p. Then $\mathbf{X}$ can be uniquely extended to a continuous and multiplicative functional in $T^{(\infty)}(\mathbb{R}^d)$ with finite p-variation.*

One of the motivation of introducing the concept of rough paths is to develop the theory of differential equations driven by rough signals.

If an $\mathbb{R}^d$-valued path X has bounded variation, we know that the Picard iteration for the following differential equation converges:

$$dY_t = V(Y_t)dX_t, \tag{5}$$

where $V = (V_1, \cdots, V_d)$ is a family of Lipschitz vector fields. Another way to consider (5) is to use the Euler scheme, which can be regarded as the Taylor expansion of functional of paths. Namely, we can write informally that

$$Y_t - Y_s \sim \sum_{n=1}^{\infty} \sum_{\alpha_1, \cdots, \alpha_n = 1}^{d} V_{\alpha_1} \cdots V_{\alpha_n} I(Y_s) \int_{s < u_1 < \cdots < u_n < t} dX_{u_1}^{\alpha_1} \cdots dX_{u_n}^{\alpha_n}. \tag{6}$$

From (6) we can see that the sequence

$$\mathbf{X}_{s,t} := (1, X_t - X_s, \int_{s<u<v<t} dX_u \otimes dX_v, \cdots, \int_{s<u_1<\cdots<u_n<t} dX_{u_1} \otimes \cdots \otimes dX_{u_n}, \cdots)$$

contains exactly all the information to determine the solution Y. On the other hand, it can be proved that $\mathbf{X}$ is multiplicative and of finite 1-variation. Since X has bounded variation, it follows from Theorem 5 that $\mathbf{X}$ is the unique enhancement of X. This is the fundamental reason why we don't need to see the higher levels when solving Eq. (5)-all information about $\mathbf{X}$, which uniquely determines the solution of (5), is incorporated in the first level.

If the driven signal is rougher, the situation becomes different. The same thing is that the information to determine the solution lies in the multiplicative structure in $T^{(\infty)}(\mathbb{R}^d)$, while the difference is that, unlike the case of paths with bounded variation, the classical path itself may not be able to determine the higher levels which are crucial to characterize the solution of a differential equation. In other words, we need to specify higher levels of the classical path in order to make sense of differential equations. According to Theorem 5, we know that the higher levels (levels above $[p]$) of a rough path $\mathbf{X}$ with roughness p are uniquely determined by $\mathbf{X}$ itself. Therefore, to establish differential equations driven by signals rougher than paths of bounded variation, we need to interpret the driven signal as a rough path with certain roughness p, that is, the driving signal should be an element in the space $\Omega_p(\mathbb{R}^d)$.

When the driving signal $\mathbf{X}$ is in some smaller space of $\Omega_p(\mathbb{R}^d)$ in which $\mathbf{X}$ can be approximated by paths of bounded variation in some sense, we are able to use a natural approximation procedure to introduce the notion of differential equations. But first we need to introduce a certain kind of topology.

Define the p-variation distance $d_p(\cdot, \cdot)$ on $\Omega_p(\mathbb{R}^d)$ by

$$d_p(\mathbf{X}, \mathbf{Y}) := \max_{1 \le i \le [p]} \sup_D (\sum_l |X^i_{t_{l-1}, t_l} - Y^i_{t_{l-1}, t_l}|^{\frac{p}{i}})^{\frac{i}{p}}.$$

Then $(\Omega_p(\mathbb{R}^d), d_p)$ is a complete metric space.

A continuous path $X \in C([0, 1]; \mathbb{R}^d)$ is called smooth if it has bounded variation. Let

$$\Omega_p^\infty(\mathbb{R}^d) := \{\mathbf{X} : \mathbf{X} \text{ is the unique enhancement of } X \text{ in } T^{([p])}(\mathbb{R}^d), \text{ for smooth } X\}$$

be the subspace of enhanced smooth paths of order $[p]$. The closure of $\Omega_p^\infty(\mathbb{R}^d)$ under the p-variation distance d_p, denoted by $G\Omega_p(\mathbb{R}^d)$, is called the space of geometric rough paths with roughness p.

The following theorem, proved by Lyons [17], which is usually known as the universal limit theorem, enables us to introduce the notion of differential equations driven by geometric rough paths.

Theorem 6 *Let $V_1, \cdots, V_d \in C_b^{[p]+1}(\mathbb{R}^d)$ be given vector fields on $\mathbb{R}^N$. For a given $y \in \mathbb{R}^d$, define the mapping*

$$F(y, \cdot) : \Omega_p^\infty(\mathbb{R}^d) \to \Omega_p^\infty(\mathbb{R}^N)$$

in the following way. For any $\mathbf{X} \in \Omega_p^\infty(\mathbb{R}^d)$, let X be the smooth path associated with $\mathbf{X}$ starting at the origin (i.e., projection of $\mathbf{X}$ onto the first level), and Y be the *unique smooth path which is the solution of the following ODE:*

$$dY_t = V_\alpha(Y_t)dX_t^\alpha$$

with $Y_0 = y$. $F(y, \mathbf{X})$ *is defined to be the enhancement of* Y *in* $\Omega_p^\infty(\mathbb{R}^d)$. *Then the mapping* $F(y, \cdot)$ *is continuous with respect to the corresponding* p-*variation distance* d_p.

According to Theorem 6, there exists a unique continuous extension of $F(y, \cdot)$ on $G\Omega_p(\mathbb{R}^d)$. The extended mapping

$$F(y, \cdot): \ G\Omega_p(\mathbb{R}^d) \to G\Omega_p(\mathbb{R}^N),$$

is called the Itô-Lyons mapping. Such a mapping defines the (unique) solution in the space $G\Omega_p(\mathbb{R}^d)$ to the following differential equation:

$$dY_t = V(Y_t)dX_t, \tag{7}$$

with initial value y. Equation (7) is called a rough differential equation driven by $\mathbf{X}$ (or simply called an RDE), and the solution $\mathbf{Y}$ is called the full solution of (7). If we are only interested in classical paths, then

$$Y_t := y + \pi_1(\mathbf{Y}), \quad t \in [0, 1],$$

is called the solution of (7) with initial value y.

3 The Euler-Maruyama Approximation for SDEs Driven by G-Brownian Motion

In this section, we are going to establish the Euler-Maruyama approximation for SDEs driven by G-Brownian motion.

This result can be used to establish the Wong-Zakai type approximation which reveals the relation between SDEs (in the sense of $L_G^2(\Omega; \mathbb{R}^N)$ by S. Peng) and RDEs (in the sense of rough paths by Lyons) driven by G-Brownian motion. In Sect. 5, the study of such relation will be our main focus. However, based on the result in the next section which reveals the rough path nature of G-Brownian motion, we are going to use the rough Taylor expansion in the theory of RDEs instead of developing the Wong-Zakai type approximation to show that the solution of an SDE solves some associated RDE with a correction term in terms of the cross variation process of multidimensional G-Brownian motion. Such approach reveals the natural of G-Brownian motion and differential equations in the sense of rough paths in a more fundamental way.

We also believe that there will be other interesting applications of the Euler-Maruyama approximation, such as in numerical analysis under G-expectation, and in mathematical finance under uncertainty.

Consider the following N-dimensional SDE driven by the canonical d-dimensional G-Brownian motion over $[0, 1]$ on the sublinear expectation space $(\Omega, L_G^2(\Omega), \mathbb{E}^G)$ which is the L_G^2-completion of the canonical path space $(\Omega, L_{ip}(\Omega), \mathbb{E}^G)$:

$$dX_t = b(X_t)dt + h_{\alpha\beta}(X_t)d\langle B^\alpha, B^\beta\rangle_t + V_\alpha(X_t)dB_t^\alpha, \tag{8}$$

with initial condition $X_0 = \xi \in \mathbb{R}^N$, where the coefficients $b^i, h_{\alpha\beta}^i, V_\alpha^i$ are bounded and uniformly Lipschitz. The existence and uniqueness of solution is studied by Peng [27].

The Euler-Maruyama approximation of the solution X_t of (8) is defined as follows.

For $n \geqslant 1$, consider the dyadic partition of the time interval $[0, 1]$, i.e.,

$$t_k^n = \frac{k}{2^n}, \; k = 0, 1, \cdots, 2^n.$$

Define X_t^n to be the approximation of X_t in the following evolutive way:

$$X_0^n = \xi,$$

and for $t \in [t_{k-1}^n, t_k^n]$,

$$(X_t^n)^i = (X_{k-1}^n)^i + V_\alpha^i(X_{k-1}^n)\Delta_k^n B^\alpha + b^i(X_{k-1}^n)\Delta t^n + h_{\alpha\beta}^i(X_{k-1}^n)\Delta_k^n\langle B^\alpha, B^\beta\rangle,$$

where

$$X_{k-1}^n := X_{t_{k-1}^n}^n, \; \Delta_k^n B^\alpha := B_{t_k^n}^\alpha - B_{t_{k-1}^n}^\alpha, \; \Delta t^n := \frac{1}{2^n},$$

$$\Delta_k^n\langle B^\alpha, B^\beta\rangle := \langle B^\alpha, B^\beta\rangle_{t_k^n} - \langle B^\alpha, B^\beta\rangle_{t_{k-1}^n}.$$

In this section, we are going to prove that X_t^n converges to the solution X_t of (8) in $L_G^2(\Omega; \mathbb{R}^N)$ with convergence rate 0.5, which coincides with the classical case when B_t reduces to a classical Brownian motion.

First of all, the following lemmas is useful for us.

Lemma 1 *Let η_t be a bounded process in $M_G^2(0, 1)$. Then for any $v \in \mathbb{R}^d$, $0 \leqslant s < t \leqslant 1$,*

$$\mathbb{E}^G[(\int_s^t \eta_u d\langle B^v\rangle_u)^2] \leqslant \overline{\sigma}_v^2(t - s)\mathbb{E}^G[\int_s^t \eta_u^2 d\langle B^v\rangle_u],$$

where $\overline{\sigma}_v^2 := 2G(v \cdot v^T)$ and $B^v := \langle v, B \rangle$, in which $\langle \cdot, \cdot \rangle$ denotes the Euclidean inner product of $\mathbb{R}^d$.

Proof By approximation, it suffices to consider

$$\eta_u = \sum_{j=1}^{k} \zeta_{j-1} 1_{[u_{j-1}, u_j)},$$

where $s = u_0 < u_1 < \cdots < u_k = t$ and $\zeta_j \in L_{ip}(\Omega_{u_j})$ are bounded. In this case, by definition

$$\int_s^t \eta_u d\langle B^v \rangle_u = \sum_{j=1}^{k} \zeta_{j-1}(\langle B^v \rangle_{u_j} - \langle B^v \rangle_{u_{j-1}}),$$

and

$$\int_s^t \eta_u^2 d\langle B^v \rangle_u = \sum_{j=1}^{k} \zeta_{j-1}^2(\langle B^v \rangle_{u_j} - \langle B^v \rangle_{u_{j-1}}),$$

which are both defined in the pathwise sense for step functions. Since $\langle B^v \rangle$ is increasing, the Cauchy-Schwarz inequality yields that

$$\left(\int_s^t \eta_u d\langle B^v \rangle_u \right)^2 \leq (\langle B^v \rangle_t - \langle B^v \rangle_s) \cdot \int_s^t \eta_u^2 d\langle B^v \rangle_u.$$

Since ζ_j are bounded, if we use M to denote an upper bound of η_u^2, it follows that for any $c \geq \overline{\sigma}_v^2$,

$$\left(\int_s^t \eta_u d\langle B^v \rangle_u \right)^2 \leq M(\langle B^v \rangle_t - \langle B^v \rangle_s - c(t - s))^+ (\langle B^v \rangle_t - \langle B^v \rangle_s)$$

$$+ c(t - s) \int_s^t \eta_u^2 d\langle B^v \rangle_u.$$

Let $\varphi(x) = (x - c(t-s))^+ x$. Since $\langle B^v \rangle_t - \langle B^v \rangle_s$ is $N([\underline{\sigma}_v^2, \overline{\sigma}_v^2] \times \{0\})$-distributed, it follows that

$$\mathbb{E}^G[\varphi(\langle B^v \rangle_t - \langle B^v \rangle_s)] = \sup_{\underline{\sigma}_v^2 \leq x \leq \overline{\sigma}_v^2} \varphi(x(t - s))$$

$$= (t - s)^2 \sup_{\underline{\sigma}_v^2 \leq x \leq \overline{\sigma}_v^2} (x - c)^+ x$$

$$= 0.$$

Therefore, by the sub-linearity of G, we have

$$\mathbb{E}^G[(\int_s^t \eta_u d\langle B^v \rangle_u)^2] \leqslant c(t-s)\mathbb{E}^G[\int_s^t \eta_u^2 d\langle B^v \rangle_u], \quad c \geqslant \bar{\sigma}_v^2.$$

Now the proof is complete.

Now we are in position to state and prove our main result of this section.

Theorem 7 *We have the following error estimate for the Euler-Maruyama approximation:*

$$\sup_{t \in [0,1]} \mathbb{E}^G[|X_t^n - X_t|^2] \leqslant C\Delta t^n,$$

where C is some positive constant only depending on d, N, G and the coefficients of (8). In particular,

$$\lim_{n \to \infty} \sup_{t \in [0,1]} \mathbb{E}^G[|X_t^n - X_t|^2] = 0.$$

Proof For $t \in [t_{k-1}^n, t_k^n]$, by construction we have

$$X_t^i - (X_t^n)^i = I_1^i + J_1^i + K_1^i + I_2^i + J_2^i + K_2^i,$$

where

$$I_1^i = \sum_{l=1}^{k-1} \int_{t_{l-1}^n}^{t_l^n} (V_\alpha^i(X_s) - V_\alpha^i(X_s^n))dB_s^\alpha + \int_{t_{k-1}^n}^t (V_\alpha^i(X_s) - V_\alpha^i(X_s^n))dB_s^\alpha,$$

$$J_1^i = \sum_{l=1}^{k-1} \int_{t_{l-1}^n}^{t_l^n} (b^i(X_s) - b^i(X_s^n))ds + \int_{t_{k-1}^n}^t (b^i(X_s) - b^i(X_s^n))ds,$$

$$K_1^i = \sum_{l=1}^{k-1} \int_{t_{l-1}^n}^{t_l^n} (h_{\alpha\beta}^i(X_s) - h_{\alpha\beta}^i(X_s^n))d\langle B^\alpha, B^\beta \rangle_s$$

$$+ \int_{t_{k-1}^n}^t (h_{\alpha\beta}^i(X_s) - h_{\alpha\beta}^i(X_s^n))d\langle B^\alpha, B^\beta \rangle_s,$$

$$I_2^i = \sum_{l=1}^{k-1} \int_{t_{l-1}^n}^{t_l^n} (V_\alpha^i(X_s^n) - V_\alpha^i(X_{l-1}^n))dB_s^\alpha + \int_{t_{k-1}^n}^t (V_\alpha^i(X_s) - V_\alpha^i(X_{l-1}^n))dB_s^\alpha,$$

$$J_2^i = \sum_{l=1}^{k-1} \int_{t_{l-1}^n}^{t_l^n} (b^i(X_s^n) - b^i(X_{l-1}^n))ds + \int_{t_{k-1}^n}^t (b^i(X_s) - b^i(X_{l-1}^n))ds,$$

$$K_2^i = \sum_{l=1}^{k-1} \int_{t_{l-1}^n}^{t_l^n} (h_{\alpha\beta}^i(X_s) - h_{\alpha\beta}^i(X_s^n)) d \langle B^\alpha, B^\beta \rangle_s$$

$$+ \int_{t_{k-1}^n}^t (h_{\alpha\beta}^i(X_s) - h_{\alpha\beta}^i(X_s^n)) d \langle B^\alpha, B^\beta \rangle_s.$$

It follows that

$$(X_t^i - (X_t^n)^i)^2 \leqslant 6((I_1^i)^2 + (J_1^i)^2 + (K_1^i)^2 + (I_2^i)^2 + (J_2^i)^2 + (K_2^i)^2). \qquad (9)$$

Throughout the rest of this section, we will always use the same notation C to denote constants only depending on d, N, G and the coefficients of (8), although they may be different from line to line.

Now the following estimates are important for further development.

1. From the G-Itô isometry, the distribution of $\langle B^\alpha \rangle$ and the Lipschitz property, we have,

$$\mathbb{E}^G[(\int_{t_{l-1}^n}^u (V_\alpha^i(X_s) - V_\alpha^i(X_s^n)) dB_s^\alpha)^2] \leqslant C \int_{t_{l-1}^n}^u \mathbb{E}^G[|X_s - X_s^n|^2] ds, \ \forall u \in [t_{l-1}^n, t_l^n].$$

2. Similarly, by Cauchy-Schwarz inequality, we have

$$\mathbb{E}^G[(\int_{t_{l-1}^n}^u (b^i(X_s) - b^i(X_s^n)) ds)^2] \leqslant C(u - t_{l-1}^n) \int_{t_{l-1}^n}^u \mathbb{E}^G[|X_s - X_s^n|^2] ds,$$

$$\forall u \in [t_{l-1}^n, t_l^n].$$

By the definition of $\langle B^\alpha, B^\beta \rangle$ and Lemma 1, we also have for any $u \in [t_{l-1}^n, t_l^n]$,

$$\mathbb{E}^G[(\int_{t_{l-1}^n}^u (h_{\alpha\beta}^i(X_s) - h_{\alpha\beta}^i(X_s^n)) d \langle B^\alpha, B^\beta \rangle_s)^2] \leqslant C(u - t_{l-1}^n) \int_{t_{l-1}^n}^u \mathbb{E}^G[|X_s - X_s^n|^2] ds.$$

3. By construction and similar arguments to (1), (2), we have

$$\mathbb{E}^G[(\int_{t_{l-1}^n}^u (V_\alpha^i(X_s^n) - V_\alpha^i(X_{l-1}^n)) dB_s^\alpha)^2] \leqslant C(u - t_{l-1}^n)^2,$$

$$\mathbb{E}^G[(\int_{t_{l-1}^n}^u (b^i(X_s^n) - b^i(X_{l-1}^n)) ds)^2] \leqslant C(u - t_{l-1}^n)^3,$$

$$\mathbb{E}^G[(\int_{t_{l-1}^n}^u (h_{\alpha\beta}^i(X_s) - h_{\alpha\beta}^i(X_s^n)) d \langle B^\alpha, B^\beta \rangle_s)^2] \leqslant C(u - t_{l-1}^n)^3,$$

for all $u \in [t_{l-1}^n, t_l^n]$.

4. By conditioning and from the properties of Itô integral with respect to G-Brownian motion, we know that the G-expectation of each "cross term" in $(I_1^i)^2$ and in $(I_2^i)^2$ is zero.

Combining (1)–(4) and applying the following elementary inequality to $(J_1^i)^2$, $(J_2^i)^2$, $(K_1^i)^2$ and $(K_2^i)^2$:

$$(a_1 + \cdots + a_m)^2 \leq m(a_1^2 + \cdots + a_m^2),$$

it is not hard to obtain that

$$\mathbb{E}^G[\|X_t - X_t^n\|^2] \leq C \int_0^t \mathbb{E}^G[\|X_s - X_s^n\|^2]ds + C(\Delta t^n), \ \forall t \in [0, 1].$$

By using Gronwall inequality, we arrive at

$$\mathbb{E}^G[\|X_t - X_t^n\|^2] \leq C(\Delta t^n),$$

which completes the proof of the theorem.

4 G-Brownian Motion as Rough Paths and RDEs Driven by G-Brownian Motion

In this section, we are going to study the nature of sample paths of G-Brownian motion under the framework of rough path theory. More precisely, we are going to show that: on the canonical path space, outside a Borel-measurable set of capacity zero, the sample paths of G-Brownian motion can be enhanced to the second level in a canonical way so that they become geometric rough paths with roughness $2 < p < 3$. As pointed out before, such a result will enable us to establish RDEs driven by G-Brownian motion in the space of geometric rough paths.

Recall that $(\Omega, L_{ip}(\Omega), \mathbb{E}^G)$ is the canonical path space associated with the function G, on which the coordinate process

$$B_t(\omega) := \omega_t, \ t \in [0, 1],$$

is a d-dimensional G-Brownian motion with continuous sample paths.

By the following moment inequality for B_t:

$$\mathbb{E}^G[|B_t - B_s|^{2q}] \leq C_q(t - s)^q, \ \forall 0 \leq s < t \leq 1, \ q > 1, \tag{10}$$

and the generalized Kolmogorov criterion (see [27] for details), we know that for quasi-surely, the sample paths of B_t are α-Hölder continuous for any $\alpha \in (0, \frac{1}{2})$. Therefore, if the sample paths of B_t can be regarded as objects in the space of geometric rough paths, the correct roughness should be $2 < p < 3$ (so we should look for the enhancement of B_t to the second level); or in other words, the right topology we should work with is the p-variation topology induced by the p-variation distance d_p on the space of geometric rough paths with roughness $2 < p < 3$. The situation here is the same as the classical Brownian motion, and the fundamental reason behind lies in the distribution of B_t (or more precisely, the moment inequality (10)), which yields the same kind of Hölder continuity for sample paths of B_t as the classical one.

From now on, we will assume that $p \in (2, 3)$ is some fixed constant.

As in the last section, for $n \geqslant 1$, $k = 0, 1, \cdots, 2^n$, let $t_k^n = \frac{k}{2^n}$ be the dyadic partition of $[0, 1]$, and let B_t^n be the piecewise linear approximation of B_t over the partition points $\{t_0^n, t_1^n, \cdots, t_{2^n}^n\}$. Since the sample paths of B_t^n are smooth, B_t^n has a unique enhancement

$$\boldsymbol{B}_{s,t}^n = (1, B_{s,t}^{n,1}, B_{s,t}^{n,2}), \ 0 \leqslant s < t \leqslant 1,$$

to the space $G\Omega_p(\mathbb{R}^d)$ of geometric rough paths with roughness p (in fact, for any $p \geqslant 1$) determined by iterated integrals.

Our goal is to show that for quasi-surely, $\boldsymbol{B}^n$ is a Cauchy sequence under the p-variation distance d_p. It follows that for quasi-surely, the sample paths of B_t can be enhanced to the second level as geometric rough paths with roughness p, which are defined as limits of $\boldsymbol{B}^n$ under d_p. Such an enhancement can be regarded as a canonical lifting by using dyadic approximations.

Throughout the rest of this section, we will use $\|\cdot\|_q$ to denote the L^q-norm under the G-expectation $\mathbb{E}^G$. Moreover, we will use the same notation C to denote constants only depending on d, G, p, although they may be different from line to line.

The following estimates are crucial for the proof of the main result of this section.

Lemma 2 *Let* $m, n \geqslant 1$, *and* $k = 1, 2, \cdots, 2^n$. *Then*

1.

$$\left\| B_{t_{k-1}^n, t_k^n}^{m,j} \right\|_{\frac{p}{j}} \leqslant \begin{cases} C(\frac{1}{2^{\frac{n}{2}}})^j, & n \leqslant m; \\ C(\frac{2^{\frac{m}{2}}}{2^n})^j, & n > m, \end{cases}$$

where $j = 1, 2$.

2.

$$\left\| B_{t_{k-1}^n, t_k^n}^{m+1,1} - B_{t_{k-1}^n, t_k^n}^{m,1} \right\|_p \leqslant \begin{cases} 0, & n \leqslant m; \\ C\frac{2^{\frac{m}{2}}}{2^n}, & n > m, \end{cases}$$

$$\left\| B_{t_{k-1}^n, t_k^n}^{m+1,2} - B_{t_{k-1}^n, t_k^n}^{m,2} \right\|_{\frac{p}{2}} \leqslant \begin{cases} C\frac{1}{2^{\frac{m}{2}} 2^{\frac{n}{2}}}, & n \leqslant m; \\ C\frac{2^m}{2^{2n}}, & n > m. \end{cases}$$

Here $\| \cdot \|_q$ denotes the L^q-norm under the G-expectation $\mathbb{E}^G$, and C is some positive constant not depending on m, n, k.

Proof 1. The first level.

If $n \leqslant m$, then

$$B^{m,1}_{t^n_{k-1}, t^n_k} = B_{t^n_k} - B_{t^n_{k-1}}.$$

It follows from the moment inequality (10) that

$$\mathbb{E}^G[|B^{m,1}_{t^n_{k-1}, t^n_k}|^p] \leqslant C \frac{1}{2^{\frac{np}{2}}},$$

and thus

$$\|B^{m,1}_{t^n_{k-1}, t^n_k}\|_p \leqslant C \frac{1}{2^{\frac{n}{2}}}.$$

Also it is trivial to see that

$$B^{m+1,1}_{t^n_{k-1}, t^n_k} - B^{m,1}_{t^n_{k-1}, t^n_k} = (B_{t^n_k} - B_{t^n_{k-1}}) - (B_{t^n_k} - B_{t^n_{k-1}}) = 0.$$

If $n > m$, then by construction we know that

$$B^{m,1}_{t^n_{k-1}, t^n_k} = \frac{2^m}{2^n}(B_{t^m_l} - B_{t^m_{l-1}}),$$

where l is the unique integer such that $[t^n_{k-1}, t^n_k] \subset [t^m_{l-1}, t^m_l]$. Therefore,

$$\|B^{m,1}_{t^n_{k-1}, t^n_k}\|_p = \frac{2^m}{2^n}\|B_{t^m_l} - B_{t^m_{l-1}}\|_p \leqslant C \frac{2^{\frac{m}{2}}}{2^n}.$$

On the other hand, if $[t^n_{k-1}, t^n_k] \subset [t^{m+1}_{2l-2}, t^{m+1}_{2l-1}]$, then

$$B^{m+1,1}_{t^n_{k-1}, t^n_k} - B^{m,1}_{t^n_{k-1}, t^n_k} = \frac{2^{m+1}}{2^n}(B_{t^{m+1}_{2l-1}} - B_{t^{m+1}_{2l-2}}) - \frac{2^m}{2^n}(B_{t^m_l} - B_{t^m_{l-1}})$$

$$= \frac{2^m}{2^n}((B_{\frac{2l-1}{2^{m+1}}} - B_{\frac{2l-2}{2^{m+1}}}) - (B_{\frac{2l}{2^{m+1}}} - B_{\frac{2l-1}{2^{m+1}}})).$$

It follows that

$$\|B^{m+1,1}_{t^n_{k-1}, t^n_k} - B^{m,1}_{t^n_{k-1}, t^n_k}\|_p \leqslant C \frac{2^{\frac{m}{2}}}{2^n}.$$

Similarly, if $[t^n_{k-1}, t^n_k] \subset [t^{m+1}_{2l-1}, t^{m+1}_{2l}]$, we will obtain the same estimate.

2. The second level.

Since $\frac{p}{2} < 2$, by monotonicity it suffices to establish the desired estimates under the L^2-norm.

First consider the term $B^{m+1,2}_{t^n_{k-1},t^n_k} - B^{m,2}_{t^n_{k-1},t^n_k}$.

If $n \leqslant m$, by the construction of $B^{m,2}_{s,t}$, we have

$$
\begin{aligned}
B^{m,2;\alpha,\beta}_{t^n_{k-1},t^n_k} &= \int_{t^n_{k-1} < u < v < t^n_k} dB^\alpha_u dB^\beta_v \\
&= \int_{t^n_{k-1}}^{t^n_k} B^{m,1;\alpha}_{t^n_{k-1},v} dB^{m,1;\beta}_v \\
&= \sum_{l=2^{(m-n)}(k-1)+1}^{2^{m-n}k} \frac{\Delta^m_l B^\beta}{\Delta t^m} \int_{t^m_{l-1}}^{t^m_l} \left(\frac{v - t^m_{l-1}}{\Delta t^m} B^\alpha_{t^m_l} + \frac{t^m_l - v}{\Delta t^m} B^\alpha_{t^m_{l-1}} - B^\alpha_{t^n_{k-1}} \right) dv \\
&= \sum_{l=2^{(m-n)}(k-1)+1}^{2^{m-n}k} \left(\frac{B^\alpha_{t^m_{l-1}} + B^\alpha_{t^m_l}}{2} - B^\alpha_{t^n_{k-1}} \right) \Delta^m_l B^\beta.
\end{aligned}
$$

Therefore,

$$
\begin{aligned}
&B^{m+1,2;\alpha,\beta}_{t^n_{k-1},t^n_k} - B^{m,2;\alpha,\beta}_{t^n_{k-1},t^n_k} \\
&= \sum_{l=2^{(m+1-n)}(k-1)+1}^{2^{m+1-n}k} \left(\frac{B^\alpha_{t^{m+1}_{l-1}} + B^\alpha_{t^{m+1}_l}}{2} - B^\alpha_{t^n_{k-1}} \right) \Delta^m_l B^\beta \\
&\quad - \sum_{l=2^{(m-n)}(k-1)+1}^{2^{m-n}k} \left(\frac{B^\alpha_{t^m_{l-1}} + B^\alpha_{t^m_l}}{2} - B^\alpha_{t^n_{k-1}} \right) \Delta^m_l B^\beta \\
&= \sum_{l=2^{(m-n)}(k-1)+1}^{2^{(m-n)}k} \left(\left(\frac{B^\alpha_{t^{m+1}_{2l-2}} + B^\alpha_{t^{m+1}_{2l-1}}}{2} - B^\alpha_{t^n_{k-1}} \right) \Delta^{m+1}_{2l-1} B^\beta + \left(\frac{B^\alpha_{t^{m+1}_{2l-1}} + B^\alpha_{t^{m+1}_{2l}}}{2} - B^\alpha_{t^n_{k-1}} \right) \right. \\
&\qquad \left. \cdot \Delta^{m+1}_{2l} B^\beta - \left(\frac{B^\alpha_{t^{m+1}_{2l-2}} + B^\alpha_{t^{m+1}_{2l}}}{2} - B^\alpha_{t^n_{k-1}} \right) (\Delta^{m+1}_{2l-1} B^\beta + \Delta^{m+1}_{2l} B^\beta) \right) \\
&= \frac{1}{2} \sum_{l=2^{m-n}(k-1)+1}^{2^{m-n}k} (\Delta^{m+1}_{2l-1} B^\alpha \Delta^{m+1}_{2l} B^\beta - \Delta^{m+1}_{2l} B^\alpha \Delta^{m+1}_{2l-1} B^\beta).
\end{aligned}
$$

By using the notation of tensor products, we have

$$
B^{m+1,2}_{t^n_{k-1},t^n_k} - B^{m,2}_{t^n_{k-1},t^n_k} = \frac{1}{2} \sum_{l=2^{m-n}(k-1)+1}^{2^{m-n}k} (\Delta^{m+1}_{2l-1} B \otimes \Delta^{m+1}_{2l} B - \Delta^{m+1}_{2l} B \otimes \Delta^{m+1}_{2l-1} B).
$$

It follows that

$$\mathbb{E}^G[|B^{m+1,2}_{t^n_{k-1},t^n_k} - B^{m,2}_{t^n_{k-1},t^n_k}|^2]$$

$$= \frac{1}{4}\mathbb{E}^G[|\sum_{l=2^{m-n}(k-1)+1}^{2^{m-n}k} (\Delta^{m+1}_{2l-1}B \otimes \Delta^{m+1}_{2l}B - \Delta^{m+1}_{2l}B \otimes \Delta^{m+1}_{2l-1}B)|^2]$$

$$\leqslant C \sum_{\substack{\alpha\neq\beta \\ \alpha,\beta=1,\cdots,d}} \mathbb{E}^G[|\sum_l (\Delta^{m+1}_{2l-1}B^\alpha \Delta^{m+1}_{2l}B^\beta - \Delta^{m+1}_{2l}B^\alpha \Delta^{m+1}_{2l-1}B^\beta)|^2]$$

$$\leqslant C \sum_{\alpha\neq\beta} \sum_{l,r} \mathbb{E}^G[(\Delta^{m+1}_{2l-1}B^\alpha \Delta^{m+1}_{2l}B^\beta - \Delta^{m+1}_{2l}B^\alpha \Delta^{m+1}_{2l-1}B^\beta)$$

$$\cdot(\Delta^{m+1}_{2r-1}B^\alpha \Delta^{m+1}_{2r}B^\beta - \Delta^{m+1}_{2r}B^\alpha \Delta^{m+1}_{2r-1}B^\beta)]$$

$$\leqslant C \sum_{\alpha\neq\beta} \sum_{l,r} (\mathbb{E}^G[\Delta^{m+1}_{2l-1}B^\alpha \Delta^{m+1}_{2r-1}B^\alpha \Delta^{m+1}_{2l}B^\beta \Delta^{m+1}_{2r}B^\beta]$$

$$+\mathbb{E}^G[\Delta^{m+1}_{2l}B^\alpha \Delta^{m+1}_{2r}B^\alpha \Delta^{m+1}_{2l-1}B^\beta \Delta^{m+1}_{2r-1}B^\beta]$$

$$+\mathbb{E}^G[-\Delta^{m+1}_{2l-1}B^\alpha \Delta^{m+1}_{2r}B^\alpha \Delta^{m+1}_{2r-1}B^\beta \Delta^{m+1}_{2l}B^\beta]$$

$$+\mathbb{E}^G[-\Delta^{m+1}_{2r-1}B^\alpha \Delta^{m+1}_{2l}B^\alpha \Delta^{m+1}_{2l-1}B^\beta \Delta^{m+1}_{2r}B^\beta]),$$

where the summation over l and r is taken from $2^{m-n}(k-1)+1$ to $2^{m-n}k$. Here we have used the sublinearity of $\mathbb{E}$. Now we study every term separately. If $l < r$, by the properties of conditional G-expectation and the distribution of B_t, we have

$$\mathbb{E}^G[\Delta^{m+1}_{2l-1}B^\alpha \Delta^{m+1}_{2l}B^\beta \Delta^{m+1}_{2r-1}B^\alpha \Delta^{m+1}_{2r}B^\beta]$$

$$= \mathbb{E}^G[\mathbb{E}[\Delta^{m+1}_{2l-1}B^\alpha \Delta^{m+1}_{2l}B^\beta \Delta^{m+1}_{2r-1}B^\alpha \Delta^{m+1}_{2r}B^\beta|\Omega_{t^{m+1}_{2r-1}}]]$$

$$= \mathbb{E}^G[\eta^+\mathbb{E}[\Delta^{m+1}_{2r}B^\beta|\Omega_{t^{m+1}_{2r-1}}] + \eta^-\mathbb{E}[-\Delta^{m+1}_{2r}B^\beta|\Omega_{t^{m+1}_{2r-1}}]]$$

$$= 0,$$

where $\eta = \Delta^{m+1}_{2l-1}B^\alpha \Delta^{m+1}_{2l}B^\beta \Delta^{m+1}_{2r-1}B^\alpha$. Similarly, we can prove that for any $l \neq r$,

$$\mathbb{E}^G[\Delta^{m+1}_{2l-1}B^\alpha \Delta^{m+1}_{2r-1}B^\beta \Delta^{m+1}_{2l}B^\beta \Delta^{m+1}_{2r}B^\beta]$$

$$= \mathbb{E}^G[(\Delta^{m+1}_{2l}B^\alpha \Delta^{m+1}_{2r-1}B^\alpha \Delta^{m+1}_{2l-1}B^\beta \Delta^{m+1}_{2r-1}B^\beta)]$$

$$= \mathbb{E}^G[(-\Delta^{m+1}_{2l-1}B^\alpha \Delta^{m+1}_{2r}B^\alpha \Delta^{m+1}_{2r-1}B^\beta \Delta^{m+1}_{2l}B^\beta)]$$

$$= \mathbb{E}^G[(-\Delta^{m+1}_{2r-1}B^\alpha \Delta^{m+1}_{2l}B^\alpha \Delta^{m+1}_{2l-1}B^\beta \Delta^{m+1}_{2r}B^\beta)]$$

$$= 0.$$

On the other hand, if $l = r$, it is straight forward that

$$\mathbb{E}^G[(\Delta_{2l-1}^{m+1}B^\alpha)^2(\Delta_{2l}^{m+1}B^\beta)^2] \leq \frac{1}{2}(\mathbb{E}^G[(\Delta_{2l-1}^{m+1}B^\alpha)^4] + \mathbb{E}^G[(\Delta_{2l}^{m+1}B^\beta)^4]) \leq C\frac{1}{2^{2m}},$$

and similarly,

$$\mathbb{E}^G(-\Delta_{2l-1}^{m+1}B^\alpha\,\Delta_{2l-1}^{m+1}B^\beta\,\Delta_{2l}^{m+1}B^\alpha\,\Delta_{2l}^{m+1}B^\beta) \leq \frac{1}{4}(\mathbb{E}^G[(\Delta_{2l-1}^{m+1}B^\alpha)^4]$$
$$+\mathbb{E}^G[(\Delta_{2l-1}^{m+1}B^\beta)^4]$$
$$+\mathbb{E}^G[(\Delta_{2l}^{m+1}B^\alpha)^4]$$
$$+\mathbb{E}^G[(\Delta_{2l}^{m+1}B^\beta)^4])$$
$$\leq C\frac{1}{2^{2m}}.$$

Combining all the estimates above, we arrive at

$$\mathbb{E}^G[|B_{t_{k-1}^n,t_k^n}^{m+1,2} - B_{t_{k-1}^n,t_k^n}^{m,2}|^2] \leq C\sum_{\alpha\neq\beta}\sum_{l=2^{m-n}(k-1)+1}^{2^{m-n}k}\frac{1}{2^{2m}} \leq C\frac{1}{2^m2^n},$$

and hence

$$\|B_{t_{k-1}^n,t_k^n}^{m+1,2} - B_{t_{k-1}^n,t_k^n}^{m,2}\|_2 \leq C\frac{1}{2^{\frac{m}{2}}2^{\frac{n}{2}}}.$$

If $n > m$, by construction we have

$$B_{t_{k-1}^n,t_k^n}^{m,2;\alpha,\beta} = \int_{t_{k-1}^n<u<v<t_k^n} d(B^m)_u^\alpha d(B^m)_v^\beta$$
$$= \int_{t_{k-1}^n}^{t_k^n} B_{t_{k-1}^n,v}^{m,1;\alpha} d(B^m)_v^\beta$$
$$= \frac{\Delta_l^m B^\alpha\,\Delta_l^m B^\beta}{(\Delta t^m)^2}\int_{t_{k-1}^n}^{t_k^n}(v - t_{k-1}^n)dv$$
$$= \frac{1}{2}2^{2(m-n)}\Delta_l^m B^\alpha\,\Delta_l^m B^\beta,$$

where l is the unique integer such that $[t_{k-1}^n, t_k^n] \subset [t_{l-1}^m, t_l^m]$. In other words, we have

$$B_{t_{k-1}^n,t_k^n}^{m,2} = \frac{1}{2}2^{2(m-n)}(\Delta_l^m B)^{\otimes 2},$$

It follows that

$$B^{m+1,2}_{t^n_{k-1},t^n_k} - B^{m,2}_{t^n_{k-1},t^n_k}$$

$$= \begin{cases} 2^{2(m-n)+1}(\Delta^{m+1}_{2l-1} B)^{\otimes 2} - 2^{2(m-n)-1}(\Delta^m_l B)^{\otimes 2}, & [t^n_{k-1}, t^n_k] \subset [t^{m+1}_{2l-2}, t^{m+1}_{2l-1}]; \\ 2^{2(m-n)+1}(\Delta^{m+1}_{2l} B)^{\otimes 2} - 2^{2(m-n)-1}(\Delta^m_l B)^{\otimes 2}, & [t^n_{k-1}, t^n_k] \subset [t^{m+1}_{2l-1}, t^{m+1}_{2l}]. \end{cases}$$

By using the Minkowski inequality, the Cauchy-Schwarz inequality and the sublinearity of $\mathbb{E}$, it is easy to obtain that

$$\| B^{m+1,2}_{t^n_{k-1},t^n_k} - B^{m,2}_{t^n_{k-1},t^n_k} \|_2 \leqslant C \frac{2^m}{2^{2n}}.$$

Now consider the term $B^{m,2}_{t^n_{k-1},t^n_k}$.
If $n \geqslant m$, by using

$$B^{m,2}_{t^n_{k-1},t^n_k} = 2^{2(m-n)-1}(\Delta^m_l B)^{\otimes 2},$$

we can proceed in the same way as before to obtain that

$$\| B^{m,2}_{t^n_{k-1},t^n_k} \|_2 \leqslant C \frac{2^m}{2^{2n}}.$$

If $n < m$, then

$$B^{m,2}_{t^n_{k-1},t^n_k} = \sum_{l=n+1}^{m} (B^{l,2}_{t^n_{k-1},t^n_k} - B^{l-1,2}_{t^n_{k-1},t^n_k}) + B^{n,2}_{t^n_{k-1},t^n_k}.$$

It follows that

$$\| B^{m,2}_{t^n_{k-1},t^n_k} \|_2 \leqslant \sum_{l=n+1}^{m} \| B^{l,2}_{t^n_{k-1},t^n_k} - B^{l-1,2}_{t^n_{k-1},t^n_k} \|_2 + \| B^{n,2}_{t^n_{k-1},t^n_k} \|_2$$

$$\leqslant C(\frac{1}{2^{\frac{n}{2}}} \sum_{l=n+1}^{\infty} \frac{1}{2^{\frac{l}{2}}} + \frac{1}{2^n})$$

$$\leqslant C \frac{1}{2^n}.$$

Now the proof is complete.

In order to study the behavior of $\boldsymbol{B}^m$ in the space $G\Omega_p(\mathbb{R}^d)$, we may need to control the p-variation distance d_p in a suitable way. For $\boldsymbol{w}, \tilde{\boldsymbol{w}} \in G\Omega(\mathbb{R}^d)$, define

$$\rho_j(\boldsymbol{w}, \tilde{\boldsymbol{w}}) := (\sum_{n=1}^{\infty} n^{\gamma} \sum_{k=1}^{2^n} |w^j_{t^n_{k-1},t^n_k} - \tilde{w}^j_{t^n_{k-1},t^n_k}|^{\frac{p}{j}})^{\frac{j}{p}}, \quad j = 1, 2, \tag{11}$$

where $\gamma > p - 1$ is some fixed universal constant. The functional ρ_j was initially introduced by Hambly and Lyons [11] to construct the stochastic area process associated with the Brownian motion on the Sierpinski gasket. We use $\rho_j(w)$ to denote $\rho_j(w, \tilde{w})$ with $\tilde{w} = (1, 0, 0)$.

The following result, which is important for us, is proved in [18].

Proposition 4 *There exists some positive constant* $R = R(p, \gamma)$, *such that for any* $w, \tilde{w} \in G\Omega(\mathbb{R}^d)$,

$$d_p(w, \tilde{w}) \leqslant R max\{\rho_1(w, \tilde{w}), \rho_1(w, \tilde{w})(\rho_1(w) + \rho_1(\tilde{w})), \rho_2(w, \tilde{w})\}.$$

Now let

$$I(w, \tilde{w}) := max\{\rho_1(w, \tilde{w}), \rho_1(w, \tilde{w})(\rho_1(w) + \rho_1(\tilde{w})), \rho_2(w, \tilde{w})\}, \tag{12}$$

and observe that

$$\{\omega : \boldsymbol{B}^m \text{ is not Cauchy under } d_p\} \subset \{\omega : \sum_{m=1}^{\infty} d_p(\boldsymbol{B}^m, \boldsymbol{B}^{m+1}) = \infty\}$$

$$\subset \limsup_{m\to\infty}\{\omega : d_p(\boldsymbol{B}^m, \boldsymbol{B}^{m+1}) > \frac{R}{2^{m\beta}}\}$$

$$\subset \limsup_{m\to\infty}\{\omega : I(\boldsymbol{B}^m, \boldsymbol{B}^{m+1}) > \frac{1}{2^{m\beta}}\}. \tag{13}$$

where β is some positive constant to be chosen. Notice that the R.H.S. of (13) is $\mathscr{B}(\Omega)$-measurable so its capacity is well-defined. Therefore, in order to prove that for quasi-surely, $\boldsymbol{B}^m$ is a Cauchy sequence under d_p, it suffices to show that the R.H.S. of (13) has capacity zero. This can be shown by using the Borel-Cantelli lemma.

According to (12), we may first need to establish estimates for

$$c(\rho_j(\boldsymbol{B}^m, \boldsymbol{B}^{m+1}) > \lambda), \ j = 1, 2,$$

and

$$c(\rho_1(\boldsymbol{B}^m) > \lambda),$$

where $m \geqslant 1$ and $\lambda > 0$. They are contained in the following lemma.

Lemma 3 *For* $m \geqslant 1$, $\lambda > 0$, *we have the following estimates.*

1.

$$c(\rho_1(\boldsymbol{B}^m) > \lambda) \leqslant C\lambda^{-p}.$$

2. *Let $\theta \in (0, \frac{p}{2} - 1)$ be some constant such that*

$$n^{\gamma+1} \leqslant C \frac{2^{n(p-1)}}{2^{n(p-\theta-1)}}, \quad \forall n \geqslant 1.$$

Then we have

$$c(\rho_j(\boldsymbol{B}^m, \boldsymbol{B}^{m+1}) > \lambda) \leqslant C\lambda^{-\frac{p}{j}} \frac{1}{2^{m(\frac{p}{2}-\theta-1)}}, \quad j = 1, 2.$$

Proof First consider

$$c(\rho_1(\boldsymbol{B}^m) > \lambda) = c(\sum_{n=1}^{\infty} n^{\gamma} \sum_{k=1}^{2^n} |B_{t_{k-1}^n, t_k^n}^{m,1}|^p > \lambda^p).$$

Define

$$A_N = \{\omega : \sum_{n=1}^{N} n^{\gamma} \sum_{k=1}^{2^n} |B_{t_{k-1}^n, t_k^n}^{m,1}|^p > \lambda^p\} \in \mathscr{B}(\Omega),$$

and

$$A = \{\omega : \sum_{n=1}^{\infty} n^{\gamma} \sum_{k=1}^{2^n} |B_{t_{k-1}^n, t_k^n}^{m,1}|^p > \lambda^p\} \in \mathscr{B}(\Omega).$$

It is obvious that $A_N \uparrow A$. By the properties of the capacity c, we have

$$c(A) = \lim_{N \to \infty} c(A_N).$$

On the other hand, by the sublinearity of $\mathbb{E}^G$, the Chebyshev inequality for the capacity c and Lemma 10, we have

$$c(A_N) \leqslant \lambda^{-p} \sum_{n=1}^{N} n^{\gamma} \sum_{k=1}^{2^n} \mathbb{E}[|B_{t_{k-1}^n, t_k^n}^{m,1}|^p]$$

$$\leqslant C\lambda^{-p}[\sum_{n=1}^{m} n^{\gamma} 2^n \frac{1}{2^{\frac{np}{2}}} + \sum_{n=m+1}^{\infty} n^{\gamma} 2^n \frac{2^{\frac{mp}{2}}}{2^{np}}]$$

$$= C\lambda^{-p}[\sum_{n=1}^{m} n^{\gamma} \frac{1}{2^{n(\frac{p}{2}-1)}} + 2^{\frac{mp}{2}} \sum_{n=m+1}^{\infty} n^{\gamma} \frac{1}{2^{n(p-1)}}]$$

$$\leqslant C\lambda^{-p}.$$

It follows that

$$c(\rho_1(\boldsymbol{B}^m) > \lambda) = c(A) \leqslant C\lambda^{-p}.$$

Now consider

$$c(\rho_1(\boldsymbol{B}^m, \boldsymbol{B}^{m+1}) > \lambda) = c\left(\sum_{n=1}^{\infty} n^{\gamma} \sum_{k=1}^{2^n} |B_{\frac{k-1}{2^n}, \frac{k}{2^n}}^{(m+1),1} - B_{\frac{k-1}{2^n}, \frac{k}{2^n}}^{(m),1}|^p > \lambda^p\right).$$

By similar reasons we will have

$$c(\rho_1(\boldsymbol{B}^m, \boldsymbol{B}^{m+1}) > \lambda) \leqslant \lambda^{-p} \sum_{n=1}^{\infty} n^{\gamma} \sum_{k=1}^{2^n} \mathbb{E}[|B_{t_{k-1}^n, t_k^n}^{m+1,1} - B_{t_{k-1}^n, t_k^n}^{m,1}|^p]$$

$$\leqslant C\lambda^{-p}\left(\sum_{n=m+1}^{\infty} n^{\gamma} 2^n \frac{2^{\frac{mp}{2}}}{2^{np}}\right)$$

$$= C\lambda^{-p} 2^{\frac{mp}{2}} \sum_{n=m+1}^{\infty} n^{\gamma} \frac{1}{2^{n(p-1)}}.$$

Since $\theta \in (0, \frac{p}{2} - 1)$ is such that

$$n^{\gamma+1} \leqslant C \frac{2^{n(p-1)}}{2^{n(p-\theta-1)}}, \quad \forall n \geqslant 1,$$

we arrive at

$$c(\rho_1(\boldsymbol{B}^m, \boldsymbol{B}^{m+1}) > \lambda) \leqslant C\lambda^{-p} \frac{1}{2^{m(\frac{p}{2}-\theta-1)}}.$$

Finally, consider the second level part. By similar reasons, we have

$$c(\rho_2(\boldsymbol{B}^m, \boldsymbol{B}^{m+1}) > \lambda) \leqslant C\lambda^{-\frac{p}{2}}\left[\sum_{n=1}^{m} n^{\gamma} 2^n \frac{1}{2^{\frac{mp}{4}} 2^{\frac{np}{4}}} + 2^{\frac{mp}{2}} \sum_{n=m+1}^{\infty} n^{\gamma} 2^n \frac{1}{2^{np}}\right]$$

$$= C\lambda^{-\frac{p}{2}}\left[\frac{1}{2^{\frac{mp}{4}}} \sum_{n=1}^{m} n^{\gamma} 2^{n(1-\frac{p}{4})} + 2^{\frac{mp}{2}} \sum_{n=m+1}^{\infty} n^{\gamma} \frac{1}{2^{n(p-1)}}\right]$$

$$\leqslant C\lambda^{-\frac{p}{2}}\left[\frac{1}{2^{\frac{mp}{4}}} m^{\gamma+1} 2^{m(1-\frac{p}{4})} + 2^{\frac{mp}{2}} \frac{1}{2^{m(p-\theta-1)}}\right]$$

$$\leqslant C\lambda^{-\frac{p}{2}} \frac{1}{2^{m(\frac{p}{2}-\theta-1)}}.$$

Now we are in position to prove the main result of this section.

Theorem 8 *Outside a $\mathcal{B}(\Omega)$-measurable set of capacity zero, $\boldsymbol{B}^m$ is a Cauchy sequence under the p-variation distance d_p. In particular, for quasi-surely, the sample paths of B_t can be enhanced to be geometric rough paths*

$$\boldsymbol{B}_{s,t} = (1, B^1_{s,t}, B^2_{s,t}), \ 0 \leqslant s < t \leqslant 1,$$

with roughness p, which are defined as the limit of sample (geometric rough) paths of $\boldsymbol{B}^m$ in $G\Omega_p(\mathbb{R}^d)$ under the p-variation distance d_p.

Proof By Lemma 3, we have

$$c(I(\boldsymbol{B}^m, \boldsymbol{B}^{m+1}) > \frac{1}{2^{m\beta}}) \leqslant \sum_{j=1}^{2} c(\rho_j(\boldsymbol{B}^m, \boldsymbol{B}^{m+1}) > \frac{1}{2^{m\beta}})$$

$$+ c(\rho_1(\boldsymbol{B}^m, \boldsymbol{B}^{m+1})(\rho_1(\boldsymbol{B}^m) + \rho_1(\boldsymbol{B}^{m+1})) > \frac{1}{2^{m\beta}})$$

$$\leqslant 2c(\rho_1(\boldsymbol{B}^m, \boldsymbol{B}^{m+1}) > \frac{1}{2^{2m\beta}}) + c(\rho_2(\boldsymbol{B}^m, \boldsymbol{B}^{m+1}) > \frac{1}{2^{m\beta}})$$

$$+ c(\rho_1(\boldsymbol{B}^m) > \frac{2^{m\beta}}{2}) + c(\rho_1(\boldsymbol{B}^{m+1}) > \frac{2^{m\beta}}{2})$$

$$\leqslant C[\frac{1}{2^{m\beta p}} + \frac{1}{2^{m(\frac{p}{2}-\theta-2\beta p-1)}} + \frac{1}{2^{m(\frac{p}{2}-\theta-\frac{\beta p}{2}-1)}}],$$

where $\theta \in (0, \frac{p}{2} - 1)$ is some fixed constant.

If we choose β such that

$$0 < \beta < \frac{p - 2\theta - 2}{4p},$$

then

$$\sum_{m=1}^{\infty} c(I(\boldsymbol{B}^m, \boldsymbol{B}^{m+1}) > \frac{1}{2^{m\beta}}) < \infty.$$

By the Borel-Cantelli lemma, we have

$$c(\limsup_{m \to \infty} \{\omega : \ I(\boldsymbol{B}^m, \boldsymbol{B}^{m+1}) > \frac{1}{2^{m\beta}}\}) = 0,$$

and the result follows from the inclusion (13). $\qquad\square$

With the help of Theorem 8 and the smoothness of $\langle B^\alpha, B^\beta \rangle_t$ (by definition the sample paths of $\langle B^\alpha, B^\beta \rangle_t$ are smooth), we are able to apply the universal limit theorem in rough path theory to define RDEs driven by G-Brownian motion in the pathwise sense. More precisely, consider the following N-dimensional RDE in the sense of rough paths:

$$dY_t = \tilde{b}(Y_t)dt + \tilde{h}_{\alpha\beta}(Y_t)d\langle B^\alpha, B^\beta\rangle_t + V_\alpha(Y_t)dB_t^\alpha, \tag{14}$$

with initial condition $Y_0 = x$, where $\tilde{b}, \tilde{h}_{\alpha\beta}, V_\alpha$ are C_b^3-vector fields on $\mathbb{R}^N$. Then outside a $\mathscr{B}(\Omega)$-measurable set of capacity zero, (14) has a unique full solution Y in $G\Omega_p(\mathbb{R}^N)$. Y is constructed as the limit of the enhancement of Y_t^n in $G\Omega_p(\mathbb{R}^N)$ under the p-variation distance, where Y_t^n is the unique classical solution of the following ordinary differential equation:

$$dY_t^n = \tilde{b}(Y_t^n)dt + \tilde{h}_{\alpha\beta}(Y_t^n)d\langle B^\alpha, B^\beta\rangle_t + V_\alpha(Y_t^n)d(B^n)_t^\alpha, \tag{15}$$

with $Y_0^n = x$, in which B_t^n is the dyadic piecewise linear approximation of B_t.

If we only consider solutions instead of full solutions (i.e., only consider the first level), then for quasi-surely, (14) has a unique solution $Y_t \in C([0, 1]; \mathbb{R}^N)$, which is constructed as the uniform limit of the solution of (15) with initial condition $Y_0^n = x$.

Before the end of this section, we are going to give an explicit description of the second level $B_{s,t}^2$ of B_t defined in Theorem 8, which reveals the nature of $B_{s,t}^2$ itself. Such result is fundamental to understand the relation between SDEs and RDEs driven by G-Brownian motion.

Lemma 4 *Assume that X_n converges to X in $L_G^2(\Omega)$ and converges to Y quasi-surely. Then for quasi-surely, $X = Y$.*

Proof By the Chebyshev inequality for the capacity, we have

$$c(|X_n - X| > \epsilon) \leqslant \frac{1}{\epsilon^2}\mathbb{E}^G[|X_n - X|^2], \ \forall\epsilon > 0.$$

Since

$$X_n \to X \quad \text{in } L_G^2(\Omega),$$

we can extract a subsequence X_{n_k}, such that for any $k \geqslant 1$,

$$\mathbb{E}^G[|X_{n_k} - X|^2] \leqslant \frac{1}{k^4}.$$

It follows that

$$c(|X_{n_k} - X| > \frac{1}{k}) \leqslant \frac{1}{k^2}, \quad \forall k \geqslant 1,$$

and

$$\sum_{k=1}^{\infty} c(|X_{n_k} - X| > \frac{1}{k}) < \infty.$$

By the Borel-Cantelli lemma for the capacity, we arrive at for quasi-surely, X_{n_k} converges to X. By assumption it follows that for quasi-surely, $X = Y$.

The following result shows the nature of the second level of B_t. In the case when B_t reduces to the classical Brownian motion, it is essentially the relation between Stratonovich and Itô integrals.

Proposition 5 *Let* $\boldsymbol{B}_{s,t} = (1, B_{s,t}^1, B_{s,t}^2)$ *be the quasi-surely defined enhancement of* B_t *in Theorem 8. Then for any* $0 \leqslant s < t \leqslant 1$, *for quasi-surely, we have*

$$B_{s,t}^{2;\alpha,\beta} = \int_s^t B_{s,u}^{\alpha} dB_u^{\beta} + \frac{1}{2} \langle B^{\alpha}, B^{\beta} \rangle_{s,t}, \tag{16}$$

where the integral on the R.H.S. of (16) is the Itô integral.

Proof We know from Theorem 8 that for quasi-surely,

$$\lim_{n \to \infty} d_p(\boldsymbol{B}^n, \boldsymbol{B}) = 0.$$

From the definition of d_p, it is straight forward that for quasi-surely, $B_{s,t}^{n,2}$ converges uniformly to $B_{s,t}^2$.

Without lost of generality, we assume that s, t are both dyadic points in $[0, 1]$. It follows that when n is large enough,

$$\begin{aligned}
B_{s,t}^{n,2;\alpha,\beta} &= \int_{s<u<v<t} d(B^n)_u^{\alpha} d(B^n)_v^{\beta} \\
&= \int_s^t (B^n)_{s,v}^{\alpha} d(B^n)_v^{\beta} \\
&= \sum_{k:[t_{k-1}^n, t_k^n] \subset [s,t]} \frac{\Delta_k^n B^{\beta}}{\Delta t^n} \int_{t_{k-1}^n}^{t_k^n} (\frac{v - t_{k-1}^n}{\Delta t^n} B_k^{\alpha} + \frac{t_k^n - v}{\Delta t^n} B_{k-1}^{\alpha} - B_s^{\alpha}) dv \\
&= \sum_{k:[t_{k-1}^n, t_k^n] \subset [s,t]} (\frac{B_{k-1}^{\alpha} + B_k^{\alpha}}{2} - B_s^{\alpha}) \Delta_k^n B^{\beta} \\
&= \sum_{k:[t_{k-1}^n, t_k^n] \subset [s,t]} (B_{k-1}^{\alpha} - B_s^{\alpha}) \Delta_k^n B^{\beta} + \frac{1}{2} \sum_k \Delta_k^n B^{\alpha} \Delta_k^n B^{\beta}.
\end{aligned}$$

From properties of Itô integral and the cross-variation $\langle B^{\alpha}, B^{\beta} \rangle_t$, we know that the R.H.S. of the above equality converges to $\int_s^t B_{s,u}^{\alpha} dB_u^{\beta} + \frac{1}{2} \langle B^{\alpha}, B^{\beta} \rangle_{s,t}$ in $L_G^2(\Omega)$.

Consequently, by Lemma 4 $B_{s,t}^2$ must coincide with $\int_s^t B_{s,u}^\alpha dB_u^\beta + \frac{1}{2}\langle B^\alpha, B^\beta\rangle_{s,t}$ quasi-surely.

5   The Fundamental Relation Between SDEs and RDEs Driven by G-Brownian Motion

So far we already know that there are two types of well-defined differential equations driven by G-Brownian motion: SDEs which are defined in the L_G^2-sense with respect to the G-expectation $\mathbb{E}^G$, and RDEs which are quasi-surely defined in the pathwise sense. This section is devoted to the study of the fundamental relation between these two types of differential equations.

Consider the following N-dimensional SDE driven by G-Brownian motion on $(\Omega, L_G^2(\Omega), \mathbb{E})$:

$$dX_t = b(X_t)dt + h_{\alpha\beta}(X_t)d\langle B^\alpha, B^\beta\rangle_t + V_\alpha(X_t)dB_t^\alpha, \tag{17}$$

with initial condition $X_0 = x \in \mathbb{R}^N$. Here we assume that $b, h_{\alpha\beta}, V_\alpha$ are C_b^3-vector fields on $\mathbb{R}^N$.

Our aim is to find the correct RDE of the form (14) whose strong solution coincides with X_t quasi-surely in the pathwise sense.

Let's first illustrate the idea in an informal way. We are going to use the rough Taylor expansion in the theory of RDEs (see Corollary 12.8 in [9]) and Proposition 5 to find the correct form of the RDE we are looking for.

Consider the following general RDE:

$$dY_t = \tilde{b}(Y_t)dt + \tilde{h}_{\alpha\beta}(Y_t)d\langle B^\alpha, B^\beta\rangle_t + \tilde{V}_\alpha(Y_t)dB_t^\alpha, \tag{18}$$

with initial condition $Y_0 = x$, where $\tilde{b}, \tilde{h}_{\alpha\beta}, \tilde{V}_\alpha$ are C_b^3-vector fields on $\mathbb{R}^N$. By the smoothness of the cross variation process $\langle B^\alpha, B^\beta\rangle$, and the roughness of B_t studied in the last section, we know from the rough Taylor expansion theorem that for quasi-surely, for some control function $\omega(s,t)$, the solution Y_t of (18) satisfies, when $\omega(s,t) \leqslant 1$,

$$|Y_{s,t} - \tilde{b}(Y_s)(t-s) - \tilde{h}_{\alpha\beta}(Y_s)\langle B^\alpha, B^\beta\rangle_{s,t} - \tilde{V}_\alpha(Y_s)B_{s,t}^{1;\alpha}$$
$$- D\tilde{V}_\beta(Y_s)\cdot\tilde{V}_\alpha(Y_s)B_{s,t}^{2;\alpha,\beta}| \leqslant C\omega(s,t)^\theta, \tag{19}$$

where $\omega(s,t)$ C and $\theta > 1$ are two constants not depending on s, t. Note that inequality (19) reveals the local behavior of the solution Y_t. It follows from Proposition 5 that for quasi-surely,

$$|Y_{s,t} - \tilde{I}_{s,t}| \leqslant C\omega(s,t)^\theta,$$

where

$$\tilde{I}_{s,t} := \tilde{b}(Y_s)(t-s) + (\tilde{h}_{\alpha\beta}(Y_s) + \frac{1}{2}D\tilde{V}_\beta(Y_s) \cdot \tilde{V}_\alpha(Y_s))d\langle B^\alpha, B^\beta \rangle_t + \tilde{V}_\alpha(Y_s)B_{s,t}^{1;\alpha}$$

$$+ D\tilde{V}_\beta(Y_s) \cdot \tilde{V}_\alpha(Y_s) \int_s^t B_{s,u}^\alpha dB_u^\beta. \tag{20}$$

Now if we consider the global behavior of Y_t, we may sum up inequality (20) over dyadic intervals $[t_{k-1}^n, t_k^n]$ and then take limit (in $L_G^2(\Omega; \mathbb{R}^N)$) to obtain that for quasi-surely,

$$Y_{s,t} = \int_s^t \tilde{b}(Y_u)du + \int_s^t (\tilde{h}_{\alpha\beta}(Y_u) + \frac{1}{2}D\tilde{V}_\beta(Y_u) \cdot \tilde{V}_\alpha(Y_u))d\langle B^\alpha, B^\beta \rangle_u$$

$$+ \int_s^t \tilde{V}_\alpha(Y_u)dB_u^\alpha + (L_G^2-) \lim_{n\to\infty} \sum_{k:[t_{k-1}^n, t_k^n] \subset [s,t]} D\tilde{V}_\alpha(Y_{t_{k-1}^n})$$

$$\cdot \tilde{V}_\beta(Y_{t_{k-1}^n}) \int_{t_{k-1}^n}^{t_k^n} B_{t_{k-1}^n, u}^\alpha dB_u^\beta, \tag{21}$$

where the integrals with respect to B_t are interpreted as Itô integrals. On the other hand, by the distribution of B_t and properties of G-Itô integral, it is not hard to prove that the L_G^2-limit in the last term of the above identity is zero. Therefore, we know that Y_t solves the SDE

$$dX_t = \tilde{b}(X_t)dt + (\tilde{h}_{\alpha\beta}(X_t) + \frac{1}{2}D\tilde{V}_\beta(X_t) \cdot \tilde{V}_\alpha(X_t))d\langle B^\alpha, B^\beta \rangle_t + \tilde{V}_\alpha(X_t)dB_t^\alpha.$$

In other words, if X_t is the solution of the SDE (17), it is natural to expect that for quasi-surely, X_t is the solution of the following RDE:

$$dY_t = b(Y_t)dt + (h_{\alpha\beta}(Y_t) - \frac{1}{2}DV_\beta(Y_t) \cdot V_\alpha(Y_t))d\langle B^\alpha, B^\beta \rangle + V_\alpha(Y_t)dB_t^\alpha, \tag{22}$$

with the same initial condition.

In the remaining of this section, we are going to prove this claim in a rigorous way.

From now on, assume that X_t is the solution of the SDE (17) and Y_t is the solution of the RDE (22) with the same initial condition $x \in \mathbb{R}^N$, where the coefficients $b, h_{\alpha\beta}, V_\alpha$ are C_b^3-vector fields on $\mathbb{R}^N$. For simplicity we will also use the same notation to denote constants only depending on d, N, G, p and the coefficients of (17), although they may be different from line to line.

The following lemma enables us to show that the L_G^2-limit in the last term of identity (21) is zero.

Lemma 5 Let $f \in C_b(\mathbb{R}^N)$, and $s < t$ be two dyadic points in $[0,1]$ (i.e., $s = t_k^m$ and $t = t_l^m$ for some m and $k < l$). Then for any $\alpha, \beta = 1, 2, \cdots, d$,

$$\lim_{n\to\infty} \mathbb{E}^G \Big[\Big(\sum_{k:[t_{k-1}^n, t_k^n] \subset [s,t]} f(Y_{t_{k-1}^n}) \int_{t_{k-1}^n}^{t_k^n} B_{t_{k-1}^n, u}^\alpha dB_u^\beta \Big)^2 \Big] = 0.$$

Proof From direct calculation, we have

$$\mathbb{E}^G \Big[\Big(\sum_{k:[t_{k-1}^n, t_k^n] \subset [s,t]} f(Y_{t_{k-1}^n}) \int_{t_{k-1}^n}^{t_k^n} B_{t_{k-1}^n, u}^\alpha dB_u^\beta \Big)^2 \Big]$$

$$\leqslant \|f\|_\infty^2 \sum_{k:[t_{k-1}^n, t_k^n] \subset [s,t]} \mathbb{E}^G \Big[\Big(\int_{t_{k-1}^n}^{t_k^n} B_{t_{k-1}^n, u}^\alpha dB_u^\beta \Big)^2 \Big]$$

$$+ 2 \sum_{\substack{k < l \\ [t_{k-1}^n, t_k^n], [t_{l-1}^n, t_l^n] \subset [s,t]}} \mathbb{E}^G \Big[f(Y_{t_{k-1}^n}) \Big(\int_{t_{k-1}^n}^{t_k^n} B_{t_{k-1}^n, u}^\alpha dB_u^\beta \Big) f(Y_{t_{l-1}^n}) \Big(\int_{t_{l-1}^n}^{t_l^n} B_{t_{l-1}^n, u}^\alpha dB_u^\beta \Big) \Big]$$

$$\leqslant C \|f\|_\infty^2 \sum_{k:[t_{k-1}^n, t_k^n] \subset [s,t]} (\Delta t^n)^2$$

$$+ 2 \sum_{\substack{k < l \\ [t_{k-1}^n, t_k^n], [t_{l-1}^n, t_l^n] \subset [s,t]}} \Big(\mathbb{E}^G \Big[\Big(f(Y_{t_{k-1}^n}) \Big(\int_{t_{k-1}^n}^{t_k^n} B_{t_{k-1}^n, u}^\alpha dB_u^\beta \Big) f(Y_{t_{l-1}^n}) \Big)^+ $$

$$\cdot \mathbb{E}^G \Big[\int_{t_{l-1}^n}^{t_l^n} B_{t_{l-1}^n, u}^\alpha dB_u^\beta | \Omega_{t_{l-1}^n} \Big] \Big] + \mathbb{E}^G \Big[\Big(f(Y_{t_{k-1}^n}) \Big(\int_{t_{k-1}^n}^{t_k^n} B_{t_{k-1}^n, u}^\alpha dB_u^\beta \Big) f(Y_{t_{l-1}^n}) \Big)^- $$

$$\cdot \mathbb{E}^G \Big[-\int_{t_{l-1}^n}^{t_l^n} B_{t_{l-1}^n, u}^\alpha dB_u^\beta | \Omega_{t_{l-1}^n} \Big] \Big] \Big)$$

$$\leqslant C \|f\|_\infty^2 \Delta t^n,$$

and the result follows easily.

Now we are in position to prove our main result of this section.

Theorem 9 *For quasi-surely,*

$$X_t = Y_t, \quad \forall t \in [0,1].$$

Proof Since the coefficients of the RDE (22) are in $C_b^3(\mathbb{R}^N)$, for quasi-surely define the following pathwise control: for $0 \leqslant s < t \leqslant 1$,

$$\omega(s,t) := (\|V\|_{2,\infty} \|\boldsymbol{B}\|_{p\text{-}var;[s,t]})^p + \|b\|_{1,\infty}(t-s)$$

$$+ \|h - \frac{1}{2}DV \cdot V\|_{1,\infty} \|\langle \boldsymbol{B}, \boldsymbol{B} \rangle\|_{1\text{-}var;[s,t]},$$

where $\|\cdot\|_{m,\infty}$ denotes the maximum of uniform norms of derivatives up to order m. It follows from the rough Taylor expansion (Corollary 12.8 [9]) that for quasi-surely, there exists some positive constant $\theta > 1$, such that for $0 \leqslant s < t \leqslant 1$, when $\omega(s,t) \leqslant 1$, we have

$$|Y_{s,t} - I_{s,t}| \leqslant C\omega(s,t)^\theta,$$

where

$$I_{s,t} = b(Y_s)(t-s) + (h_{\alpha\beta}(Y_s) - \frac{1}{2}DV_\beta(Y_s) \cdot V_\alpha(Y_s))\langle B^\alpha, B^\beta \rangle_{s,t} + V_\alpha(Y_s)B_{s,t}^{1;\alpha}$$

$$+ DV_\beta(Y_s) \cdot V_\alpha(Y_s)B_{s,t}^{2;\alpha,\beta}$$

By Proposition 5, we have for quasi-surely,

$$|Y_{s,t} - b(Y_s)(t-s) - h_{\alpha\beta}(Y_s)\langle B^\alpha, B^\beta \rangle_{s,t} - V_\alpha(Y_s)B_{s,t}^{1;\alpha}$$

$$- DV_\beta(Y_s) \cdot V_\alpha(Y_s) \int_s^t B_{s,u}^\alpha dB_u^\beta| \leqslant C\omega(s,t)^\theta. \tag{23}$$

Now consider fixed $s < t$ being two dyadic points in $[0,1]$. When n is large enough, by applying inequality (23) on each small dyadic interval $[t_{k-1}^n, t_k^n] \subset [s,t]$ and summing up through the triangle inequality, we obtain that for quasi-surely,

$$|Y_{s,t} - I_{s,t}^n| \leqslant C \sum \omega(t_{k-1}^n, t_k^n)^\theta$$

$$\leqslant C\omega(s,t) \max\{\omega(t_{k-1}^n, t_k^n)^{\theta-1} : [t_{k-1}^n, t_k^n] \subset [s,t]\},$$

where

$$I_{s,t}^n = \sum b(Y_{t_{k-1}^n})\Delta t^n + \sum h_{\alpha\beta}(Y_{t_{k-1}^n})\Delta_k^n \langle B^\alpha, B^\beta \rangle + \sum V_\alpha(Y_{t_{k-1}^n})\Delta_k^n B^\alpha$$

$$+ \sum DV_\beta(Y_{t_{k-1}^n}) \cdot V_\alpha(Y_{t_{k-1}^n}) \int_{t_{k-1}^n}^{t_k^n} B_{t_{k-1}^n,u}^\alpha dB_u^\beta,$$

and each sum is over all k such that $[t_{k-1}^n, t_k^n] \subset [s,t]$. It follows that for quasi-surely,

$$I_{s,t}^n \to Y_{s,t}, \quad n \to \infty.$$

On the other hand, the following convergence in $L_G^2(\Omega; \mathbb{R}^N)$ holds:

$$\sum b(Y_{t_{k-1}^n})\Delta t^n \rightarrow \int_s^t b(Y_u)du,$$

$$\sum h_{\alpha\beta}(Y_{t_{k-1}^n})\Delta_k^n \langle B^\alpha, B^\beta \rangle \rightarrow \int_s^t h_{\alpha\beta}(Y_u)d\langle B^\alpha, B^\beta \rangle_u,$$

$$\sum V_\alpha(Y_{t_{k-1}^n})\Delta_k^n B^\alpha \rightarrow \int_s^t V_\alpha(Y_u)dB_u^\alpha,$$

as $n \rightarrow \infty$.

The reason is the following. For simplicity we only consider the third one, as the first two are similar (and in fact easier). It is straight forward that

$$\int_0^1 |V_\alpha(Y_t) - \sum_{k=1}^{2^n} V_\alpha(Y_{t_{k-1}^n})\mathbf{1}_{[t_{k-1}^n, t_k^n)}(t)|^2 dt$$

$$= \sum_{k=1}^{2^n} \int_{t_{k-1}^n}^{t_k^n} |V_\alpha(Y_t) - V_\alpha(Y_{t_{k-1}^n})|^2 dt$$

$$\leqslant C \sum_{k=1}^{2^n} \int_{t_{k-1}^n}^{t_k^n} |Y_t - Y_{t_{k-1}^n}|^2 dt$$

$$\leqslant C \sum_{k=1}^{2^n} \|Y\|_{p-var;[t_{k-1}^n, t_k^n]}^2 \Delta t^n$$

$$\leqslant C(\sum_{k=1}^{2^n} \|Y\|_{p-var;[t_{k-1}^n, t_k^n]}^p \Delta t^n)^{\frac{2}{p}}$$

$$\leqslant C(\Delta t^n)^{\frac{2}{p}} \|Y\|_{p-var;[0,1]}^2,$$

where C depends only on V_α. Therefore, it suffices to show that $\|Y\|_{p-var;[0,1]} \in L_G^2(\Omega)$, as it will imply the G-Itô integrability of $V_\alpha(Y_t)$ and the desired convergence in $L_G^2(\Omega; \mathbb{R}^N)$ will hold. For simplicity we assume that Y_t is the solution of the following RDE

$$dY_t = V_\alpha(Y_t)dB_t^\alpha$$

with $Y_0 = \xi$ (there is no substantial difference because dt and $d\langle B^\alpha, B^\beta \rangle_t$ are more regular than dB_t), then by Theorem 10.14 in [9], we know that

$$\|Y\|_{p-var;[0,1]} \leqslant C \|\mathbf{B}\|_{p-var;[0,1]} \vee \|\mathbf{B}\|_{p-var;[0,1]}^p.$$

Therefore, we only need to show that $\|\boldsymbol{B}\|^p_{p-var;[0,1]} \in L^2_G(\Omega)$. For this purpose, we use Proposition 4 to control the p-variation norm by the functions ρ_1, ρ_2 defined in (11). It follows that

$$\|\boldsymbol{B}\|_{p-var} \leqslant C(1 + \rho_1(\boldsymbol{B})^2 + \rho_2(\boldsymbol{B})).$$

Therefore, it remains to show that $\rho_1(\boldsymbol{B})^{2p}, \rho_2(\boldsymbol{B})^p \in L^1_G(\Omega)$. First consider level one. By the distribution of B_t, we have

$$\|\sum_{n=1}^{\infty} n^{\gamma} \sum_{k=1}^{2^n} |B^1_{t^n_{k-1},t^n_k}|^p \|_2 \leqslant \sum_{n=1}^{\infty} n^{\gamma} \sum_{k=1}^{2^n} \| |B^1_{t^n_{k-1},t^n_k}|^p \|_2$$

$$\leqslant \sum_{n=1}^{\infty} n^{\gamma} (\Delta t^n)^{\frac{p}{2}-1}$$

$$< \infty,$$

and we know that $\rho_1(\boldsymbol{B})^{2p} \in L^1_G(\Omega)$. Now consider level two. By Proposition 5 and the distribution of B_t and $\langle B, B\rangle_t$, we have

$$\|\sum_{n=1}^{\infty} n^{\gamma} \sum_{k=1}^{2^n} |B^2_{t^n_{k-1},t^n_k}|^{\frac{p}{2}} \|_2 = \|\sum_{n=1}^{\infty} n^{\gamma} \sum_{k=1}^{2^n} |\int_{t^n_{k-1}}^{t^n_k} B_{t^n_{k-1},u} \otimes dB_u + \frac{1}{2}\langle B, B\rangle_{t^n_{k-1},t^n_k}|^{\frac{p}{2}} \|_2$$

$$\leqslant \sum_{n=1}^{\infty} n^{\gamma} \sum_{k=1}^{2^n} \| |\int_{t^n_{k-1}}^{t^n_k} B_{t^n_{k-1},u} \otimes dB_u + \frac{1}{2}\langle B, B\rangle_{t^n_{k-1},t^n_k}|^{\frac{p}{2}} \|_2$$

$$\leqslant C \sum_{n=1}^{\infty} n^{\gamma} (\Delta t^n)^{\frac{p}{2}-1}$$

$$< \infty.$$

It follows that $\rho_2(\boldsymbol{B})^p \in L^1_G(\Omega)$. Therefore, the desired L^2_G-convergence holds.

In addition, by Lemma 5 we also have the following L^2_G-convergence:

$$\sum DV_{\beta}(Y_{t^n_{k-1}}) \cdot V_{\alpha}(Y_{t^n_{k-1}}) \int_{t^n_{k-1}}^{t^n_k} B^{\alpha}_{t^n_{k-1},u} dB^{\beta}_u \to 0, \ n \to \infty.$$

Consequently, in $L^2_G(\Omega; \mathbb{R}^N)$,

$$I^n_{s,t} \to \int_s^t b(Y_u)du + \int_s^t h_{\alpha\beta}(Y_u)d\langle B^{\alpha}, B^{\beta}\rangle_u + \int_s^t V_{\alpha}(Y_u)dB^{\alpha}_u,$$

as $n \to \infty$.

From Lemma 4, we conclude that for quasi-surely,

$$Y_{s,t} = \int_s^t b(Y_u)du + \int_s^t h_{\alpha\beta}(Y_u)d\langle B^\alpha, B^\beta \rangle_u + \int_s^t V_\alpha(Y_u)dB_u^\alpha.$$

Since X_t and Y_t are both quasi-surely continuous, it follows that X coincides with Y quasi-surely.

Remark 2 As we mentioned at the beginning of Sect. 2, it is possible to prove Theorem 9 by establishing the Wong-Zakai type approximation. More precisely, if we let X_t^n to be the Euler-Maruyama approximation of the SDE (17) and let Y_t^n to be the unique classical solution of the following ODE:

$$dY_t^n = b(Y_t^n)dt + (h_{\alpha\beta}(Y_t^n) - \frac{1}{2}DV_\beta(Y_t^n) \cdot V_\alpha(Y_t^n))d\langle B^\alpha, B^\beta \rangle_t + V_\alpha(Y_t^n)d(B^n)_t^\alpha$$

with $X_0^n = Y_0^n = \xi$, where B_t^n is the dyadic piecewise linear approximation of B_t, then by using our main result in Sect. 2 and establishing related L_G^2-estimates, we can prove that

$$\sup_{t\in[0,1]} \mathbb{E}^G[|X_t^n - Y_t^n|^2] \leqslant C\sqrt{1+\xi^2}(\Delta t^n)^{\frac{1}{2}}.$$

In other words, Y_t^n converges to the solution X_t of the SDE (17) in the L_G^2-sense. However, we know that for quasi-surely, Y_t^n converges uniformly to the solution Y_t of the RDE (22). Again by Lemma 4 and continuity, we conclude that for quasi-sure, X coincides with Y.

From the above discussion, if we forget about the RDE (22) and only consider the L_G^2-limit of Y_t^n, it seems that there is nothing to do with rough paths at all as everything is well-defined in the classical sense. However, the fundamental point of understanding the convergence of Y_t^n in the pathwise sense lies in the crucial fact that B_t can be regarded as geometric rough paths (i.e., the enhancement defined in Sect. 3) with approximating sequence in $G\Omega_p(\mathbb{R}^d)$ being the enhancement of the natural dyadic piecewise linear approximation B_t^n. This is exactly what the universal limit theorem tells us.

Remark 3 From the RDE point of view, it is possible to reduce the regularity assumptions on the coefficients. In particular, since the regularity of t and $\langle B^\alpha, B^\beta \rangle_t$ are both "better" than B_t, the regularity assumptions on the coefficients of dt and $d\langle B^\alpha, B^\beta \rangle_t$ can be weaker than the one imposed on the coefficient of dB_t. However, we are not going to present the results under such generality. Please refer to [9] for general existence and uniqueness results of RDEs.

6 SDEs on a Differentiable Manifold Driven by *G*-Brownian Motion

Our main result in Sect. 5 can be used to establish SDEs on a differentiable manifold driven by *G*-Brownian motion, which will be the main focus of this section. The development is based on the idea in the classical case, for which one may refer to [8, 12, 13]. This part is the foundation of developing *G*-Brownian motion on a Riemannina manifold in the next section.

In classical stochastic analysis, SDEs on a manifold is established under the Stratonovich type formulation, which can be regarded as a pathwise approach. The reason of using Stratonovich type formulation instead of the Itô type one is the following. First of all, the notion of SDE can be introduced by using test functions on the manifold from an intrinsic point of view, which is consistent with ordinary differential calculus and invariant under diffeomorphisms. Moreover, when we construct solutions extrinsically, we can prove that for almost surely, the solution of the extended SDE which starts from the manifold will always live on it. This reveals the intrinsic nature of ordinary differential equations.

In the setting of *G*-expectation, we will adopt the same idea for the development. However, there is a major difficulty here. The method of constructing solutions in the classical case from the extrinsic point of view depends heavily on the localization technique, which is not available in the setting of *G*-expectation, mainly due to the reason that concepts of information flows and stopping times are not well understood. To get around with this difficulty, we will use our main result in Sect. 5 to obtain a pathwise construction. The advantage of such approach is that we can still use localization arguments but don't need to care about measurability and integrability under *G*-expectation.

Now assume that M is a differentiable manifold. For technical reasons we further assume that M is compact (it is not necessary if we impose more restrictive regularity assumptions on the generating vector fields). Let $\{b, h_{\alpha\beta}, V_{\alpha} : \alpha, \beta = 1, 2, \cdots, d\}$ be a family of C^3-vector fields on M, and let B_t be the canonical d-dimensional *G*-Brownian motion on the path space $(\Omega, L_G^2(\Omega), \mathbb{E}^G)$, where G is a function given by (2).

Consider the following symbolic Stratonovich type SDE over $[0, 1]$:

$$\begin{cases} dX_t & = b(X_t)dt + h_{\alpha\beta}(X_t)d\langle B^\alpha, B^\beta\rangle_t + V_\alpha(X_t) \circ dB_t^\alpha, \\ X_0 & = \xi \in M, \end{cases} \tag{24}$$

on M.

Definition 6 A solution X_t of the SDE (24) is an M-valued continuous stochastic process such that for any $f \in C^\infty(M)$, and any $\alpha, \beta = 1, 2, \cdots, d$,

$$\{h_{\alpha\beta} f(X_t) : t \in [0, 1]\} \in M_G^1(0, 1), \ \{V_\alpha f(X_t) : t \in [0, 1]\} \in M_G^2(0, 1),$$

and the following equality holds on $[0, 1]$:

$$f(X_t) = f(\xi) + \int_0^t bf(X_s)ds + \int_0^t h_{\alpha\beta} f(X_s)d\langle B^\alpha, B^\beta\rangle_s + \int_0^t V_\alpha f(X_s) \circ dB_s^\alpha,$$
$$(25)$$

where the last term is defined as

$$\int_0^t V_\alpha f(X_s) \circ dB_s^\alpha := \int_0^t V_\alpha f(X_s)dB_s^\alpha + \frac{1}{2}\int_0^t V_\beta V_\alpha f(X_s)d\langle B^\alpha, B^\beta\rangle_s.$$

Remark 4 Definition 6 is intrinsic. It is easy to see that Definition 6 is consistent with the Euclidean case.

Now we are going to construct the solution of (24) from the extrinsic point of view.

According to the Whitney embedding theorem (see [5]), M can be embedded into some ambient Euclidean space $\mathbb{R}^N$ as a submanifold such that the image $i(M)$ of M is closed in $\mathbb{R}^N$. We simply regard M as a subset of $\mathbb{R}^N$.

Let $F^1, \cdots, F^N \in C^\infty(M)$ be the coordinate functions on M. The following result is easy to prove. It is similar to the classical case.

Proposition 6 X_t *is a solution of (24) if and only if for any* $i = 1, 2, \cdots, N$, $\alpha, \beta = 1, 2, \cdots, d$,

$$\{h_{\alpha\beta} F^i(X_t) : t \in [0, 1]\} \in M_G^1(0, 1), \ \{V_\alpha F^i(X_t) : t \in [0, 1]\} \in M_G^2(0, 1),$$

and for any $t \in [0, 1]$,

$$F^i(X_t) = F^i(\xi) + \int_0^t bF^i(X_s)ds + \int_0^t h_{\alpha\beta} F^i(X_s)d\langle B^\alpha, B^\beta\rangle_s + \int_0^t V_\alpha F^i(X_s)\circ dB_s^\alpha.$$
$$(26)$$

Proof Necessity is obvious since $F^i \in C^\infty(M)$ for any $i = 1, 2, \cdots, N$.

Now consider sufficiency. Let $f \in C^\infty(M)$, and choose a C^∞-extension $\tilde{f}$ of f with compact support in $\mathbb{R}^N$ (it is possible since M is compact). Then for any $x \in M$,

$$f(x) = \tilde{f}(F^1(x), \cdots, F^N(x)),$$

and thus

$$f(X_t) = \tilde{f}(F^1(X_t), \cdots, F^N(X_t)), \ \forall t \in [0, 1].$$

Since M is compact and $\tilde{f}$ is smooth with compact support, it follows from the G-Itô formula that for $t \in [0, 1]$,

$$\tilde{f}(F^1(X_t), \cdots, F^N(X_t)) = f(\xi) + \int_0^t \frac{\partial \tilde{f}}{\partial y^i} (bF^i(X_s)ds + h_{\alpha\beta} F^i(X_s)d\langle B^\alpha, B^\beta \rangle_s$$
$$+ V_\alpha F^i(X_s) \circ dB_s^\alpha)$$
$$= f(\xi) + \int_0^t (bf(X_s)ds + h_{\alpha\beta} f(X_s)d\langle B^\alpha, B^\beta \rangle_s$$
$$+ V_\alpha f(X_s) \circ dB_s^\alpha),$$

where we have used the simple fact that for any C^1-vector field V on M,

$$Vf = \sum_{i=1}^N \frac{\partial \tilde{f}}{\partial y^i} VF^i.$$

By Definition 6, we know that X_t is a solution of the SDE (24).

Now we are going to prove the existence and uniqueness of (24) by using the main result of Sect. 5, namely, a pathwise approach based on the associated RDE.

Let $\tilde{b}, \tilde{h}_{\alpha\beta}, \tilde{V}_\alpha$ be C_b^3-extensions (not unique) of the vector fields $b, h_{\alpha\beta}, V_\alpha$. Consider the following Stratonovich type SDE in the ambient space $\mathbb{R}^N$:

$$dX_t = \tilde{b}(X_t)dt + \tilde{h}_{\alpha\beta}(X_t)d\langle B^\alpha, B^\beta \rangle_t + \tilde{V}_\alpha(X_t) \circ dB_t^\alpha \tag{27}$$

with $X_0 = x \in \mathbb{R}^N$, which is interpreted as the following Itô type SDE:

$$dX_t = \tilde{b}(X_t)dt + (\tilde{h}_{\alpha\beta}(X_t) + \frac{1}{2}D\tilde{V}_\alpha(X_t) \cdot \tilde{V}_\beta(X_t))d\langle B^\alpha, B^\beta \rangle_t + \tilde{V}_\alpha(X_t)dB_t^\alpha.$$

According to Sect. 5, we can alternatively interpret (27) as an RDE which is pathwisely defined. Both the SDE and the RDE has a unique solution, and according to Theorem 9 they coincide quasi-surely. Our aim is to show that for quasi-surely, the solution X_t of (27) never leaves M and it is the unique solution of (24).

The following result is important to prove the existence and uniqueness of the SDE (24) on the manifold M.

Proposition 7 *Let x_t be a path of bounded variation in $\mathbb{R}^d$. Let $W_1, \cdots, W_d$ be a family of C^1-vector fields on M and $\tilde{W}_1, \cdots, \tilde{W}_d$ be their C_b^1-extensions to $\mathbb{R}^N$. Consider the following ODE in the ambient space $\mathbb{R}^N$ over $[0, 1]$:*

$$dy_t = \tilde{W}_\alpha(y_t)dx_t^\alpha \tag{28}$$

with $y_0 = x \in M$. Then the solution $y_t \in M$ for all $t \in [0, 1]$. Moreover, y_t does not depend on extensions of the vector fields.

Proof Let $F(x) := d(x, M)^2$ be the squared distance function to the submanifold M. It follows that F is smooth in an open neighborhood of M. By using the cut-off function we may assume that $F \in C_b^\infty(M)$. Now we are able to choose an open neighborhood U of M, such that for any $x \in U$, $F(x) = 0$ if and only if $x \in M$. Moreover, since $\tilde{W}_\alpha$ ($\alpha = 1, 2, \cdots, d$) are tangent vector fields of M when restricted on M, U can be chosen such that for any $x \in U$ and $\alpha = 1, 2, \cdots, d$,

$$|\tilde{W}_\alpha F(x)| \leqslant CF(x), \tag{29}$$

for some positive constant C depending on U. The function $F(x)$ was used in [12] to construct SDEs on M driven by classical Brownian motion.

Since x_t is a path of bounded variation and $y_0 = \xi \in M$, by the change of variables formula in ordinary calculus, we have

$$F(y_t) = \int_0^t \tilde{W}_\alpha F(y_s) dx_s^\alpha, \ \forall t \in [0, 1].$$

Define $\tau := \inf\{t \in [0, 1] : y_t \notin U\}$. It follows from (29) that

$$F(y_t) \leqslant C \int_0^t F(y_s) d|x|_s, \ \forall t \in [0, \tau],$$

where $|x|_t$ is the total variation of the path x_t.

By iteration and Fubini theorem, on $[0, \tau]$ we have

$$F(y_t) \leqslant C^2 \int_0^t \left(\int_0^s F(y_u) d|x|_u \right) d|x|_s$$

$$= C^2 \int_0^t (|x|_t - |x|_s) F(y_s) d|x|_s.$$

By induction, it is easy to see that for any $k \geqslant 1$,

$$F(y_t) \leqslant C^k \int_0^t \frac{(|x|_t - |x|_s)^{k-1}}{(k-1)!} F(y_s) d|x|_s, \ \forall t \in [0, \tau].$$

Since F is bounded, we obtain further that for any $k \geqslant 1$,

$$F(y_t) \leqslant \|F\|_\infty \frac{C^k (|x|_t - |x|_0)^k}{k!}, \ \forall t \in [0, \tau].$$

By letting $k \to \infty$, it follows that $F(y_t) \equiv 0$ on $[0, \tau]$, which implies that $y_t \in M$ for any $t \in [0, \tau]$. Since y_t is continuous, the only possibility is that y_t never leaves M on $[0, 1]$.

If we rewrite the ODE (28) in its integral form:

$$y_t = \xi + \int_0^t \tilde{W}_\alpha(y_s)dx_s^\alpha, \ t \in [0, 1], \tag{30}$$

we know from previous discussion that Eq. (30) depends only on the values of $\tilde{W}_\alpha$ on M, that is, of W_α ($\alpha = 1, 2, \cdots, d$). In other words, if $\hat{W}_\alpha$ is another extension of W_α and $\hat{y}_t$ is the solution of the corresponding ODE with the same initial condition, $\hat{y}_t$ is also a solution of (28). By uniqueness, we have $y = \hat{y}$. Therefore, y_t does not depend on extensions of the vector fields.

With the help of Proposition 7, we can prove the following existence and uniqueness result.

Theorem 10 *Let* $b, h_{\alpha\beta}, V_\alpha$ *be* C^3*-vector fields on* M. *Then the Stratonovich type SDE (24) has a solution* X_t *which is unique quasi-surely.*

Proof Fix C_b^3-extensions $\tilde{b}, \tilde{h}_{\alpha\beta}, \tilde{V}_\alpha$ of $b, h_{\alpha\beta}, V_\alpha$, and let X_t be the solution of the Stratonovich type SDE (27) in $\mathbb{R}^N$ over $[0, 1]$. By Theorem 9, for quasi-surely X_t coincides with the solution of (27) when it is interpreted as an RDE. Since M is closed in $\mathbb{R}^N$, it follows from Proposition 7 and Theorem 6 (the universal limit theorem) that for quasi-surely, X_t never leaves M over $[0, 1]$. In this case, (27) is equivalent to (26), which implies from Proposition 6 that X_t is a solution of (24). On the other hand, if Y_t is another solution of (24), then it is a solution of (27) (interpreted as an SDE or an RDE). By the uniqueness of RDEs, we know that $X = Y$ quasi-surely.

Remark 5 It is possible to formulate uniqueness in the L_G^2-sense when M is regarded as a closed submanifold of $\mathbb{R}^N$. However, we use the quasi-sure formulation because the notion itself is intrinsic although the proof is developed from the extrinsic point of view.

7 G-Brownian Motion on a Compact Riemannian Manifold and the Generating Nonlinear Heat Equation

In this section, we are going to introduce the notion of G-Brownian motion on a Riemannian manifold for a wide and interesting class of G-functions, based on Eells-Elworthy-Malliavin's horizontal lifting construction (see [8, 12, 13] for the construction of Brownian motion on a Riemannian manifold and related topics). Roughly speaking, we will "roll" an Euclidean G-Brownian motion up to a Riemannian manifold "without slipping" via a proper frame bundle (for the class of G-functions we are interested in, such bundle is the orthonormal frame bundle).

In the classical case, we know that the law of a d-dimensional Brownian motion B_t is invariant under orthogonal transformations on $\mathbb{R}^d$. This is a crucial point to obtain a linear parabolic PDE (in fact, the standard heat equation associated with the Bochner horizontal Laplacian $\Delta_{\mathscr{O}(M)}$) on the orthonormal frame bundle $\mathscr{O}(M)$ over a Riemannian manifold M governing the law of the horizontal lifting ξ_t of B_t to $\mathscr{O}(M)$, which is invariant under orthogonal transformations along fibers. It is such an invariance that enables us to "project" the PDE onto the base manifold M and obtain the standard heat equation associated with the Laplace-Beltrami operator Δ_M on M. This heat equation governs the law of the development $X_t = \pi(\xi_t)$ of B_t to the Riemannian manifold M via the horizontal lifting ξ_t. As a stochastic process on M, although X_t depends on the initial orthonormal frame ξ at x as well as the initial position $x \in M$, the law of X_t depends only on the initial position x, and it is characterized by the Laplace-Beltrami operator Δ_M via the heat equation. Equivalently, it can be shown that the law of X_t is the unique solution of the martingale problem on M associated with Δ_M starting at x. X_t is called the Brownian motion on M starting at x in the sense of Eells-Elworthy-Malliavin.

It is quite natural to expect that the Brownian sample paths X_t on M will depend on the initial orthonormal frame ξ at x if we look back into the Euclidean case, in which we actually fix the standard orthonormal basis in advance and define Brownian motion in the corresponding coordinate system. If we use another orthonormal basis, we obtain a process (still a Brownian motion) which is an orthogonal transformation of the original Brownian motion. Therefore, it is the law, which is characterized by the Laplace operator on $\mathbb{R}^d$, rather than the sample paths that captures the intrinsic nature of the Brownian motion, and such nature can be developed in a Riemannian geometric setting.

It should be remarked that in a pathwise manner, we can lift B_t horizontally to the total frame bundle $\mathscr{F}(M)$ instead of $\mathscr{O}(M)$ by solving the same SDE generating by the horizontal vector fields but using a general frame instead of an orthonormal one as initial condition. Moreover, we can write down the generating heat equation on $\mathscr{F}(M)$ which takes the same form of the one on $\mathscr{O}(M)$. The key difference here is that although the horizontal lifting of B_t can be projected onto M, the heat equation on $\mathscr{F}(M)$ cannot. In other words, the heat equation is not invariant under nondegenerate linear transformations along fibers. This becomes uninteresting to us, as we are not able to obtain an intrinsic law of the development of B_t on M which is independent of initial frames. The fundamental reason of using the orthonormal frame bundle is that the Laplace operator on $\mathbb{R}^d$ is invariant exactly under orthogonal transformations.

The case of G-Brownian motion can be understood in a similar manner. From the last section we are able to solve SDEs on a differentiable manifold (in particular, on $\mathscr{F}(M)$) driven by an Euclidean G-Brownian motion B_t. By projection we obtain the development X_t of B_t to M. As we've pointed out before, such development is of no interest unless we are able to prove that the law of X_t depends only on the initial position x rather than the initial frame. In fact, if the law of X_t depends on the initial frame, we might not be able to write down the generating PDE of X_t intrinsically on M although it is possible on $\mathscr{F}(M)$. Therefore, for a given G-function, it is crucial to identify a proper frame bundle over M with a specific structure group such that parallel transport preserves fibers and the generating PDE (associated with G) of the horizontal lifting ξ_t of B_t to such frame bundle is invariant under actions by the structure group along fibers. From this, the law of X_t will be independent of initial frames in the fibre over x (x is the starting point of X_t) and we might be able to obtain the generating PDE of X_t, which is associated with G and intrinsically defined on M.

As we shall see, such idea depends on a crucial algebraic quantity associated with the G-function called the invariant group $I(G)$ of G, which will be defined later on. In this article, we are interested in the case when $I(G)$ is the orthogonal group. We will see that it contains a wide class of G-functions. In particular, one example is the generalization of the one-dimensional Barenblatt equation to higher dimensions.

The concept of the invariant group of G is motivated from the study of infinitesimal diffusive nature of SDEs driven by G-Brownian motion and their generating PDEs, which will be discussed below.

We first consider the Euclidean case.

From now on, we always assume that $G : S(d) \to \mathbb{R}$ is a given continuous, sublinear and monotonic function. Equivalently, from Sect. 2 we know that G is represented by

$$G(A) = \frac{1}{2} \sup_{B \in \Sigma} \operatorname{tr}(AB), \quad \forall A \in S(d), \tag{31}$$

where Σ is some bounded, closed and convex subset of $S_+(d)$. Let B_t be the standard d-dimensional G-Brownian motion on the path space.

Assume that $V_1, \cdots, V_d$ are C_b^3-vector fields on $\mathbb{R}^N$. Consider the following N-dimensional Stratonovich type SDE over $[0, 1]$:

$$\begin{cases} dX_{t,x} = V_\alpha(X_{t,x}) \circ dB_t^\alpha, \\ X_{0,x} = x, \end{cases} \tag{32}$$

which is either interpreted as an RDE or the associated Itô type SDE

$$\begin{cases} dX_{t,x} = V_\alpha(X_{t,x})dB_t^\alpha + \frac{1}{2}DV_\alpha(X_{t,x})V_\beta(X_{t,x})\langle B^\alpha, B^\beta \rangle_t, \\ X_{t,x} = x, \end{cases}$$

according to the main result of Sect. 5.

The following result characterizes the generator of the SDE (32) in terms of G. It describes the infinitesimal diffusive nature of (32). One might compare it with the case of linear diffusion processes.

Proposition 8 *For any* $p \in \mathbb{R}^N$, $A \in S(N)$,

$$\lim_{\delta \to 0+} \frac{1}{\delta} \mathbb{E}^G[\langle p, X_{\delta,x} - x \rangle + \frac{1}{2} \langle A(X_{\delta,x} - x), X_{\delta,x} - x \rangle]$$

$$= G((\frac{1}{2}\langle p, DV_\alpha(x)V_\beta(x) + DV_\beta(x)V_\alpha(x) \rangle + \langle AV_\alpha(x), V_\beta(x) \rangle)_{1 \le \alpha, \beta \le d}).$$

$$\tag{33}$$

Proof From the distribution of B_t we know that

$$G(A) = \frac{1}{2}\mathbb{E}^G[\langle AB_1, B_1 \rangle] = \frac{1}{2t}\mathbb{E}^G[\langle AB_t, B_t \rangle], \ \forall t > 0.$$

Therefore, the R.H.S. of (33) is equal to

$$I_\delta = \frac{1}{2\delta}\mathbb{E}^G[(\langle p, DV_\alpha(x)V_\beta(x) \rangle + \langle AV_\alpha(x), V_\beta(x) \rangle)B_\delta^\alpha B_\delta^\beta],$$

for any $\delta > 0$.

Since

$$X_{\delta,x} - x = \int_0^\delta V_\alpha(X_{s,x})dB_s^\alpha + \frac{1}{2}\int_0^\delta DV_\alpha(X_{s,x})V_\beta(X_{s,x})d\langle B^\alpha, B^\beta \rangle_s,$$

by the properties of $\mathbb{E}^G$ and the distribution of B_t, we have

$$|\frac{1}{\delta}\mathbb{E}^G[\langle p, X_{\delta,x} - x \rangle + \frac{1}{2}\langle A(X_{\delta,x} - x), X_{\delta,x} - x \rangle] - I_\delta|$$

$$\leq |\frac{1}{2\delta}\mathbb{E}^G[\int_0^\delta \langle p, DV_\alpha(X_{s,x}) \cdot V_\beta(X_{s,x}) \rangle d\langle B^\alpha, B^\beta \rangle_s$$

$$+ \langle A\int_0^\delta V_\alpha(X_{s,x})dB_s^\alpha, \int_0^\delta V_\beta(X_{s,x})dB_s^\beta \rangle]$$

$$- \frac{1}{2\delta}\mathbb{E}^G[\langle p, DV_\alpha(x) \cdot V_\beta(x) \rangle \langle B^\alpha, B^\beta \rangle_\delta$$

$$+ \langle AV_\alpha(x)B_\delta^\alpha, V_\beta(x)B_\delta^\beta \rangle]| + C\delta^{\frac{1}{2}} + C\delta$$

$$\leq \frac{1}{2\delta}(C\int_0^\delta \sqrt{\mathbb{E}^G[|X_{s,x} - x|^2]}ds + C\int_0^\delta \mathbb{E}^G[|X_{s,x} - x|^2]ds$$

$$+ C\delta^{\frac{1}{2}}\sqrt{\int_0^\delta \mathbb{E}^G[|X_{s,x} - x|^2]ds}) + C\delta^{\frac{1}{2}} + C\delta,$$

where we've also used the fact that G-Itô integrals and $B_\delta^\alpha B_\delta^\beta - \langle B^\alpha, B^\beta \rangle_\delta$ have zero mean uncertainty. Here C always denotes positive constants independent of δ.

Now the result follows easily from the fact that

$$\mathbb{E}^G[|X_{t,x} - x|^2] \leq Ct, \ \forall t \in [0, 1].$$

The infinitesimal diffusive nature of (32) characterized by Proposition 8 enables us to establish the generating PDE of (32) in terms of viscosity solutions. The understanding of this PDE, especially its intrinsic nature, is essential for the development in a geometric setting.

Theorem 11 *Let $\varphi \in C_b^{\infty}(\mathbb{R}^N)$, and define*

$$u(t, x) = \mathbb{E}^G[\varphi(X_{t,x})], \ (t, x) \in [0, 1] \times \mathbb{R}^N.$$

Then $u(t, x)$ is the unique viscosity solution of the following nonlinear parabolic PDE:

$$\begin{cases} \frac{\partial u}{\partial t} - G((\widehat{V_\alpha V_\beta}u)_{1 \leq \alpha, \beta \leq d}) = 0, \\ u(0, x) = \varphi(x), \end{cases} \tag{34}$$

where $\widehat{V_\alpha V_\beta}$ denotes the symmetrization of the second order differential operator $V_\alpha V_\beta$, that is,

$$\widehat{V_\alpha V_\beta} = \frac{1}{2}(V_\alpha V_\beta + V_\beta V_\alpha).$$

Proof The continuity of u in t and x can be shown in a standard way by using the Lipschitz continuity of φ (in fact, u is Lipchitz in x and $\frac{1}{2}$-Hölder continuous in t). Here the proof is omitted.

Fix $(t_0, x_0) \in (0, 1) \times \mathbb{R}^N$. Let $v(t, x) \in C_b^{2,3}([0, 1] \times \mathbb{R}^N)$ be a test function such that

$$u(t_0, x_0) = v(t_0, x_0)$$

and

$$u(t, x) \leq v(t, x), \ \forall (t, x) \in [0, 1] \times \mathbb{R}^N.$$

For $0 < \delta < t_0$, by the uniqueness of the SDE (32) and the fact that B_t and $\langle B^\alpha, B^\beta \rangle_t$ have independent and identically distributed increments, we know that

$$\mathbb{E}^G[\varphi(X_{t_0,x_0})|\Omega_\delta] = \mathbb{E}^G[\varphi(X_{\delta,x_0} + \int_\delta^{t_0} V_\alpha(X_{s,x_0})dB_s^\alpha$$

$$+ \frac{1}{2}\int_\delta^{t_0} DV_\alpha(X_{s,x_0}) \cdot V_\beta(X_{s,x_0})d\langle B^\alpha, B^\beta \rangle_s)|\Omega_\delta]$$

$$= \mathbb{E}^G[\varphi(X_{t_0-\delta,y})]|_{y=X_{\delta,x_0}}.$$

Therefore,

$$v(t_0, x_0) = \mathbb{E}^G[\varphi(X_{t_0,x_0})]$$

$$= \mathbb{E}^G[\mathbb{E}^G[\varphi(X_{t_0,x_0})|\Omega_\delta]]$$

$$= \mathbb{E}^G[u(t_0 - \delta, X_{\delta,x_0})]$$

$$\leq \mathbb{E}^G[v(t_0 - \delta, X_{\delta,x_0})].$$

It follows that

$$
\begin{aligned}
0 \leq\ & \mathbb{E}^G[v(t_0 - \delta, X_{\delta,x_0}) - v(t_0, x_0)] \\
=\ & \mathbb{E}^G[v(t_0 - \delta, X_{\delta,x_0}) - v(t_0, X_{\delta,x_0}) + v(t_0, X_{\delta,x_0}) - v(t_0, x_0)] \\
=\ & \mathbb{E}^G[-\delta \int_0^1 \frac{\partial v}{\partial t}(t_0 - (1-\alpha)\delta, X_{\delta,x_0})d\alpha + \langle \nabla v(t_0, x_0), X_{\delta,x_0} - x_0 \rangle \\
& + \int_0^1 \int_0^1 \langle \nabla^2 v(t_0, x_0 + \alpha\beta(X_{\delta,x_0} - x_0))(X_{\delta,x_0} - x_0), X_{\delta,x_0} - x_0 \rangle \alpha d\alpha d\beta] \\
\leq\ & -\delta \frac{\partial v}{\partial t}(t_0, x_0) + \mathbb{E}^G[\langle \nabla v(t_0, x_0), X_{\delta,x_0} - x_0 \rangle \\
& + \frac{1}{2} \langle \nabla^2 v(t_0, x_0)(X_{\delta,x_0} - x_0), X_{\delta,x_0} - x_0 \rangle] + \mathbb{E}^G[|I_\delta|] + \mathbb{E}^G[|J_\delta|],
\end{aligned}
$$

where

$$
I_\delta = -\delta \int_0^1 \left(\frac{\partial v}{\partial t}(t_0 - (1-\alpha)\delta, X_{\delta,x_0}) - \frac{\partial v}{\partial t}(t_0, x_0) \right) d\alpha,
$$

$$
\begin{aligned}
J_\delta = & \int_0^1 \int_0^1 \langle (\nabla^2 v(t_0, x_0 + \alpha\beta(X_{\delta,x_0} - x_0)) \\
& - \nabla^2 v(t_0, x_0))(X_{\delta,x_0} - x_0), X_{\delta,x_0} - x_0 \rangle \alpha d\alpha d\beta.
\end{aligned}
$$

By a standard argument one can easily show that

$$
\mathbb{E}^G[|I_\delta|] + \mathbb{E}^G[|J_\delta|] \leq C\delta^{\frac{3}{2}},
$$

where C is a positive constant independent of δ. On the other hand, the R.H.S. of (33) applying to

$$
p = \nabla v(t_0, x_0), \quad A = \nabla^2 v(t_0, x_0),
$$

is exactly the same as $G((\widehat{V_\alpha V_\beta} v(t_0, x_0))_{1 \leq \alpha, \beta \leq d})$. Therefore, by Proposition 8, we arrive at

$$
\frac{\partial v}{\partial t}(t_0, x_0) - G((\widehat{V_\alpha V_\beta} v(t_0, x_0))_{1 \leq \alpha, \beta \leq d}) \leq 0.
$$

Consequently, $u(t, x)$ is a viscosity subsolution of (34).

Similarly, one can show that $u(t, x)$ is a viscosity supersolution of (34). Therefore, $u(t, x)$ is a viscosity solution of (34).

The reason of uniqueness is the following. Define a function $F : \mathbb{R}^N \times \mathbb{R}^N \times S(N) \to \mathbb{R}$ by the R.H.S. of (33), that is,

$$F(x, p, A) = G((\frac{1}{2}\langle p, DV_\alpha(x) \cdot V_\beta(x) + DV_\beta(x) \cdot V_\alpha(x)\rangle + \langle AV_\alpha(x), V_\beta(x)\rangle)_{1 \leq \alpha, \beta \leq d}),$$

for $(x, p, A) \in \mathbb{R}^N \times \mathbb{R}^N \times S(N)$. It is easy to prove that F is sublinear in (p, A) and monotonically increasing in $S(N)$, due to the same properties held by G. Moreover, F satisfies the continuity condition (Assumption (G) in Appendix C of [27]) for the uniqueness of the associated nonlinear PDE, due to the regularity of the given vector fields V_α. In other words, all properties of G to ensure uniqueness are preserved in F, and the space dependence of F coming out are uniformly controlled. Therefore, according to the uniqueness results (see [4, 27]), the parabolic PDE has a unique viscosity solution, which is given by $u(t, x)$.

Example 1 An example which motivates the study of G-Brownian motion on a Riemannian manifold is the following.

Let $Q \in GL(d, \mathbb{R})$, where $GL(d, \mathbb{R})$ is the group of $d \times d$ real invertible matrices. Define $B_t^Q = QB_t$, and for $\varphi \in C_b^\infty(\mathbb{R}^d)$, define

$$u(t, x) = \mathbb{E}^G[\varphi(x + B_t^Q)], \quad (t, x) \in [0, 1] \times \mathbb{R}^d.$$

Then $u(t, x)$ is the unique viscosity solution of the PDE:

$$\begin{cases} \frac{\partial u}{\partial t} - G(Q^T \cdot \nabla^2 u \cdot Q) = 0, \\ u(0, x) = \varphi(x). \end{cases}$$

In fact, it follows directly from Theorem 11 if we regard $x + B_t^Q$ as the solution of the SDE over $[0, 1]$:

$$\begin{cases} dX_{t,x} = Q_\alpha \circ dB_t^\alpha, \\ X_{0,x} = x, \end{cases} \tag{35}$$

where $Q = (Q_1, \cdots, Q_d)$, and each Q_α is a constant vector field on $\mathbb{R}^d$ (so the SDE (35) coincides exactly with the Itô type one).

The result of Theorem 34 is similar to the discussion of nonlinear Feynman-Kac formula in [27], in which the solution of a forward-backward SDE is used to represent the viscosity solution of an associated nonlinear backward parabolic PDE. In our case, the intrinsic nature of (34) is fundamental and should be emphasized below in order to develop G-Brownian motion on a Riemannian manifold.

It is not hard to see that the nonlinear second order differential operator

$$G((\widehat{V_\alpha V_\beta} \cdot)_{1 \leq \alpha, \beta \leq d})$$

is intrinsically defined on $\mathbb{R}^N$, since $V_1, \cdots, V_d$ are vector fields independent of coordinates. Moreover, in local coordinates it preserves the same properties of the G-function which is defined under the standard coordinate system of $\mathbb{R}^d$. In particular, it shares the same ellipticity as G. Therefore, when the vector fields V_α are regular enough, from our results in Sect. 6, we are able to establish the generating PDE of a nonlinear diffusion process on a differentiable manifold. As in the last section, for technical simplicity we restrict ourselves to compact manifolds.

Assume that M is a compact manifold, and $V_1, \cdots, V_d$ are C^3-vector fields on M. According to Sect. 6, the Stratonovich type SDE over $[0, 1]$

$$\begin{cases} dX_{t,x} = V_\alpha(X_{t,x}) \circ dB_t^\alpha, \\ X_{0,x} = x \in M, \end{cases} \tag{36}$$

has a unique solution. The following result is immediate from Theorem 11.

Theorem 12 *Let $\varphi \in C^\infty(M)$, and define*

$$u(t, x) = \mathbb{E}^G[\varphi(X_{t,x})], \ (t, x) \in [0, 1] \times M,$$

then $u(t, x)$ is the unique viscosity solution of the following nonlinear parabolic PDE on M:

$$\begin{cases} \frac{\partial u}{\partial t} - G((\widehat{V_\alpha V_\beta}u)_{1 \leq \alpha, \beta \leq d}) = 0, \\ u(0, x) = \varphi(x), \end{cases} \tag{37}$$

where $\widehat{V_\alpha V_\beta}$ is the symmetrization of $V_\alpha V_\beta$, defined in the same way as in Theorem 11. Here the notion of viscosity solutions for the PDE (37) can be defined in the same way as in the Euclidean case by using test functions (see [1]).

Proof The result follows easily from an extrinsic point of view.

In fact, assume that M is embedded into an ambient Euclidean space $\mathbb{R}^N$ as a closed submanifold, and take a C^3-extension $\tilde{V}_\alpha$ of V_α with compact support. Consider the following Stratonovich type SDE over $[0, 1]$:

$$\begin{cases} dX_{t,x} = \tilde{V}_\alpha(X_{t,x}) \circ dB_t^\alpha, \\ X_{0,x} = x \in \mathbb{R}^N. \end{cases}$$

Let $\tilde{\varphi}$ be a C^∞-extension of φ with compact support, and define

$$\tilde{u}(t, x) = \mathbb{E}^G[\tilde{\varphi}(X_{t,x})], \ (t, x) \in [0, 1] \times \mathbb{R}^N.$$

It follows from Theorem 11 that $\tilde{u}(t, x)$ is the unique viscosity solution of the nonlinear parabolic PDE generated by the vector fields $\tilde{V}_\alpha$.

According to Sect. 6, if $x \in M$, $X_{t,x}$ will never leave M quasi-surely. Therefore, when restricted on M, $\tilde{u} = u$. In particular, we know that u is continuous. To see that u is a viscosity subsolution of (37), let $(t_0, x_0) \in (0, 1) \times M$, and $v(t, x) \in C^{2,3}([0, 1] \times M)$ be a test function such that

$$v(t_0, x_0) = u(t_0, x_0)$$

and

$$u(t, x) \leqslant v(t, x), \quad \forall (t, x) \in [0, 1] \times M.$$

Take an $C_b^{2,3}$-extension $\tilde{v}$ of v such that

$$\tilde{u}(t, x) \leqslant \tilde{v}(t, x), \quad \forall (t, x) \in [0, 1] \times \mathbb{R}^N.$$

It follows from previous discussion that

$$\frac{\partial \tilde{v}}{\partial t}(t_0, x_0) - G((\widehat{\tilde{V}_\alpha \tilde{V}_\beta \tilde{v}}(t_0, x_0))_{1 \leqslant \alpha, \beta \leqslant d}) \leqslant 0.$$

Since

$$\tilde{V}_\alpha|_M = V_\alpha, \quad \tilde{v}|_M = v,$$

from the intrinsic nature of the generating PDE, we know that

$$\frac{\partial \tilde{v}}{\partial t}(t_0, x_0) = \frac{\partial v}{\partial t}(t_0, x_0)$$

and

$$G((\widehat{\tilde{V}_\alpha \tilde{V}_\beta \tilde{v}}(t_0, x_0))_{1 \leqslant \alpha, \beta \leqslant d}) = G((\widehat{V_\alpha V_\beta v}(t_0, x_0))_{1 \leqslant \alpha, \beta \leqslant d}).$$

It follows that

$$\frac{\partial v}{\partial t}(t_0, x_0) - G((\widehat{V_\alpha V_\beta v}(t_0, x_0))_{1 \leqslant \alpha, \beta \leqslant d}) \leqslant 0.$$

Therefore, $u(t, x)$ is a viscosity subsolution of (37). Similarly we can show that it is a viscosity supersolution as well, and thus a viscosity solution.

The uniqueness of (37) follows from the same reason as in the proof of Theorem 11 once we notice that the second order differential operator $G((\widehat{V_\alpha V_\beta} \cdot)_{1 \leqslant \alpha, \beta \leqslant d})$ on M shares exactly the same properties as G (in particular, the same ellipticity), which can be seen either from an extrinsic way or via local computation. Another way to see the uniqueness is to use the results in [1] as long as we assign a complete Riemannian metric on M, which is always possible according to [20]. In this case

$$G((\widehat{\nabla_\alpha V_\beta} u)_{1\leqslant\alpha,\beta\leqslant d}) = G((\frac{1}{2}\langle\nabla u, \nabla_{V_\alpha} V_\beta + \nabla_{V_\beta} V_\alpha\rangle + \text{Hess}u(V_\alpha, V_\beta))_{1\leqslant\alpha,\beta\leqslant d}),$$

where ∇ is the Levi-Civita connection corresponding to the Riemannian metric. The uniqueness of (37) follows from Theorem 5.1 in [1] directly, as the assumptions in the theorem are verified by the properties of G. Note that we don't need the Ricci curvature condition in [1] due to the compactness of M and uniform continuity of $G((\widehat{\nabla_\alpha V_\beta}\cdot)_{1\leqslant\alpha,\beta\leqslant d})$.

Remark 6 The study of the SDE (36) as a nonlinear diffusion process on M does not require a Riemannian metric or a connection on M. The fundamental reason is that (36) is defined in the pathwise sense as an RDE generated by the vector fields V_α on M. Such an RDE only depends on the differential structure of M. The infinitesimal diffusive nature of (36) can be studied by local computation.

Now we turn to the study of G-Brownian motion on a Riemannian manifold. The Riemannian structure (the Levi-Civita connection) is used to "roll" the Euclidean G-Brownian motion up to the manifold "without slipping" by solving an SDE generated by the fundamental horizontal vector fields on a proper frame bundle (known as horizontal lifting). This is the fundamental idea of Eells-Elworthy-Malliavin on the construction of Brownian motion on a Riemannian manifold.

As is pointed out at the beginning of this section, the essential point of such development is the invariance of the generating PDE on the frame bundle under actions by the structure group along fibers. The key of capturing such invariance is Theorem 34 and Example 1, which leads to the following important concept.

Definition 7 The invariant group $I(G)$ of G is defined by

$$I(G) = \{Q \in GL(d, \mathbb{R}) : \forall A \in S(d), G(Q^T A Q) = G(A)\}.$$

It is easy to check the $I(G)$ is a group, and hence a subgroup of $GL(d, \mathbb{R})$.

By using the representation (31) of G, we have the following equivalent characterization of the invariant group $I(G)$.

Proposition 9 *Let G be represented by*

$$G(A) = \frac{1}{2}\sup_{B\in\Sigma} tr(AB), \ \forall A \in S(d),$$

where Σ is some bounded, closed and convex subset of $S_+(d)$. Then Σ is uniquely determined by G and the invariant group $I(G)$ of G is given by

$$I(G) = \{Q \in GL(d, \mathbb{R}) : Q\Sigma Q^T = \Sigma\}. \tag{38}$$

Proof It suffices to show the uniqueness of Σ, and (38) will follow immediately from the commutativity of the trace operator and the uniqueness of Σ. Note that for any $Q \in GL(d, \mathbb{R})$, $Q\Sigma Q^T$ is also a bounded, closed and convex subset of $S_+(d)$.

Introduce a symmetric bilinear form $\langle \cdot, \cdot \rangle_{\mathrm{tr}}$ on the finite dimensional vector space $S(d)$ by

$$\langle A_1, A_2 \rangle_{\mathrm{tr}} = \mathrm{tr}(A_1 A_2), \ A_1, A_2 \in S(d).$$

It is easy to check that $\langle \cdot, \cdot \rangle_{\mathrm{tr}}$ is indeed an inner product, thus $(S(d), \langle \cdot, \cdot \rangle_{\mathrm{tr}})$ is a finite dimensional Hilbert space. The form $\| \cdot \|_{tr}$ induced by $\langle \cdot, \cdot \rangle_{\mathrm{tr}}$ is equivalent to any other matrix norm on $S(d)$ since $S(d)$ is finite dimensional.

Let Σ_1, Σ_2 be two bounded, closed and convex subsets of $S_+(d)$, such that

$$\sup_{B \in \Sigma_1} \mathrm{tr}(AB) = \sup_{B \in \Sigma_2} \mathrm{tr}(AB), \ \forall A \in S(d).$$

If $\Sigma_1 \neq \Sigma_2$, without loss of generality assume that $B_0 \in \Sigma_2 \backslash \Sigma_1$. According to the Mazur separation theorem in functional analysis (see [29]), there exists a bounded linear functional $f \in S(d)^*$ and some $\alpha \in \mathbb{R}$, such that

$$f(B) < \alpha < f(B_0), \ \forall B \in \Sigma_1.$$

By the Riesz representation theorem, there exists a unique $A^* \in S(d)$, such that

$$f(B) = \langle A^*, B \rangle_{\mathrm{tr}} = \mathrm{tr}(A^* B), \ \forall B \in S(d).$$

It follows that

$$\sup_{B \in \Sigma_1} \mathrm{tr}(A^* B) \leqslant \alpha < \mathrm{tr}(A^* B_0) \leqslant \sup_{B \in \Sigma_2} \mathrm{tr}(A^* B),$$

which is a contradiction. Therefore, $\Sigma_1 = \Sigma_2$.

We list some examples for the invariant groups $I(G)$ of different G-functions.

Example 2 If $\Sigma = \{0\}$, then it is obvious that $I(G) = GL(d, \mathbb{R})$, which is a noncompact group.

Example 3 It is possible that $I(G)$ is a finite group.

Consider Σ is the set of diagonal matrices

$$\Lambda = \mathrm{diag}(\lambda_1, \cdots, \lambda_d)$$

such that each $\lambda_\alpha \in [0, 1]$, then Σ is a bounded, closed and convex subset of $S_+(d)$. We claim that

$$I(G) = \{(\pm e_{\sigma(1)}, \cdots, \pm e_{\sigma(d)}) : \ \sigma \text{ is a permutation of order } d\}, \tag{39}$$

where $\{e_1, \cdots, e_d\}$ is the standard orthonormal basis of $\mathbb{R}^d$, each e_i being regarded as a column vector.

In fact, if $Q \in GL(d, \mathbb{R})$ has the form (39), by direct computation one can show easily that

$$Q \Sigma Q^T = \Sigma. \tag{40}$$

Conversely, if Q satisfies (40), by choosing

$$\Lambda = \mathrm{diag}(1, 0, \cdots, 0),$$

we know that

$$(Q \Lambda Q^T)^\alpha_\beta = Q^\alpha_1 Q^\beta_1.$$

Therefore, if $Q \Lambda Q^T \in \Sigma$, the first column of Q must contain exactly one nonzero element q_1 such that $q_1^2 \leqslant 1$. Similarly for other columns of Q. Moreover, the corresponding nonzero elements in any two different columns of Q must be in different rows, otherwise Q will be degenerate. Consequently, Q has the form

$$Q = (q_1 e_{\sigma(1)}, \cdots, q_d e_{\sigma(d)})$$

with $q_i^2 \leqslant 1$ ($i = 1, 2, \cdots, d$). On the other hand, for the identity matrix I_d, there exists $\Lambda \in \Sigma$, such that

$$Q \Lambda Q^T = I_d.$$

By taking determinants on both sides, we have

$$q_1^2 \cdots q_d^2 \det(\Lambda) = 1,$$

which implies that $q_\alpha = \pm 1$ ($\alpha = 1, 2, \cdots, d$). Therefore, Q has the form of (39).

Note that in this case $I(G)$ is a finite subgroup of the orthogonal group $O(d)$ with order $2^d d!$. Moreover, G is given by

$$G(A) = \frac{1}{2} \sum_{\alpha=1}^{d} (A^\alpha_\alpha)^+, \ \forall A \in S(d).$$

Example 4 Now we give some examples of G such that $I(G) = O(d)$. Such case will be our main interest in this article.

1. $\Sigma = \{I_d\}$.
   Obviously (40) is equivalent to $Q \in O(d)$.

This corresponds to the case of classical Brownian motion, in which

$$G(A) = \frac{1}{2}\text{tr}(A)$$

and the generator is $\frac{1}{2}\Delta$.

2. Σ is given by the segment joining λI_d and μI_d, where $0 \leqslant \lambda < \mu$. If $Q \in GL(d, \mathbb{R})$ such that (40) holds, then

$$\mu Q Q^T = t I_d,$$

for some $t \in [\lambda, \mu]$. On the other hand, there exists some $t' \in [\lambda, \mu]$ such that

$$t' Q Q^T = \mu I_d.$$

The only possibility is that $Q Q^T = I_d$, which means $Q \in O(d)$. The converse is trivial.

In this case, G is given by

$$G(A) = \frac{1}{2}(\mu(\text{tr}A)^+ - \lambda(\text{tr}A)^-).$$

The corresponding G-heat equation can be regarded as the generalization of the one-dimensional Barenblatt equation to higher dimensions.

3. Σ is given by the subset of matrices $B \in S_+(d)$ such that the eigenvalues of B lie in the bounded interval $[\lambda, \mu]$, where $0 \leqslant \lambda < \mu$. Equivalently,

$$\Sigma = \{B \in S_+(d) : \lambda \leqslant x^T B x \leqslant \mu, \ \forall x \in \mathbb{R}^d \text{ with } |x| = 1\}.$$

It follows that Σ is a bounded, closed and convex subset of $S_+(d)$.

Since Σ is characterized by eigenvalues, and the eigenvalues of a symmetric matrix is preserved under change of orthonormal basis, it follows that for any $Q \in O(d)$, (40) holds. Conversely, let $Q \in GL(d, \mathbb{R})$ with (40). Then there exists $B_1, B_2 \in \Sigma$, such that

$$\mu Q Q^T = B_1, \ Q B_2 Q^T = \mu I_d.$$

It follows that all eigenvalues of $Q Q^T$ lie in $[\frac{\lambda}{\mu}, 1]$, and

$$\det(Q Q^T)\det(B_2) = \mu^d.$$

Therefore, the only possibility is that all eigenvalues of $Q Q^T$ are equal to 1, which implies that Q is an orthogonal matrix.

In this case G can be expressed by

$$G(A) = \frac{1}{2} \sup_{B \in \Sigma} \text{tr}(AB)$$

$$= \frac{1}{2} \sup_{P \in O(d)} \sup_{\lambda \leq c_1, \cdots, c_d \leq \mu} \text{tr}(AP^T \text{diag}(c_1, \cdots, c_d) P)$$

$$= \frac{1}{2} \sup_{P \in O(d)} \sup_{\lambda \leq c_1, \cdots, c_d \leq \mu} \text{tr}(PAP^T \text{diag}(c_1, \cdots, c_d))$$

$$= \frac{1}{2} \sup_{P \in O(d)} \sup_{\lambda \leq c_1, \cdots, c_d \leq \mu} \sum_{\alpha=1}^{d} c_\alpha (PAP^T)_\alpha^\alpha$$

$$= \frac{1}{2} \sup_{P \in O(d)} \sum_{\alpha=1}^{d} (\mu((PAP^T)_\alpha^\alpha)^+ - \lambda((PAP^T)_\alpha^\alpha)^-).$$

Similar to Example 4, for those Σ's characterized by eigenvalues, we can construct a large class of G such that $I(G) = O(d)$.

Remark 7 If Σ has at least one nondegenerate element, that is, there exists some positive definite matrix $B_0 \in \Sigma$, then $I(G)$ is a compact group. In fact, if we introduce a matrix norm $\| \cdot \|_{B_0}$ on the space $\text{Mat}(d, \mathbb{R})$ of real $d \times d$ matrices by

$$\|A\|_{B_0} = \sqrt{\text{tr}(AB_0 A^T)}, \quad A \in \text{Mat}(d, \mathbb{R}),$$

it follows that

$$\sup_{Q \in I(G)} \|Q\|_{B_0} = \sup_{Q \in I(G)} \sqrt{\text{tr}(QB_0 Q^T)} \leq \sup_{B \in \Sigma} \sqrt{\text{tr}(B)} < \infty,$$

since Σ is bounded. It is obvious that $I(G)$ is closed. Therefore, it is compact.

Now assume that (M, g) is a d-dimensional compact Riemannian manifold. If we allow explosion of a nonlinear diffusion process at some finite time, then the arguments below will carry through on a noncompact Riemannian manifold as long as the time scope is restricted from 0 up to the explosion. Here we only consider the compact case, in which explosion is not possible.

We first recall some basics about frame bundles, which is the central concept in the horizontal lifting construction. For a systematic introduction please refer to [2, 16].

Let $\mathscr{F}(M)$ be the total frame bundle over M defined by

$$\mathscr{F}(M) = \cup_{x \in M} \mathscr{F}_x(M),$$

where the fibre $\mathscr{F}_x(M)$ is the set of all frames (bases of the tangent space $T_x(M)$) at x. A frame $\xi = (\xi_1, \cdots, \xi_d) \in \mathscr{F}_x(M)$ can be equivalently regarded as a linear isomorphism from $\mathbb{R}^d$ to $T_x M$ (also denoted by ξ) if we let

$$\xi(e_\alpha) = \xi_\alpha, \ \alpha = 1, 2, \cdots, d,$$

and extend linearly to $\mathbb{R}^d$, where we always fix $\{e_1, \cdots, e_d\}$ to be the standard orthonormal basis of $\mathbb{R}^d$. $\mathscr{F}(M)$ is a principal bundle with structure group $GL(d, \mathbb{R})$ acting along fibers from the right.

Fix a frame $\xi \in \mathscr{F}_x(M)$. A vector $X \in T_\xi \mathscr{F}(M)$ is called vertical if it is tangent to the fibre $\mathscr{F}_x(M)$. The space of vertical vectors at ξ is called the vertical subspace, and it is denoted by $V_\xi \mathscr{F}(M)$. $V_\xi \mathscr{F}(M)$ is a d^2-dimensional vector space, which is independent of the Riemannian structure.

A smooth curve $\xi_t = (\xi_{1,t} \cdots, \xi_{d,t}) \in \mathscr{F}(M)$ is called horizontal if $\xi_{\alpha,t}$ is a parallel vector field along the projection curve $x_t = \pi(\xi_t)$ for each $\alpha = 1, 2, \cdots, d$. Given a smooth curve $x_t \in M$ and a frame $\xi_0 = (\xi_1, \cdots, \xi_d) \in \mathscr{F}_{x_0}(M)$, by solving a first order linear ODE, we can determine a unique parallel vector field $\xi_{\alpha,t}$ along x_t with $\xi_{\alpha,0} = \xi_\alpha$ for each $\alpha = 1, 2, \cdots, d$. The smooth curve

$$\xi_t = (\xi_{1,t}, \cdots, \xi_{d,t}) \in \mathscr{F}(M)$$

is then the unique horizontal curve with $x_t = \pi(\xi_t)$ and initial position ξ_0. ξ_t is called the horizontal lifting of x_t from ξ_0. A vector $X \in T_\xi \mathscr{F}(M)$ is called horizontal if it is tangent to a horizontal curve through ξ. The space of horizontal vectors at ξ is called the horizontal subspace, and it is denoted by $H_\xi \mathscr{F}(M)$. It is a d-dimensional vector space characterized by the Levi-Civita connection ∇.

As ξ varies, $V_\xi \mathscr{F}(M)$ (respectively, $H_\xi \mathscr{F}(M)$) determines a vertical (respectively, horizontal) subspace field on M. The following result reveals the fundamental structure of $\mathscr{F}(M)$.

Theorem 13 *The horizontal subspace field $H\mathscr{F}(M)$, which is determined by ∇, has the following properties.*

1. For each $\xi \in \mathscr{F}_x(M)$, the tangent space $T_\xi \mathscr{F}(M)$ has the decomposition

$$T_\xi \mathscr{F}(M) = H_\xi \mathscr{F}(M) \oplus V_\xi \mathscr{F}(M).$$

Moreover, $H_\xi \mathscr{F}(M)$ is isomorphic to $T_x M$ under the canonical projection π : $\mathscr{F}(M) \to M$.

2. $H\mathscr{F}(M)$ is invariant under actions by the structure group $GL(d, \mathbb{R})$. More precisely, for any $\xi \in \mathscr{F}(M)$, $Q \in GL(d, \mathbb{R})$,

$$Q_*(H_\xi \mathscr{F}(M)) = H_{\xi Q} \mathscr{F}(M).$$

It should be pointed out that given any horizontal subspace field $H\mathscr{F}(M)$ satisfying the two properties in Theorem 13, there exists an affine connection ∇^H such that $H\mathscr{F}(M)$ is the horizontal subspace field determined by ∇^H.

On $\mathscr{F}(M)$ there is a canonical way to define a frame field globally, which is not always possible on a general Riemannian manifold. This makes $\mathscr{F}(M)$ simpler than the base space M in some sense. Fix $w \in \mathbb{R}^d$. For any $\xi \in \mathscr{F}_x(M)$ regarded as a linear isomorphism $\xi : \mathbb{R}^d \to T_x M$, $\xi(w)$ is a tangent vector in $T_x M$. By Theorem 13 (1), $\xi(w)$ corresponds to a unique vector $H_w(\xi) \in H_\xi \mathscr{F}(M)$. It follows that H_w is a globally defined horizontal vector field on $\mathscr{F}(M)$. If we take $w = e_\alpha$ ($\alpha = 1, 2, \cdots, d$), then we obtain a family of horizontal vector fields $\{H_{e_1}, \cdots, H_{e_d}\}$ as a basis of the horizontal subspace $H_\xi \mathscr{F}(M)$ at each frame $\xi \in \mathscr{F}(M)$. $\{H_{e_1}, \cdots, H_{e_d}\}$ are called the fundamental horizontal fields of $\mathscr{F}(M)$, simply denoted by $\{H_1, \cdots, H_d\}$.

Now we introduce the concept of development and anti-development (see [12]), which is crucial in the construction of G-Brownian motion on M. Assume that $x_t \in M$ is a smooth curve and ξ_t is the horizontal lifting of x_t from ξ_0. Then we can determine a smooth curve

$$w_t = \int_0^t \xi_s^{-1} \dot{x}_s ds \in \mathbb{R}^d$$

starting from 0 (w_t is regarded as a column vector in $\mathbb{R}^d$). w_t is called the anti-development of x_t in $\mathbb{R}^d$ with respect to ξ_0. If ξ_t and η_t are two horizontal liftings of x_t with $\xi_0 = \eta_0 Q$ for some $Q \in GL(d, \mathbb{R})$, then the two corresponding anti-developments are related by

$$w_t^\eta = Q w_t^\xi.$$

The fundamental relation between the anti-development w_t of x_t and the horizontal lifting ξ_t is the following ODE on $\mathscr{F}(M)$:

$$d\xi_t = H_\alpha(\xi_t) dw_t^\alpha. \tag{41}$$

Conversely, given a smooth curve $w_t \in \mathbb{R}^d$ starting from 0, by solving the ODE (41) on $\mathscr{F}(M)$ with initial frame ξ_0, we obtain a horizontal curve $\xi_t \in \mathscr{F}(M)$. The projection $x_t = \pi(\xi_t)$ is called the development of w_t in M with respect to ξ_0. If we use another initial frame $\eta_0 = \xi_0 Q^{-1}$ and the driven process $v_t = Q w_t \in \mathbb{R}^d$, by solving (41) from η_0 and projection onto M we obtain the same curve x_t. In this way, we obtain a one-to-one correspondence of the Euclidean curve w_t and the manifold curve x_t via the horizontal curve ξ_t in $\mathscr{F}(M)$, which depends on the initial frame ξ_0. The procedure of getting x_t from w_t is usually known as "rolling without slipping".

A crucial point should be emphasized here is that such procedure is carried out by solving the ODE (41) in the pathwise sense, which fits well in the context of rough paths if the Euclidean curve w_t is interpreted as a rough path. In this case, (41) should be interpreted as an RDE. This is an important reason why we need to develop the notion of Stratonovich type SDEs on a differentiable manifold.

For a general Euclidean G-Brownian motion B_t, from Sect. 6 we are able to solve (41) pathwisely if the driven curve dw_t is replaced by dB_t in the Stratonovich sense (or in the RDE sense). By projecting the solution $\xi_t \in \mathscr{F}(M)$ to the manifold M, we obtain a process $X_t \in M$ pathwisely which depends on the initial position x_0 and the initial frame $\xi_0 \in \mathscr{F}_{x_0}(M)$. A disadvantage of using the total frame bundle $\mathscr{F}(M)$ is that in this way it is not possible to write down the generating PDE governing the law of X_t intrinsically on M, which does not depend on the initial frame ξ_0. Note that the generating PDE of ξ_t is well-defined on $\mathscr{F}(M)$ according to Theorem 12, which takes the form

$$\frac{\partial u}{\partial t} - G((\widehat{H_\alpha H_\beta}u)_{1 \leqslant \alpha, \beta \leqslant d}) = 0. \tag{42}$$

The main reason for such disadvantage is that the PDE (42) is not invariant under actions by $GL(d, \mathbb{R})$ along fibers, since the G-function does not have such kind of invariance.

To fix this issue, a possible way is to use the invariant group $I(G)$ of G as the structure group, so that the generating PDE will be invariant under actions by $I(G)$ along fibers due to the form (42) it takes. Therefore, we need to use a proper frame bundle (a submanifold of $\mathscr{F}(M)$ which is a principal bundle over M with structure group $I(G)$ and fibers being a suitable class of frames) instead of $\mathscr{F}(M)$. The fibers of such frame bundle should be preserved by parallel transport so the fundamental horizontal fields can be restricted on it and we are able to solve the RDE

$$d\xi_t = H_\alpha(\xi_t) \circ dB_t^\alpha$$

on the frame bundle. It will turn out that we are able to establish the generating PDE of the projection process $X_t = \pi(\xi_t)$ intrinsically on M, which does not depend on the initial frame. Therefore, although as a process the sample paths of X_t depends on the initial frame (this is not surprising since in the Euclidean case we also don't have a canonical Brownian motion if we do not fix the frame $\{e_1, \cdots, e_d\}$ in advance), the law of X_t will not. In this way we obtain a canonical PDE on M associated with the original G-function, which can be regarded as the generating PDE governing the law of X_t. The process X_t can be defined as a G-Brownian motion on M and the generating PDE will play the role of the canonical Wiener measure (the solution of the martingale problem for the operator $\frac{1}{2}\Delta_M$) on M in the nonlinear setting.

The construction of such frame bundle for a G-function with an arbitrary invariant group $I(G)$ is not clear to us at the moment. However, in the case when $I(G)$ is the orthogonal group $O(d)$, which contains a wide and interesting class of G-functions, there is a very natural frame bundle serving us well for the purpose: the orthonormal frame bundle $\mathscr{O}(M)$.

From now on, let G be given by (31) with $I(G) = O(d)$.

The orthonormal frame bundle $\mathscr{O}(M)$ over M is defined by

$$\mathscr{O}(M) = \cup_{x \in M} \mathscr{O}_x(M),$$

where the fibre $\mathscr{O}_x(M)$ is the set of orthonormal bases of $T_x M$. Since M is compact, $\mathscr{O}(M)$ is a compact submanifold of $\mathscr{F}(M)$. Moreover, since the Levi-Civita connection is compatible with the Riemannian metric g, parallel transport preserves the fibers of $\mathscr{O}(M)$. Therefore, statements about $\mathscr{F}(M)$ before on the horizontal aspect can be carried through in the case of $\mathscr{O}(M)$ directly. In particular, the fundamental horizontal fields H_α can be restricted to $\mathscr{O}(M)$. The only difference is in the vertical direction: the fibre becomes orthonormal frames, and the structure group which acts on fibers becomes the orthogonal group; the dimension in the vertical direction is reduced to $\frac{d(d-1)}{2}$.

For $\xi \in \mathscr{O}_x(M)$, according to Sect. 6, let $U_{t,\xi} \in \mathscr{O}(M)$ be the unique solution of the following RDE over $[0, 1]$:

$$\begin{cases} dU_{t,\xi} = H_\alpha(U_{t,\xi}) \circ dB_t^\alpha, \\ U_{0,\xi} = \xi. \end{cases} \tag{43}$$

Let $X_{t,\xi} = \pi(U_{t,\xi})$ be the projection of $U_{t,\xi}$ onto M.

Definition 8 $X_{t,\xi}$ is called a G-Brownian motion on the Riemannian manifold M with respect to the initial orthonormal frame $\xi \in \mathscr{O}_x(M)$, and $U_{t,\xi}$ is called a horizontal G-Brownian motion in $\mathscr{O}(M)$ starting from ξ.

For any $\varphi \in C_{Lip}(M)$ (under the Riemannian distance), define

$$u(t, \xi) = \mathbb{E}^G[\varphi(X_{t,\xi})], \ (t, \xi) \in [0, 1] \times \mathscr{O}(M).$$

Let $\hat{\varphi} = \varphi \circ \pi$ be the lifting of φ to $\mathscr{O}(M)$. It is obvious that

$$u(t, \xi) = \mathbb{E}^G[\hat{\varphi}(U_{t,\xi})].$$

By Theorem 12, we know that $u(t, \xi)$ is the unique viscosity solution of the following nonlinear parabolic PDE:

$$\begin{cases} \frac{\partial u}{\partial t} - G((\widehat{H_\alpha H_\beta u})_{1 \leqslant \alpha, \beta \leqslant d}) = 0, \\ u(0, \xi) = \hat{\varphi}(\xi), \end{cases} \tag{44}$$

on $\mathscr{O}(M)$.

The following result tells us that the law of $X_{t,\xi}$ depends only on the initial position x.

Proposition 10 *If $\xi, \eta \in \mathscr{O}_x(M)$, then*

$$u(t, \xi) = u(t, \eta).$$

Proof For any fixed orthogonal matrix $Q \in O(d)$, let $\tilde{B}_t = QB_t$, which is an orthogonal transformation of the original G-Brownian motion B_t, and let $W_{t,\xi}$ be the pathwise solution of the following RDE over $[0, 1]$:

$$\begin{cases} dW_{t,\zeta} = H_\alpha(W_{t,\zeta}) \circ d\tilde{B}_t^\alpha, \\ W_{0,\zeta} = \zeta \in \mathscr{O}(M), \end{cases} \tag{45}$$

on $\mathscr{O}(M)$. If we regard $\tilde{B}_t$ as the solution of the SDE

$$d\tilde{B}_t = Q_\alpha dB_t^\alpha$$

starting from 0 with constant coefficients, then the RDE (45) is equivalent to

$$\begin{cases} dW_{t,\zeta} = H_\beta(W_{t,\zeta})Q_\alpha^\beta \circ dB_t^\alpha, \\ W_{0,\zeta} = \zeta, \end{cases}$$

in which the generating vector fields are $H_\beta Q_\alpha^\beta$. Since the invariant group $I(G)$ of G is the orthogonal group, by Theorem 12 we know that the function

$$v(t, \zeta) = \mathbb{E}^G[\hat{\varphi}(W_{t,\zeta})], \; (t, \zeta) \in [0, 1] \times \mathscr{O}(M)$$

is the unique viscosity solution of the same PDE (44) on $\mathscr{O}(M)$. Therefore,

$$u(t, \zeta) = v(t, \zeta), \; \forall (t, \zeta) \in [0, 1] \times \mathscr{O}(M).$$

Now since $\xi, \eta \in \mathscr{O}_x(M)$, there exists some $Q \in O(d)$ such that $\xi = \eta Q$. Define $W_{t,\zeta}$ as before. By the previous discussion on the relation between different anti-developments, we know that

$$X_{t,\xi} = \pi(U_{t,\xi}) = \pi(W_{t,\eta}), \; \forall t \in [0, 1].$$

Therefore,

$$\begin{aligned} u(t, \xi) &= \mathbb{E}^G[\varphi \circ \pi(U_{t,\xi})] \\ &= \mathbb{E}^G[\varphi \circ \pi(W_{t,\eta})] \\ &= v(t, \eta) \\ &= u(t, \eta). \end{aligned}$$

From Proposition 10, we know that $u(t, \xi)$ is invariant along each fibre. Therefore, the law of $X_{t,\xi}$ depends only on the initial position $x \in M$ but not on the initial frame ξ. We use $u(t, x)$ to denote $u(t, \xi)$, where x is the base point of ξ. In this situation it is possible to establish the PDE for $u(t, x)$ intrinsically on M by "projecting down" (44), which should become the generating PDE governing the law of $X_{t,\xi}$.

For any $u \in C^\infty(M)$, take an orthonormal frame $\xi = (\xi_1, \cdots, \xi_d) \in \mathcal{O}_x(M)$, and consider the quantity

$$G((\mathrm{Hess}u(\xi_\alpha, \xi_\beta))_{1 \leq \alpha, \beta \leq d}).$$

Since $I(G) = O(d)$, it is easy to see that the above quantity is independent of the orthonormal frame $\xi \in \mathcal{O}_x(M)$. In other words, G can be regarded as a functional of the Hessian, and the nonlinear second order differential operator $G(\mathrm{Hess}(\cdot))$ is globally well-defined on M.

Now we have the following result.

Theorem 14 $u(t, x)$ *is the unique viscosity solution of the following nonlinear heat equation on M :*

$$\begin{cases} \frac{\partial u}{\partial t} - G(Hessu) = 0, \\ u(0, x) = \varphi(x). \end{cases} \tag{46}$$

Proof It suffices to show that: if $f \in C^\infty(M)$, and $\hat{f} = f \circ \pi$ is the lifting of f to $\mathcal{O}(M)$, then for any $\xi = (\xi_1, \cdots, \xi_d) \in \mathcal{O}_x(M)$,

$$\mathrm{Hess} f(\xi_\alpha, \xi_\beta)(x) = H_\alpha H_\beta \hat{f}(\xi).$$

Note that uniqueness follows from the same reason as pointed out in the proof of Theorem 12 by using results in [1].

In fact, for any $\xi = (\xi_1, \cdots, \xi_d) \in \mathcal{O}_x(M)$, let ξ_t be a horizontal curve through ξ such that $H_\beta(\xi)$ is tangent to ξ_t at $t = 0$, and let x_t be its projection onto M. It follows that the tangent vector of x_t at $t = 0$ is ξ_β, and

$$\begin{aligned} H_\beta \hat{f}(\xi) &= \frac{d\,\hat{f}(\xi_t)}{dt}\Big|_{t=0} \\ &= \frac{df(x_t)}{dt}\Big|_{t=0} \\ &= \langle \xi_\beta, \nabla f(x) \rangle_g. \end{aligned}$$

Therefore, if now assume that ξ_t is a horizontal curve through ξ with tangent vector $H_\alpha(\xi)$ at ξ and still $x_t = \pi(\xi_t)$, then

$$\begin{aligned} H_\alpha H_\beta \hat{f}(\xi) &= H_\alpha \langle \xi_\beta, \nabla f(\pi(\xi)) \rangle_g \\ &= \frac{d}{dt}\Big|_{t=0} \langle \xi_{\beta,t}, \nabla f(x_t) \rangle_g \end{aligned}$$

$$= \langle \frac{D\xi_{\beta,t}}{dt}|_{t=0}, \nabla f(x) \rangle_g + \langle \xi_\beta, \nabla_{\xi_\alpha} \nabla f(x) \rangle_g$$

$$= \text{Hess} f(\xi_\alpha, \xi_\beta)(x),$$

where we've used the fact that $\xi_{\beta,t}$ is parallel along x_t.

Since $X_{t,\xi}$ is the projection of $U_{t,\xi}$ and $U_{t,\xi}$ is the solution of the RDE (43) which is equivalent to an Itô type SDE from an extrinsic point of view, by Theorem 14 we can see that as a process on M the law of the G-Brownian motion $X_{t,\xi}$ is characterized by the nonlinear parabolic PDE (46).

Example 5 When G is given by a functional of trace, as in Example 4 (1), (2), the generating PDE (46) takes a more explicit form in terms of the Laplace-Beltrami operator Δ_M on M. This is due to the fact that

$$\Delta_M = \text{tr(Hess)}.$$

For instance, if $G(A) = \frac{1}{2}\text{tr}(A)$, then (46) becomes the classical heat equation on M:

$$\frac{\partial u}{\partial t} - \frac{1}{2}\Delta_M u = 0,$$

which governs the law of classical Brownian motion on M (see [12, 13]). If G is given by

$$G(A) = \frac{1}{2}(\mu(\text{tr}A)^+ - \lambda(\text{tr}A)^-),$$

where $0 \leqslant \lambda < \mu$, then (46) becomes

$$\frac{\partial u}{\partial t} - \frac{1}{2}(\mu(\Delta_M u)^+ - \lambda(\Delta_M u)^-) = 0.$$

It is a generalization of the one-dimensional Barenblatt equation to higher dimensions in a Riemannian geometric setting.

As pointed out before, as a process the G-Brownian motion $X_{t,\xi}$ on M depends on the initial orthonormal frame ξ and hence there is not a canonical choice of a particular one. However, if we consider the path space $W(M) = C([0, 1]; M)$, then for each $x \in M$, it is possible to define a canonical sublinear expectation $\mathbb{E}_x$ on the space $\mathscr{H}(M)$ of functionals on $W(M)$ of the form

$$f(x_{t_1}, \cdots, x_{t_n}),$$

where $0 \leqslant t_1 < \cdots < t_n \leqslant 1$ and $f \in C_{Lip}(M)$, such that under $\mathbb{E}_x$ the law of the coordinate process is characterized by the PDE (46) with $\mathbb{E}_x[\varphi(x_0)] = \varphi(x)$ for any $\varphi \in C_{Lip}(M)$.

To see this, we will define $\mathbb{E}_x$ explicitly. We use $u_\varphi(t, x)$ to denote the solution of (46), emphasizing the dependence on φ. For a functional of the form $f(x_t)$, we simply define

$$\mathbb{E}_x[f(x_t)] := u_f(t, x).$$

For a functional of the form $f(x_s, x_t)$, $\mathbb{E}_x f(x_s, x_t)$ should be defined by $\mathbb{E}^G[f(X_{s,\xi}, X_{t,\xi})]$, where $X_{t,\xi}$ is a G-Brownian motion on M with respect to an initial orthonormal frame $\xi \in \mathcal{O}_x(M)$. Similar to the proof of Theorem 11 we know that

$$\begin{aligned}
\mathbb{E}^G[f(X_{s,\xi}, X_{t,\xi})] &= \mathbb{E}^G[\mathbb{E}^G[f(X_{s,\xi}, X_{t,\xi})|\Omega_s]] \\
&= \mathbb{E}^G[\mathbb{E}^G[f(\pi(U_{s,\xi}), \pi(U_{t,\xi}))|\Omega_s]] \\
&= \mathbb{E}^G[\mathbb{E}^G[f(\pi(\eta), X_{t-s,\eta})]|_{\eta=U_{s,\xi}}].
\end{aligned}$$

But since the law of $X_{t-s,\eta}$ does not depend on the initial orthonormal frame η, we obtain that

$$\mathbb{E}^G[f(\pi(\eta), X_{t-s,\eta})]|_{\eta=U_{s,\xi}} = u_{f(X_{s,\xi},\cdot)}(t-s, X_{s,\xi}).$$

Therefore, we define

$$\mathbb{E}_x[f(x_s, x_t)] := \mathbb{E}^G[f(X_{s,\xi}, X_{t,\xi})] = u_g(s, x),$$

where

$$g(y) := u_{f(y,\cdot)}(t-s, y), \quad y \in M.$$

Inductively, assume that

$$u_f^{(n)}(t_1, \cdots, t_n, x) = \mathbb{E}_x[f(x_{t_1}, \cdots, x_{t_n})]$$

is already defined. For a functional of the form $f(x_{t_1}, \cdots, x_{t_{n+1}})$, define

$$\mathbb{E}_x[f(x_{t_1}, \cdots, x_{t_{n+1}})] := u_g(t_1, x),$$

where

$$g(y) := u_{f(y,\cdot,\cdots,\cdot)}^{(n)}(t_2 - t_1, \cdots, t_{n+1} - t_1, y), \quad y \in M.$$

Then $\mathbb{E}_x$ is the desired sublinear expectation on $\mathcal{H}(M)$.

Remark 8 As we've pointed out before, for noncompact Riemannian manifolds, the RDE (43) may possibly explode at some finite time and so may the corresponding *G*-Brownian motion as well. An interesting question is the study of explosion criterion. It might depend on the curvature and topology of the Riemannian manifold.

On the other hand, for those *G*-functions with the same invariant group, they may have some special features in common; while for those with different invariant groups, their structure should be very different. The study of classification of *G*-functions in terms of the invariant group is interesting, and it might give us some hints on generalizing our results to the case when $I(G) \neq O(d)$. We believe that in some cases it is still possible to construct a proper frame bundle with structure group $I(G)$ on which we can apply similar techniques in this section. But in some extreme cases, for instance when $I(G)$ is a finite group as in Example 3, it seems difficult to proceed along this direction unless we have a globally defined frame field over the Riemannian manifold M, which is usually not true. We might need some very different methods for those extreme cases.

Conclusion

To summarize, the motivation of the present article is to understand nonlinear diffusion processes from the viewpoint of rough path theory, and in particular, to understand the intrinsic and geometric nature of the nonlinear heat flow. Along this direction, the pathwise approach under the framework of rough paths seems to be a right tool, since it is not natural to use Itô's formulation from the geometric point of view, and localization technique from Itô's perspective is not well understood in the nonlinear situation.

In this article, we study the geometric rough path nature of sample paths of *G*-Brownian motion, which enables us to establish differential equations driven by *G*-Brownian motion in a pathwise sense. Furthermore, we establish the fundamental relation between the two types of differential equations driven by *G*-Brownian motion. Such relation enables us to develop nonlinear diffusion processes on a differentiable manifold and *G*-Brownian motion on a Riemannian manifold from the pathwise point of view. The pathwise approach seems to be quite natural if we aim at understanding the intrinsic and geometric nature of nonlinear diffusion processes. Finally, we establish the generating nonlinear heat equation of *G*-Brownian motion on a Riemannian manifold and construct the associated canonical sublinear expectation on the path space for a class of nonlinear *G*-functions whose invariant group is the orthogonal group. Although such class of *G*-functions contains the most important and interesting examples so far, it remains an open problem for the general case (in particular, the singular case) where the use of orthogonal frame bundle is no longer applicable.

Acknowledgements The authors wish to thank Professor Shige Peng for so many valuable suggestions on the present article. The authors are supported by the Oxford-Man Institute at the University of Oxford.

References

1. D. Azagra, J. Ferrera, B. Sanz, Viscosity solutions to second order partial differential equations on riemannian manifolds. J. Differ. Equ. **245**(2), 307–336 (2009)
2. S. Chern, W. Chen, K. Lam, *Lectures on Differential Geometry* (World Scientific, Singapore, 1999)
3. G. Choquet, Theory of capacities. Ann. Inst. Fourier. **5**, 87 (1953)
4. M.G. Crandall, H. Ishii, P.L. Lions, User's guide to viscosity solutions of second order partial differential equations. Bull. Am. Math. Soc. **27**(1), 1–67 (1992)
5. G. de Rham, *Riemannian Manifolds* (Springer, New York, 1984)
6. C. Dellacherie, *Capacités et Processus Stochastiques* (Springer, New York, 1972)
7. L. Denis, M. Hu, S. Peng, Function spaces and capacity related to a sublinear expectation: application to G-Brownian motion paths. Potential Anal. **34**(2), 139–161 (2011)
8. K.D. Elworthy, *Stochastic Differential Equations on Manifolds* (Springer, New York, 1998)
9. P.K. Friz, N.B. Victoir, *Multidimensional Stochastic Processes as Rough Paths: Theory and Applications* (Cambridge University Press, Cambridge, 2010)
10. F. Gao, Pathwise properties and homeomorphic flows for stochastic differential equations driven by G-Brownian motion. Stochast. Process. Appl. **119**(10), 3356–3382 (2009)
11. B.M. Hambly, T.J. Lyons, Stochastic area for Brownian motion on the sierpinski gasket. Ann. Probab. **26**(1), 132–148 (1998)
12. E.P. Hsu, *Stochastic Analysis on Manifolds* (American Mathematical Society, Providence, 2002)
13. N. Ikeda, S.Watanabe, *Stochastic Differential Equations and Diffusion Processes* (North-Holland, Amsterdam, 1989)
14. M. Kac, On distributions of certain wiener functionals. Trans. Am. Math. Soc. **65**(1), 1–13 (1949)
15. I.A. Karatzas, S.E. Shreve, *Brownian Motion and Stochastic Calculus* (Springer, New York, 1991)
16. S. Kobayashi, K. Nomizu, *Foundations of Differential Geometry, I and II* (Interscience, New York, 1963/1969)
17. T.J. Lyons, Differential equations driven by rough signals. Rev. Mat. Iberoam. **14**(2), 215–310 (1998)
18. T.J. Lyons, Z. Qian, *System Control and Rough Paths* (Oxford University Press, Oxford, 2002)
19. T.J. Lyons, C. Michael, T. Lévy, *Differential Equations Driven by Rough Paths* (Springer, Berlin, 2007)
20. K. Nomizu, H. Ozeki, The existence of complete riemannian metrics. Proc. Am. Math. Soc. **12**(6), 889–891 (1961)
21. É. Pardoux, S. Peng, Adapted solution of a backward stochastic differential equation. Syst. Control Lett. **14**(1), 55–61 (1990)
22. É. Pardoux, S. Peng, Backward stochastic differential equations and quasilinear parabolic partial differential equations, in *Stochastic Partial Differential Equations and Their Applications* (Springer, New York, 1992), pp. 200–217
23. É. Pardoux, S. Peng, Backward doubly stochastic differential equations and systems of quasilinear SPDEs. Probab. Theory Relat. Fields. **98**(2), 209–227 (1994)
24. S. Peng, Backward SDE and related g-expectation, in *Pitman Research Notes in Mathematics series*, vol. 364, ed. by N. El Karoui, L. Mazliak (1997), pp. 141–159

25. S. Peng, G-expectation, G-Brownian motion and related stochastic calculus of Itô type. Stochast. Anal. Appl. **2**, 541–567 (2007)
26. S. Peng, Multi-dimensional G-Brownian motion and related stochastic calculus under G-expectation. Stochast. Process. Appl. **118**(12), 2223–2253 (2008)
27. S. Peng, Nonlinear expectations and stochastic calculus under uncertainty. Preprint, arXiv: 1002.4546 (2010)
28. E. Wong, M. Zakai, On the relation between ordinary and stochastic differential equations. Int. J. Eng. Sci. **3**(2), 213–229 (1965)
29. K. Yosida, *Functional Analysis* (Springer, New York, 1980)

Flows Driven by Banach Space-Valued Rough Paths

Ismaël Bailleul

Abstract We show in this note how the machinery of $\mathscr{C}^1$-approximate flows devised in the work *Flows driven by rough paths*, and applied there to reprove and extend most of the results on Banach space-valued rough differential equations driven by a finite dimensional rough path can be used to deal with rough differential equations driven by an infinite dimensional Banach space-valued weak geometric Hölder p-rough paths, for any $p > 2$, giving back Lyons' theory in its full force in a simple way.

Keywords Flows • Infinite dimensional rough paths • Rough differential equations

1 Introduction

1.1 Rough Differential Equations

The theory of rough paths invented in the mid 90' by Lyons [1] has had several reformulations ever since, which have both widen its scope and made it more accessible. Of central importance in Lyons' theory is the fact that rough signals driving systems are not accurately described by the data of their increments; one needs information on what happens to the path between any two sampling times to understand the output of the system as a function of the driving signal. The nature of this additional information makes it necessary to introduce a rich stratified algebraic structure. Rough paths are paths with values in that structure, the increments of the initial rough signal being recorded in the first level of the associated rough path. A Riemann-type integral, with any rough path as integrator, can be defined in that setting, which enables to formulate differential equations driven by rough paths as fixed point problems to integral equations in the enriched algebraic structure. Gubinelli extracted in [2] the core of Lyons' machinery in the notion of controlled

I. Bailleul (✉)
IRMAR, 263 Av. du General Leclerc, 35042 Rennes, France
e-mail: ismael.bailleul@univ-rennes1.fr

© Springer International Publishing Switzerland 2014   195
C. Donati-Martin et al. (eds.), *Séminaire de Probabilités XLVI*, Lecture Notes
in Mathematics 2123, DOI 10.1007/978-3-319-11970-0_7

path. He put forward the essential fact that the class of integrands for which Riemann-type integrals can be defined in the context of rough paths is made up of time-dependent maps whose time increments are time-dependent linear transforms of the first level of the driving rough path, up to some smoother term. Gubinelli's framework was used in several works (e.g. [3–6]) to deal with a variety of subtle and hard problems in stochastic partial differential equations, and was one of the seeds of Hairer's impressive works [7, 8], where he lays the foundations of a new theoretical framework for the study of stochastic partial differential equations and solves some longstanding fundamental problems.

In parallel to this line of development, Friz and Victoir developed in [9, 10] a purely dynamical approach to rough differential equations, based on Davie's insightful work [11]. Let $F = (V_1, \ldots, V_\ell)$ be a finite collection of smooth enough vector fields on $\mathbb{R}^d$ and $\mathbf{X}$ be a rough path over $\mathbb{R}^\ell$. In Friz-Victoir's approach, a solution to the rough differential equation

$$dx_u = F(x_u) d\mathbf{X}_u$$

on $\mathbb{R}^d$, is the limit in uniform topology of the solutions to ordinary differential equations of the form

$$\dot{y}_u = \sum_{i=1}^{\ell} V_i(y_u) \dot{h}_u^i, \tag{1}$$

for some smooth control h, when the rough path canonically associated to h converges in a rough path sense to $\mathbf{X}$. The technical core of their approach takes the form of some estimates on some high order Euler expansion to the solution to Eq. (1) involving only some p-variation norm of h. However, both [9, 11] obtain dimension-dependent estimates which prevent a direct extension of their results to investigate rough differential equations driven by infinite dimensional rough paths, as can be done in Lyons or Gubinelli's framework.

A different approach to rough differential equations was proposed recently in the work "*Flows driven by rough paths*", [12]. Given some finite collection $F = (V_1, \ldots, V_\ell)$ of smooth enough vector fields on a Banach space U, a flow $(\varphi_{ts})_{0 \leqslant s \leqslant t \leqslant T}$ of maps from U to itself is said to be a solution of the equation

$$d\varphi_s = F\mathbf{X}(ds) \tag{2}$$

if its increments can accurately be described by some awaited Euler-type expansion, as in Davie or Friz-Victoir's approach. The introduction of the notion of $\mathscr{C}^1$-approximate flow, and the proof that a $\mathscr{C}^1$-approximate flow determines a unique flow close enough to it, made it possible to recover and extend most of the basic results on Banach space valued rough differential equations driven by a finite dimensional rough signal, under optimal regularity assumptions on the vector fields

V_i. Existence, well-posedness, Taylor expansion and non-explosion under linear growth conditions of the vector fields were proved in [12].

We show in the present note that the machinery of $\mathscr{C}^1$-approximate flows applies equally well to prove well-posedness results for Banach space valued rough differential equations driven by infinite dimension rough signals, extending to that setting Davie-Friz-Victoir's Euler estimates as a consequence. This infinite dimensional version of Davie's generalized estimate was proved recently in [13] by adapting the method of proof of Davie [11] and Friz-Victoir [9], and using Lyons' universal limit theorem. Our method yields both Lyons' theorem and Davie's estimate at a time.

We recall in Sect. 1.2 how to construct flows from $\mathscr{C}^1$-approximate flows [12], and construct in Sect. 2 a $\mathscr{C}^1$-approximate flow with the awaited Euler expansion. Well-posedness of the rough differential Eq. (2) on flows and Davie's estimate are direct consequences of the result on $\mathscr{C}^1$-approximate flows recalled in Sect. 1.2.

The construction of a solution flow to Eq. (2) done in Sect. 2 requires three ingredients, introduced in Sects. 1.2–1.4. Throughout that work, we denote U and V two Banach spaces. Given a non-negative real number γ, we denote by $\mathscr{C}^\gamma(U, U)$ the set of γ-Lipschitz maps from U to itself, in the sense of Stein, equipped with its natural norm (see [1] or [14]). The notation $\mathscr{C}^\gamma(U)$ stands for the set of γ-Lipschitz real valued functions on U. We use the symbol c to denote a constant whose value is unimportant and may change from place to place.

1.2 Flows and Approximate Flows

Recall that a *flow* on U is a family $\left(\varphi_{ts}\right)_{0 \leqslant s \leqslant t \leqslant T}$ of maps from U to itself such that $\varphi_{ss} = \mathrm{Id}$, for all $0 \leqslant s \leqslant T$, and $\varphi_{tu} \circ \varphi_{us} = \varphi_{ts}$, for all $0 \leqslant s \leqslant u \leqslant t \leqslant T$. The notion of $\mathscr{C}^1$-approximate flow introduced in [12] provides a convenient tool for constructing flows from families of maps which are almost flows. It is an elaboration upon Lyons' almost-multiplicative functionals, as understood by Gubinelli [2] or Feyel-de la Pradelle [15, 16], with their sewing lemmas. We state it here under a simple form which will be enough to illustrate our purpose in Sect. 2.

Definition 1 A $\mathscr{C}^1$-**approximate flow** is a family $\left(\mu_{ts}\right)_{0 \leqslant s \leqslant t \leqslant T}$ of $\mathscr{C}^2$ maps from U to itself, depending continuously on (s, t) in the topology of uniform convergence, such that

$$\left\| \mu_{ts} - Id \right\|_{\mathscr{C}^2} = o_{t-s}(1), \tag{3}$$

and there exists some positive constants c_1 and $a > 1$, such that the inequality

$$\left\| \mu_{tu} \circ \mu_{us} - \mu_{ts} \right\|_{\mathscr{C}^1} \leqslant c_1 \left| t - s \right|^a \tag{4}$$

holds for any $0 \leqslant s \leqslant u \leqslant t \leqslant T$.

Given a partition $\pi_{ts} = \{s = s_0 < s_1 < \cdots < s_{n-1} < s_n = t\}$ of $(s, t) \subset [0, T]$, set

$$\mu_{\pi_{ts}} = \mu_{s_n s_{n-1}} \circ \cdots \circ \mu_{s_1 s_0}.$$

The following statement, proved in [12], provides a flexible tool for constructing flows from $\mathscr{C}^1$-approximate flows, while generalizing the above mentioned sewing lemmas to the non-commutative setting of $\mathscr{C}^2$ maps on U.

Theorem 1 (Constructing Flows on a Banach Space) *A $\mathscr{C}^1$-approximate flow defines a unique flow $(\varphi_{ts})_{0 \leqslant s \leqslant t \leqslant T}$ on U such that the inequality*

$$\left\| \varphi_{ts} - \mu_{ts} \right\|_\infty \leqslant c |t - s|^a$$

holds for some positive constant c, for all $0 \leqslant s \leqslant t \leqslant T$ sufficiently close, say $|t - s| \leqslant \delta$; this flow satisfies the inequality

$$\left\| \varphi_{ts} - \mu_{\pi_{ts}} \right\|_\infty \leqslant c_1^2 \, T \, |\pi_{ts}|^{a-1} \tag{5}$$

for any partition π_{ts} of any interval $(s, t) \subset [0, T]$, with mesh $|\pi_{ts}| \leqslant \delta$.

A slightly more elaborated form of this theorem was used in [12] to recover and extend most of the basic results on Banach-space valued rough differential equations driven by a finite dimensional rough signal, under optimal regularity condition on the driving vector fields. Existence, well-posedness, high order Euler expansion and non-explosion under linear growth conditions on the driving vector fields were proved. It was also used in [17] to extend these results to the setting of rough differential equations driven by path-dependent vector fields.

1.3   Infinite Dimensional Rough Paths

We recall briefly in this section the setting of infinite dimensional rough paths, referring to [1, 18] for more material on this subject.

For $n \geqslant 1$, let $T_a^n(V)$ stand for the truncated algebraic tensor algebra, isomorphic to $\bigoplus_{i=0}^n V^{\otimes_a i}$, where $V^{\otimes_a i}$ stands for the algebraic tensor product of V with itself i times, and $\mathbb{R}^{\otimes 0}$ is identified to $\mathbb{R}$. Multiplication of two elements $\mathbf{e}$ and $\mathbf{e}'$ of $T^n(V)$ is simply denoted by $\mathbf{e}\mathbf{e}'$. There exists norms on $T_a^n(V)$ which satisfy the inequality

$$\|\mathbf{e}\mathbf{e}'\| \leqslant \|\mathbf{e}\| \|\mathbf{e}'\|,$$

for any pair $(\mathbf{e}, \mathbf{e}')$ of elements of $T^n(V)$, with $\|\mathbf{e}\| = |\mathbf{e}|$, if $\mathbf{e} \in V \subset T^n(V)$. Select any such norm and denote by $T^n(V)$ the completion of $T_a^n(V)$ with respect to that norm. We have $T^n(V) = \bigoplus_{i=0}^n V^{\otimes i}$, where $V^{\otimes i}$ stands for the completion of $V^{\otimes_a i}$

with respect to the restriction to $V^{\otimes_a i}$ of the norm $\|\cdot\|$, and $V^{\otimes 1} = V$, since V is complete.

Given $0 \leqslant k \leqslant n$, denote by π_k the projection map from $T^n(V)$ to $V^{\otimes k}$, and set $T_1^n(V) = \{\mathbf{e} \in T^n(V) \,;\, \pi_0(\mathbf{e}) = 1\}$, with a similar definition of $T_0^n(V)$. The exponential map is a polynomial diffeomorphism from $T_0^n(V)$ to $T_1^n(V)$, with inverse map the logarithm. Let $\mathfrak{g}^n(V)$ stand for the Lie algebra in $T^n(V)$ generated by V, and write $\mathscr{G}^n(V)$ for $\exp\big(\mathfrak{g}^n(V)\big)$.

In this setting, and for $2 \leqslant p$, a V-**valued weak geometric Hölder** p-**rough path** is a $\mathscr{G}^{[p]}(V)$-valued path $\mathbf{X}$, parametrized by some time interval $[0, T]$, such that

$$\sup_{0 \leqslant s < t \leqslant T} \frac{\left\| \pi_i \mathbf{X}_{ts} \right\|}{|t - s|^{\frac{i}{p}}} < \infty, \tag{6}$$

for all $i = 1 \ldots [p]$, where we set $\mathbf{X}_{ts} = \mathbf{X}_s^{-1} \mathbf{X}_t$. We define the norm of $\mathbf{X}$ to be

$$(\!|\mathbf{X}|\!) = \max_{i=1\ldots[p]} \sup_{0 \leqslant s < t \leqslant T} \frac{\left\| \pi_i \mathbf{X}_{ts} \right\|}{|t - s|^{\frac{i}{p}}}, \tag{7}$$

and a distance $d(\mathbf{X}, \mathbf{Y}) = (\!|\mathbf{X} - \mathbf{Y}|\!)$, on the set of V-valued weak geometric Hölder p-rough path. This definition depends on the choice of tensor norm made above.

1.4 Differential Operators

Let $\gamma > 0$ be given, and F be a continuous linear map from V to $\mathscr{C}^\gamma(U, U)$. For any $v \in V$, we identify the $\mathscr{C}^\gamma$ vector field F(v) on U with the first order differential operator

$$g \in \mathscr{C}^1(U) \mapsto (D.g)\big(\mathrm{F}(v)(\cdot)\big) \in \mathscr{C}^0(U);$$

in those terms, we recover the vector field as F(v)Id. The map F is extended to the algebraic setting of $T^{[\gamma]}(V)$ by setting $\mathrm{F}^\otimes(1) = \mathrm{Id} : \mathscr{C}^0(U) \mapsto \mathscr{C}^0(U)$, and defining $\mathrm{F}^\otimes(v_1 \otimes \cdots \otimes v_k)$, for all $1 \leqslant k \leqslant [\gamma]$ and $v_1 \otimes \cdots \otimes v_k \in V^{\otimes k}$, as the kth-order differential operator from $\mathscr{C}^k(U)$ to $\mathscr{C}^0(U)$, defined by the formula

$$\mathrm{F}^\otimes(v_1 \otimes \cdots \otimes v_k) = \mathrm{F}(v_1) \cdots \mathrm{F}(v_k),$$

and by requiring linearity. The choice of a tensor norm on $T^{[\gamma]}(V)$ ensures that this definition extends continuously to the restriction of the completed space $V^{\otimes k}$ as a continuous linear map from $V^{\otimes k}$ to the set of kth-order differential operators from $\mathscr{C}^k(U)$ to $\mathscr{C}^0(U)$, endowed with the operator norm. So, for $\mathbf{e} \in \mathfrak{g}^n(V)$ the differential operator F(**e**) is actually a first order differential operator, and we have

$$F^{\otimes}(\mathbf{e})\,F^{\otimes}(\mathbf{e}') = F^{\otimes}(\mathbf{e}\mathbf{e}'), \tag{8}$$

for all $\mathbf{e}, \mathbf{e}' \in T^n(V)$. In particular, we have

$$F^{\otimes}([v, v']) = [F(v), F(v')],$$

for any vectors v, v' of V, where the bracket on the left hand side is the bracket in $T^n(V)$ and the bracket on the right hand side is the bracket on vector fields.

2 Flows Driven by Banach Space-Valued Rough Paths

We show in this section how to associate a $\mathscr{C}^1$-approximate flow to some $\mathscr{G}^{[p]}(V)$-valued weak geometric Hölder p-rough path and some V-dependent vector field F as in Sect. 1.4, in such a way that it has the awaited Euler expansion. A solution flow to the equation

$$d\varphi_s = F\mathbf{X}(ds) \tag{9}$$

will be defined as a flow uniformly close to the $\mathscr{C}^1$-approximate flow. Well-posedness of the rough differential equation (9) on flows will follow as a direct consequence of Theorem 1.

Let $2 \leqslant p$ be given, together with a $\mathscr{G}^{[p]}(V)$-valued weak-geometric Hölder p-rough path $\mathbf{X}$, defined on some time interval $[0, T]$, and some continuous linear map F from V to $\mathscr{C}^{[p]+1}(U, U)$. For any $0 \leqslant s \leqslant t \leqslant T$, denote by $\Lambda_{ts} \in T^{[p]}(V)$ the logarithm of $\mathbf{X}_{ts}$, and let μ_{ts} stand for the well-defined time 1 map associated with the ordinary differential equation

$$\dot{y}_u = F^{\otimes}(\Lambda_{ts})(y_u), \quad 0 \leqslant u \leqslant 1. \tag{10}$$

This equation is indeed an ordinary differential equation as Λ_{ts} is an element of $\mathfrak{g}^{[p]}(V)$. It is a consequence of classical results from ordinary differential equations, and the definition of the norm on the space of weak-geometric Hölder p-rough paths, that the solution map $(r, x) \mapsto y_r$, with $y_0 = x$, depends continuously on $((s, t), \mathbf{X})$, and satisfies the following basic estimate. We have

$$\left\| y_r - \mathrm{Id} \right\|_{\mathscr{C}^1} \leqslant c \left(1 + (\!(\mathbf{X})\!)^{[p]} \right) |t - s|^{\frac{1}{p}}, \quad 0 \leqslant r \leqslant 1. \tag{11}$$

The following elementary proposition refines part of the above estimate, and is our basic step for studying flows driven by Banach space-valued weak geometric Hölder p-rough paths.

Proposition 1 *There exists a positive constant c, depending only on the data of the problem, such that the inequality*

$$\left\| f \circ \mu_{ts} - F^{\otimes}(\mathbf{X}_{ts}) f \right\|_{\infty} \leqslant c \left(1 + (\|\mathbf{X}\|)^{[p]} \right) \| f \|_{[p]+1} \, |t - s|^{\frac{[p]+1}{p}} \tag{12}$$

holds for any $f \in \mathscr{C}^{[p]+1}(U)$.

This proposition was first proved under this form in [12], following the same pattern of proof as below, for rough differential equations driven by a finite dimensional rough path. The above infinite dimensional version of it, with $f = \mathrm{Id}$, was proved independently in [13] using the same method.

The proof of this proposition and the following one are based on the elementary identity (13) below, obtained by applying repeatedly the identity

$$f(y_r) = f(x) + \int_0^r \left(F^{\otimes}(\Lambda_{ts}) f \right)(y_u) \, du, \quad 0 \leqslant r \leqslant 1,$$

together with the morphism property (8), and by separating the terms according to their size in $|t - s|$. Set $\Delta_n := \{(s_1, \ldots, s_n) \in [0, T]^n \, ; \, s_1 \leqslant \cdots \leqslant s_n\}$, for $2 \leqslant n \leqslant [p]$. The summation below is above indices k_i with range in $[\![1, [p]]\!]$; we also write ds for $ds_n \ldots ds_1$. For a γ-Lipschitz function f, we have

$$\begin{aligned}
f\left(\mu_{ts}(x)\right) = f(x) &+ \sum_{\ell=1}^{n} \frac{1}{\ell!} \sum_{k_1 + \cdots + k_\ell \leqslant [p]} \left(F^{\otimes}\left(\pi_{k_\ell}\Lambda_{ts} \cdots \pi_{k_1}\Lambda_{ts}\right) f \right)(x) \\
&+ \sum_{k_1 + \cdots + k_n \leqslant [p]} \int_{\Delta_n} \left\{ \left(F^{\otimes}\left(\pi_{k_n}\Lambda_{ts} \cdots \pi_{k_1}\Lambda_{ts}\right) f \right)(y_{s_n}) \right. \\
&\qquad\qquad\qquad\left. - \left(F^{\otimes}\left(\pi_{k_n}\Lambda_{ts} \cdots \pi_{k_1}\Lambda_{ts}\right) f \right)(x) \right\} ds \\
&+ \sum_{\ell=1}^{n} \frac{1}{\ell!} \sum_{k_1 + \cdots + k_\ell \geqslant [p]+1} \left(F^{\otimes}\left(\pi_{k_\ell}\Lambda_{ts} \cdots \pi_{k_1}\Lambda_{ts}\right) f \right)(x) \\
&+ \sum_{k_1 + \cdots + k_n \geqslant [p]+1} \int_{\Delta_n} \left\{ \left(F^{\otimes}\left(\pi_{k_n}\Lambda_{ts} \cdots \pi_{k_1}\Lambda_{ts}\right) f \right)(y_{s_n}) \right. \\
&\qquad\qquad\qquad\left. - \left(F^{\otimes}\left(\pi_{k_n}\Lambda_{ts} \cdots \pi_{k_1}\Lambda_{ts}\right) f \right)(x) \right\} ds.
\end{aligned} \tag{13}$$

We denote by $\epsilon_{ts}^{f\,;\,n}(x)$ the sum of the last two lines, made up of terms of size at least $|t - s|^{[p]+1}$. In the case where $n = [p]$, the terms in the second line involve only indices k_j with $k_j = 1$, so the elementary estimate (11) can be used to control the increment in the integral, showing that this second line is of order $|t - s|^{\frac{[p]+1}{p}}$, in infinite norm, as the maps $F^{\otimes}\left(\pi_{k_n}\boldsymbol{\Lambda}_{ts}\cdots\pi_{k_1}\boldsymbol{\Lambda}_{ts}\right)f$ are $\mathscr{C}_b^1$; we include it in the remainder $\epsilon_{ts}^{f\,;\,[p]}(x)$.

Proof (Proof of Proposition 1) Applying the above formula for $n = [p]$, together with the fact that $\exp\left(\boldsymbol{\Lambda}_{ts}\right) = \mathbf{X}_{ts}$, we get the identity

$$f\left(\mu_{ts}(x)\right) = \left(F^{\otimes}\left(\mathbf{X}_{ts}\right)f\right)(x) + \epsilon_{ts}^{f\,;\,[p]}(x).$$

It is clear on the formula for $\epsilon_{ts}^{f\,;\,[p]}(x)$ that its absolute value is bounded above by a constant multiple of $\left(1 + (\mathbf{X})^{[p]}\right)|t-s|^{\frac{[p]+1}{p}}$, for a constant depending only on the data of the problem and f as in (12).

A further look at formula (13) makes it clear that if $2 \leqslant n \leqslant [p]$ and f is $\mathscr{C}_b^{n+1}$, the estimate

$$\left\|\epsilon_{ts}^{f\,;\,n}\right\|_{\mathscr{C}^1} \leqslant c\left(1 + (\mathbf{X})^{[p]}\right)\|f\|_{\mathscr{C}^{n+1}}|t-s|^{\frac{[p]+1}{p}}, \tag{14}$$

holds as a consequence of formula (11), for a constant c depending only on the data of the problem.

Proposition 2 *The family of maps $\left(\mu_{ts}\right)_{0\leqslant s\leqslant t\leqslant T}$ is a $\mathscr{C}^1$-approximate flow.*

Proof As F is $\mathscr{C}^{[p]+1}$ as a function on U, with $[p] + 1 \geqslant 3$, the inequality $\left\|\mu_{ts} - \mathrm{Id}\right\|_{\mathscr{C}^2} = o_{t-s}(1)$, is given by classical results on ordinary differential equations; we turn to proving the $\mathscr{C}^1$-approximate flow property (4). of $\left(\mu_{ts}\right)_{0\leqslant s\leqslant t\leqslant T}$. We first use for that purpose formula (13) to write

$$\mu_{tu}\left(\mu_{us}(x)\right) = \left(F^{\otimes}\left(\mathbf{X}_{tu}\right)\mathrm{Id}\right)\left(\mu_{us}(x)\right) + \epsilon_{tu}^{\mathrm{Id}\,;\,[p]}\left(\mu_{us}(x)\right)$$

$$= \mu_{us}(x) + \sum_{m=1}^{[p]}\left(F^{\otimes}\left(\pi_m\mathbf{X}_{tu}\right)\mathrm{Id}\right)\left(\mu_{us}(x)\right) + \epsilon_{tu}^{\mathrm{Id}\,;\,[p]}\left(\mu_{us}(x)\right). \tag{15}$$

The remainder $\epsilon_{tu}^{\mathrm{Id}\,;\,[p]}\left(\mu_{us}(x)\right)$ has a $\mathscr{C}^1$-norm bounded above by $c\left(1 + (\mathbf{X})^{[p]}\right)^2|t - u|^{\frac{[p]+1}{p}}$, due (14) and inequality (11).

To deal with the term $\left(F^\otimes(\pi_m \mathbf{X}_{tu})\mathrm{Id}\right)\left(\mu_{us}(x)\right)$, we use formula (13) with $n = [p] - m$. Writing ds for $ds_{[p]-m} \ldots ds_1$, it follows that

$$
\begin{aligned}
&\left(F^\otimes(\pi_m \mathbf{X}_{tu})\mathrm{Id}\right)\left(\mu_{us}(x)\right) \\
&= \left(F^\otimes(\pi_m \mathbf{X}_{tu})\mathrm{Id}\right)(x) \\
&\quad + \sum_{\ell=1}^{[p]-m} \frac{1}{\ell!} \sum_{k_1+\cdots+k_\ell \le [p]} \left(F^\otimes(\pi_{k_\ell}\boldsymbol{\Lambda}_{us} \cdots \pi_{k_1}\boldsymbol{\Lambda}_{us}\, \pi_m \mathbf{X}_{tu})\mathrm{Id}\right)(x) \\
&\quad + \epsilon_{us}^{\bullet\,;\,p-m}(x) \\
&\quad + \sum \int_{\Delta_{[p]-m}} \Big\{ \left(F^\otimes(\pi_{k_{[p]-m}}\boldsymbol{\Lambda}_{us} \cdots \pi_{k_1}\boldsymbol{\Lambda}_{us}\, \pi_m \mathbf{X}_{tu})\mathrm{Id}\right)\left(y_{s_{[p]-m}}\right) \\
&\qquad\qquad - \left(F^\otimes(\pi_{k_\ell}\boldsymbol{\Lambda}_{us} \cdots \pi_{k_1}\boldsymbol{\Lambda}_{us}\, \pi_m \mathbf{X}_{tu})\mathrm{Id}\right)(x) \Big\}\, ds,
\end{aligned}
\tag{16}
$$

where the last sum is over the set $\{k_1 + \cdots + k_{[p]-m} \le [p]\}$ of indices. The notation $\bullet$ in the above identity stands for the $\mathscr{C}_b^{[p]+2-m}$ function $F^\otimes(\pi_m \mathbf{X}_{tu})\mathrm{Id}$ on U; it has $\mathscr{C}^1$-norm controlled by (14). It is then straightforward to use (11) to bound above the $\mathscr{C}^1$-norm of the third line in Eq. (16) by a constant multiple of $\left(1 + (\!|\mathbf{X}|\!)^{[p]}\right)|t - s|^{\frac{[p]+1}{p}}$. Writing

$$
\mu_{us}(x) = \left(F^\otimes(\mathbf{X}_{us})\mathrm{Id}\right)(x) + \epsilon_{us}^{\mathrm{Id}\,;\,[p]}(x),
$$

and using the identities $\exp\left(\boldsymbol{\Lambda}_{us}\right) = \mathbf{X}_{us}$ and $\mathbf{X}_{ts} = \mathbf{X}_{us}\mathbf{X}_{tu}$, we see that

$$
\mu_{tu}\left(\mu_{us}(x)\right) = \mu_{ts}(x) + \epsilon_{ts}(x),
$$

with a remainder

$$
\|\epsilon_{ts}\|_{\mathscr{C}^1} \le c\left(1 + (\!|\mathbf{X}|\!)^{[p]}\right)|t - s|^{\frac{[p]+1}{p}}.
\tag{17}
$$

In view of Proposition 1, the following definition is to be thought of as an analogue of Davie's definition [11] of a solution to a classical rough differential equation, in terms of Euler expansion.

Definition 2 A *flow* $(\varphi_{ts}\,;\, 0 \le s \le t \le T)$ is said to *solve the rough differential equation*

$$
d\varphi = F\,\mathbf{X}(dt)
\tag{18}
$$

if there exists a constant $a > 1$ independent of $\mathbf{X}$ and two possibly $\mathbf{X}$-dependent positive constants δ and c such that

$$\|\varphi_{ts} - \mu_{ts}\|_\infty \leq c\,|t - s|^a \tag{19}$$

holds for all $0 \leq s \leq t \leq T$ with $t - s \leq \delta$.

It is clear on this definition that if $x \in U$ and $(\varphi_{ts})_{0 \leq s \leq t \leq T}$ is a solution flow to Eq. (18), then the trajectory $(\varphi_{t0}(x))_{0 \leq t \leq T}$ is the first level of a solution to the classical rough differential equation $dx_t = \mathbf{F}d\mathbf{X}_t$, in Lyons' sense. The following well-posedness result follows directly from Theorem 1 and Proposition 2.

Theorem 2 *Suppose* $\mathbf{F}$ *is a continuous linear map from* V *to* $\mathscr{C}^{[p]+1}(U, U)$; *extend it as in Sect. 1.4 into a differential operator. Then the rough differential equation*

$$d\varphi = \mathbf{F}\,\mathbf{X}(dt)$$

has a unique solution flow; it depends continuously on $\big((s, t), \mathbf{X}\big)$.

Proof With the notations of Definition 1, identity (17) means that Eq. (4) holds with $c_1 = c\Big(1 + (\!|\mathbf{X}|\!)^{[p]}\Big)$ and $a = \frac{[p]+1}{p}$. So Theorem 1 ensures the existence of a unique flow $(\varphi_{ts})_{0 \leq s \leq t \leq T}$ close enough to $(\mu_{ts})_{0 \leq s \leq t \leq T}$; it further satisfies the inequality

$$\left\|\varphi_{ts} - \mu_{\pi_{ts}}\right\|_\infty \leq c\Big(1 + (\!|\mathbf{X}|\!)^{[p]}\Big)^2 T\,\big|\pi_{ts}\big|^{\frac{1}{p}}, \tag{20}$$

for any partition π_{ts} of $(s, t) \subset [0, T]$ of mesh $\big|\pi_{ts}\big| \leq \delta$, as a consequence of inequality (5). As these bounds are uniform in (s, t), and for $\mathbf{X}$ in a bounded set of the space of Hölder p-rough paths, and each $\mu_{\pi_{ts}}$ is a continuous function of $\big((s, t), \mathbf{X}\big)$, as a composition of continuous functions, the flow φ depends continuously on $\big((s, t), \mathbf{X}\big)$.

Davie-Friz-Victoir's estimate follows from Proposition 1 and Theorem 2 under the form

$$\left\|\varphi_{ts} - \mathbf{F}^\otimes(\mathbf{X}_{ts})\mathrm{Id}\right\|_\infty \leq c\Big(1 + (\!|\mathbf{X}|\!)^{[p]}\Big)|t - s|^{\frac{\gamma}{p}}.$$

This kind of result can be seen as a far reaching generalization of the classical works of Azencott [19], Ben Arous [20], Castell [21] and others on Stochastic Taylor expansion. A particular case of the above estimate is implicitly contained in the work [22] of Lejay, who deals with any p-rough paths, for $2 < p < 3$.

By using the refined definition of a $\mathscr{C}^1$-approximate flow given in [12] and the analogue of Theorem 1 which holds for it, the above method can be used to prove Theorem 2 under the essentially optimal condition that $\mathbf{F}$ takes values in $\mathscr{C}^\gamma(U, U)$, for any choice of $2 < p < \gamma \leq [p] + 1$. The other results proved in [12]: high order Euler expansion, non-explosion under linear growth conditions on $\mathbf{F}$, a Peano theorem on existence of a solution, can be proved in the present setting as well, by a straightforward adaptation of the notations of [12].

Acknowledgements This research was partially supported by an ANR grant "Retour post-doctorant". The author warmly thanks the UniversitÃl' de Bretagne Occidentale where part of this work was written.

References

1. T. Lyons, Differential equations driven by rough signals. Rev. Mat. Iberoam. **14**(2), 215–310 (1998)
2. M. Gubinelli, Controlling rough paths. J. Funct. Anal. **216**(1), 86–140 (2004)
3. A. Deya, M. Gubinelli, S. Tindel, Non-linear rough heat equations. Probab. Theory Relat. Fields **153**(1–2), 97–147 (2012)
4. M. Gubinelli, Rough solutions for the periodic Korteweg–de Vries equation. Commun. Pure Appl. Anal. **11**(2), 709–733 (2012)
5. F. Flandoli, M. Gubinelli, E. Priola, Well-posedness of the transport equation by stochastic perturbation. Invent. Math. **180**(1), 1–53 (2010)
6. M. Gubinelli, S. Tindel, Rough evolution equations. Ann. Probab. **38**(1), 1–75 (2010)
7. M. Hairer, Solving the KPZ equation. Ann. Maths **178**(2), 559–664 (2013)
8. M. Hairer, A theory of regularity structures. Preprint (2013)
9. P. Friz, N. Victoir, Euler estimates for rough differential equations. J. Differ. Equ. **244**(2), 388–412 (2008)
10. P. Friz, N. Victoir, *Multidimensional Stochastic Processes as Rough Paths*. CUP, Encyclopedia of Mathematics and its Applications, vol. 89. http://www.amazon.com/Multidimensional-Stochastic-Processes-Rough-Paths/dp/0521876079/ref=sr_1_1?ie=UTF8&qid=1416231458&sr=8-1&keywords=Multidimensional+Stochastic+Processes+as+Rough+Paths (2010)
11. A.M. Davie, Differential equations driven by rough paths: an approach via discrete approximation. Appl. Math. Res. Express AMRX **2**, 40 (2007)
12. I. Bailleul, Flows driven by rough paths. Preprint. arXiv:1203.0888 (2012)
13. Y. Boutaib, L.G. Gyurko, T. Lyons, D. Yang, Dimension-free estimates of rough differential equations. Preprint (2013)
14. T.J. Lyons, M. Caruana, Th. Lévy, *Differential Equations Driven by Rough Paths*. Lecture Notes in Mathematics, vol. 1908 (Springer, New York, 2007)
15. D. Feyel, A. de La Pradelle, Curvilinear integrals along enriched paths. Electron. J. Probab. **11**, 860–892 (2006)
16. D. Feyel, A. de La Pradelle, G. Mokobodzki, A non-commutative sewing lemma. Electron. Commun. Probab. **13**, 24–34 (2008)
17. I. Bailleul, Path-dependent rough differential equations. Preprint. arXiv:1309.1291 (2013)
18. T. Lyons, Z. Qian, *System Control and Rough Paths*. Oxford Mathematical Monographs. http://www.amazon.com/System-Control-Oxford-Mathematical-Monographs-ebook/dp/B000TU7360/ref=sr_1_1?ie=UTF8&qid=1416231347&sr=8-1&keywords=system+control+and+rough+paths (2002)
19. R. Azencott, Formule de Taylor stochastique et développements asymptotiques d'intégrales de Feynman. Séminaire de Probabilités, vol. XVI (1982)
20. G. Ben Arous, Flots et séries de Taylor stochastiques. Prob. Theory Relat. Fields **81**, 29–77, 1989.
21. F. Castell, Asymptotic expansion of stochastic flows. Prob. Theory Relat. Fields **96**, 225–239 (1993)
22. A. Lejay, On rough differential equations. Electron. J. Probab. **12**, 341–364 (2009)

Some Properties of Path Measures

Christian Léonard

Abstract We call any measure on a path space, a *path measure*. Some notions about path measures which appear naturally when solving the Schrödinger problem are presented and worked out in detail.

Keywords Unbounded measure • Conditional expectation • Relative entropy • Stochastic processes • Schrödinger problem

AMS classification (2010): 28A50, 60J25

1 Introduction

We call any measure on a path space, a *path measure*. Some notions about path measures which appear naturally when solving the Schrödinger problem (see (1) below and [5]) are presented and worked out in detail.

Aim of This Article

This paper is about three separate items:

1. Disintegration of an unbounded measure;
2. Basic properties of the relative entropy with respect to an unbounded measure;
3. Positive integration with respect to a Markov measure.

Although items (1) and (2) are mainly about general unbounded measures, we are motivated by their applications to path measures.

C. Léonard (✉)
Modal-X, Université Paris Ouest, Bât. G, 200 av. de la République, 92001 Nanterre, France
e-mail: christian.leonard@u-paris10.fr

© Springer International Publishing Switzerland 2014   207
C. Donati-Martin et al. (eds.), *Séminaire de Probabilités XLVI*, Lecture Notes in Mathematics 2123, DOI 10.1007/978-3-319-11970-0_8

In particular, it is shown that when Q is an unbounded path measure, some restriction must be imposed on Q for considering conditional expectations such as $Q(\cdot|X_t)$. This is the content of the notion of *conditionable path measure* which is introduced at Definition 3.

Some care is also required when working with the relative entropy with respect to an unbounded reference measure. We also give a detailed proof of the additive property of the relative entropy at Theorem 2. Indeed, we didn't find in the literature a complete proof of this well known result.

Some Notation

Let $\mathscr{X}$ be a Polish state space equipped with the corresponding Borel σ-field and $\Omega = D([0, 1], \mathscr{X})$ the space of all càdlàg (right-continuous and left-limited) paths from the unit time interval $[0, 1]$ to $\mathscr{X}$. Depending on the context, we may only consider $\Omega = C([0, 1], \mathscr{X})$, the space of all continuous paths. As usual, the σ-field on Ω is generated by the canonical process

$$X_t(\omega) := \omega_t \in \mathscr{X}, \quad \omega = (\omega_s)_{0 \leq s \leq 1} \in \Omega, \, 0 \leq t \leq 1.$$

We write $M_+(Y)$ for the set of all nonnegative measures on a space Y, and $P(Y)$ for the subset of all probability measures. Let $Q \in M_+(Y)$, the push-forward of Q by the measurable mapping $\phi : Y \to \mathscr{X}$ is denoted by $\phi_\# Q$ or $Q_\phi \in M_+(\mathscr{X})$.

Any positive measure $Q \in M_+(\Omega)$ on the path space Ω is called a *path measure*. For any subset $\mathscr{T} \subset [0, 1]$, we denote $X_{\mathscr{T}} = (X_t)_{t \in \mathscr{T}}$ and $Q_{\mathscr{T}} = (X_{\mathscr{T}})_\# Q = Q(X_{\mathscr{T}} \in \cdot) \in M_+(\Omega_{\mathscr{T}})$ the push-forward of Q by $X_{\mathscr{T}}$ on the set of positive measures on the restriction $\Omega_{\mathscr{T}}$ of Ω to $\mathscr{T}$. In particular, for each $0 \leq t \leq 1$, $Q_t = Q(X_t \in \cdot) \in M_+(\mathscr{X})$.

Motivation

Take a reference path measure $R \in M_+(\Omega)$ and consider the problem

$$H(P|R) \to \min; \qquad P \in P(\Omega) : P_0 = \mu_0, P_1 = \mu_1 \tag{1}$$

of minimizing the relative entropy

$$H(P|R) := \int_\Omega \log\left(\frac{dP}{dR}\right) dP \in (-\infty, \infty]$$

of $P \in P(\Omega)$ with respect to $R \in M_+(\Omega)$, among all path probability measures $P \in P(\Omega)$ such that the initial and final marginals P_0 and P_1 are required to equal respectively two prescribed probability measures μ_0 and $\mu_1 \in P(\mathscr{X})$ on the state space $\mathscr{X}$. This entropy minimization problem is called, since the eighties, the *Schrödinger problem*. It is described in the author's survey paper [5] where it is exemplified with R a reversible Markov process in the classical sense of [4, Sect. 1.2], for instance the *reversible Brownian motion* on $\mathbb{R}^n$.

If one wants to describe the reversible Brownian motion on $\mathbb{R}^n$ as a measure on the path space $\Omega = C([0, 1], \mathbb{R}^n)$, one has to consider an *unbounded measure*. Indeed, its reversing measure is Lebesgue measure (or any of its positive multiple), and its "law" is

$$R = \int_{\mathbb{R}^n} \mathscr{W}_x(\cdot) \, dx \in M_+(\Omega),$$

where $\mathscr{W}_x \in P(\Omega)$ stands for the Wiener measure with starting position $x \in \mathbb{R}^n$. Obviously, this path measure has the same unbounded mass as Lebesgue measure. More generally, any path measure $Q \in M_+(\Omega)$ has the same mass as its time-marginal measures $Q_t \in M_+(\mathscr{X})$ for all $t \in [0, 1]$. In particular, any reversible path measure in $M_+(\Omega)$ with an unbounded reversing measure in $M_+(\mathscr{X})$, is also unbounded.

In connection with the Schrödinger problem, the notion of (f, g)-transform of a possibly unbounded Markov measure R is introduced in [5]. It is defined by

$$P = f(X_0)g(X_1) \, R \in P(\Omega) \tag{2}$$

where f and g are measurable nonnegative functions such that $E_R(f(X_0)g(X_1)) = 1$. It is a time-symmetric extension of the usual Doob h-transform. It can be shown that the product form of the Radon-Nikodym derivative $f(X_0)g(X_1)$ implies that P is the solution to the Schrödinger problem with the correct prescribed marginals μ_0 and μ_1 given by

$$\begin{cases} \mu_0(dx) = f(x)E_R(g(X_1) \mid X_0 = x) \, R_0(dx), \\ \mu_1(dy) = E_R(f(X_0) \mid X_1 = y)g(y) \, R_1(dy). \end{cases} \tag{3}$$

Disintegration of an Unbounded Path Measure

One has to be careful when saying that the reversible Brownian motion $R \in M_+(\Omega)$ is Markov. Of course, this means that for all $0 \leq t \leq 1$, $E_R(X_{[t,1]} \in \cdot \mid X_{[0,t]}) = E_R(X_{[t,1]} \in \cdot \mid X_t)$. Similarly, we wrote (3) without hesitation. But the problem is to define properly the conditional expectation with respect to an *unbounded* measure. This will be the purpose of Sect. 2 where extensions of the conditional expectation are considered and a definition of the Markov property for an unbounded path measure is given. The general theory of conditional expectation is recalled at the appendix Sect. 5 to emphasize the role of σ-finiteness.

Relative Entropy with Respect to an Unbounded Measure

The relative entropy with respect to a probability measure is well-known. But once we have to deal with an unbounded path measure at hand, what about the relative entropy with respect to an unbounded measure and its additive property? This is the subject of Sect. 3.

Positive Integration with Respect to a Markov Measure

It is assumed in the (f, g)-transform formula (2) that $E_R(f(X_0)g(X_1)) < \infty$ with $f, g \geq 0$, while the conditional expectations $E_R(f(X_0) \mid X_1)$ and $E_R(g(X_1) \mid X_0)$ appear at (3). But the assumption that $f(X_0)g(X_1)$ is R-integrable doesn't ensure, in general, that $f(X_0)$ and $g(X_1)$ are separately R-integrable; which is a prerequisite for defining properly the conditional expectations $E_R(f(X_0) \mid X_1)$ and $E_R(g(X_1) \mid X_0)$. However, we need a general setting for the conditional expectations in (3) to be meaningful. This will be presented at Sect. 4 where we take advantage of the positivity of the functions f and g.

2   Disintegration of an Unbounded Path Measure

We often need the following notion which is a little more restrictive than the absolute continuity, but which matches with it whenever the measures are σ-finite.

Definition 1 Let R and $Q \in M_+(\Omega)$ be two positive measures on some measurable space Ω. One says that Q *admits a density with respect to* R if there exists a measurable function $\theta : \Omega \to [0, \infty)$ verifying

$$\int_\Omega f \, dQ = \int_\Omega f\theta \, dR \in [0, \infty], \quad \forall f \geq 0 \text{ measurable.}$$

We write this relation

$$Q \prec R$$

and we denote

$$\theta := \frac{dQ}{dR}$$

which is called the *Radon-Nikodym derivative* of Q with respect to R.

Thanks to the monotone convergence theorem, it is easy to check that if R is σ-finite and $\theta : \Omega \to [0, \infty)$ is a nonnegative measurable function, then

$$\theta R(A) := \int_A \theta \, dR, \quad A \in \mathscr{A},$$

defines a positive measure on the σ-field $\mathscr{A}$.

Proposition 1 *Let R and Q be two positive measures. Suppose that R is σ-finite. The following assertions are equivalent:*

(a) $Q \prec R$
(b) Q is σ-finite and $Q \ll R$.

Proof The implication $(b) \Rightarrow (a)$ is Radon-Nikodym Theorem 4. Let us show its converse $(a) \Rightarrow (b)$. The absolute continuity $Q \ll R$ is straightforward. Let us prove that Q is σ-finite. Let $(A_n)_{n \geq 1}$ be a σ-finite partition of R. Define for all $k \geq 1$, $B_k = \{k - 1 \leq dQ/dR < k\}$. The sequence $(B_k)_{k \geq 1}$ is also a measurable partition. Hence, $(A_n \cap B_k)_{n,k \geq 1}$ is a countable measurable partition. On the other hand, for any (n, k), $Q(A_n \cap B_k) = E_R(\mathbf{1}_{A_n \cap B_k} \, dQ/dR) \leq kR(A_n) < \infty$. Therefore $(A_n \cap B_k)_{n,k \geq 1}$ is a σ-finite partition of Q.

Let $Q, R \in M_+(\Omega)$ be two (possibly unbounded) positive measures on Ω. Let $\phi : \Omega \to \mathscr{X}$ be a measurable mapping from Ω to a Polish (separable, complete metric) space $\mathscr{X}$ equipped with its Borel σ-field. Although $Q \ll R$ implies that $Q_\phi \ll R_\phi$, in general we do not have $Q_\phi \prec R_\phi$ when $Q \prec R$, as the following example shows;

Example 1 The measure R is the uniform probability measure on $\Omega = [0, 1] \times [0, 1]$, Q is defined by $Q(dxdy) = 1/y \, R(dxdy)$ and we denote the canonical projections by $\phi_X(x, y) = x$, $\phi_Y(x, y) = y$, $(x, y) \in \Omega$. We observe that on the one hand R, Q and $R_{\phi_X}(dx) = \text{Leb}(dx) = dx$ are σ-finite, but on the other hand, Q_{ϕ_X} is defined by $Q_{\phi_X}(A) = \begin{cases} 0 & \text{if } \text{Leb}(A) = 0 \\ +\infty & \text{otherwise} \end{cases}$. We have $Q_{\phi_X} \ll R_{\phi_X}$, but Q_{ϕ_X} is not σ-finite. We also see that $Q_{\phi_Y}(dy) = 1/y \, dy$ is σ-finite.

An Extension of the Conditional Expectation

To extend easily results about conditional expectation with respect to a bounded measure (in particular Propositions 4 and 5) to a σ-finite measure, it is useful to rely on the following preliminary result.

Lemma 1 *Let us assume that R_ϕ is σ-finite.*

(a) Let $\gamma : \mathscr{X} \to (0, 1]$ be a measurable function such that γR_ϕ is a bounded measure. Then, $L^1(R) \subset L^1(\gamma(\phi)R)$ and for any $f \in L^1(R)$, $E_R(f \mid \phi) = E_{\gamma(\phi)R}(f \mid \phi)$, R-a.e.

(b) *There exists a function* $\gamma \in L^1(R_\phi)$ *such that* $0 < \gamma \leq 1$, R_ϕ-*a.e. In particular, the measure* $\gamma(\phi)R$ *is bounded and equivalent to* R, *i.e. for any measurable subset* A, $R(A) = 0 \iff [\gamma(\phi)R](A) = 0$.

(c) *Let* Q *be another positive measure on* Ω *such that* Q_ϕ *is* σ-*finite. Then, there exists a function* $\gamma \in L^1(R_\phi + Q_\phi)$ *such that* $0 < \gamma \leq 1$, $(R_\phi + Q_\phi)$-*a.e. In particular, the measures* $\gamma(\phi)R$ *and* $\gamma(\phi)Q$ *are bounded and respectively equivalent to* R *and* Q.

Proof • Proof of (a). Denote B_ϕ the space of all $\mathscr{A}(\phi)$-measurable and bounded functions and $\gamma B_\phi := \{h : h/\gamma(\phi) \in B_\phi\} \subset B_\phi$. For all $f \in L^1(R)$ and $h \in \gamma B_\phi$,

$$
\int_\Omega hf \, dR = \int_\Omega \frac{h}{\gamma(\phi)} f \gamma(\phi) \, dR
$$
$$
= \int_\Omega \frac{h}{\gamma(\phi)} E_{\gamma(\phi)R}(f \mid \phi) \, d(\gamma(\phi)R)
$$
$$
= \int_\Omega h E_{\gamma(\phi)R}(f \mid \phi) \, dR.
$$

On the other hand, $\int_\Omega hf \, dR = \int_\Omega h E_R(f \mid \phi) \, dR$ so that

$$
\int_\Omega h E_R(f \mid \phi) \, dR_{\mathscr{A}(\phi)} = \int_\Omega h E_{\gamma(\phi)R}(f \mid \phi) \, dR_{\mathscr{A}(\phi)}, \quad \forall h \in \gamma B_\phi.
$$

In other words, the measures $E_R(f \mid \phi)R_{\mathscr{A}(\phi)}$ and $E_{\gamma(\phi)R}(f \mid \phi)R_{\mathscr{A}(\phi)}$ match on γB_ϕ. But, since $\gamma(\phi) > 0$, the measures on $\mathscr{A}(\phi)$ are characterized by their values on γB_ϕ. Consequently, $E_R(f \mid \phi)R_{\mathscr{A}(\phi)} = E_{\gamma(\phi)R}(f \mid \phi)R_{\mathscr{A}(\phi)}$. This completes the proof of statement (1).

• Proof of (b). It is a particular instance of statement (c), taking $Q = 0$.

• Proof of (c). If R and Q are bounded, it is sufficient to take $\gamma \equiv 1$. Suppose now that $R + Q$ is unbounded. The intersection of two partitions which are respectively σ-finite with respect to R_ϕ and Q_ϕ is a partition $(\mathscr{X}_n)_{n\geq 1}$ of $\mathscr{X}$ which is simultaneously σ-finite with respect to R_ϕ and Q_ϕ. We assume without loss of generality that $(R_\phi + Q_\phi)(\mathscr{X}_n) \geq 1$ for all n. Let us define

$$
\gamma := \sum_{n\geq 1} \frac{2^{-n}}{(R_\phi + Q_\phi)(\mathscr{X}_n)} \mathbf{1}_{\mathscr{X}_n}.
$$

It is a measurable function on $\mathscr{X}$. As $\int_\Omega \gamma(\phi) \, d(R + Q) = 1$ and $0 < \gamma(\phi) \leq 1$, $(R + Q)$-a.e., $\gamma(\phi)(R + Q)$ is a probability measure that is equivalent to $R + Q$ and $L^1(R + Q) \subset L^1(\gamma(\phi)(R + Q))$.

Definition 2 (Extension of the Conditional Expectation) With Lemma 1, we see that $E_{\gamma(\phi)R}(\cdot \mid \phi)$ is an extension of $E_R(\cdot \mid \phi)$ from $L^1(R)$ to $L^1(\gamma(\phi)R)$. We denote

$$E_R(f \mid \phi) := E_{\gamma(\phi)R}(f \mid \phi), \quad f \in L^1(\gamma(\phi)R)$$

where γ is a function the existence of which is ensured by Lemma 1.

Theorem 1 *Let* $R, Q \in M_+(\Omega)$ *and* $\phi : \Omega \to \mathcal{X}$ *a measurable mapping in the Polish space* $\mathcal{X}$. *We suppose that* $Q \prec R$, *and also that* R_ϕ *are* Q_ϕ σ-*finite measures on* $\mathcal{X}$. *Then,*

(a) $E_R(\cdot \mid \phi)$ *and* $E_Q(\cdot \mid \phi)$ *admit respectively a regular conditional probability kernel* $x \in \mathcal{X} \mapsto R(\cdot \mid \phi = x) \in P(\Omega)$ *and* $x \in \mathcal{X} \mapsto Q(\cdot \mid \phi = x) \in P(\Omega)$.

(b) $Q_\phi \prec R_\phi$, $\dfrac{dQ}{dR} \in L^1(\gamma(\phi)R)$ *and*

$$\frac{dQ_\phi}{dR_\phi}(x) = E_R\left(\frac{dQ}{dR} \mid \phi = x\right), \quad \forall x \in \mathcal{X}, \ R_\phi\text{-a.e.}$$

The function γ *in the above formulas is the one whose existence is ensured by Lemma 1(c); it also appears in Definition 2.*

(c) *Moreover,* $Q(\cdot \mid \phi) \prec R(\cdot \mid \phi)$, Q-*a.e. and*

$$\frac{dQ}{dR}(\omega) = \frac{dQ_\phi}{dR_\phi}(\phi(\omega))\frac{dQ(\cdot \mid \phi = \phi(\omega))}{dR(\cdot \mid \phi = \phi(\omega))}(\omega), \quad \forall \omega \in \Omega, \ Q\text{-a.e.} \qquad (4)$$

(d) *A formula, more practical than* (4) *is the following one. For any bounded measurable function* f, *we have*

$$E_Q(f \mid \phi) = \frac{E_R\left(\frac{dQ}{dR} f \mid \phi\right)}{E_R\left(\frac{dQ}{dR} \mid \phi\right)}, \quad Q\text{-a.e.} \qquad (5)$$

where no division by zero occurs since $E_R\left(\frac{dQ}{dR} \mid \phi\right) > 0$, Q-*a.e.*

Identity (4) also writes more synthetically as

$$\frac{dQ}{dR}(\omega) = \frac{dQ_\phi}{dR_\phi}(\phi(\omega))\frac{dQ(\cdot \mid \phi)}{dR(\cdot \mid \phi)}(\omega), \quad \forall \omega \in \Omega, \ Q\text{-a.e.}$$

or more enigmatically as

$$\frac{dQ}{dR}(\omega) = \frac{dQ_\phi}{dR_\phi}(x)\frac{dQ(\cdot \mid \phi = x)}{dR(\cdot \mid \phi = x)}(\omega), \quad \forall(\omega, x), \ Q_\phi(dx)R(d\omega \mid \phi = x)\text{-a.e.}$$

since we have $\phi(\omega) = x$, $Q_\phi(dx)R(d\omega \mid \phi = x)$-almost surely.

Proof (Proof of Theorem 1) If R and Q are bounded measures, this theorem is an immediate consequence of Propositions 4 and 5.

When R_ϕ and Q_ϕ are σ-finite, we are allowed to invoke Lemma 1: $\gamma(\phi)R$ and $\gamma(\phi)Q$ are bounded measures and we can apply (i) to them. But,

$$\frac{dQ}{dR} = \frac{d(\gamma(\phi)Q)}{d(\gamma(\phi)R)} \quad \text{and} \quad \frac{dQ_\phi}{dR_\phi} = \frac{d(\gamma Q_\phi)}{d(\gamma R_\phi)}.$$

This completes the proof of the theorem.

Hilbertian Conditional Expectation

So far, we have considered the conditional expectation of a function f in $L^1(R)$. If the reference measure R is bounded, then $L^2(R) \subset L^1(R)$. But if R is unbounded, this inclusion fails and the conditional expectation which we have just built is not valid for every f in $L^2(R)$. It is immediate to extend this notion from $L^1(R) \cap L^2(R)$ to $L^2(R)$, interpreting the fundamental relation (14) in restriction to $L^2(R)$:

$$\int_\Omega hf \, dR = \int_\Omega hE_R(f \mid \mathscr{A}) \, dR, \quad \forall h \in B_\mathscr{A}, f \in L^1(R) \cap L^2(R),$$

as an Hilbertian projection. We thus define the operator

$$E_R(\cdot \mid \mathscr{A}) : L^2(R) \to L^2(R_\mathscr{A})$$

as an orthogonal projection on the Hilbertian subspace $L^2(R_\mathscr{A})$. In particular, when $\mathscr{A}$ is the σ-field generated by the measurable mapping $\phi : \Omega \to \mathscr{X}$,

$$E_R(\cdot \mid \phi) : L^2(\Omega, R) \to L^2(\mathscr{X}, R_\phi)$$

is specified for any function $f \in L^2(R)$ by

$$\int_\Omega h(\phi(\omega)) f(\omega) \, R(d\omega) = \int_\mathscr{X} h(x) E_R(f \mid \phi = x) \, R_\phi(dx), \quad \forall h \in L^2(\mathscr{X}, R_\phi).$$

Conditional Expectation of Path Measures

Now we particularize Ω to be the path space $D([0, 1], \mathscr{X})$ or $C([0, 1], \mathscr{X})$.

Lemma 2 *Let $Q \in \mathrm{M}_+(\Omega)$ be a path measure and $\mathscr{T} \subset [0, 1]$ a time subset. For $Q_\mathscr{T}$ to be a σ-finite measure, it is sufficient that there is some $t_o \in \mathscr{T}$ such that Q_{t_o} is a σ-finite measure.*

Proof Let $t_o \in \mathcal{T}$ be such that $Q_{t_o} \in M_+(\mathcal{X})$ is a σ-finite measure with $(\mathcal{X}_n)_{n \geq 1}$ an increasing sequence of measurable sets such that $Q_{t_o}(\mathcal{X}_n) < \infty$ and $\cup \mathcal{X}_n = \mathcal{X}$. Then, $Q_{\mathcal{T}}$ is also σ-finite, since $Q_{\mathcal{T}}(X_{t_o} \in \mathcal{X}_n) = Q_{t_o}(\mathcal{X}_n)$ for all n and $\cup_{n \geq 1}[\Omega_{\mathcal{T}} \cap \{X_{t_o} \in \mathcal{X}_n\}] = \Omega_{\mathcal{T}}$.

Definition 3 (Conditionable Path Measure)

1. A positive measure $Q \in M_+(\Omega)$ is called a path measure.
2. The path measure $Q \in M_+(\Omega)$ is said to be *conditionable* if for all $t \in [0, 1]$, Q_t is a σ-finite measure on $\mathcal{X}$.

With Lemma 2, for any conditionable path measure $Q \in M_+(\Omega)$, the conditional expectation $E_Q(\cdot \mid X_{\mathcal{T}})$ is well-defined for any $\mathcal{T} \subset [0, 1]$. This is the reason for this definition.

Even when $Q(\Omega) = \infty$, Proposition 4 tells us that $Q(\cdot \mid X_{\mathcal{T}})$ is a probability measure. In particular, $Q(B \mid X_{\mathcal{T}})$ and $E_Q(b \mid X_{\mathcal{T}})$ are bounded measurable functions for any measurable subset B and any measurable bounded function b.

Example 2 Let $Q \in M_+(\Omega)$ the law of the real-valued process X such that for all $0 \leq t < 1$, $X_t = X_0$ is distributed with Lebesgue measure and $X_1 = 0$, Q-almost everywhere. We see with Lemma 2 that $Q = Q_{01}$ is a σ-finite measure since Q_0 is σ-finite. But Q_1 is not a σ-finite measure. Consequently, Q is not a conditionable path measure.

Definition 4 (Markov Measure) The path measure $Q \in M_+(\Omega)$ is said to be Markov if it is conditionable in the sense of Definition 3 and if for all $0 \leq t \leq 1$

$$Q(X_{[t,1]} \in \cdot \mid X_{[0,t]}) = Q(X_{[t,1]} \in \cdot \mid X_t).$$

3 Relative Entropy with Respect to an Unbounded Measure

Let $R \in M_+(\Omega)$ be some σ-finite positive measure on some measurable space Ω. The relative entropy of the probability measure $P \in P(\Omega)$ with respect to R is loosely defined by

$$H(P|R) := \int_\Omega \log(dP/dR) \, dP \in (-\infty, \infty], \qquad P \in P(\Omega)$$

if $P \ll R$ and $H(P|R) = \infty$ otherwise.

In the special case where R is a probability measure, this definition is meaningful.

Lemma 3 *We assume that $R \in P(\Omega)$ is a probability measure.*

We have for all $P \in P(\Omega)$, $H(P|R) \in [0, \infty]$ and $H(P|R) = 0$ if and only if $P = R$.

The function $H(\cdot|R)$ is strictly convex on the convex set $P(\Omega)$.

Proof We have $H(P|R) = \int_\Omega h\left(\frac{dP}{dR}\right) dR$ with $h(a) = a \log a - a + 1$ if $a > 0$ and $h(0) = 1$. As $h \geq 0$, we see that for any $P \in P(\Omega)$ such that $P \ll R$, $H(P|R) = \int_\Omega h\left(\frac{dP}{dR}\right) dR \geq 0$. Hence $H(P|R) \in [0, \infty]$. Moreover, $h(a) = 0$ if and only if $a = 1$. Therefore, $H(P|R) = 0$ if and only if $P = R$.

The strict convexity of $H(\cdot|R)$ follows from the strict convexity of h.

If R is unbounded, one must restrict the definition of $H(\cdot|R)$ to some subset of $P(\Omega)$ as follows. As R is assumed to be σ-finite, there exists some measurable function $W : \Omega \to [0, \infty)$ such that

$$z_W := \int_\Omega e^{-W}\, dR < \infty. \tag{6}$$

Define the probability measure $R_W := z_W^{-1} e^{-W} R$ so that $\log(dP/dR) = \log(dP/dR_W) - W - \log z_W$. It follows that for any $P \in P(\Omega)$ satisfying $\int_\Omega W\, dP < \infty$, the formula

$$H(P|R) := H(P|R_W) - \int_\Omega W\, dP - \log z_W \in (-\infty, \infty]$$

is a meaningful definition of the relative entropy which is coherent in the following sense. If $\int_\Omega W'\, dP < \infty$ for another measurable function $W' : \Omega \to [0, \infty)$ such that $z_{W'} < \infty$, then $H(P|R_W) - \int_\Omega W\, dP - \log z_W = H(P|R_{W'}) - \int_\Omega W'\, dP - \log z_{W'} \in (-\infty, \infty]$.

Therefore, $H(P|R)$ is well-defined for any $P \in P(\Omega)$ such that $\int_\Omega W\, dp < \infty$ for some measurable nonnegative function W verifying (6). For any such function, let us define

$$P_W(\Omega) := \left\{ P \in P(\Omega); \int_\Omega W\, dP < \infty \right\}.$$

and $B_W(\Omega)$ the space of measurable functions $u : \Omega \to \mathbb{R}$ such that $\sup_\Omega |u|/(1 + W) < \infty$. When Ω is a topological space, we also define the space $C_W(\Omega)$ of all continuous functions on Ω such that $\sup_\Omega |u|/(1 + W) < \infty$.

Proposition 2 *Let W be some function which satisfies* (6). *For all $P \in P_W(\Omega)$,*

$$H(P|R) = \sup \left\{ \int u\, dP - \log \int e^u\, dR; u \in B_W(\Omega) \right\}$$
$$= \sup \left\{ \int u\, dP - \log \int e^u\, dR; u \in C_W(\Omega) \right\} \tag{7}$$

and for all $P \in P(\Omega)$ such that $P \ll R$,

$$H(P|R) = \sup \left\{ \int u \, dP - \log \int e^u \, dR; u : \int e^u \, dR < \infty, \int u_- \, dP < \infty \right\}$$

(8)

where $u_- = (-u) \vee 0$ and $\int u \, dP \in (-\infty, \infty]$ is well-defined for all u such that $\int u_- \, dP < \infty$.

In (7), when $C_W(\Omega)$ is invoked, it implicitly assumed that Ω is a topological space equipped with its Borel σ-field.

The proof below is mainly a rewriting of the proof of [3, Prop. B.1] in the setting where the reference measure is possibly unbounded.

Proof (Proof of Proposition 2) Once we have (8), (7) follows by standard approximation arguments.

The proof of (8) relies on Fenchel inequality for the convex function $h(t) = t \log t$:

$$st \le t \log t + e^{s-1}$$

for all $s \in [-\infty, \infty)$, $t \in [0, \infty)$, with the conventions $0 \log 0 = 0$, $e^{-\infty} = 0$ and $-\infty \times 0 = 0$ which are legitimated by limiting procedures. The equality is attained when $t = e^{s-1}$.

Taking $s = u(x)$, $t = \frac{dP}{dR}(x)$ and integrating with respect to R leads us to

$$\int u \, dP \le H(P|R) + \int e^{u-1} \, dR,$$

whose terms are meaningful with values in $(-\infty, \infty]$, provided that $\int u_- \, dP < \infty$ and $\int_\Omega e^u \, dR < \infty$. Formally, the case of equality corresponds to $\frac{dP}{dR} = e^{u-1}$. With the monotone convergence theorem, one sees that it is approached by the sequence $u_n = 1 + \log(\frac{dP}{dR} \vee e^{-n})$, as n tends to infinity. This gives us

$$H(P|R) = \sup \left\{ \int u \, dP - \int e^{u-1} \, dR; u : \int e^u \, dR < \infty, \inf u > -\infty \right\},$$

which in turn implies that

$$H(P|R) = \sup \left\{ \int u \, dP - \int e^{u-1} \, dR; u : \int e^u \, dR < \infty, \int u_- \, dP < \infty \right\}.$$

Now, we take advantage of the unit mass of $P \in P(\Omega)$:

$$\int (u + b) \, dP - \int e^{u+b-1} \, dR = \int u \, dP - e^{b-1} \int e^u \, dR + b, \quad \forall b \in \mathbb{R},$$

and we use the easy identity $\log a = \inf_{b \in \mathbb{R}}\{ae^{b-1} - b\}$ to obtain

$$\sup_{b \in \mathbb{R}}\left\{\int (u+b)\,dP - \int e^{u+b-1}\,dR\right\} = \int u\,dP - \log \int e^u\,dR.$$

Whence,

$$\sup\left\{\int u\,dP - \int e^{u-1}\,dR; u : \int e^u\,dR < \infty, \int u_-\,dP < \infty\right\}$$

$$= \sup\left\{\int (u+b)\,dP - \int e^{u+b-1}\,dR; b \in \mathbb{R}, u : \int e^u\,dR < \infty, \int u_-\,dP < \infty\right\}$$

$$= \sup\left\{\int u\,dP - \log \int e^u\,dR; u : \int e^u\,dR < \infty, \int u_-\,dP < \infty\right\}.$$

This completes the proof of (8).

Let W be a nonnegative measurable function on Ω that verifies (6). Let us introduce the space $\mathrm{M}_W(\Omega)$ of all signed measures Q on Ω such that $\int_\Omega W\,d|Q| < \infty$.

Corollary 1 *The function $H(\cdot|R)$ is convex on the vector space of all signed measures. Its effective domain* $\operatorname{dom} H(\cdot|R) := \{H(\cdot|R) < \infty\}$ *is included in* $\mathrm{P}_W(R)$

Suppose furthermore that Ω is a topological space. Then, $H(\cdot|R)$ is lower semicontinuous with respect to the topology $\sigma(\mathrm{M}_W(\Omega), C_W(\Omega))$.

As a function of its two arguments on $\mathrm{M}_W(\Omega) \times \mathrm{M}_W(\Omega)$, $H(\cdot \mid \cdot)$ is jointly convex and jointly lower semicontinuous with respect to the product topology. In particular, it is a jointly Borel function.

Proof The first statement follows from (7).

With Proposition 2, we see that $H(\cdot|R)$ is the supremum of a family of affine continuous functions: $Q \mapsto \int_\Omega u\,dQ - \log \int_\Omega e^u\,dR$ indexed by u. Hence, it is convex and lower semicontinuous. The same argument works with the joint arguments.

Let Ω and Z be two Polish spaces equipped with their Borel σ-fields. For any measurable function $\phi : \Omega \to Z$ and any measure $Q \in \mathrm{M}_+(\Omega)$ we have the disintegration formula

$$Q(\cdot) = \int_Z Q(\cdot \mid \phi = z)\,Q_\phi(dz)$$

where we write $Q_\phi := \phi_\# Q$ and $z \in Z \mapsto Q(\cdot|\phi = z) \in \mathrm{P}(\Omega)$ is measurable.

Theorem 2 (Additive Property of the Relative Entropy) *We have*

$$H(P|R) = H(P_\phi|R_\phi) + \int_Z H\Big(P(\cdot \mid \phi = z)\Big|R(\cdot \mid \phi = z)\Big)\,P_\phi(dz), \quad P \in \mathrm{P}(\Omega).$$

Proof By Theorem 1,

$$H(P|R) = \int_Z E_P\left[\log(\frac{dP}{dR}) \mid \phi = z\right] P_\phi(dz) = \int_Z \log\frac{dP_\phi}{dR_\phi}(z)\, P_\phi(dz)$$

$$+ \int_Z \left[\int_\Omega \log\frac{dP(\cdot \mid \phi = z)}{dR(\cdot \mid \phi = z)}(\omega)\, P(d\omega \mid \phi = z)\right] P_\phi(dz)$$

which is the announced result.

Remarks 1 There are serious measurability problems hidden behind this proof.

(a) The assumption that Z is Polish ensures the existence of kernels $z \mapsto P(\cdot \mid \phi = z)$ and $z \mapsto R(\cdot \mid \phi = z)$. On the other hand, we know that for any function $u \in B_W$, the mapping $z \in \mathcal{X} \mapsto E_P(u \mid \phi = z) \in \mathbb{R}$ is measurable. Therefore, the mapping $z \in Z \mapsto P(\cdot \mid \phi = z) \in P_W(\Omega)$ is measurable once $P_W(\Omega)$ is equipped with its cylindrical σ-field, i.e. generated by the mappings $Q \in P_W(\Omega) \mapsto \int_\Omega u\, dQ$ where u describes B_W. But this σ-field matches with the Borel σ-field of $\sigma(P_W(\Omega), C_W)$ when Ω is metric and separable. As H is jointly Borel (see Corollary 1), it is jointly measurable with respect to the product of the cylindrical σ-fields. Hence, $z \mapsto H\left(P(\cdot \mid \phi = z) \middle| R(\cdot \mid \phi = z)\right)$ is measurable.

   Note that in general, the Borel σ-field of $\sigma(P_W(\Omega), B_W)$ is too rich to match with the cylindrical σ-field. This is the reason why Ω is assumed to be Polish (completeness doesn't play any role here).

(b) The relative entropy $H\left(P(\cdot \mid \phi = z) \middle| R(\cdot \mid \phi = z)\right)$ inside the second integral of the additive property formula is a function of couples of probability measures. Therefore, with Lemma 3, we know that it is nonnegative in general and that it vanishes if and only if $P(\cdot \mid \phi = z) = R(\cdot \mid \phi = z)$.

(c) Together with its measurability, which was proved at Remak (a) above, this allows us to give a meaning to the integral $\int_Z H\left(P(\cdot \mid \phi = z) \middle| R(\cdot \mid \phi = z)\right) P_\phi(dz)$ in $[0, \infty]$.

Let us mention an application of this theorem in the context of the Schrödinger problem (1) where Ω is a path space, see [2,5]. For any, $R \in M_+(\Omega)$, $P \in P(\Omega)$, we have

$$H(P|R) = H(P_{01}|R_{01}) + \int_{\mathcal{X}^2} H(P^{xy}|R^{xy})\, P_{01}(dxdy)$$

where $Q_{01} := (X_0, X_1)_\# Q$ is the law of the endpoint position and $Q^{xy} := Q(\cdot|X_0 = x, X_1 = y)$ is the bridge from x to y under Q. From this additive property formula and Corollary 1, it is easily seen that the solution $\hat{P}$ of (1) (it is unique, since the entropy is *strictly* convex) satisfies

$$\hat{P}^{xy} = R^{xy}, \quad \forall (x, y) \in \mathscr{X}^2, \hat{P}_{01}\text{-a.e.}$$

and that $\hat{P}_{01}$ is the unique solution of

$$H(\pi \mid R_{01}) \to \min; \qquad \pi \in \mathrm{P}(\mathscr{X}^2) : \pi_0 = \mu_0, \pi_1 = \mu_1$$

where π_0 and $\pi_1 \in \mathrm{P}(\mathscr{X})$ are the first and second marginals of $\pi \in \mathrm{P}(\mathscr{X}^2)$.

4   Positive Integration with Respect to a Markov Measure

Integration of Nonnegative Functions

The expectation $E_R Z$ of a *nonnegative* random variable Z with respect to a positive σ-finite measure R is a well-defined notion, even when Z is not R-integrable; in which case, one sets $E_R Z = +\infty$. Indeed, with the monotone convergence theorem we have

$$E_R Z = \lim_{n \to \infty} E_R[\mathbf{1}_{\{\cup_{k \leq n} \Omega_k\}}(Z \wedge n)] \in [0, \infty]$$

where $(\Omega_k)_{k \geq 1}$ is a σ-finite partition of R.

Since $R(\cdot \mid \mathscr{A})$ is a bounded measure, we see that $E_R(Z \mid \mathscr{A})$ is well defined in $[0, \infty]$. Moreover, the fundamental formula of the conditional expectation is kept:

$$E_R[aE_R(Z \mid \mathscr{A})] = E_R(aZ)$$

for any nonnegative function $a \in \mathscr{A}$. To see this, denote $a_n = \mathbf{1}_{\{\cup_{k \leq n} \Omega_k\}}(a \wedge n)$ and $Z_n = \mathbf{1}_{\{\cup_{k \leq n} \Omega_k\}}(Z \wedge n)$. We have $E_R[a_n E_R(Z_n \mid \mathscr{A})] = E_R(a_n Z_n)$ for all $n \geq 1$. Letting n tend to infinity, we obtain the announced identity with the monotone convergence theorem.

Positive Integration with Respect to a Markov Measure

We present a technical lemma about positive integration with respect to a Markov measure $R \in M_+(\Omega)$. It is an easy result, but it is rather practical. It allows to work with (f, g)-transforms of Markov processes without assuming unnecessary integrability conditions on f and g.

Lemma 4 *Let $R \in M_+(\Omega)$ be a Markov measure.*

(a) *Let $0 \leq t \leq 1$ and α, β be nonnegative functions such that $\alpha \in \mathscr{A}_{[0,t]}$ and $\beta \in \mathscr{A}_{[t,1]}$. Then, for any ω outside an R-negligible set:*

   (i) *if $E_R(\alpha\beta \mid X_t)(\omega) = 0$, we have $E_R(\alpha \mid X_t)(\omega) = 0$ or $E_R(\beta \mid X_t)(\omega) = 0$;*

   (ii) *if $E_R(\alpha\beta \mid X_t)(\omega) > 0$, we have $E_R(\alpha \mid X_t)(\omega), E_R(\beta \mid X_t)(\omega) > 0$ and $E_R(\alpha\beta \mid X_t)(\omega) = E_R(\alpha \mid X_t)(\omega)E_R(\beta \mid X_t)(\omega) \in (0, \infty]$.*

(b) *Let $P \in M_+(\Omega)$ be a conditionable path measure such that $P \prec R$ and whose density writes as $\dfrac{dP}{dR} = \alpha\beta$ with α, β nonnegative functions such that $\alpha \in \mathscr{A}_{[0,t]}$ and $\beta \in \mathscr{A}_{[t,1]}$ for some $0 \leq t \leq 1$. Then,*

$$\begin{cases} E_R(\alpha \mid X_t), E_R(\beta \mid X_t) \in (0, \infty) \\ E_R(\alpha\beta \mid X_t) = E_R(\alpha \mid X_t)E_R(\beta \mid X_t) \in (0, \infty) \end{cases} \quad P\text{-a.e.}$$

(but not R-a.e. in general). Furthermore,

$$E_R(\alpha\beta \mid X_t) \tag{9}$$
$$= \mathbf{1}_{\{E_R(\alpha|X_t)<\infty, E_R(\beta|X_t)<\infty\}} E_R(\alpha \mid X_t)E_R(\beta \mid X_t) \in [0, \infty) \quad R\text{-a.e.}$$

As regards (9), even if $\alpha\beta$ is integrable, it is not true in general that the nonnegative functions α and β are integrable. Therefore, a priori the conditional expectations $E_R(\alpha \mid X_t)$ and $E_R(\beta \mid X_t)$ may be infinite.

Proof • Proof of (a). The measure R disintegrates with respect to the initial and final positions:

$$R = \int_{\mathscr{X}} R(\cdot \mid X_0 = x) R_0(dx) = \int_{\mathscr{X}} R(\cdot \mid X_1 = y) R_1(dy)$$

But, R_0 and R_1 are assumed to be σ-finite measures. Let $(\mathscr{X}_n^0)_{n \geq 1}$ and $(\mathscr{X}_n^1)_{n \geq 1}$ be two σ-finite partitions of R_0 and R_1, respectively. We denote $\Omega_n^0 = \{X_0 \in \cup_{k \leq n} \mathscr{X}_k^0\}$, $\Omega_n^1 = \{X_1 \in \cup_{k \leq n} \mathscr{X}_k^1\}$ and $\Omega_n = \Omega_n^0 \cap \Omega_n^1$.

As R is Markov, if the functions α and β are integrable, then $E_R(\alpha \mid X_t)$ are $E_R(\beta \mid X_t)$ well-defined and

$$E_R(\alpha\beta \mid X_t) = E_R(\alpha \mid X_t)E_R(\beta \mid X_t).$$

Letting n tend to infinity in $E_R[(\alpha \wedge n)(\beta \wedge n)\mathbf{1}_{\Omega_n} \mid X_t] = E_R((\alpha \wedge n)\mathbf{1}_{\Omega_n^0} \mid X_t)E_R((\beta \wedge n)\mathbf{1}_{\Omega_n^1} \mid X_t)$, we obtain $E_R(\alpha\beta \mid X_t) = E_R(\alpha \mid X_t)E_R(\beta \mid X_t) \in [0, \infty]$. One concludes, remarking that the sequences are increasing.

• Proof of (b). It is a consequence of the first part of the lemma. As P_t is σ-finite measure, $\frac{dP_t}{dR_t}(X_t) < \infty$, R-a.e. (hence, a fortiori P-a.e.). In addition,

$\frac{dP_t}{dR_t}(X_t) > 0$, P-a.e. (but not R-a.e. in general) and $\frac{dP_t}{dR_t}(X_t) = E_R(\alpha\beta \mid X_t)$, by Theorem 1(b). Consequently, we are allowed to apply part (ii) of (a) to obtain the identity which holds P-a.e. This identity extends R-a.e., yielding (9). To see this, remark with part (i) of (a) that when the density vanishes, the two terms of the product cannot be *simultaneously* equal to ∞ and one of them vanishes.

Analogously, one can prove the following extension.

Lemma 5 *Let $R \in \mathrm{M}_+(\Omega)$ be a Markov measure.*

1. *Let $0 \leq s \leq t \leq 1$ and two nonnegative functions α, β such that $\alpha \in \mathscr{A}_{[0,s]}$, $\beta \in \mathscr{A}_{[t,1]}$. Then, for any ω outside an R-negligible set:*

   (a) *if $E_R(\alpha\beta \mid X_{[s,t]})(\omega) = 0$, we have $E_R(\alpha \mid X_s)(\omega) = 0$ or $E_R(\beta \mid X_t)(\omega) = 0$;*

   (b) *if $E_R(\alpha\beta \mid X_{[s,t]})(\omega) > 0$, we have $E_R(\alpha \mid X_s)(\omega), E_R(\beta \mid X_t)(\omega) > 0$ and $E_R(\alpha\beta \mid X_{[s,t]})(\omega) = E_R(\alpha \mid X_s)(\omega)E_R(\beta \mid X_t)(\omega) \in (0, \infty]$.*

2. *Let $P \in \mathrm{M}_+(\Omega)$ be a conditionable path measure such that $P \prec R$ and whose density writes as $\dfrac{dP}{dR} = \alpha\zeta\beta$ with α, ζ and β nonnegative functions such that $\alpha \in \mathscr{A}_{[0,s]}, \zeta \in \mathscr{A}_{[s,t]}$ and $\beta \in \mathscr{A}_{[t,1]}$ for some $0 \leq s \leq t \leq 1$. Then,*

$$\begin{cases} E_R(\alpha \mid X_s), E_R(\beta \mid X_t) \in (0, \infty) \\ E_R(\alpha\beta \mid X_{[s,t]}) = E_R(\alpha \mid X_s)E_R(\beta \mid X_t) \in (0, \infty) \end{cases} \quad P\text{-a.e.}$$

(and not R-a.e. in general). In addition,

$$E_R(\alpha\zeta\beta \mid X_{[s,t]})$$
$$= \mathbf{1}_{\{E_R(\alpha\mid X_s)<\infty, E_R(\beta\mid X_t)<\infty\}} E_R(\alpha \mid X_s)\zeta E_R(\beta \mid X_t) \in [0, \infty) \quad R\text{-a.e.}$$

5 Conditional Expectation with Respect to an Unbounded Measure

In standard textbooks, the theory of conditional expectation is presented and developed with respect to a probability measure (or equivalently, a bounded positive measure). However, there are natural unbounded path measures, such as the reversible Brownian motion on $\mathbb{R}^n$, with respect to which a conditional expectation theory is needed. We present the details of this notion in this appendix section. From a measure theoretic viewpoint, this section is about disintegration of unbounded positive measures.

The Role of σ-Finiteness in Radon-Nikodym Theorem

The keystone of conditioning is Radon-Nikodym theorem. In order to emphasize the role of σ-finiteness, we recall a classical proof of this theorem, following von Neumann and Rudin, [6]. Let Ω be a space with its σ-field and $P, Q, R \in M_+(\Omega)$ be positive measures on Ω. One says that P is absolutely continuous with respect to R and denotes $P \ll R$, if for every measurable subset $A \subset \Omega$, $R(A) = 0 \Rightarrow P(A) = 0$. It is said to be concentrated on the measurable subset $C \subset \Omega$ if for any measurable subset $A \subset \Omega$, $P(A) = P(A \cap C)$. The measures P and Q are said to be mutually singular and one denotes $P \perp Q$, if there exist two disjoint measurable subsets $C, D \subset \Omega$ such that P is concentrated on C and Q is concentrated on D.

Theorem 3 *Let P and R be two bounded positive measures.*

(a) *There exists a unique pair (P_a, P_s) of measures such that $P = P_a + P_s$, $P_a \ll R$ and $P_s \perp R$. These measures are positive and $P_a \perp P_s$.*
(b) *There is a unique function $\theta \in L^1(R)$ such that*

$$P_a(A) = \int_A \theta \, dR, \quad \text{for any measurable subset } A.$$

Proof The uniqueness proofs are easy. Let us begin with (a). Suppose we have two Lebesgue decompositions: $P = P_a + P_s = P'_a + P'_s$. Then, $P_a - P'_a = P'_s - P_s$, $P_a - P'_a \ll R$ and $P'_s - P_s \perp R$. Hence, $P_a - P'_a = P'_s - P_s = 0$ since $Q \ll R$ and $Q \perp R$ imply that $Q = 0$. As regards (b), if we have $P_a = \theta R = \theta' R$, then $\int_A (\theta - \theta') \, dR = 0$ for any measurable $A \subset \Omega$. Therefore $\theta = \theta'$, R-a.e.

Denote $Q = P + R$. It is a bounded positive measure and for any function $f \in L^2(Q)$,

$$\left| \int_\Omega f \, dP \right| \le \int_\Omega |f| \, dQ \le \sqrt{Q(\Omega)} \|f\|_{L^2(Q)}. \tag{10}$$

It follows that $f \in L^2(Q) \mapsto \int_\Omega f \, dP \in \mathbb{R}$ is a continuous linear form on the Hilbert space $L^2(Q)$. Consequently, there exists $g \in L^2(Q)$ such that

$$\int_\Omega f \, dP = \int_\Omega fg \, dQ, \quad \forall f \in L^2(Q). \tag{11}$$

Since $0 \le P \le P + R := Q$, we obtain $0 \le g \le 1$, Q-a.e. Let us take a version of g such that $0 \le g \le 1$ everywhere. The identity (11) rewrites as

$$\int_\Omega (1 - g) f \, dP = \int_\Omega fg \, dR, \quad \forall f \in L^2(Q). \tag{12}$$

Let us set $C := \{0 \le g < 1\}$, $D = \{g = 1\}$, $P_a(\cdot) = P(\cdot \cap C)$ et $P_s(\cdot) = P(\cdot \cap D)$. Choosing $f = \mathbf{1}_D$ in (12), we obtain $R(D) = 0$ so that $P_s \perp R$.

Choosing $f = (1 + g + \cdots + g^n)1_A$ with $n \geq 1$ and A any measurable subset in (12), we obtain

$$\int_A (1 - g^{n+1})\, dP = \int_A g(1 + g + \cdots + g^n)\, dR.$$

But the sequence of functions $(1 - g^{n+1})$ increases pointwise towards 1_C. Now, by the monotone convergence theorem, we have $P(A \cap C) = \int_A 1_C g/(1-g)\, dR$. This means that $P_a = \theta R$ with $\theta = 1_{\{0 \leq g < 1\}} g/(1 - g)$.

Finally, we see that $\theta \geq 0$ is R-integrable since $\int_\Omega \theta\, dR = P_a(\Omega) \leq P(\Omega) < \infty$.

The main argument of this proof is Riesz theorem on the representation of the dual of a Hilbert space. As the continuity of the linear form is ensured by $Q(\Omega) < \infty$ at (10), we have used crucially the boundedness of the measures P and R. This can be relaxed by means of the following notion.

Definition 5 The positive measure R is said to be σ-finite if it is either bounded or if there exists a sequence $(\Omega_k)_{k\geq 1}$ of disjoint measurable subsets which partitions Ω : $\sqcup_k \Omega_k = \Omega$ and are such that $R(\Omega_k) < \infty$ for all k.

In such a case it is said that $(\Omega_k)_{k\geq 1}$ finitely partitions R or that it is a σ-finite partition of R.

Recall that an unbounded positive measure is allowed to take the value $+\infty$. For instance, the measure R which is defined on the trivial σ-field $\{\emptyset, \Omega\}$ by $R(\emptyset) = 0$ and $R(\Omega) = \infty$ is a genuine positive measure and $L^1(R) = \{0\}$. This situation may seem artificial, but in fact it is not, as can be observed with the following examples.

Examples 1

(a) The push-forward of Lebesgue measure on $\mathbb{R}$ by a function which takes finitely many values is a positive measure on the set of these values which charges at least one of them with an infinite mass. By the way this provides us with an example of a σ-finite measure whose pushed forward is not.

(b) Lebesgue measure on $\mathbb{R}^2$ is σ-finite, but its push-forward by the projection on the first coordinate assigns an infinite mass to any non-negligible Borel set.

Theorem 4 (Radon-Nikodym) *Let P and R two positive σ-finite measures such that $P \ll R$. Then, there exists a unique measurable function θ such that*

$$\int_\Omega f\, dP = \int_\Omega f\theta\, dR, \quad \forall f \in L^1(P). \tag{13}$$

Moreover, P is bounded if and only if $\theta \in L^1(R)$.

Proof Taking the intersection of two partitions which respectively finitely partition R and P, one obtains a countable measurable partition which simultaneously finitely partitions R and P. Theorem 3 applies on each subset of this partition and one obtains the desired result by recollecting the pieces. The resulting function θ

need not be integrable anymore, but it is still is locally integrable in the sense that it is integrable in restriction to each subset of the partition. We have just extended Theorem 3 when the measures P and R are σ-finite. We conclude noticing that by Theorem 3 we have: $P \ll R$ if and only if $P_s = 0$.

With respect to Radon-Nikodym theorem, making a step away from σ-finiteness seems to be hopeless, as one can guess from the following example. Take $R = \sum_{x \in [0,1]} \delta_x$: the counting measure on $\Omega = [0, 1]$, and P the Lebesgue measure on $[0, 1]$. We see that $P \ll R$, but there is no measurable function θ which satisfies (13).

Conditional Expectation with Respect to a Positive Measure

Let Ω be a space furnished with some σ-field and a sub-σ-field $\mathscr{A}$. We take a positive measure $R \in M_+(\Omega)$ on Ω and denote $R_{\mathscr{A}}$ its restriction to $\mathscr{A}$. The space of bounded measurable functions is denoted by B, while $B_{\mathscr{A}}$ is the subspace of bounded $\mathscr{A}$-measurable functions. The subspace of $L^1(R)$ consisting of the $\mathscr{A}$-measurable integrable functions is denoted by $L^1(R_{\mathscr{A}})$.

We take $g \geq 0$ in $L^1(R)$. The mapping $h \in B_{\mathscr{A}} \mapsto \int_\Omega hg \, dR := \int_\Omega h \, dR^g_{\mathscr{A}}$ defines a *finite* positive measure $R^g_{\mathscr{A}}$ on $(\Omega, \mathscr{A})$. Clearly, if $h \geq 0$ and $\int_\Omega h \, dR = \int_\Omega h \, dR_{\mathscr{A}} = 0$, then $\int_\Omega h \, dR^g_{\mathscr{A}} = 0$. This means that $R^g_{\mathscr{A}}$ is a finite measure which is absolutely continuous with respect to $R_{\mathscr{A}}$. If $R_{\mathscr{A}}$ is assumed to be σ-finite, by the Radon-Nikodym Theorem 4, there is a unique function $\theta_g \in L^1(R_{\mathscr{A}})$ such that $R^g_{\mathscr{A}} = \theta_g R_{\mathscr{A}}$. We have just obtained $\int_\Omega hg \, dR = \int_\Omega h\theta_g \, dR_{\mathscr{A}} = \int_\Omega h\theta_g \, dR$, $\forall h \in B_{\mathscr{A}}$. Now, let $f \in L^1(R)$ which might not be nonnegative. Considering its decomposition $f = f_+ - f_-$ into nonnegative and nonpositive parts: $f_+ = f \vee 0$, $f_- = (-f) \vee 0$, and setting $\theta_f = \theta_{f_+} - \theta_{f_-}$, we obtain

$$\int_\Omega hf \, dR = \int_\Omega h\theta_f \, dR_{\mathscr{A}} = \int_\Omega h\theta_f \, dR, \quad \forall h \in B_{\mathscr{A}}, f \in L^1(R). \qquad (14)$$

Definition 6 (Conditional Expectation) It is assumed that $R_{\mathscr{A}}$ is σ-finite.

For any $f \in L^1(R)$, the conditional expectation of f with respect to $\mathscr{A}$ is the unique (modulo R-a.e.-equality) function

$$E_R(f \mid \mathscr{A}) \in L^1(R_{\mathscr{A}})$$

which is integrable, $\mathscr{A}$-measurable and such that $\theta_f =: E_R(f \mid \mathscr{A})$ satisfies (14).

It is essential in this definition that $R_{\mathscr{A}}$ is assumed to be σ-finite.

Of course,

$$E_R(f \mid \mathscr{A}) = f, \quad \forall f \in L^1(R_{\mathscr{A}}) \qquad (15)$$

If, in (14), we take the function $h = \text{sign}(E_R(f|\mathscr{A}))$ which is in $B_\mathscr{A}$, we have

$$\int_\Omega |E_R(f \mid \mathscr{A})| \, dR_\mathscr{A} \le \int_\Omega |f| \, dR \qquad (16)$$

which expresses that $E_R(\cdot \mid \mathscr{A}) : L^1(R) \to L^1(R_\mathscr{A})$ is a contraction, the spaces L^1 being equipped with their usual norms $\|\cdot\|_1$. With (15), we see that the opertot norm of this contraction is 1. Therefore, $E_R(\cdot \mid \mathscr{A}) : L^1(R) \to L^1(R_\mathscr{A})$ is a continuous projection.

Taking $h = 1$ in (14), we have

$$\int_\Omega f(\omega) \, R(d\omega) = \int_\Omega E_R(f \mid \mathscr{A})(\eta) \, R(d\eta),$$

which can be written

$$E_R E_R(f \mid \mathscr{A}) = E_R(f), \qquad (17)$$

with the notation $E_R(f) := \int_\Omega f \, dR$.

Remark 1 When R is a bounded measure, the mapping $E_R(\cdot \mid \mathscr{A})$ shares the following properties.

(a) For all $f \in L^1(R) \ge 0$, $E_R(f \mid \mathscr{A}) \ge 0$, $R_\mathscr{A}$-a.e.
(b) $E_R(1 \mid \mathscr{A}) = 1$, $R_\mathscr{A}$-a.e.
(c) For all $f, g \in L^1(R)$ and $\lambda \in \mathbb{R}$, $E_R(f + \lambda g \mid \mathscr{A}) = E_R(f \mid \mathscr{A}) + \lambda E_R(g \mid \mathscr{A})$, $R_\mathscr{A}$-a.e.
(d) For any sequence $(f_n)_{n\ge 1}$ in $L^1(R)$ with $0 \le f_n \le 1$, which converges *pointwise* to 0, we have: $\lim_{n\to\infty} E_R(f_n \mid \mathscr{A}) = 0$, $R_\mathscr{A}$-a.e.

Except for the "$R_\mathscr{A}$-a.e.", these properties characterize the expectation with respect to a probability measure. They can easily be checked, using (14), as follows.

(i) For any $h \in B_\mathscr{A} \ge 0$ and $f \in L^1(R) \ge 0$, (14) implies that $\int_\Omega h E_R(f \mid \mathscr{A}) \, dR_\mathscr{A} \ge 0$, which in turns implies (a).
(ii) For any $h \in B_\mathscr{A} \ge 0$, (14) implies that $\int_\Omega h E_R(1 \mid \mathscr{A}) \, dR_\mathscr{A} = \int_\Omega h \, dR_\mathscr{A}$, whence (b).
(iii) The linearity of $f \mapsto E_R(f \mid \mathscr{A})(\eta)$ comes from the linearity of $f \mapsto \int_\Omega h f \, dR$ for all $h \in B_\mathscr{A}$. Indeed, for all $f, g \in L^1(R)$ and $\lambda \in \mathbb{R}$, we have $\int_\Omega h E_R(f + \lambda g \mid \mathscr{A}) \, dR = \int_\Omega h[E_R(f \mid \mathscr{A}) + \lambda E_R(g \mid \mathscr{A})] \, dR$, which implies (c).
(iv) For any $h \in B_\mathscr{A}$, Fatou's lemma, (14) and the dominated convergence theorem lead us to $0 \le \int_\Omega h \lim_{n\to\infty} E_R(f_n \mid \mathscr{A}) \, dR_\mathscr{A} \le \liminf_{n\to\infty} \int_\Omega h E_R(f_n \mid \mathscr{A}) \, dR_\mathscr{A} = \lim_{n\to\infty} \int_\Omega h f_n \, dR = 0$. This proves (d).

We used the boundedness of R at items (ii) and (iv), since in this case, bounded functions are integrable.

One could hope that for $R_{\mathscr{A}}$-a.e. η, there exists a probability kernel $\eta \mapsto R(\cdot \mid \mathscr{A})(\eta)$ which admits $E_R(\cdot \mid \mathscr{A})$ as its expectation. But negligible sets have to be taken into account. Indeed, the $R_{\mathscr{A}}$-negligible sets which invalidate these equalities depend on the function f, g, the real numbers λ and the sequences $(f_n)_{n \geq 1}$. Their non-countable union might not be measurable, and even in this case the measure of this union might be positive. Therefore, the σ-field on Ω must not be too rich for such a probability kernel to exist. Let us give a couple of definitions before stating at Proposition 4 that $R(\cdot \mid \mathscr{A})$ exists in a general setting.

We are looking for a conditional probability measure in the following sense.

Definition 7 (Regular Conditional Probability Kernel) The kernel $R(\cdot \mid \mathscr{A})$ is a regular conditional probability if

(a) for any $f \in L^1(R)$, $E_R(f \mid \mathscr{A})(\eta) = \int_\Omega f \, R(d\omega \mid \mathscr{A})(\eta)$ for $R_{\mathscr{A}}$-almost every η;

(b) for $R_{\mathscr{A}}$-almost every η, $R(\cdot \mid \mathscr{A})(\eta)$ is a probability measure on Ω.

Property (a) was proved at Remark 1 when R is a bounded measure. It is property (b) which requires additional work, even when R is bounded. Proposition 4 provides us with a general setting where such a regular kernel exists. When a regular conditional kernel $R(\cdot \mid \mathscr{A})$ exists, (17) is concisely expressed as a disintegration formula:

$$R(d\omega) = \int_{\{\eta \in \Omega\}} R(d\omega \mid \mathscr{A})(\eta) \, R_{\mathscr{A}}(d\eta) \tag{18}$$

Definition 8 Let $\phi : \Omega \to \mathscr{X}$ be a measurable function with values in a measurable space $\mathscr{X}$. The smallest sub-σ-field on Ω which makes ϕ a measurable function is called the *σ-field generated by ϕ*. It is denoted by $\mathscr{A}(\phi)$.

We are going to consider the conditional expectation with respect to $\mathscr{A}(\phi)$ which is denoted by

$$E(\cdot \mid \mathscr{A}(\phi)) = E(\cdot \mid \phi).$$

Proposition 3 *Let $\mathscr{B}$ be the σ-field on $\mathscr{X}$ and $\phi^{-1}(\mathscr{B}) := \{\phi^{-1}(B); B \in \mathscr{B}\}$.*

1. $\mathscr{A}(\phi) = \phi^{-1}(\mathscr{B})$.
2. Any $\mathscr{A}(\phi)$-measurable function $g : \Omega \to \mathbb{R}$ can be written as

$$g = \tilde{g} \circ \phi$$

with $\tilde{g} : \mathscr{X} \to \mathbb{R}$ a measurable function.

Proof • Proof of (1). First remark that $\mathscr{A}(\phi)$ is the smallest sub-σ-field on Ω which makes ϕ a measurable function. Consequently, it is the σ-field which is generated by $\phi^{-1}(\mathscr{B})$. But it is easy to check that $\phi^{-1}(\mathscr{B})$ is a σ-field. Hence, $\mathscr{A}(\phi) = \phi^{-1}(\mathscr{B})$.

- Proof of (2). Let $y \in g(\Omega)$. As g is $\mathscr{A}(\phi)$-measurable, $g^{-1}(y) \in \mathscr{A}(\phi)$. By (1), it follows that there exists a measurable subset $B_y \subset \mathscr{X}$ such that $\phi^{-1}(B_y) = g^{-1}(y)$. Let us set

$$\tilde{g}(x) = y, \quad \text{for all } x \in B_y.$$

For any $\omega \in g^{-1}(y)$, we have $\phi(\omega) \in B_y$, so that $g(\omega) = y = \tilde{g}(\phi(\omega))$. But $(g^{-1}(y))_{y \in g(\Omega)}$ is a partition of Ω, hence $g(\omega) = \tilde{g}(\phi(\omega))$ for all $\omega \in \Omega$.

This proposition allows us to denote

$$x \in \mathscr{X} \mapsto E_R(f \mid \phi = x) \in \mathbb{R}$$

the unique function in $L^1(\mathscr{X}, R_\phi)$ such $E_R(f \mid \phi = \phi(\eta)) = E_R(f \mid \mathscr{A}(\phi))(\eta)$, R-a.e. en η.

Proposition 4 Let $R \in M_+(\Omega)$ be a bounded positive measure on Ω and $\phi : \Omega \to \mathscr{X}$ a measurable application in the Polish (separable, complete metric) space $\mathscr{X}$ equipped with the corresponding Borel σ-field. Then, $E_R(\cdot \mid \phi)$ admits a regular conditional probability kernel $x \in \mathscr{X} \mapsto R(\cdot \mid \phi = x) \in P(\Omega)$.

Proof This well-known and technically delicate result can be found at [1, Thm. 10.2.2].

In the setting of Proposition 4, the disintegration formula (18) is

$$R(d\omega) = \int_{\mathscr{X}} R(d\omega \mid \phi = x) \, R_\phi(dx).$$

The main assumption for defining properly $E_R(f \mid \mathscr{A})$ with $f \in L^1(R)$ at Definition 6 is that $R_{\mathscr{A}}$ is σ-finite. In the special case where $\mathscr{A} = \mathscr{A}(\phi)$, it is equivalent to the following.

Assumption 1 The measure $R_\phi \in M_+(\mathscr{X})$ is σ-finite.

Remark 2 (About this Assumption) It is necessary that R is σ-finite for R_ϕ to be σ-finite too. Indeed, if $(\mathscr{X}_n)_{n \geq 1}$ is a σ-finite partition of R_ϕ, $(\phi^{-1}(\mathscr{X}_n))_{n \geq 1}$ is a countable measurable partition of Ω which satisfies $R(\phi^{-1}(\mathscr{X}_n)) = R_\phi(\mathscr{X}_n) < \infty$ for all n. This means that it finitely partitions R.

Radon-Nikodym Derivative and Conditioning

In addition to the measurable mapping $\phi : \Omega \to \mathscr{X}$ and the positive measure $R \in M_+(\Omega)$, let us introduce another positive measure $P \in M_+(\Omega)$ which admits a Radon-Nikodym derivative with respect to $R : P \prec R$.

Proposition 5 *Under the Assumption 1, let us suppose that P is bounded and $P \prec R$. Then,*

1. *We have $P_\phi \prec R_\phi$ and*

$$\frac{dP_\phi}{dR_\phi}(\phi) = E_R\left(\frac{dP}{dR} \mid \phi\right), \quad R\text{-a.e.}$$

2. *For any bounded measurable function f,*

$$E_P(f \mid \phi)E_R\left(\frac{dP}{dR} \mid \phi\right) = E_R\left(\frac{dP}{dR}f \mid \phi\right), \quad R\text{-a.e.}$$

3. *Furthermore,*

$$E_R\left(\frac{dP}{dR} \mid \phi\right) > 0, \quad P\text{-a.e.}$$

Remark 3 One might not have $E_R(\frac{dP}{dR}|\phi) > 0$, R-a.e.

Proof As P is bounded, we have

$$\frac{dP}{dR} \in L^1(R)$$

and we are allowed to consider $E_R(\frac{dP}{dR}f \mid \phi)$ for any bounded measurable function f.

- Proof of (1). For any bounded measurable function u on $\mathscr{X}$,

$$E_{P_\phi}(u) = E_P(u(\phi)) = E_R\left(\frac{dP}{dR}u(\phi)\right)$$

$$= E_R\left(u(\phi)E_R\left(\frac{dP}{dR} \mid \phi\right)\right) = E_{R_\phi}\left(uE_R\left(\frac{dP}{dR} \mid \phi = \cdot\right)\right)$$

- Proof of (2). For any bounded measurable functions f, h with $h \in \mathscr{A}(\phi)$, we have

$$E_P(hf) = E_R\left(\frac{dP}{dR}hf\right) = E_R\left(hE_R\left(\frac{dP}{dR}f \mid \phi\right)\right) \quad \text{and}$$

$$E_P(hf) = E_P(hE_P(f \mid \phi)) = E_R\left(hE_P(f \mid \phi)\frac{dP}{dR}\right)$$

$$= E_R\left[hE_P(f \mid \phi)E_R\left(\frac{dP}{dR} \mid \phi\right)\right].$$

The desired result follows by identifying the right-hand side terms of these series of equalities.

- Proof of (3). Let $A \in \mathscr{A}(\phi)$ be such that $\mathbf{1}_A E_R \left(\frac{dP}{dR} \mid \phi \right) = 0$, R-a.e. Then,

$$0 = E_R \left(\mathbf{1}_A E_R \left(\frac{dP}{dR} \mid \phi \right) \right) = E_R \left(\frac{dP}{dR} \mathbf{1}_A \right) = P(A).$$ This proves the desired result.

References

1. R.M. Dudley, *Real Analysis and Probability*. Cambridge Studies in Advanced Mathematics, vol. 74 (Cambridge University Press, Cambridge, 2002). Revised reprint of the 1989 original
2. H. Föllmer, Random fields and diffusion processes, in *École d'été de Probabilités de Saint-Flour XV-XVII-1985-87*. Lecture Notes in Mathematics, vol. 1362 (Springer, Berlin, 1988)
3. N. Gozlan, C. Léonard, Transport inequalities: a survey. Markov Process. Relat. Fields **16**, 635–736 (2010)
4. F. Kelly, *Reversibility and Stochastic Networks* (Cambridge University Press, Cambridge, 2011)
5. C. Léonard, A survey of the Schrödinger problem and some of its connections with optimal transport. Discrete Contin. Dyn. Syst. A **34**(4), 1533–1574 (2014)
6. W. Rudin, *Real and Complex Analysis* (McGraw-Hill, New York, 1987)

Semi Log-Concave Markov Diffusions

P. Cattiaux and A. Guillin

Abstract In this paper we intend to give a comprehensive approach of functional inequalities for diffusion processes under various "curvature" assumptions. One of them coincides with the usual Γ_2 curvature of Bakry and Emery in the case of a (reversible) drifted Brownian motion, but differs for more general diffusion processes. Our approach using simple coupling arguments together with classical stochastic tools, allows us to obtain new results, to recover and to extend already known results, giving in many situations explicit (though non optimal) bounds. In particular, we show new results for gradient/semigroup commutation in the log concave case. Some new convergence to equilibrium in the granular media equation is also exhibited.

Keywords Functional inequalities • Transport inequalities • Diffusion processes, Coupling • Convergence to equilibrium

AMS classification (2010): 26D10, 35K55, 39B62, 47D07, 60J60

1 Introduction and Main Results

In this paper we shall investigate some properties of time marginals (at time T finite or infinite) of Markov diffusion processes satisfying some logarithmic semi-convexity like property. The properties we are interested in are functional inequalities (Poincaré, log-Sobolev) or transportation inequalities. We shall also give some consequences for the long time behavior of such processes.

P. Cattiaux (✉)
Institut de Mathématiques de Toulouse, Université de Toulouse, CNRS UMR 5219, 118 route de Narbonne, 31062 Toulouse cedex 09, France
e-mail: cattiaux@math.univ-toulouse.fr

A. Guillin
Laboratoire de Mathématiques, CNRS UMR 6620, Université Blaise Pascal, avenue des Landais, 63177 Aubière, France
e-mail: guillin@math.univ-bpclermont.fr

© Springer International Publishing Switzerland 2014
C. Donati-Martin et al. (eds.), *Séminaire de Probabilités XLVI*, Lecture Notes in Mathematics 2123, DOI 10.1007/978-3-319-11970-0_9

Our main tools are on one hand coupling techniques and on the other hand stochastic calculus. We shall mainly use the so called "synchronous" coupling, i.e. using the same Brownian motion, but we also give some new results by using the "mirror" coupling (or coupling by reflection) introduced by Lindvall and Rogers in [34]. The main stochastic tool is (a very simple form of) Girsanov theory and h-processes.

The use of coupling techniques for obtaining analytic estimates is far to be new. It is impossible (and dangerous) to give here, even an account of the existing literature (see however [44] and references therein). The use of Girsanov theory for this goal is not new too. We shall recall later some references. The conjunction of both techniques is not usual.

The meaning of logarithmic semi-convexity will generalize the "usual" one we recall now.

Let U be a smooth (C^∞) potential defined on $\mathbb{R}^n$ and satisfying for some $K \in \mathbb{R}$,

(H.C.K)   for all (x, y), $\langle \nabla U(x) - \nabla U(y), x - y \rangle \geq K |x - y|^2$.

This property is called K-semi-convexity of U. It is clearly equivalent to the convexity of $U(x) - K|x|^2$. We denote $\Upsilon(dx) = e^{-U(x)}dx$ the Boltzmann measure associated to the potential U. If e^{-U} is dx integrable, we also introduce the normalized $\mu(dx) = \frac{1}{Z_U} e^{-U(x)} dx$ which is a probability measure. If U is semi-convex, μ is said to be semi log-concave.

Consider first the diffusion process, given by the solution of the Ito stochastic differential system

$$dX_t = dB_t - \frac{1}{2} \nabla U(X_t) \, dt;$$ (1)

$$\mathcal{L}(X_0) = \mu_0.$$

B being a standard Brownian motion. It is known that (1) has an unique non explosive strong solution, in particular on can build a solution on any probability space equipped with some Brownian motion. This is an easy consequence of Hasminski's explosion test using the Lyapunov function $x \mapsto |x|^2$.

Usual notations are in force: for a nice enough f, $P_t f(x) = \mathbb{E}(f(X_t^x))$ where X^x denotes a solution such that $\mu_0 = \delta_x$; L denotes the infinitesimal generator i.e.

$$L = \frac{1}{2} \Delta - \frac{1}{2} \langle \nabla U, \nabla \rangle,$$

and Γ denotes the carré du champ, namely here

$$\Gamma(f, g) = \frac{1}{2} \langle \nabla f, \nabla g \rangle \text{ and for simplicity } \Gamma(f) = \Gamma(f, f).$$

$\mathbb{P}_{\mu_0}$ will denote the law of the solution of (1), abridged in $\mathbb{P}_x$ when $\mu_0 = \delta_x$, i.e. $\mathbb{P}$ is defined on the usual space Ω of continuous paths; μ_t will denote the law of X_t for $t \geq 0$ and $P(t, x, .)$ denotes the law of X_t^x.

It is known that Υ is a symmetric (reversible) measure for the diffusion process, and is actually the unique invariant (stationary) measure for the process. If Υ is bounded, μ is ergodic.

In the latter case, P_t is thus a symmetric semi-group on $L^2(\mu)$. The domain $D(L)$ of its generator contains the algebra $\mathscr{A}$ generated by the constant functions and C_c^∞. In particular, if $f \in \mathscr{A}$, $\partial_t P_t f = L P_t f = P_t L f$ in $L^2(\mu)$, so that since $\partial_t - L$ is hypo-elliptic $(t, x) \mapsto P_t f(x) \in C^\infty$.

L is the basic example of generator satisfying the celebrated $C(K/2, +\infty)$ Bakry-Emery curvature condition (see [1]). Indeed if we define

$$\Gamma_2(f) = \frac{1}{2} \left(L\Gamma(f) - 2\Gamma(f, Lf) \right),$$

(H.C.K) is equivalent to $\Gamma_2(f) \geq (K/2)\,\Gamma(f)$.

This curvature condition is known to imply (and is in fact equivalent to) a lot of nice inequalities for the semi-group, in particular for all $T > 0$ and all x, a commutation between Γ and the semi group P_t holds, namely

$$\Gamma(P_T f) \leq e^{-KT} P_T \left(\sqrt{\Gamma(f)} \right)^2, \tag{2}$$

which in turn implies powerful functional inequalities such as

$$P(T, x, .)\text{satisfies a log-Sobolev inequality with constant } \frac{4}{K}(1 - e^{-KT}). \tag{3}$$

Recall that ν satisfies a (usual) log-Sobolev inequality with constant C_{LS} if

$$\text{Ent}_\nu(f) := \int f^2 \log(f^2)\,d\nu - \left(\int f^2 d\nu \right) \log \left(\int f^2 d\nu \right) \leq C_{LS} \int \Gamma(f)\,d\nu. \tag{4}$$

(3) is exactly what is (a little bit improperly) called a "local" log-Sobolev inequality in [1, Theorem 5.4.7]. For further informations and more, see the forthcoming book [5].

It is well known that a log-Sobolev inequality implies a (usual) Poincaré inequality

$$\text{Var}_\nu(f) := \int f^2\,d\nu - \left(\int f d\nu \right)^2 \leq C_P \int \Gamma(f)\,d\nu, \tag{5}$$

with $C_P = \frac{1}{2} C_{LS}$, as well as a T_2 transportation inequality

$$W_2^2(\eta, \nu) \leq C_W H(\eta|\nu), \tag{6}$$

with $C_W = C_{LS}$. Here W_2 denotes the Wasserstein distance between the probability measures η and ν, i.e.

$$W_2^2(\eta, \nu) = \frac{1}{2} \inf_\pi \int |x - y|^2 \, \pi(dx, dy),$$

where π is a coupling of η and ν (i.e. has respective marginals equal to η and ν) and

$$H(\eta|\nu) = \int \left(\frac{d\eta}{d\nu}\right) \log \left(\frac{d\eta}{d\nu}\right) d\nu,$$

denotes the Kullback-Leibler information or relative entropy of η w.r.t. ν. The latter property is due to Otto-Villani [36]. Another approach and related properties were developed by Bobkov et al. [9]. For a nice survey on transportation inequalities we refer to [26]. One can find in all these references another remarkable consequence of semi log-concavity, namely that a log-Sobolev inequality derives from a transportation inequality. This is a consequence of the following (H.W.I) inequality that holds for any nice μ density of probability h,

(**H.W.I**) If (H.C.K) holds then

$$H(h\mu|\mu) \leq \left(2 \int \frac{|\nabla h|^2}{h} \, d\mu\right)^{\frac{1}{2}} W_2(h\mu, \mu) - K \, W_2^2(h\mu, \mu).$$

As a consequence, if (H.C.K) holds for some $K \leq 0$, a T_2 transportation inequality for μ implies a log-Sobolev inequality with constant $C_{LS} \leq (4/C_W)(1 + (K/C_W))^{-2}$ provided $1 + (K/C_W) > 0$, in particular if $K = 0$.

Let us finally remark that the starting point of this approach is the Γ_2 commutation property (2) which fails however to give a direct proof of the T_2 inequality.

Our first goal is to show that functional and transportation inequalities can be derived, in the previous situation, by using coupling techniques and simple tools of stochastic calculus.

Proving (2). As a warming up, let us see how (2) can be easily derived, just using synchronous coupling, i.e. the processes X^x and X^y built with *the same* Brownian motion.

Applying Ito formula yields (almost surely)

$$e^{Kt} |X_t^x - X_t^y|^2$$

$$= |x - y|^2 + \int_0^t \left(K|X_s^x - X_s^y|^2 - \langle \nabla U(X_s^x) - \nabla U(X_s^y), X_s^x - X_s^y \rangle\right) e^{Ks} \, ds$$

$$\leq |x - y|^2,$$

thanks to (H.C.K). Hence,

$$|X_t^x - X_t^y| \le e^{-Kt/2} |x - y|, \quad \mathbb{P} \; a.s. \tag{7}$$

so that, using the mean value theorem,

$$|P_t f(x) - P_t f(y)| \le \mathbb{E}(|f(X_t^x) - f(X_t^y)|) \le e^{-Kt/2} \, \mathbb{E}(|\nabla f(z_t)| \, |x - y|)$$

for some z_t sandwiched by X_t^x and X_t^y. It remains to use the continuity (and boundedness) of ∇f and the fact that X_t^y goes almost surely to X_t^x as $y \to x$ to conclude. $\diamond$

Now consider a classical diffusion process, given by the solution of an Ito stochastic differential system

$$dX_t = \sigma(X_t) \, dB_t + b(X_t) \, dt; \tag{8}$$
$$\mathscr{L}(X_0) = \mu_0,$$

B being a standard Brownian motion. For simplicity we assume that σ is a squared matrix. Assumptions will be made ensuring again strong uniqueness and non explosion. The notations used previously are still in force.

Generalizing (2). Let us mimic what we did previously, using again the synchronous coupling and applying Ito formula. We get

$$e^{Kt} |X_t^x - X_t^y|^2 = |x - y|^2 + \int_0^t 2e^{Ks} \langle \sigma(X_s^x) - \sigma(X_s^y), X_s^x - X_s^y \rangle \, dB_s$$

$$+ \int_0^t (K|X_s^x - X_s^y|^2 + |\sigma(X_s^x) - \sigma(X_s^y)|_{HS}^2$$

$$+ 2 \langle b(X_s^x) - b(X_s^y), X_s^x - X_s^y \rangle) \, e^{Ks} \, ds.$$

This suggests to extend (H.C.K), say for Higher Convexity of order K, to this new situation

(H.C.K) $\quad \forall (x, y), \; |\sigma(x) - \sigma(y)|_{HS}^2 + 2 \langle b(x) - b(y), x - y \rangle \le -K |x - y|^2.$
$$\tag{9}$$

Indeed, if (H.C.K) holds

$$e^{Kt} |X_t^x - X_t^y|^2 \le |x - y|^2 + \int_0^t 2e^{Ks} \langle \sigma(X_s^x) - \sigma(X_s^y), X_s^x - X_s^y \rangle \, dB_s. \tag{10}$$

If the right hand side of (10) is a (true) martingale, we obtain

$$\mathbb{E}(|X_t^x - X_t^y|^2) \le e^{-Kt} |x - y|^2 . \tag{11}$$

Let $f \in \mathscr{A}$, then

$$f(X_t^x) - f(X_t^y) \le \langle \nabla f(X_t^y), X_t^x - X_t^y \rangle + C|X_t^x - X_t^y|^2$$

for some constant C, so that

$$\begin{aligned}
|P_t f(x) - P_t f(y)| &= |\mathbb{E}(f(X_t^x) - f(X_t^y))| \\
&\le |\mathbb{E}(\langle \nabla f(X_t^y), X_t^x - X_t^y \rangle)| + C\mathbb{E}(|X_t^x - X_t^y|^2) \\
&\le \left(\mathbb{E}(|\nabla f(X_t^y)|^2)\right)^{\frac{1}{2}} \left(\mathbb{E}(|X_t^x - X_t^y|^2)\right)^{\frac{1}{2}} + C\mathbb{E}(|X_t^x - X_t^y|^2) \\
&\le e^{-Kt/2} \left(P_t(|\nabla f|^2)(y)\right)^{\frac{1}{2}} |x - y| + C\, e^{-Kt} |x - y|^2 ,
\end{aligned}$$

so that, provided we know that $\nabla P_t f$ exists, we have obtained a weaker form of (2),

$$|\nabla P_t f|^2 \le e^{-Kt} P_t(|\nabla f|^2) . \tag{12}$$

We shall now explain that (12) is in general very different from (2).   ◇

Notice that if σ and b are C-Lipschitz, (H.C.K) is satisfied for $K = -(C^2 n^2 + C)$, but if σ is C-Lipschitz, (H.C.K) can be satisfied for a non-negative K provided b is sufficiently repealing. Contrary to the case of a constant diffusion coefficient, (H.C.K) is not related to the Bakry-Emery curvature condition which involves in this situation controls on derivatives of higher order of the coefficients.

For simplicity in the sequel we shall assume that $\sigma \in C_b^3$ hence is C-Lipschitz and that b is C^3, but not necessarily bounded nor with bounded derivatives. With these assumptions, once again if we assume that (H.C.K) is in force, (8) admits a unique non explosive strong solution using $x \mapsto x^2$ as a Lyapunov function for non explosion. We shall show this and other properties of the process in Sect. 3.1.

We still use the notations introduced before, but now

$$L = \frac{1}{2} \sigma \sigma^* \nabla^2 + b\nabla = \frac{1}{2} a \nabla^2 + b\nabla ,$$

and Γ the carré du champ is now

$$\Gamma(f, g) = \frac{1}{2} \langle \sigma \nabla f , \sigma \nabla g \rangle .$$

Notice that, contrary to the Bakry-Emery bounded curvature case, the previous commutation property (12) holds with the *usual gradient* and not with the *natural* one i.e. $\Gamma^{\frac{1}{2}}$.

If the proof we gave of (2) is more or less part of the folklore, the previous proof of (12) already appeared in a slightly different form in Lemma 2.2 of [40].

Starting from (2), the Γ_2 calculus of Bakry-Emery uses this commutation property and the control of the derivative of $\psi(s) = P_s(P_{t-s}f \log(P_{t-s}f))$ to get a "local" logarithmic Sobolev inequality, i.e. a logarithmic Sobolev inequality satisfied by $P(T, x, .)$ for all finite T. Note that considering rather $\psi(s) = P_s((P_{t-s}f)^2)$ leads to a local Poincaré inequality. A similar calculation can be done, at least in the uniformly elliptic situation, in the general case starting from (2), to again prove a Poincaré inequality (see e.g. [45]).

In the next two sections we propose an alternate method based on the use of h-processes. This idea is close to the one used in [14] in the stationary case, and in the paper by Djellout et al. [22, Theorem 5.6] (condition 4.5 therein being exactly (H.C.K) for $K > 0$) where these authors are looking at transportation inequalities on the path space. What we shall show is that the same scheme of proof also furnishes functional inequalities. This unified treatment of functional inequalities and transportation inequalities using an ad-hoc coupling is the novelty here.

Another interest is that the method does not require uniform ellipticity of a so that some hypo-coercive examples enter this framework as explained in Sect. 3.5. In this situation, Bakry-Emery curvature is actually equal to $-\infty$.

It also extends to the non (time-)homogeneous situation, see Sect. 5. In addition in this section we show how to directly obtain convergence to equilibrium and properties of the invariant measure for non linear diffusions of Mc Kean-Vlasov type, simplifying arguments in [35].

The case of positive curvature ($K > 0$) in (2) is important since it implies exponential decay to the equilibrium μ_∞ (or of $P_t f$ to $\int f \, d\mu_\infty$ in variance or entropy). It also provides the exponential *contraction* in W_2 distance, i.e for all initial μ_0 and ν_0,

$$W_2^2(\mu_t, \nu_t) \leq e^{-Kt} W_2^2(\mu_0, \nu_0)$$

as it clearly derives from (7). Actually as shown first by Sturm and Von Renesse [43], this contraction property is equivalent to positive curvature K. A similar statement holds under the general (H.C.K) assumption in the elliptic case (see Theorem 1).

In full generality however, (H.C.K) for a positive K, only implies the following Poincaré inequality for the invariant measure:

$$\mathrm{Var}_{\mu_\infty}(f) \leq \frac{M}{K} \int |\nabla f|^2 \, d\mu_\infty, \tag{13}$$

where M denotes the uniform norm of a. This inequality is in general strictly weaker than (5) (recall that $\Gamma(f) = |\sigma \nabla(f)|^2$), except in the uniformly elliptic case. In particular it is not sufficient to ensure the exponential decay of $\mathrm{Var}_{\mu_\infty}(P_t f)$ to 0.

If μ_∞ is not only invariant but *symmetric*, it was remarked in [17, Remark 4.9], that an exponential decay of the Wasserstein distance

$$W_2(\mu_t, \mu_\infty) \leq C \, e^{-Kt} \, W_2(\mu_0, \mu_\infty)$$

for some $C \geq 1$, implies that μ_∞ satisfies a T_2 inequality (6), and consequently a Poincaré inequality (5). This property thus holds in positive curvature.

But still in the symmetric case, one can reinforce the previous result:

Proposition 1 *Assume that for all bounded (resp. Lipschitz) density of probability h we have $W_0(P_t h\mu, \mu) \leq c_h(t)$ (reps. W_1). Then for all bounded (resp. Lipschitz and bounded) f, there exist c_f and h such that $Var_\mu(P_t f) \leq c_f\, c_h(2t)$.*

In particular if $c_h(t) = c_h\, e^{-\beta t}$, μ satisfies a Poincaré inequality (5).

The latter statement is a consequence of the following lemma one can find for instance in [19, Lemma 2.12]:

Lemma 1 *Assume that P_t is μ-symmetric. Then, if there exists $\beta > 0$ such that for all f in a dense subset of $\mathbb{L}^2(\mu)$ there exists c_f with $Var_\mu(P_t f) \leq c_f\, e^{-\beta t}$ then $Var_\mu(P_t f) \leq e^{-\beta t}\, Var_\mu(f)$ for all $f \in \mathbb{L}^2(\mu)$.*

Hence μ satisfies a Poincaré inequality with constant $C_P \leq 1/\beta$.

It is thus interesting to look at other Wasserstein distances, in particular W_1.

W_1 and Synchronous Coupling. Using again the synchronous coupling and applying Ito formula, we have

$$
e^{Kt}|X_t^x - X_t^y| = |x - y| + \int_0^t e^{Ks} \left\langle \sigma(X_s^x) - \sigma(X_s^y), \frac{X_s^x - X_s^y}{|X_s^x - X_s^y|} \right\rangle dB_s
$$

$$
+ \int_0^t K|X_s^x - X_s^y| e^{Ks} ds
$$

$$
+ \int_0^t \frac{e^{Ks}}{2|X_s^x - X_s^y|} \left(|\sigma(X_s^x) - \sigma(X_s^y)|_{HS}^2 \right.
$$

$$
\left. + 2 \langle b(X_s^x) - b(X_s^y), X_s^x - X_s^y \rangle \right) ds
$$

$$
- \int_0^t \frac{1}{2|X_s^x - X_s^y|} \left| (\sigma(X_s^x) - \sigma(X_s^y)) \left(\frac{X_s^x - X_s^y}{|X_s^x - X_s^y|} \right) \right|^2 e^{Ks} ds,
$$

almost surely for $t < T_C$, where T_C denotes the coupling time, i.e. the first time when $X_t^x = X_t^y$. After this time both processes coincide thanks to pathwise uniqueness.

The gain with respect to (H.C.K) replacing $|\sigma(x) - \sigma(y)|_{HS}^2$ by

$$
|\sigma(x) - \sigma(y)|_{HS}^2 - \left| (\sigma(x) - \sigma(y)) \left(\frac{x - y}{|x - y|} \right) \right|^2
$$

is mainly irrelevant except in the one dimensional case where the latter quantity is equal to 0.

But even in this case an exponential decay of W_1 will furnish some weak commutation property for the gradient and the semi-group namely

$$|\nabla P_t f| \leq C\, e^{-Kt}\, \|\nabla f\|_\infty, \tag{14}$$

which, nevertheless, allows us to derive weak functional inequalities.   ◇

Another approach for studying the exponential decay of W_1 was proposed by Eberle [23, 24]. Instead of synchronous coupling it uses the *mirror* (or *reflection*) coupling introduced by Lindvall and Rogers [34], and then extended by Cranston. Eberle's method allows him to look at drifted Brownian motions when the drift satisfies some "uniform convexity at infinity" property, i.e. when (H.C.K) is satisfied for some $K > 0$ but for $|x - y|$ large enough.

This situation cannot be treated by using synchronous coupling.

We recall Eberle's method and obtain some new consequences of his result in Sect. 7. In addition, up to an extra condition, we show that his result (and all the consequences we derived) can be extended to general elliptic diffusion processes. We will also use this mirror coupling to show that we may get a weak version of the commutation property in the log concave case with the "convexity at infinity" property at least in dimension one, which is the first result we know of in this direction. Still in dimension one, we will also consider using mirror coupling for non linear diffusions.

Let us come back to (1). Among potentials U satisfying Eberle's condition, one find those written as $U = V + W$ with V K-uniformly convex and W Lipschitz continuous. That $\mu(dx) = e^{-U(x)}dx$ satisfies a log-Sobolev (and a Poincaré) inequality is already known in this situation, and also when W is bounded, with a constant $C_{LS} = (4/K)\exp(\mathrm{Osc}\,W)$ where Osc denotes the oscillation of W. But, both approaches are "dimensional", i.e. furnish constants which are dimension dependent as for the celebrated double well case $|x|^4 - |x|^2$, but even for convex potentials which are not uniformly convex like $|x|^4$.

In Sect. 6, we introduce the following extension of (H.C.K).

Let α be a non decreasing function defined on $\mathbb{R}^+$. We shall say that $(\mathbf{H}.\boldsymbol{\alpha}.\mathbf{K})$ is satisfied for some $K > 0$ if for all (x, y) and all $\varepsilon > 0$,

$$\langle \nabla U(x) - \nabla U(y), x - y \rangle \geq K\,\alpha(\varepsilon)\,(|x - y|^2 - \varepsilon).$$

When $\alpha(a) = 1$, we may take $\varepsilon = 0$ and we recognize (H.C.K). In Proposition 15 we show that $U(x) = |x|^{2\beta}$ (with $\beta \geq 1$) satisfies (H.α.K_β) for $\alpha(a) = a^{\beta-1}$ and an explicit $K_\beta > 0$.

The main result of this section is then that, for suitable functions α,

if (H.α.K) holds (for $K > 0$), then μ satisfies a log-Sobolev inequality.

See Theorems 9 and 10. These theorems thus (partly) extend the Bakry-Emery criterion (3) to some non uniformly convex potentials. However, they are dealing with the invariant measure only and not with the law at time T (only incomplete results are proved in this section for these distributions). Finally they provide some explicit bound, but presumably not optimal.

Section 8 is peculiar. Using the results we have described for the Ornstein-Uhlenbeck process we show how to recover known results on the stability of functional inequalities under convolution (provided one of the terms is gaussian).

The use of stochastic calculus in deriving such inequalities is not new but only a small number of papers dealt with. One can trace back to the paper of Borell [13], who used Girsanov theory to study the propagation of log-concavity along the Schrödinger dynamics (not the Fokker-Planck one we are looking at here). In addition to [22] for transportation inequalities, and [40, 45], one can also mention [14, 15] where similar ideas are used to study hyper-boundedness. More recently, using similar arguments, Lehec [33] has studied gaussian functional inequalities and Fontbona and Jourdain [25] obtained a pathwise version of the Γ_2 theory.

Since we will meet several types of inequalities, from now on, in the whole paper, we shall say that ν satisfies a log-Sobolev inequality with constant C_{LS} if

$$\text{Ent}_\nu(f) := \int f^2 \log(f^2)\,d\nu - \left(\int f^2 d\nu\right) \log\left(\int f^2 d\nu\right) \leq C_{LS} \int |\nabla f|^2\,d\nu .$$
(15)

Similarly for the Poincaré inequality, replacing the entropy by the variance.

2 Semi Log-Concave Drifted Brownian Motion

In this first warming up section we shall look at the simplest situation given by (1) and derive the classical inequalities. Though the methods are the same in the general case, we prefer to detail the proofs first in this simpler situation. *We emphasize that the results of this section are very well known. What is new is the method of proof.*

2.1 Commutation Property

Recall the result we proved in the introduction using synchronous coupling:

Proposition 2 *In the situation of (1), assume (H.C.K). Then for all $f \in \mathscr{A}$,*

$$W_2(P_t(x,\cdot), P_t(y,\cdot)) \leq e^{-Kt/2}|x - y|,$$

$$|\nabla P_t f| \leq e^{-Kt/2} P_t|\nabla f| .$$
(16)

Remark 1 If instead of (x, y) the processes start with initial distribution π_0 the "optimal coupling" between μ_0 and ν_0 for the W_2 distance, the previous shows that $W_2^2(\mu_T, \nu_T) \leq e^{-KT} W_2^2(\mu_0, \nu_0)$. $\diamond$

2.2   h-Processes and Functional Inequalities

We now introduce the standard notion of h-process. Let $T > 0$ and h be a non-negative function such that $\int h \, d\mu_T = 1$. For simplicity, we assume for the moment that there exist c and C such that $C \geq h \geq c > 0$. We thus may define on the path-space up to time T a new probability measure

$$\frac{d\mathbb{Q}}{d\mathbb{P}_{\mu_0}}\Big|_{\mathscr{F}_T} = h(\omega_T).$$

It is immediately seen that

$$\mathbb{Q} \circ \omega_s^{-1} = P_{T-s}h \, \mu_s \quad \text{for all } 0 \leq s \leq T.$$

In this situation, it is well known (Girsanov transform theory) that one can find a progressively measurable process u_s such that

$$\frac{d\mathbb{Q}}{d\mathbb{P}_{\mu_0}}\Big|_{\mathscr{F}_T} = P_T h(\omega_0) \exp\left(\int_0^T \langle u_s, dM_s \rangle - \frac{1}{2}\int_0^T |u_s|^2 \, ds\right),$$

where ω denotes the canonical element of the path-space and M denotes the martingale part of ω under $\mathbb{P}_{\mu_0}$. In addition, it is easily seen (see e.g. [20]) that

$$H(\mathbb{Q}|\mathbb{P}_{\mu_0}) = H(h\mu_T|\mu_T) = H(P_T h\mu_0|\mu_0) + \frac{1}{2}\mathbb{E}^{\mathbb{Q}}\left(\int_0^T |u_s|^2 \, ds\right). \quad (17)$$

Actually, if $h \in \mathscr{A}$, it is immediate to check (applying Ito formula) that

$$u_s = \nabla \log P_{T-s}h(\omega_s)$$

both $\mathbb{P}_{\mu_0}$ and $\mathbb{Q}$ almost surely.

(17) thus becomes

$$H(h\mu_T|\mu_T) = H(P_T h\mu_0|\mu_0) + \frac{1}{2}\int_0^T \left(\int \frac{|\nabla P_s h|^2}{P_s h} \, d\mu_{T-s}\right) ds. \quad (18)$$

If h is smooth we may apply Proposition 2 in order to get

$$H(h\mu_T|\mu_T) \leq H(P_T h\mu_0|\mu_0) + \frac{1}{2}\int_0^T \left(\int e^{-Ks} \frac{P_s^2(|\nabla h|)}{P_s h} \, d\mu_{T-s}\right) ds$$

$$\leq H(P_T h\mu_0|\mu_0) + \frac{1}{2}\int_0^T e^{-Ks} \left(\int P_s \left(\frac{|\nabla h|^2}{h}\right) d\mu_{T-s}\right) ds$$

$$\leq H(P_T h\mu_0|\mu_0) + \frac{1}{2} \int_0^T e^{-Ks} \left(\int \frac{|\nabla h|^2}{h} d\mu_T \right) ds$$

$$\leq H(P_T h\mu_0|\mu_0) + \frac{1 - e^{-KT}}{2K} \int \frac{|\nabla h|^2}{h} d\mu_T , \qquad (19)$$

where we have used Cauchy-Schwarz inequality for the second inequality and the Markov property for the third one. The previous inequality then extends to any h in C^1 for which the right hand side makes sense, by density.

We have thus obtained the following

Proposition 3 *In the situation of (1), assume (H.C.K). If μ_0 satisfies a log-Sobolev inequality with constant $C_{LS}(0)$, μ_T satisfies a log-Sobolev inequality with constant*

$$C_{LS}(T) = e^{-KT} C_{LS}(0) + \frac{2(1 - e^{-KT})}{K} .$$

When $K = 0$ one has to replace $\frac{(1-e^{-KT})}{K}$ by T. This applies in particular to $\mu_T = P(T, x, .)$ since δ_x satisfies a log-Sobolev inequality with constant equal to 0.

Proof Apply the log-Sobolev inequality to μ_0. It furnishes (since $\int P_T h d\mu_0 = 1$),

$$H(P_T h\mu_0|\mu_0) \leq \frac{C_{LS}(0)}{4} \int \frac{|\nabla P_T h|^2}{P_T h} d\mu_0 \leq e^{-KT} \frac{C_{LS}(0)}{4} \int \frac{|\nabla h|^2}{h} d\mu_T ,$$

similarly as what we did in (19). Hence the result applying (19).

As we recalled in the introduction a log-Sobolev inequality implies a T_2 transportation inequality. It is interesting to see that one can directly obtain such an inequality for semi log-concave measures, by using the previous construction. But before to do this, just remark that the above proof using $h = 1 + \varepsilon g$ with $\int g d\mu_T = 0$ allows us to obtain a similar result replacing the log-Sobolev inequality by a Poincaré inequality i.e.

Proposition 4 *In the situation of (1), assume (H.C.K). If μ_0 satisfies a Poincaré inequality with constant $C_P(0)$, μ_T satisfies a Poincaré inequality with constant*

$$C_P(T) = e^{-KT} C_P(0) + \frac{1 - e^{-KT}}{K} .$$

When $K = 0$ one has to replace $\frac{(1-e^{-KT})}{K}$ by T. This applies in particular to $\mu_T = P(T, x, .)$ since δ_x satisfies a Poincaré inequality with constant equal to 0.

2.3 Transportation Inequalities

The existence of u_s and (17) are ensured as soon as $H(h\mu_T|\mu_T) < +\infty$ (see [20]). For our goal we do not need the explicit expression of u_s.

Indeed, Girsanov theory and Paul Lévy characterization of Brownian motion tell us that on $(\Omega, \mathbb{Q})$, there exists some standard Brownian motion w (independent of ω_0) such that, up to time T,

$$\omega_t = \omega_0 + w_t - \frac{1}{2}\int_0^t \nabla U(\omega_s)\,ds + \int_0^t u_s\,ds.$$

Since (1) has an unique *strong* solution, one can build (on $(\Omega, \mathbb{Q})$) a solution of

$$z_t = z_0 + w_t - \frac{1}{2}\int_0^t \nabla U(z_s)\,ds,$$

the law of which being given by

$$\mathbb{P}_{\nu_0} \quad \text{with} \quad \nu_0 = \mathscr{L}(z_0).$$

For instance we may choose $\nu_0 = \mu_0$ or $z_0 = \omega_0$ in which case $\nu_0 = P_T h\mu_0$. But in all situations we choose the distribution of (ω_0, z_0) in such a way that $\mathbb{E}^{\mathbb{Q}}(|\omega_0 - z_0|^2) = 2W_2^2(\nu_0, P_T h\mu_0)$ (or we take approximating sequences).

In particular

$$z_t - \omega_t = (z_0 - \omega_0) + \frac{1}{2}\int_0^t (\nabla U(\omega_s) - \nabla U(z_s))\,ds - \int_0^t u_s\,ds,$$

$\mathbb{Q}$ almost surely. Applying Ito's formula and (H.C.K) we obtain

$$\eta_t := \mathbb{E}^{\mathbb{Q}}(|z_t - \omega_t|^2) \tag{20}$$

$$\leq \mathbb{E}^{\mathbb{Q}}(|z_0 - \omega_0|^2) - K\int_0^t \eta_s\,ds + 2\int_0^t \mathbb{E}^{\mathbb{Q}}|\langle(z_s - \omega_s), u_s\rangle|\,ds$$

$$\leq \eta_0 - K\int_0^t \eta_s\,ds + 2\left(\int_0^t \eta_s\,ds\right)^{\frac{1}{2}}\left(\mathbb{E}^{\mathbb{Q}}\left(\int_0^t |u_s|^2\,ds\right)\right)^{\frac{1}{2}} \tag{21}$$

$$\leq \eta_0 - K\int_0^t \eta_s\,ds + 2\sqrt{2}\,H^{\frac{1}{2}}(h\mu_T|\mu_T)\left(\int_0^t \eta_s\,ds\right)^{\frac{1}{2}}.$$

We have then different situations depending on the sign of K.

1. First in the case where $K > 0$, one has using that $2ab \leq Ka^2 + b^2/K$

$$
\eta_t \leq \eta_0 - K \int_0^t \eta_s \, ds + 2\sqrt{2} \, H^{\frac{1}{2}}(h\mu_T|\mu_T) \left(\int_0^t \eta_s \, ds \right)^{\frac{1}{2}}
$$

$$
\leq \eta_0 + \frac{2}{K} H^{\frac{1}{2}}(h\mu_T|\mu_T)
$$

so that we recover an uniform transportation inequality when $\eta_0 = 0$, which is moreover optimal for the invariant measure, considering logarithmic Sobolev inequality and Poincaré inequality. If μ_0 satisfies some transportation inequality then one obtains that μ_T satisfies a transportation inequality with constant the sum of the initial constant plus $\frac{2}{K}$.

2. The previous simple argument has however a serious drawback in the sense that in positive curvature, μ_T does not forget the "initial measure". Let us see how to deal with this problem. Start once again from the first estimation, but using Itô's formula between t and $t + \varepsilon$ and (H.C.K)

$$
\eta_{t+\varepsilon} \leq \eta_t - K \int_t^{t+\varepsilon} \eta_s \, ds + 2 \int_t^{t+\varepsilon} \mathbb{E}^{\mathbb{Q}} |\langle (z_s - \omega_s), u_s \rangle| \, ds
$$

so that we may differentiate in time to get for all positive λ

$$
\eta_t' \leq -K\eta_t + 2\mathbb{E}^{\mathbb{Q}} |\langle (z_t - \omega_t), u_t \rangle| \, ds,
$$

$$
\leq -(K + \lambda)\eta_t + \frac{1}{\lambda} \mathbb{E}^{\mathbb{Q}} |u_t|^2.
$$

Using Gronwall's lemma, we get that

$$
\eta_T \leq e^{(-K+\lambda)T} \eta_0 + \frac{1}{\lambda} \int_0^T e^{(K-\lambda)(s-T)} \mathbb{E}^{\mathbb{Q}} |u_t|^2 \, dt.
$$

so that if $K > 0$ we get, for $\lambda < K$

$$
\eta_T \leq e^{(-K+\lambda)T} \eta_0 + \frac{1}{\lambda} \int_0^T \mathbb{E}^{\mathbb{Q}} |u_t|^2 \, dt \leq e^{(-K+\lambda)T} \eta_0 + \frac{2}{\lambda} H(h\mu_T|\mu_T).
$$

Note that this is once again optimal for the limiting measure, and captures the fact that it forgets the initial condition. When $K < 0$, we then have

$$
\eta_T \leq e^{(-K+\lambda)T} \eta_0 + \frac{2}{\lambda} e^{(-K+\lambda)T} H(h\mu_T|\mu_T).
$$

Note however the presence of the additional parameter λ.

3. Let us see how a direct approach may get rid of this additional parameter, which is particularly important in negative curvature. Define

$$a_t = e^{Kt} \int_0^t \eta_s \, ds - \frac{e^{Kt}}{K} \eta_0 .$$

We have

$$a_t' \leq 2\sqrt{2} \, e^{Kt/2} \, H^{\frac{1}{2}}(h\mu_T | \mu_T) \left(a_t + \frac{e^{Kt}}{K} \eta_0 \right)^{\frac{1}{2}} .$$

Since $\frac{e^{Kt}}{K} \leq \frac{e^{KT}}{K}$ we obtain

$$\left(a_t + \frac{e^{KT}}{K} \eta_0 \right)^{\frac{1}{2}} \leq \left(a_0 + \frac{e^{KT}}{K} \eta_0 \right)^{\frac{1}{2}} + 2\sqrt{2} \, \frac{e^{Kt/2} - 1}{K} \, H^{\frac{1}{2}}(h\mu_T | \mu_T) .$$

It follows

$$\left(\int_0^T \eta_s \, ds \right)^{\frac{1}{2}} \leq \left(\frac{1 - e^{-KT}}{K} \eta_0 \right)^{\frac{1}{2}} + 2\sqrt{2} \left(\frac{1 - e^{-KT/2}}{K} \right) H^{\frac{1}{2}}(h\mu_T | \mu_T) .$$

For $K \geq 0$, this yields, since $W_2^2(h\mu_T, \nu_T) \leq \frac{1}{2} \mathbb{E}^{\mathbb{Q}}(|z_t - \omega_t|^2)$, and using $\sqrt{a} \sqrt{b} \leq \frac{1}{2}(a + b)$,

$$W_2^2(h\mu_T, \nu_T) \leq \left(1 + \sqrt{2} \, \frac{1 - e^{-KT}}{K} \right) W_2^2(P_T h\mu_0, \nu_0)$$

$$+ \left(\frac{\sqrt{2}}{2} + \frac{4(1 - e^{-KT/2})}{K} \right) H(h\mu_T | \mu_T) . \qquad (22)$$

If $\eta_0 = 0$, (22) can be improved in

$$W_2^2(h\mu_T, \nu_T) \leq \frac{4(1 - e^{-KT/2})}{K} H(h\mu_T | \mu_T) . \qquad (23)$$

When $K \leq 0$, we obtain

$$W_2^2(h\mu_T, \nu_T) \leq \left(1 + \sqrt{2} \, \frac{1 - e^{-KT}}{K} + 2(e^{-KT} - 1) \right) W_2^2(P_T h\mu_0, \nu_0) \qquad (24)$$

$$+ \left(\frac{\sqrt{2}}{2} + 4 \frac{(1 - e^{-KT/2})}{K} - 4 \frac{(1 - e^{-KT/2})^2}{K} \right) H(h\mu_T | \mu_T) .$$

Again if $\eta_0 = 0$, (24) can be improved in

$$W_2^2 (h\mu_T, \nu_T) \leq 4 \left(\frac{(1 - e^{-KT/2})}{K} - \frac{(1 - e^{-KT/2})^2}{K} \right) H(h\mu_T | \mu_T). \quad (25)$$

The previous inequalities then extend to any non-negative h (not necessarily bounded below nor above).

If we choose $\mu_0 = \delta_x$, we have $\mu_T = P(T, x, .)$, $1 = \int h d\mu_T = P_T h(x)$ and so $\nu_0 = \delta_x$ and $\nu_T = \mu_T$. Hence

Proposition 5 *In the situation of (1), assume (H.C.K). Then $P(T, x, .)$ satisfies a T_2 transportation inequality*

$$W_2^2 (hP(T, x, .), P(T, x, .)) \leq C_T H(hP(T, x, .) | P(T, x, .)),$$

with

$$C_T = \min \left(\frac{2}{K}, \frac{4(1 - e^{-KT/2})}{K} \right)$$

when $K > 0$, $2T$ when $K = 0$ and

$$C_T = \frac{4(1 - e^{-KT/2})}{K} - 4 \frac{(1 - e^{-KT/2})^2}{K}$$

when $K \leq 0$.

If we choose $\nu_0 = \mu_0$, we may use the convexity of $t \mapsto t \log t$, i.e

$$H(P_T h\mu_0 | \mu_0) = \int P_T h \, \log P_T h \, d\mu_0 \leq \int P_T(h \, \log h) \, d\mu_0 = H(h\mu_T | \mu_T),$$

in order to get

Proposition 6 *In the situation of (1), assume (H.C.K). If μ_0 satisfies T_2 with constant $C(0)$, then μ_T satisfies T_2 with a constant $C(T)$ given,*

1. when $K > 0$, for $0 < \lambda < K$,

$$C(T) = e^{-(K-\lambda)T} C(0) + \frac{2}{\lambda},$$

2. and when $K \leq 0$, $C(T) = C_T + \frac{\sqrt{2}}{2} + B_T C(0)$ with

$$B_T = 1 + \sqrt{2} \frac{1 - e^{-KT}}{K} + 2(e^{-KT} - 1).$$

Remark 2 All what precedes holds even if Υ is not bounded (i.e. if the process is not positive recurrent), in which case of course, $K < 0$.   ◇

Remark 3 If we choose $\mu_0 = \mu$ (assuming that Υ is bounded), we have to choose $v_0 = P_T h \mu$ hence $v_T = P_{2T} h \mu_0$. After noticing that we can slightly refine the previous bound replacing $H(h\mu_T | \mu_T)$ by $H(h\mu_T | \mu_T) - H(P_T h \mu_0 | \mu_0)$ according to (17), we obtain

$$W_2(P_{2T} h \mu, h \mu) \leq \sqrt{C_T \left(H(h\mu | \mu) - H(P_T h \mu | \mu) \right)}$$

and finally

$$W_2(h\mu, \mu) \leq \sqrt{C_T \left(H(h\mu | \mu) - H(P_T h \mu | \mu) \right)} + W_2(P_{2T} h \mu, \mu). \tag{26}$$

The latter has to be compared with Remark 4.9 in [17] which shows that the inequality

$$W_2(h\mu, \mu) \leq \sqrt{T \left(H(h\mu | \mu) - H(P_T h \mu | \mu) \right)} + W_2(P_T h \mu, \mu)$$

always holds.   ◇

Remark 4 If $K > 0$ we may let T go to $+\infty$ in Proposition 3 and recover that μ satisfies a log-Sobolev inequality with constant $2/K$, hence a T_2 transportation inequality with constant $1/K$ (in particular we are loosing a factor 4 in Proposition 5).

Similarly, when $T \to +\infty$, (26) shows that if $K > 0$, μ satisfies a T_2 inequality, and since μ is log-concave, satisfies a log-Sobolev inequality. This scheme of proof does not require Proposition 2, but the (H.W.I) inequality. Unfortunately it does not furnish the optimal constant.   ◇

2.4 Transportation-Fisher Inequalities

Let us look now at another type of Transportation Information inequality recently introduced in [28], which is weaker but quite close to logarithmic Sobolev inequality (in fact equivalent under bounded curvature). For simplicity we assume here that $K \geq 0$. We are obliged to come back to the initial inequality in (20) which becomes in our new situation

$$\eta_t \leq \eta_0 - K \int_0^t \eta_s \, ds + 2 \int_0^t \mathbb{E}^{\mathbb{Q}} \left(|z_s - \omega_s| \, |\nabla \log P_{T-s} h(\omega_s)| \right) ds. \tag{27}$$

Replacing the pair $(0, t)$ by $(t, t + \varepsilon)$ we thus have

$$\eta_{t+\varepsilon} \leq \eta_t - K \int_t^{t+\varepsilon} \eta_s \, ds + 2 \int_t^{t+\varepsilon} \eta_s^{\frac{1}{2}} \left(\mathbb{E}^{\mathbb{Q}} \left(|\nabla \log P_{T-s} h(\omega_s)|^2 \right) \right)^{\frac{1}{2}} ds$$

$$\leq \eta_t - K \int_t^{t+\varepsilon} \eta_s \, ds + 2 \int_t^{t+\varepsilon} \eta_s^{\frac{1}{2}} \left(\int \frac{|\nabla P_{T-s} h|^2}{P_{T-s} h} \, d\mu_s \right)^{\frac{1}{2}} ds .$$

It follows that $t \mapsto \eta_t$ is differentiable and satisfies,

$$\eta_t' \leq - K \, \eta_t + 2 \, \eta_t^{\frac{1}{2}} \left(\int \frac{|\nabla P_{T-t} h|^2}{P_{T-t} h} \, d\mu_t \right)^{\frac{1}{2}}$$

$$\leq - K \, \eta_t + 2 \, \eta_t^{\frac{1}{2}} \left(\int \frac{P_{T-t}^2 |\nabla h|}{P_{T-t} h} \, d\mu_t \right)^{\frac{1}{2}}$$

$$\leq - K \, \eta_t + 2 \, \eta_t^{\frac{1}{2}} \left(\int P_{T-t} \left(\frac{|\nabla h|^2}{h} \right) \, d\mu_t \right)^{\frac{1}{2}}$$

$$\leq - K \, \eta_t + 2 \, \eta_t^{\frac{1}{2}} \left(\int \frac{|\nabla h|^2}{h} \, d\mu_T \right)^{\frac{1}{2}} . \tag{28}$$

(for the second inequality, recall that (H.C.0) is satisfied so that, for short, $|\nabla P_s| \leq P_s |\nabla|$.) To explore (28) we shall use the usual trick $ab \leq \lambda a^2 + \frac{1}{\lambda} b^2$ for a, b, λ positive. Hence

$$\eta_t' \leq (-K + 2\lambda) \, \eta_t + \frac{2}{\lambda} \left(\int \frac{|\nabla h|^2}{h} \, d\mu_T \right) . \tag{29}$$

We deduce, denoting $A = K - 2\lambda$,

$$W_2^2(h\mu_T, \mu_T) \leq \eta_T \leq \eta_0 \, e^{-AT} + \frac{2(1 - e^{-AT})}{A\lambda} \int \frac{|\nabla h|^2}{h} \, d\mu_T .$$

This inequality is close to what is called a $W_2 I$ inequality (see [26, Definition 10.4] or [28] for examples and details on properties of WI inequality). Here we obtain a defective $W_2 I$ inequality. However, as $T \to +\infty$, we recover the true $W_2 I$ inequality for the invariant distribution, which together with the (H.W.I) inequality allows us to recover the log-Sobolev inequality. Nevertheless, we get

Proposition 7 *Assume (H.C.K) for some $K \geq 0$. Then, for all $\lambda < K/2$, $P(T, x, \cdot)$ satisfies a WI inequality of constant $\frac{2(1-e^{-(K-2\lambda)T})}{(K-2\lambda)\lambda}$. If we suppose moreover that μ_0 satisfies a WI inequality with constant $D(0)$ then μ_T satisfies a WI inequality with constant $D(T) = e^{-(K-2\lambda)T} D(0) + \frac{2(1-e^{-(K-2\lambda)T})}{(K-2\lambda)\lambda}$.*

Remark 5 As remarked, under (H.C.K), the inequalities verified by the law μ_T depend on the inequalities verified by the initial measure, in the range between Poincaré and logarithmic Sobolev inequality. Indeed, a logarithmic Sobolev inequality implies a WI inequality, but to get the WI inequality for P_T we need only a WI inequality for the initial measure. As seen by the example of the Gaussian measure, which satisfies (H.C.K), no stronger inequalities can be obtained. ◇

3 General Diffusion Processes

We shall now extend the results of the previous section to the general situation of (8),

$$dX_t = \sigma(X_t)\, dB_t + b(X_t)\, dt. \tag{30}$$

$$\mathcal{L}(X_0) = \mu_0.$$

First of all we have to discuss some properties of the process and the associated quantities.

3.1 Some Properties of the Process

3.1.1 Non Explosion

Since we assume that $\sigma \in C_b^2$, when (H.C.K) is fulfilled, b satisfies

$$2\langle b(x) - b(y), x - y\rangle \leq -D\,|x - y|^2,$$

for some $D \in \mathbb{R}$. In particular,

$$2\langle b(x), x\rangle \leq -D\,|x|^2 + 2\,|b(0)||x|.$$

Thus, if S_k denotes the exit time from the ball $B(x,k)$, and $t_k = t \wedge S_k$ it holds

$$\mathbb{E}(|X_{t_k}^x|^2) = |x|^2 + \mathbb{E}\left(\int_0^{t_k} Trace(a(X_s^x)) + 2\langle b(X_s^x), X_s^x\rangle\, ds\right)$$

$$\leq |x|^2 + Nt + |D|\int_0^t \mathbb{E}(|X_{s_k}^x|^2)\, ds + 2|b(0)|\int_0^t \mathbb{E}(|X_{s_k}^x|)\, ds$$

$$\leq |x|^2 + (N + 2|b(0)|)t + (|D| + 2|b(0)|)\int_0^t \mathbb{E}(|X_{s_k}^x|^2)\, ds$$

where, since σ is bounded, we have defined

$$N = \| Trace(a(.)) \|_\infty,$$

and where we used $|y| \leq 1 + |y|^2$. Applying Gronwall lemma we obtain that $\mathbb{E}(|X_{t_k}^x|^2)$ is bounded independently on k, so that we may pass to the limit in k. This proves non explosion up to time t (since the explosion time is the increasing limit of the sequence S_k) for all t.

It is then easily seen that one can perform similar calculations with $g(t, x) = \exp(e^{-Ct}|x|^2)$ for a large enough C in order to kill the integrated term, i.e

Lemma 2 *There exists a large enough $C_e > 0$, such that $\mathbb{E}(\exp(e^{-C_e t}|X_t^x|^2)) \leq e^{|x|^2}$.*

It is interesting to notice that one can similarly obtain some "deviation" bound from the starting point. Indeed arguing as before, one can show the existence of constants $\alpha(T, D)$ and $\beta(T, D)$ such that for $0 \leq t \leq T$,

$$\mathbb{E}(|X_t^x - x|^2) \leq (\alpha(T, D)N + \beta(T, D)|b(x)|^2)\, t\,. \tag{31}$$

3.1.2 Properties of the Semi-group

Recall that, if (H.C.K) holds

$$e^{Kt}|X_t^x - X_t^y|^2 \leq |x - y|^2 + \int_0^t 2e^{Ks}\langle \sigma(X_s^x) - \sigma(X_s^y), X_s^x - X_s^y\rangle\, dB_s\,. \tag{32}$$

Notice that with our assumptions, the right hand side of (32) is a (true) martingale, so that

$$\mathbb{E}(|X_t^x - X_t^y|^2) \leq e^{-Kt}|x - y|^2\,. \tag{33}$$

Hence

Theorem 1 *If (H.C.K.) holds true, then*

$$W_2(P_t(x, \cdot), P_t(y, \cdot)) \leq e^{-Kt/2}|x - y|\,. \tag{34}$$

Moreover, if $\sigma\sigma^*$ is positive, (34) implies back (H.C.K.).
If we suppose moreover for some $m \geq 2$, the condition (H.C.K.m): $\forall (x, y)$

$$\frac{m}{2}|\sigma(x) - \sigma(y)|_{HS}^2 + m\langle b(x) - b(y), x - y\rangle$$

$$+m\left(\frac{m}{2} - 1\right)\frac{\|(x - y)(\sigma(x) - \sigma(y))^t\|^2}{\|x - y\|^2} \leq -K|x - y|^2$$

then

$$W_m(P_t(x, \cdot), P_t(y, \cdot)) \leq e^{-Kt/(m)}|x - y|\,. \tag{35}$$

Proof The contraction in W_2 distance inherited from (H.C.K.) has already been proved. The contraction in W_m distance is done exactly in the same way using once again synchronous coupling. The necessary part comes from [10] (or more precisely Sect. 4. in the Arxiv version 1110.3606). Let us explain the ideas of the proof. In fact, one may compute the time derivative of the Wasserstein distance: note $M : N = \sum_{i,j} M_{ij} N_{ij}$ when M and N are two matrices, then denoting ν_t and μ_t two solutions starting respectively from ν_0 and μ_0

$$\frac{d}{dt} W_2(\nu_t, \mu_t = 2\, J(\nu_t | \mu_t)$$

where if $\nu_t = \nabla \phi_t \# \mu_t$,

$$J(\nu_t, |\mu_t) = \int \Big[\frac{1}{2} \sigma \sigma^*(x) : (\nabla^2 \phi_t(x) - I)$$

$$+ \frac{1}{2} \sigma \sigma * (\nabla \phi_t(x)x) : (\nabla^2 \phi_t(x)^{-1} - I) - \langle b(\nabla \phi_t(x))$$

$$- b(x), \nabla \phi_t(x) - x \rangle \Big] d\mu_t.$$

Then the contraction property implies that at time 0 for $\nu_0 = \delta_y$ and $\mu_0 = \delta_x$

$$\frac{K}{2} |x - y|^2 \le J(\nu_0, \mu_0).$$

A clever choice of ϕ then enables to prove the result.

Let f be C-Lipschitz continuous. It holds $|f(X_t^x) - f(X_t^y)| \le C\, |X_t^x - X_t^y|$ so that, using (11), $P_t f$ is Lipschitz continuous with Lipschitz constant less than $C\, e^{-Kt/2}$.

As we said in the introduction, when $K > 0$ one deduces the existence and uniqueness of an invariant probability measure μ_∞, to which μ_T converges weakly.

In order to mimic what we have done in the previous section, we need the following: if $f \in \mathscr{A}$ (see the introduction), then $(t, x) \mapsto P_t f(x)$ is regular and satisfies (for $t > 0$)

$$\partial_t P_t f = P_t L f = L P_t f .$$

First if $f \in \mathscr{A}$, Lf is C_c^0 and we have $P_t f(x) - f(x) = \int_0^t P_s(Lf)(x)\, ds$. It follows that $\lim_{s \to 0} \frac{1}{s} (P_{t+s} f(x) - P_t f(x)) = P_t(Lf)(x)$ for all x, since $\nu \mapsto P_\nu Lf(x)$ is continuous. So $\partial_t P_t f = P_t L f$.

Now we need to know about the smoothness of $(t, x) \mapsto P_t f(x)$. But according to Theorems 40 and 49 in chapter V of Protter's book [38], $x \mapsto X_t^x$ is N times differentiable up to its explosion time, provided σ and b are $N + 1$ times differentiable. So if we assume that σ and b are C^3, and that (H.C.K) holds true, we easily get the required regularity of P_t.

The commutation of L and P_t can then be proved as in [30] (where it was assumed that the coefficients belong to C_b^∞).

3.2 Commutation Property with the Gradient

Since we know that $\nabla P_t f$ exists, the calculations made in the introduction furnished a weaker form of Proposition 2. If σ is constant, we may use the arguments of the previous section.

Proposition 8 *Assume (H.C.K) or the weaker contraction property (34). Let $f \in \mathscr{A}$. It holds*

$$|\nabla P_t f|^2 \le e^{-Kt} P_t(|\nabla f|^2).$$

If σ is a constant matrix, we have the stronger

$$|\nabla P_t f| \le e^{-Kt/2} P_t(|\nabla f|).$$

Notice that, contrary to the Bakry-Emery bounded curvature case, the previous commutation property holds with the *usual gradient* and not with the *natural* one i.e. $\Gamma^{\frac{1}{2}}$.

If Proposition 2 allowed us to obtain logarithmic Sobolev inequalities, the weaker Proposition 8 will allow us to obtain a weaker inequality, namely a Poincaré inequality.

Remark 6 It is worth mentioning here the following alternate proof of the commutation property, starting from Wasserstein contraction, as derived in the recent paper [4] following our suggestion, i.e. using Kantorovitch-Rubinstein duality we have for all bounded Lipschitz ϕ denoting the inf convolution operator $Q_t\phi(x) = \inf_y\{\phi(y) + \frac{|x-y|^2}{2t}\}$ and initial measure μ_0 and ν_0

$$\int Q_1\phi d\mu_t - \int \phi d\nu_t = \int P_t Q_1\phi d\mu_0 - \int P_t\phi d\nu_0$$
$$\le e^{-Kt} W_2^2(\mu_0, \nu_0).$$

Choose now $\mu_0 = \delta_x$, $\nu_0 = \delta_y$ to get for all y

$$P_t(Q_1\phi)(x) \le P_t\phi(y) + \frac{|x - y|^2}{2e^{Kt}}$$

which by homogeneity of the inf-convolution operator gives

$$P_t(Q_1\phi) \le Q_{e^{Kt}}(P_t\phi).$$

This assertion is in fact stronger than the gradient commutation property which can be deduced by using the fact that the inf-convolution operator is the Hopf-Lax solution of the Hamilton-Jacobi equation.   ◇

3.3   h-Processes and Functional Inequalities

We now introduce the corresponding h-process. Let $T > 0$ and $h > 0$ be such that

$$\int P_T h \, d\mu_0 = 1$$

We thus may define on the path-space up to time T a new probability measure

$$\frac{d\mathbb{Q}}{d\mathbb{P}_{\mu_0}}\Big|_{\mathscr{F}_T} = h(\omega_T).$$

Again

$$\mathbb{Q} \circ \omega_s^{-1} = P_{T-s} h \, \mu_s \quad \text{for all} \quad 0 \le s \le T.$$

For simplicity, we assume in what follows that there exist c and C such that $C \ge h \ge c > 0$. In this situation, using again Girsanov transform theory, we know that we can find a progressively measurable process u_s such that

$$\frac{d\mathbb{Q}}{d\mathbb{P}_{\mu_0}}\Big|_{\mathscr{F}_T} = P_T h(\omega_0) \exp\left(\int_0^T \langle u_s, dM_s \rangle - \frac{1}{2}\int_0^T |\sigma(\omega_s)u_s|^2 \, ds\right),$$

where ω denotes the canonical element of the path-space and M denotes the martingale part of ω under $\mathbb{P}_{\mu_0}$. In addition, it can be shown [20] that

$$H(\mathbb{Q}|\mathbb{P}_{\mu_0}) = H(h\mu_T|\mu_T) = H(P_T h\mu_0|\mu_0) + \frac{1}{2}\mathbb{E}^{\mathbb{Q}}\left(\int_0^T |\sigma(\omega_s)u_s|^2 \, ds\right), \tag{36}$$

and

$$u_s = \nabla \log P_{T-s} h(\omega_s)$$

both $\mathbb{P}_{\mu_0}$ and $\mathbb{Q}$ almost surely.

We thus have

$$H(h\mu_T|\mu_T) = H(P_T h\mu_0|\mu_0) + \frac{1}{2}\int_0^T \left(\int \frac{|\sigma \nabla P_s h|^2}{P_s h} \, d\mu_{T-s}\right) ds. \tag{37}$$

Now define

$$M = \| \, |\sigma|^2 \, \|_\infty = \sup_y \sup_{|u|=1} |\sigma(y)u|^2 .$$

If $h \in \mathscr{A}$ we may apply Proposition 8 in order to get (recall that $h \geq c$)

$$H(h\mu_T|\mu_T) \leq H(P_T h\mu_0|\mu_0) + \frac{M}{2} \int_0^T \left(\int e^{-Ks} \frac{P_s(|\nabla h|^2)}{P_s h} \, d\mu_{T-s} \right) ds$$

$$\leq H(P_T h\mu_0|\mu_0) + \frac{M}{2c} \int_0^T e^{-Ks} \left(\int |\nabla h|^2 \, d\mu_T \right) ds$$

$$\leq H(P_T h\mu_0|\mu_0) + \frac{M(1 - e^{-KT})}{2cK} \int |\nabla h|^2 \, d\mu_T , \qquad (38)$$

where we have used the Markov property for the second inequality.

Now let $g \in C_c^\infty$ be such that $\int g \, d\mu_T = \int P_T g \, d\mu_0 = 0$ and choose $h = 1 + \eta g \in \mathscr{A}$ so that $\int P_T h \, d\mu_0 = 1$ and $h > c > 0$ for η small enough. Actually we will let η go to 0 so that in the limit $c = 1$. Standard manipulations thus yield

$$\int g^2 \, d\mu_T \leq \int (P_T g)^2 \, d\mu_0 + \frac{M(1 - e^{-KT})}{K} \int |\nabla g|^2 \, d\mu_T . \qquad (39)$$

We can eventually use first the density of C_c^∞ so that, arguing as for Proposition 3 we have obtained

Proposition 9 *Assume that (H.C.K) is satisfied. Let* $M = \| \, |\sigma|^2 \, \|_\infty$.
If μ_0 *satisfies a Poincaré inequality with constant* $C_P(0)$ *then* μ_T *satisfies a Poincaré inequality with constant*

$$C_P(T) = e^{-KT} C_P(0) + \frac{M(1 - e^{-KT})}{K} .$$

This applies in particular to $P(T, x, .)$ *with* $C_P(0) = 0$.
If σ *is a constant matrix, a similar statement holds with the log-Sobolev constant instead of the Poincaré constant, replacing* M *by* $2M$.

Contrary to the log-Sobolev inequality, the Poincaré inequality does not furnish a transportation inequality, so we shall try to adapt what we did in Sect. 2.3.

3.4 Transportation Inequalities

The situation is a little bit less simple than in the previous section. Indeed the martingale term is no more a Brownian motion and we can no more use characterization tricks on $(\Omega, \mathbb{Q})$. Hence we have to consider the solution of

$$dY_t = \sigma(Y_t)\, dB_t + b(Y_t)dt + a(Y_t)\, \nabla \log P_{T-t}h(Y_t)\, dt. \tag{40}$$

As before we assume first that $h \in \mathscr{A}$, $C \geq h \geq c > 0$ so that (40) is well defined and admits a unique strong solution. We can thus build a solution with the same Brownian motion B we used in (8). Strong uniqueness follows from the local Lipschitz property of all the coefficients and non explosion (up to time T) which is ensured by construction ($\mathbb{Q}$ is a probability measure). Again we may choose in an appropriate way the distribution of the pair of initial variables.

If (H.C.K) is satisfied, it holds

$$\eta_t = \mathbb{E}(|Y_t - X_t|^2) \tag{41}$$

$$\leq \eta_0 - K \int_0^t \eta_s\, ds \tag{42}$$

$$+2 \left(\int_0^t \mathbb{E}(\langle Y_s - X_s, a(Y_s)\, \nabla \log P_{T-s}h(Y_s)\rangle) \right) ds$$

$$\leq \eta_0 - K \int_0^t \eta_s\, ds +$$

$$+2M^{\frac{1}{2}} \left(\int_0^t \eta_s\, ds \right)^{\frac{1}{2}} \left(\mathbb{E}\left(\int_0^t |\sigma(Y_s)\nabla \log P_{T-s}h(Y_s)|^2\, ds \right) \right)^{\frac{1}{2}}$$

$$\leq \eta_0 - K \int_0^t \eta_s\, ds + 2(2M)^{\frac{1}{2}} \left(\int_0^t \eta_s\, ds \right)^{\frac{1}{2}} H^{\frac{1}{2}}(h\mu_T | \mu_T). \tag{43}$$

We may thus conclude as in the previous section

Proposition 10 *Assume that (H.C.K) is satisfied. Let $M = \|\, |\sigma|^2\, \|_\infty$.*
The conclusions of Propositions 5 and 6 are still true, replacing C_T by MC_T

Of course a T_2 inequality implies a Poincaré inequality, but the constant in Proposition 9 is better (in addition we only require that μ_0 satisfies a Poincaré inequality).

Remark 7 One of the renowned consequence of such inequalities is the concentration of measure phenomenon for μ_T. In particular, under the assumptions of Proposition 10, μ_T satisfies a gaussian type concentration property. In particular $|X_T^x|^2$ has some exponential moment, fact we have already shown in Lemma 2. But this integrability does not reflect all the strength of the T_2 inequality whose tensorization property is particularly useful for statistical purposes.

When L is uniformly elliptic, this concentration property follows from gaussian estimates for the transition kernel. Here we obtain much more explicit constants (even if they are certainly far from optimality) which do not depend on the ellipticity constant. $\diamond$

Remark 8 Assume that L is uniformly elliptic, i.e.

$$e = \inf_{y} \inf_{|u|=1} |\sigma(y)u|^2 > 0.$$

Then we deduce from Proposition 9

$$P_T g^2(x) - (P_T g(x))^2 \leq \frac{2M}{e} \frac{(1 - e^{-KT})}{K} P_T(\Gamma g)(x).$$

According to [1, Proposition 5.4.1], this is equivalent to the $CD(K/2, \infty)$ condition provided $M = e$ hence when σ is constant times the identity. In the non constant diffusion case, our condition (H.C.K) seems to be really different from the Bakry-Emery curvature condition.   ◇

3.5   An Hypoelliptic Example: Kinetic Fokker-Planck Equation

We present in this section an application of the techniques developed here in an hypoelliptic example where the Bakry-Emery curvature is $-\infty$ and where (H.C.K.) may not be satisfied also.

Let (x_t, v_t) be the solution of the following SDE

$$dx_t = v_t dt$$
$$dv_t = dB_t - \nabla V(x_t)dt - v_t dt.$$

also called stochastic Hamiltonian system. The long time behavior study of such a system has been considered for a long time and have been tackled by different techniques, see for example: hypocoercivity by Villani [41] or Lyapunov function technique by Bakry et al. [3]. However, due to its hight degeneracy, the Bakry-Emery curvature is $-\infty$ so that we may not apply the Γ_2 technique. Remark also that the (H.C.K.) condition reads for all (x, v) and (y, w)

$$-\langle \nabla V(x) - \nabla V(y), v - w \rangle - |v - w|^2 \leq -K(|x - y|^2 + |v - w|^2)$$

so that it is hopeless to get $K > 0$.

Let us first remark that if ∇V is Lipshitz continuous, (H.C.K) is verified for some negative K and σ a constant diagonal matrix, so that we get that for some negative K the gradient commutation property holds

$$|\nabla P_t f| \leq e^{-Kt} P_t |\nabla f|$$

and thus the logarithmic Sobolev inequality holds for $P_t((x, v), \cdot)$. Let us remark once again that those properties are written with the usual gradient and not the Carré-du-Champ operator $\Gamma(f) = |\nabla_v f|^2$.

One may then wonder if it is possible to get the gradient commutation property with $K > 0$. In fact, using synchronous coupling and Itô's formula applied to the function $N((x, v), (y, w)) = a|x - y|^2 + b\langle x - y, v - w \rangle + |v - w|^2$, following [12], we get that if $V(x) = |x|^2 + W(x)$ where ∇W is δ-Lipschitz with δ sufficiently small there exists a, b and $K > 0$ such that N is equivalent to the euclidean norm and

$$N((x_t^x, v_t^v), (x_t^y, v_t^w)) \le e^{-Kt} N((x, v), (y, w))$$

so that we get as in Sect. 2 the commutation property for some $K > 0$ and $A > 1$

$$|\nabla P_t f| \le A e^{-Kt} P_t |\nabla f|$$

and thus a Logarithmic Sobolev inequality holds uniformly in time.

The previous method lies on the fact that one can replace the usual euclidean distance by another one, which is equivalent but more appropriate for the calculations. The same idea will be used with the mirror coupling.

It is not hard to extend the result of this simplified setting to the case where the Brownian motion in the velocity has a diffusion coefficient which is bounded and L-Lipschitz. We may then obtain a weaker gradient commutation property

$$|\nabla P_t f|^2 \le A e^{-Kt} P_t |\nabla f|^2$$

and local Poincaré type inequality or Transportation information inequality like in Propositions 9 or 10, and if L is sufficiently small uniform in time version of these inequalities (using functional N).

3.6 Interpolation of the Gradient Commutation Property and Local Beckner Inequality

We have seen here that we cannot recover a logarithmic Sobolev inequality by our technique when (H.C.K.) is in force, except when σ is a constant matrix. Remember however that we have introduced the stronger (H.C.K.m) condition which implies a contraction in Wasserstein distance W_m. It is then not hard to deduce some interpolation of the gradient commutation property

Proposition 11 *Assume (H.C.K.m) or the weaker contraction property (35). Let $f \in \mathscr{A}$, then*

$$|\nabla P_t f|^{\frac{m}{m-1}} \leq e^{-Kt/(m-1)} P_t \left(|\nabla f|^{\frac{m}{m-1}} \right). \tag{44}$$

Remark once again that this property does hold even if the diffusion coefficient is degenerate, so that variations of the hypoelliptic example of the previous subsection with a diffusion coefficient in the velocity enters into this framework. This contraction property may thus lead to a reinforcement of the Poincaré inequality to a Beckner inequality.

Proposition 12 *Assume (H.C.K.m) or the weaker (44). Let $M = \||\sigma|^2\|_\infty$. Then for all nice f, we have the following Beckner inequality*

$$P_t f^2 - P_t \left(|f|^{\frac{2m}{m+2}} \right)^{\frac{m+2}{m}} \leq M \frac{m+2}{m} \frac{1 - e^{2Kt/m}}{K} P_t |\nabla f|^2.$$

The proof is similar to the one of Wang [44] Proposition 6.3.9.

4 Convergence to Equilibrium in Positive Curvature

When $K > 0$ we already mentioned that μ_T weakly converges to the unique invariant probability measure μ_∞ (which exists).

In particular, for all smooth g (say C_b^2), $\text{Var}_{\mu_T}(g) \to \text{Var}_{\mu_\infty}(g)$ as well as $\int |\nabla g|^2 d\mu_T \to \int |\nabla g|^2 d\mu_\infty$. We deduce that if σ is uniformly elliptic,

$$\text{Var}_{\mu_\infty}(g) \leq \frac{M}{K} \int |\nabla g|^2 d\mu_\infty \leq \frac{2M}{eK} \int \Gamma(g) d\mu_\infty.$$

Summarizing all this we have obtained

Theorem 2 *Assume that σ is bounded and uniformly elliptic. Then if (H.C.K) holds for some $K > 0$, defining M and e as before, there exists an unique invariant probability measure μ_∞ and μ_∞ satisfies a Poincaré inequality with constant M/K. In addition for all $f \in \mathbb{L}^2(\mu_\infty)$ it holds*

$$\text{Var}_{\mu_\infty}(P_t f) \leq e^{-KeT/M} \text{Var}_{\mu_\infty}(f).$$

As we said, this result is not captured by the Γ_2 theory.

But we can obtain general convergence results, even in the non uniformly elliptic case. Indeed recall that in full generality

$$\text{Var}_{\mu_\infty}(P_t g) = \frac{1}{2} \int_t^{+\infty} \int |\sigma \nabla P_s g|^2 \, d\mu_\infty \, ds \,.$$

Using Proposition 8 we thus have

$$\text{Var}_{\mu_\infty}(P_t g) \leq \frac{M}{2} \int_t^{+\infty} \int |\nabla P_s g|^2 \, d\mu_\infty \, ds$$

$$\leq \frac{M}{2} \int_t^{+\infty} e^{-Ks} \int P_s(|\nabla g|^2) \, d\mu_\infty \, ds$$

$$\leq \frac{M}{2K} e^{-Kt} \int |\nabla g|^2 \, d\mu_\infty \,.$$

Hence

Theorem 3 *Assume that σ is bounded. Then if (H.C.K) holds for some $K > 0$, defining M as before, there exists an unique invariant probability measure μ_∞ and for all nice enough function g,*

$$\text{Var}_{\mu_\infty}(P_t g) \leq \frac{M}{2K} e^{-Kt} \int |\nabla g|^2 \, d\mu_\infty \,.$$

In addition if μ_∞ is symmetric (i.e. $\int fLg \, d\mu_\infty = \int gLf \, d\mu_\infty$), it holds

$$\text{Var}_{\mu_\infty}(P_t g) \leq e^{-Kt} \, \text{Var}_{\mu_\infty}(g) \,.$$

Remark once again that what is used here is the weak gradient commutation property which is a consequence of (H.C.K.). The last part of the theorem follows from Lemma 1. Of course, unless we explicitly know the invariant measure, it is not easy to see wether μ_∞ is symmetric or not.

We may further extends the previous argument to the entropic convergence to equilibrium. Let us suppose that there exists an unique invariant measure μ_∞.

Theorem 4 *Assume that σ is bounded and that the gradient commutation property*

$$|\nabla P_t f| \leq c \, e^{-Kt} P_t |\nabla f| \tag{45}$$

holds for some positive K. Then for all nice positive function f (defining M as before)

$$\text{Ent}_{\mu_\infty}(P_t f) \leq \frac{cM}{K} e^{-Kt} \int \frac{|\nabla g|^2}{g} \, d\mu_\infty \,.$$

Proof The proof is as for the L_2 decay quite standard. Indeed,

$$\text{Ent}_{\mu_\infty}(P_t g) \leq M \int_t^\infty \int \frac{|\nabla P_s g|^2}{P_s g} d\mu_\infty ds,$$

$$\leq cM \int_t^\infty e^{-Ks} \int P_s \frac{|\nabla g|^2}{g} d\mu_\infty ds,$$

$$\leq \frac{cM}{K} e^{-Kt} \int \frac{|\nabla g|^2}{g} d\mu_\infty.$$

Remark 9 One of the important point here is that we do not suppose any non-degeneracy on the diffusion coefficient, so that the result applies to the kinetic Fokker-Planck equation. It then provides an alternative to the approach by Villani [41], where he obtained such kind of convergence by completely different techniques with assumptions quite similar to the ones described in Sect. 3.5. One may then complete the approach by regularization of the Fisher Information in small time to obtain an entropic decay controlled by the initial entropy, see [41] or [27].   ◇

Remark 10 Let us point out that even in the symmetric case, such a control is not sufficient to recover a logarithmic Sobolev inequality as the analog of Lemma 1 is no more valid for the entropy. Remark however that we have shown in Sect. 2 how to recover a logarithmic Sobolev inequality for P_t using the strong commutation gradient property (45). If $K > 0$, we may then let t goes to infinity to recover a logarithmic Sobolev inequality for the invariant measure. It may be, for example, used in the context of kinetic Fokker-Planck equation with non gradient coefficient, for which the invariant measure is unknown.   ◇

Remark 11 Let us consider, as in the Poincaré case via (H.C.K.) condition, a particular class of test function g such that $g \geq \varepsilon > 0$, so that $P_t g \geq \varepsilon$. We then see adapting the preceding proof that a weak commutation of gradient property

$$|\nabla P_t f|^2 \leq c\, e^{-Kt} P_t |\nabla f|^2 \tag{46}$$

obtained for example under (H.C.K.) condition implies that

$$\text{Ent}_{\mu_\infty}(P_t g) \leq \frac{cM}{\varepsilon K} e^{-Kt} \int |\nabla g|^2 d\mu_\infty.$$

As we said in the introduction, an exponential decay of Wasserstein distances furnishes some Poincaré inequality for μ. In what follows W_0 denotes the total variation distance and W_1 is the usual 1-Wasserstein distance. We restate Proposition 1

Proposition 13 *Assume that μ is reversible. Assume that for all bounded (resp. Lipschitz) density of probability h we have $W_0(P_t h\mu, \mu) \leq c_h(t)$ (reps. W_1). Then for all bounded (resp. Lipschitz and bounded) f, there exist c_f and h such that*

$Var_\mu(P_t f) \leq c_f c_h(2t)$. In particular if $c_h(t) = c_h e^{-\beta t}$, μ satisfies a Poincaré inequality.

Proof Let f be bounded and centered, and

$$h = (f + \parallel f \parallel_\infty)/\int (f + \parallel f \parallel_\infty)\, d\mu = 1 + (f/\parallel f \parallel_\infty).$$

h is thus a density of probability with $\parallel h \parallel_\infty \leq 2$. We have

$$\mathrm{Var}_\mu(P_t f) = \parallel f \parallel_\infty^2 \, \mathrm{Var}_\mu(P_t h)$$

$$\leq \parallel f \parallel_\infty^2 \int P_t h(P_t h - 1)\, d\mu = \parallel f \parallel_\infty^2 \int h(P_{2t}h - 1)\, d\mu$$

$$\leq \parallel f \parallel_\infty^2 \parallel h \parallel_\infty \, W_0(P_{2t}h\mu, \mu) \leq 2 \parallel f \parallel_\infty^2 \, c_h(2t).$$

One can replace W_0 by W_1, just replacing $\parallel h \parallel_\infty$ by $\parallel \nabla h \parallel_\infty$ in which case

$$\mathrm{Var}_\mu(P_t f) \leq \parallel f \parallel_\infty \parallel \nabla f \parallel_\infty \, c_h(2t).$$

Remark 12 The previous result partly extends to the non symmetric situation. Indeed if we do not use the symmetry of P_t, but only the fact that $\parallel P_t \parallel \leq 1$ in $\mathbb{L}^\infty$ we obtain that provided $W_0(P_t h\mu, \mu) \leq c_h(t)$,

$$\mathrm{Var}_\mu(P_t f) \leq 2 \parallel f \parallel_\infty^2 \, c_h(t). \qquad \diamond$$

Even when the decay is not exponential, one gets a weak form of the Poincaré inequality (called a weak Poincaré inequality).

Corollary 1 *In the situation of Proposition 13, assume that $c_h(t) = c_h c(t)$ with $c(t) \to 0$ as $t \to +\infty$. Then*

$$Var_\mu(f) \leq \alpha(s) \int |\nabla f|^2\, d\mu + s\, \Psi(f),$$

for all $s > 0$ where $\Psi(f) = c_h \parallel f - \int f d\mu \parallel_\infty^2$ for W_0 and $\Psi(f) = c_h \parallel f - \int f d\mu \parallel_\infty \parallel \nabla f \parallel_\infty$ for W_1 and $\alpha(s) = s \inf_{u>0} \frac{1}{u} c^{-1}(u \exp(1 - (u/s)))$; h being defined in the proof of Proposition 13 .

Proof Once we notice that the transformation $f \mapsto \lambda f$ does not change h the result follows from [39, Theorem 2.3].

Here is an application in dimension 1, suggested by Jourdain (2013, Contraction of one-dimensional stochastic differential equations in Wasserstein distance, personal communication) who has others very interesting results in one dimension for the control of Wasserstein distances.

If $X_.$ is a solution of (8) in dimension 1, assume that

$$\langle b(x) - b(y), x - y \rangle \leq -K |x - y|^2. \tag{47}$$

As we showed in the introduction, up to the synchronous coupling time

$$e^{Kt} |X_t^x - X_t^y| \leq |x - y| + M_t$$

where M_t is a martingale term. It follows that $W_1(\mu_t, \nu_t) \leq e^{-Kt} W_1(\mu_0, \nu_0)$, so that there exists an unique invariant probability measure $\mu_\infty = e^{-V} dx$.

For μ_∞ to be reversible, we must have $2b = a' - aV'$ where $a = \sigma^2$.

5 Non Homogeneous Diffusion Processes

5.1 General Non Homogeneous Diffusion

In [21] the authors extended the Γ_2 theory to time dependent coefficients (non homogeneous diffusions). Considering the Ito system

$$dX_t = \sigma(v_t, X_t) dB_t + b(v_t, X_t) dt. \tag{48}$$

$$dv_t = dt,$$

$$\mathscr{L}(v_0, X_0) = \delta_{t_0} \otimes \mu_0,$$

we see that all what we have done can be applied to this system. Actually one can modify the "curvature" assumptions introducing for some function $K(t)$ and its derivative $K'(t)$: **(H.C.K(t))** for all (x, y), all $t \in \mathbb{R}$

$$|\sigma(t, x) - \sigma(t, y)|_{HS}^2 + 2 \langle b(t, x) - b(t, y), x - y \rangle \leq -K'(t) |x - y|^2.$$

We then have

Theorem 5 *Assume that σ and b satisfy the hypotheses of the previous section, considered as functions on $\mathbb{R} \times \mathbb{R}^n$. If (H.C.K(t)) is satisfied, then we may extend (1) (Poincaré) and (2)(log-Sobolev) replacing e^{-KT} by $e^{-K(T)}$ and $\frac{1 - e^{-KT}}{K}$ by $\int_0^T e^{-K(s)} ds$.*

Proof If f only depends on x, the proof of Proposition 2 (resp. 8) is unchanged using the process starting from $(0, x)$ and $(0, y)$ and replacing Kt by $K(t)$. To obtain the analogue of Propositions 3 and 9, it suffices to remark that $\sigma \nabla$ is equal to ∇_x, and use what precedes for h depending on x only.

For the transportation inequality we have to slightly modify the method in Sect. 2.3. With the notations therein, (20) has become,

$$\eta_t \leq \eta_0 - \int_0^t K'(s)\,\eta_s\,ds + 2\sqrt{2}\,H^{\frac{1}{2}}(h\mu_T|\mu_T)\left(\int_0^t \eta_s\,ds\right)^{\frac{1}{2}},$$

so that, as in the previous section we have to come back to

$$\eta_t \leq \eta_0 - \int_0^t K'(s)\,\eta_s\,ds + 2\int_0^t \mathbb{E}^Q\left(|z_s - \omega_s|\,|\nabla \log P_{T-s}h(\omega_s)|\right)ds. \tag{49}$$

Using as usual $(ab)^{\frac{1}{2}} \leq \lambda a + \frac{1}{\lambda}b$ we obtain (see the details of the derivation in the previous section) that for all increasing function $\lambda(t)$

$$\eta'(t) \leq (-K'(t) + \lambda'(t))\,\eta_t + \frac{4}{\lambda'(t)}I_T(h),$$

from which we deduce, provided we choose $K(0) = \lambda(0) = 0$,

$$\eta_T \leq e^{-K(T)+\lambda(T)}\,\eta_0 + 4\,e^{-K(T)+\lambda(T)}\left(\int_0^T \frac{e^{K(s)-\lambda(s)}}{\lambda'(s)}\,ds\right)I_T(h).$$

Theorem 6 *In the situation of Theorem 5. If $(H.C.K(t))$ is satisfied, then for any x and any increasing function λ, $P(T, x, .)$ satisfies a W_2I inequality*

$$W_2^2(hP(T, x, .), P(T, x, .)) \leq C(T)\int \frac{|\nabla h|^2}{h}\,d\mu_T,$$

with constant

$$C(T) \leq 4\,e^{-K(T)+\lambda(T)}\left(\int_0^T \frac{e^{K(s)-\lambda(s)}}{\lambda'(s)}\,ds\right).$$

If μ_0 satisfies a T_2 inequality with constant $C_T(0)$, then

$$W_2^2(h\mu_T, \mu_T) \leq C_T(0)\,e^{-K(T)+\lambda(T)}\,H(h\mu_T|\mu_T) + C(T)\,I_T(h).$$

The best choice of λ is not clear. If $K'(t)$ is not positive on the whole $[0, T]$, it seams that taking $\lambda(T) = \lambda T$ for some $\lambda > 0$ is enough. If $K'(t) > 0$ for all t (but not necessarily bounded from below by a positive constant), $\lambda(t) = \lambda K(t)$ seems to be natural.

Remark 13 Assume that $K(t) \to +\infty$ as $t \to +\infty$ and that $C = \int_0^{+\infty} e^{-K(s)}\,ds < +\infty$. Then, for all t, $P(t, x, .)$ (the distribution of the process starting from x and $t_0 = 0$) satisfies a Poincaré inequality (and a log-Sobolev inequality when $\sigma = Id$)

with a constant bounded by MC (or $2C$). The family $(P(t, x, .))_{t>0}$ is then tight, but we do not know whether it is weakly convergent or not. Nevertheless any weak limit satisfies the same functional inequality.

When $\sigma = Id$ we know that $|X_t^x - X_t| \leq e^{-K(t)}|x - X_0|$ for any initial random variable X_0. It follows that if a sequence $P(t_k, x, .)$ is weakly convergent to some μ, the sequence μ_{t_k} weakly converges to the same limit.

In particular if we consider $\sigma = Id$, $b(t, x) = -\frac{1}{2}(\nabla U(x) + K'(t) x)$, for some convex potential U, (H.C.K(t)) is satisfied, so that any weak limit satisfies a log-Sobolev inequality. If $d\mu = e^{-U} dx$ does not satisfy a log-Sobolev inequality, it cannot be a weak limit, even if $K'(t) \to 0$. In this situation one should expect that the "perturbation" of ∇U being smaller and smaller when t growths, the convergence to μ will still hold. This is not the case.   ◇

5.2 Application to Some Non-linear Diffusions

We shall now discuss an example that does partly enter the framework of the beginning of this section.

Following [18, 35] consider the following non-linear stochastic differential equation

$$dX_t = dB_t - \frac{1}{2} \nabla V(X_t) \, dt - \frac{1}{2} \nabla W * q_t(X_t) \, dt \tag{50}$$

$$\mathcal{L}(X_t) = q_t \, dx \, .$$

If a solution exists, q_t will solve

$$\partial_t q_t = \frac{1}{2} \nabla . (\nabla q_t + q_t \nabla V + q_t (\nabla W * q_t)) \, . \tag{51}$$

This is a non-linear diffusion of Mc Kean-Vlasov type modeling, for instance, granular media. We refer to the introduction of [18] for details and motivations. One can approximate the solution of (50) by the first coordinate of a linear large particle system with mean field interactions. This is what is done in [18,35] to study the long time behavior of X_t.

Let us see how to apply what we have just done. First, under some conditions on V and W (we later shall give some of them) existence and weak uniqueness of (50) are ensured, provided the initial law admits some big enough polynomial moment. This will imply for all x, the existence and uniqueness of q_t^x solution of (51) with initial condition δ_x. As usual for these non-linear equations, if we consider the linear time inhomogeneous S.D.E.

$$dZ_t^{x,y} = dB_t - \frac{1}{2} \nabla V(Z_t^{x,y}) \, dt - \frac{1}{2} \nabla W * q_t^x(Z_t^{x,y}) \, dt \quad Z_0^{x,y} = y \, ,$$

the pathwise unique solution (up to explosion) $Z^{x,x}$ is shown to satisfy (50) (i.e. $\mathcal{L}(Z_t^{x,x}) = q_t^x$) so that it coincides with X_t^x. So, once q_t^x and q_t^y are built, we may build our synchronous coupling (X_t^x, X_t^y) as before. We may thus state

Theorem 7 *Assume that*

H1 V, W and their first two derivatives have at most polynomial growth of order m and $W(-x) = W(x)$,
H2 V satisfies (H.C.K_V) and W satisfies (H.C.K_W).

Let $a = \max(m(m + 3), 2m^2)$. If μ_0 and ν_0 have a polynomial moment of order a, there exist an unique solution of (50) and an unique solution of (51) among the set of probability flows having a polynomial moment of order a with initial condition μ_0 or ν_0.

   Furthermore

1.

$$W_2^2(\mu_T, \nu_T) = W_2^2(q_T^{\mu_0} \, dx, q_T^{\nu_0} \, dx) \le e^{-(K_V + \min(K_W, 0))T} \, W_2^2(\mu_0, \nu_0).$$

2. If $V = 0$ and $\int x\mu_0(dx) = \int x\nu_0(dx)$ then

$$W_2^2(\mu_T, \nu_T) \le e^{-K_W T} \, W_2^2(\mu_0, \nu_0).$$

Introduce the conditions,

H'1 $K = K_V + \min(K_W, 0) > 0$.
H'2 $V = 0$, $\int x\mu_0(dx) = \int x\nu_0(dx)$ and $K_W > 0$.

If H'1 is satisfied, there exists an unique invariant distribution $\mu_\infty = q^\infty(x)dx$ of (50) and (51) satisfying the polynomial moment condition of order a, the convergence to μ_∞ in W_2 Wasserstein distance being exponential as above.

   If H'2 is satisfied the same result holds for each $A \in \mathbb{R}^n$ in the set of probability measures such that $\int x\mu(dx) = A$.

Proof The moment condition ensuring existence and uniqueness is described in [18] Sect. 2.

   Recall that we may build our synchronous coupling (X_t^x, X_t^y) as before. Now introduce an independent copy $(\bar{X}_t^x, \bar{X}_t^y)$ of (X_t^x, X_t^y).

   We have

$$\mathbb{E}\left(|X_t^x - X_t^y|^2\right)$$

$$= -\mathbb{E}\left(\int_0^t \langle \nabla V(X_s^x) - \nabla V(X_s^y), X_s^x - X_s^y \rangle ds\right) \tag{52}$$

$$-\mathbb{E}\int_0^t \int \langle \nabla W(X_s^x - z^x) - \nabla W(X_s^y - z^y), X_s^x - X_s^y \rangle q_s^x(z^x)q_s^y(z^y)dz^x dz^y ds.$$

Remark that the last term can be written

$$\int_0^t \mathbb{E}\left(\langle \nabla W(X_s^x - \bar{X}_s^x) - \nabla W(X_s^y - \bar{X}_s^y), X_s^x - X_s^y\rangle\right) ds.$$

If we assume in addition (as usual) that $W(-x) = W(x)$, and remember that $\bar{X}$ is a copy of X, it is still equal to

$$-\int_0^t \mathbb{E}\left(\langle \nabla W(X_s^x - \bar{X}_s^x) - \nabla W(X_s^y - \bar{X}_s^y), \bar{X}_s^x - \bar{X}_s^y\rangle\right) ds.$$

Hence

$$2\,\mathbb{E}\left(|X_t^x - X_t^y|^2\right)$$
$$= \mathbb{E}\left(|X_t^x - X_t^y|^2\right) + \mathbb{E}\left(|\bar{X}_t^x - \bar{X}_t^y|^2\right)$$
$$= 2|x - y|^2 - 2\,\mathbb{E}\left(\int_0^t \langle \nabla V(X_s^x) - \nabla V(X_s^y), X_s^x - X_s^y\rangle ds\right)$$
$$-\int_0^t \mathbb{E}\left(\langle \nabla W(X_s^x - \bar{X}_s^x) - \nabla W(X_s^y - \bar{X}_s^y), (X_s^x - \bar{X}_s^x) - (X_s^y - \bar{X}_s^y)\rangle\right) ds.$$

According to what precedes we have

$$\mathbb{E}\left(|X_t^x - X_t^y|^2\right) \le |x - y|^2 - K_V \int_0^t \mathbb{E}\left(|X_s^x - X_s^y|^2\right) ds$$
$$- (K_W/2) \int_0^t \mathbb{E}\left(|(X_s^x - \bar{X}_s^x) - (X_s^y - \bar{X}_s^y)|^2\right) ds$$
$$\le |x - y|^2 - K_V \int_0^t \mathbb{E}\left(|X_s^x - X_s^y|^2\right) ds$$
$$- K_W \left(\int_0^t \mathbb{E}\left(|X_s^x - X_s^y|^2\right) ds - \int_0^t |\mathbb{E}\left(X_s^x - X_s^y\right)|^2 ds\right).$$

Of course we may replace the initial δ_x and δ_y by probability distributions μ_0 and ν_0 satisfying the required moment conditions. This immediately furnishes the first assertion about the upper bound for the Wasserstein distance.

If $V = 0$ it is easily seen that $\int x q_t^{\mu_0}(x)dx = \int x \mu_0(dx)$ for all $t > 0$, hence

$$E\left(X_s^{\mu_0} - X_s^{\nu_0}\right) = 0$$

provided the same holds at time 0. This furnishes the second assertion for the upper bound.

Finally the convergence under strict positivity of our new "curvature" condition ensures the existence of the limiting measure μ_∞. To see that $\mu_\infty = q^\infty(x)dx$ is actually invariant, one can for instance use the following trick: first consider the solution q_t^∞ of (51) with initial condition q^∞. Similar bounds for the Markov non homogeneous process $Z^{q^\infty \cdot y}$ (when we replace q_t^x by q_t^∞) are obtained applying the results of the beginning of this section. Hence the law of $Z_T^{q^\infty \cdot q^\infty}$ (which is exactly μ_T starting with μ_∞ as we explained before) converges to some limiting measure $\mu_\infty^{\mu_\infty}$ which in turn is equal to μ_∞ and is invariant for $Z^{q_\infty \cdot y}$. This achieves the proof.

Remark 14 The proof of the above result is new and direct, while the result is mainly contained in [18, 35] using particle approximation. Notice that in [35] the Γ_2 approach is developed for the non homogeneous Markov diffusion Z and not for X. Also notice that some direct study of the decay to equilibrium in W_2 distance for granular media is done in [11]. $\diamond$

As said in the previous remark the Γ_2 theory does not work directly for the process X. Actually our method to control the gradient of $x \mapsto \mathbb{E}(f(X_t^x))$ should work but we do not know whether the gradient exists or not, due to the fact that we do not have any a priori regularity in the initial condition. Fortunately, if we want to obtain some properties for the time marginal distribution μ_T we may use the fact (as done by Malrieu) that this distribution coincides with the one of the non homogeneous Markov diffusion $Z_T^{x,y}$ to which we can apply the techniques of this section. In particular, in the situation of the previous theorem, when q^∞ exists we may consider the diffusion

$$dZ_t^y = dB_t - \frac{1}{2}\nabla V(Z_t^y)\,dt - \frac{1}{2}\nabla W * q^\infty(Z_t^y)\,dt \quad Z_0^y = y \,,$$

for which $\mu_\infty(dx) = q^\infty(x)dx$ the invariant probability measure. Using the results in Sect. 2 we thus have

Proposition 14 *In the situation of Theorem 7, if H'1 or H'2 are satisfied, μ_∞ satisfies a log-Sobolev inequality with constant $C_{LS} = 2/K$ or $C_{LS} = 2/K_W$.*

All what we have done extends to more general Mc Kean-Vlasov equations, with a diffusion coefficient σ and a drift b satisfying hypothesis (R). In particular, positive curvature (in the sense of (H.C.K)) will also imply existence of and convergence to an invariant probability measure. The only difference is that we have to replace log-Sobolev inequality by Poincaré inequality in the latter proposition. Let us explain quickly what kind of model we may consider. We do not aim to be optimal, but will provide a flavor of the results on contraction with some non constant diffusion term. We will not focus also on the existence of solution of such equation. Let X_t^x be solution of

$$dX_t^x = \sigma(X_t^x, \kappa * q_t(X_t^x))dB_t - \frac{1}{2}\nabla V(X_t^x)\,dt - \frac{1}{2}\nabla W * q_t^x(X_t^x)\,dt \quad (53)$$

$$X_0^x = x \quad (54)$$

$$\mathscr{L}(X_t^x) = q_t^x\,dx\,.$$

Theorem 8 *Let us suppose H1 and H2, that κ is l-Lipschitz and that*

$$|\sigma(x,y) - \sigma(x',y')|^2_{HS} \le r(|x - x'|^2 + |y - y'|^2).$$

Then (using the notation of Theorem 7)

$$W_2^2(\mu_T, \nu_T) \le e^{-(K_V - r(1+4l^2) + \min(K_W, 0))T} W_2^2(\mu_0, \nu_0).$$

Suppose moreover that $K_V - r(1 + 4l^2) + \min(K_W, 0) > 0$, then there exists an unique invariant distribution to (53), the convergence to μ_∞ in W_2 Wasserstein distance being exponential as above.

The proof follows the same line as before except that in the Itô's formula, there is the diffusion part which comes into play for which we use the Lipschitz condition of the theorem. Note that Bolley et al. [12] have considered the case of a kinetic McKean-Vlasov equation, but with a constant diffusion coefficient in speed. As before, we may obtain some functional inequality for the invariant distribution as in Proposition 14 but we have to replace log-Sobolev inequality by Poincaré inequality.

6   Extensions to Some Non Uniformly Convex Potentials

Let us come back to (1), and assume that Υ is bounded. We shall extend (H.C.K) to more general situations. The first natural extension is to replace the squared distance by some other convex functional of the distance. More precisely.

Definition 1 Let $\varphi : \mathbb{R}^+ \to \mathbb{R}^+$. We say that φ belongs to $\mathscr{C}$ if it satisfies the following conditions:

- φ is increasing and convex, with $\varphi(0) = 0$ and $\varphi(1) = 1$,
- $a \mapsto \varphi(a)/a$ is non decreasing,
- there exist a positive function ψ such that for all $a > 0$ and all $\lambda > 0$, $\varphi^{-1}(\lambda a) \le \psi(\lambda) \varphi^{-1}(a)$, where φ^{-1} denotes the inverse (reciprocal) function of φ.

Definition 2 Let $\varphi \in \mathscr{C}$. We shall say that **(H.φ.K)** is satisfied for some $K > 0$ if for all (x, y),

$$\langle \nabla U(x) - \nabla U(y), x - y \rangle \ge K \varphi(|x - y|^2).$$

On one hand, since $K > 0$ and $\varphi \ge 0$, (H.φ.K) implies that U is convex. On the other hand, if (H.φ.K) is satisfied, since U is smooth, $\varphi(a)/a$ is necessarily bounded near the origin since $\lim \sup_{a \to 0}(\varphi(a)/a) \le \inf |Hess(U)|$. Here of course if $\varphi \in \mathscr{C}$ the latter is automatically satisfied.

If $\varphi(a) = a$ this is nothing else but (H.C.K). If $\varphi(a)/a \to +\infty$ we shall say that U is super-convex. This terminology is justified by the example below.

Example 1 Let $U(x) = (|x|^2)^\beta$ for some $\beta > 1$. We shall see that (H.φ.K) is satisfied for $\varphi(a) = a^\beta$ and some K we shall estimate.

We start with the one dimensional case. In this case

$$(U'(x) - U'(y))(x - y) = 2\beta \, (sign(x)|x|^{2\beta-1} - sign(y)|y|^{2\beta-1})(x - y).$$

If $sign(x) = sign(y)$, we may assume that $|x| \geq |y|$, write $|x| = u + |y|$ for $u \geq 0$ and remark that if $2\beta - 1 \geq 1$,

$$(u + |y|)^{2\beta-1} - |y|^{2\beta-1} \geq u^{2\beta-1}$$

so that

$$(U'(x) - U'(y))(x - y) = 2\beta \, ((u + |y|)^{2\beta-1} - |y|^{2\beta-1})u \geq 2\beta u^{2\beta} = 2\beta|x - y|^{2\beta}.$$

If $sign(x) = -sign(y)$, we have, using the convexity of $x \mapsto |x|^{2\beta-1}$,

$$\begin{aligned}
(U'(x) - U'(y))(x - y) &= 2\beta \, (|x|^{2\beta-1} + |y|^{2\beta-1})(|x| + |y|) \\
&\geq 2\beta \, 2^{2-2\beta} \, (|x| + |y|)^{2\beta} \\
&= 2\beta \, 2^{2-2\beta} \, |x - y|^{2\beta}.
\end{aligned}$$

Since $\beta > 1$, we may choose $K_\beta = 2\beta \, 2^{2-2\beta}$.

The general situation is a little bit more intricate.

Pick x and y in $\mathbb{R}^n$, assume that $|x| \geq |y|$ and write $x = |x|u$ and $y = |y|(\alpha u + \gamma v)$ for unit vectors u and v such that $\langle u, v \rangle = 0$ and $\alpha^2 + \gamma^2 = 1$. Then

$$\langle \nabla U(x) - \nabla U(y), x - y \rangle = 2\beta \, ((|x|^{2\beta-1} - \alpha|y|^{2\beta-1})(|x| - \alpha|y|) + \gamma^2|y|^{2\beta}),$$

and

$$|x - y|^{2\beta} = ((|x| - \alpha|y|)^2 + \gamma^2|y|^2)^\beta \leq 2^{\beta-1} \, ((|x| - \alpha|y|)^{2\beta} + \gamma^{2\beta}|y|^{2\beta}).$$

If $\alpha \geq 0$, we write again $|x| = |y| + a$ with $a \geq 0$. Thus, since $0 \leq 1 - \alpha \leq 1$,

$$\begin{aligned}
|x|^{2\beta-1} - \alpha|y|^{2\beta-1} &= (a + |y|)^{2\beta-1} - \alpha|y|^{2\beta-1} \geq a^{2\beta-1} + (1 - \alpha)|y|^{2\beta-1} \\
&\geq a^{2\beta-1} + ((1 - \alpha)|y|)^{2\beta-1} \\
&\geq 2^{2-2\beta} \, (a + (1 - \alpha)|y|)^{2\beta-1} = 2^{2-2\beta} \, (|x| - \alpha|y|)^{2\beta-1}.
\end{aligned}$$

It follows, since $\beta \geq 1$ and $\gamma^2 \leq 1$,

$$\begin{aligned}
\langle \nabla U(x) - \nabla U(y), x - y \rangle &\geq 2\beta \, (2^{2-2\beta} \, (|x| - \alpha|y|)^{2\beta} + \gamma^2 \, |y|^{2\beta}) \\
&\geq 2\beta \, 2^{2-2\beta} \, ((|x| - \alpha|y|)^{2\beta} + \gamma^{2\beta} \, |y|^{2\beta}) \\
&\geq 2\beta \, 2^{3-3\beta} \, |x - y|^{2\beta}.
\end{aligned}$$

If $\alpha < 0$, since $|\alpha| \leq 1$, it holds

$$\langle \nabla U(x) - \nabla U(y), x - y \rangle$$
$$= 2\beta \left((|x|^{2\beta-1} + |\alpha||y|^{2\beta-1})(|x| + |\alpha||y|) + \gamma^2 |y|^{2\beta} \right)$$
$$\geq 2\beta \left((|x|^{2\beta-1} + (|\alpha||y|)^{2\beta-1})(|x| + |\alpha||y|) + \gamma^{2\beta} |y|^{2\beta} \right)$$
$$\geq 2\beta \left(2^{2-2\beta}(|x| + |\alpha||y|)^{2\beta} + \gamma^{2\beta} |y|^{2\beta} \right)$$
$$\geq 2\beta \, 2^{3-3\beta} |x - y|^{2\beta} .$$

Proposition 15 *Let $U(x) = (|x|^2)^\beta$ for some $\beta > 1$. Then $(H.\varphi.K)$ is satisfied for $\varphi(a) = a^\beta$ and $K_\beta \geq 2\beta \, 2^{3-3\beta}$. If $n = 1$ we have the better bound $K_\beta \geq 2\beta \, 2^{2-2\beta}$.* ◇

Remark 15 If $\varphi \in \mathscr{C}$, for all $a \geq 0$ and all $\varepsilon > 0$, it holds

$$\varphi(a) \geq \frac{\varphi(\varepsilon)}{\varepsilon} a - \varphi(\varepsilon) .$$

Hence $(H.\varphi.K)$ implies the following condition

$$\forall \varepsilon > 0, \forall (x, y) \quad \langle \nabla U(x) - \nabla U(y), x - y \rangle \geq K \left(\frac{\varphi(\varepsilon)}{\varepsilon} |x - y|^2 - \varphi(\varepsilon) \right) .$$
(55)

The latter appears in the study of the granular medium equation in [18] (condition(6)) for power functions φ. This formulation will be the interesting one. It can be extended in

Definition 3 Let α be a non decreasing function defined on $\mathbb{R}^+$. We shall say that **(H.α.K)** is satisfied for some $K > 0$ if for all (x, y) and all $\varepsilon > 0$,

$$\langle \nabla U(x) - \nabla U(y), x - y \rangle \geq K \alpha(\varepsilon) \left(|x - y|^2 - \varepsilon \right) .$$

$(H.\varphi.K)$ implies $(H.\alpha.K)$ with the same K and $\alpha(\varepsilon) = \varphi(\varepsilon)/\varepsilon$. In this definition we do not need that $a \mapsto a\alpha(a)$ is convex.

Now we shall see how to use $(H.\varphi.K)$.

6.1 Non Fully Convincing First Results

This subsection contains first results which are not really convincing, but have to be tested.

If we want to control the gradient $\nabla P_t f$, we may write for $t > u$,

$$|X_t^x - X_t^y|^2 = |X_u^x - X_u^y|^2 - \int_u^t \langle \nabla U(X_s^x) - \nabla U(X_s^y), X_s^x - X_s^y \rangle \, ds$$

$$\leq |X_u^x - X_u^y|^2 - K \int_u^t \varphi(|X_s^x - X_s^y|^2) \, ds.$$

Denoting $\eta_t = |X_t^x - X_t^y|^2$, we thus have $\eta_t' \leq -K \varphi(\eta_t)$. If $\varphi(a) = a^\beta$, this yields

$$|X_t^x - X_t^y|^2 \leq |x - y|^2 \left(\frac{1}{1 + K(\beta - 1)|x - y|^{2(\beta-1)} t} \right)^{1/(\beta-1)}. \tag{56}$$

This result (even after taking expectation) is not really satisfactory. Indeed, first we do not obtain any better control for $\nabla P_t f$ than the one for a general convex potential (in particular we do not obtain a rate of convergence to 0). In second place, the decay to 0 of the Wasserstein distance we obtain is desperately slow, while we expected an exponential decay (which we know to hold true for $U(x) = |x|^{2\beta}$ for $\beta \geq 1$). Notice however that we recover the exponential decay we obtained previously when $\beta \to 1$.

Remark 16 If instead of (H.φ.K) we use (H.α.K), it is not difficult to show that

$$\eta_t \leq \eta_0 e^{-K\alpha(\varepsilon)t} + \varepsilon.$$

If $\alpha(\varepsilon) = \varepsilon^{\beta-1}$, choosing $\varepsilon = \eta_0 t^{-\theta}$ for some $\theta < \beta - 1$, we get

$$W_2^2(P(t, x, .), P(t, y, .)) \leq |x - y|^2 (t^{-\theta} + e^{-K|x-y|t^{\beta-1-\theta}}).$$

The method can be extended to the Mc Kean-Vlasov situation studied in Sect. 5.2 and allows us to recover (up to the constants) Theorem 4.1 in [18] without the help of a particle approximation. However, better results in this situation are obtained in [11]. $\diamond$

Mimicking Sect. 2.3, in particular (20), do we obtain more interesting results? Using the notation therein we have

$$\eta_t := \mathbb{E}^{\mathbb{Q}}(|z_t - \omega_t|^2) \leq \eta_0 - K \int_0^t \varphi(\eta_s) \, ds + 2\sqrt{2} \, H^{\frac{1}{2}}(h\mu_T|\mu_T) \left(\int_0^t \eta_s \, ds \right)^{\frac{1}{2}}. \tag{57}$$

Using Jensen inequality we deduce

$$\varphi \left(\frac{1}{t} \int_0^t \eta_s \, ds \right) \leq \frac{1}{t} \int_0^t \varphi(\eta_s) \, ds$$

$$\leq \frac{2\sqrt{2}}{K t} H^{\frac{1}{2}}(h\mu_T | \mu_T) \left(\int_0^t \eta_s \, ds \right)^{\frac{1}{2}}$$

so that, if $v_t = \int_0^t \eta_s \, ds$,

$$v_t \leq t \, \varphi^{-1} \left(\frac{2\sqrt{2}}{K t} H^{\frac{1}{2}}(h\mu_T | \mu_T) \, v_t^{\frac{1}{2}} \right)$$

$$\leq t \, \psi \left(\frac{2\sqrt{2}}{K t} H^{\frac{1}{2}}(h\mu_T | \mu_T) \right) \, \varphi^{-1}(v_t^{1/2}).$$

If $\varphi(a) = a^\beta$, we thus obtain

$$\eta_T \leq \eta_0 + 2\sqrt{2} \, H^{\frac{1}{2}}(h\mu_T | \mu_T) \left(\int_0^T \eta_s \, ds \right)^{\frac{1}{2}}$$

$$\leq \eta_0 + (2\sqrt{2})^{\frac{\beta+1}{\beta}} \, K^{-1/\beta} \, H^{\frac{\beta+1}{2\beta}}(h\mu_T | \mu_T) \, T^{\frac{\beta-1}{2\beta-1}}. \tag{58}$$

This result is certainly not fully satisfactory too. On one hand, we get a less explosive bound in time (recall that in the general convex case the bound growths like T), but on the other hand the relative entropy appears to a power less than 1. In particular such an inequality does not imply a Poincaré inequality (which is obtained for entropies going to 0), but furnishes nice concentration properties (obtained for large entropies via Marton's argument).

6.2 An Improvement of Bakry-Emery Criterion

As we remarked at this end of Sect. 2 we may come back to the initial inequality in (20) which becomes in our new situation

$$\eta_t \leq \eta_0 - K \int_0^t \varphi(\eta_s) \, ds + 2 \int_0^t \mathbb{E}^{\mathbb{Q}} \left(|z_s - \omega_s| \, |\nabla \log P_{T-s} h(\omega_s)| \right) ds, \tag{59}$$

and yields

$$\eta_t' \leq - K \, \varphi(\eta_t) + 2 \, \eta_t^{\frac{1}{2}} \left(\int \frac{|\nabla h|^2}{h} \, d\mu_T \right)^{\frac{1}{2}}. \tag{60}$$

(here again (H.C.0) is satisfied so that, for short, $|\nabla P_s| \leq P_s|\nabla|$.) To explore (60) we shall use both the Remark 15 and the usual trick $ab \leq \lambda a^2 + \frac{1}{\lambda}b^2$ for a, b, λ positive. Hence

$$\eta_t' \leq \left(-K\frac{\varphi(\varepsilon)}{\varepsilon} + 2\lambda\right)\eta_t + \left(\frac{2}{\lambda}\left(\int \frac{|\nabla h|^2}{h}d\mu_T\right) + K\varphi(\varepsilon)\right). \tag{61}$$

We deduce, denoting $A = K\frac{\varphi(\varepsilon)}{\varepsilon} - 2\lambda$,

$$\eta_T \leq \eta_0 e^{-AT} + (1 - e^{-AT})\frac{\frac{2}{\lambda}\left(\int \frac{|\nabla h|^2}{h}d\mu_T\right) + K\varphi(\varepsilon)}{A}.$$

Choose $\lambda = (1/4)K(\varphi(\varepsilon)/\varepsilon)$ so that $A = (1/2)K(\varphi(\varepsilon)/\varepsilon) > 0$. η_T is thus bounded in time, but the bound is not tractable except for $T = +\infty$ (starting with $\mu_0 = \mu$) or if $\eta_0 = 0$. In both cases we have obtained

$$W_2^2(h\mu_T, \mu_T) \leq \varepsilon + \left(\frac{8\varepsilon^2}{K^2\varphi^2(\varepsilon)}\right)\int \frac{|\nabla h|^2}{h}d\mu_T. \tag{62}$$

It remains to optimize in ε. In full generality we choose ε such that both terms in the sum of (62) are equal (we know that we are loosing a factor less than 2). Remark that we do not use the explicit form of φ, i.e. we may replace (H.φ.K) by (H.α.K) in what we did previously. We have thus obtained

Proposition 16 *Assume that U satisfies (H.α.K) for $K > 0$. Let F be the inverse (reciprocal) function of $\varepsilon \mapsto \varepsilon\alpha^2(\varepsilon)$. Denote $\mu_T = P(T, x, .)$ and $\mu_\infty(dx) = \mu(dx) = e^{-U(x)}dx$. Then for all $0 < T \leq +\infty$, μ_T satisfies for all nice h,*

$$W_2^2(h\mu_T, \mu_T) \leq 2F\left(\frac{8}{K^2}I_T(h)\right),$$

where $I_T(h) = \int \frac{|\nabla h|^2}{h}d\mu_T$ is the Fisher information of h.

When F is equal to identity, such an inequality is called a W_2I inequality (see [26, Definition 10.4]). Here we obtained a weak form of W_2I inequality (which is clear on (62)) in the spirit of the weak Poincaré or the weak log-Sobolev inequalities.

In particular, using (H.W.I), since (H.C.0) is satisfied we obtain

Corollary 2 *Under the hypotheses of proposition 16, μ satisfies the inequality*

$$H(h\mu|\mu) \leq 2\left(I(h)F\left(\frac{8}{K^2}I(h)\right)\right)^{\frac{1}{2}}.$$

Weak logarithmic Sobolev inequalities were introduced and studied in [16]. Actually, we are not exactly here in the situation of [16] because we wrote the previous inequality in terms of a density of probability. Let $h = f^2 / \int f^2 d\mu$. We deduce from the previous corollary

$$
\int f^2 \log \left(\frac{f^2}{\int f^2 d\mu} \right) d\mu
$$
$$
\leq 4 \left(\int f^2 d\mu \right)^{\frac{1}{2}} \left(\int |\nabla f|^2 d\mu \right)^{\frac{1}{2}} F^{\frac{1}{2}} \left(\frac{32}{K^2} \frac{\int |\nabla f|^2 d\mu}{\int f^2 d\mu} \right),
$$

so that if $F(\lambda a) \leq \theta(\lambda) F(a)$,

$$
\int f^2 \log \left(\frac{f^2}{\int f^2 d\mu} \right) d\mu \leq 4 \theta^{\frac{1}{2}} \left(\frac{32}{K^2} \right) G_1 \left(\int f^2 d\mu \right) G_2 \left(\int |\nabla f|^2 d\mu \right),
$$

where $G_1(a) = a^{\frac{1}{2}} \theta^{\frac{1}{2}} (1/a)$ and $G_2(a) = a^{\frac{1}{2}} F^{\frac{1}{2}}(a)$. The previous inequality looks like the Nash inequality version of a weak log-Sobolev inequality, but with the $\mathbb{L}^2$ norm of f in place of the $\mathbb{L}^\infty$ norm of $f - \int f d\mu$. So the previous inequality is not only "weak" but also "defective".

6.2.1 Super Convex Potentials

In this sub(sub)section we assume that $\varphi(a) = a^\beta$ for some $\beta \geq 1$, so that $F(a) = a^{\frac{1}{2\beta-1}}$. We thus have

$$
\int f^2 \log \left(\frac{f^2}{\int f^2 d\mu} \right) d\mu \leq 4 \left(\frac{32}{K^2} \right)^{\frac{1}{2(2\beta-1)}} \left(\int f^2 d\mu \right)^{\frac{\beta-1}{2\beta-1}} \left(\int |\nabla f|^2 d\mu \right)^{\frac{\beta}{2\beta-1}}.
$$
$$(63)$$

Recall first that if $g \geq 0$, then $\mathrm{Var}_\mu(g) \leq \mathrm{Ent}_\mu(g)$ (see e.g. [17] (2.6)).

Next recall the following: defining $m_\mu(g)$ as a *median* of g, we have

$$
\mathrm{Var}_\mu(g) \leq 4 \int (g - m_\mu(g))^2 d\mu \leq 36 \, \mathrm{Var}_\mu(g). \tag{64}
$$

We may decompose $f - m_\mu(f) = (f - m_\mu(f))_+ - (f - m_\mu(f))_- = g_+ - g_-$ so that both g_+ and g_- are non negative with median equal to 0. In addition, if f is Lipschitz, so are g_+ and g_-, $\nabla f = \nabla g_+ + \nabla g_-$, and the product of both vanishes. Hence

$$
\mathrm{Var}_\mu(f) \leq 4 \left(\int (g_+)^2 d\mu + \int (g_-)^2 d\mu \right),
$$

while

$$\int (g_+)^2 d\mu \le 9\, \mathrm{Var}_\mu(g_+) \le 9\, \mathrm{Ent}_\mu(g_+)$$

$$\le 36 \left(\frac{32}{K^2}\right)^{\frac{1}{2(2\beta-1)}} \left(\int (g_+)^2 d\mu\right)^{\frac{\beta-1}{2\beta-1}} \left(\int |\nabla g_+|^2 d\mu\right)^{\frac{\beta}{2\beta-1}}.$$

It follows from (63)

$$\int (g_+)^2 d\mu \le (36)^{\frac{2\beta-1}{\beta}} \left(\frac{32}{K^2}\right)^{\frac{1}{2\beta}} \int |\nabla g_+|^2 d\mu,$$

similarly for g_-. We have thus obtained

Theorem 9 *Assume that U satisfies ($H.\varphi.K$) for $K > 0$ and $\varphi(a) = a^\beta$, for $\beta \ge 1$. Then, μ satisfies both a Poincaré inequality with*

$$C_P(\mu) \le C_P(K, \beta) = 4\,(36)^{\frac{2\beta-1}{\beta}} \left(\frac{32}{K^2}\right)^{\frac{1}{2\beta}},$$

and a log-Sobolev inequality with

$$C_{LS}(\mu) \le C_{LS}(K, \beta) = \left(\frac{32}{K^2}\right)^{\frac{1}{2\beta}} \left(4^{\frac{3\beta-2}{2\beta-1}} 36^{\frac{\beta-1}{\beta}} + 8 \times 36^{\frac{2\beta-1}{\beta}}\right).$$

Proof The statement on the Poincaré inequality follows from the previous discussion.

Concerning the log-Sobolev inequality, let $\tilde{f} = f - \int f d\mu$. Then, Rothaus lemma (see [1, Lemma 4.3.7]) says that

$$\mathrm{Ent}_\mu(f) \le \mathrm{Ent}_\mu(\tilde{f}) + 2\, \mathrm{Var}_\mu(f).$$

Applying (63) to $\tilde{f}$ together with the Poincaré inequality, yield the result (after some elementary calculation).

Remark 17 This theorem applies in particular to $U(x) = |x|^{2\beta}$ for $\beta \ge 1$, according to Proposition 15. The fact that μ satisfies a log-Sobolev inequality in this situation is well known, but here we obtain an explicit (though not really cute) expression for the constant that only depends on β and not on the dimension n.

Unfortunately, in this particular situation, our bounds are not optimal. Indeed, spherically symmetric log-concave probability measures are now well understood.

For the Poincaré constant, it was shown by Bobkov [8] that

$$\frac{1}{n} \mathrm{Var}_\mu(x) \le C_P(\mu) \le \frac{13}{n} \mathrm{Var}_\mu(x).$$

It is an (easy) exercise to see that $\mathrm{Var}_\mu(x) = \Gamma((n+2)/2\beta)/\Gamma(n/2\beta)$, so that $C_P(\mu) \le c(\beta) n^{\frac{1}{\beta}-1}$ which goes to 0 as $n \to +\infty$.

A famous conjecture by Kannan-Lovasz-Simonovitz is that the previous bound for spherically symmetric measures extends (up to a change of the constant 13) to any log-concave measure. If true, the KLS conjecture will presumably give a better upper bound for the Poincaré constant than ours.

Regarding the log-Sobolev constant, the work by Huet [29], furnishes a lower bound for the isoperimetric profile of μ (see Theorem 3 and the discussion p. 98 therein) which indicates a similar bound for the log-Sobolev constant as above, i.e. depending on the isotropic constant ($n^{\frac{1-\beta}{2\beta}}$) of μ. $\diamondsuit$

6.2.2 Lack of Uniform Convexity

Now choose $\alpha(a) = a^\beta$ for some $\beta \ge 1$ and $a \le 1$, and $\alpha(a) = 1$ for $a \ge 1$. (H.α.K) is less restrictive than before since it only implies a linear behavior at infinity for the gradient of the potential.

We now have $F(a) = a^{\frac{1}{2\beta+1}}$ for $a \le 1$ and $F(a) = a$ for $a \ge 1$. It follows

$$\mathrm{Ent}_\mu(f) \le 4 \left(\frac{32}{K^2}\right)^{\frac{1}{2(2\beta+1)}} \left(\int f^2 d\mu\right)^{\frac{\beta}{2\beta+1}} \left(\int |\nabla f|^2 d\mu\right)^{\frac{\beta+1}{2\beta+1}} \tag{65}$$

if $\int |\nabla f|^2 d\mu \le \frac{K^2}{32} \int f^2 d\mu$ and

$$\mathrm{Ent}_\mu(f) \le \frac{32}{K} \int |\nabla f|^2 d\mu \quad \text{otherwise.} \tag{66}$$

Proceeding as before we obtain

Theorem 10 *Assume that U satisfies (H.α.K) for $K > 0$ and $\alpha(a) = a^\beta \wedge 1$ for $\beta \ge 1$. Then, μ satisfies both a Poincaré inequality with*

$$C_P(\mu) \le C_P(K, \beta) = \max \left(\frac{32}{K}, \, 4\,(36)^{\frac{2\beta+1}{\beta+1}} \left(\frac{32}{K^2}\right)^{\frac{1}{2(\beta+1)}}\right),$$

and a log-Sobolev inequality with

$$C_{LS}(\mu) \le C_{LS}(K, \beta) = \max \left(\frac{32}{K}, \left(\frac{32}{K^2}\right)^{\frac{1}{2(\beta+1)}} \left(4^{\frac{3\beta+1}{2\beta+1}} 36^{\frac{\beta}{\beta+1}} + 8 \times 36^{\frac{2\beta+1}{\beta+1}}\right)\right).$$

Remark 18 Using the general form of the (H.W.I) inequality it is quite easy to adapt the previous proof in order to show the following result:

Let μ satisfying (H.C.K) for some $K > -\infty$. If μ satisfies a weak $W_2 I$ inequality,

$$W_2^2(h\mu, \mu) \leq C \, (I(h))^\beta$$

for some $0 < \beta \leq 1$, then μ satisfies a log-Sobolev inequality with a constant depending on C, K, β only. In particular μ satisfies a T_2 inequality.

In particular if we know that μ_T has a bounded below curvature, the previous theorems extend to μ_T. $\diamond$

7 Using Reflection Coupling

As we have seen in the previous section, the simple coupling using the same Brownian motion is not fully well suited to deal with non uniformly convex potentials.

In a recent note [23], Eberle studied the contraction property in Wasserstein distance W_1, induced by another well known coupling method: coupling by reflection (or mirror coupling) introduced in [34]. The results of the note are extended in the recent [24] which appeared more or less at the same time than the first version of the present paper.

We shall see now how to use this coupling method in the spirit of what we have done before.

7.1 Reflection Coupling for the Drifted Brownian Motion

In this section we consider X_t^x the solution starting from x of the Ito stochastic differential equation

$$dX_t = dB_t + b(X_t) \, dt, \tag{67}$$

where b is smooth enough. We introduce another formulation of the semi-convexity property, namely:

$$\kappa(r) = \inf \left\{ -2 \frac{\langle b(x) - b(y), x - y \rangle}{|x - y|^2} \, ; \, |x - y| = r \right\}, \tag{68}$$

so that, it always holds

$$2 \langle b(x) - b(y), x - y \rangle \leq -\kappa(|x - y|) \, |x - y|^2 \, .$$

We shall say that **(H.κ)** is satisfied if

$$\liminf_{r \to +\infty} \kappa(r) = \kappa_\infty > 0. \tag{69}$$

This condition is typically some "uniform convexity at infinity" condition. Indeed if $b = -\frac{1}{2} \nabla U$ where $U = U_1 + U_2$ with U_1 satisfying (H.C.κ_∞) and U_2 compactly supported, then (H.κ) is satisfied. We shall come back later to this. Notice that if (69) is satisfied, the solution of (67) is strongly unique and non explosive, using the same tools as we used before.

Now, following [23] we introduce (with some slight change of notations)

$$R_0 = \inf \{ R \geq 0 \,;\, \kappa(r) \geq 0,\, \forall r \geq R \}, \tag{70}$$

$$R_1 = \inf \{ R \geq R_0 \,;\, \kappa(r) \geq 8/(R(R - R_0)),\, \forall r \geq R \},$$

$$\varphi(r) = \exp\left(-\frac{1}{4} \int_0^r s \kappa^-(s) \, ds \right), \quad \Phi(r) = \int_0^r \varphi(s) \, ds,$$

$$g(r) = 1 - \frac{1}{2} \left(\int_0^{r \wedge R_1} \frac{\Phi(s)}{\varphi(s)} \, ds \Big/ \int_0^{R_1} \frac{\Phi(s)}{\varphi(s)} \, ds \right)$$

$$D(r) = \int_0^r \varphi(s) \, g(s) \, ds.$$

Notice that

$$\frac{1}{2} \leq g \leq 1 \text{ and } \exp\left(-\frac{1}{4} \int_0^{R_0} s \kappa^-(s) \, ds \right) = \varphi_{min} \leq \varphi \leq 1.$$

If (H.κ) is satisfied, $R_0 < +\infty$ so that $\varphi_{min} > 0$ and

$$\frac{\varphi_{min}}{2} r \leq D(r) \leq r,$$

i.e. $D(|x - y|)$ which is actually a distance, is equivalent to the euclidean distance.

Hence a consequence of Theorem 1 in [23] is the following:

Theorem 11 *Assume that (H.κ) is satisfied. Let λ be defined by*

$$\frac{1}{\lambda} = \int_0^{R_1} \frac{\Phi(s)}{\varphi(s)} \, ds \leq \frac{R_1^2}{\varphi_{min}}.$$

Then for all initial distributions ν and μ, and all t, the W_1 Wasserstein distance satisfies

$$W_1(\nu_t, \mu_t) \leq \frac{2}{\varphi_{min}} e^{-\lambda t} \, W_1(\nu, \mu).$$

In order to prove this result, Eberle adds to (67) the following Ito s.d.e.

$$dY_t = (Id - 2e_t e_t^*) dB_t + b(Y_t) dt, \ Y_0 = y, \tag{71}$$

where $e_t = (X_t - Y_t)/|X_t - Y_t|$ and e^* is the transposed of e (remark that if $n = 1$, it just changes B into $-B$). Of course one has to consider (67) and (71) together. Existence, strong uniqueness and non explosion are again easy to show. Now introduce the coupling time T_c defined by

$$T_c = \inf t \geq 0; \ X_t = Y_t,$$

and finally define

$$\bar{X}_t^y = Y_t^y \ \text{if} \ t \leq T_c, \quad \bar{X}_t^y = X_t^x \ \text{if} \ t \geq T_c.$$

It is easy to see that X^y and $\bar{X}^y$ have the same law, so that the distribution of $(X_t^x, \bar{X}_t^y)$ is a coupling of $P(t, x, .)$ and $P(t, y, .)$. Of course this extends to any initial distributions (μ, ν) and furnishes a coupling of (μ_t, ν_t).

It follows that $Z_t = X_t^x - \bar{X}_t^y$ solves

$$dZ_t = (b(X_t^x) - b(\bar{X}_t^y))dt + 2\frac{Z_t}{|Z_t|} dW_t \tag{72}$$

where $W_t = \int_0^t e_s^* dB_s$ is a one dimensional Brownian motion.

The key of the proof of Theorem 11 is then that, if $r_t = D(|X_t^x - \bar{X}_t^y|)$, r is a semi-martingale with decomposition

$$r_t = D(|x - y|) + \int_0^{t \wedge T_c} 2\varphi(r_s) g(r_s) dW_s + \int_0^{t \wedge T_c} \beta_s \, ds, \tag{73}$$

where the drift term satisfies

$$\beta_s \leq -\lambda r_s. \tag{74}$$

Taking expectation, it immediately shows that the W_D Wasserstein distance decays exponentially fast. As remarked by several authors, one can then deduce as we did previously

$$|\nabla P_t f| \leq \frac{2}{\varphi_{min}} e^{-\lambda t} \ \| \nabla f \|_\infty . \tag{75}$$

As a consequence we obtain

Theorem 12 *Assume that $b = -\frac{1}{2}\nabla U$ satisfies(H.κ) and that $\mu = (1/Z_U)e^{-U}$ is a probability measure. Then μ satisfies a Poincaré inequality with constant $C_P \leq (1/2\lambda)$.*

Proof Recall that

$$\mathrm{Var}_\mu(P_t f) = \frac{1}{2}\int_t^{+\infty}\int |\nabla P_s f|^2 \, d\mu \, ds \, .$$

According to (75), we thus have

$$\mathrm{Var}_\mu(P_t f) \leq \frac{1}{\lambda\,\varphi_{min}^2}\, e^{-2\lambda t} \parallel \nabla f \parallel_\infty^2$$

for all Lipschitz function f. According to Lemma 2.12 in [19], we deduce that $\mathrm{Var}_\mu(P_t f) \leq e^{-2\lambda t}\,\mathrm{Var}_\mu(f)$, hence the result.

Remark 19 One can see that the reflection coupling cannot furnish some information on W_2, the concavity of D near the origin being crucial. In the same negative direction, Theorem 12 cannot be extended to the log-Sobolev framework, the key Lemma 2.12 in [19] being restricted to the variance control.   ◇

Example 2 1. If (H.φ.K) is satisfied with $\varphi(a) = a^\beta$, we have $\kappa(a) = a^{2(\beta-1)}$. We thus have $R_0 = 0$, $R_1 = (8/K)^{\frac{1}{2\beta}}$, $\varphi_{min} = 1$ and finally

$$C_P \leq \frac{1}{2}\left(\frac{8}{K}\right)^{\frac{1}{\beta}}\, .$$

We recover the result in Theorem 9, i.e. a bound $C_\beta\,K^{-1/\beta}$ but with a better constant C_β.

2. If (H.α.K) is satisfied, one can choose $R_0 = \sqrt{2\varepsilon}$, $R_1 = \sqrt{2\varepsilon} + (4/\sqrt{K\alpha(\varepsilon)})$, $\varphi_{min} = \exp\left(-\frac{1}{4}\varepsilon^2\,K\alpha(\varepsilon)\right)$ and finally

$$C_P \leq \left(2\varepsilon + \frac{16}{K\alpha(\varepsilon)}\right)e^{K\varepsilon^2\alpha(\varepsilon)/4}\, .$$

3. Now assume that the potential U can be written $U = U_1 + U_2$ where U_1 satisfies (H.C.K) for some $K > 0$ and U_2 satisfies $\parallel\nabla U_2\parallel_\infty = M < +\infty$. It easily follows that (H.κ) is satisfied with $\kappa(a) = K - \frac{M}{a}$. We thus have

$$R_0 = \frac{M}{K}\, , \quad R_1 = \frac{M}{K} + \sqrt{\frac{8}{K}}\, , \quad \varphi_{min} = e^{-\frac{M^2}{8K}}\, .$$

We finally obtain

$$C_P \leq \left(\frac{\| \nabla U_2 \|_\infty}{K} + \sqrt{\frac{8}{K}} \right)^2 \exp \left(\frac{\| \nabla U_2 \|_\infty^2}{8K} \right) .$$

An old result by Miclo (unpublished but explained in [32]) indicates that such a result (without the square of the supremum of the gradient but without K in the exponential) can be obtained by using the usual Holley-Stroock perturbation argument. ◇

7.2 The Log-Concave Situation

Now consider the situation where b satisfies (H.C.0). In this situation we have $\lambda = 0$ ($R_1 = +\infty$, $\varphi = g = 1$) so that (73) becomes

$$dr_t = 2 \, dW_t + \beta_t \, dt$$

for $t \leq T_c$ with $\beta_t \leq 0$. It follows that $r_t \leq |x - y| + 2 \, W_t$ up to the first time $T_{|x-y|}$ the brownian motion W hits $- |x - y|/2$.

In particular,

$$\mathbb{P}(r_t > 0) \leq \mathbb{P}(t < T_{|x-y|}) \leq \frac{|x - y|}{\sqrt{2\pi \, t}} ,$$

since the law of $T_{|x-y|}$ is given by

$$\mathbb{P}(T_{|x-y|} \in da) = \frac{|x - y|}{2\sqrt{2\pi \, a^3}} e^{-|x-y|^2/8a} \, \mathbb{I}_{a>0} \, da .$$

As a first by-product we obtain

Proposition 17 *If b satisfies (H.C.0) then*

$$|\nabla P_t f| \leq \frac{2}{\sqrt{2\pi \, t}} \, \| f \|_\infty .$$

Actually if $b = -\frac{1}{2} \nabla U$ with U convex (i.e. in the zero curvature situation of the Γ_2 theory) the inequality $|\nabla P_t f| \leq \frac{1}{\sqrt{t}} \, \| f \|_\infty$ is well known as a consequence of what is called the reverse (local) Poincaré inequality (see [1]). The previous proposition extends this result (up to the constant) to a non-gradient drift.

Proof Recall that $X_t^x = \bar{X}_t^y$ for $t > T_c$. It follows

$$
\begin{aligned}
P_t f(x) - P_t f(y) &= \mathbb{E}\left((f(X_t^x) - f(\bar{X}_t^y))\ \mathbb{1}_{T_c > t}\right) \\
&\leq 2\ \|f\|_\infty\ \mathbb{P}(T_c > t) \leq 2\ \|f\|_\infty\ \mathbb{P}(t < T_{|x-y|}) \\
&\leq \frac{2\ \|f\|_\infty}{\sqrt{2\pi t}}\ |x - y|,
\end{aligned}
$$

hence the result.

If one wants to get a contraction bound for the gradient (in the spirit of (75) or better of Proposition 2) we cannot only use a comparison with the brownian motion for which $\nabla P_t f = P_t \nabla f$.

Remark 20 In the symmetric situation ($b = -\frac{1}{2}\nabla U$) it is known that $t \mapsto \int |\nabla P_t f|^2\, d\mu$ is non increasing. It easily follows that

$$
\|\nabla P_t f\|_{\mathbb{L}^2(\mu)} \leq \frac{\sqrt{2}}{\sqrt{t}}\ \|f\|_{\mathbb{L}^2(\mu)} .
$$

If (H.C.0) is satisfied, this remark together with Proposition 17 and Riesz-Thorin interpolation theorem show that, up to an universal constant, the same holds in all $\mathbb{L}^p(\mu)$ spaces. $\diamond$

Remark 21 If we assume that (H.κ) is satisfied, we may replace the comparison with a Brownian motion by the comparison with an Ornstein-Uhlenbeck process with parameter $\lambda/2$, according to standard comparison theorems for one dimensional Ito processes (see e.g [30, Chapter VI, Theorem 1.1]). For the O-U process, it is known (see [37]) that

$$
\mathbb{P}(T_{|x-y|} \in da) = \frac{|x - y|}{2\sqrt{2\pi}}\left(\frac{\lambda}{2\sinh(a\lambda/2)}\right)^{\frac{3}{2}} e^{\left(-\frac{\lambda|x-y|^2 e^{-a\lambda/2}}{16\sinh(a\lambda/2)} + \frac{a\lambda}{4}\right)}\, da .
$$

An explicit bound for $\mathbb{P}(t < T_{|x-y|})$ can be obtained by using the reflection principle in [46], namely

$$
\mathbb{P}(t < T_{|x-y|}) \leq \frac{\sqrt{\lambda}\, e^{-t\lambda/2}}{\sqrt{2\pi}\ \sqrt{1 - e^{-t\lambda}}}\ |x - y|,
$$

yielding

$$
|\nabla P_t f| \leq \frac{2}{\varphi_{min}}\ \frac{\sqrt{\lambda}\, e^{-t\lambda/2}}{\sqrt{2\pi}\ \sqrt{1 - e^{-t\lambda}}}\ \|f\|_\infty .
$$

These bounds are interesting as regularization bounds (from bounded to Lipschitz functions), but notice that we have lost a factor 2 in the exponential decay. $\diamond$

7.3   Reflection Coupling for General Diffusions

The case of a general diffusion process with a non constant diffusion matrix as in
Sect. 3 is more delicate to handle, as already remarked in [34, Theorem 1].

Assume that σ is a bounded and smooth square matrices field and that it is
uniformly elliptic. The quantities we need here are (notations differ from [34])

$$M = \sup_{x} \sup_{|u|=1} |\sigma(x)u|^2, \, N = \sup_{x} \sup_{|u|=1} |\sigma^{-1}(x)u|^2, \, \Lambda = \sup_{x,x'} \sup_{|u|=1} |(\sigma(x) - \sigma(x'))u|^2.$$

(76)

Recall the Lindvall-Rogers reflection coupling

$$dX_t = \sigma(X_t)dB_t + b(X_t)dt,$$

$$dX_t' = \sigma(X_t') H_t \, dB_t + b(X_t')dt,$$

where

$$H_t = Id - 2\left(\frac{\sigma^{-1}(X_t')(X_t - X_t')}{|\sigma^{-1}(X_t')(X_t - X_t')|}\right)\left(\frac{\sigma^{-1}(X_t')(X_t - X_t')}{|\sigma^{-1}(X_t')(X_t - X_t')|}\right)^*.$$

Existence and strong uniqueness can be shown as previously. Of course, as in
Sect. 7.1, we replace X_t' by X_t if $t > T_c$ the coupling time, but not to introduce
new notation we still use X'.

As in [34] define

$$Y_t = X_t - X_t', \, V_t = \frac{Y_t}{|Y_t|}, \, \alpha_t = \sigma(X_t) - \sigma(X_t')H_t, \, \beta_t = b(X_t) - b(X_t').$$

According to (15) in [34] we have

$$d(|Y_t|) = \langle V_t, \alpha_t \, dB_t\rangle + \frac{1}{2}\frac{1}{|Y_t|}\left(2\langle Y_t, \beta_t\rangle + Trace(\alpha_t \, \alpha_t^*) - |\alpha_t^* \, V_t|^2\right),$$

and a simple calculation shows that

$$Trace(\alpha_t \, \alpha_t^*) - |\alpha_t^* \, V_t|^2 =$$

$$= Trace((\sigma(X_t) - \sigma(X_t'))(\sigma(X_t) - \sigma(X_t'))^*) - |(\sigma(X_t) - \sigma(X_t'))^* \, V_t|^2,$$

while

$$|\alpha_t^* \, V_t|^2 \geq \frac{2}{N} - \Lambda.$$

Applying Ito formula we thus have for a smooth function D

$$\mathbb{E}(D(|Y_t|)) = \frac{1}{2} \mathbb{E} \left(\frac{D'(|Y_t|)}{|Y_t|} \left(2\langle Y_t, \beta_t \rangle + Trace(\alpha_t \, \alpha_t^*) - |\alpha_t^* \, V_t|^2 \right) \right.$$
$$\left. + D''(|Y_t|)|\alpha_t^* V_t|^2 \right)$$

We introduce the natural generalization of (H.κ), namely we assume that

for a κ satisfying (69), $|\sigma(x) - \sigma(y)|_{HS}^2 + 2 \langle b(x) - b(y), x - y \rangle \leq -\kappa(|x-y|)|x-y|^2.$
$$(77)$$

If D is a non decreasing, concave function we thus get, provided $(2/N) - \Lambda > 0$,

$$2 \, \mathbb{E}(D(|Y_t|)) \leq \mathbb{E} \left(-D'(|Y_t|) \kappa(|Y_t|)|Y_t| + D''(|Y_t|) \left(\frac{2}{N} - \Lambda \right) \right) .$$

Hence looking carefully at the calculations in [23], we see that, provided $(2/N) - \Lambda > 0$, the only thing we have to change in (70) is the definition of φ replacing $1/4$ by the inverse of $(2/N) - \Lambda > 0$, all other definitions being unchanged. We have thus obtained

Theorem 13 *Assume that (76) and (77) are satisfied. Assume in addition that* $(2/N) - \Lambda > 0$. *Then defining*

$$\varphi_{min} = e^{- \frac{1}{(2/N)-\Lambda} \int_0^{R_0} s\kappa^-(s) \, ds} ,$$

the conclusion of Theorem 11 is still true with $\lambda = \frac{1}{2} (\varphi_{min}/R_1^2)$.

All the consequences of Theorem 11 still hold (up to the modifications of the constants), in particular one can extend (H.C.K) to the situation of "convexity at infinity" as in Example 2(3). Details are left to the reader.

As we already said, the condition $(2/N) - \Lambda > 0$ already appears in [34] and ensures that the coupling by reflection is successful. Roughly speaking it means that the fluctuations of σ are not too big with respect to the uniform ellipticity bound.

7.4 Gradient Commutation Property and Reflection Coupling

It is of course quite disappointing at first glance that the only gradient commutation property that we get using this nice contraction results in W_1 distance, is restricted to Lipschitz function as in (75). Let us see however that we may transfer this to stronger gradient commutation properties in some cases. The main tool is the following lemma on Hölder's type inequality in Wasserstein distance.

Lemma 3 *Suppose that v and μ are two probability measures on $\mathbb{R}$, then for all $q > 1$ and p such that $\frac{1}{p} + \frac{1}{q} = 1$, we have*

$$W_2(v, \mu) \leq W_1^{\frac{1}{2q}}(v, \mu)\, W_{(2-\frac{1}{q})p}^{1-\frac{1}{2q}}(v, \mu). \tag{78}$$

Furthermore the result tensorises in the sense, that if for $i = 1, \ldots, n$, μ_i and v_i are probability measures on $\mathbb{R}$, we have for some constant $c(n)$

$$W_2(\otimes_1^n v_i, \otimes_1^n \mu_i) \leq c(n)\, W_1^{\frac{1}{2q}}(\otimes_1^n v_i, \otimes_1^n \mu_i)\, W_{(2-\frac{1}{q})p}^{1-\frac{1}{2q}}(\otimes_1^n v_i, \otimes_1^n \mu_i).$$

Proof The proof is indeed quite simple and relies mainly on Hölder's inequality. Indeed, in dimension one the optimal transport plan is the same for every convex cost (see for example Villani [42]), so that there exists a transport plan π such that

$$
\begin{aligned}
W_2^2(v, \mu) &= \int\int |x - y|^2 d\pi \\
&\leq \left(\int\int |x - y| d\pi \right)^{1/q} \left(\int\int |x - y|^{(2-\frac{1}{q})p} d\pi \right)^{1/p} \\
&= W_1^{\frac{1}{q}}(v, \mu)\, W_{(2-\frac{1}{q})p}^{2-\frac{1}{q}}(v, \mu).
\end{aligned}
$$

The case of product probability measure is deduced using the result in dimension one and the following two direct assertions

$$W_2^2(\otimes_1^n v_i, \otimes_1^n \mu_i) = \sum_{i=1}^n W_2^2(v_i, \mu_i),$$

and if v and μ have for ith marginal v_i and μ_i

$$W_p(v_i, \mu_i) \leq W_p(v, \mu).$$

Remark 22 We failed at the present time to get the general version of this lemma, i.e. does there exists a constant c only depending on the dimension such that for two probability measures on $\mathbb{R}^n$, we have

$$W_2(v, \mu) \leq c(n)\, W_1^{\frac{1}{2q}}(v, \mu)\, W_{(2-\frac{1}{q})p}^{1-\frac{1}{2q}}(v, \mu)?$$

In fact, as will be seen from our applications, even if c does depend of v and μ (in a nice way), it would be sufficient to get new gradient commutation property. $\diamond$

We are now in position to prove various gradient commutation properties in non standard cases. For simplicity, we suppose here that the diffusion coefficient is constant, i.e.

$$dX_t = dB_t + b(X_t)dt.$$

Theorem 14 *Let us suppose here that either (X_t) lives in $\mathbb{R}$ or that starting from $X_0 = x \in \mathbb{R}^n$, $X_t = (X_t^1, \ldots, X_t^n)$ is composed of independent component. Assume moreover that $(H.\kappa)$ is satisfied and that $\kappa(r) \geq -L$ then, with λ defined in Theorem 11,*

$$W_2(\mathscr{L}(X_t^x), \mathscr{L}(X_t^y)) \leq c(n) \left(\frac{2}{\phi_{min}}\right)^{\frac{1}{2q}} e^{\left[(1-\frac{1}{2q})L-\frac{\lambda}{2q}\right]t} |x-y|$$

so that the weak gradient commutation property holds

$$|\nabla P_t f|^2 \leq c(n) \left(\frac{2}{\phi_{min}}\right)^{\frac{1}{2q}} e^{\left[(1-\frac{1}{2q})L-\frac{\lambda}{2q}\right]t} P_t |\nabla f|^2$$

and thus a local Poincaré inequality holds.

Note that this theorem is the first one to give the commutation gradient property in non strictly convex case with a good behaviour at infinity.

Proof Using synchronous coupling as previously explained and the fact that $\kappa(r) \geq -L$ we have that

$$W_{(2-\frac{1}{q})p}(\mathscr{L}(X_t^x), \mathscr{L}(X_t^y)) \leq e^{Lt} |x-y|.$$

In the same time, by Theorem 11, we have that

$$W_1(\mathscr{L}(X_t^x), \mathscr{L}(X_t^y)) \leq \frac{2}{\phi_{min}} e^{-\lambda t} |x-y|.$$

We then use Lemma 78 to get the first assertion. The second one is obtained as in Proposition 8.

Example 3 Consider for example the log-concave case $b(x) = -4x^3$ for which Bakry-Emery theory enables us to get that we are in 0-curvature and thus

$$|\nabla P_t f|^2 \leq P_t |\nabla f|^2.$$

However, using Theorem 14 and this last inequality, we easily get that there exists $\lambda > 0$ such that

$$|\nabla P_t f|^2 \leq \min\left(1, \frac{2}{\phi_{min}} e^{-\lambda t}\right) P_t |\nabla f|^2,$$

which is completely new. It captures both the short time behavior equivalent to the Γ_2 0-curvature criterion and the long time behavior for which $P_t f \to \mu(f)$ and thus $\nabla P_t f \mid$ is expected to decay to 0.

Note that we may extend this example to a double well potential, in the case when the height of the well is not too large.

8 Preserving Curvature

A natural question about curvature is the following: is curvature preserved by a diffusion process? According to a result by Kolesnikov [31], the Ornstein-Uhlenbeck process is essentially the only one, among diffusion processes, preserving log-concavity (i.e. if v_0 is log-concave, so is v_T for all $T > 0$). One may also wonder if v_t may satisfy other "curvature" like inequality as HWI for example. It would have important applications on local inequalities, indeed transportation inequalities together with a HWI inequality may imply logarithmic Sobolev inequality.

In the spirit of the previous remark, consider a standard Ornstein-Uhlenbeck process $X_.$, i.e. the solution of

$$dX_t = dB_t - \frac{\lambda}{2} X_t \, dt . \tag{79}$$

The curvature K is thus equal to $\lambda \in \mathbb{R}$. If $\mathscr{L}(X_0) = v$, it is known that the law v_T of X_T is the same as the law of

$$e^{-\lambda T/2} \left(Z + \sqrt{\frac{e^{\lambda T} - 1}{\lambda}} \, G \right) ,$$

where G and Z are independent random variables, G being a standard gaussian variable and Z having distribution v. Hence

$$C_{LS}(v_T) \le e^{-\lambda T} C_{LS}(v) + \frac{2(1 - e^{-\lambda T})}{\lambda} ,$$

or, if we use the notation $C_{LS}(Y) = C_{LS}(\eta)$ for a random variable Y with distribution η,

$$C_{LS} \left(e^{-\lambda T/2} \left(Z + \sqrt{\frac{e^{\lambda T} - 1}{\lambda}} \, G \right) \right) \le e^{-\lambda T} C_{LS}(Z) + \frac{2(1 - e^{-\lambda T})}{\lambda} . \tag{80}$$

But if A is a random variable it is clear that $C_{LS}(\lambda A) = \lambda^2 C_{LS}(A)$. It follows

$$C_{LS}\left(Z + \sqrt{\frac{e^{\lambda T} - 1}{\lambda}}\, G\right) \le C_{LS}(Z) + \frac{2(e^{\lambda T} - 1)}{\lambda}. \tag{81}$$

This is not surprising since a more general result can be obtained directly (extending a similar result for the Poincaré inequality in [6]):

Proposition 18 *Let X and Y be independent random variables and $\lambda \in [0,1]$ then,*

$$C_{LS}(\sqrt{\lambda}X + \sqrt{1-\lambda}Y) \le \lambda C_{LS}(X) + (1-\lambda)C_{LS}(Y)\,, \text{ the same holds with } C_P\,.$$

Proof The first result for C_P is proved in [6] proposition 1. For C_{LS} the proof is very similar. Let f be smooth. Then

$$\mathbb{E}\left((f^2 \log f^2)(\sqrt{\lambda}X + \sqrt{1-\lambda}Y)\right)$$

$$\le \int \left(\int f^2(\sqrt{\lambda}x + \sqrt{1-\lambda}y)\, d\mathbb{P}_X(x)\right)$$

$$\log\left(\int f^2(\sqrt{\lambda}x + \sqrt{1-\lambda}y)\, d\mathbb{P}_X(x)\right) dP_Y(y)$$

$$+ \int \left(C_{LS}(X) \int \lambda |\nabla f|^2(\sqrt{\lambda}x + \sqrt{1-\lambda}y)\, d\mathbb{P}_X(x)\right) dP_Y(y)$$

$$\le \left(\int f^2(\sqrt{\lambda}x + \sqrt{1-\lambda}y)\, d\mathbb{P}_X(x)\, d\mathbb{P}_Y(y)\right)$$

$$\log\left(\int f^2(\sqrt{\lambda}x + \sqrt{1-\lambda}y)\, d\mathbb{P}_X(x)\, d\mathbb{P}_Y(y)\right)$$

$$+ C_{LS}(Y) \int \left|\nabla \sqrt{\int f^2(\sqrt{\lambda}x + \sqrt{1-\lambda}y)\, d\mathbb{P}_X(x)}\right|^2 d\mathbb{P}_Y(y)$$

$$+ \lambda\, C_{LS}(X) \int |\nabla f|^2(\sqrt{\lambda}x + \sqrt{1-\lambda}y)\, d\mathbb{P}_X(x)\, dP_Y(y)$$

$$\le \mathbb{E}(f^2(\sqrt{\lambda}X + \sqrt{1-\lambda}Y)) \log\left(\mathbb{E}(f^2(\sqrt{\lambda}X + \sqrt{1-\lambda}Y))\right)$$

$$+ \lambda\, C_{LS}(X) \int |\nabla f|^2(\sqrt{\lambda}x + \sqrt{1-\lambda}y)\, d\mathbb{P}_X(x)\, dP_Y(y)$$

$$+ (1-\lambda)\, C_{LS}(Y) \int \left(\frac{\int |\nabla f^2|(\sqrt{\lambda}x + \sqrt{1-\lambda}y)\, d\mathbb{P}_X(x)}{2\sqrt{\int f^2(\sqrt{\lambda}x + \sqrt{1-\lambda}y)\, d\mathbb{P}_X(x)}}\right)^2 d\mathbb{P}_Y(y).$$

Since $\nabla f^2 = 2f \nabla f$, we may use Cauchy-Schwarz inequality in order to bound the last term in the latter sum. This yields exactly the desired result.

Remark 23 It is known that any log-concave probability measure satisfies some Poincaré inequality. The result is due to Bobkov [7] (a short proof is contained in [2]). If Z is a log-concave random variable, $Z + \alpha G$ is still log-concave according to the Prekopa-Leindler theorem.

Here is an amusing proof of the consequence of Prekopa's result when one variable is gaussian.

Let X (resp. Y) be a random variable with law $e^{-V(x)}dx$ (resp. a standard gaussian variable). We assume that X and Y are independent. The density of $X + \sqrt{\lambda}\, Y$ is thus given by

$$q(x) = (2\pi\lambda)^{-n/2} \int e^{-V(u)} e^{-\frac{|x-u|^2}{2\lambda}} du = (2\pi\lambda)^{-n/2} p(x).$$

Let $H(x)$ be the hessian matrix of $\log p$. Then

$$p^2(x) \langle \xi, H(x)\xi \rangle = p(x) \frac{1}{\lambda^2} \left(\int \langle \xi, (x-u) \rangle^2 e^{-V(u)} e^{-\frac{|x-u|^2}{2\lambda}} du \right) - \frac{1}{\lambda} p^2(x)|\xi|^2$$

$$- \frac{1}{\lambda^2} \left(\int \langle \xi, (x-u) \rangle e^{-V(u)} e^{-\frac{|x-u|^2}{2\lambda}} du \right)^2.$$

Hence

$$\langle \xi, H(x)\xi \rangle = -\frac{1}{\lambda}|\xi|^2 + \frac{1}{\lambda^2} \left(\int \langle \xi, (x-u) \rangle^2 e^{-V(u)} e^{-\frac{|x-u|^2}{2\lambda}} \frac{du}{p(x)} \right)$$

$$- \frac{1}{\lambda^2} \left(\int \langle \xi, (x-u) \rangle e^{-V(u)} e^{-\frac{|x-u|^2}{2\lambda}} \frac{du}{p(x)} \right)^2.$$

Now assume that V satisfies (H.C.K) for some $K \in \mathbb{R}$. The probability measure

$$e^{-V(u)} e^{-\frac{|x-u|^2}{2\lambda}} \frac{du}{p(x)}$$

(or if one prefers its potential) satisfies (H.C.$K + (1/\lambda)$). If $K + (1/\lambda) > 0$, it thus satisfies a Poincaré inequality with constant $\lambda/(1 + K\lambda)$. Applying this Poincaré inequality to the function $u \mapsto \langle \xi, (x-u) \rangle$, we obtain

$$\langle \xi, H(x)\xi \rangle \leq -\frac{K}{1 + K\lambda} |\xi|^2.$$

Thanks to simple scales we may thus state

Proposition 19 *Let X be a random variable with law $e^{-V(x)}dx$ and Y a standard gaussian variable independent of X. If V satisfies (H.C.K) for $K \in \mathbb{R}$, then for $0 \le \lambda \le 1$, the distribution of $\sqrt{\lambda}\, X + \sqrt{1 - \lambda}\, Y$ satisfies $(H.C.\frac{K}{\lambda + K(1-\lambda)})$ as soon as $\lambda + K(1 - \lambda) > 0$.*
In particular if X is log-concave $(K = 0)$ so is $\sqrt{\lambda}\, X + \sqrt{1 - \lambda}\, Y$.

References

1. C. Ané, S. Blachère, D. Chafaï, P. Fougères, I. Gentil, F. Malrieu, C. Roberto, G. Scheffer, *Sur les inégalités de Sobolev logarithmiques*. Panoramas et Synthèses, vol. 10 (Société Mathématique de France, Paris, 2000)
2. D. Bakry, F. Barthe, P. Cattiaux, A. Guillin, A simple proof of the Poincaré inequality for a large class of probability measures. Electon. Commun. Probab. **13**, 60–66 (2008)
3. D. Bakry, P. Cattiaux, A. Guillin, Rate of convergence for ergodic continuous Markov processes: Lyapunov versus Poincaré. J. Funct. Anal. **254**, 727–759 (2008)
4. D. Bakry, I. Gentil, L. Ledoux, On Harnack inequalities and optimal transportation. Preprint, available on ArXiv (2012)
5. D. Bakry, I. Gentil, L. Ledoux, Analysis and Geometry of Markov diffusion operators, Springer, Grundlehren der mathematischen Wissenschaften, Vol. 348 (2014)
6. K. Ball, F. Barthe, A. Naor, Entropy jumps in the presence of a spectral gap. Duke Math. J. **119**, 41–63 (2003)
7. S.G. Bobkov, Isoperimetric and analytic inequalities for log-concave probability measures. Ann. Probab. **27**(4), 1903–1921 (1999)
8. S.G. Bobkov, Spectral gap and concentration for some spherically symmetric probability measures, in *Geometric Aspects of Functional Analysis, Israel Seminar 2000–2001*. Lecture Notes in Mathematics, vol. 1807 (Springer, Berlin, 2003), pp. 37–43
9. S.G. Bobkov, I. Gentil, M. Ledoux, Hypercontractivity of Hamilton-Jacobi equations. J. Math. Pure Appl. **80**(7), 669–696 (2001)
10. F. Bolley, I. Gentil, A. Guillin, Convergence to equilibrium in Wasserstein distance for Fokker-Planck equation. J. Funct. Anal. **263**(8), 2430–2457 (2012)
11. F. Bolley, I. Gentil, A. Guillin, Uniform convergence to equilibrium for granular media. Arch. Ration. Mech. Anal. **208**(2), 429–445 (2013)
12. F. Bolley, A. Guillin, F. Malrieu, Trend to equilibrium and particle approximation for a weakly selfconsistent Vlasov-Fokker-Planck equation. Math. Model. Numer. Anal. **44**(5), 867–884 (2010)
13. C. Borell, Diffusion equations and geometric inequalities. Potential Anal. **12**, 49–71 (2000)
14. P. Cattiaux, A pathwise approach of some classical inequalities. Potential Anal. **20**, 361–394 (2004)
15. P. Cattiaux, Hypercontractivity for perturbed diffusion semi-groups. Ann. Fac. des Sc. de Toulouse **14**(4), 609–628 (2005)
16. P. Cattiaux, I. Gentil, A. Guillin, Weak logarithmic-Sobolev inequalities and entropic convergence. Probab. Theory Relat. Fields **139**, 563–603 (2007)
17. P. Cattiaux, A. Guillin, On quadratic transportation cost inequalities. J. Math. Pures Appl. **88**(4), 341–361 (2006)
18. P. Cattiaux, A. Guillin, F. Malrieu, Probabilistic approach for granular media equations in the non uniformly convex case. Probab. Theory Relat. Fields **140**, 19–40 (2008)
19. P. Cattiaux, A. Guillin, P.A. Zitt, Poincaré inequalities and hitting times. Ann. Inst. Henri Poincaré. Probab. Stat. **49**(1), 95–118 (2013)

20. P. Cattiaux, C. Léonard, Minimization of the Kullback information of diffusion processes. Ann. Inst. Henri Poincaré. Prob. Stat. **30**(1), 83–132 (1994); and correction in Ann. Inst. Henri Poincaré **31**, 705–707 (1995)
21. J.F. Collet, F. Malrieu, Logarithmic Sobolev inequalities for inhomogeneous semigroups. ESAIM Probab. Stat. **12**, 492–504 (2008)
22. H. Djellout, A. Guillin, L. Wu, Transportation cost information inequalities for random dynamical systems and diffusions. Ann. Probab. **334**, 1025–1028 (2002)
23. A. Eberle, Reflection coupling and Wasserstein contractivity without convexity. C. R. Acad. Sci. Paris Ser. I **349**, 1101–1104 (2011)
24. A. Eberle, Couplings, distances and contractivity for diffusion processes revisited. Available on Math. arXiv:1305.1233 [math.PR] (2013)
25. J. Fontbona, B. Jourdain, A trajectorial interpretation of the dissipations of entropy and Fisher information for stochastic differential equations. Available on Math. arXiv:1107.3300 [math.PR] (2011)
26. N. Gozlan, C. Léonard, Transport inequalities—a survey. Markov Process. Relat. Fields **16**, 635–736 (2010)
27. A. Guillin, F.-Y. Wang, Degenerate Fokker-Planck equations: Bismut formula, gradient estimate and Harnack inequality. J. Differ. Equ. **253**(1), 20–40 (2012)
28. A. Guillin, C. Léonard, L. Wu, N. Yao, Transportation-information inequalities for Markov processes. Probab. Theory Relat. Fields **144**(3–4), 669–695 (2009)
29. N. Huet, Isoperimetry for spherically symmetric log-concave probability measures. Rev. Mat. Iberoam. **27**(1), 93–122 (2011)
30. N. Ikeda, S. Watanabe, *Stochastic Differential Equations and Diffusion Processes*, 2nd edn. (North-Holland, Amsterdam, 1988)
31. A.V. Kolesnikov, On diffusion semigroups preserving the log-concavity. J. Funct. Anal. **186**(1), 196–205 (2001)
32. M. Ledoux, Logarithmic Sobolev inequalities for unbounded spin systems revisited, in *Séminaire de Probabilités XXXV*. Lecture Notes in Mathematics, vol. 1755 (Springer, New York, 2001), pp. 167–194
33. J. Lehec, Representation formula for the entropy and functional inequalities. Ann. Inst. Henri Poincaré. Prob. Stat. **49**(3), 885–899 (2013)
34. T. Lindvall, L.C.G. Rogers, Coupling of multidimensional diffusions by reflection. Ann. Probab. **14**, 860–872 (1986)
35. F. Malrieu, Logarithmic Sobolev inequalities for some nonlinear PDE's. Stoch. Process. Appl. **95**(1), 109–132 (2001)
36. F. Otto, C. Villani, Generalization of an inequality by Talagrand and links with the logarithmic Sobolev inequality. J. Funct. Anal. **173**, 361–400 (2000)
37. J. Pitman, M. Yor, Bessel processes and infinitely divisible laws, in *Stochastic Integrals*. Lecture Notes in Mathematics, vol. 851 (Springer, New York, 1980), pp. 285–370
38. P.E. Protter, *Stochastic Integration and Differential Equations*. Stochastic Modelling and Applied Probability, vol. 21 (Springer, Berlin, 2005)
39. M. Röckner, F.Y. Wang, Weak Poincaré inequalities and L^2-convergence rates of Markov semigroups. J. Funct. Anal. **185**(2), 564–603 (2001)
40. M. Röckner, F.Y. Wang, Log-Harnack inequality for stochastic differential equations in Hilbert spaces and its consequences. Anal. Quant. Probab. Relat. Top. **13**, 27–37 (2010)
41. C. Villani, Hypocoercivity. Mem. Am. Math. Soc. **202**(950), iv+141 (2009)
42. C. Villani, *Optimal Transport*. Grundlehren der Mathematischen Wissenschaften [Fundamental Principles of Mathematical Sciences], vol. 338 (Springer, Berlin, 2009)
43. M. Von Renesse, K.T. Sturm, Transport inequalities, gradient estimates, entropy, and Ricci curvature. Commun. Pure Appl. Math. **58**(7), 923–940 (2005)

44. F.Y. Wang, Functional Inequalities, Markov Processes and Spectral Theory (Science Press, Beijing, 2004)
45. F.Y. Wang, T. Zhang, Log-Harnack inequality for mild solutions of SPDEs with strongly multiplicative noise. Available on Math. arXiv:1210.6416 [math.PR] (2012)
46. C. Yi, On the first passage time distribution of an Ornstein-Uhlenbeck process. Quant. Fin. **10**(9), 957–960 (2010)

On Maximal Inequalities for Purely Discontinuous Martingales in Infinite Dimensions

Carlo Marinelli and Michael Röckner

Abstract The purpose of this paper is to give a survey of a class of maximal inequalities for purely discontinuous martingales, as well as for stochastic integral and convolutions with respect to Poisson measures, in infinite dimensional spaces. Such maximal inequalities are important in the study of stochastic partial differential equations with noise of jump type.

1 Introduction

The purpose of this work is to collect several proofs, in part revisited and extended, of a class of maximal inequalities for stochastic integrals with respect to compensated random measures, including Poissonian integrals as a special case. The precise formulation of these inequalities can be found in Sects. 3–5 below. Their main advantage over the maximal inequalities of Burkholder, Davis and Gundy is that their right-hand side is expressed in terms of predictable "ingredients", rather than in terms of the quadratic variation. Since our main motivation is the application to stochastic partial differential equations (SPDE), in particular to questions of existence, uniqueness, and regularity of solutions (cf. [25–29]), we focus on processes in continuous time taking values in infinite-dimensional spaces. Corresponding estimates for finite-dimensional processes have been used in many areas, for instance in connection to Malliavin calculus for processes with jumps, flow properties of solutions to SDEs, and numerical schemes for Lévy-driven SDEs (see e.g. [2, 16, 18]). Very recent extensions to vector-valued settings have been used to develop the theory of stochastic integration with jumps in (certain) Banach spaces (see [8] and references therein).

We have tried to reconstruct the historical developments around this class of inequalities (an investigation which led us to quite a few surprises), providing

C. Marinelli (✉)
Department of Mathematics, University College London, Gower Street, London WC1E 6BT, UK
e-mail: c.marinelli@ucl.ac.uk

M. Röckner
Fakultät für Mathematik, Universität Bielefeld, Bielefeld, Germany

© Springer International Publishing Switzerland 2014
C. Donati-Martin et al. (eds.), *Séminaire de Probabilités XLVI*, Lecture Notes
in Mathematics 2123, DOI 10.1007/978-3-319-11970-0_10

relevant references, and we hope that our account could at least serve to correct
some terminology that seems, from an historical point of view, not appropriate. In
fact, while we refer to Sect. 6 below for details, it seems important to remark already
at this stage that the estimates which we termed "Bichteler-Jacod's inequalities" in
our previous article [27] should have probably more rightfully been baptized as
"Novikov's inequalities", in recognition of the contribution [33].

Let us conclude this introductory section with a brief outline of the remaining
content: after fixing some notation and collecting a few elementary (but useful)
results in Sect. 2, we state and prove several upper and lower bounds for purely
discontinuous Hilbert-space-valued continuous-time martingales in Sect. 3. We
actually present several proofs, adapting, simplifying, and extending arguments of
the existing literature. The proofs in Sects. 3.2 and 3.3 might be, at least in part, new.
On the issue of who proved what and when, however, we refer to the (hopefully)
comprehensive discussion in Sect. 6. Section 4 deals with L_q-valued processes that
can be written as stochastic integrals with respect to compensated Poisson random
measures. Unfortunately, to keep this survey within a reasonable length, it has not
been possible to reproduce the proof, for which we refer to the original contribution
[8]. The (partial) extension to the case of stochastic convolutions is discussed in
Sect. 5.

2 Preliminaries

Let $(\Omega, \mathscr{F}, \mathbb{F} = (\mathscr{F}_t)_{t\geq 0}, \mathbb{P})$ be a filtered probability space satisfying the "usual"
conditions, on which all random elements will be defined, and H a real (separable)
Hilbert space with norm $\|\cdot\|$. If ξ is an E-valued random variable, with E a normed
space, and $p > 0$, we shall use the notation[1]

$$\|\xi\|_{\mathbb{L}_p(E)} := \left(\mathbb{E}\|\xi\|_E^p\right)^{1/p}.$$

Let μ be a random measure on a measurable space $(Z, \mathscr{Z})$, with dual predictable
projection (compensator) ν. We shall use throughout the paper the symbol M to
denote a martingale of the type $M = g \star \bar{\mu}$, where $\bar{\mu} := \mu - \nu$ and g is a vector-
valued (predictable) integrand such that the stochastic integral

$$(g \star \bar{\mu})_t := \int_{(0,t]}\int_Z g(s,z)\,\bar{\mu}(ds,dz)$$

is well defined. We shall deal only with the case that g (hence M) takes values in
H or in an L_q space. Integrals with respect to μ, ν and $\bar{\mu}$ will often be written in

---

[1]Just to avoid (unlikely) confusion, we note that $\mathbb{E}(\cdots)^\alpha$ always stands for the expectation of
$(\cdots)^\alpha$, and not for $[\mathbb{E}(\cdots)]^\alpha$.

abbreviated form, e.g. $\int_0^t g \, d\bar{\mu} := (g \star \bar{\mu})_t$ and $\int g \, d\bar{\mu} := (g \star \bar{\mu})_\infty$. If M is H-valued, the following well-known identities hold for the quadratic variation $[M, M]$ and the Meyer process $\langle M, M \rangle$:

$$[M, M]_T = \sum_{s \leq T} \|\Delta M_s\|^2 = \int_0^T \|g\|^2 \, d\mu, \qquad \langle M, M \rangle_T = \int_0^T \|g\|^2 \, d\nu$$

for any stopping time T. Moreover, we shall need the fundamental Burkholder-Davis-Gundy's (BDG) inequality:

$$\left\| M_\infty^* \right\|_{\mathbb{L}_p} \eqsim \left\| [M, M]_\infty^{1/2} \right\|_{\mathbb{L}_p} \qquad \forall p \in [1, \infty[,$$

where $M_\infty^* := \sup_{t \geq 0} \|M_t\|$. An expression of the type $a \lesssim b$ means that there exists a (positive) constant N such that $a \leq Nb$. If N depends on the parameters $p_1, \ldots, p_n$, we shall write $a \lesssim_{p_1, \ldots, p_n} b$. Moreover, if $a \lesssim b$ and $b \lesssim a$, we shall write $a \eqsim b$.

The following lemma about (Fréchet) differentiability of powers of the norm of a Hilbert space is elementary and its proof is omitted.

Lemma 1 *Let* $\phi : H \to \mathbb{R}$ *be defined as* $\phi : x \mapsto \|x\|^p$, *with* $p > 0$. *Then* $\phi \in C^\infty(H \setminus \{0\})$, *with first and second Fréchet derivatives*

$$\phi'(x) : \eta \mapsto p\|x\|^{p-2} \langle x, \eta \rangle, \tag{1}$$

$$\phi''(x) : (\eta, \zeta) \mapsto p(p-2)\|x\|^{p-4} \langle x, \eta \rangle \langle x, \zeta \rangle + p\|x\|^{p-2} \langle \eta, \zeta \rangle. \tag{2}$$

In particular, $\phi \in C^1(H)$ *if* $p > 1$, *and* $\phi \in C^2(H)$ *if* $p > 2$.

It should be noted that, here and in the following, for $p \in [1, 2[$ and $p \in [2, 4[$, the linear form $\|x\|^{p-2} \langle x, \cdot \rangle$ and the bilinear form $\|x\|^{p-4} \langle x, \cdot \rangle \langle x, \cdot \rangle$, respectively, have to be interpreted as the zero form if $x = 0$.

The estimate contained in the following lemma is simple but perhaps not entirely trivial.

Lemma 2 *Let* $1 \leq p \leq 2$. *One has, for any* $x, y \in H$,

$$0 \leq \|x + y\|^p - \|x\|^p - p\|x\|^{p-2} \langle x, y \rangle \lesssim_p \|y\|^p. \tag{3}$$

Proof We can clearly assume $x, y \neq 0$, otherwise (3) trivially holds. Since the function $\phi : x \mapsto \|x\|^p$ is convex and Fréchet differentiable on $H \setminus \{0\}$ for all $p \geq 1$, one has

$$\phi(x + y) - \phi(x) \geq \langle \nabla\phi(x), y \rangle,$$

hence, by (1),

$$\|x + y\|^p - \|x\|^p - p\|x\|^{p-2}\langle x, y\rangle \geq 0.$$

To prove the upper bound we distinguish two cases: if $\|x\| \leq 2\|y\|$, it is immediately seen that (3) is true; if $\|x\| > 2\|y\|$, Taylor's formula applied to the function $[0, 1] \ni t \mapsto \|x + ty\|^p$ implies

$$\|x + y\|^p - \|x\|^p - p\|x\|^{p-2}\langle x, y\rangle \lesssim_p \|x + \theta y\|^{p-2}\|y\|^2$$

for some $\theta \in]0, 1[$ (in particular $x + \theta y \neq 0$). Moreover, we have

$$\|x + \theta y\| \geq \|x\| - \|y\| > 2\|y\| - \|y\| = \|y\|,$$

hence, since $p - 2 \leq 0$, $\|x + \theta y\|^{p-2} \leq \|y\|^{p-2}$.   □

For the purposes of the following lemma only, let $(X, \mathscr{A}, m)$ be a measure space, and denote $L_p(X, \mathscr{A}, m)$ simply by L_p.

Lemma 3 *Let* $1 < q < p$. *For any* $\alpha \geq 0$, *one has*

$$\|f\|_{L_q}^\alpha \leq \|f\|_{L_1}^\alpha + \|f\|_{L_p}^\alpha$$

Proof By a well-known consequence of Hölder's inequality one has

$$\|f\|_{L_q} \leq \|f\|_{L_1}^r \|f\|_{L_p}^{1-r},$$

for some $0 < r < 1$. Raising to the power α and applying Young's inequality (see e.g. [13, Sect. 4.8]) with conjugate exponents $s := 1/r$ and $s' := 1/(1 - r)$ yields

$$\|f\|_{L_q}^\alpha \leq \|f\|_{L_1}^{r\alpha} \|f\|_{L_p}^{(1-r)\alpha} \leq r\|f\|_{L_1}^\alpha + (1-r)\|f\|_{L_p}^\alpha \leq \|f\|_{L_1}^\alpha + \|f\|_{L_p}^\alpha.$$

□

3 Inequalities for Martingales with Values in Hilbert Spaces

The following domination inequality, due to Lenglart [21], will be used several times.

Lemma 4 *Let* X *and* A *be a positive adapted right-continuous process and an increasing predictable process, respectively, such that* $\mathbb{E}[X_T|\mathscr{F}_0] \leq \mathbb{E}[A_T|\mathscr{F}_0]$ *for any bounded stopping time. Then one has*

$$\mathbb{E}(X_\infty^*)^p \lesssim_p \mathbb{E}A_\infty^p \qquad \forall p \in]0, 1[.$$

Theorem 1 *Let $\alpha \in [1, 2]$. One has*

$$
\mathbb{E}(M_\infty^*)^p \lesssim_{\alpha,p}
\begin{cases}
\mathbb{E}\left(\int \|g\|^\alpha \, dv \right)^{p/\alpha} & \forall p \in \,]0, \alpha] \, , \\[2ex]
\mathbb{E}\left(\int \|g\|^\alpha \, dv \right)^{p/\alpha} + \mathbb{E} \int \|g\|^p \, dv & \forall p \in [\alpha, \infty[\, ,
\end{cases}
\tag{BJ}
$$

and

$$
\mathbb{E}(M_\infty^*)^p \gtrsim_p \mathbb{E}\left(\int \|g\|^2 \, dv \right)^{p/2} + \mathbb{E} \int \|g\|^p \, dv \qquad \forall p \in [2, \infty[\, .
\tag{4}
$$

Sometimes we shall use the notation $\mathsf{BJ}_{\alpha,p}$ to denote the inequality BJ with parameters α and p.

Several proofs of BJ will be given below. Before doing that, a few remarks are in order. Choosing $\alpha = 2$ and $\alpha = p$, respectively, one obtains the probably more familiar expressions

$$
\mathbb{E}(M_\infty^*)^p \lesssim_p
\begin{cases}
\mathbb{E}\left(\int \|g\|^2 \, dv \right)^{p/2} & \forall p \in \,]0, 2] \, , \\[2ex]
\mathbb{E} \int \|g\|^p \, dv & \forall p \in [1, 2] \, , \\[2ex]
\mathbb{E}\left(\int \|g\|^2 \, dv \right)^{p/2} + \mathbb{E} \int \|g\|^p \, dv & \forall p \in [2, \infty[\, .
\end{cases}
$$

In more compact notation, BJ may equivalently be written as

$$
\|M_\infty^*\|_{\mathbb{L}_p} \lesssim_{\alpha,p}
\begin{cases}
\|g\|_{\mathbb{L}_p(L_\alpha(v))} & \forall p \in \,]0, \alpha] \, , \\[2ex]
\|g\|_{\mathbb{L}_p(L_\alpha(v))} + \|g\|_{\mathbb{L}_p(L_p(v))} & \forall p \in [\alpha, \infty[\, ,
\end{cases}
$$

where

$$
\|g\|_{\mathbb{L}_p(L_\alpha(v))} := \big\| \|g\|_{L_\alpha(v)} \big\|_{\mathbb{L}_p}, \qquad \|g\|_{L_\alpha(v)} := \left(\int \|g\|^\alpha \, dv \right)^{1/\alpha}.
$$

This notation is convenient but slightly abusive, as it is not standard (nor clear how) to define L_p spaces with respect to a random measure. However, if μ is a Poisson measure, then v is "deterministic" (i.e. it does not depend on $\omega \in \Omega$), and the above notation is thus perfectly lawful. In particular, if v is deterministic, it is rather straightforward to see that the above estimates imply

$$
\|M_\infty^*\|_{\mathbb{L}_p} \lesssim_p \inf_{g_1 + g_2 = g} \|g_1\|_{\mathbb{L}_p(L_2(v))} + \|g_2\|_{\mathbb{L}_p(L_p(v))}
$$

$$
=: \|g\|_{\mathbb{L}_p(L_2(v)) + \mathbb{L}_p(L_p(v))}, \qquad 1 \le p \le 2,
$$

as well as

$$\left\| M_\infty^* \right\|_{\mathbb{L}_p} \lesssim_p \max\left(\|g\|_{\mathbb{L}_p(L_2(\nu))}, \|g\|_{\mathbb{L}_p(L_p(\nu))} \right) =: \|g\|_{\mathbb{L}_p(L_2(\nu)) \cap \mathbb{L}_p(L_p(\nu))}, \quad p \geq 2$$

(for the notions of sum and intersection of Banach spaces see e.g. [17]). Moreover, since the dual space of $\mathbb{L}_p(L_2(\nu)) \cap \mathbb{L}_p(L_p(\nu))$ is $\mathbb{L}_{p'}(L_2(\nu)) + \mathbb{L}_{p'}(L_{p'}(\nu))$ for any $p \in [1, \infty[$, where $1/p + 1/p' = 1$, by a duality argument one can obtain the lower bound

$$\left\| M_\infty^* \right\|_{\mathbb{L}_p} \gtrsim \|g\|_{\mathbb{L}_p(L_2(\nu)) + \mathbb{L}_p(L_p(\nu))} \qquad \forall p \in]1, 2].$$

One thus has

$$\left\| M_\infty^* \right\|_{\mathbb{L}_p} \eqsim_p \begin{cases} \|g\|_{\mathbb{L}_p(L_2(\nu)) + \mathbb{L}_p(L_p(\nu))} & \forall p \in]1, 2], \\ \|g\|_{\mathbb{L}_p(L_2(\nu)) \cap \mathbb{L}_p(L_p(\nu))} & \forall p \in [2, \infty[. \end{cases}$$

By virtue of the Lévy-Itô decomposition and of the BDG inequality for stochastic integrals with respect to Wiener processes, the above maximal inequalities admit corresponding versions for stochastic integrals with respect to Lévy processes (cf. [15, 25]). We do not dwell on details here.

3.1 Proofs

We first prove the lower bound (4). The proof is taken from [24] (we recently learned, however, cf. Sect. 6 below, that the same argument already appeared in [10]).

Proof (Proof of (4)) Since $p/2 > 1$, one has

$$\mathbb{E}[M, M]_\infty^{p/2} = \mathbb{E}\left(\sum \|\Delta M\|^2 \right)^{p/2} \geq \mathbb{E} \sum \|\Delta M\|^p = \mathbb{E} \int \|g\|^p \, d\mu = \mathbb{E} \int \|g\|^p \, d\nu,$$

as well as, since $x \mapsto x^{p/2}$ is convex,

$$\mathbb{E}[M, M]_\infty^{p/2} \geq \mathbb{E}\langle M, M \rangle_\infty^{p/2} = \mathbb{E}\left(\int \|g\|^2 \, d\nu \right)^{p/2},$$

see e.g. [22]. Therefore, recalling the BDG inequality,

$$\mathbb{E}(M_\infty^*)^p \gtrsim \mathbb{E}[M, M]_\infty^{p/2} \gtrsim \mathbb{E}\left(\int \|g\|^2 \, d\nu \right)^{p/2} + \mathbb{E} \int \|g\|^p \, d\nu. \qquad \square$$

We now give several alternative arguments for the upper bounds.

The first proof we present is based on Itô's formula and Lenglart's domination inequality. It does not rely, in particular, on the BDG inequality, and it is probably, in this sense, the most elementary.

Proof (First Proof of BJ*)* Let $\alpha \in]1, 2]$, and $\phi : H \ni x \mapsto \|x\|^\alpha = h(\|x\|^2)$, with $h : y \mapsto y^{\alpha/2}$. Furthermore, let $(h_n)_{n \in \mathbb{N}}$ be a sequence of functions of class $C_c^\infty(\mathbb{R})$ such that $h_n \to h$ pointwise, and define $\phi_n : x \mapsto h_n(\|x\|^2)$, so that $\phi_n \in C_b^2(H)$.[2] Itô's formula (see e.g. [31]) then yields

$$\phi_n(M_\infty) = \int_0^\infty \phi_n'(M_-) \, dM + \sum \left(\phi_n(M_- + \Delta M) - \phi_n(M_-) - \phi_n'(M_-) \Delta M \right).$$

Taking expectation and passing to the limit as $n \to \infty$, one has, by estimate (3) and the dominated convergence theorem,

$$\mathbb{E}\|M_\infty\|^\alpha \leq \mathbb{E} \sum \left(\|M_- + \Delta M\|^\alpha - \|M_-\|^\alpha - \alpha \|M_-\|^{\alpha-2} \langle M_-, \Delta M \rangle \right)$$

$$\lesssim_\alpha \mathbb{E} \sum \|\Delta M\|^\alpha = \mathbb{E} \int \|g\|^\alpha \, d\mu = \mathbb{E} \int \|g\|^\alpha \, d\nu,$$

which implies, by Doob's inequality,

$$\mathbb{E}(M_\infty^*)^\alpha \lesssim_\alpha \mathbb{E} \int \|g\|^\alpha \, d\nu.$$

If $\alpha = 1$ we cannot use Doob's inequality, but we can argue by a direct calculation:

$$\mathbb{E}M_\infty^* = \mathbb{E} \sup_{t \geq 0} \left\| \int_0^t g \, d\bar{\mu} \right\| \leq \mathbb{E} \sup_{t \geq 0} \left\| \int_0^t g \, d\mu \right\| + \mathbb{E} \sup_{t \geq 0} \left\| \int_0^t g \, d\nu \right\|$$

$$\leq \mathbb{E} \sup_{t \geq 0} \int_0^t \|g\| \, d\mu + \mathbb{E} \sup_{t \geq 0} \int_0^t \|g\| \, d\nu$$

$$\leq 2\mathbb{E} \int \|g\| \, d\nu.$$

An application of Lenglart's domination inequality finishes the proof of the case $\alpha \in [1, 2], \ p \in]0, \alpha]$.

Let us now consider the case $\alpha = 2, \ p > 2$. We apply Itô's formula to a C_b^2 approximation of $x \mapsto \|x\|^p$, as in the first part of the proof, then take expectation and pass to the limit, obtaining

$$\mathbb{E}\|M_\infty\|^p \leq \mathbb{E} \sum \left(\|M_- + \Delta M\|^p - \|M_-\|^p - p \|M_-\|^{p-2} \langle M_-, \Delta M_s \rangle \right).$$

---

[2]The subscript $\cdot_c$ means "with compact support", and $C_b^2(H)$ denotes the set of twice continuously differentiable functions $\varphi : H \to \mathbb{R}$ such that φ, φ' and φ'' are bounded.

Applying Taylor's formula to the function $t \mapsto \|x + ty\|$ we obtain, in view of (2),

$$
\begin{aligned}
\|M_- + \Delta M\|^p &- \|M_-\|^p - p\|M_-\|^{p-2}\langle M_-, \Delta M\rangle \\
&= \frac{1}{2}p(p-2)\|M_- + \theta\Delta M\|^{p-4}\langle M_- + \theta\Delta M, \Delta M\rangle^2 \\
&\quad + \frac{1}{2}p\|M_- + \theta\Delta M\|^{p-2}\|\Delta M\|^2 \\
&\leq \frac{1}{2}p(p-1)\|M_- + \theta\Delta M\|^{p-2}\|\Delta M\|^2,
\end{aligned}
$$

where $\theta \equiv \theta_s \in \,]0,1[$. Since $\|M_- + \theta\Delta M\| \leq \|M_-\| + \|\Delta M\|$, we also have

$$
\|M_- + \theta\Delta M\|^{p-2} \lesssim_p \|M_-\|^{p-2} + \|\Delta M\|^{p-2} \leq (M_-^*)^{p-2} + \|\Delta M\|^{p-2}.
$$

Appealing to Doob's inequality, one thus obtains

$$
\begin{aligned}
\mathbb{E}(M_\infty^*)^p &\lesssim_p \mathbb{E}\|M_\infty\|^p \lesssim_p \mathbb{E}\sum\left((M_-^*)^{p-2}\|\Delta M\|^2 + \|\Delta M\|^p\right) \\
&= \mathbb{E}\int\left((M_-^*)^{p-2}\|g\|^2 + \|g\|^p\right)d\mu \\
&= \mathbb{E}\int\left((M_-^*)^{p-2}\|g\|^2 + \|g\|^p\right)d\nu \\
&\leq \mathbb{E}(M_\infty^*)^{p-2}\int\|g\|^2\,d\nu + \mathbb{E}\int\|g\|^p\,d\nu.
\end{aligned}
$$

By Young's inequality in the form

$$
ab \leq \varepsilon a^{\frac{p}{p-2}} + N(\varepsilon)b^{p/2},
$$

we are left with

$$
\mathbb{E}(M_\infty^*)^p \leq \varepsilon N(p)\mathbb{E}(M_\infty^*)^p + N(\varepsilon, p)\mathbb{E}\left(\int\|g\|^2\,d\nu\right)^{p/2} + \mathbb{E}\int\|g\|^p\,d\nu.
$$

The proof of the case $p > \alpha = 2$ is completed choosing ε small enough.

We are thus left with the case $\alpha \in [1, 2[$, $p > \alpha$. Note that, by Lemma 3,

$$
\|\cdot\|_{L_2(\nu)} \leq \|\cdot\|_{L_2(\nu)} + \|\cdot\|_{L_p(\nu)} \lesssim \|\cdot\|_{L_\alpha(\nu)} + \|\cdot\|_{L_p(\nu)},
$$

hence the desired result follows immediately by the cases with $\alpha = 2$ proved above.

$\square$

Remark 1 The proof of $\mathsf{BJ}_{2,p}$, $p \geq 2$, just given is a (minor) adaptation of the proof in [27], while the other cases are taken from [24]. However (cf. Sect. 6 below), essentially the same result with a very similar proof was already given by Novikov [33]. In the latter paper the author treats the finite-dimensional case, but the constants are explicitly dimension-free. Moreover, he deduces the case $p < \alpha$ from the case $p = \alpha$ using the extrapolation principle of Burkholder and Gundy [6], where we used instead Lenglart's domination inequality. However, the proof of the latter is based on the former.

Proof (Second Proof of $\mathsf{BJ}_{\alpha,p}$ ($p \leq \alpha$)) An application of the BDG inequality to M, taking into account that $\alpha/2 \leq 1$, yields

$$\mathbb{E}(M_T^*)^\alpha \lesssim_\alpha \mathbb{E}\Big(\sum_{\leq T}\|\Delta M\|^2\Big)^{\alpha/2}$$

$$\leq \mathbb{E}\sum_{\leq T}\|\Delta M\|^\alpha = \mathbb{E}\big(\|g\|^\alpha \star \mu\big)_T = \mathbb{E}\big(\|g\|^\alpha \star \nu\big)_T$$

for any stopping time T. The result then follows by Lenglart's domination inequality. □

We are now going to present several proofs for the case $p > \alpha$. As seen at the end of the first proof of BJ, it suffices to consider the case $p > \alpha = 2$.

Proof (Second Proof of $\mathsf{BJ}_{2,p}$ ($p > 2$)) Let us show that $\mathsf{BJ}_{2,2p}$ holds if $\mathsf{BJ}_{2,p}$ does: the identity

$$[M, M] = \|g\|^2 \star \mu = \|g\|^2 \star \bar{\mu} + \|g\|^2 \star \nu,$$

the BDG inequality, and $\mathsf{BJ}_{2,p}$ imply

$$\mathbb{E}(M_\infty^*)^{2p} \lesssim_p \mathbb{E}[M, M]_\infty^p \lesssim \mathbb{E}\big|\big(\|g\|^2 \star \bar{\mu}\big)_\infty\big|^p + \mathbb{E}\big(\|g\|^2 \star \nu\big)_\infty^p$$

$$\lesssim_p \mathbb{E}\int \|g\|^{2p}\,d\nu + \mathbb{E}\Big(\int \|g\|^4\,d\nu\Big)^{p/2} + \mathbb{E}\Big(\int \|g\|^2\,d\nu\Big)^{\frac{1}{2}2p} \tag{5}$$

$$= \|g\|_{L_{2p}(\nu)}^{2p} + \|g\|_{L_4(\nu)}^{2p} + \|g\|_{L_2(\nu)}^{2p}$$

Since $2 < 4 < 2p$, one has, by Lemma 3,

$$\|g\|_{L_4(\nu)}^{2p} \leq \|g\|_{L_{2p}(\nu)}^{2p} + \|g\|_{L_2(\nu)}^{2p},$$

which immediately implies that $\mathsf{BJ}_{2,2p}$ holds true. Let us now show that $\mathsf{BJ}_{2,p}$ implies $\mathsf{BJ}_{2,2p}$ also for any $p \in [1, 2]$. Recalling that $\mathsf{BJ}_{2,p}$ does indeed hold for $p \in [1, 2]$, this proves that $\mathsf{BJ}_{2,p}$ holds for all $p \in [2, 4]$, hence for all $p \geq 2$, thus completing the proof. In fact, completely similarly as above, one has, for any $p \in [1, 2]$,

$$\mathbb{E}(M_\infty^*)^{2p} \lesssim_p \mathbb{E}\big|\big(\|g\|^2 \star \bar{\mu}\big)_\infty\big|^p + \mathbb{E}\big(\|g\|^2 \star \nu\big)_\infty^p$$

$$\lesssim_p \mathbb{E} \int \|g\|^{2p} \, d\nu + \mathbb{E}\bigg(\int \|g\|^2 \, d\nu\bigg)^{\frac{1}{2} \, 2p}. \qquad \square$$

Remark 2 The above proof, with $p > 2$, is adapted from [3], where the authors assume $H = \mathbb{R}$ and $p = 2^n$, $n \in \mathbb{N}$, mentioning that the extension to any $p \geq 2$ can be obtained by an interpolation argument.

Proof (Third Proof of $\mathsf{BJ}_{2,p}$ *($p > 2$))* Let $k \in \mathbb{N}$ be such that $2^k \leq p < 2^{k+1}$. Applying the BDG inequality twice, one has

$$\mathbb{E}\big\|(g \star \bar{\mu})_\infty\big\|^p \lesssim_p \mathbb{E}\big(\|g\|^2 \star \mu\big)_\infty^{p/2} \lesssim_p \mathbb{E}\big|\big(\|g\|^2 \star \bar{\mu}\big)_\infty\big|^{p/2} + \mathbb{E}\big(\|g\|^2 \star \nu\big)_\infty^{p/2},$$

where

$$\mathbb{E}\big|\big(\|g\|^2 \star \bar{\mu}\big)_\infty\big|^{p/2} \lesssim_p \mathbb{E}\big(\|g\|^2 \star \mu\big)_\infty^{p/4} \lesssim_p \mathbb{E}\big|\big(\|g\|^4 \star \bar{\mu}\big)_\infty\big|^{p/4} + \mathbb{E}\big(\|g\|^4 \star \nu\big)_\infty^{p/4}.$$

Iterating we are left with

$$\mathbb{E}\big\|(g \star \bar{\mu})_\infty\big\|^p \lesssim_p \mathbb{E}\big(\|g\|^{2^{k+1}} \star \mu\big)_\infty^{p/2^{k+1}} + \sum_{i=1}^k \mathbb{E}\bigg(\int \|g\|^{2^i} \, d\nu\bigg)^{p/2^i},$$

where, recalling that $p/2^{k+1} < 1$,

$$\mathbb{E}\big(\|g\|^{2^{k+1}} \star \mu\big)_\infty^{p/2^{k+1}} = \mathbb{E}\Big(\sum \|\Delta M\|^{2^{k+1}}\Big)^{p/2^{k+1}}$$

$$\leq \mathbb{E} \sum \|\Delta M\|^p = \mathbb{E} \int \|g\|^p \, d\mu = \mathbb{E} \int \|g\|^p \, d\nu.$$

The proof is completed observing that, since $2 \leq 2^i \leq p$ for all $1 \leq i \leq k$, one has, by Lemma 3,

$$\mathbb{E}\bigg(\int \|g\|^{2^i} \, d\nu\bigg)^{p/2^i} = \mathbb{E}\|g\|_{L_{2^i}(\nu)}^p \leq \mathbb{E}\|g\|_{L_2(\nu)}^p + \mathbb{E}\|g\|_{L_p(\nu)}^p$$

$$= \mathbb{E}\bigg(\int \|g\|^2 \, d\nu\bigg)^{p/2} + \mathbb{E} \int \|g\|^p \, d\nu. \qquad \square$$

Remark 3 The above proof, which can be seen as a variation of the previous one, is adapted from [37, Lemma 4.1] (which was translated to the H-valued case in [25]). In [37] the interpolation step at the end of the proof is obtained in a rather tortuous (but interesting way), which is not reproduced here.

The next proof is adapted from [16].

Proof (Fourth Proof of $\mathsf{BJ}_{2,p}$ $(p > 2))$ Let us start again from the BDG inequality:

$$\mathbb{E}(M_\infty^*)^p \lesssim_p \mathbb{E}[M,M]_\infty^{p/2}.$$

Since $[M,M]$ is a real, positive, increasing, purely discontinuous process with $\Delta[M,M] = \|\Delta M\|^2$, one has

$$[M,M]_\infty^{p/2} = \sum([M,M]^{p/2} - [M,M]_-^{p/2})$$

$$= \sum\left(([M,M]_- + \|\Delta M\|^2)^{p/2} - [M,M]_-^{p/2}\right).$$

For any $a, b \geq 0$, the mean value theorem applied to the function $x \mapsto x^{p/2}$ yields the inequality

$$(a+b)^{p/2} - a^{p/2} = (p/2)\xi^{p/2-1}b \leq (p/2)(a+b)^{p/2-1}b$$

$$\leq (p/2)2^{p/2-1}(a^{p/2-1}b + b^{p/2}),$$

where $\xi \in \,]a,b[$, hence also

$$([M,M]_- + \|\Delta M\|^2)^{p/2} - [M,M]_-^{p/2} \lesssim_p [M,M]_-^{p/2-1}\|\Delta M\|^2 + \|\Delta M\|^p.$$

This in turn implies

$$\mathbb{E}[M,M]_\infty^{p/2} \lesssim_p \sum\left([M,M]_-^{p/2-1}\|\Delta M\|^2 + \|\Delta M\|^p\right)$$

$$= \mathbb{E}\int\left([M,M]_-^{p/2-1}\|g\|^2 + \|g\|^p\right)d\mu$$

$$= \mathbb{E}\int\left([M,M]_-^{p/2-1}\|g\|^2 + \|g\|^p\right)d\nu$$

$$\leq \mathbb{E}[M,M]_\infty^{p/2-1}\int\|g\|^2\,d\nu + \mathbb{E}\int\|g\|^p\,d\nu.$$

By Young's inequality in the form

$$a^{p/2-1}b \leq \varepsilon a^{p/2} + N(\varepsilon)b^{p/2}, \qquad a, b \geq 0,$$

one easily infers

$$\mathbb{E}[M, M]_\infty^{p/2} \lesssim_p \mathbb{E}\left(\int \|g\|^2 \, dv\right)^{p/2} + \mathbb{E}\int \|g\|^p \, dv, \qquad \square$$

thus concluding the proof.

3.2 A (Too?) Sophisticated Proof

In this subsection we prove a maximal inequality valid for any H-valued local martingale M (that is, we do not assume that M is purely discontinuous), from which $\mathsf{BJ}_{2,p}$, $p > 2$, follows immediately.

Theorem 2 Let M be any local martingale with values in H. One has, for any $p \geq 2$,

$$\mathbb{E}(M_\infty^*)^p \lesssim_p \mathbb{E}\langle M, M\rangle_\infty^{p/2} + \mathbb{E}\big((\Delta M)_\infty^*\big)^p.$$

Proof We are going to use Davis' decomposition (see [32] for a very concise proof in the case of real martingales, a detailed "transliteration" of which to the case of Hilbert-space-valued martingales can be found in [30]): setting $S := (\Delta M)^*$, one has $M = L + K$, where L and K are martingales satisfying the following properties:

(i) $\|\Delta L\| \lesssim S_-$;
(ii) K has integrable variation and $K = K^1 + \widetilde{K^1}$, where $\widetilde{K^1}$ is the predictable compensator of K^1 and $\int |dK^1| \lesssim S_\infty$.

Since $M^* \leq L^* + K^*$, we have

$$\|M_\infty^*\|_{\mathbb{L}_p} \leq \|L_\infty^*\|_{\mathbb{L}_p} + \|K_\infty^*\|_{\mathbb{L}_p},$$

where, by the BDG inequality, $\|K_\infty^*\|_{\mathbb{L}_p} \lesssim_p \|[K, K]^{1/2}\|_{\mathbb{L}_p}$. Moreover, by the maximal inequality for martingales with predictably bounded jumps in [22, p. 37][3] and the elementary estimate $\langle L, L\rangle^{1/2} \leq \langle M, M\rangle^{1/2} + \langle K, K\rangle^{1/2}$, one has

$$\|L_\infty^*\|_{\mathbb{L}_p} \lesssim_p \|\langle L, L\rangle_\infty^{1/2}\|_{\mathbb{L}_p} + \|S_\infty\|_{\mathbb{L}_p}$$

$$\leq \|\langle M, M\rangle_\infty^{1/2}\|_{\mathbb{L}_p} + \|\langle K, K\rangle_\infty^{1/2}\|_{\mathbb{L}_p} + \|(\Delta M)_\infty^*\|_{\mathbb{L}_p}.$$

---

[3]One can verify that the proof in [22] goes through without any change also for Hilbert-space-valued martingales.

Since $p \geq 2$, the inequality between moments of a process and of its dual predictable projection in [22, Theoreme 4.1] yields $\|\langle K, K\rangle^{1/2}\|_{\mathbb{L}_p} \lesssim_p \|[K, K]^{1/2}\|_{\mathbb{L}_p}$. In particular, we are left with

$$\|M^*_\infty\|_{\mathbb{L}_p} \lesssim_p \|\langle M, M\rangle_\infty^{1/2}\|_{\mathbb{L}_p} + \|(\Delta M)^*_\infty\|_{\mathbb{L}_p} + \|[K, K]_\infty^{1/2}\|_{\mathbb{L}_p}.$$

Furthermore, applying a version of Stein's inequality between moments of a process and of its predictable projection (see e.g. [30], and [39, p. 103] for the original formulation), one has, for $p \geq 2$,

$$\|[\widetilde{K^1}, \widetilde{K^1}]^{1/2}\|_{\mathbb{L}_p} \lesssim_p \|[K^1, K^1]^{1/2}\|_{\mathbb{L}_p},$$

hence, recalling property (ii) above and that the quadratic variation of a process is bounded by its first variation, we are left with

$$\|[K, K]^{1/2}\|_{\mathbb{L}_p} \leq \|[K^1, K^1]^{1/2}\|_{\mathbb{L}_p} + \|[\widetilde{K^1}, \widetilde{K^1}]^{1/2}\|_{\mathbb{L}_p}$$

$$\lesssim_p \|[K^1, K^1]^{1/2}\|_{\mathbb{L}_p} \leq \left\|\int |dK^1|\right\|_{\mathbb{L}_p} \lesssim \|(\Delta M)^*_\infty\|_{\mathbb{L}_p}. \qquad \square$$

It is easily seen that Theorem 2 implies $\mathsf{BJ}_{2,p}$ (for $p \geq 2$): in fact, one has

$$\mathbb{E}\left((\Delta M)^*\right)^p \leq \mathbb{E}\sum \|\Delta M\|^p = \mathbb{E}\int \|g\|^p \, d\nu.$$

Remark 4 The above proof is a simplified version of an argument from [24]. As we recently learned, however, a similar argument was given in [10]. As a matter of fact, their proof is somewhat shorter than ours, as they claim that $[K, L] = 0$. Unfortunately, we have not been able to prove this claim.

3.3 A Conditional Proof

The purpose of this subsection is to show that if $\mathsf{BJ}_{2,p}$, $p \geq 2$, holds for real (local) martingales, then it also holds for (local) martingales with values in H. For this we are going to use Khinchine's inequality: let $x \in H$, and $\{e_k\}_{k \in \mathbb{N}}$ be an orthonormal basis of H. Setting $x_k := \langle x, e_k\rangle$. Then one has

$$\|x\| = \left(\sum_k x_k^2\right)^{1/2} = \left\|\sum_k x_k \varepsilon_k\right\|_{L_2(\tilde{\Omega})} \approx \left\|\sum_k x_k \varepsilon_k\right\|_{L_p(\tilde{\Omega})},$$

where $(\bar{\Omega}, \bar{\mathscr{F}}, \bar{\mathbb{P}})$ is an auxiliary probability space, on which a sequence (ε_k) of i.i.d Rademacher random variables are defined.

Writing

$$M_k := \langle M, e_k \rangle = g_k \star \bar{\mu}, \qquad g_k := \langle g, e_k \rangle,$$

one has $\sum_k M_k \varepsilon_k = \left(\sum_k g_k \varepsilon_k\right) \star \bar{\mu}$, hence Khinchine's inequality, Tonelli's theorem, and Theorem 1 for real martingales yield

$$\mathbb{E}\|M\|^p \eqsim \mathbb{E}\left\|\left(\sum g_k \varepsilon_k\right) \star \bar{\mu}\right\|_{L_p(\bar{\Omega})}^p = \bar{\mathbb{E}}\,\mathbb{E}\left|\left(\sum g_k \varepsilon_k\right) \star \bar{\mu}\right|^p$$

$$\lesssim_p \bar{\mathbb{E}}\,\mathbb{E}\left(\int \left|\sum g_k \varepsilon_k\right|^2 d\nu\right)^{p/2} + \bar{\mathbb{E}}\,\mathbb{E}\int \left|\sum g_k \varepsilon_k\right|^p d\nu$$

$$=: I_1 + I_2.$$

Tonelli's theorem, together with Minkowski's and Khinchine's inequalities, yield

$$I_1 = \mathbb{E}\,\bar{\mathbb{E}}\left(\int \left|\sum g_k \varepsilon_k\right|^2 d\nu\right)^{p/2} = \mathbb{E}\left\|\int \left|\sum g_k \varepsilon_k\right|^2 d\nu\right\|_{L_{p/2}(\bar{\Omega})}^{p/2}$$

$$\leq \mathbb{E}\left(\int \left\|\left|\sum g_k \varepsilon_k\right|^2\right\|_{L_{p/2}(\bar{\Omega})} d\nu\right)^{p/2}$$

$$= \mathbb{E}\left(\int \left\|\sum g_k \varepsilon_k\right\|_{L_p(\bar{\Omega})}^2 d\nu\right)^{p/2} \eqsim \left(\int \|g\|^2 d\nu\right)^{p/2}.$$

Similarly, one has

$$I_2 = \mathbb{E}\int \left\|\sum g_k \varepsilon_k\right\|_{L_p(\bar{\Omega})}^p d\nu \eqsim \mathbb{E}\int \|g\|^p d\nu. \qquad \square$$

The proof is completed by appealing to Doob's inequality.

Remark 5 This conditional proof has probably not appeared in published form, although the idea is contained in [36].

4 Inequalities for Poisson Stochastic Integrals with Values in L_q Spaces

Even though there exist in the literature some maximal inequalities for stochastic integrals with respect to compensated Poisson random measures and Banach-space-valued integrands, here we limit ourselves to reporting about (very recent) two-sided estimates in the case of L_q-valued integrands. Throughout this section we assume that μ is a Poisson random measure, so that its compensator ν is of the form Leb $\otimes$ ν_0, where Leb stands for the one-dimensional Lebesgue measure and ν_0 is a (non-random) σ-finite measure on Z. Let $(X, \mathscr{A}, n)$ be a measure space, and denote L_q spaces on X simply by L_q, for any $q \geq 1$. Moreover, let us introduce the following spaces, where $p_1, p_2, p_3 \in [1, \infty[$:

$$L_{p_1, p_2, p_3} := \mathbb{L}_{p_1}(L_{p_2}(\mathbb{R}_+ \times Z \to L_{p_3}(X))), \quad \tilde{L}_{p_1, p_2} := \mathbb{L}_{p_1}(L_{p_2}(X \to L_2(\mathbb{R}_+ \times Z))).$$

Then one has the following result, due to Dirksen [8]:

$$\left(\mathbb{E} \sup_{t \geq 0} \|(g \star \bar{\mu})_t\|_{L_q}^p\right)^{1/p} \eqsim_{p,q} \|g\|_{\mathscr{I}_{p,q}}, \tag{6}$$

where

$$\mathscr{I}_{p,q} := \begin{cases} L_{p,p,q} + L_{p,q,q} + \tilde{L}_{p,q}, & 1 < p \leq q \leq 2, \\ (L_{p,p,q} \cap L_{p,q,q}) + \tilde{L}_{p,q}, & 1 < q \leq p \leq 2, \\ L_{p,p,q} \cap (L_{p,q,q} + \tilde{L}_{p,q}), & 1 < q < 2 \leq p, \\ L_{p,p,q} + (L_{p,q,q} \cap \tilde{L}_{p,q}), & 1 < p < 2 \leq q, \\ (L_{p,p,q} + L_{p,q,q}) \cap \tilde{L}_{p,q}, & 2 \leq p \leq q, \\ L_{p,p,q} \cap L_{p,q,q} \cap \tilde{L}_{p,q}, & 2 \leq q \leq p. \end{cases} \tag{7}$$

The proof of this result is too long to be included here. We limit instead ourselves to briefly recalling what the main "ingredients" are: the core of the argument is to establish suitable extensions of the classical Rosenthal inequality

$$\mathbb{E}\left|\sum \xi_k\right|^p \lesssim_p \max\left(\mathbb{E}\sum |\xi_k|^p, \left(\mathbb{E}\sum |\xi_k|^2\right)^{p/2}\right),$$

where $p \geq 2$ and $\xi = (\xi_k)_k$ is any (finite) sequence of centered independent real random variables (see [38]). In particular, if $\xi = (\xi_k)_k$ is a finite sequence of independent centered random variables taking values in L_q, one has

$$\left(\mathbb{E}\left\|\sum \xi_k\right\|_{L_q}^p\right)^{1/p} \eqsim_{p,q} \|\xi\|_{s_{p,q}} \qquad \forall p, q \in]1, \infty[, \tag{8}$$

where the space $s_{p,q}$ is defined replacing in the above definition of $\mathscr{I}_{p,q}$ the spaces $L_{p,q,q}$, $L_{p,p,q}$ and $\tilde{L}_{p,q}$ by the spaces $D_{q,q}$, $D_{p,q}$ and $S_{p,q}$, respectively, with

$$\|\xi\|_{D_{p,q}} := \left(\sum \mathbb{E}\|\xi_k\|_{L_q}^p\right)^{1/p}, \qquad \|\xi\|_{S_{p,q}} := \left\|\left(\sum \mathbb{E}|\xi_k|^2\right)^{1/2}\right\|_{L_q}.$$

The proof of the vector-valued Rosenthal inequality (8) combines in a clever and elegant way classical inequalities for sums of independent random variables in Banach spaces (see e.g. [7, 20]) with geometric properties of L_q spaces (in particular in connection with the notions of type and cotype). An important role is also played by the duality of sums and intersections of Banach spaces. The maximal inequalities (6) for stochastic integrals of step processes with respect to compensated Poisson random measures are then implied by (8), via a simple argument based on decoupling techniques and on Doob's inequality (for the decoupling approach to stochastic integration cf. [19]).

5 Inequalities for Stochastic Convolutions

In this section we show how one can extend, under certain assumptions, maximal inequalities from stochastic integrals to stochastic convolutions using dilations of semigroups. As is well known, stochastic convolutions are in general not semimartingales, hence establishing maximal inequalities for them is, in general, not an easy task. Usually one tries to approximate stochastic convolutions by processes which can be written as solutions to stochastic differential equations in either a Hilbert or a Banach space, for which one can (try to) obtain estimates using tools of stochastic calculus. As a final step, one tries to show that such estimates can be transferred to stochastic convolutions as well, based on establishing suitable convergence properties. At present it does not seem possible to claim that any of the two methods is superior to the other (cf., e.g., the discussion in [41]). We choose to concentrate on the dilation technique for its simplicity and elegance.

We shall say that a linear operator A on a Banach space E, such that $-A$ is the infinitesimal generator of a strongly continuous semigroup S, is of class D if there exist a Banach space $\bar{E}$, an isomorphic embedding $\iota : E \to \bar{E}$, a projection $\pi : \bar{E} \to \iota(E)$, and a strongly continuous bounded group $(U(t))_{t \in \mathbb{R}}$ on $\bar{E}$ such that the following diagram commutes for all $t > 0$:

$$
\begin{array}{ccc}
E & \xrightarrow{\ S(t)\ } & E \\
{\scriptstyle\iota}\downarrow & & \uparrow{\scriptstyle \iota^{-1}\circ\pi} \\
\bar{E} & \xrightarrow[\ U(t)\ ]{} & \bar{E}
\end{array}
$$

As far as we know there is no general characterization of operators of class D.[4] Several sufficient conditions, however, are known.

We begin with the classical dilation theorem by Sz.-Nagy (see e.g. [40]).

---

[4]The definition of class D is not standard and it is introduced just for the sake of concision.

Proposition 1 *Let A be a linear m-accretive operator on a Hilbert space H. Then A is of class D.*

The next result, due to Fendler [11], is analogous to Sz.-Nagy's dilation theorem in the context of L_q spaces, although it requires an extra positivity assumption. Here and in the following X stands for a measure space and m for a measure on it.

Proposition 2 *Let $E = L_q(X,m)$, with $q \in \,]1, \infty[$. Assume that A is a linear densely defined m-accretive operator on E such that $S(t) := e^{-tA}$ is positivity preserving for all $t > 0$. Then A is of class D, with $\bar{E} = L_q(Y)$, where Y is another measure space.*

The following very recent result, due to Fröhlich and Weis [12], allows one to consider classes of operators that are not necessarily accretive (for many interesting examples, see e.g. [41]). For all unexplained notions of functional calculus for operators we refer to, e.g., [42].

Proposition 3 *Let $E = L_q(X,m)$, with $q \in \,]1, \infty[$, and assume that A is sectorial and admits a bounded H^∞-calculus with $\omega_{H^\infty}(A) < \pi/2$. Then A is of class D, and one can choose $\bar{E} = L_q([0,1] \times X, Leb \otimes m)$.*

We are now going to show how certain maximal estimates for stochastic integrals yield maximal estimates for convolutions involving the semigroup generated by an operator of class D. As mentioned at the beginning of the section, the problem is that stochastic convolutions are not martingales, hence maximal inequalities for the latter class of processes cannot be directly used. The workaround presented here is, roughly speaking, based on the idea of embedding the stochastic convolution in the larger space $\bar{E}$, where the semigroup S can be replaced by the group U, to the effect that inequalities for martingales can be applied. In particular, note that, since U is a strongly continuous group of contractions and the operator norms of π is less than or equal to one, we have

$$\mathbb{E} \sup_{t \geq 0} \left\| \int_0^t \!\!\! \int_Z S(t-s) g(s,z) \, \bar{\mu}(ds, dz) \right\|_E^p$$

$$= \mathbb{E} \sup_{t \geq 0} \left\| \pi \int_0^t \!\!\! \int_Z U(t-s) \iota(g(s,z)) \, \bar{\mu}(ds, dz) \right\|_{\bar{E}}^p$$

$$= \mathbb{E} \sup_{t \geq 0} \left\| \pi U(t) \int_0^t \!\!\! \int_Z U(-s) \iota(g(s,z)) \, \bar{\mu}(ds, dz) \right\|_{\bar{E}}^p$$

$$\leq \|\pi\|_\infty^p \, \sup_{t \geq 0} \|U(t)\|_\infty^p \, \mathbb{E} \sup_{t \geq 0} \left\| \int_0^t \!\!\! \int_Z U(-s) \iota(g(s,z)) \, \bar{\mu}(ds, dz) \right\|_{\bar{E}}^p$$

$$\leq \mathbb{E} \sup_{t \geq 0} \left\| \int_0^t \!\!\! \int_Z U(-s) \iota(g(s,z)) \, \bar{\mu}(ds, dz) \right\|_{\bar{E}}^p, \tag{9}$$

where $\|\cdot\|_\infty$ denotes the operator norm. We have thus reduced the problem to finding a maximal estimate for a stochastic integral, although involving a different integrand and on a larger space.

If E is a Hilbert space we can proceed rather easily.

Proposition 4 *Let A be of class D on a Hilbert space E. Then one has, for any $\alpha \in [1, 2]$,*

$$\mathbb{E}\sup_{t\geq 0}\left\|\int_0^t S(t-\cdot)g\,d\bar\mu\right\|_E^p \lesssim_{\alpha,p} \begin{cases} \mathbb{E}\left(\int \|g\|^\alpha\,d\nu\right)^{p/\alpha} & \forall p \in\,]0,\alpha]\,, \\ \mathbb{E}\left(\int \|g\|^\alpha\,d\nu\right)^{p/\alpha} + \mathbb{E}\int \|g\|^p\,d\nu & \forall p \in [\alpha,\infty[\,. \end{cases}$$

Proof We consider only the case $p > \alpha$, as the other one is actually simpler. The estimate $\mathsf{BJ}_{\alpha,p}$ and (9) yield

$$\mathbb{E}\sup_{t\geq 0}\left\|\int_0^t S(t-\cdot)g\,d\bar\mu\right\|_E^p$$

$$\lesssim_{\alpha,p} \mathbb{E}\int \|U(-\cdot)\iota\circ g\|_{\bar E}^p\,d\nu + \mathbb{E}\left(\int \|U(-\cdot)\iota\circ g\|_{\bar E}^\alpha\,d\nu\right)^{p/\alpha}$$

$$\leq \mathbb{E}\int \|g\|_E^p\,d\nu + \mathbb{E}\left(\int \|g\|_E^\alpha\,d\nu\right)^{p/\alpha},$$

because U is a unitary group and the embedding ι is isometric.   □

If $E = L_q(X)$, the transposition of maximal inequalities from stochastic integrals to stochastic convolution is not so straightforward. In particular, (9) implies that the corresponding upper bounds will be functions of the norms of $U(-\cdot)\iota\circ g$ in three spaces of the type $L_{p,p,q}$, $L_{p,q,q}$ and $\tilde L_{p,q}$ (with X replaced by a different measure space Y, so that $\bar E = L_q(Y)$). In analogy to the previous proposition, it is not difficult to see that

$$\left\|U(-\cdot)\iota\circ g\right\|_{\mathbb{L}_{p_1}L_{p_2}(\mathbb{R}_+\times Z\to L_{p_3}(Y))} \leq \|g\|_{\mathbb{L}_{p_1}L_{p_2}(\mathbb{R}_+\times Z\to L_{p_3}(X))}. \tag{10}$$

However, estimating the norm of $U(-\cdot)\iota\circ g$ in $\tilde L_{p,q}(Y)$ in terms of the norm of g in $\tilde L_{p,q}$ does not seem to be possible without further assumptions. Nonetheless, the following sub-optimal estimates can be obtained.

Proposition 5 *Let A be of class D on $E = L_q := L_q(X)$ and μ a Poisson random measure. Then one has*

$$\mathbb{E}\sup_{t\geq 0}\left\|\int_0^t S(t-\cdot)g\,d\bar\mu\right\|_{L_q}^p \lesssim_{p,q} \|g\|_{\mathscr{I}_{p,q}},$$

where

$$\mathscr{I}_{p,q} := \begin{cases} L_{p,p,q} + L_{p,q,q}, & 1 < p \le q \le 2, \\ L_{p,p,q} \cap L_{p,q,q}, & 1 < q \le p \le 2, \\ L_{p,p,q} \cap L_{p,q,q}, & 1 < q < 2 \le p, \\ L_{p,p,q} + (L_{p,q,q} \cap L_{p,2,q}), & 1 < p < 2 \le q, \\ (L_{p,p,q} + L_{p,q,q}) \cap L_{p,2,q}, & 2 \le p \le q, \\ L_{p,p,q} \cap L_{p,q,q} \cap L_{p,2,q}, & 2 \le q \le p. \end{cases}$$

Proof Note that, if $q < 2$, one has, by definition, $\|\cdot\|_{\mathscr{I}_{p,q}} \le \|\cdot\|_{\mathscr{J}_{p,q}}$ [where the spaces $\mathscr{I}_{p,q}$ have been defined in (7)]; if $q \ge 2$, by Minkowski's inequality,

$$\left\| \left(\int |g|^2 \, dv \right)^{1/2} \right\|_{L_q} = \left\| \int |g|^2 \, dv \right\|_{L_{q/2}}^{1/2} \le \left(\int \|g\|_{L_q}^2 \, dv \right)^{1/2},$$

that is, $\|\cdot\|_{\tilde{L}_{p,q}} \le \|\cdot\|_{L_{p,2,q}}$. This implies $\|\cdot\|_{\mathscr{I}_{p,q}} \le \|\cdot\|_{\mathscr{J}_{p,q}}$ for all $q \ge 2$, hence for all (admissible) values of p and q. Therefore (9) and the maximal estimate (7) yield the desired result. □

Remark 6 The above maximal inequalities for stochastic convolutions continue to hold if A is only quasi-m-accretive and g has compact support in time. In this case the inequality sign $\lesssim_{p,q}$ has to be replaced by $\lesssim_{p,q,\eta,T}$, where T is a finite time horizon. One simply has to repeat the same arguments using the m-accretive operator $A + \eta I$, for some $\eta > 0$.

6   Historical and Bibliographical Remarks

In this section we try to reconstruct, at least in part, the historical developments around the maximal inequalities presented above. Before doing that, however, let us briefly explain how we became interested in the class of maximal inequalities: the first-named author used in [25] a Hilbert-space version of a maximal inequality in [37] to prove well-posedness for a Lévy-driven SPDE arising in the modeling of the term structure of interest rates. The second-named author pointed out that such an inequality, possibly adapted to the more general case of integrals with respect to compensated Poisson random measures (rather than with respect to Lévy processes), was needed to solve a problem he was interested in, namely to establish regularity of solutions to SPDEs with jumps with respect to initial conditions: our joint efforts led to the results in [27], where we proved a slightly less general version of the inequality $\mathsf{BJ}_{2,p}$, $p \ge 2$, using an argument involving only Itô's formula. At the time of writing [27] we did not realize that, as demonstrated in the present paper, it would have been possible to obtain the same result adapting one of the two arguments (for Lévy-driven integrals) we were aware of, i.e. those in [3] and [37].

The version in [27] of the inequality $\mathsf{BJ}_{2,p}$, $p \geq 2$, was called in that paper "Bichteler-Jacod inequality", as we believed it appeared (in dimension one) for the first time in [3]. This is still what we believed until a few days ago (this explains the label BJ), when, after this paper as well as the first drafts of [23] and [24] were completed, we found a reference to [33] in [44]. This is one of the surprises we alluded to in the introduction. Namely, Novikov proved (in 1975, hence well before Bichteler and Jacod, not to mention how long before ourselves) the upper bound $\mathsf{BJ}_{\alpha,p}$ for all values of α and p, assuming $H = \mathbb{R}^n$, but with constants that are independent of the dimension. For this reason it seems that, if one wants to give a name (as we do) to the inequality BJ and its extensions, they should be called Novikov's inequalities.[5] Unfortunately Novikov's paper [33] was probably not known also to Kunita, who proved in [18] (in 2004) a slightly weaker version of $\mathsf{BJ}_{2,p}$, $p \geq 2$, in $H = \mathbb{R}^n$, also using Itô's formula. Moreover, Applebaum [1] calls these inequalities "Kunita's estimates", but, again, they are just a version of what we called (and are going to call) Novikov's inequality.

Even though the proofs in [2, 3] are only concerned with the real-valued case, the authors explicitly say that they knew how to get the constant independent of the dimension (see, in particular, [2, Lemma 5.1 and remark 5.2]). The proofs in [16, 37] are actually concerned with integrals with respect to Lévy processes, but the adaptation to the more general case presented here is not difficult. Moreover, the inequalities in [2, 3, 16, 37] are of the type

$$\mathbb{E} \sup_{t \leq T} \|(g \star \bar{\mu})_t\|^p \lesssim_{p,d,T} \mathbb{E} \int_0^T \left(\int_Z \|g((s,\cdot)\|^2 \, dm \right)^{p/2} ds + \mathbb{E} \int_0^T \int_Z \|g((s,\cdot)\|^p \, d\nu_0 \, ds,$$

where μ is a Poisson random measure with compensator $\nu = \mathrm{Leb} \otimes \nu_0$. Our proofs show that all their arguments can be improved to yield a constant depending only on p and that the first term on the right-hand side can be replaced by $\mathbb{E}\big(\|g\|^2 \star \nu\big)_T^{p/2}$.

Again through [44] we also became aware of the Novikov-like inequality by Dzhaparidze and Valkeila [10], where Theorem 2 in proved with $H = \mathbb{R}$. It should be observed that the inequality in the latter theorem is apparently more general than, but actually equivalent to BJ (cf. [24]).

Another method to obtain Novikov-type inequalities, also in vector-valued settings, goes through their analogs in discrete time, i.e. the Burkholder-Rosenthal inequality. We have not touched upon this method, as we are rather interested in

---

[5]It should be mentioned that there are discrete-time real-valued analogs of $\mathsf{BJ}_{2,p}$, $p \geq 2$, that go under the name of Burkholder-Rosenthal (in alphabetical but reverse chronological order: Rosenthal [38] proved it for sequences of independent random variables in 1970, then Burkholder [5] extended it to discrete-time (real) martingales in 1973), and some authors speak of continuous-time Burkholder-Rosenthal inequalities. One may then also propose to use the expression Burkholder-Rosenthal-Novikov inequality, that, however, seems too long.

"direct" methods in continuous time. We refer the interested reader to the very recent preprints [8, 9], as well as to [35, 43] and references therein.

The idea of using dilation theorems to extend results from stochastic integrals to stochastic convolutions has been introduced, to the best of our knowledge, in [14]. This method has then been generalized in various directions, see e.g. [15, 25, 41]. In this respect, it should be mentioned that the "classical" direct approach, which goes through approximations by regular processes and avoid dilations (here "classical" stands for equations on Hilbert spaces driven by Wiener process), has been (partially) extended to Banach-space valued stochastic convolutions with jumps in [4]. The former and the latter methods are complementary, in the sense that none is more general than the other. Furthermore, it is well known (see e.g. [34]) that the factorization method breaks down when applied to stochastic convolutions with respect to jump processes.

Acknowledgements A large part of the work for this paper was carried out during visits of the first-named author to the Interdisziplinäres Zentrum für Komplexe Systeme, Universität Bonn, invited by S. Albeverio. The second-named author is supported by the DFG through the SFB 701.

References

1. D. Applebaum, *Lévy Processes and Stochastic Calculus*, 2nd edn. (Cambridge University Press, Cambridge, 2009). MR 2512800 (2010m:60002)
2. K. Bichteler, J.-B. Gravereaux, J. Jacod, *Malliavin Calculus for Processes with Jumps* (Gordon and Breach Science Publishers, New York, 1987). MR MR1008471 (90h:60056)
3. K. Bichteler, J. Jacod, Calcul de Malliavin pour les diffusions avec sauts: existence d'une densité dans le cas unidimensionnel, in *Seminar on Probability, XVII*. Lecture Notes in Math., vol. 986 (Springer, Berlin, 1983), pp. 132–157. MR 770406 (86f:60070)
4. Z. Brzeźniak, E. Hausenblas, J. Zhu, *Maximal inequality of stochastic convolution driven by compensated Poisson random measures in Banach spaces*, arXiv:1005.1600 (2010)
5. D.L. Burkholder, Distribution function inequalities for martingales. Ann. Probab. **1**, 19–42 (1973). MR 0365692 (51 #1944)
6. D.L. Burkholder, R.F. Gundy, Extrapolation and interpolation of quasi-linear operators on martingales. Acta Math. **124**, 249–304 (1970). MR 0440695 (55 #13567)
7. V.H. de la Peña, E. Giné, *Decoupling* (Springer, New York, 1999). MR 1666908 (99k:60044)
8. S. Dirksen, Itô isomorphisms for L^p-valued Poisson stochastic integrals. Ann. Probab. **42**(6), 2595–2643 (2014). doi:10.1214/13-AOP906
9. S. Dirksen, J. Maas, and J. van Neerven, Poisson stochastic integration in Banach spaces, Electron. J. Probab. **18**(100), 28 pp. (2013)
10. K. Dzhaparidze, E. Valkeila, On the Hellinger type distances for filtered experiments. Probab. Theory Relat. Fields **85**(1), 105–117 (1990). MR 1044303 (91d:60102)
11. G. Fendler, Dilations of one parameter semigroups of positive contractions on L^p spaces. Can. J. Math. **49**(4), 736–748 (1997). MR MR1471054 (98i:47035)
12. A.M. Fröhlich, L. Weis, H^∞ calculus and dilations. Bull. Soc. Math. France **134**(4), 487–508 (2006). MR 2364942 (2009a:47091)
13. G.H. Hardy, J.E. Littlewood, G.Pólya, *Inequalities*, 2nd edn. (Cambridge University Press, Cambridge, 1988). MR 0046395 (13,727e)
14. E. Hausenblas, J. Seidler, A note on maximal inequality for stochastic convolutions. Czech. Math. J. **51(126)**(4), 785–790 (2001). MR MR1864042 (2002j:60092)

15. E. Hausenblas, J. Seidler, Stochastic convolutions driven by martingales: maximal inequalities and exponential integrability. Stoch. Anal. Appl. **26**(1), 98–119 (2008). MR 2378512 (2009a:60066)
16. J. Jacod, Th.G. Kurtz, S. Méléard, Ph. Protter, The approximate Euler method for Lévy driven stochastic differential equations. Ann. Inst. H. Poincaré Probab. Stat. **41**(3), 523–558 (2005). MR MR2139032 (2005m:60149)
17. S.G. Kreĭn, Yu.Ī. Petunīn, E.M. Semënov, *Interpolation of Linear Operators*. Translations of Mathematical Monographs, vol. 54 (American Mathematical Society, Providence, 1982). MR 649411 (84j:46103)
18. H. Kunita, Stochastic differential equations based on Lévy processes and stochastic flows of diffeomorphisms, in *Real and Stochastic Analysis* (Birkhäuser Boston, Boston, 2004), pp. 305–373. MR 2090755 (2005h:60169)
19. S. Kwapień, W.A. Woyczyński, *Random Series and Stochastic Integrals: Single and Multiple* (Birkhäuser, Boston, 1992). MR 1167198 (94k:60074)
20. M. Ledoux, M. Talagrand, *Probability in Banach Spaces* (Springer, Berlin, 1991). MR 1102015 (93c:60001)
21. E. Lenglart, Relation de domination entre deux processus. Ann. Inst. H. Poincaré Sect. B (N.S.) **13**(2), 171–179 (1977). MR 0471069 (57 #10810)
22. E. Lenglart, D. Lépingle, M. Pratelli, Présentation unifiée de certaines inégalités de la théorie des martingales, in *Séminaire de Probabilités, XIV (Paris, 1978/1979)*. Lecture Notes in Math., vol. 784 (Springer, Berlin, 1980), pp. 26–52. MR 580107 (82d:60087)
23. C. Marinelli, *On maximal inequalities for purely discontinuous L_q-valued martingales*. Arxiv:1311.7120v1 (2013)
24. C. Marinelli, *On regular dependence on parameters of stochastic evolution equations*, in preparation
25. C. Marinelli, Local well-posedness of Musiela's SPDE with Lévy noise. Math. Finance **20**(3), 341–363 (2010). MR 2667893
26. C. Marinelli, Approximation and convergence of solutions to semilinear stochastic evolution equations with jumps. J. Funct. Anal. **264**(12), 2784–2816 (2013). MR 3045642
27. C. Marinelli, C. Prévôt, M.Röckner, Regular dependence on initial data for stochastic evolution equations with multiplicative Poisson noise. J. Funct. Anal. **258**(2), 616–649 (2010). MR MR2557949
28. C. Marinelli, M. Röckner, On uniqueness of mild solutions for dissipative stochastic evolution equations. Infinite Dimens. Anal. Quantum Probab. Relat. Top. **13**(3), 363–376 (2010). MR 2729590 (2011k:60220)
29. C. Marinelli, M. Röckner, Well-posedness and asymptotic behavior for stochastic reaction-diffusion equations with multiplicative Poisson noise. Electron. J. Probab. **15**(49), 1528–1555 (2010). MR 2727320
30. C. Marinelli, M. Röckner, *On the maximal inequalities of Burkholder, Davis and Gundy*, arXiv preprint (2013)
31. M. Métivier, *Semimartingales* (Walter de Gruyter & Co., Berlin, 1982). MR MR688144 (84i:60002)
32. P.A. Meyer, *Le dual de H^1 est BMO (cas continu)*, Séminaire de Probabilités, VII (Univ. Strasbourg). Lecture Notes in Math., vol. 321 (Springer, Berlin, 1973), pp. 136–145. MR 0410910 (53 #14652a)
33. A.A. Novikov, Discontinuous martingales. Teor. Verojatnost. i Primemen. **20**, 13–28 (1975). MR 0394861 (52 #15660)
34. Sz. Peszat, J. Zabczyk, *Stochastic Partial Differential Equations with Lévy Noise* (Cambridge University Press, Cambridge, 2007). MR MR2356959
35. Io. Pinelis, Optimum bounds for the distributions of martingales in Banach spaces. Ann. Probab. **22**(4), 1679–1706 (1994). MR 1331198 (96b:60010)
36. C. Prévôt (Knoche), Mild solutions of SPDE's driven by Poisson noise in infinite dimensions and their dependence on initial conditions, Ph.D. thesis, Universität Bielefeld, 2005

37. Ph. Protter, D. Talay, The Euler scheme for Lévy driven stochastic differential equations. Ann. Probab. **25**(1), 393–423 (1997). MR MR1428514 (98c:60063)
38. H.P. Rosenthal, On the subspaces of L^p ($p > 2$) spanned by sequences of independent random variables. Isr. J. Math. **8**, 273–303 (1970). MR 0271721 (42 #6602)
39. E.M. Stein, *Topics in Harmonic Analysis Related to the Littlewood-Paley Theory* (Princeton University Press, Princeton, 1970). MR 0252961 (40 #6176)
40. B. Sz.-Nagy, C. Foias, H. Bercovici, L. Kérchy, *Harmonic analysis of operators on Hilbert space*, 2nd edn. (Springer, New York, 2010). MR 2760647 (2012b:47001)
41. M. Veraar, L. Weis, A note on maximal estimates for stochastic convolutions. Czech. Math. J. **61**(**136**)(3), 743–758 (2011). MR 2853088
42. L. Weis, The H^∞ holomorphic functional calculus for sectorial operators—a survey, in *Partial Differential Equations and Functional Analysis* (Birkhäuser, Basel, 2006), pp. 263–294. MR 2240065 (2007c:47018)
43. A.T.A. Wood, Rosenthal's inequality for point process martingales. Stoch. Process. Appl. **81**(2), 231–246 (1999). MR 1694561 (2000f:60073)
44. A.T.A. Wood, Acknowledgement of priority. Comment on: Rosenthal's inequality for point process martingales. Stoch. Process. Appl. **81**(2), 231–246 (1999). MR1694561 (2000f:60073); Stoch. Process. Appl. **93**(2), 349 (2001). MR 1828780

Admissible Trading Strategies Under Transaction Costs

Walter Schachermayer

Abstract A well known result in stochastic analysis reads as follows: for an $\mathbb{R}$-valued super-martingale $X = (X_t)_{0 \leq t \leq T}$ such that the terminal value X_T is non-negative, we have that the entire process X is non-negative. An analogous result holds true in the no arbitrage theory of mathematical finance: under the assumption of no arbitrage, an admissible portfolio process $x + (H \cdot S)$ verifying $x + (H \cdot S)_T \geq 0$ also satisfies $x + (H \cdot S)_t \geq 0$, for all $0 \leq t \leq T$.

In the present paper we derive an analogous result in the presence of transaction costs. In fact, we give two versions: one with a numéraire-based, and one with a numéraire-free notion of admissibility. It turns out that this distinction on the primal side perfectly corresponds to the difference between local martingales and true martingales on the dual side.

A counter-example reveals that the consideration of transaction costs makes things more delicate than in the frictionless setting.

1 A Theorem on Admissibility

We consider a stock price process $S = (S_t)_{0 \leq t \leq T}$ in continuous time with a fixed horizon T. This stochastic process is assumed to be based on a filtered probability space $(\Omega, \mathscr{F}, (\mathscr{F}_t)_{0 \leq t \leq T}, \mathbb{P})$, satisfying the usual conditions of completeness and right continuity. We assume that S is adapted and has càdlàg (right continuous, left limits), and strictly positive trajectories, i.e. the function $t \to S_t(\omega)$ is càdlàg and strictly positive, for almost each $\omega \in \Omega$.

In mathematical finance a key assumption is that the process S is *free of arbitrage*. The Fundamental Theorem of Asset Pricing states that this property is

Partially supported by the Austrian Science Fund (FWF) under grant P25815, the European Research Council (ERC) under grant FA506041 and by the Vienna Science and Technology Fund (WWTF) under grant MA09-003.

W. Schachermayer (✉)
Fakultät für Mathematik, Universität Wien, Nordbergstrasse 15, 1090 Wien, Austria
e-mail: walter.schachermayer@univie.ac.at

© Springer International Publishing Switzerland 2014
C. Donati-Martin et al. (eds.), *Séminaire de Probabilités XLVI*, Lecture Notes in Mathematics 2123, DOI 10.1007/978-3-319-11970-0_11

essentially equivalent to the property that S admits an equivalent local martingale measure (see, [4, 9], or the books [5, 13]).

Definition 1 The process S admits an equivalent local martingale measure, if there is a probability measure $Q \sim \mathbb{P}$ such that S is a local martingale under Q.

Fix a process S satisfying the above assumption and note that Definition 1 implies in particular that S is a semi-martingale as this property is invariant under equivalent changes of measure. Turning to the theme of the paper, we now consider *trading strategies*, i.e. S-integrable predictable processes $H = (H_t)_{0 \le t \le T}$. We call H *admissible* if there is $M > 0$ such that

$$(H \cdot S)_t \ge -M, \qquad \mathbb{P} - a.s. \quad \text{for} \quad 0 \le t \le T. \tag{1}$$

The stochastic integral

$$(H \cdot S)_t = \int_0^t H_u dS_u, \qquad 0 \le t \le T, \tag{2}$$

then is a local Q-martingale by a result of Ansel-Stricker under each equivalent local martingale measure Q (see [1] and [16]). Assumption (1) also implies that the local martingale $H \cdot S$ is a *super-martingale* (see [5], Prop. 7.2.7) under each equivalent local martingale measure Q. We thus infer from the easy result mentioned in the first line of the abstract that $(H \cdot S)_T \ge -x$ almost surely implies that $(H \cdot S)_t \ge -x$ almost surely under Q (and therefore also under $\mathbb{P}$), for all $0 \le t \le T$. In fact, we may replace the deterministic time t by a $[0, T]$-valued stopping time τ.

We resume our findings in the subsequent well-known Proposition (compare [14, Prop. 4.1]).

Proposition 1 *Let the process S admit an equivalent local martingale measure, let H be admissible, and suppose that there is $x \in \mathbb{R}_+$ such that*

$$x + (H \cdot S)_T \ge 0, \qquad \mathbb{P} - a.s. \tag{3}$$

Then

$$x + (H \cdot S)_\tau \ge 0, \qquad \mathbb{P} - a.s. \tag{4}$$

for every $[0, T]$-valued stopping time τ.

We now introduce transaction costs: fix $0 \le \lambda < 1$. We define the *bid-ask spread* as the interval $[(1 - \lambda)S, S]$. The interpretation is that an agent can buy the stock at price S, but sell it only at price $(1 - \lambda)S$. Of course, the case $\lambda = 0$ corresponds to the usual frictionless theory.

In the setting of transaction costs the notion of *consistent price systems*, which goes back to [3, 10], plays a role analogous to the notion of equivalent martingale measures in the frictionless theory (Definition 1).

Definition 2 Fix $1 > \lambda \geq 0$. A process $S = (S_t)_{0 \leq t \leq T}$ satisfies the condition (CPS^λ) of having a consistent price system under transaction costs λ if there is a process $\tilde{S} = (\tilde{S}_t)_{0 \leq t \leq T}$, such that

$$(1 - \lambda)S_t \leq \tilde{S}_t \leq S_t, \qquad 0 \leq t \leq T, \qquad (5)$$

as well as a probability measure Q on $\mathscr{F}$, equivalent to $\mathbb{P}$, such that $(\tilde{S}_t)_{0 \leq t \leq T}$ is a local martingale under Q.

We say that S admits consistent price systems for arbitrarily small transaction costs if (CPS^λ) is satisfied, for all $1 > \lambda > 0$.

For continuous process S, in [8] the condition of *admitting consistent price systems* for arbitrarily small transaction costs has been related to the condition of *no arbitrage* under arbitrarily small transaction costs, thus proving a version of the Fundamental Theorem of Asset Pricing under small transaction costs (compare [11] for a large amount of related material).

It is important to note that we *do not assume* that S is a semi-martingale as one is forced to do in the frictionless theory [4, Theorem 7.2]. Only the process $\tilde{S}$ appearing in Definition 2 has to be a semi-martingale, as it becomes a local martingale after passing to an equivalent measure Q.

To formulate a result analogous to Proposition 1 in the setting of transaction costs we have to define the notion of $\mathbb{R}^2$-valued *self-financing trading strategies*.

Definition 3 Fix a strictly positive stock price process $S = (S_t)_{0 \leq t \leq T}$ with càdlàg paths, as well as transaction costs $1 > \lambda > 0$.

A self-financing trading strategy starting with zero endowment is a pair of predictable, finite variation processes $(\varphi_t^0, \varphi_t^1)_{0 \leq t \leq T}$ such that

(*i*) $\varphi_0^0 = \varphi_0^1 = 0$,

(*ii*) denoting by $\varphi_t^0 = \varphi_t^{0,\uparrow} - \varphi_t^{0,\downarrow}$ and $\varphi_t^1 = \varphi_t^{1,\uparrow} - \varphi_t^{1,\downarrow}$, the canonical decompositions of φ^0 and φ^1 into the difference of two increasing processes, starting at $\varphi_0^{0,\uparrow} = \varphi_0^{0,\downarrow} = \varphi_0^{1,\uparrow} = \varphi_0^{1,\downarrow} = 0$, these processes satisfy

$$d\varphi_t^{0,\uparrow} \leq (1 - \lambda)S_t d\varphi_t^{1,\downarrow}, \quad d\varphi_t^{0,\downarrow} \geq S_t d\varphi_t^{1,\uparrow}, \quad 0 \leq t \leq T. \qquad (6)$$

The trading strategy $\varphi = (\varphi^0, \varphi^1)$ is called admissible if there is $M > 0$ such that the liquidation value V_t^{liq} satisfies

$$V_\tau^{liq}(\varphi^0, \varphi^1) := \varphi_\tau^0 + (\varphi_\tau^1)^+(1 - \lambda)S_\tau - (\varphi_\tau^1)^- S_\tau \geq -M, \qquad (7)$$

a.s., for all $[0, T]$-valued stopping times τ.

The processes φ_t^0 and φ_t^1 model the holdings at time t in units of bond and stock respectively. We normalize the bond price by $B_t \equiv 1$. The differential notation in (6) needs some explanation. If φ is continuous, then (6) has to be understood as the integral requirement.

$$\int_\sigma^\tau ((1-\lambda)S_t \, d\varphi_t^{1,\downarrow} - d\varphi_t^{0,\uparrow}) \geq 0, \qquad a.s. \qquad (8)$$

for all stopping times $0 \leq \sigma \leq \tau \leq T$, and analogously for the second differential inequality in (6). The above integral makes pathwise sense as Riemann-Stieltjes intregral, as φ is continuous and of finite variation and S is càdlàg. Things become more delicate when we also consider jumps of φ: note that, for every stopping time τ the left and right limits $\varphi_{\tau-}$ and $\varphi_{\tau+}$ exist as φ is of bounded variation. But the three values $\varphi_{\tau-}, \varphi_\tau$ and $\varphi_{\tau+}$ may very well be different. As in [2] we denote the increments by

$$\Delta\varphi_\tau = \varphi_\tau - \varphi_{\tau-}, \qquad\qquad \Delta_+\varphi_\tau = \varphi_{\tau+} - \varphi_\tau. \qquad (9)$$

For totally inaccessible stopping times τ, the predictability of φ implies that $\Delta\varphi_\tau = 0$ almost surely, while for accessible stopping times τ it may happen that $\Delta\varphi_\tau \neq 0$ as well as $\Delta_+\varphi_\tau \neq 0$.

To the assumption that (8) has to hold true for the continuous part of φ the following requirements therefore have to be added to take care of the jumps of φ.

$$\Delta\varphi_\tau^{0,\uparrow} \leq (1-\lambda)S_{\tau-}\Delta\varphi_\tau^{1,\downarrow}, \qquad\qquad \Delta\varphi_\tau^{0,\downarrow} \geq S_{\tau-}\Delta\varphi_\tau^{1,\uparrow} \qquad (10)$$

and in the case of right jumps

$$\Delta_+\varphi_\tau^{0,\uparrow} \leq (1-\lambda)S_\tau\Delta_+\varphi_\tau^{1,\downarrow}, \qquad\qquad \Delta_+\varphi_\tau^{0,\downarrow} \geq S_\tau\Delta_+\varphi_\tau^{1,\uparrow}, \qquad (11)$$

holding true a.s. for all $[0,T]$-valued stopping times τ. Let us give an economic interpretation of the significance of (10) and (11). For simplicity we let $\lambda = 0$. Think of a predictable time τ, say the time τ of a speech of the chairman of the Fed. The speech does not come as a surprise. It was announced some time before which—mathematically speaking—corresponds to the predictability of τ. It is to be expected that this speech will have a sudden effect on the price of a stock S, say a possible jump from $S_{\tau-}(\omega) = 100$ to $S_\tau(\omega) = 110$ (recall that S is assumed to be càdlàg). A trader may want to follow the following strategy: she holds a position of $\varphi_{\tau-}^1(\omega)$ stocks until "immediately before the speech". Then, one second before the speech starts, she changes the position from $\varphi_{\tau-}^1(\omega)$ to $\varphi_\tau^1(\omega)$ causing an increment of $\Delta\varphi_\tau^1(\omega)$. Of course, the price $S_{\tau-}(\omega)$ still applies, corresponding to (10). Subsequently, the speech starts and the jump $\Delta S_\tau(\omega) = S_\tau(\omega) - S_{\tau-}(\omega)$ is revealed. The agent may now decide "immediately after learning the size of $\Delta S_\tau(\omega)$" to change her position from $\varphi_\tau^1(\omega)$ to $\varphi_{\tau+}^1(\omega)$ on the base of the price $S_\tau(\omega)$ which corresponds to (11).

We have chosen to define the trading strategy φ by explicitly specifying both accounts, the holdings in bond φ^0 as well as the holdings in stock φ^1. It would be sufficient to only specify φ^1 similarly as in the frictionless theory where we usually only specify the process H in (1) which corresponds to φ^1 in the present notation. Given a predictable finite variation process $\varphi^1 = (\varphi^1_t)_{0 \le t \le T}$ starting at $\varphi^1_0 = 0$, which we canonically decompose into the difference $\varphi^1 = \varphi^{1,\uparrow} - \varphi^{1,\downarrow}$, we may define the process φ^0 by

$$d\varphi^0_t = (1 - \lambda) S_t d\varphi^{1,\downarrow}_t - S_t d\varphi^{1,\uparrow}_t.$$

The resulting pair (φ^0, φ^1) obviously satisfies (6) with equality holding true rather than inequality. Not withstanding, it is convenient in (6) to consider trading strategies (φ^0, φ^1) which allow for an inequality in (6), i.e. for "throwing away money". But it is clear from the preceding argument that we may always pass to a dominating pair (φ^0, φ^1) where equality holds true in (6).

In the theory of financial markets under transaction costs the super-martingale property of the value process is formulated in Proposition 2 below. First we have to recall a definition from [6] which extends the notion of a super-martingale beyond the framework of càdlàg processes.

Definition 4 An optional process $X = (X_t)_{0 \le t \le T}$ is called an optional strong super-martingale if, for all stopping times $0 \le \sigma \le \tau \le T$ we have

$$\mathbb{E}[X_\tau \mid \mathscr{F}_\sigma] \le X_\sigma, \tag{12}$$

where we impose that X_τ is integrable.

An optional strong super-martingale can be decomposed in the style of Doob-Meyer which is known under the name of Mertens decomposition (see [6]). X is an optional strong super-martingale if and only if it can be decomposed into

$$X = M - A, \tag{13}$$

where M is a local martingale (and therefore càdlàg) as well as a super-martingale, and A an increasing predictable process (which is làdlàg but has no reason to be càglàd or càdlàg). This decomposition then is unique.

One may also define the notion of a local optional strong supermartingale in an obvious way. In this case the process M in (15) only is required to be a local martingale and not necessarily a super-martingale, while the requirements on A remain unchanged.

Proposition 2 *Fix S, transaction costs $1 > \lambda > 0$, and an admissible self-financing trading strategy $\varphi = (\varphi^0, \varphi^1)$ as above. Suppose that $(\tilde{S}, Q)$ is a consistent price system under transaction costs λ. Then the process*

$$\tilde{V}_t := \varphi^0_t + \varphi^1_t \tilde{S}_t, \qquad\qquad 0 \le t \le T,$$

satisfies $\tilde{V} \geq V^{liq}$ *almost surely and is an optional strong super-martingale under* Q.

Proof The assertion $\tilde{V} \geq V^{liq}$ is an obvious consequence of $\tilde{S} \in [(1 - \lambda)S, S]$.

We have to show that $\tilde{V}$ decomposes as in (13). Arguing formally, we may apply the product rule to obtain

$$d\tilde{V}_t = (d\varphi_t^0 + \tilde{S}_t d\varphi_t^1) + \varphi_t^1 d\tilde{S}_t \qquad (14)$$

so that

$$\tilde{V}_t = \int_0^t (d\varphi_u^0 + \tilde{S}_u d\varphi_u^1) + \int_0^t \varphi_u^1 d\tilde{S}_u. \qquad (15)$$

The first term in (15) is decreasing by (6) and the fact that $\tilde{S} \in [(1 - \lambda)S, S]$. The second term defines, at least formally speaking, a local Q-martingale as $\tilde{S}$ is so. Hence the sum of the two integrals should be an (optional strong) super-martingale.

The justification of the above formal reasoning deserves some care (compare the proof of Lemma 8, in [2]). Suppose first that φ is continuous. In this case φ is a semi-martingale so that we are allowed to apply Itô calculus to $\tilde{V}$. Formula (15) therefore makes perfect sense as an Itô integral, bearing in mind that φ has finite variation, which coincides with the pointwise interpretation of the integral via partial integration. The first integral in (15) is a well-defined decreasing predictable process. As regards the second integral, note that by the admissibility of φ it is uniformly bounded from below. Hence by a result of Ansel-Stricker ([1], see also [16]) it is a local Q-martingale as well as a super-martingale. Hence $\tilde{V}$ is indeed a super-martingale under Q (in the classical càdlàg sense).

Passing to the case when φ is allowed to have jumps, the process $\tilde{V}$ need not be càdlàg anymore. It still is an optional process and we have to verify that it decomposes as in (13). Assume first that φ is of the form

$$\varphi_t = (f^0, f^1)\mathbb{1}_{\rrbracket\tau, T\rrbracket}(t), \qquad (16)$$

where $(f^0, f^1) = \Delta_+(\varphi_\tau^0, \varphi_\tau^1)$ are $\mathscr{F}_\tau$-measurable bounded random variables verifying (11) and τ is a $[0, T]$-stopping time. We obtain

$$\tilde{V}_t = [\Delta_+\varphi_\tau^0 + (\Delta_+\varphi_\tau^1)\tilde{S}_t]\mathbb{1}_{\rrbracket\tau, T\rrbracket}(t)$$
$$= [\Delta_+\varphi_\tau^0 + (\Delta_+\varphi_\tau^1)\tilde{S}_\tau]\mathbb{1}_{\rrbracket\tau, T\rrbracket}(t) + (\Delta_+\varphi_\tau^1)(\tilde{S}_t - \tilde{S}_\tau)\mathbb{1}_{\rrbracket\tau, T\rrbracket}(t). \qquad (17)$$

Again, the first term is a decreasing predictable process and the second term is a local martingale under Q.

Next assume that φ is of the form

$$\varphi_t = (f^0, f^1)\mathbb{1}_{\llbracket\tau, T\rrbracket}(t), \qquad (18)$$

where τ is a predictable stopping time, and $(f^0, f^1) = \Delta(\varphi_\tau^0, \varphi_\tau^1)$ are bounded $\mathscr{F}_{\tau-}$-measurable random variables verifying (10). Similarly as in (17) we obtain

$$\tilde{V}_t = [\Delta\varphi_\tau^0 + (\Delta\varphi_\tau^1)\tilde{S}_t]\mathbb{1}_{[\![\tau,T]\!]}(t)$$
$$= [\Delta\varphi_\tau^0 + (\Delta\varphi_\tau^1)\tilde{S}_{\tau-}]\mathbb{1}_{[\![\tau,T]\!]}(t) + (\Delta\varphi_\tau^1)(\tilde{S}_t - \tilde{S}_{\tau-})\mathbb{1}_{[\![\tau,T]\!]}(t). \qquad (19)$$

Once more, the first term is a decreasing predictable process (this time it is even càdlàg) and the second term is a local martingale under Q.

Finally we have to deal with a general admissible self-financing trading strategy φ. To show that $\tilde{V}$ is of the form (13) we first assume that the total variation of φ is uniformly bounded. We decompose φ into its continuous and purely discontinuous part $\varphi = \varphi^c + \varphi^{pd}$. We also may find a sequence $(\tau_n)_{n=1}^\infty$ of $[0, T] \cup \{\infty\}$-valued stopping times such that the supports $([\![\tau_n]\!])_{n=1}^\infty$ are mutually disjoint and $\bigcup_{n=1}^\infty [\![\tau_n]\!]$ exhausts the right jumps of φ. Similarly, we may find a sequence $(\tau_n^p)_{n=1}^\infty$ of predictable stopping times such that their supports $([\![\tau_n^p]\!])_{n=1}^\infty$ are mutually disjoint and $\bigcup_{n=1}^\infty [\![\tau_n^p]\!]$ exhausts the left jumps of φ. We apply the above argument to φ^c, and to each $(\tau_n, \Delta_+\varphi_{\tau_n})$ and $(\tau_n^p, \Delta\varphi_{\tau_n^p})$, and sum up the corresponding terms in (15), (17) and (19). This sum converges to $\tilde{V} = M - A$, where M is a local Q-martingale and A an increasing process, as we have assumed that the total variation of φ is bounded (compare [12] and the proof of Lemma 8 in [2]). By the boundedness from below we conclude that M is also a super-martingale.

Passing to the case where φ has only finite instead of uniformly bounded variation, we use the predictability of φ to find a localizing sequence $(\sigma_k)_{k=1}^\infty$ such that each stopped process φ^{σ_k} has uniformly bounded variation. Apply the above argument to each φ^{σ_k} to obtain the same conclusion for φ.

Summing up, we have shown that $\tilde{V}$ admits a Mertens decomposition (13) and therefore is an optional strong super-martingale.

We can now state the analogous result to Proposition 1 in the presence of transaction costs.

Theorem 1 *Fix the càdlàg, adapted process S and $1 > \lambda > 0$ as above, and suppose that S satisfies $(CPS^{\lambda'})$, for each $1 > \lambda' > 0$.*

Let $\varphi = (\varphi_t^0, \varphi_t^1)_{0 \le t \le T}$ be an admissible, self-financing trading strategy under transaction costs λ, starting with zero endowment, and suppose that there is $x > 0$ s.t. for the terminal liquidation value V_T^{liq} we have a.s.

$$V_T^{liq}(\varphi^0, \varphi^1) = \varphi_T^0 + (\varphi_T^1)^+(1 - \lambda)S_T - (\varphi_T^1)^- S_T \ge -x. \qquad (20)$$

We then also have that

$$V_\tau^{liq}(\varphi^0, \varphi^1) = \varphi_\tau^0 + (\varphi_\tau^1)^+(1 - \lambda)S_\tau - (\varphi_\tau^1)^- S_\tau \ge -x, \qquad (21)$$

a.s., for every stopping time $0 \leq \tau \leq T$.

Proof Supposing that (21) fails, we may find $\frac{\lambda}{2} > \alpha > 0$, and a stopping time $0 \leq \tau \leq T$, such that either $A = A_+$ or $A = A_-$ satisfies $\mathbb{P}[A] > 0$, where

$$A_+ = \{\varphi_\tau^1 \geq 0, \varphi_\tau^0 + \varphi_\tau^1 \frac{1-\lambda}{1-\alpha} S_\tau < -x\}, \tag{22}$$

$$A_- = \{\varphi_\tau^1 \leq 0, \varphi_\tau^0 + \varphi_\tau^1 (1-\alpha)^2 S_\tau < -x\}. \tag{23}$$

Indeed, focusing on (22) and denoting by $A_+(\alpha)$ the set in (22) we have $\cup_{\alpha>0} A_+(\alpha) = \{\varphi_\tau^1 \geq 0, \varphi_\tau^0 + \varphi_\tau^1 (1-\lambda) S_\tau < -x\}$, showing that the failure of (21) implies the existence of $\alpha > 0$ such that $\mathbb{P}[A] > 0$.

Choose $0 < \lambda' < \alpha$ and a λ'-consistent price system $(\tilde{S}, Q)$. As $\tilde{S}$ takes values in $[(1-\lambda')S, S]$, we have that $(1-\alpha)\tilde{S}$ as well as $\frac{1-\lambda}{1-\alpha}\tilde{S}$ take values in $[(1-\lambda)S, S]$ as $(1-\lambda')(1-\lambda) > (1-\lambda)$ and $(1-\lambda')\frac{1-\lambda}{1-\alpha} > 1 - \lambda$. It follows that $((1-\alpha)\tilde{S}, Q)$ as well as $(\frac{1-\lambda}{1-\alpha}\tilde{S}, Q)$ are consistent price systems under transaction costs λ. By Proposition 2 we obtain that

$$\left(\varphi_t^0 + \varphi_t^1 (1-\alpha)\tilde{S}_t\right)_{0 \leq t \leq T} \quad \text{and} \quad \left(\varphi_t^0 + \varphi_t^1 \frac{1-\lambda}{1-\alpha}\tilde{S}_t\right)_{0 \leq t \leq T}$$

are optional strong Q-super-martingales. Arguing with the second process using $\tilde{S} \leq S$, we obtain from (22) the inequality

$$\mathbb{E}_Q[V_T^{liq} \mid A_+] \leq \mathbb{E}_Q\left[\varphi_T^0 + \varphi_T^1 \frac{1-\lambda}{1-\alpha}\tilde{S}_T \Big| A_+\right]$$

$$\leq \mathbb{E}_Q\left[\varphi_\tau^0 + \varphi_\tau^1 \frac{1-\lambda}{1-\alpha}\tilde{S}_\tau \Big| A_+\right]$$

$$\leq \mathbb{E}_Q\left[\varphi_\tau^0 + \varphi_\tau^1 \frac{1-\lambda}{1-\alpha}S_\tau \Big| A_+\right] < -x.$$

Arguing with the first process and using that $\tilde{S} \geq (1 - \lambda')S \geq (1 - \alpha)S$ (which implies that $\varphi_\tau^1(1-\alpha)\tilde{S}_\tau \leq \varphi_\tau^1(1-\alpha)^2 S_\tau$ on A_-) we obtain from (23) the inequality

$$\mathbb{E}_Q[V_T^{liq} \mid A_-] \leq \mathbb{E}_Q\left[\varphi_T^0 + \varphi_T^1(1-\alpha)\tilde{S}_T | A_-\right]$$

$$\leq \mathbb{E}_Q\left[\varphi_\tau^0 + \varphi_\tau^1(1-\alpha)\tilde{S}_\tau | A_-\right]$$

$$\leq \mathbb{E}_Q\left[\varphi_\tau^0 + \varphi_\tau^1(1-\alpha)^2 S_\tau | A_-\right] < -x.$$

Either A_+ or A_- has strictly positive probability; hence we arrive at a contradiction to $V_T^{liq} \geq -x$ almost surely.

2   The Numéraire-Free Setting

In this section we derive results analogous to Proposition 2 and Theorem 1 in a numéraire-free setting. This is inspired by the discussion of the numéraire-based versus numéraire-free setting in [8] and [15] (compare also [7, 11, 17, 18]).

   We complement the above notions of admissibility and consistent price systems by the following numéraire-free variants.

Definition 5 In the setting of Definition 3 we call a self-financing strategy φ *admissible in a numéraire-free sense* if there is $M > 0$ such that

$$V_\tau^{liq}(\varphi^0, \varphi^1) := \varphi_\tau^0 + (\varphi_\tau^1)^+ (1 - \lambda) S_\tau - (\varphi_\tau^1)^- S_\tau \geq -M(1 + S_\tau), \qquad \text{a.s.,} \qquad (24)$$

for each $[0, T]$-valued stopping time τ.

   While the control of the portfolio process φ in (7) is in terms of M units of bond (which is considered as numéraire), the present condition (24) stipulates that the risk involved by the trading strategy φ can be super-hedged by holding M units of bond plus $\frac{M}{1-\lambda}$ units of stock.

Definition 6 Fix $1 > \lambda \geq 0$. In the setting of Definition 2 we call a pair $(\tilde{S}, Q) = ((\tilde{S}_t)_{0 \leq t \leq T, Q})$ satisfying (5) a consistent price process *in the non-local sense* if $\tilde{S}$ is a true martingale under Q, *not only a local martingale*.

   The passage from the numéraire-based to numéraire-free admissibility for the primal objects, i.e. the trading strategies φ, perfectly corresponds to the passage from local martingales to martingales in Definition 6 for the dual objects, i.e. the consistent price systems. This is the message of the two subsequent results (compare also [15]).

Proposition 3 *In the setting of Proposition 2 fix a self-financing trading strategy* $\varphi = (\varphi^0, \varphi^1)$ *which we now assume to be admissible in the numéraire-free sense. Also fix $(\tilde{S}, Q)$ which we now assume to be a λ-consistent price system in the non-local sense, i.e. $\tilde{S}$ is a true Q-martingale. We again may conclude that the process*

$$\tilde{V}_t := \varphi_t^0 + \varphi_t^1 \tilde{S}_t, \qquad\qquad 0 \leq t \leq T,$$

satisfies $\tilde{V} \geq V^{liq}$ almost surely and is an optional strong super-martingale under Q.

Proof We closely follow the proof of Proposition 2 which carries over verbatim, also under the present weaker assumption of numéraire-free admissibility. Again, we conclude that the second integral in (15) is a local Q-martingale from the fact that $\tilde{S}$ is a local Q-martingale and φ^1 is predictable and of finite variation. The only subtlety is the following: contrary to the setting of Proposition 2 we now may only deduce the obvious implication that $\tilde{V} = (\tilde{V}_t)_{0 \leq t \leq T}$ is a *local* optional strong super-martingale under Q.

What needs extra work is an additional argument which finally shows that the word *local* may be dropped, i.e. that $\tilde{V}$ again is an optional strong super-martingale under Q.

By the numéraire-free admissibility condition we know that there is some $M > 0$ such that, for all $[0, T]$-valued stopping times τ,

$$\tilde{V}_\tau \geq V_\tau^{liq} \geq -M(1 + S_\tau), \qquad \text{a.s.} \tag{25}$$

We also know that $\tilde{S}$ is a uniformly integrable martingale under Q. Hence the family of random variables $\tilde{S}_\tau$ as well as that of S_τ (note that $S_\tau \leq \frac{\tilde{S}_\tau}{1-\lambda}$), where τ ranges through the $[0, T]$-valued stopping times, is uniformly integrable.

We have to show that, for all stopping times $0 \leq \rho \leq \sigma \leq T$ we have

$$\mathbb{E}_Q[\tilde{V}_\sigma | \mathscr{F}_\rho] \leq \tilde{V}_\rho. \tag{26}$$

We know that $\tilde{V}$ is a local optional strong super-martingale under Q, so that there is a localizing sequence $(\tau_n)_{n=1}^\infty$ of stopping times such that

$$\mathbb{E}_Q[\tilde{V}_{\sigma \wedge \tau_n} | \mathscr{F}_{\rho \wedge \tau_n}] \leq \tilde{V}_{\rho \wedge \tau_n}, \quad n \geq 1. \tag{27}$$

Using (25) we may deduce (26) from (27) by the (conditional version of the) following well-known variant of Fatou's lemma: Let $(f_n)_{n=1}^\infty$ be a sequence of random variables on $(\Omega, \mathscr{F}, \mathbb{R})$ converging almost surely to f_0 and such that the negative parts $(f_n^-)_{n=1}^\infty$ are uniformly Q-integrable. Then

$$\mathbb{E}_Q[f_0] \leq \liminf_{n \to \infty} \mathbb{E}_Q[f_n].$$

Remark 1 We have assumed in Proposition 2 as well as in the above Proposition 3 that Q is equivalent to $\mathbb{P}$. In fact, we may also assume that Z_T^0 vanishes on a non-trivial set so that Q is only absolutely continuous w.r. to $\mathbb{P}$. The assertions of the two propositions still remain valid for $\mathbb{P}$-absolutely continuous Q, provided that we replace the requirements *almost surely* by *Q-almost surely*.

We now state and prove the numéraire-free version of Theorem 1.

Theorem 2 *In the setting of Theorem 1 suppose now that S satisfies $(CPS^{\lambda'})$ in the non-local sense, for each $1 > \lambda' > 0$. As in Theorem 1, let φ be admissible, but now in the numéraire-free sense, and let $x > 0$ such that*

$$V_T^{liq}(\varphi^0, \varphi^1) \geq -x. \tag{28}$$

We then also have

$$V_\tau^{liq}(\varphi^0, \varphi^1) \geq -x, \tag{29}$$

a.s., for every stopping time $0 \leq \tau \leq T$.

Proof The proof of Theorem 1 carries over verbatim to the present setting, replacing the application of Proposition 2 by an application of its numéraire-free version Proposition 3.

3 A Counter-Example

The assumption $(CPS^{\lambda'})$, for *each* $\lambda' > 0$, cannot be dropped in Proposition 1 as shown by the example presented in the next lemma.

Lemma 1 *Fix* $1 > \lambda \geq \lambda' > 0$ *and* $C > 1$. *There is a continuous process* $S = (S_t)_{0 \leq t \leq 1}$ *satisfying* $(CPS^{\lambda'})$, *and a* λ-*self-financing, admissible trading strategy* $(\varphi^0, \varphi^1) = (\varphi^0_t, \varphi^1_t)_{0 \leq t \leq 1}$ *such that*

$$V^{liq}_1(\varphi^0, \varphi^1) \geq -1, \qquad\qquad a.s. \qquad (30)$$

while

$$\mathbb{P}\left[V^{liq}_{\frac{1}{2}}(\varphi^0, \varphi^1) \leq -C \right] > 0. \qquad (31)$$

Proof In order to focus on the central (and easy) idea of the construction we first show the assertion for the constant $C = 2 - \lambda$ and under the assumption $\lambda = \lambda'$. In this case we can give a deterministic example, i.e. S, φ^0 and φ^1 will not depend on the random element $\omega \in \Omega$.

Define $S_0 = S_1 = 1$, and $S_{\frac{1}{2}} = 1 - \lambda$ where we fix $T = 1$.

To make $S = (S_t)_{0 \leq t \leq T}$ continuous, we interpolate linearly, i.e.

$$S_t = 1 - 2t\lambda, \qquad 0 \leq t \leq \tfrac{1}{2}, \qquad (32)$$

$$S_t = 1 - 2(1 - t)\lambda, \qquad \tfrac{1}{2} \leq t \leq 1. \qquad (33)$$

Note that condition (CPS^{λ}) is satisfied, as the constant process $\tilde{S}_t \equiv (1 - \lambda)$ defines a λ-consistent price system: it trivially is a martingale (under any probability measure) and takes values in $[(1 - \lambda)S, S]$.

Starting from the initial endowment $(\varphi^0_0, \varphi^1_0) = (0, 0)$, we might invest, at time $t = 0$, the maximal amount into the stock so that at time $t = 1$ condition (30) holds true. In other words, we let $\varphi^1_{0+} = -\varphi^0_{0+}$ be the biggest number such that

$$(1 - \lambda)\varphi^1_{0+} + \varphi^0_{0+} \geq -1,$$

which clearly gives $\varphi^1_{0+} = \tfrac{1}{\lambda}$. Hence $(\varphi^0_t, \varphi^1_t) = (-\tfrac{1}{\lambda}, \tfrac{1}{\lambda})$, for all $0 < t \leq T$, is a self-financing strategy, starting at $(\varphi^0_0, \varphi^1_0) = (0, 0)$ for which (30) is satisfied.

Looking at (31) we calculate

$$V_{\frac{1}{2}}(\varphi^0, \varphi^1) = (1 - \lambda) \cdot (1 - \lambda) \cdot \frac{1}{\lambda} - \frac{1}{\lambda} = -2 + \lambda.$$

In order to replace $\lambda' = \lambda$ by an arbitrarily small constant $\lambda' > 0$, and $C = 2 - \lambda$ by an arbitrarily large constant $C > 1$, we make the following observation: if the initial endowment $(\varphi_0^0, \varphi_0^1) = (0, 0)$ were replaced by $(\varphi_0^0, \varphi_0^1) = (M, 0)$, for some large M, the agent could play the above game on a larger scale: she could choose $(\varphi_t^0, \varphi_t^1) = (M - \frac{M+1}{\lambda}, \frac{M+1}{\lambda})$, for $0 < t \le 1$, to still satisfy (30):

$$V_1(\varphi^0, \varphi^1) = M - \frac{M+1}{\lambda} + (1 - \lambda)\frac{M+1}{\lambda} = -1.$$

As regards the liquidation value $V_{\frac{1}{2}}^{liq}$, we now assume $S_{\frac{1}{2}} = 1 - \lambda'$ [instead of $S_{\frac{1}{2}} = 1 - \lambda$ in (32) and (33)] to make sure that $(CPS^{\lambda'})$ holds true. The liquidation value at time $t = \frac{1}{2}$ then becomes

$$V_{\frac{1}{2}}^{liq}(\varphi^0, \varphi^1) = M - \frac{M+1}{\lambda} + (1 - \lambda)(1 - \lambda')\frac{M+1}{\lambda}$$

$$= M - (M + 1)[1 + \lambda'(\frac{1}{\lambda} - 1)]$$

which tends to $-\infty$, as $M \to \infty$ in view of $0 < \lambda' \le \lambda < 1$.

Turning back to the original endowment $(\varphi_0^0, \varphi_0^1) = (0, 0)$, the idea is that, during the time interval $[0, \frac{1}{4}]$, the price process S provides the agent with the opportunity to become rich with positive probability, i.e. $\mathbb{P}[(\varphi_{\frac{1}{4}}^0, \varphi_{\frac{1}{4}}^1) = (M, 0)] > 0$. We then play the above game, conditionally on the event $\{(\varphi_{\frac{1}{4}}^0, \varphi_{\frac{1}{4}}^1) = (M, 0)\}$ and with $[0, 1]$ replaced by $[\frac{1}{4}, 1]$.

The subsequent construction makes this idea concrete. Let $(\mathscr{F}_t)_{0 \le t \le 1}$ be generated by a Brownian motion $(W_t)_{0 \le t \le 1}$. Fix disjoint sets A_+ and A_- in $\mathscr{F}_{\frac{1}{8}}$ such that $\mathbb{P}[A_+] = \frac{1}{2\tilde{M}-1}$ and $\mathbb{P}[A_-] = 1 - \mathbb{P}[A_+]$, where $\tilde{M} > 1$ is defined by $M = -1 + \tilde{M}(1 - \lambda')$. The set A_+ is split into two sets A_{++} and A_{+-} such that A_{++} and A_{+-} are in $\mathscr{F}_{\frac{1}{4}}$ and

$$\mathbb{P}\left[A_{++} \Big| \mathscr{F}_{\frac{1}{8}}\right] = \mathbb{P}\left[A_{+-} \Big| \mathscr{F}_{\frac{1}{8}}\right] = \frac{1}{2}\mathbb{1}_{A_+}.$$

We define $S_{\frac{1}{4}}$ by

$$
S_{\frac{1}{4}} = \begin{cases} 2\tilde{M} - 1 & \text{on } A_{++} \\ 1 & \text{on } A_{+-} \\ \frac{1}{2} & \text{on } A_{-} \end{cases}
$$

and

$$
S_t = \mathbb{E}\left[S_{\frac{1}{4}} \big| \mathscr{F}_t \right], \qquad\qquad 0 \le t \le \frac{1}{4}, \tag{34}
$$

so that $(S_t)_{0 \le t \le \frac{1}{4}}$ is a continuous $\mathbb{P}$-martingale. The numbers above were designed in such a way that

$$
S_0 = 1,
$$

and

$$
S_{\frac{1}{8}} = \begin{cases} \tilde{M} & \text{on } A_{+} \\ \frac{1}{2} & \text{on } A_{-} \end{cases}
$$

To define S_t also for $\frac{1}{4} < t \le 1$ we simply let $S_t = S_{\frac{1}{4}}$ on $A_{++} \cup A_{-}$ while, conditionally on A_{+-}, we repeat the above deterministic construction on $[\frac{1}{4}, 1]$:

$$
\begin{aligned}
S_t &= 1 - 4(t - \tfrac{1}{4})\lambda', & \tfrac{1}{4} \le t \le \tfrac{1}{2}, \\
S_t &= 1 - 2(1 - t)\lambda', & \tfrac{1}{2} \le t \le 1.
\end{aligned}
$$

This defines the process S. Condition $(CPS^{\lambda'})$ is satisfied as $(\tilde{S}_t)_{0 \le t \le 1} := ((1 - \lambda')S_{t \wedge \frac{1}{4}})_{0 \le t \le 1}$ is a $\mathbb{P}$-martingale taking values in the bid-ask spread $[(1 - \lambda')S_t, S_t]_{0 \le t \le 1}$.

Let us now define the strategy (φ^0, φ^1) : starting with $(\varphi_0^0, \varphi_0^1) = (0, 0)$ we define $(\varphi_t^0, \varphi_t^1) = (-1, 1)$, for $0 < t \le \frac{1}{8}$. In prose: the agent buys one stock at time $t = 0$ and holds it until time $t = \frac{1}{8}$. At time $t = \frac{1}{8}$ she sells the stock again, so that $(\varphi_{\frac{1}{8}}^0, \varphi_{\frac{1}{8}}^1) = (-1 + \frac{(1-\lambda)}{2}, 0)$ on A_{-}, while $(\varphi_{\frac{1}{8}+}^0, \varphi_{\frac{1}{8}+}^1) = (-1 + \tilde{M}(1 - \lambda'), 0) = (M, 0)$ on A_{+}.

On A_- we simply define $(\varphi_t^0, \varphi_t^1) = (-1 + \frac{1-\lambda}{2}, 0)$, for all $\frac{1}{8} < t \le 1$ and note that (30) is satisfied on A_-.

On A_+ we define $(\varphi_t^0, \varphi_t^1) = (M, 0)$, for $\frac{1}{8} < t \le \frac{1}{4}$. In prose: during $]\frac{1}{8}, \frac{1}{4}]$ the agent does not invest into the stock and is happy about the M bonds in her portfolio. At time $t = \frac{1}{4}$ we distinguish two cases: on A_{++} we continue to define $(\varphi_t^0, \varphi_t^1) = (M, 0)$, also for $\frac{1}{4} < t \le 1$. On A_{+-} we let $(\varphi_t^0, \varphi_t^1) = (M - \frac{M+1}{\lambda}, \frac{M+1}{\lambda})$, for $\frac{1}{4} < t \le 1$. As discussed above, inequality (30) then holds true almost surely, while $V_{\frac{1}{2}}(\varphi^0, \varphi^1)$ attains the value $M - (M + 1)[1 + \lambda'(\frac{1}{\lambda} - 1)]$ which tends to $-\infty$ as M tends to ∞. This happens with positive probability $\mathbb{P}[A_{+-}] > 0$.

The construction of the example now is complete.

Acknowledgements I warmly thank Irene Klein without whose encouragement this note would not have been written and who strongly contributed to its shaping. Thanks go also to Christoph Czichowsky for his advice on some of the subtle technicalities of this note. I thank an anonymous referee for careful reading and for pointing out a number of inaccuracies.

References

1. J.P. Ansel, C. Stricker, Couverture des actifs contingents et prix maximum. Annales de l'Institut Henri Poincaré – Probabilités et Statistiques **30**, 303–315 (1994)
2. L. Campi, W. Schachermayer, A super-replication theorem in Kabanov's model of transaction costs. Finance Stoch. **10**, 579–596 (2006)
3. J. Cvitanić, I. Karatzas, Hedging and portfolio optimization under transaction costs: a martingale approach. Math. Finance **6**(2), 133–165 (1996)
4. F. Delbaen, W. Schachermayer, A general version of the fundamental theorem of asset pricing. Mathematische Annalen **300**(1), 463–520 (1994)
5. F. Delbaen, W. Schachermayer, *The Mathematics of Arbitrage* (Springer, Berlin, 2006)
6. C. Dellacherie, P.A. Meyer, *Probabilities and Potential B. Theory of Martingales* (North-Holland, Amsterdam, 1982)
7. F. Delbaen, W. Schachermayer, The No-Arbitrage Property under a change of numéraire. Stoch. Stoch. Rep. **53b**, 213–226 (1995)
8. P. Guasoni, M. Rásonyi, W. Schachermayer, The fundamental theorem of asset pricing for continuous processes under small transaction costs. Ann. Finance **6**(2), 157–191 (2008)
9. J.M. Harrison, D.M. Kreps, Martingales and arbitrage in multiperiod securities markets. J. Econ. Theory **20**, 381–408 (1979)
10. E. Jouini, H. Kallal, Martingales and arbitrage in securities markets with transaction costs. J. Econ. Theory **66**, 178–197 (1995)
11. Y.M. Kabanov, M. Safarian, *Markets with Transaction Costs: Mathematical Theory*, Springer Finance (Springer, Berlin, 2009)
12. Yu.M. Kabanov, Ch. Stricker, Hedging of contingent claims under transaction costs, in *Advances in Finance and Stochastics*, ed. by K. Sandmann, Ph. Schönbucher. Essays in Honour of Dieter Sondermann (Springer, Berlin, 2002), pp. 125–136
13. I. Karatzas, S.E. Shreve, *Methods of Mathematical Finance* (Springer, New York, 1998)
14. W. Schachermayer, Martingale Measures for discrete-time processes with infinite horizon. Math. Finance **4**(1), 25–55 (1994)
15. W. Schachermayer, *The super-replication theorem under proportional transaction costs revisited*, To appear in Mathematics and Financial Economics (2014)

16. E. Strasser, Necessary and sufficient conditions for the supermartingale property of a stochastic integral with respect to a local martingale, in *Séminaire de Probabilités XXXVII*. Springer Lecture Notes in Mathematics, vol. 1832 (Springer, Berlin, 2003), pp. 385–393
17. J.A. Yan, A new look at the fundamental theorem of asset pricing. J. Korean Math. Soc. **35**, 659–673 (1998), World Scientific Publishers
18. J.A. Yan, A Numéraire-free and original probability based framework for financial markets. In: *Proceedings of the ICM 2002*, **III** (World Scientific Publishers, Beijing, 2005), pp. 861–874

Potentials of Stable Processes

A.E. Kyprianou and A.R. Watson

Abstract For a stable process, we give an explicit formula for the potential measure of the process killed outside a bounded interval and the joint law of the overshoot, undershoot and undershoot from the maximum at exit from a bounded interval. We obtain the equivalent quantities for a stable process reflected in its infimum. The results are obtained by exploiting a simple connection with the Lamperti representation and exit problems of stable processes.

Keywords Lévy processes • Stable processes • Reflected stable processes • Hitting times • Positive self-similar Markov processes • Lamperti representation • Potential measures • Resolvent measures

AMS classification (2000): 60G52, 60G18, 60G51

1 Introduction and Results

For a Lévy process X, the measure

$$U^A(x, \mathrm{d}y) = \mathrm{E}_x \int_0^\infty \mathbb{1}\left[X_t \in \mathrm{d}y\right] \mathbb{1}\left[\forall s \le t : X_s \in A\right] \mathrm{d}t,$$

Part of this work was done while the second author was at the University of Bath, UK, and at CIMAT, Mexico.

A.E. Kyprianou
University of Bath, Bath, UK
e-mail: a.kyprianou@bath.ac.uk

A.R. Watson (✉)
University of Zürich, Zürich, Switzerland
e-mail: alexander.watson@math.uzh.ch

© Springer International Publishing Switzerland 2014   333
C. Donati-Martin et al. (eds.), *Séminaire de Probabilités XLVI*, Lecture Notes
in Mathematics 2123, DOI 10.1007/978-3-319-11970-0_12

called the *potential* (or *resolvent*) *measure of X killed outside A*, is a quantity of great interest, and is related to exit problems.

The main cases where the potential measure can be computed explicitly are as follows. If X is a Lévy process with known Wiener–Hopf factors, it can be obtained when A is half-line or $\mathbb{R}$; see [2, Theorem VI.20]. When X is a totally asymmetric Lévy process with known scale functions, it can be obtained for A a bounded interval, a half-line or $\mathbb{R}$; see [9, Section 8.4]. Finally, [1] details a technique to obtain a potential measure for a reflected Lévy process killed outside a bounded interval from the same quantity for the unreflected process.

In this note, we consider the case where X is a stable process and A is a bounded interval. We compute the measure $U^{[0,1]}$, from which U^A may be obtained for any bounded interval A via spatial homogeneity and scaling; and from this we compute the joint law at first exit of $[0, 1]$ of the overshoot, undershoot and undershoot from the maximum. Furthermore, we give the potential measure and triple law also for the process reflected in its infimum.

The potential measure has been previously been computed when X is symmetric; see [4, Corollary 4] and references therein, as well as [1]. We extend these results to asymmetric stable processes with jumps on both sides. The essential observation is that a potential for X with killing outside a bounded interval may be converted into a potential for the *Lamperti transform of X*, say ξ, with killing outside a half-line. To compute this potential in a half-line, it is enough to know the killing rate of ξ and the solution of certain exit problems for X. The results for the reflected process are obtained via the work of [1].

We now give our results. Some facts we will rely on are summarised in Sect. 2, and proofs are given in Sect. 3.

We work with the (strictly) stable process X with scaling parameter α and positivity parameter ρ, which is defined as follows. For (α, ρ) in the set

$$\mathscr{A} = \{(\alpha, \rho) : \alpha \in (0, 1), \ \rho \in (0, 1)\} \cup \{(\alpha, \rho) = (1, 1/2)\}$$

$$\cup \{(\alpha, \rho) : \alpha \in (1, 2), \ \rho \in (1 - 1/\alpha, 1/\alpha)\},$$

let X, with probability laws $(\mathrm{P}_x)_{x \in \mathbb{R}}$, be the Lévy process with characteristic exponent

$$\Psi(\theta) = \begin{cases} c|\theta|^\alpha (1 - i\beta \tan \frac{\pi\alpha}{2} \operatorname{sgn} \theta) & \alpha \in (0, 2) \setminus \{1\}, \\ c|\theta| & \alpha = 1, \end{cases} \qquad \theta \in \mathbb{R},$$

where $c = \cos(\pi\alpha(\rho - 1/2))$ and $\beta = \tan(\pi\alpha(\rho - 1/2))/\tan(\pi\alpha/2)$. This Lévy process has absolutely continuous Lévy measure with density

$$c_+ x^{-(\alpha+1)} \mathbb{1}[x > 0] + c_- |x|^{-(\alpha+1)} \mathbb{1}[x < 0], \qquad x \in \mathbb{R},$$

where

$$c_+ = \frac{\Gamma(\alpha+1)}{\Gamma(\alpha\rho)\Gamma(1-\alpha\rho)}, \qquad c_- = \frac{\Gamma(\alpha+1)}{\Gamma(\alpha\hat{\rho})\Gamma(1-\alpha\hat{\rho})}$$

and $\hat{\rho} = 1 - \rho$.

The parameter set $\mathscr{A}$ and the characteristic exponent Ψ represent, up a multiplicative constant in Ψ, all (strictly) stable processes which jump in both directions, except for Brownian motion and the symmetric Cauchy processes with non-zero drift. The normalisation is the same as that in [8], and when X is symmetric, that is when $\rho = 1/2$, the normalisation agrees with that of [4]. We remark that the quantities we are interested in can also be derived in cases of one-sided jumps: either X is a subordinator, in which case the results are trivial, or X is a spectrally one-sided Lévy process, in which case the potentials in question may be assembled using the theory of scale functions; see [9, Theorem 8.7 and Exercise 8.2].

The choice α and ρ as parameters is explained as follows. X satisfies the α-*scaling property*, that

$$\text{under } P_x, \text{ the law of } (cX_{tc^{-\alpha}})_{t\geq 0} \text{ is } P_{cx}, \tag{1}$$

for all $x \in \mathbb{R}$, $c > 0$. The second parameter satisfies $\rho = P_0(X_t > 0)$.
Having defined the stable process, we proceed to our results. Let

$$\sigma^{[0,1]} = \inf\{t \geq 0 : X_t \notin [0,1]\},$$

and define the killed potential measure and potential density

$$U_1(x, \mathrm{d}y) := U^{[0,1]}(x, \mathrm{d}y) = E_x \int_0^{\sigma^{[0,1]}} \mathbb{1}[X_t \in \mathrm{d}y]\,\mathrm{d}t = u_1(x,y)\,\mathrm{d}y,$$

provided the density u_1 exists.

Theorem 1 *For $0 < x, y < 1$,*

$$u_1(x,y) = \begin{cases} \dfrac{1}{\Gamma(\alpha\rho)\Gamma(\alpha\hat{\rho})}(x-y)^{\alpha-1}\displaystyle\int_0^{\frac{y(1-x)}{x-y}} s^{\alpha\rho-1}(s+1)^{\alpha\hat{\rho}-1}\,\mathrm{d}s, & y < x, \\[3mm] \dfrac{1}{\Gamma(\alpha\rho)\Gamma(\alpha\hat{\rho})}(y-x)^{\alpha-1}\displaystyle\int_0^{\frac{x(1-y)}{y-x}} s^{\alpha\hat{\rho}-1}(s+1)^{\alpha\rho-1}\,\mathrm{d}s, & x < y. \end{cases}$$

Part of the claim of this theorem is that $u_1(x,y)$ exists and is finite on the domain given; this will also be the case in the coming results, and so we will not remark on it again. When X is symmetric, the theorem reduces, by spatial homogeneity and scaling of X and substituting in the integral, to [4, Corollary 4].

With very little extra work, Theorem 1 yields an apparently stronger result. Let

$$\tau_0^- = \inf\{t \geq 0 : X_t < 0\}; \qquad \overline{X}_t = \sup_{s \leq t} X_s, \qquad t \geq 0,$$

and write

$$\mathbb{E}_x \int_0^{\tau_0^-} \mathbb{1}\left[X_t \in \mathrm{d}y, \, \overline{X}_t \in \mathrm{d}z\right] \mathrm{d}t = u(x, y, z) \, \mathrm{d}y \, \mathrm{d}z,$$

if the right-hand side exists. Then we have the following.

Corollary 2 *For $x > 0$, $y \in [0, z)$, $z > x$,*

$$u(x, y, z) = \frac{1}{\Gamma(\alpha\rho)\Gamma(\alpha\hat{\rho})} x^{\alpha\hat{\rho}} y^{\alpha\rho} \frac{(z - x)^{\alpha\rho-1}(z - y)^{\alpha\hat{\rho}-1}}{z^\alpha} \, \mathrm{d}y \, \mathrm{d}z. \qquad (2)$$

Proof Rescaling, we obtain

$$\mathbb{E}_x \int_0^{\tau_0^-} \mathbb{1}\left[X_t \in \mathrm{d}y, \, \overline{X}_t \leq z\right] \mathrm{d}t = z^{\alpha-1} u_1(x/z, y/z),$$

and the density is found by differentiating the right-hand side in z.

From this density, one may recover the following hitting distribution, which originally appeared in [11, Corollary 15]. Let

$$\tau_1^+ = \inf\{t \geq 0 : X_t > 1\}.$$

Corollary 3 *For $u \in [0, 1 - x)$, $v \in (u, 1]$, $y \geq 0$,*

$$\mathbb{P}_x(1 - \overline{X}_{\tau_1^+-} \in \mathrm{d}u, \, 1 - X_{\tau_1^+-} \in \mathrm{d}v, \, X_{\tau_1^+} - 1 \in \mathrm{d}y, \, \tau_1^+ < \tau_0^-)$$

$$= \frac{\Gamma(\alpha + 1)}{\Gamma(\alpha\hat{\rho})\Gamma(1 - \alpha\rho)} \frac{x^{\alpha\hat{\rho}}(1 - v)^{\alpha\rho}(1 - u - x)^{\alpha\rho-1}(v - u)^{\alpha\hat{\rho}-1}}{(1 - u)^\alpha (v + y)^{\alpha+1}} \, \mathrm{d}u \, \mathrm{d}v \, \mathrm{d}y. \quad (3)$$

Proof Following the proof of [2, Proposition III.2], one may show that the left-hand side of (3) is equal to $u(x, 1 - v, 1 - u)\pi(v + y)$, where π is the Lévy density of X.   □

Remark 4 The proof of Corollary 3 suggests an alternative derivation of Theorem 1. Since the identity (3) is known, one may deduce $u(x, y, z)$ from it by following the proof backwards. The potential $u_1(x, y)$ without $\overline{X}$ may then be obtained via integration. However, in Sect. 3 we offer instead a self-contained proof based on well-known hitting distributions for the stable process.

Now let Y denote the stable process X reflected in its infimum, that is,

$$Y_t = X_t - \underline{X}_t, \qquad t \geq 0,$$

where $\underline{X}_t = \inf\{X_s, 0 \leq s \leq t\} \wedge 0$ for $t \geq 0$. Y is a self-similar Markov process.

Let $T_1^+ = \inf\{t > 0 : Y_t > 1\}$ denote the first passage time of Y above the level 1, and define

$$R_1(x, dy) = E_x \int_0^{T_1^+} \mathbb{1}\left[Y_t \in dy\right] dt = r_1(x, y) \, dy,$$

where the density r_1 exists by [1, Theorem 4.1]. Note that, as Y is self-similar, R_1 suffices to deduce the potential of Y killed at first passage above any level.

Theorem 5 *For $0 < y < 1$,*

$$r_1(0, y) = \frac{1}{\Gamma(\alpha)} y^{\alpha\rho - 1} (1 - y)^{\alpha\hat{\rho}}.$$

Hence, for $0 < x, \, y < 1$,

$$r_1(x, y) = \begin{cases} \frac{1}{\Gamma(\alpha\rho)\Gamma(\alpha\hat{\rho})} \left[(x - y)^{\alpha - 1} \int_0^{\frac{y(1-x)}{x-y}} s^{\alpha\rho - 1}(s + 1)^{\alpha\hat{\rho} - 1} \, ds \right. \\ \qquad\qquad \left. + y^{\alpha\rho - 1}(1 - y)^{\alpha\hat{\rho}} \int_0^{1-x} t^{\alpha\rho - 1}(1 - t)^{\alpha\hat{\rho} - 1} \, dt \right], \quad y < x, \\[2ex] \frac{1}{\Gamma(\alpha\rho)\Gamma(\alpha\hat{\rho})} \left[(y - x)^{\alpha - 1} \int_0^{\frac{x(1-y)}{y-x}} s^{\alpha\hat{\rho} - 1}(s + 1)^{\alpha\rho - 1} \, ds \right. \\ \qquad\qquad \left. + y^{\alpha\rho - 1}(1 - y)^{\alpha\hat{\rho}} \int_0^{1-x} t^{\alpha\rho - 1}(1 - t)^{\alpha\hat{\rho} - 1} \, dt \right], \quad x < y. \end{cases}$$

Writing

$$E_x \int_0^\infty \mathbb{1}\left[Y_t \in dy, \, \overline{Y}_t \in dz\right] dt = r(x, y, z) \, dy \, dz,$$

where $\overline{Y}_t$ is the supremum of Y up to time t, we obtain the following corollary, much as we had for X.

Corollary 6 *For $y \in (0, z), \, z \geq 0$,*

$$r(0, y, z) = \frac{\alpha\hat{\rho}}{\Gamma(\alpha)} y^{\alpha\rho - 1}(z - y)^{\alpha\hat{\rho} - 1},$$

and for $x > 0$, $y \in (0, z)$, $z \geq x$,

$$r(x, y, z) = \frac{1}{\Gamma(\alpha\rho)\Gamma(\alpha\hat{\rho})} y^{\alpha\rho-1}(z - y)^{\alpha\hat{\rho}-1} \left[x^{\alpha\hat{\rho}}(z - x)^{\alpha\rho-1}z^{1-\alpha} \right.$$
$$\left. + \alpha\hat{\rho} \int_0^{1-\frac{x}{z}} t^{\alpha\rho-1}(1 - t)^{\alpha\hat{\rho}-1} \, dt \right].$$

We also have the following corollary, which is the analogue of Corollary 3.

Corollary 7 *For $u \in (0, 1]$, $v \in (u, 1)$, $y \geq 0$,*

$$\mathbb{P}_0(1 - \overline{Y}_{T_1^+} \in du, \ 1 - Y_{T_1^+} \in dv, \ Y_{T_1^+} - 1 \in dy)$$
$$= \frac{\alpha \cdot \alpha\hat{\rho}}{\Gamma(\alpha\rho)\Gamma(1 - \alpha\rho)} \frac{(1 - v)^{\alpha\rho-1}(v - u)^{\alpha\hat{\rho}-1}}{(v + y)^{\alpha+1}} du \, dv \, dy,$$

and for $x \geq 0$, $u \in [0, 1 - x)$, $v \in (u, 1)$, $y \geq 0$,

$$\mathbb{P}_x(1 - \overline{Y}_{T_1^+} \in du, \ 1 - Y_{T_1^+} \in dv, \ Y_{T_1^+} - 1 \in dy)$$
$$= \frac{\Gamma(\alpha + 1)}{\Gamma(\alpha\hat{\rho})\Gamma(1 - \alpha\rho)} \frac{(1 - v)^{\alpha\rho-1}(v - u)^{\alpha\hat{\rho}-1}}{(v + y)^{\alpha+1}}$$
$$\times \left[x^{\alpha\hat{\rho}}(1 - u - x)^{\alpha\rho-1}(1 - u)^{1-\alpha} + \alpha\hat{\rho} \int_0^{1-\frac{x}{1-u}} t^{\alpha\rho-1}(1 - t)^{\alpha\hat{\rho}-1} \, dt \right] du \, dv \, dy.$$

The marginal in $dv \, dy$ appears in [1, Corollary 3.5] for the case where X is symmetric and $x = 0$. The marginal in dy is given in [10] for the process reflected in the supremum; this corresponds to swapping ρ and $\hat{\rho}$. However, unless $x = 0$, it appears to be difficult to integrate in Corollary 7 and obtain the expression found in [10].

Finally, one may integrate in Theorem 5 and obtain the expected first passage time for the reflected process.

Corollary 8 *For $x \geq 0$,*

$$\mathbb{E}_x[T_1^+] = \frac{1}{\Gamma(\alpha + 1)} \left[x^{\alpha\hat{\rho}}(1 - x)^{\alpha\rho} + \alpha\hat{\rho} \int_0^{1-x} t^{\alpha\rho-1}(1 - t)^{\alpha\hat{\rho}-1} \, dt \right].$$

In particular,

$$\mathbb{E}_0[T_1^+] = \frac{1}{\Gamma(\alpha)} \frac{\Gamma(\alpha\rho)\Gamma(\alpha\hat{\rho} + 1)}{\Gamma(\alpha + 1)}.$$

2 The Lamperti Representation

We will calculate potentials related to X by appealing to the Lamperti transform [12, 15]. Recall that a process Y with probability measures $(P_x)_{x>0}$ is a *positive self-similar Markov process (pssMp)* if it is a standard Markov process (in the sense of [3]) with state space $[0, \infty)$ which has zero as an absorbing state and satisfies the scaling property:

$$\text{under } P_x, \text{ the law of } (cY_{tc^{-\alpha}})_{t\geq0} \text{ is } P_{cx},$$

for all $x, c > 0$.

The Lamperti transform gives a correspondence between pssMps and killed Lévy processes, as follows. Let $S(t) = \int_0^t (Y_u)^{-\alpha} \, du$; this process is continuous and strictly increasing until Y reaches zero. Let T be its inverse. Then, the process

$$\xi_s = \log Y_{T(s)}, \qquad s \geq 0$$

is a Lévy process, possibly killed at an independent exponential time, and termed the *Lamperti transform* of Y. Note that $\xi_0 = \log x$ when $Y_0 = x$, and one may easily see from the definition of S that $e^{\alpha\xi_{S(t)}} \, dS(t) = dt$.

A simple example of the Lamperti transform in action is given by considering the process X. Define

$$\tau_0^- = \inf\{t \geq 0 : X_t < 0\},$$

and let

$$P_x^*(X_t \in \cdot) = P_x(X_t \in \cdot, \, t < \tau_0^-), \qquad t \geq 0, \, x > 0.$$

The process X with laws $(P_x^*)_{x>0}$ is a pssMp. Caballero and Chaumont [5] gives explicitly the generator of its Lamperti transform, whose laws we denote $(\mathbb{P}_y^*)_{y\in\mathbb{R}}$, finding in particular that it is killed at rate

$$q := c_-/\alpha = \frac{\Gamma(\alpha)}{\Gamma(\alpha\hat{\rho})\Gamma(1 - \alpha\hat{\rho})}. \tag{4}$$

3 Proofs

To avoid the proliferation of symbols, we generally distinguish processes only by the measures associated with them; the exception is that self-similar processes will be distinguished from processes obtained by Lamperti transform. Thus, the time

$$\tau_1^+ = \inf\{t \geq 0 : X_t > 0\}$$

always refers to the canonical process of the measure it appears under, and will be used for self-similar processes; and

$$S_0^+ = \inf\{s \geq 0 : \xi_s > 0\}, \quad \text{and} \quad S_0^- = \inf\{s \geq 0 : \xi_s < 0\}$$

will likewise be used for processes obtained by Lamperti transform.

Proof of Theorem 1 Our proof makes use of the pssMp (X, P^*) and its Lamperti transform $(\xi, \mathbb{P}^*)$, both defined in Sect. 2. Let $0 < x, y < 1$. Then

$$\begin{aligned}
U_1(x, \mathrm{d}y) &= \mathrm{E}_x \int_0^{\sigma^{[0,1]}} \mathbb{1}\,[X_t \in \mathrm{d}y]\,\mathrm{d}t \\
&= \mathrm{E}_x^* \int_0^{\tau_1^+} \mathbb{1}\,[X_t \in \mathrm{d}y]\,\mathrm{d}t,
\end{aligned}$$

using nothing more than the definition of (X, P^*). We now use the Lamperti representation to relate this to $(\xi, \mathbb{P}^*)$. This process is killed at the rate q given in (4), and so it may be represented as an unkilled Lévy process $(\xi, \mathbb{P})$ which is sent to some cemetery state at the independent exponential time $\mathbf{e}_q$. The time-changes in this representation are denoted S and T, and we recall that $\mathrm{d}t = e^{\alpha S(t)}\mathrm{d}S(t)$. In the following calculation, we first use the fact that τ_1^+ (the first passage of X) under P_x is equal to $T(S_0^+)$ (the time-change of the first-passage of ξ) under $\mathbb{P}_{\log x}$. The second line follows by a time substitution in the integral (see, for example, [14, Sect. A4])

$$\begin{aligned}
U_1(x, \mathrm{d}y) &= \mathrm{E}_{\log(x)}^* \int_0^{T(S_0^+)} \mathbb{1}\left[e^{\xi_{S(t)}} \in \mathrm{d}y\right] e^{\alpha \xi_{S(t)}}\,\mathrm{d}S(t) \\
&= y^\alpha \mathrm{E}_{\log(x)} \int_0^{S_0^+} \mathbb{1}\left[e^{\xi_s} \in \mathrm{d}y\right] \mathbb{1}\left[\mathbf{e}_q > s\right]\,\mathrm{d}s \\
&= y^\alpha \hat{\mathbb{E}}_{\log(1/x)} \int_0^{S_0^-} \mathbb{1}\,[\xi_s \in \log(1/\mathrm{d}y)]\,e^{-qs}\,\mathrm{d}s,
\end{aligned}$$

where $\hat{\mathbb{E}}$ refers to the dual Lévy process $-\xi$. Examining the proof of Theorem VI.20 in [2] reveals that, for any $a > 0$,

$$\begin{aligned}
&\hat{\mathbb{E}}_a \int_0^{S_0^-} \mathbb{1}\,[\xi_s \in \cdot]\,e^{-qs}\,\mathrm{d}s \\
&= \frac{1}{q} \int_{[0,\infty)} \hat{\mathbb{P}}_0\left(\overline{\xi}_{\mathbf{e}_q} \in \mathrm{d}w\right) \int_{[0,a]} \hat{\mathbb{P}}_0\left(-\underline{\xi}_{\mathbf{e}_q} \in \mathrm{d}z\right) \mathbb{1}\,[a + w - z \in \cdot],
\end{aligned}$$

where for each $t \geq 0, \overline{\xi}_t = \sup\{\xi_s : s \leq t\}$ and $\underline{\xi}_t = \inf\{\xi_s : s \leq t\}$. Then, provided that the measures $\hat{\mathbb{P}}_0(\overline{\xi}_{\mathbf{e}_q} \in \cdot)$ and $\hat{\mathbb{P}}_0(\underline{\xi}_{\mathbf{e}_q} \in \cdot)$ possess respective densities g_S and g_I (as we will shortly see they do), it follows that for $a > 0$,

$$\hat{\mathbb{E}}_a \int_0^{S_0^-} \mathbb{1}\left[\xi_s \in \mathrm{d}v\right] e^{-qs} \, \mathrm{d}s = \frac{\mathrm{d}v}{q} \int_{(a-v)\vee 0}^a \mathrm{d}z \, g_I(-z) g_S(v - a + z).$$

We may apply this result to our potential measure U_1 in order to find its density, giving

$$u_1(x, y) = \frac{1}{q} y^{\alpha-1} \int_{\frac{y}{x} \vee 1}^{\frac{1}{x}} t^{-1} g_I(\log t^{-1}) g_S(\log(tx/y)) \, \mathrm{d}t. \tag{5}$$

It remains to determine the densities g_S and g_I of the measures $\hat{\mathbb{P}}_0(\overline{\xi}_{\mathbf{e}_q} \in \cdot)$ and $\hat{\mathbb{P}}_0(\underline{\xi}_{\mathbf{e}_q} \in \cdot)$. These can be related to functionals of X by the Lamperti transform:

$$\hat{\mathbb{P}}_0(\overline{\xi}_{\mathbf{e}_q} \in \cdot) = \mathbb{P}_0(-\underline{\xi}_{\mathbf{e}_q} \in \cdot) = \mathbb{P}_1(-\log \underline{X}_{\tau_0^-} \in \cdot)$$

$$\hat{\mathbb{P}}_0(\underline{\xi}_{\mathbf{e}_q} \in \cdot) = \mathbb{P}_0(-\overline{\xi}_{\mathbf{e}_q} \in \cdot) = \mathbb{P}_1(-\log \overline{X}_{\tau_0^-} \in \cdot). \tag{6}$$

The laws of the rightmost random variables in (6) are available explicitly, as we now show. For the law of $\underline{X}_{\tau_0^-}$, we transform it into an overshoot problem and make use of Example 7 in Doney and Kyprianou [6], as follows. We omit the calculation of the integral, which uses [7, 8.380.1].

$$\mathbb{P}_1(\underline{X}_{\tau_0^-} \in \mathrm{d}y) = \hat{\mathbb{P}}_0(1 - \overline{X}_{\tau_1^+} \in \mathrm{d}y)$$

$$= K \int_y^\infty \mathrm{d}v \int_0^\infty \mathrm{d}u \, (v - y)^{\alpha\rho-1}(v + u)^{-(\alpha+1)}(1 - y)^{\alpha\hat{\rho}-1} \, \mathrm{d}y$$

$$= \frac{\sin(\pi\alpha\hat{\rho})}{\pi} y^{-\alpha\hat{\rho}}(1 - y)^{\alpha\hat{\rho}-1} \, \mathrm{d}y, \qquad y \in [0, 1].$$

$$\tag{7}$$

For the law of $\overline{X}_{\tau_0^-}$, consider the following calculation.

$$\mathbb{P}_1(\overline{X}_{\tau_0^-} \geq y) = \mathbb{P}_1(\tau_y^+ < \tau_0^-) = \mathbb{P}_{1/y}(\tau_1^+ < \tau_0^-).$$

This final quantity depends on the solution of the two-sided exit problem for the stable process; it is computed in [13], where it is denoted $f_1(1/y, \infty)$. Note that [13] contains a typographical error: in Lemma 3 of that work and the discussion after it, the roles of q (which is ρ in our notation) and $1 - q$ should be swapped. In the corrected form, we have

$$P_1(\overline{X}_{\tau_0^-} \geq y) = \frac{\Gamma(\alpha)}{\Gamma(\alpha\rho)\Gamma(\alpha\hat{\rho})} \int_0^{1/y} u^{\alpha\hat{\rho}-1}(1-u)^{\alpha\rho-1}\, du$$

$$= \frac{\Gamma(\alpha)}{\Gamma(\alpha\rho)\Gamma(\alpha\hat{\rho})} \int_y^\infty t^{-\alpha}(t-1)^{\alpha\rho-1}\, dt,$$

which gives us the density for $y \geq 1$.

Since g_S and g_I possess densities on their whole support, we may substitute (7) and (3) into (5) and obtain

$$u_1(x, y) = \frac{1}{\Gamma(\alpha\rho)\Gamma(\alpha\hat{\rho})} x^{\alpha\hat{\rho}-1} y^{\alpha\rho} \int_{\frac{y}{x}\vee 1}^{\frac{1}{x}} t^{-\alpha}(t-1)^{\alpha\rho-1}\left(t - \frac{y}{x}\right)^{\alpha\hat{\rho}-1} dt,$$

for $x, y \in (0, 1)$. The expression in the statement follows by a short manipulation of this integral.

Proof of Theorem 5 According to [1, Theorem 4.1], since X is regular upwards, we have the following formula for $r_1(0, y)$:

$$r_1(0, y) = \lim_{z \downarrow 0} \frac{u_1(z, y)}{P_z(\tau_1^+ < \tau_0^-)}.$$

We have found u_1 above, and as we already mentioned, we have from [13] that

$$P_x(\tau_1^+ < \tau_0^-) = \frac{\Gamma(\alpha)}{\Gamma(\alpha\rho)\Gamma(\alpha\hat{\rho})} \int_0^x t^{\alpha\hat{\rho}-1}(1-t)^{\alpha\rho-1}\, dt.$$

We may then make the following calculation, using l'Hôpital's rule on the second line since the integrals converge,

$$r_1(0, y) = \frac{1}{\Gamma(\alpha)} y^{\alpha-1} \lim_{z \downarrow 0} \frac{\displaystyle\int_0^{\frac{z(1-y)}{y-z}} s^{\alpha\hat{\rho}-1}(s+1)^{\alpha\rho-1}\, ds}{\displaystyle\int_0^z t^{\alpha\hat{\rho}-1}(1-t)^{\alpha\rho-1}\, dt}$$

$$= \frac{1}{\Gamma(\alpha)} y^{\alpha-1} \lim_{z \downarrow 0} \frac{z^{\alpha\hat{\rho}-1}(1-y)^{\alpha\hat{\rho}-1}(y-z)^{1-\alpha\hat{\rho}}\frac{\partial}{\partial z}\left[\frac{z(1-y)}{y-z}\right]}{z^{\alpha\hat{\rho}-1}\frac{\partial}{\partial z}[z]}$$

$$= \frac{1}{\Gamma(\alpha)} y^{\alpha\rho-1}(1-y)^{\alpha\hat{\rho}}.$$

Finally, the full potential density $r_1(x, y)$ follows simply by substituting in the following formula, from the same theorem in [1]:

$$r_1(x, y) = u_1(x, y) + P_x(\tau_0^- < \tau_1^+)r_1(0, y). \qquad \Box$$

Acknowledgements We would like to thank the referee for his careful reading of this paper.

References

1. E.J. Baurdoux, Some excursion calculations for reflected Lévy processes. ALEA Lat. Am. J. Probab. Math. Stat. **6**, 149–162 (2009)
2. J. Bertoin, *Lévy Processes*. Cambridge Tracts in Mathematics, vol. 121 (Cambridge University Press, Cambridge, 1996)
3. R.M. Blumenthal, R.K. Getoor, *Markov Processes and Potential Theory*. Pure and Applied Mathematics, vol. 29 (Academic, New York, 1968)
4. R.M. Blumenthal, R.K. Getoor, D.B. Ray, On the distribution of first hits for the symmetric stable processes. Trans. Am. Math. Soc. **99**, 540–554 (1961)
5. M.E. Caballero, L. Chaumont, Conditioned stable Lévy processes and the Lamperti representation. J. Appl. Probab. **43**(4), 967–983 (2006). doi:10.1239/jap/1165505201
6. R.A. Doney, A.E. Kyprianou, Overshoots and undershoots of Lévy processes. Ann. Appl. Probab. **16**(1), 91–106 (2006). doi:10.1214/105051605000000647
7. I.S. Gradshteyn, I.M. Ryzhik, Table of Integrals, Series, and Products, 7th edn. (Elsevier/Academic, Amsterdam, 2007). Translated from the Russian. Translation edited and with a preface by Alan Jeffrey and Daniel Zwillinger
8. A. Kuznetsov, J.C. Pardo, Fluctuations of stable processes and exponential functionals of hypergeometric Lévy processes. Acta Appl. Math. **123**, 113–139 (2013). doi:10.1007/s10440-012-9718-y
9. A.E. Kyprianou, *Introductory Lectures on Fluctuations of Lévy Processes with Applications*. Universitext (Springer, Berlin, 2006)
10. A.E. Kyprianou, First passage of reflected strictly stable processes. ALEA Lat. Am. J. Probab. Math. Stat. **2**, 119–123 (2006)
11. A.E. Kyprianou, J.C. Pardo, V. Rivero, Exact and asymptotic n-tuple laws at first and last passage. Ann. Appl. Probab. **20**(2), 522–564 (2010). doi:10.1214/09-AAP626
12. J. Lamperti, Semi-stable Markov processes. I. Z. Wahrscheinlichkeitstheorie und Verw. Gebiete **22**, 205–225 (1972)
13. B.A. Rogozin, The distribution of the first hit for stable and asymptotically stable walks on an interval. Theory Probab. Appl. **17**(2), 332–338 (1972). doi:10.1137/1117035
14. M. Sharpe, *General Theory of Markov Processes*. Pure and Applied Mathematics, vol. 133 (Academic, Boston, 1988)
15. J. Vuolle-Apiala, Itô excursion theory for self-similar Markov processes. Ann. Probab. **22**(2), 546–565 (1994)

Unimodality of Hitting Times for Stable Processes

Julien Letemplier and Thomas Simon

Abstract We show that the hitting times for points of real α-stable Lévy processes ($1 < \alpha \leq 2$) are unimodal random variables. The argument relies on strong unimodality and several recent multiplicative identities in law. In the symmetric case we use a factorization of Yano et al. (Sémin Probab XLII:187–227, 2009), whereas in the completely asymmetric case we apply an identity of the second author (Simon, Stochastics 83(2):203–214, 2011). The method extends to the general case thanks to a fractional moment evaluation due to Kuznetsov et al. (Electr. J. Probab. 19:30, 1–26, 2014), for which we also provide a short independent proof.

Keywords Hitting time • Kanter random variable • Self-decomposability • Size-bias • Stable Lévy process • Unimodality

1 Introduction and Statement of the Result

A real random variable X is said to be unimodal if there exists $a \in \mathbb{R}$ such that its distribution function $\mathbb{P}[X \leq x]$ is convex on $(-\infty, a)$ and concave on $(a, +\infty)$. When X is absolutely continuous, this means that its density is non-decreasing on $(-\infty, a]$ and non-increasing on $[a, +\infty)$. The number a is called a mode of X and might not be unique. A random variable with a single mode is called strictly unimodal. The problem of unimodality has been intensively studied for infinitely divisible random variables and we refer to Chapter 10 in [10] for details. This problem has also been settled in the framework of hitting times of processes and

J. Letemplier (✉)
Laboratoire Paul Painlevé, Université Lille 1, Cité Scientifique, F-59655 Villeneuve d'Ascq Cedex, France
e-mail: ju.letemplier@math.univ-lille1.fr

T. Simon
Laboratoire Paul Painlevé, Université Lille 1, Cité Scientifique, F-59655 Villeneuve d'Ascq Cedex, France

Laboratoire de physique théorique et modèles statistiques, Université Paris Sud, Bâtiment 100, 15 avenue Georges Clémenceau, F-91405 Orsay Cedex, France
e-mail: simon@math.univ-lille1.fr

© Springer International Publishing Switzerland 2014
C. Donati-Martin et al. (eds.), *Séminaire de Probabilités XLVI*, Lecture Notes in Mathematics 2123, DOI 10.1007/978-3-319-11970-0_13

Rösler—see Theorem 1.2 in [9]—showed that hitting times for points of real-valued diffusions are always unimodal. However, much less is known when the underlying process has jumps, for example when it is a Lévy process.

In this paper we consider a real strictly α-stable process ($1 < \alpha \leq 2$), which is a Lévy process $\{X_t, \ t \geq 0\}$ starting from zero and having characteristic exponent

$$\log[\mathbb{E}[e^{i\lambda X_1}]] \ = \ -(i\lambda)^\alpha e^{-i\pi\alpha\rho\,\mathrm{sgn}(\lambda)}, \quad \lambda \in \mathbb{R},$$

where $\rho \in [1 - 1/\alpha, 1/\alpha]$ is the positivity parameter of $\{X_t, \ t \geq 0\}$ that is $\rho = \mathbb{P}[X_1 \geq 0]$. We refer to [17] and to Chapter 3 in [10] for an account on stable laws and processes. In particular, comparing the parametrisations (B) and (C) in the introduction of [17] shows that the characteristic exponent of $\{X_t, \ t \geq 0\}$ takes the more familiar form

$$c\,|\lambda|^\alpha(1 - i\theta \tan(\frac{\pi\alpha}{2})\,\mathrm{sgn}(\lambda))$$

with $\rho = 1/2 + (1/\pi\alpha)\tan^{-1}(\theta \tan(\pi\alpha/2))$ and $c = \cos(\pi\alpha(\rho - 1/2))$. The constant c is a scaling parameter which could take any arbitrary positive value without changing our purposes below. We are interested in the hitting times for points of $\{X_t, \ t \geq 0\}$:

$$\tau_x \ = \ \inf\{t > 0, \ X_t = x\}, \quad x \in \mathbb{R}.$$

It is known [7] that τ_x is a proper random variable which is also absolutely continuous. Recall also—see e.g. Example 43.22 in [10]—that points are polar for strictly α-stable Lévy process with $\alpha \leq 1$, so that $\tau_x = +\infty$ a.s. in this situation. In the following we will focus on the random variable $\tau = \tau_1$. Again, this does not cause any loss of generality since by self-similarity one has

$$\tau_x \overset{d}{=} x^\alpha \tau_1 \qquad \text{and} \qquad \tau_{-x} \overset{d}{=} x^\alpha \tau_{-1}$$

for any $x \geq 0$, and because the law of τ_{-1} can be deduced from that of τ_1 in considering the dual process $\{-X_t, \ t \geq 0\}$. We show the

Theorem 1 *The random variable τ is unimodal.*

In the spectrally negative case $\rho = 1/\alpha$, the result is plain because τ is then a positive stable random variable of order $1/\alpha$, which is known to be unimodal—see e.g. Theorem 53.1 in [10]. We will implicitly exclude this situation in the sequel and focus on the case with positive jumps. To proceed with this non-trivial situation we use several facts from the recent literature, in order to show that τ factorizes into the product of a certain unimodal random variable and a product of powers of Gamma random variables. The crucial property that the latter product preserves unimodality by independent multiplication [2] allows to conclude. Our argument follows that of [12], where a new proof of Yamazato's theorem for the unimodality of stable densities was established, but it is more involved. In passing we obtain a self-decomposability property for the Kanter random variable, which is interesting in itself and extends the main result of [8].

For the sake of clarity we divide the proof into three parts. We first consider the symmetric case $\rho = 1/2$, appealing to a factorization of τ in terms of generalized Rayleigh and Beta random variables which was discovered in [16]. Second, we deal with the spectrally positive case $\rho = 1 - 1/\alpha$, with the help of a multiplicative identity in law for τ involving positive stable and shifted Cauchy random variables which was obtained in [11]. In the third part, we observe the remarkable fact that this latter identity extends to the general case $\rho \in (1 - 1/\alpha, 1/\alpha)$, thanks to the evaluation of the Mellin transform of τ which was performed in [6], and which can actually be obtained very easily—see Sect. 2.4. This identity allows also to show that the density of τ is real-analytic and that τ is strictly unimodal—see the final remark (a).

2 Proof of the Theorem

2.1 The Symmetric Case

Formula (5.12) and Lemma 2.17 in [16] yield together with the normalization (4.1) therein, which is the same as ours, the following independent factorization

$$\tau \overset{d}{=} 2^{-\alpha} \mathbf{L}^{-\frac{\alpha}{2}} \times \left(\mathbf{Z}_{\frac{\alpha}{2}}^{(-\frac{1}{2})} \right)^{-\frac{\alpha}{2}} \times \mathbf{B}_{1-\frac{1}{\alpha}, \frac{1}{\alpha}}^{-1} \tag{1}$$

with the following notation, which will be used throughout the text:

- $\mathbf{L}$ is the unit exponential random variable.
- $\mathbf{B}_{a,b}$ is the Beta random variable $(a, b > 0)$ with density

$$\frac{\Gamma(a+b)}{\Gamma(a)\Gamma(b)} x^{a-1}(1-x)^{b-1}\mathbf{1}_{(0,1)}(x)$$

- $\mathbf{Z}_c$ is the positive c-stable random variable $(0 < c \leq 1)$, normalized such that

$$\mathbb{E}\left[e^{-\lambda \mathbf{Z}_c}\right] = e^{-\lambda^c}, \qquad \lambda \geq 0.$$

- For every $t \in \mathbb{R}$ and every positive random variable X such that $\mathbb{E}[X^t] < +\infty$, the random variable $X^{(t)}$ is the size-biased sampling of X at order t, which is defined by

$$\mathbb{E}\left[f(X^{(t)})\right] = \frac{\mathbb{E}\left[X^t f(X)\right]}{\mathbb{E}[X^t]}$$

for every $f : \mathbb{R}^+ \to \mathbb{R}$ bounded continuous.

Notice that the above random variable $\mathbf{Z}_{\frac{\alpha}{2}}^{(-\frac{1}{2})}$ makes sense from the closed expression of the fractional moments of $\mathbf{Z}_c$:

$$\mathbb{E}[\mathbf{Z}_c^s] = \frac{\Gamma(1-s/c)}{\Gamma(1-s)}, \qquad s < c, \tag{2}$$

which ensures $\mathbb{E}[\mathbf{Z}_{\frac{\alpha}{2}}^{-\frac{1}{2}}] < +\infty$. Observe also that for every $t > 0$ one has $\mathbf{L}^{(t-1)} \overset{d}{=} \boldsymbol{\Gamma}_t$ where $\boldsymbol{\Gamma}_t$ is the Gamma random variable with density

$$\frac{x^{t-1}e^{-x}}{\Gamma(t)} \mathbf{1}_{(0,+\infty)}(x).$$

It is easy to see that

$$X^{(t)} \times Y^{(t)} \overset{d}{=} (X \times Y)^{(t)} \quad \text{and} \quad \left(X^{(t)}\right)^p \overset{d}{=} (X^p)^{(\frac{t}{p})} \tag{3}$$

for every $t, p \in \mathbb{R}$ such that the involved random variables exist, and where the products in the first identity are supposed to be independent. In particular, one has

$$(\kappa X)^{(t)} \overset{d}{=} \kappa X^{(t)}$$

for every positive constant κ. Combined with (1) the second identity in (3) entails

$$\tau \overset{d}{=} 2^{-\alpha} \mathbf{L}^{-\frac{\alpha}{2}} \times \left(\mathbf{Z}_{\frac{\alpha}{2}}^{-\frac{\alpha}{2}}\right)^{(\frac{1}{\alpha})} \times \mathbf{B}_{1-\frac{1}{\alpha},\frac{1}{\alpha}}^{-1}.$$

On the other hand, Kanter's factorization—see Corollary 4.1 in [4]—reads

$$\mathbf{Z}_{\frac{\alpha}{2}}^{-\frac{\alpha}{2}} \overset{d}{=} \mathbf{L}^{1-\frac{\alpha}{2}} \times b_{\frac{\alpha}{2}}(\mathbf{U}) \tag{4}$$

where $\mathbf{U}$ is uniform on $(0, 1)$ and

$$b_c(u) = \frac{\sin(\pi u)}{\sin^c(\pi c u) \sin^{1-c}(\pi(1-c)u)}$$

for all $u, c \in (0, 1)$. Since b_c is a decreasing function from $\kappa_c = c^{-c}(1-c)^{c-1}$ to 0—see the proof of Theorem 4.1 in [4] for this fact, let us finally notice that the support of the random variable

$$\mathbf{K}_c = \kappa_c^{-1} b_c(\mathbf{U})$$

is $[0, 1]$. Putting everything together shows that

$$\tau \overset{d}{=} 2^{-\alpha} \kappa_{\frac{\alpha}{2}} \, \mathbf{L}^{-\frac{\alpha}{2}} \times \left(\mathbf{L}^{1-\frac{\alpha}{2}} \right)^{\left(\frac{1}{\alpha} \right)} \times \mathbf{K}_{\frac{\alpha}{2}}^{\left(\frac{1}{\alpha} \right)} \times \mathbf{B}^{-1}_{1-\frac{1}{\alpha}, \frac{1}{\alpha}}.$$

By Theorem 3.7. and Corollary 3.14. in [2], the random variable

$$\mathbf{X}_\alpha = 2^{-\alpha} \kappa_{\frac{\alpha}{2}} \, \mathbf{L}^{-\frac{\alpha}{2}} \times \left(\mathbf{L}^{1-\frac{\alpha}{2}} \right)^{\left(\frac{1}{\alpha} \right)}$$

$$\overset{d}{=} 2^{-\alpha} \kappa_{\frac{\alpha}{2}} \, \mathbf{L}^{-\frac{\alpha}{2}} \times \left(\mathbf{L}^{\left(\frac{1}{\alpha} - \frac{1}{2} \right)} \right)^{1-\frac{\alpha}{2}} \overset{d}{=} 2^{-\alpha} \kappa_{\frac{\alpha}{2}} \, \mathbf{L}^{-\frac{\alpha}{2}} \times \Gamma^{1-\frac{\alpha}{2}}_{\frac{1}{\alpha} + \frac{1}{2}}$$

is multiplicatively strongly unimodal, that is its independent product with any uni-modal random variable remains unimodal. Indeed, a straightforward computation shows that the random variables $\log(\mathbf{L})$ and $\log(\Gamma_{\frac{1}{\alpha} + \frac{1}{2}})$ have a log-concave density, and the same is true for $\log(\mathbf{X}_\alpha)$ by Prékopa's theorem. All in all, we are reduced to show the

Proposition 1 *With the above notation, the random variable* $\mathbf{K}_{\frac{\alpha}{2}}^{\left(\frac{1}{\alpha} \right)} \times \mathbf{B}^{-1}_{1-\frac{1}{\alpha}, \frac{1}{\alpha}}$ *is unimodal.*

Proof A computation shows that the density of $\mathbf{B}^{-1}_{1-\frac{1}{\alpha}, \frac{1}{\alpha}}$ decreases on $(1, +\infty)$. Hence, there exists $F_\alpha : (0, 1) \mapsto (1, +\infty)$ increasing and convex such that

$$\mathbf{B}^{-1}_{1-\frac{1}{\alpha}, \frac{1}{\alpha}} \overset{d}{=} F_\alpha(\mathbf{U}).$$

On the other hand, up to normalization the density of $\mathbf{K}_{\frac{\alpha}{2}}^{\left(\frac{1}{\alpha} \right)}$ writes

$$g_\alpha(x) = x^{\frac{1}{\alpha}} f_\alpha(x)$$

on $(0, 1)$, where f_α is the density of $\mathbf{K}_{\frac{\alpha}{2}}$. It follows from Lemma 2.1 in [13] that f_α increases on $(0, 1)$, so that the density of g_α also increases on $(0, 1)$ and that there exists $G_\alpha : (0, 1) \mapsto (0, 1)$ increasing and concave such that

$$\mathbf{K}_{\frac{\alpha}{2}}^{\left(\frac{1}{\alpha} \right)} \overset{d}{=} G_\alpha(\mathbf{U}).$$

We can now conclude by the lemma in [12]. $\qquad \square$

Remark 1 The lemma in [12] shows that the mode of $\mathbf{K}_{\frac{\alpha}{2}}^{\left(\frac{1}{\alpha} \right)} \times \mathbf{B}^{-1}_{1-\frac{1}{\alpha}, \frac{1}{\alpha}}$ is actually 1. However, this does not give any information on the mode of τ.

2.2   The Spectrally Positive Case

This situation corresponds to the value $\rho = 1 - 1/\alpha$ of the positivity parameter. The characteristic exponent of $\{X_t, \ t \geq 0\}$ can be extended to the negative half-plane, taking the simple form

$$\log[\mathbb{E}[e^{-\lambda X_1}]] = -\lambda^\alpha, \qquad \lambda \geq 0.$$

With this normalization, we will use the following independent factorization which was obtained in [11]:

$$\tau \stackrel{d}{=} \mathbf{U}_\alpha \times \mathbf{Z}_{\frac{1}{\alpha}}$$

where $\mathbf{U}_\alpha$ is a random variable with density

$$f_{\mathbf{U}_\alpha}(t) = \frac{-(\sin \pi\alpha) t^{1/\alpha}}{\pi(t^2 - 2t \cos \pi\alpha + 1)}.$$

It is easy to see that $\mathbf{U}_\alpha$ is multiplicatively strongly unimodal for $\alpha \leq 3/2$ and this was used in Proposition 8 of [11] to deduce the unimodality of τ in this situation. To deal with the general case $\alpha \in (1, 2)$ we proceed via a different method. First, it is well-known and easy to see—solve e.g. Exercise 4.21 (3) in [1]—that the independent quotient

$$\left(\frac{\mathbf{Z}_{\alpha-1}}{\mathbf{Z}_{\alpha-1}} \right)^{\alpha-1}$$

has the density

$$\frac{-\sin(\pi\alpha)}{\pi(t^2 - 2t \cos \pi\alpha + 1)}$$

over $\mathbb{R}^+$, whence we deduce

$$\tau \stackrel{d}{=} \left(\frac{\mathbf{Z}_{\alpha-1}^{\alpha-1}}{\mathbf{Z}_{\alpha-1}^{\alpha-1}} \right)^{(\frac{1}{\alpha})} \times \mathbf{Z}_{\frac{1}{\alpha}}$$

$$\stackrel{d}{=} \kappa_{\frac{1}{\alpha}}^{-\alpha} \left(\frac{\mathbf{L}^{2-\alpha}}{\mathbf{L}^{2-\alpha}} \right)^{(\frac{1}{\alpha})} \times \mathbf{L}^{1-\alpha} \times \mathbf{K}_{\alpha-1}^{(\frac{1}{\alpha})} \times (\mathbf{K}_{\alpha-1}^{-1})^{(\frac{1}{\alpha})} \times \mathbf{K}_{\frac{1}{\alpha}}^{-\alpha}$$

with the above notation. Similarly as above, the first product with the three exponential random variables is multiplicatively strongly unimodal, whereas the random variable $\mathbf{K}_{\alpha-1}^{(\frac{1}{\alpha})}$ has an increasing density on $(0, 1)$. Hence, reasoning as in

Proposition 1 it is enough to show that the random variable $(\mathbf{K}_{\alpha-1}^{-1})^{(\frac{1}{\alpha})} \times \mathbf{K}_{\frac{1}{\alpha}}^{-\alpha}$ has a decreasing density on $(1, +\infty)$. We show the more general

Proposition 2 *With the above notation, the random variable*

$$(\mathbf{K}_\beta^{-r})^{(t)} \times \mathbf{K}_\gamma^{-s}$$

has a decreasing density on $(1, +\infty)$ *for every* $r, s > 0$ *and* β, γ, t *in* $(0, 1)$.

The proof of the proposition uses the notion of self-decomposability—see Chapter 3 in [10] for an account. Recall that a positive random variable X is self-decomposable if its Laplace transform reads

$$\mathbb{E}[e^{-\lambda X}] = \exp - \left[a_X \lambda + \int_0^\infty (1 - e^{-\lambda x}) \frac{\varphi_X(x)}{x} dx \right], \quad \lambda \geq 0,$$

for some $a_X \geq 0$ which is called the drift coefficient of X, and some non-increasing function $\varphi_X : (0, +\infty) \rightarrow \mathbb{R}^+$ which will be henceforth referred to as the spectral function of X. Introduce the following random variable

$$\mathbf{W}_\beta = -\log(\mathbf{K}_\beta)$$

and notice that its support is $\mathbb{R}^+$, thanks to our normalization for $\mathbf{K}_\beta$. A key-observation is the following

Lemma 1 *The random variable* $\mathbf{W}_\beta$ *is self-decomposable, without drift and with a spectral function taking the value* $1/2$ *at* $0+$.

Proof Combining (2), (4) and the classical formula for the Gamma function

$$\Gamma(1 - u) = \exp \left[\gamma u + \int_0^\infty (e^{ux} - 1 - ux) \frac{dx}{x(e^x - 1)} \right], \quad u < 1,$$

(where γ is Euler's constant) yields the following expression for the Laplace transform of $\mathbf{W}_\beta$—see (3.5) in [8]:

$$\mathbb{E}[e^{-\lambda \mathbf{W}_\beta}] = \mathbb{E}[\mathbf{K}_\beta^\lambda] = \exp - \left[\int_0^\infty (1 - e^{-\lambda x}) \frac{\varphi_\beta(x)}{x} dx \right], \quad \lambda \geq 0,$$

with

$$\varphi_\beta(x) = \frac{e^{-x}}{1 - e^{-x}} - \frac{e^{-x/\beta}}{1 - e^{-x/\beta}} - \frac{e^{-x/(1-\beta)}}{1 - e^{-x/(1-\beta)}}, \quad x > 0.$$

It was shown in Lemma 3 of [8] that the function φ_β is non-negative and an asymptotic expansion at order 2 yields $\varphi_\beta(0+) = 1/2$.

We finally show that φ_β is non-increasing on $(0, +\infty)$. Following the proof and the notation of Lemma 3 in [8] this amounts to the fact that the function $x \mapsto x\psi_\beta'(x)$ therein is non-decreasing on $(0, 1)$. Decomposing

$$\varphi_\beta(x) = \beta \frac{e^{-x}}{1 - e^{-x}} - \frac{e^{-x/\beta}}{1 - e^{-x/\beta}} + (1 - \beta) \frac{e^{-x}}{1 - e^{-x}} - \frac{e^{-x/(1-\beta)}}{1 - e^{-x/(1-\beta)}},$$

the non-decreasing property is now a clear consequence of the following claim:

$$t \mapsto \log(1 - e^t) - \log(1 - e^{rt}) \quad \text{is convex on } \mathbb{R}^- \text{ for every } r \in (0, 1). \tag{5}$$

Let us show the claim. Differentiating twice, we see that we are reduced to prove that

$$\frac{r^2 x^{r-1}(1 - x)^2}{(1 - x^r)^2} \geq 1, \qquad 0 < x, r < 1.$$

The limit of the quantity on the left-hand side is 1 when $x \to 1-$, whereas its logarithmic derivative equals

$$\frac{(r - 1)(1 - x^{r+1}) + (r + 1)(x^r - x)}{x(1 - x)(1 - x^r)}.$$

The latter fraction is negative for all $0 < x, r < 1$ because its numerator is concave as a function of $x \in (0, 1)$ which vanishes together with its derivative at $x = 1-$. This shows (5) and finishes the proof of the lemma. □

Proof of Proposition 2 Set $f_{\beta,r}$ resp. $f_{\gamma,s}$ for the density of $r\mathbf{W}_\beta$ resp. $s\mathbf{W}_\gamma$. By multiplicative convolution, the density of $(\mathbf{K}_\beta^{-r})^{(t)} \times \mathbf{K}_\gamma^{-s}$ writes

$$\int_1^x (xy^{-1})^t f_{\mathbf{K}_\beta^{-r}}(xy^{-1}) f_{\mathbf{K}_\gamma^{-s}}(y) \frac{dy}{y}$$

on $(1, +\infty)$, up to some normalization constant. This transforms into

$$x^{t-1} \int_1^x f_{\beta,r}(\log(x) - \log(y)) f_{\gamma,s}(\log(y)) \frac{dy}{y^{t+1}}$$

$$= x^{t-1} \int_0^{\log(x)} f_{\beta,r}(\log(x) - u) e^{-tu} f_{\gamma,s}(u) \, du$$

and since $t \in (0, 1)$ it is enough to prove that the function

$$v \mapsto \int_0^v f_{\beta,r}(v - u) e^{-tu} f_{\gamma,s}(u) \, du \tag{6}$$

is non-increasing on $(0, +\infty)$. Lemma 1 and a change of variable show that $f_{\beta,r}(u)$ resp. $e^{-tu} f_{\gamma,s}(u)$ is up to normalization the density of a positive self-decomposable random variable without drift and with spectral function $\varphi_\beta(xr^{-1})$ resp. $e^{-tx}\varphi_\gamma(xs^{-1})$, with the notation of the proof of Lemma 1. By additive convolution this entails that the function in (6) is the constant multiple of the density of a positive self-decomposable random variable without drift and with spectral function

$$\varphi_\beta(xr^{-1}) + e^{-tx}\varphi_\gamma(xs^{-1}).$$

By Lemma 1 this latter function takes the value 1 at $0+$, and we can conclude by Theorem 53.4 (ii) in [10]. □

Remark 2 By Theorem 4 in [3] we know that $\mathbf{W}_\beta$ has also a completely monotone density, in other words—see Theorem 51.12 in [10]—that its spectral function writes

$$\varphi_\beta(x) = x \int_0^\infty e^{-tx}\theta_\alpha(t)dt, \quad x \geq 0,$$

for some function $\theta_\alpha(t)$ valued in $[0, 1]$ and such that $t^{-1}\theta_\alpha(t)$ is integrable at $0+$. This entails that $\mathbf{K}_\beta^{-r}$ has a completely monotone density as well for every $r > 0$—see Corollary 3 in [3]. However, this latter property does not seem true in general for $(\mathbf{K}_\beta^{-r})^{(t)} \times \mathbf{K}_\gamma^{-s}$.

2.3   The General Case

We now suppose $\rho \in (1 - 1/\alpha, 1/\alpha)$, which means that our stable Lévy process has jumps of both signs. The symmetric case was dealt with previously but it can also be handled with the present argument. Theorem 3.10 in [6] computes the fractional moments of τ in closed form:

$$\mathbb{E}[\tau^s] = \frac{\sin(\frac{\pi}{\alpha})\sin(\pi\rho\alpha(s + \frac{1}{\alpha}))}{\sin(\pi\rho)\sin(\pi(s + \frac{1}{\alpha}))} \times \frac{\Gamma(1 - \alpha s)}{\Gamma(1 - s)} \tag{7}$$

for $-1 - 1/\alpha < s < 1 - 1/\alpha$ (the initial normalization of [6] is the same as ours—see the introduction therein—but beware that with their notation our τ has the law of T_0 under $\mathbf{P}_{-1}$). On the other hand, it is easy to see from (2) and the complement formula for the Gamma function that

$$\mathbb{E}\left[\left(\frac{\mathbf{Z}_{\rho\alpha}^{\rho\alpha}}{\mathbf{Z}_{\rho\alpha}^{\rho\alpha}}\right)^s\right] = \frac{\sin(\pi\rho\alpha s)}{\rho\alpha\sin(\pi s)}, \quad -1 < s < 1.$$

Hence, a fractional moment identification entails

$$\tau \stackrel{d}{=} \left(\frac{\mathbf{Z}_{\rho\alpha}^{\rho\alpha}}{\mathbf{Z}_{\rho\alpha}^{\rho\alpha}} \right)^{(\frac{1}{\alpha})} \times \mathbf{Z}_{\frac{1}{\alpha}}. \tag{8}$$

Making the same manipulations as in the spectrally positive case, we obtain

$$\tau \stackrel{d}{=} \kappa_{\frac{1}{\alpha}}^{-\alpha} \left(\frac{\mathbf{L}^{1-\rho\alpha}}{\mathbf{L}^{1-\rho\alpha}} \right)^{(\frac{1}{\alpha})} \times \mathbf{L}^{1-\alpha} \times \mathbf{K}_{\rho\alpha}^{(\frac{1}{\alpha})} \times (\mathbf{K}_{\rho\alpha}^{-1})^{(\frac{1}{\alpha})} \times \mathbf{K}_{\frac{1}{\alpha}}^{-\alpha} \tag{9}$$

and we can conclude because Proposition 2 applies here as well. $\qquad\square$

2.4   A Short Proof of (7)

In this paragraph we give an independent proof of the fractional moment evaluation (7), which is short and standard. This method relying on the following potential formula was suggested to us by L. Chaumont and P. Patie and we would like to thank them for reminding us this formula. Setting f_{X_t} for the density of X_t, by Theorem 43.3 in [10] one has

$$\mathbb{E}[e^{-q\tau}] = \frac{u^q(1)}{u^q(0)}$$

where by self-similarity

$$u^q(1) = \int_0^\infty e^{-qt} f_{X_t}(1)\, dt = \int_0^\infty e^{-qt} t^{-1/\alpha} f_{X_1}(t^{-1/\alpha})\, dt,$$

and

$$u^q(0) = \int_0^\infty e^{-qt} f_{X_t}(0)\, dt = \frac{1}{\varphi(q)}$$

with φ the Laplace exponent of the inverse local time at zero of $\{X_t,\, t \geq 0\}$. It is well-known and easy to see by self-similarity—see e.g. Theorem 2 in [14]—that

$$\varphi(q) = \kappa\, q^{\frac{\alpha-1}{\alpha}},$$

where $\kappa > 0$ is a normalizing constant to be determined later. Making a change of variable, we deduce

$$\mathbb{E}[e^{-q\tau}] = \tilde{\kappa}\, q^{\frac{\alpha-1}{\alpha}} \mathbb{E}[Z_1 e^{-qZ_1}]$$

with $\tilde{\kappa} = \alpha\rho\kappa$ and $Z_1 = (X_1|X_1 \geq 0)^{-\alpha}$. For every $s \in (0, 1)$ one has

$$\mathbb{E}[\tau^{-s}] = \frac{1}{\Gamma(s)} \int_0^\infty \mathbb{E}[e^{-q\tau}]q^{s-1}dq$$

$$= \frac{\tilde{\kappa}}{\Gamma(s)}\mathbb{E}\left[Z_1 \int_0^\infty e^{-qZ_1}q^{\frac{\alpha-1}{\alpha}+s-1}dq\right]$$

$$= \tilde{\kappa}\frac{\Gamma(1-1/\alpha+s)}{\Gamma(s)}\mathbb{E}\left[Z_1^{\frac{1}{\alpha}-s}\right]$$

$$= \tilde{\kappa}\frac{\Gamma(1-1/\alpha+s)}{\Gamma(s)}\mathbb{E}\left[(X_1|X_1 \geq 0)^{\alpha s-1}\right].$$

On the other hand, by the formula (2.6.20) in [17] one has

$$\mathbb{E}\left[(X_1|X_1 \geq 0)^{\alpha s-1}\right] = \frac{\Gamma(\alpha s)\Gamma(1+1/\alpha-s)}{\Gamma(1-\rho+\rho\alpha s)\Gamma(1+\rho-\rho\alpha s)}$$

and putting everything together entails

$$\mathbb{E}[\tau^{-s}] = \frac{\tilde{\kappa}\Gamma(\alpha s)\Gamma(1-1/\alpha+s)\Gamma(1+1/\alpha-s)}{\Gamma(s)\Gamma(1-\rho+\rho\alpha s)\Gamma(1+\rho-\rho\alpha s)}$$

$$= \frac{\sin(\frac{\pi}{\alpha})\sin(\pi\rho\alpha(-s+\frac{1}{\alpha}))\Gamma(1+\alpha s)}{\sin(\pi\rho)\sin(\pi(-s+\frac{1}{\alpha}))\Gamma(1+s)}$$

for every $s \in (0, 1)$, where we used standard properties of the Gamma function and the identification of the constant comes from $\mathbb{E}[\tau^0] = 1$. This completes the proof of (7) for $s \in (-1, 0)$, and hence for $s \in (-1 - 1/\alpha, 1 - 1/\alpha)$ by analytic continuation. $\qquad\square$

Remark 3 The above computation shows that the normalizing constant for the inverse local time at zero reads

$$\kappa = \frac{\alpha\sin(\frac{\pi}{\alpha})}{\sin(\pi\rho)}.$$

One can check that this constant is the same as the one computed by Fourier inversion in [14] p. 636.

2.5 Final Remarks

(a) The identity in law (8) shows that the density of τ is the multiplicative convolution of two densities which are real-analytic on $(0, +\infty)$. Indeed, it is well-known—see e.g. [17] Theorem 2.4.1—that the density of $\mathbf{Z}_{\frac{1}{\alpha}}$ is real-analytic on $(0, +\infty)$, whereas the density of the first factor in (8) reads

$$\frac{\sin(\pi\rho\alpha)\sin(\frac{\pi}{\alpha})t^{\frac{1}{\alpha}}}{\pi\sin(\pi\rho)(t^2 + 2t\cos(\pi\rho\alpha) + 1)}.$$

Hence, the density of τ is itself real-analytic on $(0, +\infty)$ and a combination of our main result and the principle of isolated zeroes entails that τ is strictly unimodal. Besides, its mode is positive since we know from Theorem 3.15 (iii) in [6] in the spectrally two-sided case, and from Proposition 2 in [11] in the spectrally positive case, that the density of τ always vanishes at $0+$ (with an infinite first derivative). The strict unimodality of τ can also be obtained in analyzing more sharply the factors in (9) and using Step 6 p. 212 in [2].

(b) The identity in law (8) can be extended in order to encompass the whole set of admissible parameters $\{\alpha \in (1, 2], 1 - 1/\alpha \le \rho \le 1/\alpha\}$ of strictly α-stable Lévy processes that hit points in finite time a.s. Using the Legendre-Gauss multiplication formula and a fractional moment identification, one can also deduce from (8) with $\rho = 1/2$ the formula (5.12) of [16]. It is possible to derive a factorization of τ with the same inverse Beta factor $\mathbf{B}^{-1}_{1-\frac{1}{\alpha}, \frac{1}{\alpha}}$ for $\alpha \le n/(n-1)$ and $\rho = 1/n$, but this kind of identity in law does not seem to be true in general.

(c) When $\rho\alpha \ge 1/2$, the formula (8) shows that the law of τ is closely related to that of the positive branch of a real stable random variable with scaling parameter $1/\alpha$ and positivity parameter $\rho\alpha$. Indeed, Bochner's subordination—see e.g. Chapter 6 in [10] or Section 3.2 in [17] for details—shows that the latter random variable decomposes into the independent product

$$\left(\frac{\mathbf{Z}^{\rho\alpha}_{\rho\alpha}}{\mathbf{Z}^{\rho\alpha}_{\rho\alpha}}\right) \times \mathbf{Z}_{\frac{1}{\alpha}}.$$

(d) The identity (8) is attractive in its simplicity. Compare with the distribution of first passage times of stable Lévy processes, whose fractional moments can be computed in certain situations—see Theorem 3 in [5] and the references therein for other recent results in the same vein—but with complicated formulæ apparently not leading to tractable multiplicative identities in law. In the framework of hitting times, it is natural to ask whether (8) could not help to investigate further distributional properties of τ, in the spirit of [15]. This will be the matter of further research.

Acknowledgements Ce travail a bénéficié d'une aide de l'Agence Nationale de la Recherche portant la référence ANR-09-BLAN-0084-01.

References

1. L. Chaumont, M. Yor, *Exercises in Probability* (Cambridge University Press, Cambridge, 2003)
2. I. Cuculescu, R. Theodorescu, Multiplicative strong unimodality. Aust. New Zeal. J. Stat. **40**(2), 205–214 (1998)
3. W. Jedidi, T. Simon, Further examples of GGC and HCM densities. Bernoulli **19**, 1818–1838 (2013)
4. M. Kanter, Stable densities under change of scale and total variation inequalities. Ann. Probab. **3**, 697–707 (1975)
5. A. Kuznetsov. On the density of the supremum of a stable process. Stoch. Proc. Appl. **123**(3), 986–1003 (2013)
6. A. Kuznetsov, A.E. Kyprianou, J.C. Millan, A.R. Watson, The hitting time of zero for a stable process. Electr. J. Probab. **19**, Paper 30, 1–26 (2014)
7. D. Monrad. Lévy processes: Absolute continuity of hitting times for points. Z. Wahrsch. verw. Gebiete **37**, 43–49 (1976)
8. D. Pestana, D.N. Shanbhag, M. Sreehari. Some further results in infinite divisibility. Math. Proc. Camb. Phil. Soc. **82**, 289–295 (1977)
9. U. Rösler, Unimodality of passage times for one-dimensional strong Markov processes. Ann. Probab. **8**(4), 853–859 (1980)
10. K. Sato, *Lévy Processes and Infinitely Divisible Distributions* (Cambridge University Press, Cambridge, 1999)
11. T. Simon, Hitting densities for spectrally positive stable processes. Stochastics **83**(2), 203–214 (2011)
12. T. Simon, A multiplicative short proof for the unimodality of stable densities. Elec. Comm. Probab. **16**, 623–629 (2011)
13. T. Simon, On the unimodality of power transformations of positive stable densities. Math. Nachr. **285**(4), 497–506 (2012)
14. C. Stone, The set of zeros of a semi-stable process. Ill. J. Math. **7**, 631–637 (1963)
15. M. Yamazato, Hitting time distributions of single points for 1-dimensional generalized diffusion processes. Nagoya Math. J. **119**, 143–172 (1990)
16. K. Yano, Y. Yano, M. Yor, On the laws of first hitting times of points for one-dimensional symmetric stable Lévy processes. Sémin. Probab. **XLII**, 187–227 (2009)
17. V.M. Zolotarev. *One-Dimensional Stable Distributions* (Nauka, Moskva, 1983)

On the Law of a Triplet Associated with the Pseudo-Brownian Bridge

Mathieu Rosenbaum and Marc Yor

Abstract We identify the distribution of a natural triplet associated with the pseudo-Brownian bridge. In particular, for B a Brownian motion and T_1 its first hitting time of the level one, this remarkable law allows us to understand some properties of the process $(B_{uT_1}/\sqrt{T_1},\ u \leq 1)$ under uniform random sampling, a study started in (Elie, Rosenbaum, and Yor, On the expectation of normalized Brownian functionals up to first hitting times, Preprint, arXiv:1310.1181, 2013).

Keywords Brownian motion • Pseudo-Brownian bridge • Bessel process • Local time • Hitting times • Scaling • Uniform sampling • Mellin transform

1 Introduction and Main Results

Let $(B_t,\ t \geq 0)$ be a standard Brownian motion and $(T_a,\ a > 0)$ its first hitting times process:

$$T_a = \inf\{t,\ B_t > a\}.$$

Let U denote a uniform random variable on $[0, 1]$, independent of B. In [2], the very interesting distribution of the random variable

$$\frac{B_{UT_a}}{\sqrt{T_a}}$$

is described. In particular, it is shown that this law does not depend on a, admits moments of any order and is centered. Furthermore, its density is quite remarkable, see [2] for details. In this work, our goal is to extend this study by giving some general properties of the rescaled Brownian motion up to its first hitting time of

M. Rosenbaum (✉) • M. Yor
University Pierre et Marie Curie (Paris 6), LPMA, Case courrier 188, 4 Place Jussieu, 75252 Paris cedex 05, France
e-mail: mathieu.rosenbaum@upmc.fr

© Springer International Publishing Switzerland 2014
C. Donati-Martin et al. (eds.), *Séminaire de Probabilités XLVI*, Lecture Notes in Mathematics 2123, DOI 10.1007/978-3-319-11970-0_14

level 1, that is the process $(\alpha_u, \ u \le 1)$ defined by

$$\alpha_u = \frac{B_{uT_1}}{\sqrt{T_1}}.$$

Here also, we will focus on the case of a uniform random sampling. Indeed, it is a very natural sampling scheme, which is known to lead to deep properties, see the seminal paper [4].

Let $(L_t, \ t \ge 0)$ be the local time of the Brownian motion at point 0 and $(\tau_l, \ l > 0)$ be the inverse local time process:

$$\tau_l = \inf\{t, \ L_t > l\}.$$

It turns out that our results on the process (α_u) will be deduced from properties obtained on the pseudo-Brownian bridge. This process was introduced in [1] and is defined by

$$(\frac{B_{u\tau_1}}{\sqrt{\tau_1}}, \ u \le 1).$$

The pseudo-Brownian bridge is equal to 0 at time 0 and time 1 and has the same quadratic variation as the Brownian motion. Thus, it shares some similarities with the Brownian bridge, which explains its name. We refer to [1] for more on this process. More precisely, we consider in this paper the triplet

$$(\frac{B_{U\tau_1}}{\sqrt{\tau_1}}, \frac{1}{\sqrt{\tau_1}}, L_{U\tau_1}),$$

where U is a uniform random variable on $[0, 1]$, independent of B. Our main theorem, whose proof is given in Sect. 2 and where $\underset{\mathscr{L}}{=}$ denotes equality in law, is the following.

Theorem 1 *The following identity in law holds:*

$$(\frac{B_{U\tau_1}}{\sqrt{\tau_1}}, \frac{1}{\sqrt{\tau_1}}, L_{U\tau_1}) \underset{\mathscr{L}}{=} (\frac{1}{2}B_1, L_1, \Lambda),$$

with Λ a uniform random variable on $[0, 1]$, independent of (B_1, L_1).

This result is quite remarkable in the sense that the distribution of the triplet is surprisingly simple. In particular, the marginal laws of the variables in the triplet are respectively Gaussian, absolute Gaussian and uniform distributions.

Let $(M_t, \ t \ge 0)$ be the one sided supremum of B and $(R_t, \ t \ge 0)$ be a three dimensional Bessel process starting from 0. We define the random variable γ by

$$\gamma = \sup\{t \ge 0, \ R_t = 1\}$$

and the process $(J_u, \ u \geq 0)$ by

$$J_u = \inf_{t \geq u} R_t.$$

Using Lévy's characterization of the reflecting Brownian motion and Pitman's representation of the three dimensional Bessel process, see [3] and for example [5], the aficionados of Brownian motion will easily deduce the variants of Theorem 1 stated in the following corollary, where U is a uniform random variable independent of B and R.

Corollary 1 *We have*

$$\left(\frac{B_{UT_1}}{\sqrt{T_1}}, \frac{1}{\sqrt{T_1}}, M_{UT_1}\right) \underset{\mathscr{L}}{=} \left(\Lambda L_1 - \frac{1}{2}|B_1|, L_1, \Lambda\right),$$

$$\left(\frac{R_{U\gamma}}{\sqrt{\gamma}}, \frac{1}{\sqrt{\gamma}}, J_{U\gamma}\right) \underset{\mathscr{L}}{=} \left(\Lambda L_1 + \frac{1}{2}|B_1|, L_1, \Lambda\right),$$

with Λ a uniform random variable on $[0, 1]$, independent of (B_1, L_1).

We clearly see that the distribution of the couple (B_1, L_1) plays an essential role in the description of the laws of the triplets in Theorem 1 and Corollary 1. Recall that for $s > 0$, the density of (B_s, L_s) at point $(x, l) \in \mathbb{R} \times \mathbb{R}^+$ is given by

$$\frac{1}{\sqrt{2\pi s^3}}(|x| + l)\exp\left(-\frac{(l + |x|)^2}{2s}\right). \tag{1}$$

From this expression, we deduce for fixed time $s \geq 0$ the useful and well-known factorization:

$$(|B_s|, L_s) \underset{\mathscr{L}}{=} R_s(1 - U, U).$$

Using elementary computations, this last expression enables to obtain interesting consequences of Theorem 1 and Corollary 1. For example, we give in the following corollary unexpected equalities in law and independence properties for some Brownian functionals.

Corollary 2 *The three triplets*

$$\left(|B_{U\tau_1}|, L_{U\tau_1}, \frac{1 + 2|B_{U\tau_1}|}{\sqrt{\tau_1}}\right),$$

$$\left(M_{UT_1} - B_{UT_1}, M_{UT_1}, \frac{1 + 2(M_{UT_1} - B_{UT_1})}{\sqrt{T_1}}\right),$$

$$\left(R_{U\gamma} - J_{U\gamma}, J_{U\gamma}, \frac{1 + 2(R_{U\gamma} - J_{U\gamma})}{\sqrt{\gamma}}\right)$$

are identically distributed and their common law is that of the triplet of independent variables

$$\left(\frac{1}{2}\left(\frac{1}{U} - 1\right), \Lambda, R_1\right).$$

As a general comment, we should mention that we in fact derived several ways to obtain the preceding results. We present here the approach that we think is the most easily accessible.

The rest of the paper is organized as follows. Section 2 contains the proof of Theorem 1. The law of the random variable $B_{UT_1}/\sqrt{T_1}$ is investigated in Sect. 3 and some applications of Theorem 1 and Corollary 1 can be found in Sect. 4. Some additional remarks are gathered in an appendix.

2 Proof of Theorem 1

In this section, we give the proof of our main result, Theorem 1.

2.1 Step 1: Introducing a Mellin Type Transform

First remark that using the symmetry of the Brownian motion, Theorem 1 is equivalent to the equality in law:

$$\left(\frac{|B_{U\tau_1}|}{\sqrt{\tau_1}}, \frac{1}{\sqrt{\tau_1}}, L_{U\tau_1}\right) \underset{\mathscr{L}}{=} \left(\frac{1}{2}|B_1|, L_1, \Lambda\right). \tag{2}$$

We now introduce the function $\mathscr{M}_l(a, c)$ which is defined for $a \geq 0$, $c \geq 0$ and $l \leq 1$ by

$$\mathscr{M}_l(a, c) = \mathbb{E}\left[\left(\frac{|B_{U\tau_1}|}{\sqrt{\tau_1}}\right)^a \left(\frac{1}{\sqrt{\tau_1}}\right)^c 1_{\{L_{U\tau_1} \leq l\}}\right].$$

Using properties of the Mellin transform together with a monotone class argument, we easily get that proving (2) is equivalent to show the following equality:

$$\mathscr{M}_l(a, c) = l\mathbb{E}\left[|\frac{1}{2}B_1|^a(L_1)^c\right]. \tag{3}$$

2.2 Step 2: The Mellin Transform of (B_1, L_1)

In this second step, we give the Mellin transform of the couple (B_1, L_1). Having an expression for this functional is of course helpful in order to prove Equality (3).

2.2.1 The Result

The Mellin transform of the couple $(|B_1|, L_1)$ is given in the following proposition.

Proposition 1 *For $a > 0$ and $c > 0$, we have*

$$\mathbb{E}[|B_1|^a (L_1)^c] = \frac{\Gamma(1+a)\Gamma(1+c)}{2^{\frac{a+c}{2}}\Gamma(1 + \frac{a+c}{2})},$$

where Γ denotes the classical gamma function.

2.2.2 Technical Lemma

Before starting the proof of Proposition 1, we give a useful lemma.

Lemma 1 *Let $\mathscr{E}$ and $\mathscr{E}'$ be two independent standard exponential variables, independent of B. The following equality in law holds:*

$$(|B_{2\mathscr{E}}|, L_{2\mathscr{E}}) \stackrel{\mathscr{L}}{=} (\mathscr{E}, \mathscr{E}').$$

This result is in fact quite well known. However, for sake of completeness we give its proof here.

Proof To prove Lemma 1, we compute the Mellin transform of the couple $(|B_{2\mathscr{E}}|, L_{2\mathscr{E}})$. Let $a > 0$ and $c > 0$. From the law of the couple $(|B_t|, L_t)$ obtained from (1), we get

$$\mathbb{E}[|B_{2\mathscr{E}}|^a (L_{2\mathscr{E}})^c] = \frac{1}{2}\int_0^{+\infty} dt\, e^{-t/2}\mathbb{E}[|B_t|^a(L_t)^c]$$

$$= \frac{1}{2}\int_0^{+\infty} dt\, e^{-t/2}\int_0^{+\infty} dx \int_0^{+\infty} dl\,\sqrt{\frac{2}{\pi t^3}}x^a l^c(x+l)e^{-\frac{(l+x)^2}{2t}}$$

$$= \int_0^{+\infty} dx \int_0^{+\infty} dl x^a l^c \int_0^{+\infty} \frac{dt}{\sqrt{2\pi t^3}}e^{-t/2}(x+l)e^{-\frac{(l+x)^2}{2t}}.$$

Using the expression of the density of an inverse Gaussian random variable, it is easily seen that the integral in (dt) is equal to $\exp(-(x+l))$. Therefore, the remaining double integral is equal to $\Gamma(1+a)\Gamma(1+c)$. This gives the result since

$$\mathbb{E}[(\mathscr{E})^a(\mathscr{E}')^c] = \Gamma(1+a)\Gamma(1+c).\qquad\square$$

2.2.3 Proof of Proposition 1

We now give the proof of Proposition 1. First remark that, by scaling, we have

$$\sqrt{2\mathscr{E}}(|B_1|, L_1) \underset{\mathscr{L}}{=} (|B_{2\mathscr{E}}|, L_{2\mathscr{E}}).$$

Thus, using Lemma 1, we get

$$\mathbb{E}[(2\mathscr{E})^{\frac{a+c}{2}}]\mathbb{E}[|B_1|^a(L_1)^c] = \mathbb{E}[(\mathscr{E})^a(\mathscr{E}')^c] = \Gamma(1+a)\Gamma(1+c).$$

We eventually obtain the result since

$$\mathbb{E}[(2\mathscr{E})^{\frac{a+c}{2}}] = 2^{\frac{a+c}{2}}\Gamma(1 + \frac{a+c}{2}).$$

2.3 Step 3: End of the Proof of Theorem 1

According to Proposition 1, in order to prove Theorem 1, it suffices to show that

$$\mathscr{M}_l(a,c) = l\frac{\Gamma(1+a)\Gamma(1+c)}{2^{\frac{a+c}{2}}\Gamma(1 + \frac{a+c}{2})2^a}. \tag{4}$$

To this purpose, let us write $\mathscr{M}_l(a,c)$ under the form

$$\mathscr{M}_l(a,c) = \mathbb{E}\Big[\frac{1}{\tau_1^{\frac{a+c}{2}+1}}\int_0^{\tau_1} ds|B_s|^a 1_{\{L_s \leq l\}}\Big]$$

$$= \mathbb{E}\Big[\frac{1}{\tau_1^{\frac{a+c}{2}+1}}\int_0^{\tau_l} ds|B_s|^a\Big].$$

Then, using the definition of the Gamma function and a change of variable, we easily get

$$\mathscr{M}_l(a,c) = \frac{1}{2^{\frac{a+c}{2}}\Gamma(1 + \frac{a+c}{2})}\int_0^{+\infty} d\mu\mu^{1+a+c}\mathbb{E}\Big[\int_0^{\tau_l} ds|B_s|^a e^{-\frac{\mu^2}{2}\tau_1}\Big]. \tag{5}$$

Now recall that the process τ_l is a subordinator with Laplace exponent at point $\lambda \in \mathbb{R}^{+*}$ equal to $\sqrt{2\lambda}$, see [5]. Therefore,

$$\mathbb{E}\Big[\int_0^{\tau_l} ds|B_s|^a e^{-\frac{\mu^2}{2}\tau_1}\Big] = \mathbb{E}\Big[\int_0^{\tau_l} ds|B_s|^a e^{-\frac{\mu^2}{2}\tau_l - \mu(1-l)}\Big]$$

$$= e^{-\mu}\mathbb{E}\Big[\int_0^{\tau_l} ds|B_s|^a e^{-\frac{\mu^2}{2}\tau_l - \mu(|B_{\tau_l}| - L_{\tau_l})}\Big].$$

Then, using the optional stopping theorem for the martingale $e^{\mu(L_s-|B_s|)-\frac{\mu^2 s}{2}}$, see [5], we get

$$\mathbb{E}\Big[\int_0^{\tau_l} ds |B_s|^a e^{-\frac{\mu^2}{2}\tau_1}\Big] = e^{-\mu}\mathbb{E}\Big[\int_0^{+\infty} ds |B_s|^a e^{\mu(L_s-|B_s|)-\frac{\mu^2 s}{2}} 1_{\{L_s \le l\}}\Big].$$

With the help of the joint law of $(|B_s|, L_s)$ given in (1), we obtain that this last quantity is also equal to

$$e^{-\mu}\int_0^{+\infty} dx \int_0^l dm x^a e^{\mu(m-x)} 2\int_0^{+\infty} \frac{ds}{\sqrt{2\pi s^3}}(m+x)e^{-\frac{(m+x)^2}{2s}}e^{-\frac{\mu^2 s}{2}}.$$

Using again the expression of the density of an inverse Gaussian random variable, we see that the integral in (ds) is equal to $\exp(-\mu(x+m))$. Consequently, we obtain

$$\mathbb{E}\Big[\int_0^{\tau_l} ds |B_s|^a e^{-\frac{\mu^2}{2}\tau_1}\Big] = 2e^{-\mu}\int_0^{+\infty} dx \int_0^l dm x^a e^{-2\mu x}$$

$$= 2l e^{-\mu}\frac{\Gamma(1+a)}{(2\mu)^{a+1}}.$$

Plugging this equality into Eq. (5) gives

$$\mathcal{M}_l(a,c) = \frac{l}{2^{\frac{a+c}{2}}\Gamma(1+\frac{a+c}{2})}\frac{\Gamma(1+a)\Gamma(1+c)}{2^a},$$

which is the desired identity (4).

3 The Properties of the Law of α Revisited

In this section, we focus on the random variable

$$\alpha = \frac{B_{UT_1}}{\sqrt{T_1}},$$

with U a uniform random variable on $[0, 1]$ independent of B.

3.1 Two Equivalent Characterizations of the Distribution of α

In [2], we describe in term of its density the law of the variable α. In particular, we show that this variable is centered. This characterization of the distribution of

α is obtained thanks to the computation of the Mellin transform of the positive and negative parts of α. More precisely, we have for $m > 0$

$$\mathbb{E}[(\alpha_+)^m] = \mathbb{E}[|N|^m]2\int_0^1 dz\frac{z^{1+m}}{(1+2z)}$$

$$\mathbb{E}[(\alpha_-)^m] = \mathbb{E}[|\frac{N}{2}|^m](\frac{\log(3)}{2}).$$

This leads to the following description of the law of α, which is of course equivalent to that given in [2].

Proposition 2 *Let N and Z be two independent random variables with N standard Gaussian and Z with density*

$$\frac{2z}{(1-\frac{\log(3)}{2})(1+2z)}1_{\{0<z<1\}}.$$

We have the following equalities in law:

$$(\alpha|\alpha > 0) \underset{\mathscr{L}}{=} |N|Z$$

$$(-\alpha|\alpha < 0) \underset{\mathscr{L}}{=} \frac{1}{2}|N|$$

$$\mathbb{P}[\alpha > 0] = 1 - \frac{\log(3)}{2}.$$

We wish to compare this characterization with the one obtained from Corollary 1, namely

$$\alpha \underset{\mathscr{L}}{=} \Lambda L_1 - \frac{1}{2}|B_1|. \tag{6}$$

We first note that the centering property is easily recovered since

$$\mathbb{E}[\Lambda L_1 - \frac{1}{2}|B_1|] = \frac{1}{2}\mathbb{E}[L_1 - |B_1|] = 0.$$

Moreover, the second moment can also be computed without difficulty. Indeed,

$$\mathbb{E}[(\Lambda L_1 - \frac{1}{2}|B_1|)^2] = \frac{1}{3}\mathbb{E}[L_1^2] - \frac{1}{2}\mathbb{E}[L_1|B_1|] + \frac{1}{4}.$$

Then, using for example the formula for the Mellin transform of the couple $(|B_1|, L_1)$ given in Proposition 1, we get

$$\mathbb{E}[(\Lambda L_1 - \frac{1}{2}|B_1|)^2] = \frac{1}{3}.$$

We now show that Proposition 2 and Eq. (6) match. This is deduced from the following elementary description of the random variable A defined by

$$A = \Lambda U - \frac{1}{2}(1 - U),$$

with Λ and U two independent uniform random variables on $[0, 1]$.

Proposition 3 *Let Z be the random variable defined in Proposition 2 and V a uniform variable on $[0, 1]$, independent of Z. We have*

$$(A|A > 0) \underset{\mathscr{L}}{=} VZ$$

$$(-A|A < 0) \underset{\mathscr{L}}{=} \frac{1}{2}V$$

$$\mathbb{P}[A > 0] = 1 - \frac{\log(3)}{2}.$$

Proof Let f be a positive measurable function. The density of A can be computed directly as follows. We have

$$\mathbb{E}[f(A)] = \int_0^1 d\lambda \mathbb{E}\left[f\left(\lambda U - \frac{1}{2}(1 - U)\right)\right]$$

$$= \int_{\mathbb{R}} dx f(x) u(x),$$

with

$$u(x) = \mathbb{E}\left[\frac{1}{U} 1_{\{\frac{1+2x}{3} \le U \le 1+2x\}}\right].$$

Now, note that on the one hand

– if $-\frac{1}{2} \le x \le 0$,

$$u(x) = \int_{\frac{1+2x}{3}}^{1+2x} \frac{du}{u} = \log(3),$$

– if $0 \le x \le 1$,

$$u(x) = \int_{\frac{1+2x}{3}}^{1} \frac{du}{u} = \log\left(\frac{3}{1 + 2x}\right),$$

– if $x \le -\frac{1}{2}$ or $x > 1$,

$$u(x) = 0.$$

On the other hand, it is easily seen that the density of VZ is

$$\frac{1}{1 - \frac{\log(3)}{2}} \log\left(\frac{3}{1 + 2x}\right).$$

This ends the proof of Proposition 3.   □

Let us now start from Eq. (6) and recover Proposition 2. From Eq. (6), with our usual notation, we get

$$\alpha \underset{\mathscr{L}}{=} R_1 A.$$

Thus,

$$(\alpha|\alpha > 0) \underset{\mathscr{L}}{=} (R_1 A|A > 0).$$

From Proposition 3, we obtain

$$(R_1 A|A > 0) \underset{\mathscr{L}}{=} R_1 VZ \underset{\mathscr{L}}{=} |N|Z,$$

which gives the first result in Proposition 2. The two other results are proved similarly, with the help of Proposition 3.

3.2 A Warning and Some Developments Around Proposition 2

Since $1/\sqrt{T_1}$ is distributed as $|N|$, from Proposition 2, it may be tempting to think that

$$(B_{UT_1}|B_{UT_1} > 0)$$

is distributed as Z and is independent of T_1. However, this is wrong. This incited us to look at the joint law of $1/\sqrt{T_1}$ and B_{UT_1}. Indeed, although encoded in Corollary 1, it may deserve an explicit presentation which we give in the following theorem.

Theorem 2

* Let $p \geq 0$. For ϕ a positive measurable function, the following formulas hold:

$$\mathbb{E}\left[\frac{1}{(\sqrt{T_1})^p} \phi(B_{UT_1}) 1_{\{B_{UT_1} > 0\}}\right] = c_p \int_0^1 db \phi(b)\left(1 - \frac{1}{(3 - 2b)^{p+1}}\right)$$

$$\mathbb{E}\left[\frac{1}{(\sqrt{T_1})^p} \phi(B_{UT_1}) 1_{\{B_{UT_1} < 0\}}\right] = c_p \int_{-\infty}^0 dx \phi(x)\left(\frac{1}{(1 - 2x)^{p+1}} - \frac{1}{(3 - 2x)^{p+1}}\right),$$

where $c_p = \mathbb{E}[|N|^p] = \frac{\Gamma(1+p)}{2^{p/2}\Gamma(1+p/2)}$.

- *The joint law of $(1/\sqrt{T_1}, B_{UT_1})$ admits the density h defined on $\mathbb{R}^{+*} \times (-\infty, 1]$ by*

$$h(z, x) = \sqrt{\frac{2}{\pi}} (e^{-z^2/2} - e^{-(3-2x)^2 z^2/2}) 1_{\{z>0,\, 0<x<1\}}$$

$$+ \sqrt{\frac{2}{\pi}} (e^{-(1-2x)^2 z^2/2} - e^{-(3-2x)^2 z^2/2}) 1_{\{z>0,\, x<0\}}.$$

- *The law of B_{UT_1} admits the density k defined on $(-\infty, 1)$ by*

$$k(x) = \frac{2(1-x)}{3-2x} 1_{\{0<x<1\}} + \frac{2}{(1-2x)(3-2x)} 1_{\{x<0\}}.$$

Remark that from Theorem 2, we get that

$$(1 - B_{UT_1} | B_{UT_1} > 0)$$

is distributed as Z. We now give the proof of Theorem 2.

Proof • To prove the first part of Theorem 2, we use the fact that

$$\mathbb{E}\left[\frac{1}{(\sqrt{T_1})^p} \phi(B_{UT_1}) 1_{\{B_{UT_1}>0\}}\right]$$

is equal to

$$\frac{1}{2^{p/2} \Gamma(1 + p/2)} \int_0^{+\infty} d\mu \mu^{1+p} \mathbb{E}\left[\int_0^{T_1} ds \phi(B_s) 1_{\{B_s>0\}} e^{-\frac{\mu^2}{2} T_1}\right].$$

From Proposition 3.1 in [2] (or the computation of I_μ in the same paper), the above expectation is equal to

$$\int_0^1 db \phi(b) \frac{1}{\mu} (e^{-\mu} - e^{-\mu(3-2b)}).$$

Then, using Fubini's theorem and integrating in μ yields the first formula. The second formula is obtained in the same manner.

- We now show that Part 1 of Theorem 2 may also be obtained without relying upon [2], but only on the first identity of Corollary 1. Let

$$\phi_p = \mathbb{E}\left[\frac{1}{(\sqrt{T_1})^p} \phi(B_{UT_1}) 1_{\{B_{UT_1}>0\}}\right].$$

From Corollary 1, we have

$$
\phi_p = \mathbb{E}\Big[(L_1)^p \phi\Big(\Lambda - \frac{1}{2}\big(\frac{1}{U} - 1\big)\big)1_{\{\Lambda > \frac{1}{2}(\frac{1}{U}-1)\}}\Big]
$$

$$
= \mathbb{E}[(R_1)^p]\mathbb{E}\Big[U^p \phi\Big(\Lambda - \frac{1}{2}\big(\frac{1}{U} - 1\big)\big)1_{\{\Lambda > \frac{1}{2}(\frac{1}{U}-1)\}}\Big].
$$

We note that, as a consequence of the identity in law

$$
L_1 \underset{\mathscr{L}}{=} R_1 U,
$$

we get $\mathbb{E}[(R_1)^p] = c_p(p + 1)$. Hence, with the help of Fubini's theorem and simple changes of variables,

$$
\phi_p = c_p(p + 1) \int_0^1 db\phi(b) \int_{\frac{1}{3-2b}}^1 du u^p = c_p \int_0^1 db\phi(b)\Big(1 - \frac{1}{(3 - 2b)^{p+1}}\Big).
$$

The second formula of Part 1 may be obtained in the same manner.

- For the proof of Part 2 in Theorem 2, in order to obtain the density at point (z, x) in $\mathbb{R}^{+*} \times (0, 1)$, $h(z, x)$, we note that from Part 1 of Theorem 2, the formula

$$
\mathbb{E}\Big[f\big(\frac{1}{\sqrt{T_1}}, B_{UT_1}\big)1_{\{B_{UT_1} > 0\}}\Big] = \mathbb{E}[f(|N|, V)] - \int_0^1 \frac{db}{3 - 2b}\mathbb{E}\Big[f\big(\frac{|N|}{3 - 2b}, b\big)\Big],
$$

with V a uniform variable on $[0, 1]$ independent of N, holds for every function f of the form

$$
f(z, b) = z^p \phi(b),
$$

with ϕ some positive measurable function. An application of the monotone class theorem yields the validity of the above formula for every positive measurable function f. Then, a simple change of variables gives the first formula in Part 2. The second formula is proved likewise.

- To prove Part 3 in Theorem 2, it suffices to take $p = 0$ in Part 1.   □

4 Applications

In this section, we give two applications of Theorem 1 and Corollary 1.

4.1 A Family of Centered Brownian Functionals

In [2], we established that the variable H defined by

$$H = \frac{1}{T_1^{3/2}} \int_0^{T_1} ds\, B_s$$

admits moments of all orders and is centered. This centering property is equivalent to that of the random variable

$$\alpha = \frac{B_{UT_1}}{\sqrt{T_1}},$$

where U is a uniform random variable on $[0, 1]$, independent of B. In fact, Theorem 1 and Corollary 1 enable to build families of centered functionals involving the Brownian motion and its first hitting time of level 1, the local time at point 0 and its inverse process, and the running maximum. We have the following theorem, in which H_1 is equal to H.

Theorem 3 *For any $p \geq 1$, the random variables H_p and H_p' are centered, with*

$$H_p = \frac{1}{T_1^{p/2+1}} \int_0^{T_1} ds\Big(\big(\frac{p+1}{2p^2} - 1\big) M_s^p + B_s M_s^{p-1}\Big)$$

and

$$H_p' = \frac{1}{\tau_1^{p/2+1}} \int_0^{\tau_1} ds\Big(\frac{p+1}{2p^2} L_s^p - |B_s| L_s^{p-1}\Big).$$

Proof First remark that from Lévy's equivalence theorem, H_p and H_p' have the same law. Then, obviously the expectation of H_p' is that of

$$\big(\frac{p+1}{2p^2}\big)\frac{L_{U\tau_1}^p}{(\sqrt{\tau_1})^p} - \frac{|B_{U\tau_1}|(L_{U\tau_1})^{p-1}}{(\sqrt{\tau_1})^p}.$$

From Theorem 1, this random variable has the same law as

$$\big(\frac{p+1}{2p^2}\big)\Lambda^p L_1^p - \frac{1}{2}|B_1|\Lambda^{p-1} L_1^{p-1}.$$

The expectation of this last quantity is equal to

$$\frac{1}{2p^2}\mathbb{E}[L_1^p] - \frac{1}{2p}\mathbb{E}[|B_1| L_1^{p-1}].$$

Using for example the Mellin transform of the couple $(|B_1|, L_1)$ given in Proposition 1, it is easily seen that this expression is equal to zero (one may also use the martingale property of $\frac{1}{p}L_t^p - |B_t|L_t^{p-1}$, see [5]).   □

4.2   On the Law of $R_{U\gamma}/\sqrt{\gamma}$

We now focus on the distribution of $R_{U\gamma}/\sqrt{\gamma}$. From Corollary 1, we know that

$$\frac{R_{U\gamma}}{\sqrt{\gamma}} \underset{\mathscr{L}}{=} \Lambda L_1 + \frac{1}{2}|B_1| \underset{\mathscr{L}}{=} R_1 A',$$

with

$$A' = \Lambda U + \frac{1}{2}(1 - U).$$

There is the following description of the laws of A' and $R_{U\gamma}/\sqrt{\gamma}$.

Proposition 4 *The law of A' admits the density l given by*

$$l(a) = \log\left(\frac{1}{|2a - 1|}\right)1_{\{0<a<1\}}.$$

Consequently, the law of $R_{U\gamma}/\sqrt{\gamma}$ admits the following density:

$$\sqrt{\frac{2}{\pi}}x^2 \int_1^{+\infty} dy\, y\exp\left(-\frac{x^2y^2}{2}\right)l\left(\frac{1}{y}\right).$$

Proof The density of A' is obtained thanks to straightforward computations. Let f be a positive measurable function. Using the density of a three dimensional Bessel variable R_1, see [5], we have

$$\mathbb{E}[f(R_1 A')] = \sqrt{\frac{2}{\pi}} \int_0^{+\infty} dr\, r^2 e^{-r^2/2}\mathbb{E}[f(rA')].$$

Now, using the density of A', we get

$$\mathbb{E}[f(R_1 A')] = \sqrt{\frac{2}{\pi}} \int_0^{+\infty} dr\, r^2 e^{-r^2/2} \int_0^r \frac{dx}{r}l\left(\frac{x}{r}\right)f(x)$$

$$= \int_0^{+\infty} dx f(x)\sqrt{\frac{2}{\pi}} \int_x^{+\infty} dr\, re^{-r^2/2}l\left(\frac{x}{r}\right)$$

$$= \int_0^{+\infty} dx f(x)\sqrt{\frac{2}{\pi}}x^2 \int_1^{+\infty} dy\, ye^{-x^2y^2/2}l\left(\frac{1}{y}\right). \qquad □$$

Conclusion and Future Work

In this paper, we establish the law of a triplet associated with the pseudo-Brownian bridge. This process has been introduced in [1] and is defined as

$$\left(\frac{B_{u\tau_1}}{\sqrt{\tau_1}}, u \leq 1\right).$$

In particular, this enables us to understand in depth some properties of the random variable

$$\alpha = \frac{B_{UT_1}}{\sqrt{T_1}}$$

studied in [2]. In a forthcoming work, we intend to develop some consequences of the obtained results for the Brownian bridge and the Brownian meander.

Acknowledgements We thank the referee for a thorough reading of our paper.

A Appendix

A.1 A Simple Proof for the Joint Law of $(1/\sqrt{\tau_1}, L_{U\tau_1})$

The fact that

$$\left(\frac{1}{\sqrt{\tau_1}}, L_{U\tau_1}\right) \underset{\mathscr{L}}{=} (L_1, \Lambda)$$

can obviously be deduced from Theorem 1. However, interestingly, we can give a simple proof for this equality in law. Indeed, for $\lambda \geq 0$ and $l < 1$, we have

$$\mathbb{E}[e^{-\lambda\tau_1}1_{\{L_{U\tau_1}\leq l\}}] = \mathbb{E}[\frac{1}{\tau_1}\int_0^{\tau_1} ds 1_{\{L_s \leq l\}}e^{-\lambda\tau_1}]$$

$$= \mathbb{E}[\frac{\tau_l}{\tau_1}e^{-\lambda\tau_1}].$$

Now, consider in general (τ_l) a subordinator and denote by ψ its Laplace exponent. Thus, we have

$$\mathbb{E}[\frac{\tau_l}{\tau_1}e^{-\lambda\tau_1}] = \mathbb{E}[\tau_l \int_0^{+\infty} dt e^{-(t+\lambda)\tau_1}]$$

$$= \int_0^{+\infty} dt \mathbb{E}[\tau_l e^{-(t+\lambda)\tau_l}]e^{-(1-l)\psi(t+\lambda)}.$$

Using the fact that the Laplace exponent is differentiable on $\mathbb{R}^{+*}$, we get

$$\mathbb{E}[\frac{\tau_l}{\tau_1}e^{-\lambda\tau_1}] = \int_0^{+\infty} dt l\psi'(t+\lambda)e^{-l\psi(t+\lambda)}e^{-(1-l)\psi(t+\lambda)}$$

$$= le^{-\psi(\lambda)}.$$

This proves the independence of $1/\sqrt{\tau_1}$ and $L_{U\tau_1}$ and the fact that $L_{U\tau_1}$ is uniformly distributed. The equality in law

$$\frac{1}{\sqrt{\tau_1}} \underset{\mathscr{L}}{=} L_1$$

is easily obtained by scaling.

A.2 On a One Parameter Family of Random Variables Including α

In this section of the appendix, we consider the family of variables defined for $0 < c \leq 1$ by

$$\alpha_c = \Lambda L_1 - c|B_1|,$$

as an extension of our study of

$$\alpha \underset{\mathscr{L}}{=} \alpha_{1/2} \underset{\mathscr{L}}{=} B_{UT_1}/\sqrt{T_1}.$$

The variables α_c, although less natural than $\alpha_{1/2}$, enjoy some similar remarkable properties. Indeed, Propositions 3 and 2 admit the following extensions.

Proposition 5 *Let $0 < c \leq 1$ and $C = 1/c$. Let Λ and U be two independent uniform variables on $[0, 1]$ and*

$$A_c = \Lambda U - c(1 - U).$$

We have

$$(A_c|A_c > 0) \underset{\mathscr{L}}{=} VZ_C$$

$$(-A_c|A_c < 0) \underset{\mathscr{L}}{=} cV$$

$$\mathbb{P}[A_c > 0] = 1 - c\log(1 + C),$$

where V and Z_C are independent, with V uniform on $[0, 1]$ and Z_C a random variable with density given by

$$\frac{C}{1 - c\log(1 + C)} \frac{d\,zz}{(1 + C\,z)} 1_{\{0 < z < 1\}}.$$

Proposition 6 *Let $0 < c \leq 1$. The following equalities in law hold.*

$$(\alpha_c | \alpha_c > 0) \underset{\mathscr{L}}{=} |N| Z_C$$

$$(-\alpha_c | \alpha_c < 0) \underset{\mathscr{L}}{=} c|N|$$

$$\mathbb{P}[\alpha_c > 0] = 1 - c\log(1 + C).$$

Proof To establish Proposition 5, we simply compute the density of A_c. Proposition 6 ensues since

$$\alpha_c \underset{\mathscr{L}}{=} R_1 A_c$$

and

$$R_1 V \underset{\mathscr{L}}{=} |N|,$$

with the same notation as previously. $\square$

References

1. P. Biane, J.-F. Le Gall, M. Yor, Un processus qui ressemble au pont brownien. In: *Séminaire de Probabilités XXI* (Springer, New York, 1987), pp. 270–275
2. R. Elie, M. Rosenbaum, M. Yor, On the expectation of normalized brownian functionals up to first hitting times. arXiv preprint arXiv:1310.1181 (2013)
3. J.W. Pitman, One-dimensional brownian motion and the three-dimensional bessel process. Adv. Appl. Probab. **7**, 511–526 (1975)
4. J.W. Pitman, Brownian motion, bridge, excursion, and meander characterized by sampling at independent uniform times. Electron. J. Probab. **4**(11), 1–33 (1999)
5. D. Revuz, M. Yor, *Continuous Martingales and Brownian Motion*, vol. 293 (Springer, New York, 1999)

Skew-Product Decomposition of Planar Brownian Motion and Complementability

Jean Brossard, Michel Émery, and Christophe Leuridan

Abstract Let Z be a complex Brownian motion starting at 0 and W the complex Brownian motion defined by

$$W_t = \int_0^t \frac{\overline{Z}_s}{|Z_s|} \, \mathrm{d}Z_s \; .$$

The natural filtration $\mathcal{F}^W$ of W is the filtration generated by Z up to an arbitrary rotation. We show that given any two different matrices Q_1 and Q_2 in $O_2(\mathbf{R})$, there exists an $\mathcal{F}^Z$-previsible process H taking values in $\{Q_1, Q_2\}$ such that the Brownian motion $\int H \cdot \mathrm{d}W$ generates the whole filtration $\mathcal{F}^Z$. As a consequence, for all a and b in $\mathbf{R}$ such that $a^2 + b^2 = 1$, the Brownian motion $a \, \Re(W) + b \, \Im(W)$ is complementable in $\mathcal{F}^Z$.

Keywords Brownian filtrations • Complementability • Planar Brownian motion • Skew-product decomposition

AMS classification (2010): 60J65, 60H20

1 Introduction

Brownian filtrations constitute a rich topic, where innocent-looking questions sometimes turn out to be quite tricky. How can one recognize if a given filtration is Brownian (that is, generated by some Brownian motion)? Few characterizations

J. Brossard • C. Leuridan (✉)
Institut Fourier, Université Joseph Fourier et CNRS, BP 74, 38 402 Saint-Martin-d'Hères Cedex, France
e-mail: jean.brossard@ujf-grenoble.fr; christophe.leuridan@ujf-grenoble.fr

M. Émery
IRMA, Université Unique de Strasbourg et CNRS, 7 rue René Descartes, 67 084 Strasbourg Cedex, France
e-mail: emery@math.unistra.fr

© Springer International Publishing Switzerland 2014
C. Donati-Martin et al. (eds.), *Séminaire de Probabilités XLVI*, Lecture Notes in Mathematics 2123, DOI 10.1007/978-3-319-11970-0_15

are known, and most of the time one has to exhibit a generating Brownian motion.

In 1980, Stroock and Yor [6] raised the following question.

(Q1): *If a filtration $\mathcal{F}$ has the previsible representation property (PRP) w.r.t. some Brownian motion, is $\mathcal{F}$ necessarily Brownian?*

Not until 15 years later was the matter settled, with two very different counter-examples provided by Dubins, Feldman, Smorodinsky and Tsirelson [3] and by Tsirelson [8]. The former shows the existence on Wiener space of a probability Q equivalent to the Wiener measure P, such that no Q-BM generates the filtration (even though the filtration must have the PRP w.r.t. the Q-BM obtained as the Girsanov transform of P). And the latter counter-example asserts that the Walsh BM on three or more branches generates a non-Brownian filtration.

In view of the new light shed by these examples, an updated version of Stroock and Yor's original question can be asked.

(Q2): *If a filtration $\mathcal{F}$ is immersed[1] in some (possibly infinite-dimensional) Brownian filtration and has the PRP w.r.t. some $\mathcal{F}$-Brownian motion, is $\mathcal{F}$ necessarily Brownian?*

The general case remains elusive, although a positive answer is obtained in [4] in the particular case of filtrations which are Brownian on every interval $[\varepsilon, \infty[$ with $\varepsilon > 0$; under this very strong additional hypothesis, Brownianness or non Brownianness is a germ property at time $0+$. Natural examples of such filtrations are provided by quotienting the filtration of a d-dimensional Brownian motion by some subgroup of the orthogonal group. These filtrations are already proved to be Brownian by Malric in [5], who explicitly constructs Brownian motions generating them.

(Q2) draws attention to the various ways a Brownian filtration can be immersed in another Brownian filtration. For instance, given Z a d-dimensional BM and given $k \in [\![1, d-1]\!]$, for each k-dimensional linear subspace S of $\mathbf{R}^d$ the orthogonal projection of Z on S is a BM whose filtration is immersed in the filtration $\mathcal{Z}$ of Z, and all these immersions are clearly isomorphic to each other.[2] Moreover, given a k-dimensional $\mathcal{Z}$-BM B, its filtration is immersed in $\mathcal{Z}$, and this immersion is isomorphic to all previous ones if and only if there exists a $(d-k)$-dimensional $\mathcal{Z}$-BM B' independent of B such that the d-dimensional $\mathcal{Z}$-BM (B, B') generates the full filtration $\mathcal{Z}$. This property was introduced in [2] and called *complementability;* it is the simplest way a k-dimensional Brownian filtration can be immersed in a d-dimensional one.

---

[1]A filtration $\mathcal{F}$ is said to be immersed in a filtration $\mathcal{G}$ when every $\mathcal{F}$-martingale is a $\mathcal{G}$-martingale.

[2]Given four filtrations $\mathcal{F}$, $\mathcal{G}$, $\mathcal{F}'$ and $\mathcal{G}'$ with $\mathcal{F}$ immersed in $\mathcal{G}$ and $\mathcal{F}'$ immersed in $\mathcal{G}'$, *the immersion of $\mathcal{F}'$ in $\mathcal{G}'$ is isomorphic to the immersion of $\mathcal{F}$ in $\mathcal{G}$ if $\mathcal{G}$ and $\mathcal{G}'$ are in correspondence by some isomorphism which maps $\mathcal{F}$ onto $\mathcal{F}'$.*

Another property, maximality,[3] is defined in [2] and shown there to be necessary for complementability. This leads to the question of the sufficiency.

(Q3): *Conversely, does maximality imply complementability?*

Even in the simplest case, when $k = 1$ and $d = 2$, question (Q3) is still open. Although there is no direct mathematical relation between (Q2) and (Q3), they turn out to be similar, at two levels. First, each of them asks if some BM can be constructed, so as to generate (alone for (Q2), or together with B for (Q3)) a given filtration. Second, in all instances so far that such a BM has successfully been constructed, the methods are similar, be it for (Q2) or (Q3); they rely on coupling arguments. It thus appears that studying (Q3) might indirectly contribute to progress on (Q2). At this stage, both (Q2) and (Q3) seem difficult: we have in view no strategy of proof, nor any candidate for a counter-example; but one of (Q2) and (Q3) may eventually turn out to be less difficult than the other.

The present work describes a family of real BM shown to be complementable in the filtration of a complex BM (Corollary 2); although these real BM are very simply defined, our proof of their complementability (by explicit construction of a complement) is rather involved.

From now on, we fix a complex Brownian motion $Z = X + iY$ starting at 0. Almost surely, the Bessel process $R = |Z|$ never returns to 0. We shall focus on the independent Brownian motions U and V given by

$$U_t = \int_0^t \frac{X_s \, dX_s + Y_s \, dY_s}{R_s} \quad \text{and} \quad V_t = \int_0^t \frac{X_s \, dY_s - Y_s \, dX_s}{R_s} \, .$$

The complex Brownian motion $W = U + iV = \int (\overline{Z}/R) \, dZ$ is known to generate the quotient filtration generated by Z up to an arbitrary rotation.

Actually, W is not the generating Brownian motion constructed by Malric in [5]. But among all Brownian motions generating the quotient filtration of Z modulo SO_2, the complex Brownian motion $W = U + iV$ is somehow the most natural one, since its real part U governs the Bessel process R whereas its imaginary part V governs the increments of the angular part of Z. These facts are due to Stroock and Yor [7] and recalled in Proposition 1.

The complementability of U is proved in [2] by exhibiting an independent complement. Is V complementable too? In [1], V is shown to be maximal, and its complementability is announced, without proof, at the end of the introduction. A proof will be given in the present paper; we shall actually establish a more general result as follows.

---

[3]A k-dimensional Z-BM B is called maximal if no other k-dimensional Z-BM generates a strictly bigger filtration than B.

Theorem 1 *Let Q_1 and Q_2 be any two different matrices in $O_2(\mathbf{R})$. There exists an $\mathcal{F}^Z$-previsible process H taking values in $\{Q_1, Q_2\}$ such that the Brownian motion $\int H \cdot \mathrm{d}W$ generates the whole filtration $\mathcal{F}^Z$.*

Here, for every $Q \in O_2(\mathbf{R})$ and $z \in \mathbf{C}$, we denote by $Q \cdot z$ the complex number provided by the usual action of the 2×2 real matrices on $\mathbf{C}$ identified with $\mathbf{R}^2$. More precisely, set

$$Q = \begin{pmatrix} a & -\sigma b \\ b & \sigma a \end{pmatrix} \tag{1}$$

with a and b in $\mathbf{R}$, $a^2 + b^2 = 1$, $\sigma \in \{-1, 1\}$, and put $c = a + ib$. Then $Q \cdot z = cz$ or $Q \cdot z = c\bar{z}$ according to σ being equal to 1 or -1.

Observe that for each $Q \in O_2(\mathbf{R})$, the Brownian motion $\int (QH) \cdot \mathrm{d}W = \int Q \cdot \mathrm{d}(\int H \cdot \mathrm{d}W)$ generates the same filtration as $\int H \cdot \mathrm{d}W$. Hence, assuming that $Q_2 = I_2$ in the proof of the theorem is no restriction. Moreover, Theorem 1 is completely contained in its two corollaries below.

Corollary 1 *For every complex number $c \neq 1$ with modulus 1, there exists an $\mathcal{F}^Z$-previsible process η taking values in $\{1, c\}$ such that the Brownian motion $\int \eta \, \mathrm{d}W$ generates the whole filtration $\mathcal{F}^Z$.*

Corollary 2 *For all a and b in $\mathbf{R}$ such that $a^2 + b^2 = 1$, the Brownian motion $aU + bV$ is complementable in $\mathcal{F}^Z$.*

Corollary 1 directly follows from Theorem 1 applied to the matrices Q_1 and Q_2 such that $Q_1 \cdot z = z$ and $Q_2 \cdot z = c z$. Similarly, Corollary 2 follows from Theorem 1 applied to the matrices

$$Q_1 = \begin{pmatrix} a & b \\ b & -a \end{pmatrix} \quad \text{and} \quad Q_2 = \begin{pmatrix} a & b \\ -b & a \end{pmatrix}. \tag{2}$$

Indeed, since $\Re(Q_1 \cdot z) = \Re(Q_2 \cdot z) = a \Re(z) + b \Im(z)$, the complex Brownian motion $\int H \cdot \mathrm{d}W$ provided by Theorem 1 has real part $\Re(\int H \cdot \mathrm{d}W) = aU + bV$; so $aU + bV$ is complemented by the imaginary part $\Im(\int H \cdot \mathrm{d}W)$.

Note that the complementability of U, already proved in [2], corresponds to the choice $a = 1$ and $b = 0$ in formulas (2). Thus, what remains to be proved is Theorem 1 when $Q_2 = I_2$ and $Q_1 = Q$ given by formula (1) with $(a, b) \neq (1, 0)$ and $\sigma \in \{-1, 1\}$.

Actually, the proof given below for the case $(a, b) \neq (1, 0)$ does not work any longer when $(a, b) = (1, 0)$ and $\sigma = -1$ and must then be modified. We will explain why in Sect. 3.4. Observe that the case when $(a, b) = (1, 0)$ and $\sigma = 1$ is not to be considered since it corresponds to $Q = I_2$.

The first section provides preliminary results whereas the second section is devoted to the proof.

2 Notations and Tools

In the sequel, we fix a complex-valued Brownian motion $Z = X + iY$ started from 0, and U, V, W and R are defined as above. The filtration $\mathcal{F}^Z$ generated by Z will be the ambient filtration: unless otherwise specified, martingales, Brownian motions, stopping times are always relative to $\mathcal{F}^Z$.

We fix an orthogonal matrix

$$Q = \begin{pmatrix} a & -\sigma b \\ b & \sigma a \end{pmatrix} ,$$

with a and b in $\mathbf{R}$, $a^2 + b^2 = 1$, $\sigma \in \{-1, 1\}$ and $a \neq 1$. Our aim is to construct a previsible process H with values in $\{I_2, Q\}$ such that the complex Brownian motion $\widehat{W} = \int H \cdot dW$ generates $\mathcal{F}^Z$.

Before introducing the tools for the construction of H, we recall some well-known facts (see Proposition 3.1 and Theorem 3.4 of Stroock-Yor [7]).

Proposition 1 (Classical Properties of U and V)

- *The process R is the unique and strong solution of the stochastic differential equation*

$$dR_s = dU_s + \frac{ds}{2R_s} ; \qquad R_0 = 0 .$$

  In particular, the processes R and U generate the same filtration.
- *For $t \geq s > 0$,*

$$\frac{Z_t}{R_t} = \frac{Z_s}{R_s} \exp\left(i \int_s^t \frac{dV_r}{R_r}\right) .$$

- *Hence, given $t \geq s > 0$, the value Z_t can be recovered from Z_s and from the increments of W on the time-interval $[s, t]$.*
- *For each $t > 0$, one has $\mathcal{F}_t^Z = \mathcal{F}_t^W \vee \sigma(Z_t / R_t)$, and the r.v. Z_t / R_t is independent of W, with uniform law on the unit circle.*

The fourth point of Proposition 1 describes the information missing in W to recover Z, whereas the third point shows that the loss of information occurs only at time $0+$. This is a key fact in the proof of the next lemma.

Lemma 1 *Let $(C_t)_{0 \leq t \leq 1}$ be a complex $\mathcal{F}^Z$-Brownian motion such that the r.v. Z_1 is measurable in the σ-field $\mathcal{F}_1^C$. For $t > 1$ set $C_t = C_1 + W_t - W_1$. Then $(C_t)_{t \geq 0}$ is a Brownian motion which generates the whole filtration $\mathcal{F}^Z$.*

Proof By the third point of Proposition 1, since C has the same increments as W after time 1, it suffices to show that C generates on the time-interval $[0, 1]$ the same filtration as Z. We only need to check that for every $t \in [0, 1]$, Z_t is $\mathcal{F}_t^C$-measurable.

Consider the $\mathcal{F}^C$-martingale M given by $M_t = \mathbf{E}[Z_1 \mid \mathcal{F}_t^C]$, whose final value M_1 is equal to Z_1 by our measurability hypothesis. As C is an $\mathcal{F}^Z$-Brownian motion, every $\mathcal{F}^C$-martingale is an $\mathcal{F}^Z$-martingale (by the previsible representation property); so M is on $[0, 1]$ the $\mathcal{F}^Z$-martingale $(\mathbf{E}[Z_1 \mid \mathcal{F}_s^Z])_{s \in [0,1]}$, namely, $M = Z$ on $[0, 1]$. Thus Z_t is $\mathcal{F}_t^C$-measurable for $t \in [0, 1]$.   □

Given a previsible process $(H_t)_{0<t\leq 1}$ with values in $\{I_2, Q\}$, one may apply Lemma 1 to the Brownian motion $C = \widehat{W} = \int H \cdot dW$. To prove Theorem 1, it is thus sufficient to construct a previsible process $(H_t)_{0<t\leq 1}$ with values in $\{I_2, Q\}$ such that the r.v. Z_1 can be recovered from the Brownian motion $(\widehat{W}_t)_{0\leq t\leq 1}$.

2.1 Solution of a SDE Governed by a Complex Brownian Motion C

Let $C = A + iB$ be any complex $\mathcal{F}^Z$-Brownian motion. For $t \geq s > 0$ and ζ any r.v. measurable in $\mathcal{F}_s^Z$ and valued in $\mathbf{C}^*$, we denote by $Sol(C, s, t, \zeta)$ the value at time t of the solution on the time interval $[s, +\infty[$ of the stochastic differential equation

$$dZ_t' = \frac{Z_t'}{|Z_t'|} \, dC_t \qquad \text{with initial condition } Z_s' = \zeta .$$

A priori, the solution to this SDE is only defined on some interval $[s, \gamma[$, where γ is the hitting time of 0 by Z'. But setting $dZ' = dC$ on $[\gamma, \infty[$, one has $dZ' = H \, dC$ with H previsible and $|H| = 1$; so Z' is a complex BM, and γ, the hitting time of 0 by a complex BM, must therefore be a.s. infinite. This shows that the above equation was well posed, with solution Z' some complex BM started at time s from the value ζ. We have obtained the following statement.

Lemma 2 *The process* $\big(Sol(C, s, t, \zeta)\big)_{t \geq s}$ *is a complex Brownian motion starting from* ζ. *Moreover, the random variable* $Sol(C, s, t, \zeta)$ *depends only on* ζ *and on the increments of* C *on the time interval* $[s, t]$.

According to Proposition 1, the process $Z_t' = Sol(C, s, t, z)$ can be obtained by first solving on the time interval $[s, +\infty[$ the stochastic differential equation

$$dR_t' = dA_t + \frac{dt}{2R_t'} \quad \text{with initial condition } R_s' = |\zeta| , \tag{3}$$

and by then setting, for every $t \geq s$

$$\frac{Z_t'}{R_t'} = \frac{\zeta}{|\zeta|} \exp\Big(i \int_s^t \frac{dB_r}{R_r'}\Big) . \tag{4}$$

Formulas (3) and (4) show that two solutions associated to two initial conditions which have the same modulus evolve parallelly. This statement is made formal in the next lemma, which will repeatedly be used.

Lemma 3 (Parallel Evolution) *Given* $s > 0$, *let* $C = A + iB$ *be any complex* $\mathcal{F}_s^Z$-*Brownian motion and* ζ' *and* ζ'' *two* $\mathcal{F}_s^Z$-*measurable r.v. valued in* $\mathbf{C}^*$. *Then, on the event* $\left[|\zeta'| = |\zeta''| \right]$, *one almost surely has*

$$\text{for all } t \geq s, \qquad \frac{\text{Sol}(C, s, t, \zeta'')}{\text{Sol}(C, s, t, \zeta')} = \frac{\zeta''}{\zeta'} \,.$$

2.2 Metrics on $\mathbf{C}^*$ and on $\mathbf{C}^*$-Valued Random Variables

Lemma 3 suggests that it is relevant to introduce the distance d defined on the set $\mathbf{C}^*$ by

$$d(z, z') = 2\pi \qquad\qquad \text{if } |z| \neq |z'|$$
$$d(z, z') = |\arg(z'/z)| \qquad \text{if } |z| = |z'|,$$

where $\arg(z'/z)$ is chosen in $[-\pi, \pi]$. Notice that $d(z, z') \leq \pi \Leftrightarrow |z| = |z'|$. With this notation, the next result follows immediately from Lemma 3.

Corollary 3 *Let* s, C, ζ' *and* ζ'' *be as in Lemma 3. Almost surely, one has for all* $t \geq s$

$$d\left(\text{Sol}(C, s, t, \zeta'), \ \text{Sol}(C, s, t, \zeta'')\right) \leq d(\zeta', \zeta'') \,.$$

Proof On the event $\left[|\zeta'| \neq |\zeta''| \right]$ the right-hand side is 2π, the maximal value of d; on $\left[|\zeta'| = |\zeta''| \right]$, Lemma 3 says that $d\left(\text{Sol}(C, s, t, \zeta'), \text{Sol}(C, s, t, \zeta'')\right)$ remains constant in time. □

It is not difficult to see that equality always holds in the conclusion of Corollary 3; but except in Sect. 3.4 where the case that $(a, b) = (1, 0)$ is discussed, we shall only need the majoration.

We come back to properties of d. For $\varepsilon < 2\pi$, the relation $d(z, z') \leq \varepsilon$ implies $|z - z'| \leq \varepsilon |z|$; consequently, the d-topology is finer than the usual topology.

We shall also use the corresponding distance on random variables: since the metric d is bounded, the formula $D(\zeta', \zeta'') = \mathbf{E}[d(\zeta', \zeta'')]$ defines a distance on the set of all random variables valued in $\mathbf{C}^*$ and defined up to almost sure equality. The topology associated to D is nothing but the topology of convergence in probability for the metric d, and the comparison with the usual topology easily extends to random variables.

Lemma 4 *Let $(\zeta_n)_{n \geq 0}$ and ζ be random variables valued in $\mathbf{C}^*$, defined on $(\Omega, \mathcal{A}, \mathbf{P})$. If $\zeta_n \to \zeta$ in D-distance, this convergence also holds in probability for the usual metric.*

Proof For z and z' in $\mathbf{C}^*$, if $|z| \neq |z'|$, one has $d(z, z') = 2\pi$; and if $|z| = |z'|$, one has $|z' - z| \leq |z| \, d(z, z')$. So, for all z and z' in $\mathbf{C}^*$,

$$|z' - z| \wedge 2\pi \leq (1 + |z|) d(z, z') .$$

This estimate gives

$$|\zeta_n - \zeta| \wedge 2\pi \leq (1 + |\zeta|) d(\zeta, \zeta_n).$$

Hence if $d(\zeta_n, \zeta)$ tends to zero in probability, so does $|\zeta_n - \zeta|$. $\square$

When $\varepsilon \in [0, \pi]$, we denote by $z \underset{\varepsilon}{\simeq} z'$ the relation $d(z, z') \leq \varepsilon$, which means that $z' = z e^{i\theta}$ for some $\theta \in [-\varepsilon, \varepsilon]$.

Given $\varepsilon \in \,]0, \pi]$, set $m_\varepsilon = \lceil 2\pi/\varepsilon \rceil = \min\{n \in \mathbf{N} : \pi/n \leq \varepsilon/2\}$. For every $z \in \mathbf{C}^*$, the definition

$$z^\varepsilon = |z| e^{i2k\pi/m_\varepsilon} \text{ if } z = |z| e^{i\theta} \text{ with } (2k - 1)\pi/m_\varepsilon < \theta \leq (2k + 1)\pi/m_\varepsilon$$

provides an approximation z^ε of z such that $z \underset{\varepsilon/2}{\simeq} z^\varepsilon$; the modulus of z^ε is the same as the modulus of z, and the argument of z^ε is the argument of the closest m_ε-th root of unity. When z belongs to the set

$$\Delta_\varepsilon = \bigcup_{k=0}^{m_\varepsilon - 1} \mathbf{R}_+ e^{i(2k+1)\pi/m_\varepsilon} ,$$

which consists of m_ε rays separating the roots of unity, the root of unity closest to z is not unique, and an arbitrary choice has been made. The map $z \mapsto z^\varepsilon$ is continuous on $\mathbf{C} \setminus \Delta_\varepsilon$ for the usual distance, and locally constant on $\mathbf{C} \setminus \Delta_\varepsilon$ for the distance d.

2.3 Modifying the Increments Before a Stopping Time

For every stopping time τ and every complex Brownian motion C, we define two new complex Brownian motions $C^{Q,\tau}$ and $C^{Q^{-1},\tau}$ by

$$dC^{Q,\tau} = (\mathbf{1}_{[0,\tau]} Q + \mathbf{1}_{]\tau,+\infty[} I_2) \cdot dC ;$$

$$dC^{Q^{-1},\tau} = (\mathbf{1}_{[0,\tau]} Q^{-1} + \mathbf{1}_{]\tau,+\infty[} I_2) \cdot dC .$$

Given $s > 0$ and $\varepsilon > 0$, we introduce the stopping time

$$\tau_{s,\varepsilon} = \inf\{t \geq s : \mathrm{Sol}(W, s, t, Z_s^\varepsilon) \underset{\varepsilon/2}{\simeq} \mathrm{Sol}(Q \cdot W, s, t, \sqrt{s})\}.$$

In this formula, s is a time, but $\sqrt{s}$ (considered as a complex number) is a spatial position; due to Brownian scaling invariance,

$$\tau_{s,\varepsilon} \quad \text{has the same law as} \quad s\,\tau_{1,\varepsilon}\,. \tag{5}$$

Note that the definition of $\tau_{s,\varepsilon}$ involves Z_s^ε but not Z_s; since almost surely $Z_s \notin \Delta_\varepsilon$, a small perturbation of the argument of Z_s does not change Z_s^ε, $\tau_{s,\varepsilon}$ and $W^{Q,\tau_{s,\varepsilon}}$. Moreover, parallel evolution of

$$Z = \mathrm{Sol}(W, s, \cdot, Z_s) \quad \text{and} \quad \mathrm{Sol}(W, s, \cdot, Z_s^\varepsilon) \quad \text{on } [s, +\infty[$$

and parallel evolution of

$$\mathrm{Sol}(W, s, \cdot, Z_s^\varepsilon) \quad \text{and} \quad \mathrm{Sol}(W^{Q,\tau_{s,\varepsilon}}, s, \cdot, \sqrt{s}) \quad \text{on the interval } [\tau_{s,\varepsilon}, +\infty[$$

show that

$$Z \underset{\varepsilon/2}{\simeq} \mathrm{Sol}(W, s, \cdot, Z_s^\varepsilon) \underset{\varepsilon/2}{\simeq} \mathrm{Sol}(W^{Q,\tau_{s,\varepsilon}}, s, \cdot, \sqrt{s}) \quad \text{on } [\tau_{s,\varepsilon}, +\infty[\,.$$

This establishes the next lemma.

Lemma 5 *For $s > 0$ and $\varepsilon > 0$, the process $W^{Q,\tau_{s,\varepsilon}}$ is an $\mathcal{F}^Z$-Brownian motion. Moreover,*

$$Z \underset{\varepsilon}{\simeq} \mathrm{Sol}(W^{Q,\tau_{s,\varepsilon}}, s, \cdot, \sqrt{s}) \quad \text{on the interval } [\tau_{s,\varepsilon}, +\infty[\,.$$

We will see later that $\tau_{s,\varepsilon}$ is a.s. finite; the scaling property (5) will then show that $\tau_{s,\varepsilon} \to 0$ in probability when $s \to 0$. Thus, when s is small enough, knowing the increments after s of the Brownian motion $W^{Q,\tau_{s,\varepsilon}}$ is sufficient to approach Z_t.

In order to get a Brownian motion $\widehat{W} = \int H \cdot dW$ on the time interval $[0, 1]$ such that Z_1 can be recovered from $(\widehat{W}_t)_{0 \leq t \leq 1}$, we shall concatenate pieces of Brownian motions $W^{Q,\tau_{s_n,\varepsilon_n}}$ for some decreasing sequences $(s_n)_{n \geq 0}$ and $(\varepsilon_n)_{n \geq 0}$ tending to 0.

Assume that, for some $s > 0$, we have constructed an approximation of Z_s using the Brownian motion $W^{Q,\tau_{r,\delta}}$ for some $s \in\,]0, r[$. How can we now utilize this approximation and the Brownian motion $W^{Q,\tau_{s,\varepsilon}}$ to approach Z_t for $t \geq s$? This is the role of the random maps $F_{C,s,t}^\varepsilon$ that we now introduce.

2.4 The Random Maps $F_{C,s,t}^\varepsilon$

Let C be any $\mathcal{F}^Z$-complex Brownian motion. For $s > 0$, $\varepsilon > 0$ and ζ a $\mathbf{C}^*$-valued, $\mathcal{F}_s^Z$-measurable r.v., we define the stopping time

$$\tau_{C,s,\zeta}^\varepsilon = \inf\{t \geq s : \mathrm{Sol}(Q^{-1} \cdot C, s, t, \zeta^\varepsilon) \underset{\varepsilon/2}{\simeq} \mathrm{Sol}(C, s, t, \sqrt{s})\},$$

and for $t \geq s$ we set

$$F_{C,s,t}^\varepsilon(\zeta) = F^\varepsilon(C, s, t, \zeta) = \mathrm{Sol}(C^{Q^{-1}, \tau_{C,s,\zeta}^\varepsilon}, s, t, \zeta).$$

In particular, taking ζ a constant r.v. z, we get an almost surely well defined random map $z \mapsto F_{C,s,t}^\varepsilon(z)$ from $\mathbf{C}^*$ to $\mathbf{C}^*$, and $F_{C,s,t}^\varepsilon(\zeta)$ can be identified with the compound $F_{C,s,t}^\varepsilon \circ \zeta$.

As a consequence of Lemma 2,

$$F_{C,s,t}^\varepsilon(\zeta) \text{ is measurable for } \sigma\big((C_r - C_s)_{r \in [s,t]}, \zeta\big); \tag{6}$$

this will be useful later.

When $C = W^{Q, \tau_{s,\varepsilon}}$, the Brownian motions C and $Q \cdot W$ coincide up to time $\tau_{s,\varepsilon}$, so $\tau_{C,s,Z_s}^\varepsilon = \tau_{s,\varepsilon}$ and $C^{Q^{-1}, \tau_{C,s,Z_s}^\varepsilon} = W$. Thus

$$F^\varepsilon(W^{Q, \tau_{s,\varepsilon}}, s, t, Z_s) = \mathrm{Sol}(W, s, t, Z_s) = Z_t. \tag{7}$$

It will also be useful to know that if ζ is a complex random variable close to Z_s, the random variable $F^\varepsilon(W^{Q, \tau_{s,\varepsilon}}, s, t, \zeta)$ is close to Z_t. We now establish this continuity property of the map $F_{W^{Q, \tau_{s,\varepsilon}}}^\varepsilon$.

Lemma 6 Fix $\varepsilon > 0$, $s > 0$ and C a complex Brownian motion in $\mathcal{F}^Z$; let also ζ_0 be any complex, $\mathcal{F}_s^Z$-measurable r.v such that $\mathbf{P}[\zeta_0 \in \Delta_\varepsilon] = 0$. The map $\zeta \mapsto F_{C,s,t}^\varepsilon(\zeta)$, defined on the set of all $\mathbf{C}^*$-valued, $\mathcal{F}_s^Z$-measurable random variables, is continuous at ζ_0 in the D-distance. In other terms, $D\big(F_{C,s,t}^\varepsilon(\zeta), F_{C,s,t}^\varepsilon(\zeta_0)\big) \to 0$ when $D(\zeta, \zeta_0) \to 0$.

Proof On the event $[d(\zeta, \zeta_0) < d(\zeta_0, \Delta_\varepsilon)]$, one has $d(\zeta, \zeta_0) < 2\pi$, whence $|\zeta| = |\zeta_0|$; moreover, ζ and ζ_0 are in the same connected component of Δ_ε^c, so $\zeta_0^\varepsilon = \zeta^\varepsilon$. This implies $\tau_{C,s,\zeta}^\varepsilon = \tau_{C,s,\zeta_0}^\varepsilon$, and consequently also $C^{Q, \tau_{C,s,\zeta}^\varepsilon} = C^{Q, \tau_{C,s,\zeta_0}^\varepsilon}$; by Corollary 3, $d\big(F_{C,s,t}^\varepsilon(\zeta), F_{C,s,t}^\varepsilon(\zeta_0)\big) \leq d(\zeta, \zeta_0)$.

On the complementary event $[d(\zeta, \zeta_0) \geq d(\zeta_0, \Delta_\varepsilon)]$, we simply majorize $d\big(F_{C,s,t}^\varepsilon(\zeta), F_{C,s,t}^\varepsilon(\zeta_0)\big)$ by 2π.

All in all,

$$d\big(F_{C,s,t}^\varepsilon(\zeta), F_{C,s,t}^\varepsilon(\zeta_0)\big) \leq d(\zeta, \zeta_0) + 2\pi \, \mathbf{1}_{d(\zeta,\zeta_0) \geq d(\zeta_0, \Delta_\varepsilon)};$$

and taking expectations gives

$$D\big(F^\varepsilon_{C,s,t}(\zeta), F^\varepsilon_{C,s,t}(\zeta_0)\big) \le \mathbf{E}[d(\zeta, \zeta_0)] + 2\pi\, \mathbf{P}[d(\zeta, \zeta_0) \ge d(\zeta_0, \Delta_\varepsilon)]\;.$$

By hypothesis, the r.v. $d(\zeta_0, \Delta_\varepsilon)$ is a.s. > 0; so when $d(\zeta, \zeta_0)$ tends to zero in probability, the right-hand side tends to zero. $\qquad\square$

3 Construction of $\widehat{W} = \int H \cdot dW$ on the Time Interval $[0, 1]$

The construction will rest on the continuity properties of the maps $F^\varepsilon_{C,s,t}$ (Lemma 6) and on the next statement, which will be admitted in Sect. 3.1, and proved in Sect. 3.2 as a consequence of the convergence $P[\tau_{s,\varepsilon} \le t] = P[\tau_{1,\varepsilon} \le t/s] \to 1$ as $s \to 0$.

Lemma 7 *For all $t \ge s > 0$ and $\varepsilon > 0$,*

$$F^\varepsilon_{W^{Q,\tau_{s,\varepsilon}},s,t}(\sqrt{s}) = \mathrm{Sol}(W^{Q,\tau_{s,\varepsilon}}, s, t, \sqrt{s}).$$

Moreover, given $t > 0$ and $\varepsilon > 0$, $D\big(Z_t, F^\varepsilon_{W^{Q,\tau_{s,\varepsilon}},s,t}(\sqrt{s})\big) < 2\varepsilon$ provided s is small enough.

3.1 The Construction

First, we recursively define two decreasing sequences $(s_n)_{n\ge0}$ and $(\varepsilon_n)_{n\ge0}$ such that for every $n \ge 1$, $D\big(Z_1, G_n(\sqrt{s_n})\big) \le 1/n$, with

$$G_n = F^{\varepsilon_1}_{W^{Q,\tau_{s_1,\varepsilon_1}},s_1,s_0} \circ \cdots \circ F^{\varepsilon_n}_{W^{Q,\tau_{s_n,\varepsilon_n}},s_n,s_{n-1}}.$$

The construction begins with $s_0 = 1$ and $\varepsilon_0 = 1$.

Assume that, for some $n \ge 1$, $s_0 > \ldots > s_{n-1} > 0$ and $\varepsilon_0 > \ldots > \varepsilon_{n-1} > 0$ have already been constructed. Property (7) says that $F^\varepsilon_{W^{Q,\tau_{s,\varepsilon}},s,t}(Z_s) = Z_t$ for $t \ge s$, so $G_{n-1}(Z_{s_{n-1}}) = Z_{s_0} = Z_1$; and Lemma 6 says that the map $\zeta \mapsto G_{n-1}(\zeta)$ is D-continuous at $Z_{s_{n-1}}$. This provides some $\varepsilon_n \in \,]0, \varepsilon_{n-1}[$ such that, for each ζ measurable in $\mathcal{F}^Z_{s_{n-1}}$,

$$D(Z_{s_{n-1}}, \zeta) \le 2\varepsilon_n \quad\Longrightarrow\quad D\big(Z_1, G_{n-1}(\zeta)\big) \le 1/n\;. \tag{8}$$

Then Lemma 7 (provisionally admitted) provides some $s_n \in \,]0, s_{n-1}/2[$ such that

$$D\big(Z_{s_{n-1}}, F^{\varepsilon_n}_{W^{Q,\tau_{s_n,\varepsilon_n}},s_n,s_{n-1}}(\sqrt{s_n})\big) < 2\varepsilon_n\;,$$

and taking $\zeta = F^{\varepsilon_n}_{W^{Q,\tau_{s_n,\varepsilon_n}},s_n,s_{n-1}}(\sqrt{s_n})$ in (8) yields $D\big(Z_1, G_n(\sqrt{s_n})\big) \le 1/n$, which completes the recursion.

Let now H be the process defined on the interval $]0, 1]$ by

$$H_s = Q\mathbf{1}_{[s \le \tau_{s_n,\varepsilon_n}]} + I_2\mathbf{1}_{[s > \tau_{s_n,\varepsilon_n}]} \quad \text{on the interval} \quad]s_n, s_{n-1}] .$$

As H takes its values in $\{I_2, Q\}$, the process $\widehat{W} = \int H \cdot dW$ is a complex Brownian motion. By construction, $\widehat{W}$ has the same increments as $W^{Q,\tau_{s_n,\varepsilon_n}}$ on the interval $]s_n, s_{n-1}]$. By Lemma 1, we only have to show that the r.v. Z_1 can be recovered from the observation of $\widehat{W}$ on $[0, 1]$.

Define a family $(\zeta_{n,k})_{1 \le k \le n}$ of **C**-valued random variables by $\zeta_{n,n} = \sqrt{s_n}$ and, for k decreasing from n to 1,

$$\zeta_{n,k-1} = F^{\varepsilon_k}_{\widehat{W},s_k,s_{k-1}}(\zeta_{n,k}) = F^{\varepsilon_k}_{W^{Q,\tau_{s_k,\varepsilon_k}},s_k,s_{k-1}}(\zeta_{n,k}) .$$

Property (6) shows by recursion that $\zeta_{n,k}$ is $\mathcal{F}^{\widehat{W}}_{s_k}$-measurable; in particular $\zeta_{n,1}$ is $\mathcal{F}^{\widehat{W}}_1$-measurable. By definition of G_n, one has $\zeta_{n,1} = G_n(\sqrt{s_n})$, wherefrom $D(Z_1, \zeta_{n,1}) \le 1/n$. By Lemma 4, $\zeta_{n,1} \to Z_1$ in probability, and Z_1 too is $\mathcal{F}^{\widehat{W}}_1$-measurable. $\qquad\qquad\square$

3.2 Proof of Lemma 7

The last missing step in the proof is Lemma 7, to be proved now. A key point in this proof is the following coupling property.

Lemma 8 *The orthogonal matrix*

$$Q = \begin{pmatrix} a & -\sigma b \\ b & \sigma a \end{pmatrix}$$

*is still fixed, with a and b in **R**, $a^2 + b^2 = 1$ and $\sigma \in \{-1, 1\}$. Assume that $(a, b) \ne (1, 0)$. Then, given $\varepsilon > 0$, the stopping time*

$$\tau_{1,\varepsilon} = \inf\{t \ge 1 : \mathrm{Sol}(W, 1, t, Z^\varepsilon_1) \underset{\varepsilon/2}{\simeq} \mathrm{Sol}(Q \cdot W, 1, t, 1)\}$$

is almost surely finite.

We stress the fact that the conclusion turns out to be false if one takes $(a, b) = (1, 0)$. This case will be discussed in Sect. 3.4.

Proof of Lemma 7, assuming Lemma 8 Fix $t \ge s > 0$, and let C be any complex $\mathcal{F}^Z$-Brownian motion. Since the stopping time $\tau^\varepsilon_{C,s,\zeta}$ defined in Sect. 2.4 is equal to s when $\zeta = \sqrt{s}$, one has $C^{Q^{-1} \cdot \tau^\varepsilon_{C,s} \cdot \sqrt{s}} = C$ and $F^\varepsilon_{C,s,t}(\sqrt{s}) = \mathrm{Sol}(C, s, t, \sqrt{s})$. Taking in particular $C = W^{Q,\tau_{s,\varepsilon}}$ yields the first part of Lemma 7.

By Lemma 5, $Z_t \underset{\varepsilon}{\simeq} \mathrm{Sol}(W^{Q,\tau_{s,\varepsilon}}, s, t, \sqrt{s})$ on the event $[\tau_{s,\varepsilon} \le t]$, so

$$D\big(Z_t, \mathrm{Sol}(W^{Q,\tau_{s,\varepsilon}}, s, t, \sqrt{s})\big) \le \varepsilon\, P[\tau_{s,\varepsilon} \le t] + 2\pi\, P[\tau_{s,\varepsilon} > t]$$

$$\le \varepsilon + 2\pi\, P[\tau_{s,\varepsilon} > t]\,.$$

Now, by Lemma 8 and by the scaling property (5),

$$P[\tau_{s,\varepsilon} > t] = P[\tau_{1,\varepsilon} > t/s] \to 0 \quad \text{as} \quad s \to 0\,;$$

hence $\limsup\limits_{s \to 0} D\big(Z_t, \mathrm{Sol}(W^{Q,\tau_{s,\varepsilon}}, s, t, \sqrt{s})\big) \le \varepsilon.$ $\qquad\qquad\qquad\square$

3.3 Proof of Lemma 8

By parallel evolution, for every $t \ge 1$, $\mathrm{Sol}(W, 1, t, Z_1^\varepsilon) = (Z_1^\varepsilon/Z_1) \times Z_t$. Set $Z'_t = \mathrm{Sol}(Q \cdot W, 1, t, Z_1/Z_1^\varepsilon) = (Z_1/Z_1^\varepsilon) \times \mathrm{Sol}(Q \cdot W, 1, t, 1)$. Then

$$\tau_{1,\varepsilon} = \inf\{t \ge 1 : Z_t \underset{\varepsilon/2}{\simeq} Z'_t)\}.$$

For $t \ge 1$, set $Z_t = R_t\, e^{i\Theta_t}$ and $Z'_t = R'_t\, e^{i\Theta'_t}$, where Θ and Θ' are continuous and $\mathcal{F}^Z$-adapted determinations of the arguments of Z and Z' on the time-interval $[1, +\infty[$, and put

$$L_t = \ln R_t - \ln R'_t\,, \qquad M_t = \Theta_t - \Theta'_t\,.$$

With this notation, $\tau_{1,\varepsilon}$ becomes

$$\tau_{1,\varepsilon} = \inf\{t \ge 1 : L_t = 0 \text{ and } M_t \in [-\varepsilon/2, \varepsilon/2] + 2\pi \mathbf{Z}\}\,.$$

Set $c = a + ib$. By assumption, c has modulus 1 and $c \ne 1$. Two cases will be distinguished according to the value of σ.

1. Case Where $\sigma = 1$

If $\sigma = 1$, then $Q \cdot W = cW$, and stochastic calculus yields

$$d(L_t + iM_t) = \frac{dZ_t}{Z_t} - \frac{dZ'_t}{Z'_t} = \left(\frac{1}{R_t} - \frac{c}{R'_t}\right) dW_t\,;$$

so $L+\mathrm{i}M$ is a conformal local martingale. Now, $\left|(1/R_t)-(c/R'_t)\right|$ can be minorated by the imaginary part $|b|/R'_t$; or, if $b=0$ and $c=-1$, by $1/R'_t$. This minoration entails that

$$\langle L\rangle_t = \langle M\rangle_t = \int_1^t \left|\frac{1}{R_s}-\frac{c}{R'_s}\right|^2 \mathrm{d}s \to +\infty \qquad \text{as} \quad t\to +\infty ,$$

since the time spent by R' below any level is infinite. Consequently, $L+\mathrm{i}M$ is a time-changed complex Brownian motion, and it almost surely visits the set $\mathrm{i}\bigl([-\varepsilon/2,\varepsilon/2]+2\pi\mathbf{Z}\bigr)$; so $\tau_{1,\varepsilon}$ is finite.

2. Case Where $\sigma = -1$

If $\sigma=-1$, then $Q\cdot W=c\overline{W}$, and one has

$$\mathrm{d}(L_t+\mathrm{i}M_t) = \frac{\mathrm{d}Z_t}{Z_t}-\frac{\mathrm{d}Z'_t}{Z'_t} = \frac{\mathrm{d}W_t}{R_t}-\frac{c\,\mathrm{d}\overline{W}_t}{R'_t} .$$

We can define a real local martingale $(N_t)_{t\ge 1}$ by $N_1=0$ and

$$\mathrm{i}\,\mathrm{d}(N_t-M_t) = \frac{1}{R'_t}\left(c\,\mathrm{d}\overline{W}_t-\overline{c}\,\mathrm{d}W_t\right) .$$

Observe that

$$\mathrm{d}(L_t+\mathrm{i}N_t) = \left(\frac{1}{R_t}-\frac{\overline{c}}{R'_t}\right)\mathrm{d}W_t = \frac{R'_t-\overline{c}R_t}{R_t R'_t}\,\mathrm{d}W_t ,$$

so $L+\mathrm{i}N$ is a conformal local martingale and

$$\frac{c\,\mathrm{d}\overline{W}_t}{R'_t} = \frac{cR_t}{R'_t-cR_t}\,\mathrm{d}(L_t-\mathrm{i}N_t) .$$

But using $c=a+\mathrm{i}b,\, c\overline{c}=1$ and $R_t/R'_t=\mathrm{e}^{L_t}$, one gets

$$\frac{cR_t}{R'_t-cR_t} = \frac{cR_t(R'_t-\overline{c}R_t)}{R'^2_t-2aR_tR'_t+R^2_t} = \frac{c-\mathrm{e}^{L_t}}{\mathrm{e}^{-L_t}-2a+\mathrm{e}^{L_t}} = \frac{c-\mathrm{e}^{L_t}}{2(\cosh(L_t)-a)} .$$

Hence

$$\mathrm{d}(N_t-M_t) = 2\Im\left(\frac{c\,\mathrm{d}\overline{W}_t}{R'_t}\right) = \frac{b\,\mathrm{d}L_t-a\,\mathrm{d}N_t+\mathrm{e}^{L_t}\,\mathrm{d}N_t}{\cosh(L_t)-a} ,$$

or equivalently,

$$dM_t = \frac{-b\,dL_t - \sinh(L_t)\,dN_t}{\cosh(L_t) - a}.$$ (9)

Moreover, $\langle L, N \rangle_t = 0$ because $L + iN$ is conformal, and

$$\langle L \rangle_t = \langle N \rangle_t = \int_1^t \left| \frac{1}{R_s} - \frac{\bar{c}}{R'_s} \right|^2 ds.$$

As in the first case, the continuous, strictly increasing process $\langle L \rangle = \langle N \rangle$ starts from 0 at time 1 and goes to $+\infty$. Hence, on the time-interval $[1, +\infty[$, the local martingale (L, N) is a time-changed two-dimensional Brownian motion.

Call $(\alpha_s)_{s \geq 0}$ the inverse of the process $\langle L \rangle = \langle N \rangle$, and set

$$\lambda_s = L_{\alpha_s}, \ \mu_s = M_{\alpha_s} \text{ and } \nu_s = N_{\alpha_s}.$$

Then the processes λ and ν are independent one-dimensional Brownian motions and (9) becomes

$$d\mu_s = -\frac{b}{\cosh(\lambda_s) - a}\,d\lambda_s - \frac{\sinh(\lambda_s)}{\cosh(\lambda_s) - a}\,d\nu_s.$$

To prove that $\tau_{1,\varepsilon}$ is (almost surely) finite, we have to show that the process (λ, μ) almost surely visits the set $\{0\} \times ([-\varepsilon/2, \varepsilon/2] + 2\pi\mathbf{Z})$.

Changing time again, call ϱ the right-continuous inverse of the local time of λ at 0. Observing that $\lambda_{\varrho_\ell} = 0$, it now suffices to verify that the process $(\mu_{\varrho_\ell})_{\ell \geq 0}$ visits $[-\varepsilon/2, \varepsilon/2] + 2\pi\mathbf{Z}$.

For every $\ell \geq 0$, set

$$S_\ell = \mu_0 - \mu_{\varrho_\ell} = \int_0^{\varrho_\ell} \frac{b}{\cosh(\lambda_s) - a}\,d\lambda_s + \int_0^{\varrho_\ell} \frac{\sinh(\lambda_s)}{\cosh(\lambda_s) - a}\,d\nu_s;$$

The Markov property of Brownian motion and the independence of λ and ν imply that, conditionally on λ, S is a Gaussian process with independent increments, and the variance of S_ℓ is

$$\int_0^{\varrho_\ell} \left(\frac{\sinh(\lambda_s)}{\cosh(\lambda_s) - a} \right)^2 ds.$$

This variance tends to infinity with ℓ because the time spent by λ outside $[-1, 1]$ is infinite. Hence, the conclusion follows from the next lemma. □

Lemma 9 Let $(S_t)_{t \geq 0}$ be a Gaussian process with independent increments, such that $\mathrm{Var}(S_t) \to +\infty$ as $t \to +\infty$. Then almost surely, for every interval I with positive length, $(S_t)_{t \geq 0}$ visits $I + 2\pi\mathbf{Z}$.

Proof First, notice that if G is a Gaussian random variable with law $\mathcal{N}(m, v)$, then the law of $G - m$ is $\mathcal{N}(0, v)$; hence, calling g_v the density of $\mathcal{N}(0, v)$, one has, for every interval I with length $\leq 2\pi$,

$$P[G \in I + 2\pi\mathbf{Z}] = \sum_{k \in \mathbf{Z}} P[G \in I + 2\pi k] = \int_{I-m} \left(\sum_{k \in \mathbf{Z}} g_v(x - k2\pi) \right) dx .$$

Now, the Poisson summation formula yields, for every $x \in \mathbf{R}$,

$$\sum_{k \in \mathbf{Z}} g_v(x - k2\pi) = \frac{1}{2\pi} \sum_{n \in \mathbf{Z}} e^{-vn^2/2} \, e^{inx} .$$

If $v \geq 2$,

$$2\pi \sum_{k \in \mathbf{Z}} g_v(x - k2\pi) \geq 1 - 2\sum_{n \geq 1} e^{-vn^2/2} \geq 1 - 2\sum_{n \geq 1} e^{-n^2} \geq \frac{1}{5} ,$$

hence for every interval I with length $\leq 2\pi$,

$$P[G \in I + 2\pi\mathbf{Z}] \geq \frac{1}{10\pi}|I|.$$

For every $t \geq 0$, set $V(t) = \mathrm{Var}(S_t)$. Since $V(t) \to +\infty$ as $t \to +\infty$, one can recursively construct an increasing sequence $(t_n)_{n \geq 1}$ such that $V(t_1) \geq 2$ and $V(t_n) \geq V(t_{n-1}) + 2$ for every $n \geq 2$. The random variables S_{t_1} and $(S_{t_n} - S_{t_{n-1}})_{n \geq 2}$ are independent, Gaussian with variance ≥ 2, so for every $n \geq 1$,

$$P\Big[S_{t_n} \in I + 2\pi\mathbf{Z} \,\big|\, \sigma(S_{t_1}, \ldots, S_{t_{n-1}}) \Big] \geq \frac{|I|}{10\pi} ;$$

and by recursion,

$$P\Big[\forall k \in [1 \ldots n], \ S_{t_k} \notin I + 2\pi\mathbf{Z} \Big] \leq \left(1 - \frac{|I|}{10\pi} \right)^n .$$

Hence

$$P\Big[\forall k \geq 1, \ S_{t_k} \notin I + 2\pi\mathbf{Z}\Big] = 0 .$$

This provides the results for a given interval I. Letting the bounds of the interval vary in $\mathbf{Q}$ yields the result. $\square$

3.4   The Case When $(a, b) = (1, 0)$

Assume now that $(a, b) = (1, 0)$. First, we explain why the conclusion of Lemma 8 does not hold any longer.

The case when $\sigma = 1$ is not interesting in view of Theorem 1 since it corresponds to the choice $Q = I_2$. In that case, both processes $\mathrm{Sol}(W, 1, t, Z_1^\varepsilon)$ and $\mathrm{Sol}(Q \cdot W, 1, t, 1)$ evolve parallelly, and by the remark after Corollary 3, the stopping time $\tau_{1,\varepsilon}$ is finite only on the null event $[Z_1 \underset{\varepsilon/2}{\simeq} 1]$.

The case when $\sigma = -1$ is much more interesting since it corresponds to the complementability of U. In that case, Q is diagonal with diagonal $(1, -1)$, and $Q \cdot W = \overline{W} = U - iV$. Thus the processes R and R', introduced in the proof of Lemma 8, are solutions of stochastic differential equations governed by the same Brownian motion U, namely,

$$\mathrm{d}R_s = \mathrm{d}U_s + \frac{\mathrm{d}s}{2R_s} \qquad \text{and} \qquad \mathrm{d}R'_s = \mathrm{d}U_s + \frac{\mathrm{d}s}{2R'_s}$$

with initial conditions $R_1 = |Z_1|$ and $R'_1 = 1$. Almost surely, $|Z_1| \neq 1$; hence, by noticing that the quantity

$$\ln|R_t - R'_t| = \ln|R_1 - R'_1| - \int_1^t \frac{\mathrm{d}s}{2R_s R'_s}$$

does not explode in finite time, or by applying classical results on flows of solutions of SDEs, one sees that the processes R and R' never meet, so the stopping time $\tau_{1,\varepsilon}$ is infinite.

Still, the proof of Theorem 1 can be adapted to the case when $(a, b) = (1, 0)$ and $\sigma = -1$; it even becomes simpler, and, in fact, reduces to the proof given in [2]. Here are the changes to be made in the preceding proof.

First, the stopping times $\tau_{s,\varepsilon}$ must be replaced with the stopping times

$$\tau_s = \inf\{t \geq s : \mathrm{Sol}(W, s, t, Z_s) = \mathrm{Sol}(Q \cdot W, s, t, R_s)\}.$$

Note the modifications in the definition:

- the $\varepsilon/2$-almost equality becomes a true equality;
- the initial position Z_s^ε in $\mathrm{Sol}(W, s, t, Z_s^\varepsilon)$ is replaced by Z_s;
- the initial position $\sqrt{s}$ in $\mathrm{Sol}(Q \cdot W, s, t, \sqrt{s})$ is replaced by R_s.

Choosing R_s as the initial position will not cause any difficulty since for every previsible process H with values in $\{I_2, Q\}$, the real part of the Brownian motion $\widehat{W} = \int H \cdot \mathrm{d}W$ is U, so the process R is adapted to $\mathcal{F}^{\widehat{W}}$.

Almost sure finiteness is much more simply proved for τ_1 than for $\tau_{1,\varepsilon}$. Indeed, setting $Z_t = \mathrm{Sol}(W, 1, t, Z_1) = R_t \mathrm{e}^{\mathrm{i}\Theta_t}$ for $t \geq 1$ as before, one gets $\mathrm{Sol}(Q \cdot W, 1, t, R_1) = R_t \mathrm{e}^{\mathrm{i}(\Theta_0 - \Theta_t)}$, so

$$\tau_1 = \inf\{t \geq 1 : 2\Theta_t \in \Theta_0 + 2\pi \mathbf{Z}\}.$$

As Θ is a time-changed Brownian motion, τ_1 is almost surely finite.

Then we only need to choose any decreasing sequence $(s_n)_{n \geq 0}$ of positive real numbers such that $s_n / s_{n-1} \to 0$ as $n \to \infty$ (this ensures that $P[\tau_{s_n} \leq s_{n-1}] = P[\tau_1 \leq s_{n-1}/s_n] \to 1$), and to define the Brownian motion $\widehat{W} = \int H \cdot \mathrm{d}W$ on the interval $[0, 1]$ as before. In the construction proposed in Sect. 3.1, the starting value $\zeta_{n,n}$ of the sequence $(\zeta_{n,k})_{k \in \{1,\dots,n\}}$ also has to be modified; a possible choice is $\zeta_{n,n} = R_{s_n}$.

Acknowledgements The second author gratefully acknowledges the support of the ANR programme ProbaGeo.

References

1. J. Brossard, M. Émery, C. Leuridan, Maximal Brownian motions. Ann. de l'Institut Henri Poincaré Probab. Stat. **45**(3), 876–886 (2009)
2. J. Brossard, C. Leuridan, Filtrations browniennes et compléments indépendants. Séminaire de Probabilités XLI. Lect. Notes Math. **1934**, 265–278 (2008). Springer
3. L. Dubins, J. Feldman, M. Smorodinsky, B. Tsirelson, Decreasing sequences of σ-fields and a measure change for Brownian motion. Ann. Probab. **24**(2), 882–904 (1996)
4. M. Émery, On certain almost Brownian filtrations. Ann. de l'Institut Henri Poincaré Probab. Stat. **41**(3), 285–305 (2005)
5. M. Malric, Filtrations quotients de la filtration brownienne. Séminaire de Probabilités XXXV. Lect. Notes Math. **1755**, 260–264 (2001). Springer
6. D. Stroock, M. Yor, On extremal solutions of martingale problems. Ann. Scientifiques de l'École Normale Supérieure **13**(1), 95–164 (1980)
7. D.W. Stroock, M. Yor, Some remarkable martingales. Séminaire de Probabilités XV. Lect. Notes Math. **850**, 590–603 (1981). Springer
8. B. Tsirelson, Triple points: from non-Brownian filtrations to harmonic measures. Geomet. Funct. Anal. **7**(6), 1096–1142 (1997)

On the Exactness of the Lévy-Transformation

Vilmos Prokaj

Abstract In a recent paper we gave a sufficient condition for the strong mixing property of the Lévy-transformation. In this note we show that it actually implies a much stronger property, namely exactness.

1 Introduction

Our aim in this short note, to supplement the result of [1]. In that work we obtained a condition which implies the strong mixing property, hence the ergodicity of the Lévy-transformation. We reformulate this condition, see (3) below, and show that it actually implies a stronger property called exactness. That is, we deduce that the tail σ-algebra of the Lévy transformation is trivial provided that condition (3) holds.

2 Summary of the Results of [1]

First, we fix some notations. $\mathbf{W} = C[0, \infty)$ is the space of continuous function defined on $[0, \infty)$, $\mathbf{P}$ is the Wiener measure on the Borel σ-field of $\mathbf{W}$, and β is the canonical process on $\mathbf{W}$. Finally T is a $\mathbf{P}$ almost everywhere defined transformation of $\mathbf{W}$ defined by the formula

$$(T\beta) = \int h(s, \beta)d\beta_s \tag{1}$$

where h is a progressively measurable function on $[0, \infty) \times \mathbf{W}$ taking values in $\{-1, 1\}$. We use the notation $\beta^{(n)}$ for $T^n\beta$ and $(\mathcal{F}_t^{(n)})_{t \geq 0}$ for the filtration generated by $\beta^{(n)}$ and $h_s^{(n)} = \prod_{k=0}^{n-1} h(s, \beta^{(k)})$.

V. Prokaj (✉)
Department of Probability Theory and Statistics, Eötvös Loránd University, 1117 Budapest, Pázmány P. sétány 1/C, Hungary
e-mail: prokaj@cs.elte.hu

© Springer International Publishing Switzerland 2014

395

C. Donati-Martin et al. (eds.), *Séminaire de Probabilités XLVI*, Lecture Notes in Mathematics 2123, DOI 10.1007/978-3-319-11970-0_16

The transformation T is called *exact*, whenever $\bigcap_n \mathcal{F}_\infty^{(n)}$ is trivial.

The Lévy transformation is obtained by the choice $h(s, \beta) = \text{sign}(\beta_s)$ and denoted by **T**. The rest of this section is devoted to this special case.

The main observation of [1] was that the existence of certain stopping times makes it possible to estimate the covariance of $h_s^{(n)}$ and $h_1^{(n)}$, which is the key to prove the strong mixing property of **T**. More precisely, for $r \in (0, 1)$ and $C > 0$ let

$$\tau_{r,C} = \inf \left\{ s > r : \exists n, \ \beta_s^{(n)} = 0, \ \min_{0 \le k < n} |\beta_s^{(k)}| > C \sqrt{(1-s)_+} \right\}.$$

That is $\tau_{r,C}$ is the first time after r when for some n the first n iterated paths are relatively far away from the origin while $\beta^{(n)}$ is zero.

Then it was proved that

$$\limsup_{n \to \infty} \left| \mathbf{E} \left(h_r^{(n)} h_1^{(n)} \right) \right| \le \mathbf{P}(\tau_{r,C} = 1) + \mathbf{P} \left(\sup_{0 \le s \le 1} |\beta_s| > C \right). \tag{2}$$

It was stated without the first term on the right, under the assumption that this term is zero. The proof of this inequality used the coupling of the shadow path $\tilde{\beta}$, reflected after $\tau_{r,C}$ and the original path β. This argument actually yields the following form of (2)

$$\lim_{n \to \infty} \left| \mathbf{E} \left(h_1^{(n)} \mid \mathcal{F}_1^{(n)} \vee \mathcal{F}_r^{(0)} \right) \right| \le \mathbf{P}(\tau_{r,C} = 1) + \mathbf{P} \left(\sup_{0 \le s \le 1} |\beta_s| > C \right). \tag{3}$$

Note that the limit on the left hand side exists as $\left| \mathbf{E} \left(h_1^{(n)} \mid \mathcal{F}_1^{(n)} \vee \mathcal{F}_r^{(0)} \right) \right|$ is a reversed submartingale.

By virtue of the estimates in (2) and (3) a sufficient condition for the strong mixing of the Lévy transformation is that

$$\tau_{r,C} < 1, \quad \text{almost surely, for all } r \in (0, 1), C > 0. \tag{4}$$

The main result of this paper is the following theorem.

Theorem 1 *If* (4) *holds then the Lévy transformation is exact.*

The proof is based on the estimate (3) and is given in the next section where we do not assume the special form of the Lévy transformation. That is, we prove the next statement from which Theorem 1 follows.

Proposition 1 *Let T be the transformation of the Wiener-space as in* (1). *If*

$$\lim_{n \to \infty} \mathbf{E} \left(h_t^{(n)} \mid \mathcal{F}_{rt}^{(0)} \vee \mathcal{F}_t^{(n)} \right) = 0, \quad \text{for almost all } t > 0 \text{ and } r \in [0, 1) \tag{5}$$

then T is exact.

3 Proof of Proposition 1

For a deterministic function $f \in L^2([0, \infty))$ we will use the notation $\mathcal{E}(f)$ for

$$\mathcal{E}(f) = \exp \int_0^\infty f(s) d\beta_s - \frac{1}{2} \int_0^\infty f^2(s) ds.$$

Since the linear hull of the set of $\{\mathcal{E}(f) : f \in L^2([0, \infty))\}$ is dense in $L^2(\mathbf{W})$ the following statement is obvious.

Proposition 2 $\bigcap_n \mathcal{F}_\infty^{(n)}$ *is trivial if and only if* $\mathbf{E}\left(\mathcal{E}(f) \mid \mathcal{F}_\infty^{(n)}\right) \to 1$ *for each* $f \in L^2([0, \infty))$.

To express $\mathbf{E}\left(\mathcal{E}(f) \mid \mathcal{F}_\infty^{(n)}\right)$ we use the next proposition.

Lemma 1 *Assume that* ξ *is a measurable and* $\mathcal{F}^{(0)}$-*adapted process satisfying* $\mathbf{E}\left(\int_0^\infty \xi_s^2 ds\right) < \infty$. *Then*

$$\mathbf{E}\left(\int_0^\infty \xi_s d\beta_s^{(0)} \mid \mathcal{F}_\infty^{(n)}\right) = \int_0^\infty mathbf{E}\left(\xi_s h_s^{(n)} \mid \mathcal{F}_s^{(n)}\right) d\beta_s^{(n)}$$

Proof First observe that both sides of the equation makes sense.

Denote by V the left hand side of the equation and by V' the right hand side. Besides let $U \in L^2(\mathcal{F}_\infty^{(n)})$ and write it, using that $\mathcal{F}^{(n)}$ is generated by the Brownian motion $\beta^{(n)}$, as $U = c + \int_0^\infty u_s d\beta_s^{(n)}$ with some $c \in \mathbb{R}$ and $\mathcal{F}^{(n)}$-predictable u. Then

$$\mathbf{E}(UV) = \mathbf{E}\left(\int_0^\infty \xi_s h_s^{(n)} u_s ds\right)$$

$$= \mathbf{E}\left(\int_0^\infty \mathbf{E}\left(\xi_s h_s^{(n)} \mid \mathcal{F}_s^{(n)}\right) u_s ds\right) = \mathbf{E}(UV').$$

This proves that $V = V'$ which is the claim.

In the proof of the next statement we call a probability measure $Q \sim \mathbf{P}$ simple when it is in the form $dQ = \mathcal{E}(f) d\mathbf{P}$ with some $f \in L^2([0, \infty))$.

Proposition 3 $\bigcap_n \mathcal{F}_\infty^{(n)}$ *is trivial if and only if for all* $Q \sim \mathbf{P}$

$$\mathbf{P}_Q\left(h_s^{(n)} \mid \mathcal{F}_s^{(n)}\right) \to 0, \quad \mathbf{P}\text{-almost surely, for almost all } s > 0. \tag{6}$$

Proof In the proof we mostly work with simple equivalent measures, and obtain the conclusion of the "only if" part by approximation.

First we get a formula for $\mathbf{E}\left(\mathcal{E}(f) \mid \mathcal{F}_\infty^{(n)}\right)$ when $f \in L^2([0, \infty))$ and then we apply Proposition 2.

So for the simple equivalent measure $dQ = \mathcal{E}(f)d\mathbf{P}$, let the density process be denoted by $Z_t = \mathbf{E}\left(\mathcal{E}(f) \mid \mathcal{F}_t\right)$. Then $dZ_t = Z_t f(t)d\beta_t$ and by Lemma 1

$$
\begin{aligned}
Z_\infty^{(n)} = \mathbf{E}\left(\mathcal{E}(f) \mid \mathcal{F}_\infty^{(n)}\right) &= \mathbf{E}\left(1 + \int_0^\infty Z_t f(t)d\beta_t \,\middle|\, \mathcal{F}_\infty^{(n)}\right) \\
&= 1 + \int_0^\infty f(t)\mathbf{E}\left(Z_t h_t^{(n)} \mid \mathcal{F}_t^{(n)}\right) d\beta_t^{(n)}.
\end{aligned}
$$

By the Bayes rule $\mathbf{E}\left(Z_t h_t^{(n)} \mid \mathcal{F}_t^{(n)}\right) = \mathbf{P}_Q\left(h_t^{(n)} \mid \mathcal{F}_t^{(n)}\right) Z_t^{(n)}$. That is, with $\xi_t^{(n)} = \mathbf{P}_Q\left(h_t^{(n)} \mid \mathcal{F}_t^{(n)}\right)$ and $M^{(n)} = \int \xi_s^{(n)} f(s)d\beta_s^{(n)}$ we can write

$$
\mathbf{E}\left(\mathcal{E}(f) \mid \mathcal{F}_\infty^{(n)}\right) = \exp M_\infty^{(n)} - \frac{1}{2}\langle M^{(n)}\rangle_\infty.
$$

When (6) holds then $\langle M^{(n)}\rangle_\infty \to 0$ in $L^1(\mathbf{P})$, $M_\infty^{(n)} \to 0$ in $L^2(\mathbf{P})$, hence $\ln Z_\infty^{(n)} \to 0$ in $L^1(\mathbf{P})$. Since $Z_\infty^{(n)} = \mathbf{E}\left(\mathcal{E}(f) \mid \mathcal{F}_\infty^{(n)}\right)$ converges almost surely we get that its limit is 1. This is true for all $f \in L^2[0,\infty)$ and by Proposition 2 we obtain that the tail σ-field $\bigcap_n \mathcal{F}_\infty^{(n)}$ is trivial.

For the converse we prove below that when $\bigcap_n \mathcal{F}_\infty^{(n)}$ is trivial then for each $f \in L^2[0,\infty)$

$$
f(s)\mathbf{E}\left(\mathcal{E}(f)h_s^{(n)} \mid \mathcal{F}_s^{(n)}\right) \to 0, \quad \text{almost surely, for almost all } s > 0. \tag{7}
$$

Then we consider

$$
\mathcal{H}_s = \left\{\xi \in L^1(\mathbf{P}) : \mathbf{E}\left(\xi h_s^{(n)} \mid \mathcal{F}_s^{(n)}\right) \to 0 \text{ in } L^1(\mathbf{P})\right\}, \quad s > 0.
$$

$\mathcal{H}_s$ is obviously a closed subspace of $L^1(\mathbf{P})$. It is possible to choose $D = \{f_1, f_2, \dots\} \subset L^2([0,\infty))$, a countable set of deterministic, nowhere vanishing functions, such that the linear hull of $\{\mathcal{E}(f) : f \in D\}$ is dense in $L^1(\mathbf{P})$. Finally let

$$
\mathcal{T} = \left\{s > 0 : \forall f \in D, \, \mathbf{E}\left(\mathcal{E}(f)h_s^{(n)} \mid \mathcal{F}_s^{(n)}\right) \to 0\right\}.
$$

Then $\mathcal{T}$ has full Lebesgue measure within $[0,\infty)$ and for $s \in \mathcal{T}$ we obviously have $\mathcal{H}_s = L^1(\mathbf{P})$. For $s \in \mathcal{T}$ (6) follows, by considering $\xi = dQ/d\mathbf{P}$.

It remains to show that

$$
\bigcap \mathcal{F}_\infty^{(n)} \quad \text{is trivial} \tag{8}
$$

implies (7). So we fix f and use the notation Q, $\xi^{(n)}$, $M^{(n)}$ introduced at the beginning of the proof. Note that $(|\xi_s^{(n)}|, \mathcal{F}_s^{(n)})_{n \geq 0}$ is a reversed Q-submartingale for each fixed s. Hence $|\xi_s^{(n)}|$ is convergent almost surely (both under $\mathbf{P}$ and Q by their equivalence) and the limit is $\bigcap_n \mathcal{F}_s^{(n)} \subset \bigcap_n \mathcal{F}_\infty^{(n)}$ measurable. Since $\bigcap_n \mathcal{F}_\infty^{(n)}$ is trivial there is a deterministic function g such that $|\xi_s^{(n)}| \to g(s)$ almost surely for almost all s. Obviously $0 \leq g(s) \leq 1$.

Another implication of (8) is that

$$\ln \mathbf{E}\left(\mathcal{E}(f) \mid \mathcal{F}_\infty^{(n)}\right) = M_\infty^{(n)} - \frac{1}{2}\langle M^{(n)}\rangle_\infty \to 0, \quad \text{almost surely.} \tag{9}$$

Here

$$\langle M^{(n)}\rangle_\infty \to \sigma^2 = \int_0^\infty (f(s)g(s))^2 ds, \quad \text{almost surely}$$

and we will see that $M_\infty^{(n)}$ has normal limit with expectation zero and variance σ^2. Then (9) can only hold if $\sigma^2 = 0$ which obviously implies (7).

To finish the proof we write $M_\infty^{(n)}$ as

$$M_\infty^{(n)} = \int_0^\infty f(s)g(s)\operatorname{sign}(\xi_s^{(n)})d\beta_s^{(n)} + \int_0^\infty f(s)(\xi_s^{(n)} - g(s)\operatorname{sign}(\xi_s^{(n)}))d\beta_s^{(n)}.$$

Here the law of the first term is normal $N(0, \sigma^2)$ not depending on n, while the second term goes to zero in $L^2(\mathbf{P})$.

To finish the proof of Proposition 1 assume that (5) holds, that is

$$\lim_{n\to\infty} \mathbf{E}\left(h_t^{(n)} \mid \mathcal{F}_{rt}^{(0)} \vee \mathcal{F}_t^{(n)}\right) = 0, \quad \text{for almost all } t > 0 \text{ and } r \in [0, 1).$$

Fix a $Q \sim \mathbf{P}$ and denote by $Z_t = \dfrac{dQ|_{\mathcal{F}_t^{(0)}}}{d\mathbf{P}|_{\mathcal{F}_t^{(0)}}}$ the density process. By the Bayes formula it is enough to show that

$$\mathbf{E}\left(Z_t h_t^{(n)} \mid \mathcal{F}_t^{(n)}\right) \to 0.$$

Since $|h^{(n)}| \leq 1$ we have the next estimate

$$\left\|\mathbf{E}\left(Z_t h_t^{(n)} \mid \mathcal{F}_{rt}^{(0)} \vee \mathcal{F}_t^{(n)}\right) - \mathbf{E}\left(Z_{rt} h_t^{(n)} \mid \mathcal{F}_{rt}^{(0)} \vee \mathcal{F}_t^{(n)}\right)\right\|_{L^1} \leq \|Z_t - Z_{rt}\|_{L^1},$$

and by (5)

$$\mathbf{E}\left(Z_{rt} h_t^{(n)} \mid \mathcal{F}_{rt}^{(0)} \vee \mathcal{F}_t^{(n)}\right) = Z_{rt} \mathbf{E}\left(h_t^{(n)} \mid \mathcal{F}_{rt}^{(0)} \vee \mathcal{F}_t^{(n)}\right) \to 0$$

almost surely and in L^1. That is,

$$\limsup_{n\to\infty}\left\|\mathbf{E}\left(Z_t h_t^{(n)} \mid \mathcal{F}_t^{(n)}\right)\right\|_{L^1} \le \inf_{r\in[0,1)}\|Z_t - Z_{rt}\|_{L^1} = 0.$$

This means that the limit of the reversed submartingale $\left|\mathbf{P}_Q\left(h_t^{(n)} \mid \mathcal{F}_t^{(n)}\right)\right|$ is zero and T is exact by Proposition 3. This completes the proof of Proposition 1.

Acknowledgements The author thanks Michel Emery for reading the first version of this note and offering helpful comments, and the referee for suggesting a simplification in the proof of Proposition 3.

Reference

1. V. Prokaj, *Some Sufficient Conditions for the Ergodicity of the Lévy-Transformation*, ed. by C. Donati-Martin, A. Lejay, A. Rouault. Séminaire de Probabilités, XLV (Springer, New York, 2013), pp. 93–121. Doi: 10.1007/978-3-319-00321-4_2, arxiv:1206.2485

Multi-Occupation Field Generates the Borel-Sigma-Field of Loops

Yinshan Chang

Abstract In this article, we consider the space of càdlàg loops on a Polish space S. The loop space can be equipped with a "Skorokhod" metric. Moreover, it is Polish under this metric. Our main result is to prove that the Borel-σ-field on the space of loops is generated by a class of loop functionals: the multi-occupation field. This result generalizes the result in the discrete case, see (Le Jan, Markov Paths, Loops and Fields, vol. 2026, Springer, Heidelberg, 2011).

1 Introduction

The Markovian loops have been studied by Le Jan [3] and Sznitman [5]. Under reasonable assumptions of the state space, as an application of Blackwell's theorem, we would like to prove that multi-occupation field generates the Borel-σ-field on the space of loops, see Theorem 1. This generalizes the result in [3], see the paragraph below Proposition 10 in Chapter 2 of [3]. For self-containedness, we introduce several necessary definitions and notations in the following paragraphs.

Let (S, d_S) be a Polish space with the Borel-σ-field. As usual, denote by $D_S([0, a])$ the Skorokhod space, i.e. the space of càdlàg[1]-paths over time interval $[0, a]$ which is also left-continuous at time a. We equip it with the Skorokhod metric and the corresponding Borel-σ-field.

Definition 1 (Based Loop) A based loop is an element $l \in D_S([0, t])$ for some $t > 0$ such that $l(0) = l(t)$. We call t the duration of the based loop and denote it by $|l|$.

Definition 2 (Loop) We say two based loops are equivalent iff they are identical up to some circular translation. A loop is defined as an equivalence class of based loops. For a based loop l, we denote by l^o its equivalence class.

---

[1]The terminology "càdlàg" is short for right-continuous with left hand limits.

Y. Chang (✉)
Max Planck Institute for Mathematics in the Sciences, 04103 Leipzig, Germany
e-mail: ychang@mis.mpg.de

© Springer International Publishing Switzerland 2014
C. Donati-Martin et al. (eds.), *Séminaire de Probabilités XLVI*, Lecture Notes in Mathematics 2123, DOI 10.1007/978-3-319-11970-0_17

Definition 3 (Multi-Occupation Field/Time) Define the rotation operator r_j as follows: $r_j(z^1, \ldots, z^n) = (z^{1+j}, \ldots, z^n, z^1, \ldots, z^j)$. For any $f : S^n \to \mathbb{R}$ measurable, define the multi-occupation field of based loop l of time duration t as

$$\langle l, f \rangle = \sum_{j=0}^{n-1} \int_{0 < s^1 < \cdots < s^n < t} f \circ r_j(l(s^1), \ldots, l(s^n)) \, ds^1 \cdots ds^n.$$

If l_1 and l_2 are two equivalent based loops, they correspond to the same multi-occupation field. Therefore, the multi-occupation field is well-defined for loops. For discrete S, define the multi-occupation time $\hat{l}^{x_1, \ldots, x_n}$ of a (based) loop to be $\langle l, 1_{(x_1, \ldots, x_n)}(\cdot) \rangle$ where

$$1_{(x_1, \ldots, x_n)}((y_1, \ldots, y_n)) = \begin{cases} 1 & \text{if } (y_1, \ldots, y_n) = (x_1, \ldots, x_n) \\ 0 & \text{otherwise.} \end{cases}$$

The following idea for defining the distance of loops is due to Titus Lupu. Given two based loops l_1 and l_2, they can be normalized to have duration 1 by linear time scaling. Denote them $l_1^{\text{normalized}}$ and $l_2^{\text{normalized}}$. As S is Polish, by Theorem 5.6 in [2], the Skorokhod space $(D_S([0, 1]), d)$ is also Polish under the following metric:

$$d(l_1, l_2) \overset{\text{def}}{=} \inf_{\lambda} \left(\sup_{s < t} \left| \log \frac{\lambda(t) - \lambda(s)}{t - s} \right| + \sup_{u \in [0,1]} d_S(l_1(\lambda(u)), l_2(u)) \right) \tag{1}$$

where the infimum is taken over all increasing bijections $\lambda : [0, 1] \to [0, 1]$. Then, it is straightforward to see that the space of based loops under the following metric D is also Polish:

$$D(l_1, l_2) \overset{\text{def}}{=} \left| |l_1| - |l_2| \right| + d(l_1^{\text{normalized}}, l_2^{\text{normalized}}).$$

Definition 4 (Distance on Loops) Define the distance D^o of two loops l_1^o and l_2^o by

$$D^o(l_1^o, l_2^o) \overset{\text{def}}{=} \inf\{D(l, l') : l \in l_1^o \text{ and } l' \in l_2^o\}.$$

Remark 1 This is not the standard way to define a pseudo metric on quotient space. In general, the above definition does not satisfy the triangular inequality. In this special situation, the distance D is in fact invariant under suitable circular translation which guarantees the triangular inequality, as is stated below.

We provide the proofs of the following three propositions in Appendix.

Proposition 1 *The D^o indeed defines a distance.*

Proposition 2 *The loop space is Polish under the metric D^o.*

Then, we equip the loop space with the Borel-σ-field. The next proposition states the measurability of the multi-occupation field.

Proposition 3 *Fix any bounded Borel measurable function f on S^n, the following map is Borel measurable functional on the loop space:*

$$l \rightarrow \langle l, f \rangle.$$

Our main result is the following theorem.

Theorem 1 *The Borel-σ-field on the loops is generated by the multi-occupation field if (S, d_S) is Polish.*

2 Proof of Theorem 1

We will prove the main theorem in this section as an application of the following Blackwell's theorem.

Theorem 2 (Blackwell's Theorem, Theorem 26, Chapter III of [1]) *Suppose $(E, \mathcal{E})$ is a Blackwell space, $\mathcal{S}, \mathcal{F}$ are sub-σ-field of $\mathcal{E}$ and $\mathcal{S}$ is separable. Then $\mathcal{F} \subset \mathcal{S}$ iff every atom of $\mathcal{F}$ is a union of atoms of $\mathcal{S}$.*

As a consequence, we have the following lemma.

Lemma 1 *Suppose $(E, \mathcal{B}(E))$ is a Polish space with the Borel-σ-field. Let $\{f_i, i \in \mathbb{N}\}$ be measurable functions and denote $\mathcal{F} = \sigma(f_i, i \in \mathbb{N})$. Then, $\mathcal{F} = \mathcal{B}(E)$ iff for all $x \neq y \in E$, there exists f_i such that $f_i(x) \neq f_i(y)$.*

Proof Since E is Polish, $\mathcal{B}(E)$ is separable and $(E, \mathcal{B}(E))$ is a Blackwell space. The atoms of $\mathcal{B}(E)$ are all the one point sets. Obviously, $\mathcal{F} \subset \mathcal{B}(E)$ and $\mathcal{F}$ is separable. By Blackwell's theorem, $\mathcal{F} = \mathcal{B}(E)$ iff the atoms of $\mathcal{F}$ are all the one point sets which is equivalent to the following: for all $x \neq y \in E$, there exists f_i such that $f_i(x) \neq f_i(y)$.

Then, we are ready for the proof of the main theorem.

From the definition of the multi-occupation field, any loop defines a finite measure on S^n for all $n \in \mathbb{N}_+$. Let $\mathfrak{B} = (B_i, i \in \mathbb{N})$ be a countable topological basis of S. The σ-field generated by the multi-occupation field must equal to the σ-field generated by the following countable[2] functionals $\{\langle \cdot, 1_B \rangle : B \in \bigcup_{k=1}^{\infty} \mathfrak{B}^k\}$.

---

[2]The countability is required by Lemma 1.

In fact, if two loop l_1 and l_2 are the same under these countable loop functionals $\{\langle \cdot, 1_B \rangle : B \in \bigcup_{k=1}^{\infty} \mathfrak{B}^k\}$, they must agree on all the functionals of the form $\langle \cdot, f \rangle$. By Lemma 1, it remains to check that two loops with the same occupation field are the same loop.

Suppose loops l_1^o and l_2^o have the same occupation field, i.e. $\langle l_1^o, f \rangle = \langle l_2^o, f \rangle$ for all positive f on some S^n ($n \in \mathbb{N}_+$). Recall that a loop is an equivalence class of based loops. Take two based loops l_1, l_2 in the equivalence class l_1^o, l_2^o respectively. Then, $\langle l_1, f \rangle = \langle l_1^o, f \rangle = \langle l_2^o, f \rangle = \langle l_2, f \rangle$. Define $m_1(A) = \langle l_1, 1_A \rangle$ and $m_2(A) = \langle l_2, 1_A \rangle$ for $A \in \mathcal{B}(S)$. Then, we have $m_1 = m_2$ which means that the time spent in some Borel measurable set is the same for these two loops. In particular, the two (based) loops have the same time duration, say t. For simplicity of the notations, we will use m instead of m_1 and m_2. Now, we are ready to show that $l_1^o = l_2^o$ in three steps. Let us present the sketch of the proof before providing the details. We first decompose the space into an approximate partition which is used in [4]. Next, we replace arcs of trajectory in each part by a single point with corresponding holding times. In this way, we get two loops in the same discrete space. By the construction of these discrete loops, their multi-occupation fields coincide. It is known that Theorem 1 is true for loops in discrete space. Thus, these two modified loops in the discrete space are exactly the same. Moreover, when the rough partition is small enough, these modified loops are actually good approximation of the original loops l_1^o and l_2^o in the sense of Skorokhod. For that reason, we conclude in the last step that $l_1^o = l_2^o$.

I. For all $\epsilon > 0$ fixed, we choose a collection of open sets U_i^{ϵ} satisfying the following properties:

  – their boundaries are negligible with respect to m,
  – they have positive distances from each other,
  – the complement of their union has mass smaller than 2ϵ with respect to m,
  – their diameters are smaller than ϵ.

Let U^{ϵ} be the union of $(U_i^{\epsilon})_i$.

Actually, the rough partitions $(U_i^{\epsilon})_i$ are chosen in the following way. It is well-known that every finite measure on the Borel-σ-field of a Polish space is regular. Therefore, for $\epsilon > 0$, we can find some compact set K_{ϵ} such that $m(K_{\epsilon}^c) < \epsilon$ where K_{ϵ}^c is the complement of K_{ϵ}. Let $\mathcal{D} = \{x_1, \ldots, x_n, \ldots\}$ be a countable dense subset of S. Fix any $x \in S$, except for countable many $r \in \mathbb{R}_+$, the measure m does not charge the boundary $\partial(B(x, r))$ of the ball $B(x, r)$. Then, for any $\epsilon > 0$, there exists a collection of open ball $B(x_i, r_i)$ with radius smaller than ϵ such that their boundaries are negligible with respect to m. Then, they cover the compact set K_{ϵ} as $\mathcal{D}$ is dense in S. Therefore, we can extract a finite open covering $\{B(y_1, r_1), \ldots, B(y_k, r_k)\}$. The boundaries of these open balls cut the whole space S into a finite partition of the space $S \setminus \bigcup_i \partial(B(y_i, r_i))$: q open sets $P_0, \ldots, P_q$ with $P_0 = (\bigcup_i \overline{B(y_i, r_i)})^c$ and

$q \geq k$. Let $U_{i,\delta}^{\epsilon} = \{y \in S : d_S(y, P_i^c) > \delta\}$ which is contained in P_i. In fact, one can always choose some δ_0 small enough and good enough such that the boundary sets $\{y \in S : d_S(y, P_i^c) = \delta_0\}$ of these open sets are negligible under m. Moreover, $m(S \setminus (\bigcup_i U_{i,\delta_0}^{\epsilon} \cap K_{\epsilon})) < 2\epsilon$. Set $U_i^{\epsilon} = U_{i,\delta_0}^{\epsilon}, i = 1, \ldots, q$.

Then, they satisfy the desired properties stated above.

II. From the based loop l_j $(j = 1, 2)$, we will construct two piecewise-constant based loops l_j^{ϵ} $(j = 1, 2)$ with finitely many jumps such that l_1^{ϵ} and l_2^{ϵ} are the same in the sense of loop and that they are quite close to the trace of l_1 and l_2 on U^{ϵ} respectively.

To be more precise, define $A_{j,u}^{\epsilon} = \int_0^u 1_{\{l_j(s) \in U^{\epsilon}\}} ds$ with the convention that $l_j(s + kt) = l_j(s)$ for $s \in [0, t]$ and $k \in \mathbb{Z}$ where t is the time duration of the based loops. Then, $(A_{j,u}^{\epsilon}, u \in \mathbb{R}_+)$ is right-continuous and increasing for $j = 1, 2$. Let $(\sigma_{j,s}^{\epsilon}, s \in \mathbb{R}_+)$ be the right-continuous inverse of $(A_{j,u}^{\epsilon}, u \in \mathbb{R}_+)$ for $j = 1, 2$ respectively. More precisely,

$$\sigma_{j,s}^{\epsilon} = \inf\{s \in \mathbb{R}_+ : A_{j,u}^{\epsilon} > s\}.$$

Let $t_{\epsilon} = A_{1,t}^{\epsilon} = A_{2,t}^{\epsilon} = m(U^{\epsilon})$ be the total occupation time of the loops within U^{ϵ}. Then, $A_{j,u+kt}^{\epsilon} = kt_{\epsilon} + A_{j,u}^{\epsilon}$ and $A_{j,u}^{\epsilon} \leq u$ for $u \in \mathbb{R}_+$. Thus, $\sigma_{j,s+kt_{\epsilon}}^{\epsilon} = \sigma_{j,s}^{\epsilon} + kt$ for $k \in \mathbb{N}, s \in [0, t_{\epsilon}[$ and $\sigma_{j,s}^{\epsilon} \geq s$ for $s \in \mathbb{R}_+$. Moreover, as $\epsilon \downarrow 0$, $(\sigma_{j,s}^{\epsilon}, s \in \mathbb{R}_+)$ decreases to $(s, s \in \mathbb{R}_+)$ uniformly on any compact of $\mathbb{R}_+$. We know that $l_j(\sigma_{j,s}^{\epsilon}) \in \bigcup_i \overline{U_i^{\epsilon}}$. We choose a point y_i in each U_i^{ϵ} and define $l_j^{\epsilon}(s) = y_i$ iff $l_j(\sigma_{j,s}^{\epsilon}) \in \overline{U_i^{\epsilon}}$ for $j = 1, 2$. Then, as the diameters of $(U_i^{\epsilon})_i$ are less than ϵ, $\sup_s d_S(l_j^{\epsilon}(s), l_j(\sigma_{j,s}^{\epsilon})) \leq \epsilon$ for $j = 1, 2$. Moreover, $s \to l_j^{\epsilon}(s)$ is càdlàg for $j = 1, 2$. Since the based loops l_1 and l_2 are càdlàg and all the $\overline{U_i^{\epsilon}}$ have a positive distance from each other, $s \to l_j^{\epsilon}(s)$ has finitely many jumps in any finite time interval for $j = 1, 2$. Then, $(l_j^{\epsilon}(s), s \in [0, t_{\epsilon}])^o$ is a loop on the same finite state space for $j = 1, 2$ respectively. Since the boundary of U_i^{ϵ} is negligible with respect to m, by Lebesgue's change of measure formula,

$$\left((l_j^{\epsilon}(s), s \in [0, t_{\epsilon}])^o\right)^{y_{i_1}, \ldots, y_{i_n}} = \left\langle l_j, 1_{U_{i_1}^{\epsilon}} \cdots 1_{U_{i_n}^{\epsilon}} \right\rangle \text{ for } j = 1, 2.$$

Thus, $(l_1^{\epsilon}(s), s \in [0, t_{\epsilon}])^o$ and $(l_2^{\epsilon}(s), s \in [0, t_{\epsilon}])^o$ have the same multi-occupation field. It is known that Theorem 1 is true for loops in finite discrete space, see the paragraph below Proposition 10 in Chapter 2 of [3]. Thus,

$$(l_1^{\epsilon}(s), s \in [0, t_{\epsilon}])^o = (l_2^{\epsilon}(s), s \in [0, t_{\epsilon}])^o.$$

Consequently, there exists some $T_2(\epsilon) \in [0, t_{\epsilon}[$ such that $l_2^{\epsilon}(s + T_2) = l_1^{\epsilon}(s)$ for $s \geq 0$.

III. Take the limit on a subsequence and use the right-continuity of the path to conclude $l_1 = l_2$ up to circular translation. To be precise, we can find a sequence $(\epsilon_k)_k$ with $\lim\limits_{k\to\infty} \epsilon_k = 0$ such that $T_2(\epsilon_k)$ converges to $T \in [0, t]$ as $k \to \infty$. Then, $\lim\limits_{k\to\infty} \sigma^{\epsilon_k}_{2,s+T_2} = s + T$ for fixed $s \geq 0$. Accordingly,

$$\lim_{k\to\infty} \min\{d_S(l_2(\sigma^{\epsilon_k}_{2,s+T_2}), l_2(s+T)), d_S(l_2(\sigma^{\epsilon_k}_{2,s+T_2}), l_2((s+T)-))\} = 0. \tag{2}$$

On the other hand, we have $\lim\limits_{k\to\infty} \sigma^{\epsilon_k}_{2,s} = s$ and $\sigma^{\epsilon_k}_{2,s} \geq s$ for all $k \in \mathbb{N}$. Therefore, by the right continuity of l_1,

$$\lim_{k\to\infty} d_S(l_1(\sigma^{\epsilon_k}_{1,s}), l_1(s)) = d_S(l_1(s+), l_1(s)) = 0. \tag{3}$$

From the constructions of l_1^ϵ and l_2^ϵ and the argument in part II, we see that

$$\sup_s d_S(l_2(\sigma^{\epsilon_k}_{2,s+T_2}), l_1(\sigma^{\epsilon_k}_{1,s})) \leq \sup_s d_S(l_2^\epsilon(s+T_2), l_2(\sigma^{\epsilon_k}_{2,s+T_2}))$$

$$+ \sup_s d_S(l_1^\epsilon(s), l_1(\sigma^{\epsilon_k}_{1,s}))$$

$$\leq 2\epsilon_k. \tag{4}$$

As a result, by (2)+(3)+(4), for any $s \geq 0$, either $d_S(l_2(s+T), l_1(s)) = 0$ or $d_S(l_2((s+T)-), l_1(s)) = 0$. Finally, by right-continuity of the paths l_1 and l_2,

$$l_2(s+T) = l_1(s)$$

and the proof is complete.

Appendix

As promised, we give the proofs for Propositions 1, 2 and 3 in this section. For that reason, we prepare several notations and lemmas in the following.

Definition 5 Suppose $\lambda : [0, 1] \to [0, 1]$ is a increasing bijection. For $t \in [0, 1[$, define

$$\theta_t \lambda(s) = \begin{cases} \lambda(t+s) - \lambda(t) & \text{for } s \in [0, 1-t] \\ 1 - \lambda(t) + \lambda(t+s-1) & \text{for } s \in [1-t, 1]. \end{cases}$$

In fact, we cut the graph of λ at the time t, exchange the first part of the graph with the second part and then glue them together to get an increasing bijection over $[0, 1]$.

Lemma 2

$$\sup_{s<t} \left| \log \frac{\theta_r \lambda(t) - \theta_r \lambda(s)}{t - s} \right| = \sup_{s<t} \left| \log \frac{\lambda(t) - \lambda(s)}{t - s} \right|.$$

Proof Denote by $\phi(\lambda, s, t)$ the quantity $|\log \frac{\lambda(t)-\lambda(s)}{t-s}|$. We see that for $a < b < c$,

$$\max(\phi(\lambda, a, b), \phi(\lambda, b, c)) \geq \phi(\lambda, a, c).$$

Thus, $\sup_{s<t} \phi(\lambda, s, t) = \sup_{s<t, t-s \text{ is small}} \phi(\lambda, s, t)$. As a result, $\sup_{s<t} |\log \frac{\lambda(t)-\lambda(s)}{t-s}|$ is a function of λ which is invariant under θ_t.

Definition 6 For a based loop l of time duration t and $r \in [0, t[$, denote by Θ_r the circular translation of l:

$$\Theta_r(l)(u) = \begin{cases} l(u+r) & \text{for } u \in [0, t-r] \\ l(u+r-t) & \text{for } u \in [t-r, t]. \end{cases}$$

Then, we can extend Θ_r for all $r \in \mathbb{R}$ by periodical extension.

Notice that $\Theta_r(l)$ is a based loop iff the periodical extension of l is continuous at time r. Nevertheless, we define the distance $D(\Theta_r l, l)$ in the same way. The next lemma shows the continuity of $r \to \Theta_r l$ at time r when the based loop l is continuous at r.

Lemma 3 *Suppose l is a based loop. Then,* $\lim_{h \to 0} D(\Theta_h l, l) = 0.$

Proof Without loss of generality, we can assume l has time duration 1. By definition, we have that

$$D(\Theta_h(l), l) = d(\Theta_h(l), l)$$

$$= \inf \left\{ \sup_{s<t} \left| \log \frac{\lambda(t) - \lambda(s)}{t - s} \right| + \sup_{u \in [0,1]} d_S(l(\lambda(u)), \Theta_h(l)(u)) : \right.$$

$$\left. \lambda \text{ increasing bijection on } [0, 1] \right\}.$$

Fix $0 < a < b < 1$, take $\lambda(0) = 0, \lambda(a) = a + h, \lambda(b) = b + h, \lambda(1) = 1$ and linearly interpolate λ elsewhere. Then,

$$D(\Theta_h(l), l) \leq \max \left(\left| \log \frac{a+h}{a} \right|, \left| \log \frac{1-b-h}{1-b} \right| \right)$$

$$+ 2 \sup_{u,v \in [0, a+|h|] \cup [b-|h|, 1]} |l(u) - l(v)|.$$

Thus, for any $0 < a < b < 1$,

$$\limsup_{h \to 0} D(\Theta_h(l), l) \le 2 \sup_{u,v \in [0,a] \cup [b,1]} |l(u) - l(v)|.$$

Since l is a based loop, $\inf_{a,b} (\sup_{u,v \in [0,a] \cup [b,1]} |l(u) - l(v)|) = 0$. Therefore,

$$\lim_{h \to 0} D(\Theta_h l, l) = 0.$$

Lemma 4 *Suppose l_1 is a based loop with time duration t and l_1 is continuous at time $r \in [0, t[$. Then,*

$$\inf\{D(l_1, l) : l \in l_2^o\} = \inf\{D(\Theta_r(l_1), l) : l \in l_2^o\}.$$

Proof Recall that $D(l_1, l) = \Big| |l| - |l_1| \Big| + d(l_1^{\text{normalized}}, l^{\text{normalized}})$ where

$$d(l_1^{\text{normalized}}, l^{\text{normalized}}) = \inf \Big\{ \sup_{u \in [0,1]} d_S(l_1^{\text{normalized}}(u), l^{\text{normalized}}(\lambda(u)))$$

$$+ \sup_{s < t} \Big| \log \frac{\lambda(t) - \lambda(s)}{t - s} \Big| : \lambda \text{ increasing bijection over } [0, 1] \Big\}.$$

Then, for $\epsilon > 0$, there exists $l \in l_2^o$ and λ such that

$$\sup_{s < t} \Big| \log \frac{\lambda(t) - \lambda(s)}{t - s} \Big| + \sup_{u \in [0,1]} d_S(l_1^{\text{normalized}}(u), l^{\text{normalized}}(\lambda(u)))$$

$$< \inf\{D(l_1, l) : l \in l_2^o\} + \epsilon. \qquad (5)$$

Since the paths are càdlàg, for fixed l_1 and l, the following set is at most countable:

$$\{a : l_1 \text{ jumps at time } a \text{ or } l \text{ jumps at } |l|\lambda(a/|l_1|)\}.$$

Thus, we can find a sequence $(r_n)_n$ such that

- $r_n \downarrow r$ as $n \to \infty$,
- $\Theta_{r_n}(l_1)$ and $\Theta_{|l|\lambda(r_n/|l_1|)}(l)$ are both based loops.

By Lemma 2, we have that

$$\sup_{s < t} \Big| \log \frac{\lambda(t) - \lambda(s)}{t - s} \Big| = \sup_{s < t} \Big| \log \frac{\theta_{r_n/|l_1|}\lambda(t) - \theta_{r_n/|l_1|}\lambda(s)}{t - s} \Big|. \qquad (6)$$

Meanwhile, we have that

$$\sup_{u \in [0,1]} d_S(l_1^{\text{normalized}}(u), l^{\text{normalized}}(\lambda(u)))$$

$$= \sup_{u \in [0,1]} d_S\left((\Theta_{r_n} l_1)^{\text{normalized}}(u), (\Theta_{|l|\lambda(r_n/|l_1|)}l)^{\text{normalized}}(\theta_{r_n/|l_1|}\lambda(u))\right). \tag{7}$$

Notice that $\Theta_{|l|\lambda(r_n/|l_1|)}l \in l_2^o$. Thus, by (5)+(6)+(7), for any $\epsilon > 0$, there exists $(r_n)_n$ with decreasing limit r such that

$$\inf\{D(\Theta_{r_n}l_1, l) : l \in l_2^o\} < \inf\{D(l_1, l) : l \in l_2^o\} + \epsilon. \tag{8}$$

By triangular inequality of D,

$$D(\Theta_r l_1, l) \le D(\Theta_{r_n}l_1, \Theta_r l_1) + D(\Theta_{r_n}l_1, l).$$

We take the infimum on both sides, then

$$\inf\{D(\Theta_r l_1, l) : l \in l_2^o\} \le D(\Theta_{r_n}l_1, \Theta_r l_1) + \inf\{D(\Theta_{r_n}l_1, l) : l \in l_2^o\}.$$

By (8),

$$\inf\{D(\Theta_r l_1, l) : l \in l_2^o\} \le D(\Theta_{r_n}l_1, \Theta_r l_1) + \inf\{D(l_1, l) : l \in l_2^o\} + \epsilon. \tag{9}$$

By Lemma 3, for the based loop l_1, $\lim_{n \to \infty} D(\Theta_{r_n}l_1, \Theta_r l_1) = 0$. By taking $n \to \infty$ in (9), we see that

$$\inf\{D(\Theta_r l_1, l) : l \in l_2^o\} \le \inf\{D(l_1, l) : l \in l_2^o\} + \epsilon \text{ for all } \epsilon > 0.$$

Therefore,

$$\inf\{D(\Theta_r l_1, l) : l \in l_2^o\} \le \inf\{D(l_1, l) : l \in l_2^o\}.$$

If we replace r by $|l_1| - r$ and l_1 by $\Theta_r l_1$, we have the inequality in opposite direction:

$$\inf\{D(\Theta_r l_1, l) : l \in l_2^o\} \ge \inf\{D(l_1, l) : l \in l_2^o\}.$$

Then, we turn to prove Propositions 1, 2 and 3.

Proof (Proof of Proposition 1)

- Reflexivity: straightforward from the definition.
- Triangular inequality: directly from Lemma 4.

- $D^o(l_1^o, l_2^o) = 0 \implies l_1^o = l_2^o$: by Lemma 4, it is enough to show that

$$\inf\{D(l_1, l) : l \in l_2^o\} = 0 \implies l_1 \in l_2^o.$$

Suppose $\inf\{D(l_1, l) : l \in l_2^o\} = 0$. Then, we can find a sequence $(r_n)_n$ with limit r such that $\lim_{n\to\infty} D(\Theta_{r_n} l_2, l_1) = 0$. Since $l_1(|l_1|-) = l_1(0)$, l_2 must be continuous at r and $\lim_{n\to\infty} \Theta_{r_n} l_2 = \Theta_r l_2$ by Lemma 3. Thus, $l_1 = \Theta_r l_2$.

Proof (Proof of Proposition 2)

- Completeness: given a Cauchy sequence $(l_n^o)_n$, one can always extract a subsequence $(l_{n_k}^o)_k$ such that $D^o(l_{n_k}^o, l_{n_{k+1}}^o) < 2^{-k}$. By Lemma 4, one can find in each equivalence class $l_{n_k}^o$ a based loop L_k such that $D(L_k, L_{k+1}) < 2^{-k}$. By the completeness of D, there exists a based loop L such that $\lim_{k\to\infty} L_k = L$. Thus, $\lim_{k\to\infty} l_{n_k}^o = L^o$. So it is the same for $(l_n^o)_n$.
- Separability: the based loop space is separable. Then, as a continuous image, the loop space is separable.

Proof (Proof of Proposition 3) For any bounded continuous function $f : S^n \to \mathbb{R}$, $l \to \langle l, f \rangle$ is continuous in l. In particular, it is measurable. By monotone class theorem for functions, $l \to \langle l, f \rangle$ is measurable for all bounded measurable $f : S^n \to \mathbb{R}$.

Acknowledgements The author is grateful to Professor Y. Le Jan for inspiring discussions, valuable suggestions and great help in preparation of the manuscript. The author also thanks Titus Lupu for inspiring discussions.

References

1. C. Dellacherie, P-A. Meyer, *Probabilities and Potential*, vol. 29 of *North-Holland Mathematics Studies* (North-Holland Publishing, Amsterdam, 1978)
2. S.N. Ethier, T.G. Kurtz, *Markov Processes*. Wiley Series in Probability and Mathematical Statistics: Probability and Mathematical Statistics (Wiley, New York, 1986). Characterization and convergence
3. Y. Le Jan, *Markov Paths, Loops and Fields*, vol. 2026 of *Lecture Notes in Mathematics* (Springer, Heidelberg, 2011) Lectures from the 38th Probability Summer School held in Saint-Flour, 2008, École d'Été de Probabilités de Saint-Flour [Saint-Flour Probability Summer School]
4. Y. Le Jan, Z. Qian, Stratonovich's signatures of Brownian motion determine Brownian sample paths. Probab. Theory Relat. Fields **157**(1–2), 209–223 (2013)
5. A-S. Sznitman, *Topics in Occupation Times and Gaussian Free Fields*. Zurich Lectures in Advanced Mathematics (European Mathematical Society (EMS), Zürich, 2012)

Ergodicity, Decisions, and Partial Information

Ramon van Handel

Abstract In the simplest sequential decision problem for an ergodic stochastic process X, at each time n a decision u_n is made as a function of past observations $X_0, \ldots, X_{n-1}$, and a loss $l(u_n, X_n)$ is incurred. In this setting, it is known that one may choose (under a mild integrability assumption) a decision strategy whose pathwise time-average loss is asymptotically smaller than that of any other strategy. The corresponding problem in the case of partial information proves to be much more delicate, however: if the process X is not observable, but decisions must be based on the observation of a different process Y, the existence of pathwise optimal strategies is not guaranteed. The aim of this paper is to exhibit connections between pathwise optimal strategies and notions from ergodic theory. The sequential decision problem is developed in the general setting of an ergodic dynamical system $(\Omega, \mathcal{B}, \mathbf{P}, T)$ with partial information $\mathcal{Y} \subseteq \mathcal{B}$. The existence of pathwise optimal strategies grounded in two basic properties: the conditional ergodic theory of the dynamical system, and the complexity of the loss function. When the loss function is not too complex, a general sufficient condition for the existence of pathwise optimal strategies is that the dynamical system is a conditional K-automorphism relative to the past observations $\bigvee_{n \geq 0} T^n \mathcal{Y}$. If the conditional ergodicity assumption is strengthened, the complexity assumption can be weakened. Several examples demonstrate the interplay between complexity and ergodicity, which does not arise in the case of full information. Our results also yield a decision-theoretic characterization of weak mixing in ergodic theory, and establish pathwise optimality of ergodic nonlinear filters.

1 Introduction

Let $X = (X_k)_{k \in \mathbb{Z}}$ be a stationary and ergodic stochastic process. A decision maker must select at the beginning of each day k a decision u_k depending on the past observations $X_0, \ldots, X_{k-1}$. At the end of the day, a loss $l(u_k, X_k)$ is incurred. The

R. van Handel (✉)
Sherrerd Hall 227, Princeton University, Princeton, NJ 08544, USA
e-mail: rvan@princeton.edu

© Springer International Publishing Switzerland 2014   411
C. Donati-Martin et al. (eds.), *Séminaire de Probabilités XLVI*, Lecture Notes in Mathematics 2123, DOI 10.1007/978-3-319-11970-0_18

decision maker would like to minimize her time-average loss

$$L_T(\mathbf{u}) = \frac{1}{T} \sum_{k=1}^{T} l(u_k, X_k).$$

How should she go about selecting a decision strategy $\mathbf{u} = (u_k)_{k \geq 1}$?

There is a rather trivial answer to this question. Taking the expectation of the time-average loss, we obtain for any strategy $\mathbf{u}$ using the tower property

$$\mathbf{E}[L_T(\mathbf{u})] = \mathbf{E}\left[\frac{1}{T} \sum_{k=1}^{T} \mathbf{E}[l(u_k, X_k) | X_0, \ldots, X_{k-1}] \right]$$

$$\geq \mathbf{E}\left[\frac{1}{T} \sum_{k=1}^{T} \min_u \mathbf{E}[l(u, X_k) | X_0, \ldots, X_{k-1}] \right] = \mathbf{E}[L_T(\tilde{\mathbf{u}})],$$

where $\tilde{\mathbf{u}} = (\tilde{u}_k)_{k \geq 1}$ is defined as $\tilde{u}_k = \arg\min_u \mathbf{E}[l(u, X_k) | X_0, \ldots, X_{k-1}]$ (we disregard for the moment integrability and measurability issues, existence of minima, and the like; such issues will be properly addressed in our results). Therefore, the strategy $\tilde{\mathbf{u}}$ minimizes the *mean* time-average loss $\mathbf{E}[L_T(\mathbf{u})]$.

However, there are conceptual reasons to be dissatisfied with this obvious solution. In many decision problems, one only observes a single sample path of the process X. For example, if X_k is the return of a financial market in day k and $L_T(\mathbf{u})$ is the loss of an investment strategy $\mathbf{u}$, only one sample path of the model is ever realized: we do not have the luxury of averaging our investment loss over multiple "alternative histories". The choice of a strategy for which the mean loss is small does not guarantee, a priori, that it will perform well on the one and only realization that happens to be chosen by nature. Similarly, if X_k models the state of the atmosphere and $L_T(\mathbf{u})$ is the error of a weather prediction strategy, we face a similar conundrum. In such situations, the use of stochastic models could be justified by some sort of ergodic theorem, which states that the mean behavior of the model with respect to different realizations captures its time-average behavior over a single sample path. Such an ergodic theorem for sequential decisions was obtained by Algoet [1, Theorem 2] under a mild integrability assumption.

Theorem 1.1 (Algoet [1]) *Suppose that* $|l(u, x)| \leq \Lambda(x)$ *with* $\Lambda \in L \log L$. *Then*

$$\liminf_{T \to \infty} \{ L_T(\mathbf{u}) - L_T(\tilde{\mathbf{u}}) \} \geq 0 \quad a.s.$$

for every strategy $\mathbf{u}$: *that is, the mean-optimal strategy* $\tilde{\mathbf{u}}$ *is pathwise optimal.*

The proof of this result follows from a simple martingale argument. What is remarkable is that the details of the model do not enter the picture at all: nothing is assumed on the properties of X or l beyond some integrability (ergodicity is not needed, and a similar result holds even in the absence of stationarity, cf. [1,

Theorem 3]). This provides a universal justification for optimizing the mean loss: the much stronger pathwise optimality property is obtained "for free."

In the proof of Theorem 1.1, it is essential that the decision maker has full information on the history $X_0, \ldots, X_{k-1}$ of the process X. However, the derivation of the mean-optimal strategy can be done in precisely the same manner in the more general setting where only partial or noisy information is available. To formalize this idea, let $Y = (Y_k)_{k \in \mathbb{Z}}$ be the stochastic process observable by the decision maker, and suppose that the pair (X, Y) is stationary and ergodic. The loss incurred at time k is still $l(u_k, X_k)$, but now u_k may depend on the observed data $Y_0, \ldots, Y_{k-1}$ only. It is easily seen that in this setting, the mean-optimal strategy $\tilde{\mathbf{u}}$ is given by $\tilde{u}_k = \arg\min_u \mathbf{E}[l(u, X_k)|Y_0, \ldots, Y_{k-1}]$, and it is tempting to assume that $\tilde{\mathbf{u}}$ is also pathwise optimal. Surprisingly, this is very far from being the case.

Example 1.2 (Weissman and Merhav [32]) Let $X_0 \sim$ Bernoulli(1/2) and let $X_k = 1 - X_{k-1}$ and $Y_k = 0$ for all k. Then (X, Y) is stationary and ergodic: $Y_k = 0$ indicates that we are in the setting of *no* information (that is, we must make blind decisions). Consider the loss $l(u, x) = (u - x)^2$. Then the mean-optimal strategy $\tilde{u}_k = 1/2$ satisfies $L_T(\tilde{\mathbf{u}}) = 1/4$ for all T. However, the strategy $u_k = k \bmod 2$ satisfies $L_T(\mathbf{u}) = 0$ for all T with probability 1/2. Therefore, $\tilde{\mathbf{u}}$ is not pathwise optimal. In fact, it is easily seen that no pathwise optimal strategy exists.

Example 1.2 illustrates precisely the type of conundrum that was so fortuitously ruled out in the full information setting by Theorem 1.1. Indeed, it would be hard to argue that either $\mathbf{u}$ or $\tilde{\mathbf{u}}$ in Example 1.2 is superior: a gambler placing blind bets u_k on a sequence of games with loss $l(u_k, X_k)$ may prefer either strategy depending on his demeanor. The example may seem somewhat artificial, however, as the hidden process X has infinitely long memory; the gambler can therefore beat the mean-optimal strategy by simply guessing the outcome of the first game. But precisely the same phenomenon can appear when (X, Y) is nearly memoryless.

Example 1.3 Let $(\xi_k)_{k \in \mathbb{Z}}$ be i.i.d. Bernoulli(1/2), and let $X_k = (\xi_{k-1}, \xi_k)$ and $Y_k = |\xi_k - \xi_{k-1}|$ for all k. Then (X, Y) is a stationary 1-dependent sequence: $(X_k, Y_k)_{k \leq n}$ and $(X_k, Y_k)_{k \geq n+2}$ are independent for every k. We consider the loss $l(u, x) = (u - x_1)^2$. It is easily seen that X_k is independent of $Y_1, \ldots, Y_{k-1}$, so that the mean-optimal strategy $\tilde{u}_k = 1/2$ satisfies $L_T(\tilde{\mathbf{u}}) = 1/4$ for all T. On the other hand, note that $\xi_{k-1} = (\xi_0 + Y_1 + \cdots + Y_{k-1}) \bmod 2$. It follows that the strategy $u_k = (Y_1 + \cdots + Y_{k-1}) \bmod 2$ satisfies $L_T(\mathbf{u}) = 0$ for all T with probability 1/2.

Evidently, pathwise optimality cannot be taken for granted in the partial information setting even in the simplest of examples: in contrast to the full information setting, the existence of pathwise optimal strategies depends both on specific ergodicity properties of the model (X, Y) and (as will be seen later) on the complexity on the loss l. What mechanism is responsible for pathwise optimality under partial information is not very well understood. Weissman and Merhav [32], who initiated the study of this problem, give a strong sufficient condition in the binary setting. Little is known beyond their result, beside one particularly special case of quadratic loss and additive noise considered by Nobel [24].

The aim of this paper is twofold. On the one hand, we will give general conditions for pathwise optimality under partial information, and explore some tradeoffs inherent in this setting. On the other hand, we aim to exhibit some connections between the pathwise optimality problem and certain notions and problems in ergodic theory, such as conditional mixing and individual ergodic theorems for subsequences. To make such connections in their most natural setting, we begin by rephrasing the decision problem in the general setting of ergodic dynamical systems.

1.1 The Dynamical System Setting

Let T be an invertible measure-preserving transformation of a probability space $(\Omega, \mathcal{B}, \mathbf{P})$. T defines the time evolution of the dynamical system $(\Omega, \mathcal{B}, \mathbf{P}, T)$: if the system is initially in state $\omega \in \Omega$, then at time k the system is in the state $T^k \omega$. The state of the system is not directly observable, however. To model the available information, we fix a σ-field $\mathcal{Y} \subseteq \mathcal{B}$ of events that can be observed at a single time. Therefore, if we have observed the system in the time interval $[m, n]$, the information contained in the observations is given by the σ-field $\mathcal{Y}_{m,n} = \bigvee_{k \in [m,n]} T^{-k} \mathcal{Y}$.

In this general setting, the decision problem is defined as follows. Let $\ell : U \times \Omega \to \mathbb{R}$ be a given loss function, where U is the set of possible decisions. At each time k, a decision u_k is made and a loss $\ell_k(u_k) := \ell(u_k, T^k \omega)$ is incurred. The decision can only depend on the observations: that is, a strategy $\mathbf{u} = (u_k)_{k \geq 1}$ is admissible if u_k is $\mathcal{Y}_{0,k}$-measurable for every k. The time-average loss is given by

$$L_T(\mathbf{u}) := \frac{1}{T} \sum_{k=1}^{T} \ell_k(u_k).$$

The basic question we aim to answer is whether there exists a pathwise optimal strategy, that is, a strategy $\mathbf{u}^\star$ such that for every admissible strategy $\mathbf{u}$

$$\liminf_{T \to \infty} \{L_T(\mathbf{u}) - L_T(\mathbf{u}^\star)\} \geq 0 \quad \text{a.s.}$$

The stochastic process setting discussed above can be recovered as a special case.

Example 1.4 Let (X, Y) be a stationary and ergodic stochastic process, where X_k takes values in the measurable space $(E, \mathcal{E})$ and Y_k takes values in the measurable space $(F, \mathcal{F})$. We can realize (X, Y) as the coordinate process on the canonical path space $(\Omega, \mathcal{B}, \mathbf{P})$ where $\Omega = E^{\mathbb{Z}} \times F^{\mathbb{Z}}$, $\mathcal{B} = \mathcal{E}^{\mathbb{Z}} \otimes \mathcal{F}^{\mathbb{Z}}$, and $\mathbf{P}$ is the law of (X, Y). Let $T : \Omega \to \Omega$ be the canonical shift $(T(x, y))_n = (x_{n+1}, y_{n+1})$. Then $(\Omega, \mathcal{B}, \mathbf{P}, T)$ is an ergodic dynamical system. If we choose the observation σ-field $\mathcal{Y} = \sigma\{Y_0\}$ and the loss $\ell(u, \omega) = l(u, X_1(\omega))$, we recover the decision problem with partial information for the stochastic process (X, Y) as it was introduced above.

More generally, we could let the loss depend arbitrarily on future or past values of (X, Y).

Let us briefly discuss the connection between pathwise optimal strategies and classical ergodic theorems. The key observation in the derivation of the mean-optimal strategy $\tilde{u}_k = \arg\min_u \mathbf{E}[\ell_k(u)|\mathcal{Y}_{0,k}]$ is that by the tower property

$$\mathbf{E}\left[\frac{1}{T}\sum_{k=1}^{T}\ell_k(u_k)\right] = \mathbf{E}\left[\frac{1}{T}\sum_{k=1}^{T}\mathbf{E}[\ell_k(u_k)|\mathcal{Y}_{0,k}]\right].$$

As the summands on the right-hand side depend only on the observed information, we can minimize inside the sum to obtain the mean-optimal strategy $\tilde{u}$. Precisely the same considerations would show that $\tilde{u}$ is pathwise optimal if we could prove the ergodic counterpart of the tower property of conditional expectations

$$\frac{1}{T}\sum_{k=1}^{T}\{\ell_k(u_k) - \mathbf{E}[\ell_k(u_k)|\mathcal{Y}_{0,k}]\} \xrightarrow{T\to\infty} 0 \quad \text{a.s.} \quad ?$$

The validity of such a statement is far from obvious, however.

In the special case of blind decisions (that is, $\mathcal{Y}$ is the trivial σ-field) the "ergodic tower property" reduces to the question of whether, given $f_k(\omega) := \ell(u_k, \omega)$,

$$\frac{1}{T}\sum_{k=1}^{T}\{f_k - \mathbf{E}[f_k]\} \circ T^k \xrightarrow{T\to\infty} 0 \quad \text{a.s.} \quad ?$$

If the functions f_k do not depend on k, this is precisely the individual ergodic theorem. However, an individual ergodic theorem need not hold for arbitrary sequences f_k. Special cases of this problem have long been investigated in ergodic theory. For example, if $f_k = a_k f$ for some fixed function f and bounded sequence $(a_k) \subset \mathbb{R}$, the problem reduces to a weighted individual ergodic theorem, see [2] and the references therein. If $a_k \in \{0, 1\}$ for all k, the problem reduces further to an individual ergodic theorem along a subsequence (at least if the sequence has positive density), cf. [2, 6] and the references therein. A general characterization of such ergodic properties does not appear to exist, which suggests that it is probably very difficult to obtain necessary and sufficient conditions for pathwise optimality. The situation is better for mean (rather than individual) ergodic theorems, cf. [3] and the references therein, and we will also obtain more complete results in a weaker setting.

The more interesting case where the information $\mathcal{Y}$ is nontrivial provides additional complications. In this situation, the "ergodic tower property" could be viewed as a type of *conditional* ergodic theorem, in between the individual ergodic theorem and Algoet's result [1]. Our proofs are based on an elaboration of this idea.

1.2 Some Representative Results

The essence of our results is that, when the loss ℓ is not too complex, pathwise optimal strategies exist under suitable conditional mixing assumptions on the ergodic dynamical system $(\Omega, \mathcal{B}, \mathbf{P}, T)$. To this end, we introduce conditional variants of two standard notions in ergodic theory: weak mixing and K-automorphisms.

Definition 1.5 An invertible dynamical system $(\Omega, \mathcal{B}, \mathbf{P}, T)$ is said to be *conditionally weak mixing* relative to a σ-field $\mathcal{Z}$ if for every $A, B \in \mathcal{B}$

$$\frac{1}{T} \sum_{k=1}^{T} |\mathbf{P}[A \cap T^k B | \mathcal{Z}] - \mathbf{P}[A | \mathcal{Z}] \, \mathbf{P}[T^k B | \mathcal{Z}]| \xrightarrow{T \to \infty} 0 \quad \text{in } L^1.$$

Definition 1.6 An invertible dynamical system $(\Omega, \mathcal{B}, \mathbf{P}, T)$ is called a *conditional K-automorphism* relative to a σ-field $\mathcal{Z} \subset \mathcal{B}$ if there is a σ-field $\mathcal{X} \subset \mathcal{B}$ such that

1. $\mathcal{X} \subset T^{-1}\mathcal{X}$.
2. $\bigvee_{k=1}^{\infty} T^{-k}\mathcal{X} = \mathcal{B} \mod \mathbf{P}$.
3. $\bigcap_{k=1}^{\infty} (\mathcal{Z} \vee T^k \mathcal{X}) = \mathcal{Z} \mod \mathbf{P}$.

When the σ-field $\mathcal{Z}$ is trivial, these definitions reduce[1] to the usual notions of weak mixing and K-automorphism, cf. [31]. Similar conditional mixing conditions also appear in the ergodic theory literature, see [26] and the references therein.

An easily stated consequence of our main results, for example, is the following.

Theorem 1.7 *Suppose that $(\Omega, \mathcal{B}, \mathbf{P}, T)$ is a conditional K-automorphism relative to $\mathcal{Y}_{-\infty,0}$. Then the mean-optimal strategy $\tilde{\mathbf{u}}$ is pathwise optimal for every loss function $\ell : U \times \Omega \to \mathbb{R}$ such that U is finite and $|\ell(u, \omega)| \leq \Lambda(\omega)$ with $\Lambda \in L^1$.*

This result gives a general sufficient condition for pathwise optimality when the decision space U is finite. In the stochastic process setting (Example 1.4), the conditional K-property would follow from the validity of the σ-field identity

$$\bigcap_{k=1}^{\infty} (\mathcal{Y}_{-\infty,0} \vee \mathcal{X}_{-\infty,-k}) = \mathcal{Y}_{-\infty,0} \mod \mathbf{P},$$

where $\mathcal{X}_{-\infty,k} = \sigma\{X_i : i \leq k\}$ (choose $\mathcal{X} := \mathcal{X}_{-\infty,0} \vee \mathcal{Y}_{-\infty,0}$ in Definition 1.6). In the Markovian setting, this is a familiar identity in filtering theory: it is precisely the necessary and sufficient condition for the optimal filter to be ergodic, see Sect. 3.3 below. Our results therefore lead to a new pathwise optimality property of nonlinear

---

[1]To be precise, our definitions are time-reversed with respect to the textbook definitions; however, T is a K-automorphism if and only if T^{-1} is a K-automorphism [31, p. 110], and the corresponding statement for weak mixing is trivial. Therefore, our definitions are equivalent to those in [31].

filters. Conversely, results from filtering theory yield a broad class of (even non-Markovian) models for which the conditional K-property can be verified [14,27]. It is interesting to note that despite the apparent similarity between the conditions for filter ergodicity and pathwise optimality, there appears to be no direct connection between these phenomena, and their proofs are entirely distinct. Let us also note that, in the full information setting ($Y_k = X_k$) the conditional K-property holds trivially, which explains the deceptive simplicity of Algoet's result.

While the conditional ergodicity assumption of Theorem 1.7 is quite general, the requirement that the decision space U is finite is a severe restriction on the complexity of the loss function ℓ. We have stated Theorem 1.7 here in order to highlight the basic ingredients for the existence of a pathwise optimal strategy. The assumption that U is finite will be replaced by various complexity assumptions on the loss ℓ; such extensions will be developed in the sequel. While some complexity assumption on the loss is needed in the partial information setting, there is a tradeoff between the complexity and ergodicity: if the notion of conditional ergodicity is strengthened, then the complexity assumption on the loss can be weakened.

All our pathwise optimality results are corollaries of a general master theorem, Theorem 2.6 below, that ensures the existence of a pathwise optimal strategy under a certain uniform version of the K-automorphism property. However, in the absence of further assumptions, this theorem does not ensure that the mean-optimal strategy $\tilde{\mathbf{u}}$ is in fact pathwise optimal: the pathwise optimal strategy constructed in the proof may be difficult to compute. We do not know, in general, whether it is possible that a pathwise optimal strategy exists, while the mean-optimal strategy fails to be pathwise optimal. In order to gain further insight into such questions, we introduce another notion of optimality that is intermediate between pathwise and mean optimality. A strategy $\mathbf{u}^\star$ is said to be weakly pathwise optimal if

$$\mathbf{P}[L_T(\mathbf{u}) - L_T(\mathbf{u}^\star) \geq -\varepsilon] \xrightarrow{T \to \infty} 1 \quad \text{for every } \varepsilon > 0.$$

It is not difficult to show that if a weakly pathwise optimal strategy exists, then the mean-optimal strategy $\tilde{\mathbf{u}}$ must also be weakly pathwise optimal. However, the notion of weak pathwise optimality is distinctly weaker than pathwise optimality. For example, we will prove the following counterpart to Theorem 1.7.

Theorem 1.8 *Suppose that* $(\Omega, \mathcal{B}, \mathbf{P}, T)$ *is conditionally weak mixing relative to* $\mathcal{Y}_{-\infty,0}$. *Then the mean-optimal strategy* $\tilde{\mathbf{u}}$ *is weakly pathwise optimal for every loss function* $\ell : U \times \Omega \to \mathbb{R}$ *such that* U *is finite and* $|\ell(u,\omega)| \leq \Lambda(\omega)$ *with* $\Lambda \in L^1$.

There is a genuine gap between Theorems 1.8 and 1.7: in fact, a result of Conze [6] on individual ergodic theorems for subsequences shows that there is a loss function ℓ such that for a generic (in the weak topology) weak mixing system, a mean-optimal blind strategy $\tilde{\mathbf{u}}$ fails to be pathwise optimal.

While weak pathwise optimality may not be as conceptually appealing as pathwise optimality, the weak pathwise optimality property is easier to characterize. In particular, we will show that the conditional weak mixing assumption in

Theorem 1.8 is not only sufficient, but also necessary, in the special case that $\mathcal{Y}$ is an invariant σ-field (that is, $\mathcal{Y} = T^{-1}\mathcal{Y}$). Invariance of $\mathcal{Y}$ is somewhat unnatural in decision problems, as it implies that no additional information is gained over time as more observations are accumulated. On the other hand, invariance of $\mathcal{Z}$ in Definitions 1.5 and 1.6 is precisely the situation of interest in applications of conditional mixing in ergodic theory (e.g., [26]). The interest of this result is therefore that it provides a decision-theoretic characterization of the (conditional) weak mixing property.

1.3   Further Questions

This paper is chiefly concerned with the existence of pathwise optimal strategies under partial information. There are, however, a number of other questions that may be of significant interest in this setting. While such questions are beyond the scope of this paper, we briefly highlight two potential areas for further investigation.

First, it should be emphasized that the pathwise optimal strategies that are obtained in our main results rely fundamentally on the assumption that we know the distribution of the underlying model (X, Y). In practical situations, however, the latter might not be known to the decision maker. To this end, the papers [1, 24, 32] have investigated also a separate question: even if a pathwise optimal strategy exists, it possible to attain the optimal asymptotic loss by means of a *universal* decision scheme that does not require any knowledge of the law of the model? Such strategies must "learn" the law of the model on the fly from the observed data. In the setting of partial information, such universal schemes evidently cannot exist without further assumptions: for example, in the blind setting (cf. Example 1.2) there is no information at all, so there is no hope to learn anything about the underlying model. However, if some a priori information is available on the structure of the model, one might still be able to construct strategies that are universal within certain model classes: this is done in [24, 32], for example. Precisely what conditions are required on the model structure in order to develop such results is an interesting question that appears to be quite distinct from the problem of existence that we investigate here.

Second, it should be noted that the notions of optimality investigated in this paper are purely qualitative in nature. One could aim to investigate more quantitative aspects of the sequential decision problem. Particularly the notion of weak pathwise optimality may lend itself to quantitative analysis under assumptions on the conditional mixing rate of the model, for example, see Sect. 3.4.2 below. The systematic investigation of such quantitative questions is beyond the scope of this paper.

1.4   Organization of This Paper

The remainder of the paper is organized as follows. In Sect. 2, we state and discuss the main results of this paper. We also give a number of examples that illustrate various aspects of our results. Our main results require two types of assumptions: conditional mixing assumptions on the dynamical system, and complexity assumptions on the loss. In Sect. 3 we discuss various methods to verify these assumptions, as well as further examples and consequences (such as pathwise optimality of nonlinear filters). Finally, the proofs of our main results are given in Sect. 4.

2   Main Results

2.1   Basic Setup and Notation

Throughout this paper, we will consider the following setting:

- $(\Omega, \mathcal{B}, \mathbf{P})$ is a probability space.
- $\mathcal{Y} \subseteq \mathcal{B}$ is a sub-σ-field.
- $T : \Omega \to \Omega$ is an invertible measure-preserving ergodic transformation.
- $(U, \mathcal{U})$ is a measurable space.

As explained in the introduction, we aim to make sequential decisions in the ergodic dynamical system $(\Omega, \mathcal{B}, \mathbf{P}, T)$. The decisions take values in the decision space U, and the σ-field $\mathcal{Y}$ represents the observable part of the system. We define

$$\mathcal{Y}_{m,n} = \bigvee_{k=m}^{n} T^{-k} \mathcal{Y} \qquad \text{for } -\infty \le m \le n \le \infty,$$

that is, $\mathcal{Y}_{m,n}$ is the σ-field generated by the observations in the time interval $[m, n]$. An admissible decision strategy must depend causally on the observed data.

Definition 2.1   A strategy $\mathbf{u} = (u_k)_{k \ge 1}$ is called *admissible* if it is $\mathcal{Y}_{0,k}$-adapted, that is, $u_k : \Omega \to U$ is $\mathcal{Y}_{0,k}$-measurable for every $k \ge 1$.

It will be convenient to introduce the following notation. For every $m \le n$, define

$$\mathbb{U}_{m,n} = \{u : \Omega \to U : u \text{ is } \mathcal{Y}_{m,n}\text{-measurable}\}, \qquad \mathbb{U}_n = \bigcup_{-\infty < m \le n} \mathbb{U}_{m,n}.$$

Thus a strategy $\mathbf{u}$ is admissible whenever $u_k \in \mathbb{U}_{0,k}$ for all k. Note that $\mathbb{U}_n \subsetneq \mathbb{U}_{-\infty,n}$: this distinction will be essential for the validity of our results.

To describe the loss of a decision strategy, we introduce a loss function ℓ.

- $\ell : U \times \Omega \to \mathbb{R}$ is a measurable function and $|\ell(u, \omega)| \le \Lambda(\omega)$ with $\Lambda \in L^1$.

If $|\ell(u, \omega)| \le \Lambda(\omega)$ with $\Lambda \in L^p$, the loss is said to be dominated in L^p. As indicated above, we will always assume[2] that our loss functions are dominated in L^1.

The loss function $\ell(u, \omega)$ represents the cost incurred by the decision u when the system is in state ω. In particular, the cost of the decision u_k at time k is given by $\ell(u_k, T^k\omega) = \ell_k(u_k)$, where we define for notational simplicity

$$\ell_n(u) : \Omega \to \mathbb{R}, \qquad \ell_n(u)(\omega) = \ell(u, T^n\omega).$$

Our aim is to select an admissible strategy $\mathbf{u}$ that minimizes the time-average loss

$$L_T(\mathbf{u}) = \frac{1}{T} \sum_{k=1}^{T} \ell_k(u_k)$$

in a suitable sense.

Definition 2.2 An admissible strategy $\mathbf{u}^\star$ is *pathwise optimal* if

$$\liminf_{T \to \infty} \{L_T(\mathbf{u}) - L_T(\mathbf{u}^\star)\} \ge 0 \quad \text{a.s.}$$

for every admissible strategy $\mathbf{u}$.

Definition 2.3 An admissible strategy $\mathbf{u}^\star$ is *weakly pathwise optimal* if

$$\mathbf{P}[L_T(\mathbf{u}) - L_T(\mathbf{u}^\star) \ge -\varepsilon] \xrightarrow{T \to \infty} 1 \quad \text{for every } \varepsilon > 0$$

for every admissible strategy $\mathbf{u}$.

Definition 2.4 An admissible strategy $\mathbf{u}^\star$ is *mean optimal* if

$$\liminf_{T \to \infty} \{\mathbf{E}[L_T(\mathbf{u})] - \mathbf{E}[L_T(\mathbf{u}^\star)]\} \ge 0$$

for every admissible strategy $\mathbf{u}$.

These notions of optimality are progressively weaker: a pathwise optimal strategy is clearly weakly pathwise optimal, and a weakly pathwise optimal strategy is mean optimal (as the loss function is assumed to be dominated in L^1).

In the introduction, it was stated that $\tilde{u}_k = \arg\min_{u \in U} \mathbf{E}[\ell_k(u)|\mathcal{Y}_{0,k}]$ defines a mean-optimal strategy. This disregards some technical issues, as the arg min may not exist or be measurable. It suffices, however, to consider a slight reformulation.

---

[2]Non-dominated loss functions may also be of significant interest, see [24] for example. We will restrict attention to dominated loss functions, however, which suffice in many cases of interest.

Lemma 2.5 *There exists an admissible strategy* $\tilde{\mathbf{u}}$ *such that*

$$\mathbf{E}[\ell_k(\tilde{u}_k)|\mathcal{Y}_{0,k}] \le \operatorname*{ess\,inf}_{u \in \mathbb{U}_{0,k}} \mathbf{E}[\ell_k(u)|\mathcal{Y}_{0,k}] + k^{-1} \quad a.s.$$

for every $k \ge 1$. *In particular,* $\tilde{\mathbf{u}}$ *is mean-optimal.*

Proof It follows from the construction of the essential supremum [25, p. 49] that there exists a countable family $(U^n)_{n \in \mathbb{N}} \subseteq \mathbb{U}_{0,k}$ such that

$$\operatorname*{ess\,inf}_{u \in \mathbb{U}_{0,k}} \mathbf{E}[\ell_k(u)|\mathcal{Y}_{0,k}] = \inf_{n \in \mathbb{N}} \mathbf{E}[\ell_k(U^n)|\mathcal{Y}_{0,k}].$$

Define the random variable

$$\tau = \inf\left\{n : \mathbf{E}[\ell_k(U^n)|\mathcal{Y}_{0,k}] \le \operatorname*{ess\,inf}_{u \in \mathbb{U}_{0,k}} \mathbf{E}[\ell_k(u)|\mathcal{Y}_{0,k}] + k^{-1}\right\}.$$

Note that $\tau < \infty$ a.s. as $\operatorname*{ess\,inf}_{u \in \mathbb{U}_{0,k}} \mathbf{E}[\ell_k(u)|\mathcal{Y}_{0,k}] \ge -\mathbf{E}[\Lambda \circ T^k|\mathcal{Y}_{0,k}] > -\infty$ a.s. We therefore define $\tilde{u}_k = U^\tau$. To show that $\tilde{\mathbf{u}}$ is mean optimal, it suffices to note that

$$\mathbf{E}[L_T(\mathbf{u})] - \mathbf{E}[L_T(\tilde{\mathbf{u}})] = \frac{1}{T} \sum_{k=1}^{T} \mathbf{E}\left[\mathbf{E}[\ell_k(u_k)|\mathcal{Y}_{0,k}] - \mathbf{E}[\ell_k(\tilde{u}_k)|\mathcal{Y}_{0,k}]\right] \ge -\frac{1}{T} \sum_{k=1}^{T} k^{-1}$$

for any admissible strategy $\mathbf{u}$ and $T \ge 1$. $\square$

In particular, we emphasize that a mean-optimal strategy $\tilde{\mathbf{u}}$ always exists. In the remainder of this paper, we will fix a mean-optimal strategy $\tilde{\mathbf{u}}$ as in Lemma 2.5.

2.2 Pathwise Optimality

Our results on the existence of pathwise optimal strategies are all consequences of one general result, Theorem 2.6, that will be stated presently. The essential assumption of this general result is that the properties of the conditional K-automorphism (Definition 1.6) hold uniformly with respect to the loss function ℓ. Note that, in principle, the assumptions of this result do not imply that $(\Omega, \mathcal{B}, \mathbf{P}, T)$ is a conditional K-automorphism, though this will frequently be the case.

Theorem 2.6 (Pathwise Optimality) *Suppose that for some σ-field $\mathfrak{X} \subset \mathcal{B}$*

1. $\mathfrak{X} \subset T^{-1}\mathfrak{X}$.

2. The following martingales converge uniformly:

$$\operatorname*{ess\,sup}_{u \in \mathbb{U}_0} \left| \mathbf{E}[\ell_0(u)|\mathcal{Y}_{-\infty,0} \vee T^{-n}\mathfrak{X}] - \ell_0(u) \right| \xrightarrow{n \to \infty} 0 \quad in \ L^1,$$

$$\operatorname*{ess\,sup}_{u \in \mathbb{U}_0} \left| \mathbf{E}[\ell_0(u)|\mathcal{Y}_{-\infty,0} \vee T^n\mathfrak{X}] - \mathbf{E}[\ell_0(u)|\bigcap_{k=1}^{\infty}(\mathcal{Y}_{-\infty,0} \vee T^k\mathfrak{X})] \right| \xrightarrow{n \to \infty} 0 \quad in \ L^1.$$

3. The remote past does not affect the asymptotic loss:

$$L^\star := \mathbf{E}\left[\operatorname*{ess\,inf}_{u \in \mathbb{U}_0} \mathbf{E}[\ell_0(u)|\mathcal{Y}_{-\infty,0}] \right] = \mathbf{E}\left[\operatorname*{ess\,inf}_{u \in \mathbb{U}_0} \mathbf{E}[\ell_0(u)|\bigcap_{k=1}^{\infty}(\mathcal{Y}_{-\infty,0} \vee T^k\mathfrak{X})] \right].$$

Then there exists an admissible strategy $\mathbf{u}^\star$ such that for every admissible strategy $\mathbf{u}$

$$\liminf_{T \to \infty}\{L_T(\mathbf{u}) - L_T(\mathbf{u}^\star)\} \geq 0 \quad a.s., \qquad \lim_{T \to \infty} L_T(\mathbf{u}^\star) = L^\star \quad a.s.,$$

that is, $\mathbf{u}^\star$ is pathwise optimal and $L^\star$ is the optimal long time-average loss.

The proof of this result will be given in Sect. 4.1 below.

Before going further, let us discuss the conceptual nature of the assumptions of Theorem 2.6. The assumptions encode two separate requirements:

1. Assumption 3 of Theorem 2.6 should be viewed as a mixing assumption on the dynamical system $(\Omega, \mathcal{B}, \mathbf{P}, T)$ that is tailored to the decision problem. Indeed, $\mathcal{Y}_{-\infty,0}$ represents the information contained in the observations, while $\bigcap_{k=1}^{\infty}(\mathcal{Y}_{-\infty,0} \vee T^k\mathfrak{X})$ includes in addition the remote past of the generating σ-field $\mathfrak{X}$. The assumption states that knowledge of the remote past of the unobserved part of the model cannot be used to improve our present decisions.

2. Assumption 2 of Theorem 2.6 should be viewed as a complexity assumption on the loss function ℓ. Indeed, in the absence of the essential suprema, these statements hold automatically by the martingale convergence theorem. The assumption requires that the convergence is in fact uniform in $u \in \mathbb{U}_0$. This will be the case when the loss function is not too complex.

The assumptions of Theorem 2.6 can be verified in many cases of interest. In Sect. 3 below, we will discuss various methods that can be used to verify both the conditional mixing and the complexity assumptions of Theorem 2.6.

In general, neither the conditional mixing nor the complexity assumption can be dispensed with in the presence of partial information.

Example 2.7 (Assumption 3 is Essential) We have seen in Examples 1.2 and 1.3 in the introduction that no pathwise optimal strategy exists. In both these examples Assumption 2 is satisfied, that is, the loss function is not too complex (this will follow from general complexity results, cf. Example 3.6 in Sect. 3 below). On the other hand, it is easily seen that the conditional mixing Assumption 3 is violated.

Example 2.8 (Assumption 2 is Essential) Let $X = (X_k)_{k \in \mathbb{Z}}$ be the stationary Markov chain in $[0, 1]$ defined by $X_{k+1} = (X_k + \varepsilon_{k+1})/2$ for all k, where $(\varepsilon_k)_{k \in \mathbb{Z}}$ is an i.i.d. sequence of Bernoulli$(1/2)$ random variables. We consider the setting of blind decisions with the loss function $\ell_k(u) = \lfloor 2^u X_k \rfloor \bmod 2$, $u \in U = \mathbb{N}$. Note that

$$X_k = \sum_{i=0}^{\infty} 2^{-i-1} \varepsilon_{k-i}, \qquad \ell_k(u) = \varepsilon_{k-u+1}.$$

We claim that no pathwise optimal strategy can exist. Indeed, consider for fixed $r \geq 0$ the strategy $\mathbf{u}^r$ such that $u_k^r = k + r$. Then $\ell_k(u_k^r) = \varepsilon_{1-r}$ for all k. Therefore,

$$\varepsilon_{1-r} - \limsup_{T \to \infty} L_T(\mathbf{u}^\star) = \liminf_{T \to \infty} \{L_T(\mathbf{u}^r) - L_T(\mathbf{u}^\star)\} \geq 0 \quad \text{a.s.} \quad \text{for all } r \geq 0$$

for every pathwise optimal strategy $\mathbf{u}^\star$. In particular,

$$0 = \inf_{r \geq 0} \varepsilon_{1-r} \geq \limsup_{T \to \infty} L_T(\mathbf{u}^\star) \geq \liminf_{T \to \infty} L_T(\mathbf{u}^\star) \geq 0 \quad \text{a.s.}$$

As $|L_T(\mathbf{u}^\star)| \leq 1$ for all T, it follows by dominated convergence that a pathwise optimal strategy $\mathbf{u}^\star$ must satisfy $\mathbf{E}[L_T(\mathbf{u}^\star)] \to 0$ as $T \to \infty$. But clearly $\mathbf{E}[L_T(\mathbf{u})] = 1/2$ for every T and strategy $\mathbf{u}$, which entails a contradiction.

Nonetheless, in this example the dynamical system is a K-automorphism (even a Bernoulli shift), so that Assumption 3 is easily satisfied. As no pathwise optimal strategy exists, this must be caused by the failure of Assumption 2. For example, for the natural choice $\mathcal{X} = \sigma\{X_k : k \leq 0\}$, Assumption 3 holds as $\bigcap_k T^k \mathcal{X}$ is trivial by the Kolmogorov zero-one law, but it is easily seen that the second equation of Assumption 2 fails. Note that the function $l(u, x) = \lfloor 2^u x \rfloor \bmod 2$ becomes increasingly oscillatory as $u \to \infty$; this is precisely the type of behavior that obstructs uniform convergence in Assumption 2 (akin to "overfitting" in statistics).

Example 2.9 (Assumption 2 is Essential, continued) In the previous example, pathwise optimality fails due to failure of the second equation of Assumption 2. We now give a variant of this example where the first equation of Assumption 2 fails.

Let $X = (X_k)_{k \in \mathbb{Z}}$ be an i.i.d. sequence of Bernoulli$(1/2)$ random variables. We consider the setting of blind decisions with the loss function $\ell_k(u) = X_{k+u}$, $u \in U = \mathbb{N}$. We claim that no pathwise optimal strategy can exist. Indeed, consider for $r = 0, 1$ the strategy $\mathbf{u}^r$ defined by $u_k = 2^{r+n+1} - k$ for $2^n \leq k < 2^{n+1}$, $n \geq 0$. Then

$$L_{2^n - 1}(\mathbf{u}^r) = \frac{1}{2^n - 1} \sum_{m=0}^{n-1} \sum_{k=2^m}^{2^{m+1}-1} X_{k+u_k} = \frac{2^n}{2^n - 1} \sum_{m=0}^{n-1} 2^{-(n-m)} X_{2^{r+m+1}}.$$

Suppose that $\mathbf{u}^\star$ is pathwise optimal. Then

$$\liminf_{T\to\infty} \mathbf{E}[L_T(\mathbf{u}^0) \wedge L_T(\mathbf{u}^1) - L_T(\mathbf{u}^\star)] \geq \mathbf{E}\left[\liminf_{T\to\infty}\{L_T(\mathbf{u}^0) \wedge L_T(\mathbf{u}^1) - L_T(\mathbf{u}^\star)\}\right] \geq 0.$$

But a simple computation shows that $\mathbf{E}[L_{2^n-1}(\mathbf{u}^0) \wedge L_{2^n-1}(\mathbf{u}^1)]$ converges as $n \to \infty$ to a quantity strictly less than $1/2 = \mathbf{E}[L_T(\mathbf{u}^\star)]$, so that we have a contradiction.

Nonetheless, in this example Assumption 3 and the second line of Assumption 2 are easily satisfied, e.g., for the natural choice $\mathcal{X} = \sigma\{X_k : k \leq 0\}$. However, the first line of Assumption 2 fails, and indeed no pathwise optimal strategy exists.

It is evident from the previous examples that an assumption on both conditional mixing and on complexity of the loss function is needed, in general, to ensure existence of a pathwise optimal strategy. In this light, the complete absence of any such assumptions in the full information case is surprising. The explanation is simple, however: all assumptions of Theorem 2.6 are automatically satisfied in this case.

Example 2.10 (Full Information) Let $X = (X_k)_{k\in\mathbb{Z}}$ be any stationary ergodic process, and consider the case of full information: that is, we choose the observation σ-field $\mathcal{Y} = \sigma\{X_0\}$ and the loss $\ell(u, \omega) = l(u, X_1(\omega))$. Then all assumptions of Theorem 2.6 are satisfied: indeed, if we choose $\mathcal{X} = \sigma\{X_k : k \leq 0\}$, then $\mathcal{Y}_{-\infty,0} = \mathcal{Y}_{-\infty,0} \vee T^k\mathcal{X}$ for all $k \geq 0$, so that Assumption 3 and the second line of Assumption 2 hold trivially. Moreover, $\ell_0(u)$ is $T^{-k}\mathcal{X}$-measurable for every $u \in \mathbb{U}_0$ and $k \geq 1$, and thus the first line of Assumption 2 holds trivially. It follows that in the full information setting, a pathwise optimal strategy always exists.

In a sense, Theorem 2.6 provides additional insight even in the full information setting: it provides an explanation as to why the case of full information is so much simpler than the partial information setting. Moreover, Theorem 2.6 provides an explicit expression for the optimal asymptotic loss $L^\star$, which is not given in [1].[3]

However, it should be emphasized that Theorem 2.6 does not state that the mean-optimal strategy $\tilde{\mathbf{u}}$ is pathwise optimal; it only guarantees the existence of some pathwise optimal strategy $\mathbf{u}^\star$. In contrast, in the full information setting, Theorem 1.1 ensures pathwise optimality of the mean-optimal strategy. This is of practical importance, as the mean-optimal strategy can in many cases be computed explicitly or by efficient numerical methods, while the pathwise optimal strategy constructed in the proof of Theorem 2.6 may be difficult to compute. We do not know whether it is possible in the general setting of Theorem 2.6 that a pathwise

---

[3]In [1, Appendix II.B] it is shown that under a continuity assumption on the loss function l, the optimal asymptotic loss in the full information setting is given by $\mathbf{E}[\inf_u \mathbf{E}[l(u, X_1)|X_0, X_{-1}, \ldots]]$. However, a counterexample is given of a discontinuous loss function for which this expression does not yield the optimal asymptotic loss. The key difference with the expression for $L^\star$ given in Theorem 2.6 is that in the latter the essential infimum runs over $u \in \mathbb{U}_0$, while it is implicit in [1] that the infimum in the above expression is an essential infimum over $u \in \mathbb{U}_{-\infty,0}$. As the counterexample in [1] shows, these quantities need not coincide in the absence of continuity assumptions.

optimal strategy exists, but that the mean-optimal strategy $\tilde{\mathbf{u}}$ is not pathwise optimal. Pathwise optimality of the mean-optimal strategy $\tilde{\mathbf{u}}$ can be shown, however, under somewhat stronger assumptions. The following corollary is proved in Sect. 4.2 below.

Corollary 2.11 *Suppose that for some σ-field $\mathfrak{X} \subset \mathcal{B}$*

1. $\mathfrak{X} \subset T^{-1}\mathfrak{X}$.
2. The following martingales converge uniformly:

$$\underset{u \in \mathbb{U}_0}{\mathrm{ess\,sup}} \left| \mathbf{E}[\ell_0(u) | \mathcal{Y}_{-\infty,0} \vee T^{-n}\mathfrak{X}] - \ell_0(u) \right| \xrightarrow{n \to \infty} 0 \quad in \ L^1,$$

$$\underset{u \in \mathbb{U}_0}{\mathrm{ess\,sup}} \left| \mathbf{E}[\ell_0(u) | \mathcal{Y}_{-\infty,0} \vee T^n\mathfrak{X}] - \mathbf{E}[\ell_0(u) | \bigcap_{k=1}^{\infty}(\mathcal{Y}_{-\infty,0} \vee T^k\mathfrak{X})] \right| \xrightarrow{n \to \infty} 0 \quad in \ L^1,$$

$$\underset{u \in \mathbb{U}_{-n,0}}{\mathrm{ess\,sup}} \left| \mathbf{E}[\ell_0(u) | \mathcal{Y}_{-n,0}] - \mathbf{E}[\ell_0(u) | \mathcal{Y}_{-\infty,0}] \right| \xrightarrow{n \to \infty} 0 \quad a.s.$$

3. The remote past does not affect the present:

$$\mathbf{E}[\ell_0(u) | \mathcal{Y}_{-\infty,0}] = \mathbf{E}[\ell_0(u) | \bigcap_{k=1}^{\infty}(\mathcal{Y}_{-\infty,0} \vee T^k\mathfrak{X})] \qquad for \ all \ u \in \mathbb{U}_0.$$

Then the mean-optimal strategy $\tilde{\mathbf{u}}$ (Lemma 2.5) satisfies $L_T(\tilde{\mathbf{u}}) \to L^\star$ a.s. as $T \to \infty$. In particular, it follows from Theorem 2.6 that $\tilde{\mathbf{u}}$ is pathwise optimal.

The assumptions of Corollary 2.11 are stronger than those of Theorem 2.6 in two respects. First, Assumption 3 is slightly strengthened; however, this is a very mild requirement. More importantly, a third martingale is assumed to converge uniformly (pathwise!) in Assumption 2. The latter is not an innocuous requirement: while the assumption holds in many cases of interest, substantial regularity of the loss function is needed (see Sect. 3.1 for further discussion). In particular, this requirement is not automatically satisfied in the case of full information, and Theorem 1.1 therefore does not follow in its entirety from our results. It remains an open question whether it is possible to establish pathwise optimality of the mean-optimal strategy $\tilde{\mathbf{u}}$ under a substantial weakening of the assumptions of Corollary 2.11.

A particularly simple regularity assumption on the loss is that the decision space U is finite. In this case uniform convergence is immediate, so that the assumptions of Corollary 2.11 reduce essentially to the $\mathcal{Y}_{-\infty,0}$-conditional K-property. Therefore, evidently Corollary 2.11 implies Theorem 1.7. More general conditions that ensure the validity of the requisite assumptions will be discussed in Sect. 3.

2.3 Weak Pathwise Optimality

In the previous section, we have seen that a pathwise optimal strategy $\mathbf{u}^\star$ exists under general assumptions. However, unlike in the full information case, it is not clear whether in general (without a nontrivial complexity assumption) the mean-optimal strategy $\tilde{\mathbf{u}}$ is pathwise optimal. In the present section, we will aim to obtain some additional insight into this issue by considering the notion of weak pathwise optimality (Definition 2.3) that is intermediate between pathwise optimality and mean optimality. This notion is more regularly behaved than pathwise optimality; in particular, it is straightforward to prove the following simple result.

Lemma 2.12 *Suppose that a weakly pathwise optimal strategy* $\mathbf{u}^\star$ *exists. Then the mean-optimal strategy* $\tilde{\mathbf{u}}$ *is also weakly pathwise optimal.*

Proof Let $\Lambda_T = \frac{1}{T} \sum_{k=1}^{T} \Lambda \circ T^k$. As $|L_T(\mathbf{u})| \leq \Lambda_T$ for any strategy $\mathbf{u}$, we have

$$\mathbf{E}[(L_T(\tilde{\mathbf{u}}) - L_T(\mathbf{u}^\star))_-] \leq \varepsilon \, \mathbf{P}[L_T(\tilde{\mathbf{u}}) - L_T(\mathbf{u}^\star) \geq -\varepsilon] + \mathbf{E}[2\Lambda_T \, \mathbf{1}_{L_T(\tilde{\mathbf{u}}) - L_T(\mathbf{u}^\star) < -\varepsilon}]$$

for any $\varepsilon > 0$. Note that the sequence $(\Lambda_T)_{T \geq 1}$ is uniformly integrable as $\Lambda_T \to \mathbf{E}[\Lambda]$ in L^1 by the ergodic theorem. Therefore, using weak pathwise optimality of $\mathbf{u}^\star$, it follows that $\mathbf{E}[(L_T(\tilde{\mathbf{u}}) - L_T(\mathbf{u}^\star))_-] \to 0$ as $T \to \infty$. We therefore have

$$\limsup_{T \to \infty} \mathbf{E}[|L_T(\tilde{\mathbf{u}}) - L_T(\mathbf{u}^\star)|] = -\liminf_{T \to \infty}\{\mathbf{E}[L_T(\mathbf{u}^\star)] - \mathbf{E}[L_T(\tilde{\mathbf{u}})]\} \leq 0$$

by mean-optimality of $\tilde{\mathbf{u}}$. It follows easily that $\tilde{\mathbf{u}}$ is also pathwise optimal. □

While Theorem 2.6 does not ensure that the mean-optimal strategy $\tilde{\mathbf{u}}$ is pathwise optimal, the previous lemma guarantees that $\tilde{\mathbf{u}}$ is at least weakly pathwise optimal. However, we will presently show that the latter conclusion may follow under considerably weaker assumptions than those of Theorem 2.6. Indeed, just as pathwise optimality was established for conditional K-automorphisms, we will establish weak optimality for conditionally weakly mixing automorphisms.

Let us begin by developing a general result on weak pathwise optimality, Theorem 2.13 below, that plays the role of Theorem 2.6 in the present setting. The essential assumption of this general result is that the conditional weak mixing property (Definition 1.5) holds uniformly with respect to the loss function ℓ. For simplicity of notation, let us define as in Theorem 2.6 the optimal asymptotic loss

$$L^\star := \mathbf{E}\left[\underset{u \in \mathbb{U}_0}{\mathrm{ess\,inf}} \, \mathbf{E}[\ell_0(u)|\mathcal{Y}_{-\infty,0}] \right]$$

(let us emphasize, however, that the Assumption 3 of Theorem 2.6 need not hold in the present setting!) In addition, let us define the modified loss functions

$$\bar{\ell}_0(u) := \ell_0(u) - \mathbf{E}[\ell_0(u)|\mathcal{Y}_{-\infty,0}], \quad \bar{\ell}_0^M(u) := \ell_0(u)\mathbf{1}_{\Lambda \leq M} - \mathbf{E}[\ell_0(u)\mathbf{1}_{\Lambda \leq M}|\mathcal{Y}_{-\infty,0}].$$

The proof of the following theorem will be given in Sect. 4.3.

Theorem 2.13 *Suppose that the uniform conditional mixing assumption*

$$\lim_{M\to\infty} \limsup_{T\to\infty} \left\| \frac{1}{T} \sum_{k=1}^{T} \operatorname{ess\,sup}_{u,u'\in\mathbb{U}_0} |\mathbf{E}[\{\bar{\ell}_0^M(u) \circ T^{-k}\} \, \bar{\ell}_0^M(u') | \mathcal{Y}_{-\infty,0}]| \right\|_1 = 0$$

holds. Then the mean-optimal strategy $\tilde{u}$ is weakly pathwise optimal, and the optimal long time-average loss satisfies the ergodic theorem $L_T(\tilde{u}) \to L^\star$ in L^1.

Remark 2.14 We have assumed throughout that the loss function ℓ is dominated in L^1. If the loss is in fact dominated in L^2, that is, $|\ell(u,\omega)| \le \Lambda(\omega)$ with $\Lambda \in L^2$, then the assumption of Theorem 2.13 is evidently implied by the natural assumption

$$\frac{1}{T} \sum_{k=1}^{T} \operatorname{ess\,sup}_{u,u'\in\mathbb{U}_0} |\mathbf{E}[\{\bar{\ell}_0(u) \circ T^{-k}\} \, \bar{\ell}_0(u') | \mathcal{Y}_{-\infty,0}]| \xrightarrow{T\to\infty} 0 \quad \text{in } L^1,$$

and in this case $L_T(\tilde{u}) \to L^\star$ in L^2 (by dominated convergence). The additional truncation in Theorem 2.13 is included only to obtain a result that holds in L^1.

Conceptually, as in Theorem 2.6, the assumption of Theorem 2.13 combines a conditional mixing assumption and a complexity assumption. Indeed, the conditional weak mixing property relative to $\mathcal{Y}_{-\infty,0}$ (Definition 1.5) implies that

$$\frac{1}{T} \sum_{k=1}^{T} |\mathbf{E}[\{f \circ T^{-k}\} \, g | \mathcal{Y}_{-\infty,0}] - \mathbf{E}[f \circ T^{-k} | \mathcal{Y}_{-\infty,0}] \, \mathbf{E}[g | \mathcal{Y}_{-\infty,0}]| \xrightarrow{T\to\infty} 0 \quad \text{in } L^1$$

for every $f, g \in L^2$ (indeed, for simple functions f, g this follows directly from the definition, and the claim for general f, g follows by approximation in L^2). Therefore, in the absence of the essential supremum, the assumption of Theorem 2.13 reduces essentially to the assumption that the dynamical system $(\Omega, \mathcal{B}, \mathbf{P}, T)$ is conditionally weak mixing relative to $\mathcal{Y}_{-\infty,0}$. However, Theorem 2.13 requires in addition that the convergence in the definition of the conditional weak mixing property holds uniformly with respect to the possible decisions $u \in \mathbb{U}_0$. This will be the case when the loss function ℓ is not too complex (cf. Sect. 3). For example, in the extreme case where the decision space U is finite, uniformity is automatic, and thus Theorem 1.8 in the introduction follows immediately from Theorem 2.13.

Recall that a pathwise optimal strategy is necessarily weakly pathwise optimal. This is reflected, for example, in Theorems 1.7 and 1.8: indeed, note that

$$\|\mathbf{P}[A \cap T^k B | \mathcal{Z}] - \mathbf{P}[A | \mathcal{Z}] \, \mathbf{P}[T^k B | \mathcal{Z}]\|_1$$

$$= \|\mathbf{E}[\{\mathbf{1}_A - \mathbf{P}[A | \mathcal{Z}]\} \, \mathbf{1}_{T^k B} | \mathcal{Z}]\|_1$$

$$\le \|\mathbf{E}[\{\mathbf{1}_A - \mathbf{P}[A | \mathcal{Z}]\} \, \mathbf{P}[T^k B | T^{k-n}\mathcal{X}] | \mathcal{Z}]\|_1 + \|\mathbf{1}_{T^k B} - \mathbf{P}[T^k B | T^{k-n}\mathcal{X}]\|_1$$

$$\le \|\mathbf{P}[A | \mathcal{Z} \vee T^{k-n}\mathcal{X}] - \mathbf{P}[A | \mathcal{Z}]\|_1 + \|\mathbf{1}_B - \mathbf{P}[B | T^{-n}\mathcal{X}]\|_1$$

for any n, k, so that the conditional K-property implies the conditional weak mixing property (relative to any σ-field $\mathcal{Z}$) by letting $k \to \infty$, then $n \to \infty$. Along the same lines, one can show that a slight variation of the assumptions of Theorem 2.6 imply the assumption of Theorem 2.13 (modulo minor issues of truncation, which could have been absorbed in Theorem 2.6 also at the expense of heavier notation). It is not entirely obvious, at first sight, how far apart the conclusions of our main results really are. For example, in the setting of full information, cf. Example 2.10, the assumption of Theorem 2.13 holds automatically (as then $\bar{\ell}_0^M (u) \circ T^{-k}$ is $\mathcal{Y}_{-\infty,0}$-measurable for every $u \in \mathbb{U}_0$ and $k \geq 1$). Moreover, the reader can easily verify that in all the examples we have given where no pathwise optimal strategy exists (Examples 1.2, 1.3, 2.8, 2.9), even the existence of a weakly pathwise optimal strategy fails. It is therefore tempting to assume that in a typical situation where a weakly pathwise optimal strategy exists, there will likely also be a pathwise optimal strategy. The following example, which is a manifestation of a rather surprising result in ergodic theory due to Conze [6], provides some evidence to the contrary.

Example 2.15 (Generic Transformations) In this example, we fix the probability space $(\Omega, \mathcal{B}, \mathbf{P})$, where $\Omega = [0, 1]$ with its Borel σ-field $\mathcal{B}$ and the Lebesgue measure $\mathbf{P}$. We consider the decision space $U = \{0, 1\}$ and loss function ℓ defined as

$$\ell(u, \omega) = -u \left(\mathbf{1}_{[0,1/2]}(\omega) - 1/2 \right) \qquad \text{for } (u, \omega) \in U \times \Omega.$$

Moreover, we will consider the setting of blind decisions, that is, $\mathcal{Y}$ is trivial.

We have not yet defined a transformation T. Our aim is to prove the following: *for a generic invertible measure-preserving transformation T, there is a mean-optimal strategy $\tilde{\mathbf{u}}$ that is weakly pathwise optimal but not pathwise optimal*. This shows not only that there can be a substantial gap between Theorems 1.7 and 1.8, but that this is in fact the typical situation (at least in the sense of weak topology).

Let us recall some basic notions. Denote by $\mathcal{T}$ the set of all invertible measure-preserving transformations of $(\Omega, \mathcal{B}, \mathbf{P})$. The weak topology on $\mathcal{T}$ is the topology generated by the basic neighborhoods $B(T_0, B, \varepsilon) = \{T \in \mathcal{T} : \mathbf{P}[TB \triangle T_0B] < \varepsilon\}$ for all $T_0 \in \mathcal{T}$, $B \in \mathcal{B}$, $\varepsilon > 0$. A property is said to hold for a *generic* transformation if it holds for every transformation T in a dense G_δ subset of $\mathcal{T}$. A well-known result of Halmos [13] states that a generic transformation is weak mixing. Therefore, for a generic transformation, any mean-optimal strategy $\tilde{\mathbf{u}}$ is weakly pathwise optimal by Theorem 1.8. This proves the first part of our statement.

Of course, in the present setting, $\mathbf{E}[\ell_k(u)|\mathcal{Y}_{0,k}] = \mathbf{E}[\ell_k(u)] = 0$ for every decision $u \in U$. Therefore, *every* admissible strategy $\mathbf{u}$ is mean-optimal, and the optimal mean loss is given by $L^\star = 0$, regardless of the choice of transformation $T \in \mathcal{T}$. It is natural to choose a stationary strategy $\tilde{\mathbf{u}}$ (for example, $\tilde{u}_k = 1$ for all k) so that $\lim_{T \to \infty} L_T(\tilde{\mathbf{u}}) = L^\star$ a.s. We will show that for a generic transformation, the strategy $\tilde{\mathbf{u}}$ is not pathwise optimal. To this end, it evidently suffices to find another strategy $\mathbf{u}$ such that $\liminf_{T \to \infty} L_T(\mathbf{u}) < L^\star$ with positive probability.

To this end, we use the following result of Conze that can be read off from the proof of [6, Theorem 5]: there exists a sequence $n_k \uparrow \infty$ with $k/n_k \to 1/2$ such that for every $0 < \alpha < 1$ and $1/2 < \lambda < 1$, a generic transformation T satisfies

$$\mathbf{P}\left[\limsup_{N\to\infty} \frac{1}{N} \sum_{k=1}^{N} \mathbf{1}_{[0,1/2]} \circ T^{n_k} \geq \lambda\right] \geq 1 - \alpha.$$

Define the strategy $\mathbf{u}$ such that $u_n = 1$ if $n = n_k$ for some k, and $u_n = 0$ otherwise. Then, for a generic transformation T, we have with probability at least $1 - \alpha$

$$\liminf_{T\to\infty} L_{n_T}(\mathbf{u}) = -\limsup_{T\to\infty} \frac{1}{n_T} \sum_{k=1}^{T} (\mathbf{1}_{[0,1/2]} \circ T^{n_k} - 1/2) \leq -\frac{2\lambda - 1}{4}.$$

In words, we have shown that for a generic transformation T, the time-average loss of the mean-optimal strategy $\tilde{\mathbf{u}}$ exceeds that of the strategy $\mathbf{u}$ infinitely often by almost $1/4$ with almost unit probability. Thus the mean-optimal strategy $\tilde{\mathbf{u}}$ fails to be pathwise optimal in a very strong sense, and our claim is established.

Example 2.15 only shows that there is a mean-optimal strategy $\tilde{\mathbf{u}}$ that is weakly pathwise optimal but not pathwise optimal. It does not make any statement about whether or not a pathwise optimal strategy $\mathbf{u}^\star$ actually exists. However, we do not know of any mechanism that might lead to pathwise optimality in such a setting. We therefore conjecture that for a generic transformation a pathwise optimal strategy in fact fails to exist at all, so that (unlike in the full information setting) pathwise optimality and weak pathwise optimality are really distinct notions.

The result of Conze used in Example 2.15 originates from a deep problem in ergodic theory that aims to understand the validity of individual ergodic theorems for subsequences, cf. [2, 6] and the references therein. A general characterization of such ergodic properties does not appear to exist, which suggests that the pathwise optimality property may be difficult to characterize beyond general sufficient conditions such as Theorem 2.6. In contrast, the weak pathwise optimality property is much more regularly behaved. The following theorem, which will be proved in Sect. 4.4 below, provides a complete characterization of weak pathwise optimality in the special case that the observation field $\mathcal{Y}$ is invariant.

Theorem 2.16 *Let $(\Omega, \mathcal{B}, \mathbf{P}, T)$ be an ergodic dynamical system, and suppose that $(\Omega, \mathcal{B}, \mathbf{P})$ is a standard probability space and that $\mathcal{Y} \subseteq \mathcal{B}$ is an invariant σ-field (that is, $\mathcal{Y} = T^{-1}\mathcal{Y}$). Then the following are equivalent:*

1. *$(\Omega, \mathcal{B}, \mathbf{P}, T)$ conditionally weak mixing relative to $\mathcal{Y}$.*
2. *For every bounded loss function $\ell : U \times \Omega \to \mathbb{R}$ with finite decision space card $U < \infty$, there exists a weakly pathwise optimal strategy.*

The invariance of $\mathcal{Y}$ is automatic in the setting of blind decisions (as $\mathcal{Y}$ is trivial), in which case Theorem 2.16 yields a decision-theoretic characterization of the weak mixing property. In more general observation models, invariance of $\mathcal{Y}$ may be an unnatural requirement from the point of view of decisions under partial information, as it implies that there is no information gain over time. On the other hand, applications of the notion of conditional weak mixing relative to a σ-field $\mathcal{Z}$ in ergodic theory almost always assume that $\mathcal{Z}$ is invariant (e.g., [26]). Theorem 2.16 yields a decision-theoretic interpretation of this property by choosing $\mathcal{Y} = \mathcal{Z}$.

3 Complexity and Conditional Ergodicity

3.1 Universal Complexity Assumptions

The goal of this section is to develop complexity assumptions on the loss function ℓ that ensure that the uniform convergence assumptions in our main results hold regardless of any properties of the transformation T or observations $\mathcal{Y}$. While such universal complexity assumptions are not always necessary (for example, in the full information setting uniform convergence holds regardless of the loss function), they frequently hold in practice and provide easily verifiable conditions that ensure that our results hold in a broad class of decision problems with partial information.

The simplest assumption is Grothendieck's notion of equimeasurability [12].

Definition 3.1 The loss function $\ell : U \times \Omega \to \mathbb{R}$ on the probability space $(\Omega, \mathcal{B}, \mathbf{P})$ is said to be *equimeasurable* if for every $\varepsilon > 0$, there exists $\Omega_\varepsilon \in \mathcal{B}$ with $\mathbf{P}[\Omega_\varepsilon] \geq 1 - \varepsilon$ such that the class of functions $\{\ell_0(u)\mathbf{1}_{\Omega_\varepsilon} : u \in U\}$ is totally bounded in $L^\infty(\mathbf{P})$.

The beauty of this simple notion is that it ensures uniform convergence of almost anything. In particular, we obtain the following results.

Lemma 3.2 *Suppose that the loss function ℓ is equimeasurable. Then Assumption 2 of Corollary 2.11 holds, and thus Assumption 2 of Theorem 2.6 holds as well, provided that $\mathcal{X}$ is a generating σ-field (that is, $\bigvee_n T^{-n}\mathcal{X} = \mathcal{B}$).*

Proof Let us establish the first line of Assumption 2. Fix $\varepsilon > 0$ and Ω_ε as in Definition 3.1. Then there exist $N < \infty$ measurable functions $l_1, \ldots, l_N : \Omega \to \mathbb{R}$ such that for every $u \in U$, there exists $k(u) \in \{1, \ldots, N\}$ such that

$$\|\ell_0(u)\mathbf{1}_{\Omega_\varepsilon} - l_{k(u)}\mathbf{1}_{\Omega_\varepsilon}\|_\infty \leq \varepsilon$$

(and $u \mapsto k(u)$ can clearly be chosen to be measurable). It follows that

$$\operatorname*{ess\,sup}_{u \in \mathbb{U}_0} \left|\mathbf{E}[\ell_0(u)|\mathcal{Y}_{-\infty,0} \vee T^{-n}\mathcal{X}] - \ell_0(u)\right| \leq \max_{1 \leq k \leq N} \left|\mathbf{E}[l_k \mathbf{1}_{\Omega_\varepsilon}|\mathcal{Y}_{-\infty,0} \vee T^{-n}\mathcal{X}] - l_k \mathbf{1}_{\Omega_\varepsilon}\right|$$

$$+ 2\varepsilon + \mathbf{E}[\Lambda \mathbf{1}_{\Omega_\varepsilon^c}|\mathcal{Y}_{-\infty,0} \vee T^{-n}\mathcal{X}] + \Lambda \mathbf{1}_{\Omega_\varepsilon^c}.$$

As $\mathcal{X}$ is generating, the martingale convergence theorem gives

$$\limsup_{n\to\infty} \left\| \operatorname*{ess\,sup}_{u\in U_0} \left| \mathbf{E}[\ell_0(u)|\mathcal{Y}_{-\infty,0} \vee T^{-n}\mathcal{X}] - \ell_0(u) \right| \right\|_1 \le 2\varepsilon + \mathbf{E}[2\Lambda \mathbf{1}_{\Omega_\varepsilon^c}].$$

Letting $\varepsilon \downarrow 0$ yields the first line of Assumption 2. The remaining statements of Assumption 2 follow by an essentially identical argument. $\qquad\square$

Lemma 3.3 *Suppose that the following conditional mixing assumption holds:*

$$\lim_{M\to\infty} \limsup_{T\to\infty} \left\| \frac{1}{T} \sum_{k=1}^{T} |\mathbf{E}[\{\bar{\ell}_0^M(u)\circ T^{-k}\}\, \bar{\ell}_0^M(u')|\mathcal{Y}_{-\infty,0}]| \right\|_1 = 0 \quad \text{for every } u, u' \in U.$$

If the loss function ℓ is equimeasurable, then the assumption of Theorem 2.13 holds.

Proof The proof is very similar to that of Lemma 3.2 and is therefore omitted. $\qquad\square$

As an immediate consequence of these lemmas, we have:

Corollary 3.4 *The conclusions of Theorems 1.7 and 1.8 remain in force if the assumption that U is finite is replaced by the assumption that ℓ is equimeasurable.*

We now give a simple condition for equimeasurability that suffices in many cases. It is closely related to a result of Mokobodzki (cf. [9, Theorem IX.19]).

Lemma 3.5 *Suppose that U is a compact metric space and that $u \mapsto \ell(u, \omega)$ is continuous for a.e. $\omega \in \Omega$. Then ℓ is equimeasurable.*

Proof As U is a compact metric space (with metric d), it is certainly separable. Let $U_0 \subseteq U$ be a countable dense set, and define the functions

$$b_n = \sup_{u,u'\in U_0 : d(u,u')\le n^{-1}} |\ell_0(u) - \ell_0(u')|.$$

b_n is measurable, as it is the supremum of countably many random variables. Moreover, for almost every ω, the function $u \mapsto \ell(u, \omega)$ is uniformly continuous (being continuous on a compact metric space). Therefore, $b_n \downarrow 0$ a.s. as $n \to \infty$.

By Egorov's theorem, there exists for every $\varepsilon > 0$ a set Ω_ε with $\mathbf{P}[\Omega_\varepsilon] \ge 1 - \varepsilon$ such that $\|b_n \mathbf{1}_{\Omega_\varepsilon}\|_\infty \downarrow 0$. We claim that $\{\ell_0(u)\mathbf{1}_{\Omega_\varepsilon} : u \in U\}$ is compact in L^∞. Indeed, for any sequence $(u_n)_{n\ge1} \subseteq U$ we may choose a subsequence $(u_{n_k})_{k\ge1}$ that converges to $u_\infty \in U$. Then for every r, we have $|\ell_0(u_{n_k}) - \ell_0(u_\infty)| \le b_r$ for all k sufficiently large, and therefore $\|\ell_0(u_{n_k})\mathbf{1}_{\Omega_\varepsilon} - \ell_0(u_\infty)\mathbf{1}_{\Omega_\varepsilon}\|_\infty \to 0$. $\qquad\square$

Let us give two standard examples of decision problems (cf. [1, 24]).

Example 3.6 (ℓ_p-Prediction) Consider the stochastic process setting (X, Y), and let f be a bounded function. The aim is, at each time k, to choose a predictor u_k of $f(X_{k+1})$ on the basis of the observation history $Y_0, \ldots, Y_k$. We aim to minimize the pathwise time-average ℓ_p-prediction loss $\frac{1}{T}\sum_{k=1}^{T} |u_k - f(X_{k+1})|^p$

($p \geq 1$). This is a particular decision problem with partial information, where the loss function is given by $\ell_0(u) = |u - f(X_1)|^p$ and the decision space is $U = [\inf_x f(x), \sup_x f(x)]$. It is immediate that ℓ is equimeasurable by Lemma 3.5.

Example 3.7 (Log-Optimal Portfolios) Consider a market with d securities (e.g., $d - 1$ stocks and one bond) whose returns in day k are given by the random variable X_k with values in $\mathbb{R}_+^d$. The decision space $U = \{p \in \mathbb{R}_+^d : \sum_{i=1}^d p_i = 1\}$ is the simplex: u_k^i represents the fraction of wealth invested in the ith security in day k. The total wealth at time T is therefore given by $\prod_{k=1}^T \langle u_k, X_k \rangle$. We only have access to partial information Y_k in day k, e.g., from news reports. We aim to choose an investment strategy on the basis of the available information that maximizes the wealth, or, equivalently, its growth $\frac{1}{T} \sum_{k=1}^T \log\langle u_k, X_k \rangle$. This corresponds to a decision problem with partial information for the loss function $\ell_0(u) = -\log\langle u, X_0 \rangle$.

In order for the loss to be dominated in L^1, we impose the mild assumption $\mathbf{E}[\Lambda] < \infty$ with $\Lambda = \sum_{i=1}^d |\log X_0^i|$. We claim that the loss ℓ is then also equimeasurable. Indeed, as $\mathbf{E}[\Lambda] < \infty$, the returns must satisfy $X_0^i > 0$ a.s. for every i. Therefore, equimeasurability follows directly from Lemma 3.5.

As we have seen above, equimeasurability follows easily when the loss function possesses some mild pointwise continuity properties. However, there are situations when this may not be the case. In particular, suppose that $\ell(u, \omega)$ only takes the values 0 and 1, that is, our decisions are sets (as may be the case, for example, in predicting the shape of an oil spill or in sequential classification problems). In such a case, equimeasurability will rarely hold, and it is of interest to investigate alternative complexity assumptions. As we will presently explain, equimeasurability is almost necessary to obtain a universal complexity assumption for Corollary 2.11; however, in the setting of Theorem 2.6, the assumption can be weakened considerably.

The simplicity of the equimeasurability assumption hides the fact that there are two distinct uniformity assumptions in Corollary 2.11: we require uniform convergence of both martingales and reverse martingales, which are quite distinct phenomena (cf. [17,18]). The uniform convergence of martingales can be restrictive.

Example 3.8 (Uniform Martingale Convergence) Let $(\mathcal{G}_n)_{n \geq 1}$ be a filtration such that each $\mathcal{G}_n = \sigma\{\pi_n\}$ is generated by a finite measurable partition π_n of the probability space $(\Omega, \mathcal{B}, \mathbf{P})$. Let $L : \mathbb{N} \times \Omega \to \mathbb{R}$ a bounded function such that $L(u, \cdot)$ is $\mathcal{G}_\infty$-measurable for every $u \in \mathbb{N}$. Then $\mathbf{E}[L(u, \cdot)|\mathcal{G}_n] \to L(u, \cdot)$ a.s. for every u. We claim that if this martingale convergence is in fact uniform, that is,

$$\sup_{u \in \mathbb{N}} |\mathbf{E}[L(u, \cdot)|\mathcal{G}_n] - L(u, \cdot)| \xrightarrow{n \to \infty} 0 \quad \text{in } L^1,$$

then L must necessarily be equimeasurable. To see this, let us first extract a subsequence $n_k \uparrow \infty$ along which the uniform martingale convergence holds a.s. Fix $\varepsilon > 0$. By Egorov's theorem, there exists a set Ω_ε with $\mathbf{P}[\Omega_\varepsilon] \geq 1 - \varepsilon$ such that

$$\sup_{u \in \mathbb{N}} \|\mathbf{E}[L(u, \cdot)|\mathcal{G}_{n_k}]\mathbf{1}_{\Omega_\varepsilon} - L(u, \cdot)\mathbf{1}_{\Omega_\varepsilon}\|_\infty \xrightarrow{k \to \infty} 0.$$

Therefore, for every $\alpha > 0$, there exists k such that

$$\sup_{u\in\mathbb{N}} \|\alpha\lfloor\alpha^{-1}\mathbf{E}[L(u,\cdot)|\mathcal{G}_{n_k}]\mathbf{1}_{\Omega_\varepsilon}\rfloor - L(u,\cdot)\mathbf{1}_{\Omega_\varepsilon}\|_\infty \le 2\alpha.$$

But as $\mathcal{G}_n$ is finitely generated, we can write

$$\mathbf{E}[L(u,\cdot)|\mathcal{G}_n]\mathbf{1}_{\Omega_\varepsilon} = \sum_{P\in\pi_n} L_{n,u,P}\mathbf{1}_{P\cap\Omega_\varepsilon},$$

with $|L_{n,u,P}| \le \|L\|_\infty$ for all n, u, P. In particular, $\{\alpha\lfloor\alpha^{-1}\mathbf{E}[L(u,\cdot)|\mathcal{G}_n]\mathbf{1}_{\Omega_\varepsilon}\rfloor :$ $u \in \mathbb{N}\}$ is a finite family of random variables for every n. We have therefore established that the family $\{L(u,\cdot)\mathbf{1}_{\Omega_\varepsilon} : u \in \mathbb{N}\}$ is totally bounded in L^∞.

In the context of Corollary 2.11, the previous example can be interpreted as follows. Suppose that the observations are finite-valued, that is, $\mathcal{Y}$ is a finitely generated σ-field. Let us suppose, for simplicity, that the decision space U is countable (the same conclusion holds for general U modulo some measurability issues). Then, if the third line of Assumption 2 in Corollary 2.11 holds, then the conditioned loss $\mathbf{E}[\ell_0(u)|\mathcal{Y}_{-\infty,0}]$ is necessarily equimeasurable. While it is possible that the conditioned loss is equimeasurable even when the loss ℓ is not (e.g., in the case of blind decisions), this is somewhat unlikely to be the case given a nontrivial observation structure. Therefore, it appears that equimeasurability is almost necessary to obtain universal complexity assumptions in the setting of Corollary 2.11.

The situation is much better in the setting of Theorem 2.6, however. While the first line of Assumption 2 in Theorem 2.6 is still a uniform martingale convergence property, the σ-field $\mathcal{X}$ cannot be finitely generated except in trivial cases. In fact, in many cases the loss ℓ will be $T^{-n}\mathcal{X}$-measurable for some $n < \infty$, in which case the first line of Assumption 2 is automatically satisfied (in particular, in the stochastic process setting, this will be the case for *finitary* loss $\ell_0(u) = l(u, X_{n_1}, \ldots, X_{n_k})$ if we choose $\mathcal{X} = \sigma\{X_k, Y_k : k \le 0\}$). The remainder of Assumption 2 is a uniform reverse martingale convergence property, which holds under much weaker assumptions.

Definition 3.9 The loss $\ell : U \times \Omega \to \mathbb{R}$ on $(\Omega, \mathcal{B})$ is said to be *universally bracketing* if for every probability measure $\mathbf{P}$ and $\varepsilon, M > 0$, the family $\{\ell_0(u)\mathbf{1}_{\Lambda \le M} : u \in U\}$ can be covered by finitely many brackets $\{f : g \le f \le h\}$ with $\|g - h\|_{L^1(\mathbf{P})} \le \varepsilon$.

Lemma 3.10 *Let $(\Omega, \mathcal{B})$ be a standard space, and let $\mathcal{X}, \mathcal{Y}$ be countably generated. Suppose the loss ℓ is universally bracketing and finitary (that is, for some $n \in \mathbb{Z}$, $\ell_0(u)$ is $T^{-n}\mathcal{X}$-measurable for all $u \in U$). Then Assumption 2 of Theorem 2.6 holds.*

Proof The finitary assumption trivially implies the first line of Assumption 2. The second line follows along the lines of the proof of [17, Corollary 1.4(2⇒7)].[4]   □

To show that universal bracketing can be much weaker than equimeasurability, we give a simple example in the context of set estimation.

Example 3.11 (Confidence Intervals) Consider the stochastic process setting (X, Y) where X takes values in the set $[-1, 1]$, and fix $\varepsilon > 0$. We would like to pin down the value of X_k up to precision ε; that is, we want to choose $u_k \in [-1, 1]$ as a function of the observations $Y_0, \ldots, Y_k$ such that $u_k \leq X_k < u_k + \varepsilon$ as often as possible. This is a partial information decision problem with loss function $\ell_0(u) = \mathbf{1}_{\mathbb{R} \setminus [u, u+\varepsilon[}(X_0)$.

The proof of the universal bracketing property of ℓ is standard. Given **P** and $\varepsilon > 0$, we choose $-1 = a_0 < a_1 < \cdots < a_n = 1$ (for some finite n) in such a way that $\mathbf{P}[a_i < X_0 < a_{i+1}] \leq \varepsilon$ for all i (note that every atom of X_0 with probability greater than ε is one of the values a_i). Put each function $\ell_0(u)$ such that $u = a_i$ or $u + \varepsilon = a_i$ for some i in its own bracket, and consider the additional brackets $\{f : \mathbf{1}_{\mathbb{R} \setminus]a_{i-1}, a_{j+1}[} \leq f \leq \mathbf{1}_{\mathbb{R} \setminus [a_i, a_j]}\}$ for all $1 \leq i \leq j < n$. Then evidently each of the brackets has diameter not exceeding 2ε, and for every $u \in U$ the function $\ell_0(u)$ is included in one of the brackets thus constructed.

On the other hand, whenever the law of X_0 is not purely atomic, the loss ℓ cannot be equimeasurable. Indeed, as $\|\ell_0(u)\mathbf{1}_{\Omega_\varepsilon} - \ell_0(u')\mathbf{1}_{\Omega_\varepsilon}\|_\infty = 1$ whenever $\ell_0(u)\mathbf{1}_{\Omega_\varepsilon} \neq \ell_0(u')\mathbf{1}_{\Omega_\varepsilon}$, it is impossible for $\{\ell_0(u)\mathbf{1}_{\Omega_\varepsilon} : u \in U\}$ to be totally bounded in L^∞ for any infinite set Ω_ε (and therefore for any set of sufficiently large measure).

In [17] a detailed characterization is given of the universal bracketing property. In particular, it is shown that a uniformly bounded, separable loss ℓ on a standard measurable space is universally bracketing if and only if $\{\ell_0(u) : u \in U\}$ is a universal Glivenko-Cantelli class, that is, a class of functions for which the law of large numbers always holds uniformly. Many useful methods have been developed in empirical process theory to verify this property, cf. [10, 29]. In particular, for a separable $\{0, 1\}$-valued loss, a very useful sufficient condition is that $\{\ell_0(u) : u \in U\}$ is a Vapnik-Chervonenkis class. We refer to [10, 17, 29] for further details.

3.2   *Conditional Absolute Regularity*

In the previous section, we have developed universal complexity assumptions that are applicable regardless of other details of the model. In the present section, we will in some sense take the opposite approach: we will develop a sufficient condition for

---

[4]The pointwise separability assumption in [17, Corollary 1.4(2⇒7)] is not needed here, as the essential supremum can be reduced to a countable supremum as in the proof of Lemma 2.5.

a stronger version of the conditional K-property (in the stochastic process setting) under which no complexity assumptions are needed. This shows that there is a tradeoff between mixing and complexity; if the mixing assumption is strengthened, then the complexity assumption can be weakened. An additional advantage of the sufficient condition to be presented is that it is in practice one of the most easily verifiable conditions that ensures the conditional K-property.

In the remainder of this section, we will work in the stochastic process setting. Let (X, Y) be a stationary ergodic process taking values in the Polish space $E \times F$. We define $\mathcal{Y}_{n,m} = \sigma\{Y_k : n \le k \le m\}$ and $\mathcal{X}_{n,m} = \sigma\{X_k : n \le k \le m\}$ for $n \le m$, and we consider the observation and generating fields $\mathcal{Y} = \sigma\{Y_0\}$, $\mathcal{X} = \mathcal{X}_{-\infty,0} \vee \mathcal{Y}_{-\infty,0}$. In this setting, the conditional K-property relative to $\mathcal{Y}_{-\infty,0}$ reduces to

$$\bigcap_{k=1}^{\infty}(\mathcal{Y}_{-\infty,0} \vee \mathcal{X}_{-\infty,-k}) = \mathcal{Y}_{-\infty,0} \quad \mathrm{mod}\,\mathbf{P}.$$

If $\mathcal{Y}$ is trivial (that is, the observations Y are noninformative), this reduces to the statement that X has a trivial past tail σ-field, that is, X is regular (or purely nondeterministic) in the sense of Kolmogorov. This property is often fairly easy to check: for example, any Markov chain whose law converges weakly to a unique invariant measure is regular (cf. [28, Prop. 3]). When $\mathcal{Y}$ is nontrivial, the conditional K-property is generally not so easy to check, however. We therefore give a condition, arising from filtering theory [27], that allows to deduce conditional mixing properties from their more easily verifiable unconditional counterparts.

We will require two assumptions. The first assumption states that the pair (X, Y) is absolutely regular in the sense of Volkonskiĭ and Rozanov [30] (this property is also known as β-mixing). Absolute regularity is a strengthening of the regularity property; assuming regularity of (X, Y) is not sufficient for what follows [16]. Many techniques have been developed to verify the absolute regularity property; for example, any Harris recurrent and aperiodic Markov chain is absolutely regular [22].

Definition 3.12 The process (X, Y) is said to be *absolutely regular* if

$$\left\| \mathbf{P}[(X_k, Y_k)_{k \ge n} \in \cdot \,|\, \mathcal{X}_{-\infty,0} \vee \mathcal{Y}_{-\infty,0}] - \mathbf{P}[(X_k, Y_k)_{k \ge n} \in \cdot\,] \right\|_{\mathrm{TV}} \xrightarrow{n \to \infty} 0 \quad \text{in } L^1.$$

By itself, however, absolute regularity of (X, Y) is not sufficient for the conditional K-property, as can be seen in Example 1.3. In this example, the relation between the processes X and Y is very singular, so that things go wrong when we condition. The following nondegeneracy assumption rules out this possibility.

Definition 3.13 The process (X, Y) is said to be *nondegenerate* if

$$\mathbf{P}[Y_1, \dots, Y_m \in \cdot \,|\, \mathcal{Z}_{-\infty,0} \vee \mathcal{Z}_{m+1,\infty}] \sim \mathbf{P}[Y_1, \dots, Y_m \in \cdot \,|\, \mathcal{Y}_{-\infty,0} \vee \mathcal{Y}_{m+1,\infty}] \quad \text{a.s.}$$

for every $1 \le m < \infty$, where $\mathcal{Z}_{n,m} := \mathcal{X}_{n,m} \vee \mathcal{Y}_{n,m}$.

The nondegeneracy assumption ensures that the null sets of the law of the observations Y do not depend too much on the unobserved process X. The assumption is often easily verified. For example, if $Y_k = f(X_k) + \eta_k$ where η_k is an i.i.d. sequence of random variables with strictly positive density, then the conditional distributions in Definition 3.13 have strictly positive densities and are therefore equivalent a.s.

Theorem 3.14 ([27]) *If (X, Y) is absolutely regular and nondegenerate, then*

$$\bigcap_{k=1}^{\infty} (\mathcal{Y}_{-\infty,0} \vee \mathcal{X}_{-\infty,-k}) = \mathcal{Y}_{-\infty,0} \quad \text{mod } \mathbf{P}.$$

Theorem 3.14 provides a practical method to check the conditional K-property. However, the proof of Theorem 3.14 actually yields a much stronger statement. It is shown in [27, Theorem 3.5] that if (X, Y) is absolutely regular and nondegenerate, then X is *conditionally* absolutely regular relative to $\mathcal{Y}_{-\infty,\infty}$ in the sense that

$$\left\| \mathbf{P}[(X_k)_{k \geq n} \in \cdot \,|\, \mathcal{X}_{-\infty,0} \vee \mathcal{Y}_{-\infty,\infty}] - \mathbf{P}[(X_k)_{k \geq n} \in \cdot \,|\, \mathcal{Y}_{-\infty,\infty}] \right\|_{\mathrm{TV}} \xrightarrow{n \to \infty} 0 \quad \text{in } L^1.$$

Moreover, it is shown[5] that under the same assumptions [27, Proposition 3.9]

$$\mathbf{P}[(X_k)_{k \leq 0} \in \cdot \,|\, \mathcal{Y}_{-\infty,0}] \sim \mathbf{P}[(X_k)_{k \leq 0} \in \cdot \,|\, \mathcal{Y}_{-\infty,\infty}] \quad \text{a.s.}$$

From these properties, we can deduce the following result.

Theorem 3.15 *In the setting of the present section, suppose that (X, Y) is absolutely regular and nondegenerate, and consider a loss function of the form $\ell_0(u) = l(u, X_0)$. Then the conclusions of Theorem 2.6 hold.*

The key point about Theorem 3.15 is that no complexity assumption is imposed: the loss function $l(u, x)$ may be an arbitrary measurable function (as long as it is dominated in L^1 in accordance with our standing assumption). The explanation for this is that the conditional absolute regularity property is so strong that the regular conditional probabilities $\mathbf{P}[X_0 \in \cdot \,|\, \mathcal{Y}_{-\infty,\infty} \vee \mathcal{X}_{-\infty,-n}]$ converge in total variation. Therefore, the corresponding reverse martingales converge uniformly over any dominated family of measurable functions. The strength of the conditional mixing property therefore eliminates the need for any additional complexity assumptions. In contrast, we may certainly have pathwise optimal strategies when absolute regularity fails, but then a complexity assumption is essential (cf. Example 2.8).

The proof of Theorem 3.15 will be given in Sect. 4.5. The proof is a straightforward adaptation of Theorem 2.6; unfortunately, the fact that the conditional absolute

---

[5]Some of the statements in [27] are time-reversed as compared to their counterparts stated here. However, as both the absolute regularity and the nondegeneracy assumptions are invariant under time reversal (cf. [30] for the former; the latter is trivial), the present statements follow immediately.

regularity property is relative to $\mathcal{Y}_{-\infty,\infty}$ rather than $\mathcal{Y}_{-\infty,0}$ complicates a direct verification of the assumptions of Theorem 2.6 (while this should be possible along the lines of [27], we will follow the simpler route here). The results of [27] could also be used to obtain the conclusion of Corollary 2.11 in the setting of Theorem 3.15 under somewhat stronger nondegeneracy assumptions.

3.3   Hidden Markov Models and Nonlinear Filters

The goal of the present section is to explore some implications of our results to filtering theory. For simplicity of exposition, we will restrict attention to the classical setting of (general state space) hidden Markov models (see, e.g., [4]).

We adopt the stochastic process setting and notations of the previous section. In addition, we assume that (X, Y) is a hidden Markov model, that is, a Markov chain whose transition kernel can be factored as $\tilde{P}(x, y, dx', dy') = P(x, dx')\, \Phi(x', dy')$. This implies that the process X is a Markov chain in its own right, and that the observations Y are conditionally independent given X. In the following, we will assume that the observation kernel Φ has a density, that is, $\Phi(x, dy) = g(x, y)\, \varphi(dy)$ for some measurable function g and reference measure φ.

A fundamental object in this theory is the nonlinear filter Π_k, defined as

$$\Pi_k := \mathbf{P}[X_k \in \cdot\, | Y_0, \ldots, Y_k].$$

The measure-valued process $\Pi = (\Pi_k)_{k \geq 0}$ is itself a (nonstationary) Markov chain [16] with transition kernel $\mathscr{P}$. To study the stationary behavior of the filter, which is of substantial interest in applications (see, for example, [15] and the references therein), one must understand the relationship between the ergodic properties of X and Π. The following result, proved in [16], is essentially due to Kunita [20].

Theorem 3.16 *Suppose that the transition kernel P possesses a unique invariant measure (that is, X is uniquely ergodic). Then the filter transition kernel $\mathscr{P}$ possesses a unique invariant measure (that is, Π is uniquely ergodic) if and only if*

$$\bigcap_{k=1}^{\infty} (\mathcal{Y}_{-\infty,0} \vee \mathcal{X}_{-\infty,-k}) = \mathcal{Y}_{-\infty,0} \quad \mathrm{mod}\, \mathbf{P}.$$

Evidently, ergodicity of the filter is closely related to the conditional K-property. We will exploit this fact to prove a new optimality property of nonlinear filters.

The usual interpretation of the filter Π_k is that one aims to track to current location X_k of the unobserved process on the basis of the observation history $Y_0, \ldots, Y_k$.

By the elementary property of conditional expectations, $\Pi_k(f)$ provides, for any bounded test function f, an optimal mean-square error estimate of $f(X_k)$:

$$\mathbf{E}\big[\{f(X_k) - \Pi_k(f)\}^2\big] \leq \mathbf{E}\big[\{f(X_k) - \hat{f}_k(Y_0, \dots, Y_k)\}^2\big] \quad \text{for any measurable } \hat{f}_k.$$

This interpretation may not be satisfying, however, if only one sample path of the observations is available (recall Examples 1.2 and 1.3): one would rather show that

$$\liminf_{T \to \infty} \left[\frac{1}{T} \sum_{k=1}^{T} \{f(X_k) - \hat{f}_k(Y_0, \dots, Y_k)\}^2 - \frac{1}{T} \sum_{k=1}^{T} \{f(X_k) - \Pi_k(f)\}^2 \right] \geq 0 \quad \text{a.s.}$$

for any alternative sequence of estimators $(\hat{f}_k)_{k \geq 0}$. If this property holds for any bounded test function f, the filter will be said to be *pathwise optimal*.

Corollary 3.17 *Suppose that the filtering process Π is uniquely ergodic. Then the filter is both mean-square optimal and pathwise optimal.*

Proof Note that the filter $\Pi_k(f)$ is the mean-optimal policy for the partial information decision problem with loss $\ell_0(u) = \{f(X_0) - u\}^2$. As the latter is equimeasurable, the result follows directly from Theorem 3.16 and Corollary 2.11.
□

The interaction between our main results and the ergodic theory of nonlinear filters is therefore twofold. On the one hand, our main results imply that ergodic nonlinear filters are always pathwise optimal. Conversely, Theorem 3.16 shows that ergodicity of the filter is a sufficient condition for our main results to hold in the context of hidden Markov models with equimeasurable loss. This provides another route to establishing the conditional K-property: the filtering literature provides a variety of methods to verify ergodicity of the filter [5, 7, 14, 27]. It should be noted, however, that ergodicity of the filter is not necessary for the conditional K-property to hold, even in the setting of hidden Markov models.

Example 3.18 Consider the hidden Markov model (X, Y) where X is the stationary Markov chain such that $X_0 \sim \mathrm{Uniform}([0, 1])$ and $X_{k+1} = 2X_k \bmod 1$, $Y_k = 0$ for all $k \in \mathbb{Z}$ (that is, we have noninformative observations). Clearly the tail σ-field $\bigcap_n \mathcal{X}_{-\infty,n}$ is nontrivial, and thus the filter fails to be ergodic by Theorem 3.16. Nonetheless, we claim that the conditional K-property holds, so that our main results apply for any equimeasurable loss; in particular, the filter is pathwise optimal.

The key point is that, even in the hidden Markov model setting, one need not choose the "canonical" generating σ-field $\mathcal{X} = \mathcal{X}_{-\infty,0}$ in Definition 1.6. In the present example, we choose instead $\mathcal{X} = \sigma\{\mathbf{1}_{X_k > 1/2} : k \leq 0\}$. To verify the conditional K-property, note that $(\mathbf{1}_{X_k > 1/2})_{k \in \mathbb{Z}}$ are i.i.d. Bernoulli$(1/2)$ and

$$X_k = \sum_{\ell=0}^{\infty} 2^{-\ell-1} \mathbf{1}_{X_{k+\ell} > 1/2} \quad \text{a.s.} \quad \text{for all } k \in \mathbb{Z}.$$

Thus $\mathcal{X} \subset T^{-1}\mathcal{X}$ by construction, $\bigvee_k T^{-k}\mathcal{X} = \sigma\{X_n : n \in \mathbb{Z}\}$ is a generating σ-field, and $\bigcap_k T^k\mathcal{X}$ is trivial by the Kolmogorov zero-one law.

Let us now consider the decision problem in the setting of a hidden Markov model with equimeasurable loss function $\ell_0(u) = l(u, X_0)$. If the filter is ergodic, then Corollary 3.4 ensures that the mean-optimal strategy $\tilde{\mathbf{u}}$ is pathwise optimal. In this setting, the mean-optimal strategy can be expressed in terms of the filter:

$$\tilde{u}_k = \arg\min_{u \in U} \mathbf{E}[l(u, X_k)|Y_0, \dots, Y_k] = \arg\min_{u \in U} \int l(u, x)\, \Pi_k(dx).$$

When X_k takes values in a finite set $E = \{1, \dots, d\}$, the filter can be recursively computed in a straightforward manner [4]. In this case, the mean-optimal strategy $\tilde{\mathbf{u}}$ can be implemented directly. On the other hand, when E is a continuous space, the conditional measure Π_k is an infinite-dimensional object which cannot be computed exactly except in special cases. However, Π_k can often be approximated very efficiently by recursive Monte Carlo approximations $\Pi_k^N = \frac{1}{N} \sum_{i=1}^{N} \delta_{Z_k^N(i)}$, known as particle filters [4], that converge to the true filter Π_k as the number of particles increases $N \to \infty$. This suggests to approximate the mean-optimal strategy $\tilde{\mathbf{u}}$ by

$$\tilde{u}_k \approx \tilde{u}_k^N := \arg\min_{u \in U} \int l(u, x)\, \Pi_k^N(dx) = \arg\min_{u \in U} \frac{1}{N} \sum_{i=1}^{N} l(u, Z_k^N(i)).$$

The strategy $\tilde{\mathbf{u}}^N$ is a type of sequential stochastic programming algorithm to approximate the mean-optimal strategy. In this setting, it is of interest to establish whether the strategy $\tilde{\mathbf{u}}^N$ is in fact approximately pathwise optimal, at least in the weak sense. To this end, we prove the following approximation lemma.

Lemma 3.19 *In the hidden Markov model setting with equimeasurable loss $\ell_0(u) = l(u, X_0)$, suppose that the filter is ergodic, and let Π_k^N be an approximation of Π_k. If*

$$\lim_{N \to \infty} \limsup_{T \to \infty} \mathbf{E}\left[\frac{1}{T} \sum_{k=1}^{T} \operatorname*{ess\,sup}_{u \in \mathbb{U}_{0,k}} |\Pi_k^N(l(u, \cdot)) - \Pi_k(l(u, \cdot))| \right] = 0,$$

then the strategy $\tilde{\mathbf{u}}^N$ is approximately weakly pathwise optimal in the sense that

$$\lim_{N \to \infty} \liminf_{T \to \infty} \mathbf{P}[L_T(\mathbf{u}) - L_T(\tilde{\mathbf{u}}^N) \geq -\varepsilon] = 1 \quad \textit{for every } \varepsilon > 0$$

holds for every admissible strategy $\mathbf{u}$.

Proof We begin by noting that

$$\mathbf{P}[L_T(\mathbf{u}) - L_T(\tilde{\mathbf{u}}^N) < -\varepsilon] \leq \mathbf{P}[L_T(\mathbf{u}) - L_T(\tilde{\mathbf{u}}) < -\varepsilon/2] + \mathbf{P}[L_T(\tilde{\mathbf{u}}^N) - L_T(\tilde{\mathbf{u}}) > \varepsilon/2].$$

Under the present assumptions, the mean-optimal strategy $\tilde{\mathbf{u}}$ is (weakly) pathwise optimal. It follows[6] as in the proof of Lemma 2.12 that $\mathbf{E}[(L_T(\tilde{\mathbf{u}}^N) - L_T(\tilde{\mathbf{u}}))_-] \to 0$ as $T \to \infty$, and we obtain for any admissible strategy $\mathbf{u}$ and $\varepsilon > 0$

$$\limsup_{T\to\infty} \mathbf{P}[L_T(\mathbf{u}) - L_T(\tilde{\mathbf{u}}^N) < -\varepsilon] \le \frac{2}{\varepsilon} \limsup_{T\to\infty} \mathbf{E}[L_T(\tilde{\mathbf{u}}^N) - L_T(\tilde{\mathbf{u}})].$$

To proceed, we estimate

$$\mathbf{E}[L_T(\tilde{\mathbf{u}}^N) - L_T(\tilde{\mathbf{u}})] = \mathbf{E}\left[\frac{1}{T}\sum_{k=1}^{T} \int \{l(\tilde{u}_k^N, x) - l(\tilde{u}_k, x)\} \, \Pi_k(dx)\right]$$

$$\le \mathbf{E}\left[\frac{1}{T}\sum_{k=1}^{T} \int \{l(\tilde{u}_k^N, x) - l(\tilde{u}_k, x)\} \, \Pi_k^N(dx)\right]$$

$$+ 2\,\mathbf{E}\left[\frac{1}{T}\sum_{k=1}^{T} \operatorname*{ess\,sup}_{u\in\mathbb{U}_{0,k}} |\Pi_k^N(l(u,\cdot)) - \Pi_k(l(u,\cdot))|\right].$$

But note that by the definition of $\tilde{\mathbf{u}}^N$

$$\int \{l(\tilde{u}_k^N, x) - l(\tilde{u}_k, x)\} \, \Pi_k^N(dx) = \inf_{u\in U} \int l(u,x) \, \Pi_k^N(dx) - \int l(\tilde{u}_k, x) \, \Pi_k^N(dx) \le 0.$$

The proof is therefore easily completed by applying the assumption.   □

Evidently, the key difficulty in this problem is to control the time-average error of the filter approximation (in a norm determined by the loss function l) uniformly over the time horizon. This problem is intimately related with the ergodic theory of nonlinear filters. The requisite property follows from the results in [15] under reasonable ergodicity assumptions but under very stringent complexity assumptions on the loss (essentially that $\{l(u,\cdot) : u \in U\}$ is uniformly Lipschitz). Alternatively, one can apply the results in [8], which require exceedingly strong ergodicity assumptions but weaker complexity assumptions. Let us note that one could similarly obtain a pathwise version of Lemma 3.19, but the requisite pathwise approximation property of particle filters has not been investigated in the literature.

---

[6]As particle filters employ a random sampling mechanism, the strategy $\tilde{\mathbf{u}}^N$ is technically speaking not admissible in the sense of this paper: Π_k^N (and therefore $\tilde{u}_k^N$) depends also on auxiliary sampling variables $\xi_0, \ldots, \xi_k$ that are independent of $Y_0, \ldots, Y_k$. However, it is easily seen that all our results still hold when such *randomized* strategies are considered. Indeed, it suffices to condition on $(\xi_k)_{k\ge0}$, so that all our results apply immediately under the conditional distribution.

3.4 The Conditions of Algoet, Weissman, Merhav, and Nobel

The aim of this section is to briefly discuss the assumptions imposed in previous work on pathwise optimality due to Algoet [1], Weissman and Merhav [32], and Nobel [24]. Let us emphasize that, while our results cover a much broader range of decision problems, none of these previous results follow in their entirety from our general results. This highlights once more that our results are, unfortunately, nowhere close to a complete characterization of the pathwise optimality property.

3.4.1 Algoet

Algoet's results [1], which cover the full information setting only, were already discussed at length in the introduction and in Sect. 2.2. The existence of a pathwise optimal strategy can be obtained in this setting under no additional assumptions from Theorem 2.6, which even goes beyond Algoet's result in that it gives an explicit expression for the optimal asymptotic loss. However, Algoet establishes that in fact the mean-optimal strategy $\tilde{\mathbf{u}}$ is pathwise optimal in this setting, while our general Corollary 2.11 can only establish this under an additional complexity assumption. We do not know whether this complexity assumption can be weakened in general.

3.4.2 Weissman and Merhav

Weissman and Merhav [32] consider the stochastic process setting (X, Y), where X_k takes values in $\{0, 1\}$ and Y_k takes values in $\mathbb{R}$ for all $k \in \mathbb{Z}$, and where the loss function takes the form $\ell_0(u) = l(u, X_1)$ and is assumed to be uniformly bounded. As X is binary-valued, it is immediate that any loss function l is equimeasurable. Therefore, our results show that the mean-optimal strategy $\tilde{\mathbf{u}}$ is pathwise optimal whenever the model is a conditional K-automorphism relative to $\mathcal{Y}_{-\infty,0}$.

The assumption imposed by Weissman and Merhav in [32] is as follows:

$$\sum_{k=1}^{\infty} \sup_{r \geq 1} \mathbf{E}[|\mathbf{P}[X_{r+k} = a | X_r = a, \mathcal{Y}_{0,r+k-1}] - \mathbf{P}[X_{r+k} = a | \mathcal{Y}_{0,r+k-1}]|] < \infty \text{ for } a = 0, 1.$$

Using stationarity, this condition is equivalent to

$$\sum_{k=0}^{\infty} \sup_{r \geq 1} \mathbf{E}[|\mathbf{P}[X_1 = a | X_{-k} = a, \mathcal{Y}_{-r-k,0}] - \mathbf{P}[X_1 = a | \mathcal{Y}_{-r-k,0}]|] < \infty \text{ for } a = 0, 1,$$

which readily implies

$$\sum_{k=0}^{\infty} \mathbf{E}[|\mathbf{P}[X_1 = a | \sigma\{X_{-k}\} \vee \mathcal{Y}_{-\infty,0}] - \mathbf{P}[X_1 = a | \mathcal{Y}_{-\infty,0}]|] < \infty.$$

If the σ-field $\sigma\{X_{-k}\} \vee \mathcal{Y}_{-\infty,0}$ could be replaced by the larger σ-field $\mathcal{X}_{-\infty,-k} \vee \mathcal{Y}_{-\infty,0}$ in this expression, then Assumption 3 of Corollary 2.11 would follow immediately. However, the smaller σ-field appears to yield a slightly better variant of the assumption imposed in [32]. This is possible because the result is restricted to the special choice of loss $\ell_0(u) = l(u, X_1)$ that depends on X_1 only. On the other hand, it is to be expected that in most cases the assumption of [32] is much more stringent than that of Corollary 2.11. Note that Assumption 3 of Corollary 2.11 is purely qualitative in nature: it states, roughly speaking, that two σ-fields coincide. This is a structural property of the model. On the other hand, the assumption of [32] is inherently *quantitative* in nature: it requires that a certain mixing property holds at a sufficiently fast rate (the mixing coefficients must be summable). A quantitative bound on the mixing rate is both much more restrictive and much harder to verify, in general, as compared to a purely structural property.

In a sense, the approach of Weissman and Merhav is much closer in spirit to the weak pathwise optimality results in this paper than it is to the pathwise optimality results. Indeed, if we replace the weak pathwise optimality property

$$\mathbf{P}[L_T(\mathbf{u}) - L_T(\mathbf{u}^\star) < -\varepsilon] \xrightarrow{T \to \infty} 0 \quad \text{for every } \varepsilon > 0$$

by its quantitative counterpart

$$\sum_{T=1}^{\infty} \mathbf{P}[L_T(\mathbf{u}) - L_T(\mathbf{u}^\star) < -\varepsilon] < \infty \quad \text{for every } \varepsilon > 0,$$

then pathwise optimality will automatically follow from the Borel-Cantelli lemma. In the same spirit, if in Theorem 2.13 we replace the uniform conditional mixing assumption by the corresponding quantitative counterpart

$$\sum_{k=1}^{T} \mathbf{E}\left[\operatorname*{ess\,sup}_{u,u' \in \mathbb{U}_0} |\mathbf{E}[\{\bar{\ell}_0^M(u) \circ T^{-k}\} \bar{\ell}_0^M(u')|\mathcal{Y}_{-\infty,0}]| \right] = O(T^\alpha)$$

for some $\alpha < 1$ (that may depend on M), then we easily obtain a pathwise version of Lemma 4.7 below (using Etemadi's well-known device [11]), and consequently the conclusion of Theorem 2.13 is replaced by that of Theorem 2.6. It is unclear whether such quantitative mixing conditions provide a distinct mechanism for pathwise optimality as compared to qualitative structural conditions as in our main results.

3.4.3 Nobel

Nobel [24] considers the stochastic process setting (X, Y) with observations of the additive form $Y_k = X_k + N_k$, where $N = (N_k)_{k \in \mathbb{Z}}$ is an L^2-martingale difference sequence independent of X. The loss function considered is the mean-square loss

$\ell_0(u) = (u - X_1)^2$. This very special scenario is essential for the result given in [24]; on the other hand, it is not assumed that (X, Y) is even stationary or that the decision space U is a compact set (when $U = \mathbb{R}$, the quadratic loss is not dominated). In order to compare with our general results, we will additionally assume that (X, Y) is stationary and ergodic and that X_k are uniformly bounded random variables (so that we may choose $U = [-\|X_1\|_\infty, \|X_1\|_\infty]$ without loss of generality).

While this is certainly a decision problem with partial information, the key observation is that this special problem is in fact a decision problem with full information in disguise. Indeed, note that we can write for any strategy $\mathbf{u}$

$$L_T(\mathbf{u}) = \frac{1}{T} \sum_{k=1}^{T} (u_k - Y_{k+1})^2 + \frac{1}{T} \sum_{k=1}^{T} \{X_{k+1}^2 - Y_{k+1}^2\} + \frac{1}{T} \sum_{k=1}^{T} 2u_k N_{k+1}.$$

The last term of this expression converges to zero a.s. as $T \to \infty$ for any admissible strategy $\mathbf{u}$ by the martingale law of large numbers, as $(u_k N_{k+1})_{k \in \mathbb{Z}}$ is an L^2-martingale difference sequence. On the other hand, the second to last term of this expression does not depend on the strategy $\mathbf{u}$ at all. Therefore,

$$\liminf_{T \to \infty} \{L_T(\mathbf{u}) - L_T(\tilde{\mathbf{u}})\} = \liminf_{T \to \infty} \left\{ \frac{1}{T} \sum_{k=1}^{T} (u_k - Y_{k+1})^2 - \frac{1}{T} \sum_{k=1}^{T} (\tilde{u}_k - Y_{k+1})^2 \right\} \quad \text{a.s.,}$$

which corresponds to the decision problem with the full information loss $\ell_0(u) = (u - Y_1)^2$. Thus pathwise optimality of the mean-optimal strategy $\tilde{\mathbf{u}}$ follows from Algoet's result. (The main difficulty in [24] is to introduce suitable truncations to deal with the lack of boundedness, which we avoided here.)

Of course, we could deduce the result from our general theory in the same manner: reduce first to a full information decision problem as above, and then invoke Corollary 2.11 in the full information setting. However, a more relevant test of our general theory might be to ask whether one can deduce the result directly from Corollary 2.11, without first reducing to the full information setting. Unfortunately, it is not clear whether it is possible, in general, to find a generating σ-field $\mathfrak{X}$ such that Assumption 3 of Corollary 2.11 holds.

One might interpret the additive noise model as a type of "informative" observations: while X cannot be reconstructed from the observations Y, the law of X can certainly be reconstructed from the law of Y even if the former were not known a priori (this idea is exploited in [24, 32] to devise universal prediction strategies that do not require prior knowledge of the law of X). In the hidden Markov model setting, there is in fact a connection between "informative" observations and the conditional K-property. In particular, if (X, Y) is a hidden Markov model where X_k takes a finite number of values, and $Y_k = X_k + \xi_k$ where ξ_k are i.i.d. and independent of X, then the conditional K-property holds, and we therefore have pathwise optimal strategies for *any* dominated loss. This follows from observability conditions in the Markov setting, cf. [5, section 6.2] and the references therein.

However, the ideas that lead to this result do not appear to extend to more general situations.

4 Proofs

4.1 Proof of Theorem 2.6

Throughout the proof, we fix a generating σ-field $\mathfrak{X}$ that satisfies the conditions of Theorem 2.6. In the following, we define the σ-fields

$$\mathfrak{G}_k^n = \mathcal{Y}_{-\infty,k} \vee T^{n-k}\mathfrak{X}, \qquad \mathfrak{G}_k^\infty = \bigcap_n \mathfrak{G}_k^n.$$

Note that $\mathfrak{G}_k^n$ is decreasing in n and increasing in k.

We begin by establishing the following lemma.

Lemma 4.1 *For any admissible strategy* $\mathbf{u}$ *and any* $m, n \in \mathbb{Z}$

$$\frac{1}{T} \sum_{k=1}^{T} \{\mathbf{E}[\ell_k(u_k)|\mathfrak{G}_k^m] - \mathbf{E}[\ell_k(u_k)|\mathfrak{G}_k^n]\} \xrightarrow{T\to\infty} 0 \quad a.s.$$

Proof Assume $m < n$ without loss of generality. Fix $r < \infty$, and define

$$\Delta_k^j = \mathbf{E}[\ell_k(u_k)\mathbf{1}_{\Lambda\circ T^k \le r}|\mathfrak{G}_k^j] - \mathbf{E}[\ell_k(u_k)\mathbf{1}_{\Lambda\circ T^k \le r}|\mathfrak{G}_k^{j+1}]$$

for $m \le j < n$. Then it is easily seen that we have the inequality

$$\left|\frac{1}{T} \sum_{k=1}^{T} \{\mathbf{E}[\ell_k(u_k)|\mathfrak{G}_k^m] - \mathbf{E}[\ell_k(u_k)|\mathfrak{G}_k^n]\}\right| \le \sum_{j=m}^{n-1} \left|\frac{1}{T} \sum_{k=1}^{T} \Delta_k^j\right| +$$

$$\frac{1}{T} \sum_{k=1}^{T} \{\mathbf{E}[\Lambda\mathbf{1}_{\Lambda>r}|\mathfrak{G}_0^m] + \mathbf{E}[\Lambda\mathbf{1}_{\Lambda>r}|\mathfrak{G}_0^n]\} \circ T^k.$$

By the ergodic theorem, the second term on the right converges to $\kappa(r) := \mathbf{E}[2\Lambda\mathbf{1}_{\Lambda>r}]$ a.s. as $T \to \infty$. It remains to consider the first term.

To this end, note the inclusions $\mathfrak{G}_k^{j+1} \subseteq \mathfrak{G}_k^j \subseteq \mathfrak{G}_{k+1}^{j+1}$. It follows that

$$\Delta_k^j \text{ is } \mathfrak{G}_{k+1}^{j+1}\text{-measurable}, \quad \mathbf{E}[\Delta_k^j|\mathfrak{G}_k^{j+1}] = 0, \quad \text{and } |\Delta_k^j| \le 2r$$

for $0 \leq j < n$. Thus $(\Delta_k^j)_{k\geq 1}$ is a uniformly bounded martingale difference sequence with respect to the filtration $(\mathcal{G}_{k+1}^{j+1})_{k\geq 1}$, and we consequently have

$$\frac{1}{T}\sum_{k=1}^T \Delta_k^j \xrightarrow{T\to\infty} 0 \quad \text{a.s.}$$

by the simplest form of the martingale law of large numbers (indeed, it is easily seen that $M_n = \sum_{k=1}^n \Delta_k^j/k$ is an L^2-bounded martingale, so that the result follows from the martingale convergence theorem and Kronecker's lemma).

Putting together these results, we obtain

$$\limsup_{T\to\infty}\left|\frac{1}{T}\sum_{k=1}^T\{\mathbf{E}[\ell_k(u_k)|\mathcal{G}_k^m]-\mathbf{E}[\ell_k(u_k)|\mathcal{G}_k^n]\}\right| \leq \kappa(r) \quad \text{a.s.}$$

for arbitrary $r < \infty$. Letting $r \to \infty$ completes the proof. $\qquad\square$

We can now establish a lower bound on the loss of any strategy.

Corollary 4.2 *Under the assumptions of Theorem 2.6, we have*

$$\frac{1}{T}\sum_{k=1}^T\{\ell_k(u_k)-\mathbf{E}[\ell_k(u_k)|\mathcal{G}_k^\infty]\} \xrightarrow{T\to\infty} 0 \quad \text{a.s.}$$

for any admissible strategy **u**. *In particular,*

$$\liminf_{T\to\infty} L_T(\mathbf{u}) \geq \mathbf{E}\left[\operatorname*{ess\,inf}_{u\in\mathbb{U}_0}\mathbf{E}[\ell_0(u)|\mathcal{G}_0^\infty]\right] = L^\star \quad \text{a.s.}$$

Proof We begin by noting that

$$\left|\frac{1}{T}\sum_{k=1}^T\{\mathbf{E}[\ell_k(u_k)|\mathcal{G}_k^n]-\mathbf{E}[\ell_k(u_k)|\mathcal{G}_k^\infty]\}\right| \leq \frac{1}{T}\sum_{k=1}^T\operatorname*{ess\,sup}_{u\in\mathbb{U}_k}|\mathbf{E}[\ell_k(u)|\mathcal{G}_k^n]-\mathbf{E}[\ell_k(u)|\mathcal{G}_k^\infty]|$$

$$\xrightarrow{T\to\infty}\mathbf{E}\left[\operatorname*{ess\,sup}_{u\in\mathbb{U}_0}|\mathbf{E}[\ell_0(u)|\mathcal{G}_0^n]-\mathbf{E}[\ell_0(u)|\mathcal{G}_0^\infty]|\right] \quad \text{a.s.}$$

by the ergodic theorem. Similarly,

$$\left|\frac{1}{T}\sum_{k=1}^T\{\mathbf{E}[\ell_k(u_k)|\mathcal{G}_k^m]-\ell_k(u_k)\}\right| \leq \frac{1}{T}\sum_{k=1}^T\operatorname*{ess\,sup}_{u\in\mathbb{U}_k}|\mathbf{E}[\ell_k(u)|\mathcal{G}_k^m]-\ell_k(u)|$$

$$\xrightarrow{T\to\infty}\mathbf{E}\left[\operatorname*{ess\,sup}_{u\in\mathbb{U}_0}|\mathbf{E}[\ell_0(u)|\mathcal{G}_0^m]-\ell_0(u)|\right] \quad \text{a.s.}$$

Therefore, using Lemma 4.1 and Assumption 2 of Theorem 2.6, the first statement of the Corollary follows by letting $n \to \infty$ and $m \to -\infty$.

For the second statement, it suffices to note that

$$\frac{1}{T} \sum_{k=1}^{T} \mathbf{E}[\ell_k(u_k)|\mathcal{G}_k^\infty] \geq \frac{1}{T} \sum_{k=1}^{T} \operatorname*{ess\,inf}_{u \in \mathbb{U}_k} \mathbf{E}[\ell_k(u)|\mathcal{G}_k^\infty] \xrightarrow{T \to \infty} L^\star \quad \text{a.s.}$$

by the ergodic theorem and Assumption 3 of Theorem 2.6. $\qquad\qquad\square$

As was explained in the introduction, a pathwise optimal strategy could easily be obtained if one can prove "ergodic tower property" of the form

$$\frac{1}{T} \sum_{k=1}^{T} \{\ell_k(u_k) - \mathbf{E}[\ell_k(u_k)|\mathcal{Y}_{0,k}]\} \xrightarrow{T \to \infty} 0 \quad \text{a.s.} \quad ?$$

Corollary 4.2 establishes just such a property, but where the σ-field $\mathcal{Y}_{0,k}$ is replaced by the larger σ-field $\mathcal{G}_k^\infty$. This yields a lower bound on the asymptotic loss, but it is far from clear that one can choose a $\mathcal{Y}_{0,k}$-adapted strategy that attains this bound.

Therefore, what remains is to show that there exists an admissible strategy $\mathbf{u}^\star$ that attains the lower bound in Corollary 4.2. A promising candidate is the mean-optimal strategy $\tilde{\mathbf{u}}$. Unfortunately, we are not able to prove pathwise optimality of the mean-optimal strategy in the general setting of Theorem 2.6. However, we will obtain a pathwise optimal strategy $\mathbf{u}^\star$ by a judicious modification of the mean-optimal strategy $\tilde{\mathbf{u}}$. The key idea is the following "uniform" version of the martingale convergence theorem, which we prove following Neveu [23, Lemma V-2-9].

Lemma 4.3 *The following holds:*

$$\operatorname*{ess\,inf}_{u \in \mathbb{U}_{-k,0}} \mathbf{E}[\ell_0(u)|\mathcal{Y}_{-k,0}] \xrightarrow{k \to \infty} \operatorname*{ess\,inf}_{u \in \mathbb{U}_0} \mathbf{E}[\ell_0(u)|\mathcal{Y}_{-\infty,0}] \quad \textit{a.s. and in } L^1.$$

Proof Using the construction of the essential supremum as in the proof of Lemma 2.5, we can choose for each $0 \leq k < \infty$ a countable family $\mathbb{U}_{-k,0}^c \subset \mathbb{U}_{-k,0}$ such that

$$\operatorname*{ess\,inf}_{u \in \mathbb{U}_{-k,0}} \mathbf{E}[\ell_0(u)|\mathcal{Y}_{-k,0}] = \inf_{u \in \mathbb{U}_{-k,0}^c} \mathbf{E}[\ell_0(u)|\mathcal{Y}_{-k,0}] \quad \text{a.s.,}$$

and a countable family $\mathbb{U}_0^c \subset \mathbb{U}_0$ such that

$$\operatorname*{ess\,inf}_{u \in \mathbb{U}_0} \mathbf{E}[\ell_0(u)|\mathcal{Y}_{-\infty,0}] = \inf_{u \in \mathbb{U}_0^c} \mathbf{E}[\ell_0(u)|\mathcal{Y}_{-\infty,0}] \quad \text{a.s.}$$

For every $0 \leq k < \infty$, choose an arbitrary ordering $(U_k^n)_{n \in \mathbb{N}}$ of the elements of the countable set $\mathbb{U}_{-k,0}^c \cup (\mathbb{U}_0^c \cap \mathbb{U}_{-k,0})$. Then we clearly have

$$M_k := \operatorname*{ess\,inf}_{u \in \mathbb{U}_{-k,0}} \mathbf{E}[\ell_0(u)|\mathcal{Y}_{-k,0}] = \min_{0 \leq l \leq k} \inf_{n \in \mathbb{N}} \mathbf{E}[\ell_0(U_l^n)|\mathcal{Y}_{-k,0}] \quad \text{a.s.}$$

and

$$M := \operatorname*{ess\,inf}_{u \in \mathbb{U}_0} \mathbf{E}[\ell_0(u)|\mathcal{Y}_{-\infty,0}] = \inf_{0 \leq l < \infty} \inf_{n \in \mathbb{N}} \mathbf{E}[\ell_0(U_l^n)|\mathcal{Y}_{-\infty,0}] \quad \text{a.s.}$$

Our aim is to prove that $M_k \to M$ a.s. and in L^1 as $k \to \infty$.

We begin by noting that $|M_k| \leq \mathbf{E}[\Lambda|\mathcal{Y}_{-k,0}]$. Therefore, the sequence $(M_k)_{k \geq 0}$ is uniformly integrable. Moreover, $(M_k)_{k \geq 0}$ is a supermartingale with respect to the filtration $(\mathcal{Y}_{-k,0})_{k \geq 0}$: indeed, we can easily compute

$$\mathbf{E}[M_{k+1}|\mathcal{Y}_{-k,0}] \leq \mathbf{E}\left[\min_{0 \leq l \leq k} \inf_{n \in \mathbb{N}} \mathbf{E}[\ell_0(U_l^n)|\mathcal{Y}_{-k-1,0}]\Big|\mathcal{Y}_{-k,0}\right] \leq M_k.$$

Thus $M_k \to M_\infty$ a.s. and in L^1 by the martingale convergence theorem for some random variable M_∞. We must now show that $M_\infty = M$ a.s. Note that

$$M_\infty = \lim_{k \to \infty} M_k \leq \lim_{k \to \infty} \mathbf{E}[\ell_0(U_l^n)|\mathcal{Y}_{-k,0}] = \mathbf{E}[\ell_0(U_l^n)|\mathcal{Y}_{-\infty,0}] \quad \text{a.s.}$$

for every $n \in \mathbb{N}$ and $0 \leq l < \infty$, so $M_\infty \leq M$ a.s. To complete the proof, it therefore suffices to show that $\mathbf{E}[M_\infty] = \mathbf{E}[M]$.

To this end, define for $N \in \mathbb{N}$ and $0 \leq k \leq \infty$

$$M_k^N = \min_{l \leq N \wedge k} \min_{n \leq N} \mathbf{E}[\ell_0(U_l^n)|\mathcal{Y}_{-k,0}].$$

As $(M_k^N)_{k \geq 0}$ is again a supermartingale, clearly $\mathbf{E}[M_k^N]$ is doubly nonincreasing in k and N. The exchange of limits is therefore permitted, so that

$$\mathbf{E}[M_\infty] = \lim_{k \to \infty} \lim_{N \to \infty} \mathbf{E}[M_k^N] = \lim_{N \to \infty} \lim_{k \to \infty} \mathbf{E}[M_k^N] = \mathbf{E}[M].$$

This completes the proof. □

Corollary 4.4 *Suppose that Assumption 3 of Theorem 2.6 holds. Then*

$$\mathbf{E}[\ell_k(\tilde{u}_k)|\mathcal{G}_k^\infty] \circ T^{-k} \xrightarrow{k \to \infty} \operatorname*{ess\,inf}_{u \in \mathbb{U}_0} \mathbf{E}[\ell_0(u)|\mathcal{G}_0^\infty] \quad in \ L^1.$$

Proof Define $\hat{u}_k = \tilde{u}_k \circ T^{-k} \in \mathbb{U}_{-k,0}$, so that

$$\mathbf{E}[\ell_k(\tilde{u}_k)|\mathcal{G}_k^\infty] \circ T^{-k} = \mathbf{E}[\ell_0(\hat{u}_k)|\mathcal{G}_0^\infty].$$

By stationarity and the definition of $\tilde{\mathbf{u}}$, we have

$$\mathbf{E}[\mathbf{E}[\ell_0(\hat{u}_k)|\mathcal{G}_0^\infty]] = \mathbf{E}[\mathbf{E}[\ell_0(\hat{u}_k)|\mathcal{Y}_{-k,0}]]$$
$$\leq \mathbf{E}\left[\operatorname*{ess\,inf}_{u\in\mathbb{U}_{-k,0}} \mathbf{E}[\ell_0(u)|\mathcal{Y}_{-k,0}]\right] + k^{-1}.$$

Therefore, by Lemma 4.3, we have

$$\limsup_{k\to\infty} \mathbf{E}[\mathbf{E}[\ell_0(\hat{u}_k)|\mathcal{G}_0^\infty]] \leq \mathbf{E}\left[\operatorname*{ess\,inf}_{u\in\mathbb{U}_0} \mathbf{E}[\ell_0(u)|\mathcal{Y}_{-\infty,0}]\right] = L^\star.$$

On the other hand, note that

$$\mathbf{E}[\ell_0(\hat{u}_k)|\mathcal{G}_0^\infty] \geq \operatorname*{ess\,inf}_{u\in\mathbb{U}_0} \mathbf{E}[\ell_0(u)|\mathcal{G}_0^\infty] \quad \text{a.s.}$$

Using Assumption 3, we therefore have

$$\limsup_{k\to\infty} \left\| \mathbf{E}[\ell_0(\hat{u}_k)|\mathcal{G}_0^\infty] - \operatorname*{ess\,inf}_{u\in\mathbb{U}_0} \mathbf{E}[\ell_0(u)|\mathcal{G}_0^\infty] \right\|_1 \leq 0.$$

This completes the proof. $\square$

We are now in the position to construct the pathwise optimal strategy $\mathbf{u}^\star$. By Corollary 4.4, we can choose a (nonrandom) sequence $k_n \uparrow \infty$ such that

$$\mathbf{E}[\ell_{k_n}(\tilde{u}_{k_n})|\mathcal{G}_{k_n}^\infty] \circ T^{-k_n} \xrightarrow{n\to\infty} \operatorname*{ess\,inf}_{u\in\mathbb{U}_0} \mathbf{E}[\ell_0(u)|\mathcal{G}_0^\infty] \quad \text{a.s.}$$

Let us define

$$u_k^\star = \tilde{u}_{k_n} \circ T^{k-k_n} \quad \text{for } k_n \leq k < k_{n+1}, \, n \in \mathbb{N}.$$

Then clearly $\mathbf{u}^\star = (u_k^\star)_{k\geq 1}$ is an admissible strategy.

Lemma 4.5 *Suppose that the assumptions of Theorem 2.6 hold. Then*

$$\lim_{T\to\infty} L_T(\mathbf{u}^\star) = L^\star \quad \text{a.s.}$$

Proof By construction,

$$\mathbf{E}[\ell_k(u_k^\star)|\mathcal{G}_k^\infty] \circ T^{-k} \xrightarrow{k\to\infty} \operatorname*{ess\,inf}_{u\in\mathbb{U}_0} \mathbf{E}[\ell_0(u)|\mathcal{G}_0^\infty] \quad \text{a.s.}$$

Moreover,

$$\sup_{k\geq 1} \left| \mathbf{E}[\ell_k(u_k^\star)|\mathcal{G}_k^\infty] \circ T^{-k} \right| \leq \mathbf{E}[\Lambda|\mathcal{G}_0^\infty] \in L^1.$$

Therefore, by Maker's generalized ergodic theorem [19, Corollary 10.8]

$$\frac{1}{T} \sum_{k=1}^{T} \mathbf{E}[\ell_k(u_k^\star)|\mathcal{G}_k^\infty] \xrightarrow{T\to\infty} \mathbf{E}\left[\operatorname*{ess\,inf}_{u\in\mathbb{U}_0} \mathbf{E}[\ell_0(u)|\mathcal{G}_0^\infty] \right] = L^\star \quad \text{a.s.}$$

Thus $L_T(\mathbf{u}^\star) \to L^\star$ a.s. as $T \to \infty$ by Corollary 4.2. □

The proof of Theorem 2.6 is now complete. Indeed, if $\mathbf{u}$ is admissible, then

$$\liminf_{T\to\infty}\{L_T(\mathbf{u}) - L_T(\mathbf{u}^\star)\} = \liminf_{T\to\infty} L_T(\mathbf{u}) - L^\star \geq 0 \quad \text{a.s.}$$

by Lemma 4.5 and Corollary 4.2, so $\mathbf{u}^\star$ is pathwise optimal.

4.2 Proof of Corollary 2.11

To prove pathwise optimality, it suffices to show $L_T(\tilde{\mathbf{u}}) \to L^\star$ a.s.

Lemma 4.6 *Under the assumptions of Corollary 2.11, the mean-optimal strategy $\tilde{\mathbf{u}}$ (Lemma 2.5) satisfies $L_T(\tilde{\mathbf{u}}) \to L^\star$ a.s. as $T \to \infty$.*

Proof By the definition of $\tilde{\mathbf{u}}$ and Lemma 4.3,

$$\mathbf{E}[\ell_k(\tilde{u}_k)|\mathcal{Y}_{0,k}] \circ T^{-k} \xrightarrow{k\to\infty} \operatorname*{ess\,inf}_{u\in\mathbb{U}_0} \mathbf{E}[\ell_0(u)|\mathcal{Y}_{-\infty,0}] \quad \text{a.s.}$$

Therefore, the third part of Assumption 2 of Corollary 2.11 implies that

$$\mathbf{E}[\ell_k(\tilde{u}_k)|\mathcal{Y}_{-\infty,k}] \circ T^{-k} \xrightarrow{k\to\infty} \operatorname*{ess\,inf}_{u\in\mathbb{U}_0} \mathbf{E}[\ell_0(u)|\mathcal{Y}_{-\infty,0}] \quad \text{a.s.}$$

But by Assumption 3 of Corollary 2.11 and stationarity, we obtain

$$\mathbf{E}[\ell_k(\tilde{u}_k)|\mathcal{G}_k^\infty] \circ T^{-k} \xrightarrow{k\to\infty} \operatorname*{ess\,inf}_{u\in\mathbb{U}_0} \mathbf{E}[\ell_0(u)|\mathcal{Y}_{-\infty,0}] \quad \text{a.s.}$$

Moreover, we have

$$\sup_{k\geq 1} \left| \mathbf{E}[\ell_k(\tilde{u}_k)|\mathcal{G}_k^\infty] \circ T^{-k} \right| \leq \mathbf{E}[\Lambda|\mathcal{G}_0^\infty] \in L^1.$$

Maker's generalized ergodic theorem [19, Corollary 10.8] therefore yields

$$\frac{1}{T} \sum_{k=1}^{T} \mathbf{E}[\ell_k(\tilde{u}_k)|\mathcal{G}_k^\infty] \xrightarrow{T\to\infty} \mathbf{E}\left[\operatorname*{ess\,inf}_{u\in\mathbb{U}_0} \mathbf{E}[\ell_0(u)|\mathcal{Y}_{-\infty,0}] \right] = L^\star \quad \text{a.s.}$$

As the assumptions of Corollary 2.11 imply those of Theorem 2.6, the result as well as pathwise optimality of $\tilde{\mathbf{u}}$ now follow from Corollary 4.2. ☐

4.3 Proof of Theorem 2.13

The proof of the Theorem is once again based on a variant of the "ergodic tower property" described in the introduction. In the present setting, the result follows rather easily from the conditional weak mixing assumption.

Lemma 4.7 *Suppose that the assumption of Theorem 2.13 holds. Then*

$$\frac{1}{T}\sum_{k=1}^{T}\{\ell_k(u_k) - \mathbf{E}[\ell_k(u_k)|\mathcal{Y}_{-\infty,k}]\} \xrightarrow{T\to\infty} 0 \quad in\ L^1$$

for every admissible strategy **u**.

Proof Define $\bar{\ell}_k^M(u) = \bar{\ell}_0^M(u) \circ T^k$ for $u \in U$. We begin by noting that

$$\mathbf{E}\left[\left(\frac{1}{T}\sum_{k=1}^{T}\bar{\ell}_k^M(u_k)\right)^2\right] = \frac{1}{T^2}\sum_{n,m=1}^{T}\mathbf{E}[\bar{\ell}_n^M(u_n)\bar{\ell}_m^M(u_m)].$$

Suppose that $m \leq n$. Then by stationarity and as **u** is admissible

$$\mathbf{E}[\bar{\ell}_n^M(u_n)\bar{\ell}_m^M(u_m)] = \mathbf{E}[\bar{\ell}_0^M(u_n \circ T^{-n})\{\bar{\ell}_0^M(u_m \circ T^{-m}) \circ T^{-(n-m)}\}]$$

$$\leq \mathbf{E}\left[\operatorname*{ess\,sup}_{u,u'\in\mathbb{U}_0}|\mathbf{E}[\bar{\ell}_0^M(u')\{\bar{\ell}_0^M(u)\circ T^{-(n-m)}\}|\mathcal{Y}_{-\infty,0}]|\right].$$

We can therefore estimate

$$\mathbf{E}\left[\left(\frac{1}{T}\sum_{k=1}^{T}\bar{\ell}_k^M(u_k)\right)^2\right] \leq \frac{2}{T^2}\sum_{n=1}^{T}\sum_{k=0}^{n-1}\mathbf{E}\left[\operatorname*{ess\,sup}_{u,u'\in\mathbb{U}_0}|\mathbf{E}[\bar{\ell}_0^M(u')\{\bar{\ell}_0^M(u)\circ T^{-k}\}|\mathcal{Y}_{-\infty,0}]|\right]$$

$$= \frac{2}{T^2}\sum_{k=0}^{T-1}(T-k)\,\mathbf{E}\left[\operatorname*{ess\,sup}_{u,u'\in\mathbb{U}_0}|\mathbf{E}[\bar{\ell}_0^M(u')\{\bar{\ell}_0^M(u)\circ T^{-k}\}|\mathcal{Y}_{-\infty,0}]|\right]$$

$$\leq \frac{2}{T}\sum_{k=0}^{T-1}\mathbf{E}\left[\operatorname*{ess\,sup}_{u,u'\in\mathbb{U}_0}|\mathbf{E}[\bar{\ell}_0^M(u')\{\bar{\ell}_0^M(u)\circ T^{-k}\}|\mathcal{Y}_{-\infty,0}]|\right].$$

By the uniform conditional mixing assumption, it follows that

$$\lim_{M\to\infty}\limsup_{T\to\infty}\left\|\frac{1}{T}\sum_{k=1}^{T}\bar{\ell}_k^M(u_k)\right\|_2 = 0.$$

On the other hand, note that

$$\sup_{T\geq 1}\left\|\frac{1}{T}\sum_{k=1}^{T}\{\ell_k(u_k)-\mathbf{E}[\ell_k(u_k)|\mathcal{Y}_{-\infty,k}]\}-\frac{1}{T}\sum_{k=1}^{T}\bar{\ell}_k^M(u_k)\right\|_1 \leq \mathbf{E}[2\Lambda 1_{\Lambda>M}] \xrightarrow{M\to\infty} 0.$$

The result now follows by applying the triangle inequality. $\square$

Corollary 4.8 *Under the assumption of Theorem 2.13, we have*

$$\mathbf{P}\Big[L_T(\mathbf{u}) - L^\star \leq -\varepsilon\Big] \xrightarrow{T\to\infty} 0 \quad \text{for every } \varepsilon > 0$$

for every admissible strategy $\mathbf{u}$.

Proof Let $\mathbf{u}$ be any admissible strategy. Then by Lemma 4.7

$$L_T(\mathbf{u}) - \frac{1}{T}\sum_{k=1}^{T}\mathbf{E}[\ell_k(u_k)|\mathcal{Y}_{-\infty,k}] \xrightarrow{T\to\infty} 0 \quad \text{in } L^1.$$

On the other hand, note that

$$\frac{1}{T}\sum_{k=1}^{T}\mathbf{E}[\ell_k(u_k)|\mathcal{Y}_{-\infty,k}] \geq \frac{1}{T}\sum_{k=1}^{T}\operatorname*{ess\,inf}_{u\in\mathbb{U}_k}\mathbf{E}[\ell_k(u)|\mathcal{Y}_{-\infty,k}] \xrightarrow{T\to\infty} L^\star \quad \text{in } L^1$$

by the ergodic theorem. The result follows directly. $\square$

In view of Corollary 4.8, in order to establish weak pathwise optimality of $\tilde{\mathbf{u}}$ it evidently suffices to prove that $\tilde{\mathbf{u}}$ satisfies the ergodic theorem $L_T(\tilde{\mathbf{u}}) \to L^\star$ in L^1. However, most of the work was already done in the proof of Theorem 2.6.

Lemma 4.9 *Under the assumption of Theorem 2.13, $L_T(\tilde{\mathbf{u}}) \to L^\star$ in L^1.*

Proof By the definition of $\tilde{\mathbf{u}}$, we have

$$\mathbf{E}[\ell_k(\tilde{u}_k)|\mathcal{Y}_{0,k}] \circ T^{-k} \leq \operatorname*{ess\,inf}_{u\in\mathbb{U}_{-k,0}}\mathbf{E}[\ell_0(u)|\mathcal{Y}_{-k,0}] + k^{-1} \quad \text{a.s.}$$

Therefore, by Lemma 4.3, we obtain

$$\limsup_{k\to\infty}\mathbf{E}[\mathbf{E}[\ell_k(\tilde{u}_k)|\mathcal{Y}_{-\infty,k}] \circ T^{-k}] \leq \mathbf{E}\Big[\operatorname*{ess\,inf}_{u\in\mathbb{U}_0}\mathbf{E}[\ell_0(u)|\mathcal{Y}_{-\infty,0}]\Big] = L^\star.$$

On the other hand,

$$\mathbf{E}[\ell_k(\tilde{u}_k)|\mathcal{Y}_{-\infty,k}] \circ T^{-k} \geq \operatorname*{ess\,inf}_{u\in\mathbb{U}_0}\mathbf{E}[\ell_0(u)|\mathcal{Y}_{-\infty,0}] \quad \text{a.s.}$$

for all $k \in \mathbb{N}$. It follows that

$$
\limsup_{k \to \infty} \left\| \mathbf{E}[\ell_k(\tilde{u}_k)|\mathcal{Y}_{-\infty,k}] \circ T^{-k} - \operatorname*{ess\,inf}_{u \in \mathbb{U}_0} \mathbf{E}[\ell_0(u)|\mathcal{Y}_{-\infty,0}] \right\|_1 =
$$

$$
\limsup_{k \to \infty} \mathbf{E}[\mathbf{E}[\ell_k(\tilde{u}_k)|\mathcal{Y}_{-\infty,k}] \circ T^{-k}] - \mathbf{E}\left[\operatorname*{ess\,inf}_{u \in \mathbb{U}_0} \mathbf{E}[\ell_0(u)|\mathcal{Y}_{-\infty,0}] \right] \le 0.
$$

Therefore, by Maker's generalized ergodic theorem [19, Corollary 10.8]

$$
\frac{1}{T} \sum_{k=1}^{T} \mathbf{E}[\ell_k(\tilde{u}_k)|\mathcal{Y}_{-\infty,k}] \xrightarrow{T \to \infty} L^\star \quad \text{in } L^1.
$$

The result now follows using Lemma 4.7. $\qquad\square$

Combining Corollary 4.8 and Lemma 4.9 completes the proof of Theorem 2.13.

4.4 Proof of Theorem 2.16

The implication $1 \Rightarrow 2$ of Theorem 2.16 follows immediately from Theorem 1.8. In the following, we will prove the converse implication $2 \Rightarrow 1$: that is, we will show that if $(\Omega, \mathcal{B}, \mathbf{P}, T)$ is *not* conditionally weak mixing relative to $\mathcal{Y}$, then we can construct a bounded loss function ℓ with some finite decision space U for which there exists no weakly pathwise optimal strategy.

We begin by providing a "diagonal" characterization of conditional weak mixing.

Lemma 4.10 $(\Omega, \mathcal{B}, \mathbf{P}, T)$ *is conditionally weak mixing relative to* $\mathcal{Z}$ *if and only if*

$$
\frac{1}{T} \sum_{k=1}^{T} |\mathbf{E}[\{h \circ T^{-k}\} h|\mathcal{Z}] - \mathbf{E}[h \circ T^{-k}|\mathcal{Z}] \mathbf{E}[h|\mathcal{Z}]| \xrightarrow{T \to \infty} 0 \quad \text{in } L^1
$$

for every $h \in L^2$, *provided that* $\mathcal{Z} \subseteq T^{-1}\mathcal{Z}$.

Proof It suffices to show that if the equation display in the lemma holds, then $(\Omega, \mathcal{B}, \mathbf{P}, T)$ is conditionally weak mixing relative to $\mathcal{Z}$. To this end, let us fix $h \in L^2$ and denote by $\mathscr{A}$ the class of all functions $g \in L^2$ such that

$$
\frac{1}{T} \sum_{k=1}^{T} |\mathbf{E}[\{g \circ T^{-k}\} h|\mathcal{Z}] - \mathbf{E}[g \circ T^{-k}|\mathcal{Z}] \mathbf{E}[h|\mathcal{Z}]| \xrightarrow{T \to \infty} 0 \quad \text{in } L^1.
$$

Clearly $\mathscr{A}$ is closed linear subspace of L^2. Note that $\mathscr{A}$ certainly contains every random variable of the form $h\mathbf{1}_B \circ T^m$ or $\mathbf{1}_B \circ T^m$ for $m \in \mathbb{Z}$ and $B \in \mathcal{Z}$. Therefore, the closed linear span K of all such random variables is included in $\mathscr{A}$. On the other hand, suppose that $g \in K^\perp$. Then for every $k \in \mathbb{Z}$, we have

$$\mathbf{E}[\mathbf{E}[\{g \circ T^{-k}\} h | \mathbb{Z}] \mathbf{1}_B] = \mathbf{E}[g \{h\mathbf{1}_B \circ T^k\}] = 0$$

for all $B \in \mathbb{Z}$. It follows that $\mathbf{E}[\{g \circ T^{-k}\} h | \mathbb{Z}] = 0$ a.s. for all $k \in \mathbb{N}$. Similarly, we find that $\mathbf{E}[g \circ T^{-k} | \mathbb{Z}] = 0$ a.s. for all $k \in \mathbb{N}$. Thus evidently $K^\perp \subseteq \mathscr{A}$ also. Therefore, $\mathscr{A}$ contains $K \oplus K^\perp = L^2$, and the proof is complete. $\square$

In the remainder of this section, we suppose that $(\Omega, \mathcal{B}, \mathbf{P}, T)$ is not conditionally weakly mixing relative to $\mathcal{Y}$. By Lemma 4.10, there is a function $h \in L^2$ such that

$$\limsup_{T \to \infty} \mathbf{E}\left[\frac{1}{T} \sum_{k=1}^{T} |\mathbf{E}[\{H \circ T^{-k}\} H | \mathcal{Y}]| \right] \geq \varepsilon > 0$$

where $H := h - \mathbf{E}[h|\mathcal{Y}]$. By approximation in L^2, we may clearly assume without loss of generality that h takes values in $[0, 1]$, so that H takes values in $[-1, 1]$. We will fix such a function in the sequel, and consider the loss function

$$\ell(u, \omega) = u \, H(\omega)$$

where we initially choose decisions $u \in [-1, 1]$ (the decision space will be discretized at the end of the proof as required by Theorem 2.16). We claim that for the loss function ℓ there exists no weakly pathwise optimal strategy. This will be proved by a randomization procedure that will be explained presently.

In the following $([0, 1], \mathcal{I})$ denotes the unit interval with its Borel σ-field.

Lemma 4.11 *Suppose that $(\Omega, \mathcal{B}, \mathbf{P})$ is a standard probability space. Then there exists a $(\mathcal{Y} \otimes \mathcal{I})$-measurable map $\iota : \Omega \times [0, 1] \to \Omega$ such that*

$$\mathbf{E}[X|\mathcal{Y}](\omega) = \int_0^1 X(\iota(\omega, \lambda)) \, d\lambda \qquad \mathbf{P}\text{-a.e. } \omega \in \Omega.$$

for any bounded $(\mathcal{B}\text{-})$measurable function $X : \Omega \to \mathbb{R}$.

Proof As $(\Omega, \mathcal{B}, \mathbf{P})$ is a standard probability space, this is [19, Lemma 3.22] together with the existence of regular conditional probabilities [19, Theorem 6.3]. $\square$

Consider the quantity

$$A_T^\lambda(\omega) = \frac{1}{T} \sum_{k=1}^{T} H(T^k \iota(\omega, \lambda)) \, H(T^k \omega).$$

Then we can compute

$$\int_0^1 (A_T^\lambda)^2 \, d\lambda = \frac{1}{T^2} \sum_{m,n=1}^T H(T^m \omega) \, H(T^n \omega) \int_0^1 H(T^m \iota(\omega, \lambda)) \, H(T^n \iota(\omega, \lambda)) \, d\lambda$$

$$= \frac{1}{T^2} \sum_{m,n=1}^T H(T^m \omega) \, H(T^n \omega) \, \mathbf{E}[\{H \circ T^m\}\{H \circ T^n\} | \mathcal{Y}](\omega).$$

In particular, using the invariance of $\mathcal{Y}$, we have

$$\left[\int_0^1 \mathbf{E}[(A_T^\lambda)^2] \, d\lambda \right]^{1/2} = \mathbf{E}\left[\frac{1}{T^2} \sum_{m,n=1}^T \mathbf{E}[\{H \circ T^m\}\{H \circ T^n\} | \mathcal{Y}]^2 \right]^{1/2}$$

$$\geq \mathbf{E}\left[\frac{1}{T^2} \sum_{m,n=1}^T |\mathbf{E}[\{H \circ T^m\}\{H \circ T^n\} | \mathcal{Y}]| \right]$$

$$\geq \mathbf{E}\left[\frac{1}{T^2} \sum_{n=1}^T \sum_{m=1}^n |\mathbf{E}[\{H \circ T^{m-n}\} \, H | \mathcal{Y}]| \right]$$

$$= \mathbf{E}\left[\frac{1}{T^2} \sum_{k=0}^{T-1} (T-k) |\mathbf{E}[\{H \circ T^{-k}\} \, H | \mathcal{Y}]| \right]$$

$$\geq \mathbf{E}\left[\frac{1}{2T} \sum_{k=0}^{\lfloor T/2 \rfloor} |\mathbf{E}[\{H \circ T^{-k}\} \, H | \mathcal{Y}]| \right].$$

By our choice of H, it follows that

$$\limsup_{T \to \infty} \mathbf{E}[(A_T^\lambda)^2] \geq \frac{\varepsilon^2}{16}$$

for some $\lambda = \lambda_0 \in [0, 1]$. Define

$$u_k(\omega) = H(T^k \iota(\omega, \lambda_0)).$$

Then u_k is $\mathcal{Y}$-measurable for all k (and is therefore admissible if we choose, for the time being, the continuous decision space $U = [-1, 1]$), and $L_T(\mathbf{u}) = A_T^{\lambda_0}$. Moreover,

$$\frac{\varepsilon^2}{16} \leq \limsup_{T \to \infty} \mathbf{E}[(L_T(\mathbf{u}))^2] \leq \frac{\varepsilon^2}{64} + \limsup_{T \to \infty} \mathbf{P}\left[L_T(\mathbf{u}) > \frac{\varepsilon}{8} \right] + \limsup_{T \to \infty} \mathbf{P}\left[L_T(\mathbf{u}) < -\frac{\varepsilon}{8} \right]$$

implies that we may assume without loss of generality that

$$\limsup_{T\to\infty} \mathbf{P}\left[L_T(\mathbf{u}) < -\frac{\varepsilon}{8}\right] > 0$$

(if this is not the case, simply substitute $-\mathbf{u}$ for $\mathbf{u}$ in the following). But note that the strategy $\tilde{\mathbf{u}}$ defined by $\tilde{u}_k = 0$ for all k is mean-optimal (indeed, $\mathbf{E}[\ell_k(u)|\mathcal{Y}] = u\,\mathbf{E}[H|\mathcal{Y}]\circ T^k = 0$ for all u by construction). Thus evidently

$$\limsup_{T\to\infty} \mathbf{P}\left[L_T(\mathbf{u}) - L_T(\tilde{\mathbf{u}}) < -\frac{\varepsilon}{8}\right] > 0,$$

so $\tilde{\mathbf{u}}$ is not weakly pathwise optimal. It follows from Lemma 2.12 that no weakly pathwise optimal strategy can exist if we choose the decision space $U = [-1, 1]$.

To complete the proof of Theorem 2.16, it remains to show that this conclusion remains valid if we replace $U = [-1, 1]$ by some finite set. This is easily attained by discretization, however. Indeed, let $U = \{k\varepsilon/16 : k = -\lfloor 16/\varepsilon\rfloor,\ldots,\lfloor 16/\varepsilon\rfloor\}$, and construct a new strategy $\mathbf{u}'$ such that u'_k equals the value of u_k (which takes values in $[-1, 1]$) rounded to the nearest element of U. Clearly $\tilde{\mathbf{u}}$ and $\mathbf{u}'$ both take values in the finite set U, and we have $|L_T(\mathbf{u}) - L_T(\mathbf{u}')| \le \varepsilon/16$. Therefore,

$$\limsup_{T\to\infty} \mathbf{P}\left[L_T(\mathbf{u}') - L_T(\tilde{\mathbf{u}}) < -\frac{\varepsilon}{16}\right] > 0,$$

and it follows again by Lemma 2.12 that no weakly pathwise optimal strategy exists.

4.5 Proof of Theorem 3.15

By stationarity, we can rewrite the conditional absolute regularity property as

$$\left\|\mathbf{P}[(X_k)_{k\ge0} \in \cdot\,|\mathcal{X}_{-\infty,-n} \vee \mathcal{Y}_{-\infty,\infty}] - \mathbf{P}[(X_k)_{k\ge0} \in \cdot\,|\mathcal{Y}_{-\infty,\infty}]\right\|_{\mathrm{TV}} \xrightarrow{n\to\infty} 0 \quad \text{in } L^1.$$

Using a simple truncation argument (as the loss is dominated in L^1), this implies

$$\operatorname*{ess\,sup}_{u\in U_0} \left|\mathbf{E}[l(u, X_0)|\mathcal{X}_{-\infty,-n} \vee \mathcal{Y}_{-\infty,\infty}] - \mathbf{E}[l(u, X_0)|\mathcal{Y}_{-\infty,\infty}]\right| \xrightarrow{n\to\infty} 0 \quad \text{in } L^1.$$

If only we could replace $\mathcal{Y}_{-\infty,\infty}$ by $\mathcal{Y}_{-\infty,0}$ in this expression, all the assumptions of Theorem 2.6 would follow immediately. Unfortunately, it is not immediately obvious whether this replacement is possible without additional assumptions.

Remark 4.12 In general, it is not clear whether a conditional K-automorphism relative to $\mathcal{Y}_{-\infty,\infty}$ is necessarily a conditional K-automorphism relative to $\mathcal{Y}_{-\infty,0}$.

In this context, it is interesting to note that the corresponding property does hold for conditional weak mixing. We briefly sketch the proof. Suppose that $(\Omega, \mathcal{B}, \mathbf{P}, T)$ is conditionally weakly mixing relative to $\mathcal{Y}_{-\infty,\infty}$. We claim that then also

$$\frac{1}{T}\sum_{k=1}^{T} |\mathbf{E}[\{f \circ T^{-k}\}\, g | \mathcal{Y}_{-\infty,0}] - \mathbf{E}[f \circ T^{-k} | \mathcal{Y}_{-\infty,0}]\,\mathbf{E}[g | \mathcal{Y}_{-\infty,0}]| \xrightarrow{T\to\infty} 0 \quad \text{in } L^1$$

for every $f, g \in L^2$. Indeed, the conclusion is clearly true whenever f is $\mathcal{Y}_{-\infty,n}$-measurable for some $n \in \mathbb{Z}$. By approximation in L^2, the conclusion holds whenever f is $\mathcal{Y}_{-\infty,\infty}$-measurable, and it therefore suffices to consider $f \in L^2(\mathcal{Y}_{-\infty,\infty})^\perp$. But in this case we have $\mathbf{E}[f \circ T^{-k} | \mathcal{Y}_{-\infty,\infty}] = \mathbf{E}[f \circ T^{-k} | \mathcal{Y}_{-\infty,0}] = 0$ for all k, and

$$\left\| \frac{1}{T}\sum_{k=1}^{T} |\mathbf{E}[\{f \circ T^{-k}\}\, g | \mathcal{Y}_{-\infty,0}]| \right\|_1$$

$$\leq \left\| \frac{1}{T}\sum_{k=1}^{T} |\mathbf{E}[\{f \circ T^{-k}\}\, g | \mathcal{Y}_{-\infty,\infty}]| \right\|_1 \xrightarrow{T\to\infty} 0 \quad \text{in } L^1$$

by Jensen's inequality and the conditional weak mixing property relative to $\mathcal{Y}_{-\infty,\infty}$.

As we cannot directly replace $\mathcal{Y}_{-\infty,\infty}$ by $\mathcal{Y}_{-\infty,0}$, we take an alternative approach. We begin by noting that, using the conditional absolute regularity property as described above, we obtain the following trivial adaptation of Corollary 4.2.

Lemma 4.13 *Under the assumptions of Theorem 3.15, we have*

$$\frac{1}{T}\sum_{k=1}^{T}\{l(u_k, X_k) - \mathbf{E}[l(u_k, X_k) | \mathcal{Y}_{-\infty,\infty}]\} \xrightarrow{T\to\infty} 0 \quad a.s.$$

for any admissible strategy **u**.

We will now proceed to replace $\mathcal{Y}_{-\infty,\infty}$ by $\mathcal{Y}_{-\infty,k}$ in Lemma 4.13. To this end, we use the additional property established in [27, Proposition 3.9]:

$$\mathbf{P}[(X_k)_{k\leq 0} \in \cdot\, | \mathcal{Y}_{-\infty,0}] \sim \mathbf{P}[(X_k)_{k\leq 0} \in \cdot\, | \mathcal{Y}_{-\infty,\infty}] \quad \text{a.s.}$$

Theorem 3.14 implies that the past tail σ-field $\bigcap_n \mathcal{X}_{-\infty,n}$ is $\mathbf{P}[\cdot\, | \mathcal{Y}_{-\infty,0}]$-trivial a.s. (cf. [33]). Thus a standard argument [21, Theorem III.14.10] yields

$$\left\| \mathbf{P}[(X_k)_{k\leq n} \in \cdot\, | \mathcal{Y}_{-\infty,0}] - \mathbf{P}[(X_k)_{k\leq n} \in \cdot\, | \mathcal{Y}_{-\infty,\infty}] \right\|_{\mathrm{TV}} \xrightarrow{n\to-\infty} 0 \quad \text{in } L^1.$$

Therefore, by stationarity and a simple truncation argument, we have

$$\operatorname*{ess\,sup}_{u\in\mathbb{U}_0}\left|\mathbf{E}[l(u,X_0)|\mathcal{Y}_{-\infty,n}]-\mathbf{E}[l(u,X_0)|\mathcal{Y}_{-\infty,\infty}]\right|\xrightarrow{n\to\infty}0\quad\text{in }L^1.$$

This yields the following consequence.

Corollary 4.14 *Under the assumptions of Theorem 3.15, we have*

$$\frac{1}{T}\sum_{k=1}^{T}\{l(u_k,X_k)-\mathbf{E}[l(u_k,X_k)|\mathcal{Y}_{-\infty,k}]\}\xrightarrow{T\to\infty}0\quad a.s.$$

for any admissible strategy **u**. *In particular,*

$$\liminf_{T\to\infty}L_T(\mathbf{u})\geq\mathbf{E}\left[\operatorname*{ess\,inf}_{u\in\mathbb{U}_0}\mathbf{E}[\ell_0(u)|\mathcal{Y}_{-\infty,0}]\right]=L^\star\quad a.s.$$

Proof (Sketch) Following almost verbatim the proof of Lemma 4.1, one can prove

$$\frac{1}{T}\sum_{k=1}^{T}\{\mathbf{E}[l(u_k,X_k)|\mathcal{Y}_{-\infty,k}]-\mathbf{E}[l(u_k,X_k)|\mathcal{Y}_{-\infty,k+r}]\}\xrightarrow{T\to\infty}0\quad a.s.$$

for any $r\in\mathbb{N}$. On the other hand, we have

$$\limsup_{T\to\infty}\left|\frac{1}{T}\sum_{k=1}^{T}\{\mathbf{E}[l(u_k,X_k)|\mathcal{Y}_{-\infty,k+r}]-\mathbf{E}[l(u_k,X_k)|\mathcal{Y}_{-\infty,\infty}]\}\right|$$

$$\leq\lim_{T\to\infty}\frac{1}{T}\sum_{k=1}^{T}\operatorname*{ess\,sup}_{u\in\mathbb{U}_k}\left|\mathbf{E}[l(u,X_k)|\mathcal{Y}_{-\infty,k+r}]-\mathbf{E}[l(u,X_k)|\mathcal{Y}_{-\infty,\infty}]\right|$$

$$=\mathbf{E}\left[\operatorname*{ess\,sup}_{u\in\mathbb{U}_0}\left|\mathbf{E}[l(u,X_0)|\mathcal{Y}_{-\infty,r}]-\mathbf{E}[l(u,X_0)|\mathcal{Y}_{-\infty,\infty}]\right|\right]\quad a.s.$$

by the ergodic theorem. It was shown above that the latter quantity converges to zero as $r\to\infty$, and the result now follows using Lemma 4.13. □

The remainder of the proof of Theorem 3.15 is identical to that of Theorem 2.6 modulo trivial modifications, and is therefore omitted.

Acknowledgements This work was partially supported by NSF grant DMS-1005575.

References

1. P.H. Algoet, The strong law of large numbers for sequential decisions under uncertainty. IEEE Trans. Inform. Theory **40**(3), 609–633 (1994)
2. A. Bellow, V. Losert, The weighted pointwise ergodic theorem and the individual ergodic theorem along subsequences. Trans. Am. Math. Soc. **288**(1), 307–345 (1985)
3. D. Berend, V. Bergelson, Mixing sequences in Hilbert spaces. Proc. Am. Math. Soc. **98**(2), 239–246 (1986)
4. O. Cappé, E. Moulines, T. Rydén, *Inference in Hidden Markov Models* (Springer, New York, 2005)
5. P. Chigansky, R. van Handel, A complete solution to Blackwell's unique ergodicity problem for hidden Markov chains. Ann. Appl. Probab. **20**(6), 2318–2345 (2010)
6. J.P. Conze, *Convergence des moyennes ergodiques pour des sous-suites*. In: Contributions au calcul des probabilités. Bull. Soc. Math. France, Mém. No. 35 (Soc. Math. France, Paris, 1973), pp. 7–15
7. D. Crisan, B. Rozovskiĭ (eds.), *The Oxford Handbook of Nonlinear Filtering* (Oxford University Press, Oxford, 2011)
8. P. Del Moral, M. Ledoux, Convergence of empirical processes for interacting particle systems with applications to nonlinear filtering. J. Theor. Probab. **13**(1), 225–257 (2000)
9. C. Dellacherie, P.A. Meyer, *Probabilities and Potential. C* (North-Holland, Amsterdam, 1988)
10. R.M. Dudley, *Uniform Central Limit Theorems* (Cambridge University Press, Cambridge, 1999)
11. N. Etemadi, An elementary proof of the strong law of large numbers. Z. Wahrsch. Verw. Gebiete **55**(1), 119–122 (1981)
12. A. Grothendieck, Produits tensoriels topologiques et espaces nucléaires. Mem. Am. Math. Soc. **1955**(16), 140 (1955)
13. P.R. Halmos, In general a measure preserving transformation is mixing. Ann. Math. (2) **45**, 786–792 (1944)
14. R. van Handel, The stability of conditional Markov processes and Markov chains in random environments. Ann. Probab. **37**(5), 1876–1925 (2009)
15. R. van Handel, Uniform time average consistency of Monte Carlo particle filters. Stoch. Process. Appl. **119**(11), 3835–3861 (2009)
16. R. van Handel, On the exchange of intersection and supremum of σ-fields in filtering theory. Isr. J. Math. **192**, 763–784 (2012)
17. van Handel, R.: The universal Glivenko-Cantelli property. Probab. Theor. Relat. Fields **155**(3–4), 911–934 (2013)
18. J. Hoffmann-Jørgensen, *Uniform Convergence of Martingales*. In: Probability in Banach spaces, 7 (Oberwolfach, 1988), *Progr. Probab.*, vol. 21 (Birkhäuser Boston, Boston, 1990), pp. 127–137
19. O. Kallenberg, *Foundations of Modern Probability*, 2nd edn. (Springer, New York, 2002)
20. H. Kunita, Asymptotic behavior of the nonlinear filtering errors of Markov processes. J. Multivariate Anal. **1**, 365–393 (1971)
21. T. Lindvall, *Lectures on the Coupling Method* (Dover Publications, Mineola, 2002). Corrected reprint of the 1992 original
22. S. Meyn, R.L. Tweedie, *Markov Chains and Stochastic Stability*, 2nd edn. (Cambridge University Press, Cambridge, 2009)
23. J. Neveu, *Discrete-Parameter Martingales* (North-Holland, Amsterdam, 1975)
24. A.B. Nobel, On optimal sequential prediction for general processes. IEEE Trans. Inform. Theory **49**(1), 83–98 (2003)
25. D. Pollard, *A User'S Guide to Measure Theoretic Probability* (Cambridge University Press, Cambridge, 2002)
26. D.J. Rudolph, Pointwise and L^1 mixing relative to a sub-sigma algebra. Ill. J. Math. **48**(2), 505–517 (2004)

27. X.T. Tong, R. van Handel, Conditional ergodicity in infinite dimension (2012). Preprint
28. H. Totoki, On a class of special flows. Z. Wahrscheinlichkeitstheorie und Verw. Gebiete **15**, 157–167 (1970)
29. A.W. van der Vaart, J.A. Wellner, Weak Convergence and Empirical Processes (Springer, New York, 1996)
30. V.A. Volkonskiĭ, Y.A. Rozanov, Some limit theorems for random functions. I. Theor. Probab. Appl. **4**, 178–197 (1959)
31. P. Walters, *An Introduction to Ergodic Theory* (Springer, New York, 1982)
32. T. Weissman, N. Merhav, Universal prediction of random binary sequences in a noisy environment. Ann. Appl. Probab. **14**(1), 54–89 (2004)
33. H. von Weizsäcker, Exchanging the order of taking suprema and countable intersections of σ-algebras. Ann. Inst. H. Poincaré Sect. B (N.S.) **19**(1), 91–100 (1983)

Invariance Principle for the Random Walk Conditioned to Have Few Zeros

Laurent Serlet

Abstract We consider a nearest neighbor random walk on $\mathbb{Z}$ starting at zero, conditioned to return at zero at time $2n$ and to have a number z_n of zeros on $(0, 2n]$. As $n \to +\infty$, if $z_n = o(\sqrt{n})$, we show that the rescaled random walk converges toward the Brownian excursion normalized to have unit duration. This generalizes the classical result for the case $z_n \equiv 1$.

Keywords Conditioned random walk • Excursions • Invariance principle

1 Introduction and Statement of the Results

For $n \geq 1$, let $(X_j)_{0 \leq j \leq 2n}$ be, under a probability $\mathbb{P}$, a standard nearest neighbor random walk on $\mathbb{Z}$ starting from 0. This means that $X_0 = 0$ and for $0 < j \leq 2n$, the random variables $X_j - X_{j-1}$ are independent and uniform over $\{-1, +1\}$. See for instance [9] as a general reference on random walks.

We define the conditional probability $\mathbb{P}_n = \mathbb{P}(\cdot \mid X_{2n} = 0)$ so that, under this probability, $(X_j)_{0 \leq j \leq 2n}$ becomes a "discrete random bridge". Let us introduce the usual rescaling of this walk by defining, for $0 \leq t \leq 1$,

$$X_t^n = \frac{1}{\sqrt{2n}} \left(X_{[2nt]} + (2nt - [2nt])(X_{[2nt]+1} - X_{[2nt]}) \right) \tag{1}$$

where $[\cdot]$ is the integer part. It is well known by Donsker's famous theorem that, under $\mathbb{P}_n$, the law of $(X_t^n)_{t \in [0,1]}$ converges weakly to the law of the Brownian bridge. Here and below, weak convergence of laws on the space of continuous functions is understood, as usual, with this space of continuous functions being equipped with the uniform norm. See [1] for generalities on weak convergence and application to Donsker's Theorem in an unconditioned setting and [7] for the conditioned case.

L. Serlet (✉)
Clermont Université, Université Blaise Pascal, Laboratoire de Mathématiques (CNRS umr 6620),
Complexe scientifique des Cézeaux, BP 80026, 63171 Aubière, France
e-mail: Laurent.Serlet@math.univ-bpclermont.fr

© Springer International Publishing Switzerland 2014   461
C. Donati-Martin et al. (eds.), *Séminaire de Probabilités XLVI*, Lecture Notes
in Mathematics 2123, DOI 10.1007/978-3-319-11970-0_19

Under $\mathbb{P}_n$, let Z be the number of zeros of the walk on $(0, 2n]$. The law of $\frac{1}{\sqrt{n}} Z$ converges weakly to the law of the local time at time 1 and level 0 of a Brownian bridge, see [8] for the original statement. As a consequence, under $\mathbb{P}_n$, Z is typically of order $\sqrt{n}$ as $n \to +\infty$. The subject of this paper is to examine the asymptotic behavior in law of the random walk under $\mathbb{P}_n$, if we condition the number of zeros Z to be equal to z_n where this quantity is asymptotically smaller than $\sqrt{n}$. To this end we set $\mathbb{P}_n^z(\cdot) = \mathbb{P}_n(\cdot \mid Z = z)$.

Kaigh [5, Theorem 2.6] has given an invariance principle in the case $z_n \equiv 1$; in equivalent terms, Kaigh conditions the walk to be positive between the endpoints and vanish at the endpoints. The result is that the law of $(X_t^n)_{t \in [0,1]}$ under $\mathbb{P}_n^1$ converges weakly to the Brownian excursion normalized to have duration 1. See for instance [10] for the definition and properties of the normalized Brownian excursion. Conditioning walks to be positive has been studied in many papers, see [2] for a recent contribution giving many references. Several recent papers have been motivated by random polymer models.

We want to know whether a result similar to Kaigh's Theorem holds under $\mathbb{P}_n^{z_n}$ when (z_n) is any sequence of positive integers such that $z_n = o(\sqrt{n})$ i.e. $z_n / \sqrt{n} \to 0$. Note that if we condition the random walk to have zeros at prescribed sites — for instance a zero at n — the invariance principle is of course different. But if the positions of the zeros are not imposed, they are distributed according to the conditional law and could accumulate (in the asymptotic scale) near the endpoints of the interval. This is precisely what happens.

Proposition 1 *Suppose* $z_n = o(\sqrt{n})$. *Then, there exists a sequence* (ε_n) *converging toward zero as* $n \to +\infty$ *such that*

$$\lim_{n \to +\infty} \mathbb{P}_n^{z_n}[\exists k \in [n\varepsilon_n, (2 - \varepsilon_n) n], \ X_k = 0] = 0. \tag{2}$$

A consequence of this proposition is that Kaigh's invariance principle can be extended with the same conclusion.

Theorem 1 *Suppose* $z_n = o(\sqrt{n})$ *as before. Then, under* $\mathbb{P}_n^{z_n}$, *the law of the process* $(X_t^n)_{t \in [0,1]}$ *converges weakly to the law of the signed Brownian excursion normalized to have duration 1.*

In this statement, by signed normalized Brownian excursion we mean the process $(\sigma W_t)_{t \in [0,1]}$ where $(W_t)_{t \in [0,1]}$ is the nonnegative Brownian excursion normalized to unit duration and σ is, independently, a uniform random variable in $\{-1, +1\}$. The theorem is also valid for a nonnegative reflecting random walk and in that case, the limit is the nonnegative normalized excursion.

Let us give examples of application. First it can apply to a polymer whose profile is modeled as the graph of a (one-dimensional) random walk. This is the very simple model of directed polymer. See [4] Chap. 1 for the description of many polymer models and motivations. In our setting, the conditioning we are studying corresponds to impose a prescribed number z_n of pining on the first axis.

Theorem 1 gives the asymptotic profile of a directed polymer with a large number of constituting molecules under the condition of a certain number of pining.

In another context, a nonnegative reflecting random walk is, under $\mathbb{P}_n^{z_n}$, the contour process of a Galton-Watson random forest with geometric offspring law, conditioned to have z_n trees and n as total number of nodes. In this context Theorem 1 can be reformulated, in an informal way, as the fact that, as $n \to +\infty$, this specified Galton-Watson random forest converges to a single continuous random tree (CRT) "coded by a Brownian excursion". We do not go further in this direction since it is not the subject of the present paper but we refer to [6] for a survey of the concepts and vast literature on random trees.

The rest of the paper is devoted to proofs. We start with a lemma that will be essential.

2 The Key Lemma

Let us recall and introduce some notation. Under $\mathbb{P}_n$, the random walk $(X_j)_{0 \leq j \leq 2n}$ has $Z + 1$ total number of zeros that we denote in increasing order by $Y_0 = 0 < Y_1 < \cdots < Y_{Z-1} < 2n = Y_Z$. There are Z excursions out of zero which are the pieces of trajectory between two consecutive times Y_{i-1} and Y_i ($1 \leq i \leq Z$).

Lemma 1 *For any $A > 0$, there exists a constant c_1 such that, for any $n \geq 1$ and any $1 \leq z \leq \min(n, A\sqrt{n})$,*

$$\mathbb{P}_n(Z = z) \leq c_1 \frac{z}{n}. \tag{3}$$

Moreover, if $1 \leq z_n \leq n$ and $z_n = o(\sqrt{n})$ as $n \to +\infty$ then there exists a constant $c_2 > 0$ such that, for every n,

$$\mathbb{P}_n(Z = z_n) \geq c_2 \frac{z_n}{n}. \tag{4}$$

Proof Under $\mathbb{P}_n$ let us denote by $L_1, \ldots, L_Z$ the successive excursion lengths that is $L_i = Y_i - Y_{i-1}$ for $i \leq Z$. The study of these excursion lengths has been made in several papers, see [3] and the references therein. A "deterministic nonnegative excursion" of length 2ℓ is simply a finite sequence $0 = u_0, u_1, \ldots, u_{2\ell} = 0$ with $|u_{i+1} - u_i| = 1$ and $u_1, \ldots, u_{2\ell-1} > 0$. The number $e(\ell)$ of such deterministic excursions is the $(\ell - 1)$-th Catalan number

$$e(\ell) = \frac{1}{2(2\ell - 1)} \binom{2\ell}{\ell} = \frac{(2\ell - 2)!}{\ell! \, (\ell - 1)!} \tag{5}$$

as proved for instance in [11, 6.2.3 (iv)]. It follows from elementary counting that, for positive integers $\ell_1, \ldots, \ell_z$ such that $\ell_1 + \cdots + \ell_z = n$,

$$\mathbb{P}_n(Z = z, L_1 = 2\,\ell_1, \ldots, L_z = 2\,\ell_z) = \frac{2^z}{\binom{2n}{n}}\, e(l_1) \ldots e(l_z),$$

the factor 2^z coming from the choice of the signs of the excursions. We obtain the formula

$$\mathbb{P}_n(Z = z) = \frac{2^z}{\binom{2n}{n}} \sum_{\ell_1 + \cdots + \ell_z = n} e(l_1) \ldots e(l_z). \tag{6}$$

But, a well known property of Catalan numbers [11, 6.2 p.178] is that, for $u \in (-1/4, 1/4)$,

$$\sum_{\ell=1}^{+\infty} e(\ell)\, u^\ell = \frac{1 - \sqrt{1 - 4u}}{2}. \tag{7}$$

Combining with (6), we deduce that, for $u \in (-1, 1)$,

$$\sum_{n=z}^{+\infty} 4^{-n} \binom{2n}{n} \mathbb{P}_n(Z = z)\, u^n = \left(1 - \sqrt{1 - u}\right)^z. \tag{8}$$

Let us set

$$\alpha_n = 4^{-n} \binom{2n}{n} = \mathbb{P}(X_{2n} = 0) \tag{9}$$

recalling that $\mathbb{P}$ is the probability under which $(X_j)_{0 \leq j \leq 2n}$ is the standard (unconditioned) random walk on $\mathbb{Z}$ starting from 0.

Let us take $\rho \in (0, 1)$ and denote by $S(\rho)$ the circle of radius ρ and center 0 in the complex plane $\mathbb{C}$. It follows from Eq. (8) that, for $n \geq z$,

$$2\pi i\, \alpha_n\, \mathbb{P}_n(Z = z) = \int_{S(\rho)} \frac{\left(1 - \sqrt{1 - u}\right)^z}{u^{n+1}}\, du \tag{10}$$

where we use the standard extension of the square root function to the complex domain $\mathbb{C} \setminus \mathbb{R}_-$. For the integral on the right hand side we expand the term $\left(1 - \sqrt{1 - u}\right)^z$ according to Newton's formula and note that, for $j \in \{0, \ldots, z\}$ even and $n \geq z$,

$$\int_{S(\rho)} \frac{\left(\sqrt{1 - u}\right)^j}{u^{n+1}}\, du = 0.$$

so only the contribution of the odd powers remains. We get

$$
2\pi i\, \alpha_n\, \mathbb{P}_n(Z = z) = -\sum_{j=0}^{[(z-1)/2]} \binom{z}{1+2j} \int_{S(\rho)} \frac{\sqrt{1-u}\,(1-u)^j}{u^{n+1}}\, du
$$

$$
= -\sum_{j=0}^{[(z-1)/2]} \binom{z}{1+2j} \sum_{r=0}^{j} \binom{j}{r}(-1)^r \int_{S(\rho)} \frac{\sqrt{1-u}\,u^r}{u^{n+1}}\, du.
$$

But, it follows from (7) that, for any integer $q \geq 0$,

$$
-\int_{S(\rho)} \frac{\sqrt{1-u}}{u^{q+1}}\, du = 2\pi\, i\beta_q
$$

where $\beta_0 = -1$ and for $q \geq 1$, $\beta_q = 2\, 4^{-q}\, e(q)$. Taking this fact into account, we get

$$
\alpha_n\, \mathbb{P}_n(Z = z) = \sum_{j=0}^{[(z-1)/2]} \binom{z}{1+2j} \sum_{r=0}^{j} \binom{j}{r}(-1)^r\, \beta_{n-r}. \tag{11}
$$

We introduce the notation δ for the shift operator on sequences –that is $\delta(\beta)_n = \beta_{n-1}$. For the validity of this definition one can extend the definition of β by $\beta_k = \beta_0$ for $k < 0$ but this plays no role in the sequel. We denote I the identity operator on sequences, so that the Formula (11) can be rewritten as

$$
\alpha_n\, \mathbb{P}_n(Z = z) = \sum_{j=0}^{[(z-1)/2]} \binom{z}{1+2j} (I - \delta)^j\, (\beta)_n
$$

for all $n \geq z \geq 1$. But for $n \geq 1$, one easily checks that

$$
\beta_n - \beta_{n-1} = -\frac{3}{2n-3}\, \beta_n
$$

from which we easily deduce by induction on j that

$$
(I - \delta)^j\, (\beta)_n = (-1)^j \left(\prod_{i=1}^{j} \frac{2i+1}{2n-2i-1} \right) \beta_n
$$

for $n \geq j$. We obtain

$$
\frac{\alpha_n}{z\, \beta_n}\, \mathbb{P}_n(Z = z) = 1 + R(n, z) \tag{12}
$$

with a remainder term

$$R(n,z) = \sum_{j=1}^{[(z-1)/2]} \frac{1}{z}\binom{z}{1+2j}(-1)^j \left(\prod_{i=1}^{j} \frac{2i+1}{2n-2i-1}\right). \tag{13}$$

This term is easily shown to be bounded by

$$|R(n,z)| \leq \sum_{j=1}^{[(z-1)/2]} \frac{1}{2^j\,j!}\left(\frac{z^2}{2n-2j-1}\right)^j.$$

For $n \geq 1$ and $1 \leq z \leq A\sqrt{n}$, it is clear that $|R(n,z)|$ stays bounded. But $\alpha_n/\beta_n = 2n - 1$ so that (12) gives (3). Moreover the assumption $z_n = o(\sqrt{n})$ entails $R(n,z_n) \to 0$ and (4) follows. The proof of Lemma 1 is complete. Another technical lemma needed in the sequel is the following.

Lemma 2 *There exists a constant c_3 such that, for $1 \leq g^2 < n/2$,*

$$\sum_{\substack{2g^2 \leq 2k \leq n \\ n < 2j \leq 2n}} (j-k)^{-3/2}\,k^{-3/2} \leq \frac{c_3}{\sqrt{n}\,g}. \tag{14}$$

Proof By taking out of the sum the contribution of the value $j = [n/2]+1$, we see that the left hand side in (14) is bounded from above by

$$\frac{1}{n}\sum_{k=g^2}^{[n/2]} \sum_{j=[n/2]+2}^{n} \frac{1}{n^2}\left(\frac{k}{n}\right)^{-3/2}\left(\frac{j-k}{n}\right)^{-3/2} \tag{15}$$

$$+\ \frac{1}{n^3}\sum_{k=g^2}^{[n/2]} \left(\frac{k}{n}\right)^{-3/2}\left(\frac{[n/2]+1-k}{n}\right)^{-3/2}. \tag{16}$$

But, when $g \geq 2$, the quantity in (15) is lower than

$$\frac{1}{n}\sum_{k=g^2}^{[n/2]} \sum_{j=[n/2]+2}^{n} \int_{[\frac{k-1}{n},\frac{k}{n}]\times[\frac{j-2}{n},\frac{j-1}{n}]} x^{-3/2}\,(y-x)^{-3/2}\,dx\,dy$$

$$\leq \frac{1}{n}\int_{\frac{g^2-1}{n}}^{\gamma} \frac{dx}{x^{3/2}}\int_{\gamma}^{1} \frac{dy}{(y-x)^{3/2}}$$

where $\gamma = \frac{1}{n}\left[\frac{n}{2}\right]$. The double integral above is easily bounded by $c\left(\frac{g^2-1}{n}\right)^{-1/2}$ so the quantity (15) is effectively at most of order $\frac{c}{\sqrt{n}\,g}$. The case $g = 1$ is similar. Also, the reader can check that the quantity in (16) is at most of order $c\,n^{-3/2}$ which is lower than $\frac{c}{\sqrt{n}\,g}$. This completes the proof.

3 Proof of Proposition 1

We recall that, under $\mathbb{P}_n^z$, we denote by $Y_0 = 0 < Y_1 < \cdots < Y_{z-1} < 2n = Y_z$ the increasing sequence of times $k \in \{0, \ldots, 2n\}$ such that $X_k = 0$. Let us denote by $G \in \{0, \ldots, z-1\}$ the index of the largest zero lower than n i.e. $Y_G \leq n < Y_{G+1}$.

Let $(\varepsilon_n)_{n \geq 1}$ be a sequence of real numbers in $(0, 1)$ converging to 0 and such that

$$\lim_{n \to 0} \frac{z_n}{\sqrt{n} \, \varepsilon_n^{3/2}} = 0. \tag{17}$$

Note that $n \, \varepsilon_n \longrightarrow +\infty$ as we will use it later. To describe different configurations where there is a zero in the interval $[n \, \varepsilon_n, (2-\varepsilon_n) \, n]$, we introduce four (non disjoint) sets : $\Lambda_1, \Lambda_2, \Lambda_3, \Lambda_4$ as follows. For $q \in \{1, 2, 3, 4\}$, we denote by Λ_q the set of triples of integers (g, j, k) such that $1 \leq g \leq z_n - 2$ and moreover

$$2g^2 \leq 2k \leq n < 2j \leq (2 - \varepsilon_n) \, n \text{ in the case } q = 1,$$

$$2g \leq 2k \leq 2g^2 \leq n < 2j \leq (2 - \varepsilon_n) \, n \text{ in the case } q = 2,$$

$$\varepsilon_n \, n \leq 2k \leq n < 2j \leq 2n - 2(z_n - g - 1)^2 \text{ in the case } q = 3,$$

$$\varepsilon_n \, n \leq 2k \leq n < 2n - 2(z_n - g - 1)^2 \leq 2j \leq 2n - 2(z_n - g - 1) \text{ in the case } q = 4.$$

Proposition 1 will be established once we prove the following facts

$$\lim_{n \to +\infty} \sum_{(g,j,k) \in \Lambda_q} \mathbb{P}_n^{z_n} \left[G = g, \, Y_g = 2k, \, Y_{g+1} = 2j \right] = 0 \tag{18}$$

for $q \in \{1, 2, 3, 4\}$,

$$\lim_{n \to +\infty} \sum_{n < 2k \leq (2-\varepsilon_n)n} \mathbb{P}_n^{z_n} \left[G = 0, \, Y_1 = 2k \right] = 0 \tag{19}$$

and

$$\lim_{n \to +\infty} \sum_{\varepsilon_n n \leq 2j \leq n} \mathbb{P}_n^{z_n} \left[G = z_n - 1, \, Y_{z_n - 1} = 2j \right] = 0. \tag{20}$$

We will only prove (18) for $q \in \{1, 2\}$. The cases $q \in \{3, 4\}$ are almost exactly symmetrical. Moreover (19) and (20) are obtained by the same techniques and are easier.

Let us set $Z(a, b] = \#(\{k; \ a < k \le b, \ X_k = 0\})$ so that, under $\mathbb{P}$ or $\mathbb{P}_n$, the notation Z previously introduced is $Z(0, 2n]$. We first re-express the probability

$$\mathbb{P}_n^{z_n} \left[G = g, \ Y_g = 2k, \ Y_{g+1} = 2j \right]$$
$$= \frac{\mathbb{P}\left(G = g, \ Y_g = 2k, \ Y_{g+1} = 2j, \ X_{2n} = 0, \ Z = z_n \right)}{\mathbb{P}(X_{2n} = 0) \, \mathbb{P}_n(Z = z_n)}$$

and note that the event appearing at the numerator is

$$Z(0, 2k] = g, X_{2k} = 0, Z(2k, 2j] = 1, X_{2j} = 0, Z(2j, 2n] = z_n - g - 1, X_{2n} = 0.$$

Using the Markov property under $\mathbb{P}$ and recalling the notation α_n introduced in (9), we obtain, for $2k \le n < 2j$,

$$\mathbb{P}_n^{z_n} \left[G = g, \ Y_g = 2k, \ Y_{g+1} = 2j \right]$$
$$= \frac{\mathbb{P}_k(Z = g) \, \alpha_k \, \mathbb{P}_{j-k}(Z = 1) \, \alpha_{j-k} \, \mathbb{P}_{n-j}(Z = z_n - g - 1) \, \alpha_{n-j}}{\alpha_n \, \mathbb{P}_n(Z = z_n)} \quad (21)$$

and we know that

$$\mathbb{P}_{j-k}(Z = 1) = \frac{1}{2(j - k) - 1}.$$

Note also that, as $n \to +\infty$, Stirling's Formula easily implies that $\alpha_n \sim 1/\sqrt{\pi n}$ and consequently, α_i is also bounded by $c/\sqrt{i}$ for every i.

Let us prove (18) in the case $q = 1$. Note that, for $(g, k, j) \in \Lambda_1$, the ratio $g/\sqrt{k}$ is trivially bounded from above and so is the ratio $(z_n - g - 1)/\sqrt{n - j}$ thanks to the hypothesis on (ε_n). As a consequence we can use the bound (3) of Lemma 1 twice in the numerator of (21) and we can also use (4) for the denominator. We get that, for $(g, k, j) \in \Lambda_1$, the probability $\mathbb{P}_n^{z_n} \left[G = g, \ Y_g = 2k, \ Y_{g+1} = 2j \right]$ is bounded from above, up to a multiplicative constant, by

$$\frac{g \ (z_n - g - 1) \ n^{3/2}}{z_n \ (j - k)^{3/2} \ k^{3/2} \ (n - j)^{3/2}}.$$

We observe that, in the expression above, $(n - j)^{-3/2} \le (\varepsilon_n n/2)^{-3/2}$. We bound $z_n - g - 1/z_n$ by 1. Then we sum over k, j with $2g^2 \le 2k < n$ and $n < 2j \le (2 - \varepsilon_n)n$ and bound this sum using Lemma 2. We sum over g with $1 \le g \le z_n - 2$. Finally we get

$$\sum_{(g,k,j) \in \Lambda_1} \mathbb{P}_n^{z_n} \left[G = g, \ Y_g = 2k, \ Y_{g+1} = 2j \right] \le \frac{c}{\varepsilon_n^{3/2}} \frac{z_n}{\sqrt{n}}$$

and this last quantity converges to 0 by the assumption (17) on (ε_n).

Now let us pass to the case $q = 2$. We follow the same lines as in the case $q = 1$, bounding the terms in (21) in the same way except that $\mathbb{P}_k(Z = g)$ cannot be bounded using (3) of Lemma 1. We obtain

$$\mathbb{P}_n^{z_n}\left[G = g,\ Y_g = 2k,\ Y_{g+1} = 2j\right]$$

$$\leq c\ \mathbb{P}_k(Z = g)\ \frac{z_n - g - 1}{z_n\ (j - k)^{3/2}\ \sqrt{k}}\ \varepsilon_n^{-3/2}$$

and we have to show the convergence toward 0 of the sum of these quantities over $(g, j, k) \in \Lambda_2$. We remark that $j - k \geq c\,n$ for large n. We bound $z_n - g - 1/z_n$ by 1, as before. We sum over j and then over g. We finally get that

$$\sum_{(g,k,j)\in\Lambda_2} \mathbb{P}_n^{z_n}\left[G = g,\ Y_g = 2k,\ Y_{g+1} = 2j\right]$$

$$\leq \frac{c}{\sqrt{n}}\ \varepsilon_n^{-3/2} \sum_{k=1}^{z_n^2} \frac{1}{\sqrt{k}} \leq c\ \frac{z_n}{\sqrt{n}}\varepsilon_n^{-3/2} \to 0.$$

This completes the proof of (18) for the case $q = 2$ and, as said before, we leave the other cases—which use similar arguments—to the reader.

4   Proof of Theorem 1

As stated in Proposition 1 we can choose a sequence (ε_n) such that, as $n \to +\infty$, we have $\varepsilon_n \to 0$, $n\,\varepsilon_n \to +\infty$ and $\mathbb{P}_n^{z_n}(A_n) \to 0$ where

$$A_n = \{\exists k \in [\varepsilon_n\,n, (2 - \varepsilon_n)\,n],\ X_k = 0\}.$$

It follows that we can show the desired convergence in law of Theorem 1 with the supplementary condition that A_n^C is realized. Let us set $\hat{\mathbb{P}}_n^{z_n} = \mathbb{P}_n^{z_n}(\cdot\,|A_n^C)$. Under $\hat{\mathbb{P}}_n^{z_n}$, it is clear that, for large n, the longest excursion of $(X_k)_{0\leq k\leq 2n}$ out of zero is the one straddling n and more precisely it is $(X_k)_{g_n\leq k\leq d_n}$ where

$$d_n = \min\{k \geq n,\ X_k = 0\} \text{ and } g_n = \max\{k \leq n,\ X_k = 0\}.$$

Thus $2\tilde{L}_1^n := d_n - g_n$ is the length of this longest excursion. We define the re-indexed excursion $(U_k)_{0\leq k\leq 2n}$ by $U_k = X_{(g_n+k)\wedge d_n}$.

We define $(V_k)_{0\leq k\leq 2n}$ as the walk that remains when the former "central" excursion is removed, that is $V_k = X_k$ for $k \leq g_n$ and $V_k = X_{(d_n+k-g_n)\wedge 2n}$ for $g_n < k \leq 2n$. Note that $(V_k)_{0\leq k\leq 2n}$ is stopped at the random time $2n - d_n + g_n = 2(n - \tilde{L}_1^n)$ and, conditionally on $\tilde{L}_1^n = \ell$, $(V_k)_{0\leq k\leq 2(n-\ell)}$ follows the law $\mathbb{P}_{n-\ell}^{z_n-1}$.

We introduce the processes $(U_t^n)_{t\in[0,1]}$ and $(V_t^n)_{t\in[0,1]}$ which are obtained from $(U_k)_{0\le k\le 2n}$ and $(V_k)_{0\le k\le 2n}$ respectively by rescaling and linear interpolation, as $(X_t^n)_{t\in[0,1]}$ is obtained from $(X_k)_{0\le k\le 2n}$ according to (1). To prove Theorem 1, it suffices to prove that, under $\hat{\mathbb{P}}_n^{z_n}$,

(i) $(U_t^n)_{t\in[0,1]}$ converges in law to $I^{(1)}$, the law of the Brownian excursion normalized to have unit length.

(ii) $(V_t^n)_{t\in[0,1]}$ converges in law to 0.

Indeed, if these convergences in law hold, we apply Skorohod representation Theorem to construct the following objects on a certain probability space:

- $(\tilde{U}_t^n)_{t\in[0,1]}$ having the same law as $(U_t^n)_{t\in[0,1]}$
- $(\tilde{V}_t^n)_{t\in[0,1]}$ having the same law as $(V_t^n)_{t\in[0,1]}$
- $(U_t^\infty)_{t\in[0,1]}$ distributed as $I^{(1)}$

such that, almost surely, $(\tilde{U}_t^n)_{t\in[0,1]}$ converges to $(U_t^\infty)_{t\in[0,1]}$ and $(\tilde{V}_t^n)_{t\in[0,1]}$ converges to zero, both with respect to the uniform norm. From this argument it is possible to find $(\tilde{X}_t^n)_{t\in[0,1]}$ having the same law as $(X_t^n)_{t\in[0,1]}$ under $\hat{\mathbb{P}}_n^{z_n}$ and which converges almost surely to $(U_t^\infty)_{t\in[0,1]}$, which distributed as $I^{(1)}$. Hence Theorem 1 will follow.

Point (ii) claimed above is easily seen since, conditionally on A_n^C, the process $(V_t^n)_{t\in[0,1]}$ vanishes on $[\varepsilon_n, 1]$ with $\varepsilon_n \downarrow 0$ and is, on $[0, \varepsilon_n]$, a rescaled random walk with a prescribed number of zeros. More precisely, for any $\eta > 0$, we have

$$\hat{\mathbb{P}}_n^{z_n}\left(\sup_{t\in[0,1]} |V_t^n| \ge \eta\right) = \int_0^{n\varepsilon_n} \mathbb{P}_q^{z_n-1}\left(\sup_{k\le 2q} |X_k| \ge \sqrt{2n}\,\eta\right) \hat{\mathbb{P}}_n^{z_n}\left(n - \tilde{L}_1^n \in dq\right)$$

$$\le \sup_{q\le n\varepsilon_n} \mathbb{P}_q^1\left(\sup_{k\le 2q} |X_k| \ge \sqrt{2n}\,\eta\right)$$

$$\le \mathbb{P}_{[n\varepsilon_n]}^1\left(\sup_{k\le 2[n\varepsilon_n]} |X_k| \ge \sqrt{2n}\,\eta\right)$$

$$\le \mathbb{P}_{[n\varepsilon_n]}^1\left(\sup_{s\in[0,1]} |X_s^{[n\varepsilon_n]}| \ge \frac{\eta}{\sqrt{\varepsilon_n}}\right). \tag{22}$$

As proved by Kaigh [5], the law of $(X_s^{[n\varepsilon_n]})_{s\in[0,1]}$ under $\mathbb{P}_{[n\varepsilon_n]}^1$ converges (weakly) to the law of the normalized Brownian excursion. Since moreover $\eta/\sqrt{\varepsilon_n} \to +\infty$, we deduce that the quantity in (22) tends to 0. Hence the convergence toward 0 claimed in (ii) holds in probability hence in law.

Point (i) is a slight generalization of the same result of Kaigh [5]. It consists in proving that, for any bounded continuous functional F on the space of continuous functions,

$$\lim_{n\to+\infty} \hat{\mathbb{E}}_n^{z_n}\left[F\big((U_t^n)_{t\in[0,1]}\big)\right] = I^{(1)}[F]. \tag{23}$$

But

$$\hat{\mathbb{E}}_n^{z_n} \left[F\left((U_t^n)_{t\in[0,1]} \right) \right] = \int_{(1-\varepsilon_n)n}^{n} \mathbb{E}_q^1 \left[F\left((X_t^n)_{t\in[0,1]} \right) \right] \hat{\mathbb{P}}_n^{z_n} \left(\tilde{L}_1^n \in dq \right). \tag{24}$$

In this formula $\mathbb{E}_q^1$ is the expectation operator associated to $\mathbb{P}_q^1$ in the sense that the process $(X_t^n)_{t\in[0,1]}$ is obtained from $(X_k)_{0\le k\le 2n}$ according to (1) when $(X_k)_{0\le k\le 2n}$ is a standard random walk excursion of length $2q$, and then stopped at zero. To obtain (23) using (24) it suffices to see that, for any sequence (q_n) such that $q_n/n \to 1$, we have the limit

$$\mathbb{E}_{q_n}^1 \left[F\left((X_t^n)_{t\in[0,1]} \right) \right] \to I^{(1)}(F).$$

Kaigh proved the result for $q_n \equiv n$ but the same proof can give this slightly generalized version. This completes the proof of Theorem 1.

5 Remark and Open Question

Remark Equations (12) and (13) give the simplified form :

$$\frac{2n-1}{z} \mathbb{P}_n(Z = z) = 1 + \sum_{j=1}^{[(z-1)/2]} \frac{(-1)^j}{2^j \, j!} \prod_{i=1}^{j} \frac{(z-2i+1)(z-2i)}{2n-2i-1}.$$

In particular, for $x \neq 0$, we get

$$\sqrt{n} \, \mathbb{P}_n(Z = [x\sqrt{n}]) = \frac{\sqrt{n}[x\sqrt{n}]}{2n-1} \sum_{j=0}^{\left[\frac{[x\sqrt{n}]-1}{2}\right]} \frac{(-1)^j}{j! 2^j} \prod_{i=1}^{j} \frac{([x\sqrt{n}]-2i+1/2)^2 - 1/4}{2n-2i-1}$$

from which we deduce

$$\lim_{n\to+\infty} \sqrt{n} \, \mathbb{P}_n(Z = [x\sqrt{n}]) = \frac{x}{2} \, e^{-\frac{x^2}{4}}.$$

This is an elementary way to obtain the convergence of the law of $Z/\sqrt{n}$ under $\mathbb{P}_n$ to the law of density $\frac{x}{2} e^{-\frac{x^2}{4}}$ (Rayleigh's law). The local form above even implies via Scheffé's usual argument that the convergence of the laws holds in total variation. Of course, a more sophisticated argument is to use the convergence of the number of zeros of the discrete random bridge toward the local time at level 0 of a Brownian bridge and identify the law of this local time as Rayleigh's law (see [10, Exercise VI.2.35]).

Open question: under $\mathbb{P}_n^{z_n}$ with $z_n = o(\sqrt{n})$ as before, it would be interesting to describe the asymptotic behavior of Y_G and Y_{G+1} for instance via $\max(Y_G, 2n - Y_{G+1})$ which is the maximal distance of a zero from the extremities. Starting from (21), it would however require a new version of Lemma 1, freed from the assumptions on $z/\sqrt{n}$.

References

1. P. Billingsley, *Convergence of Probability Measures* (Wiley, New York, 1968)
2. F. Caravenna, L. Chaumont, An invariance principle for random walk bridges conditioned to stay positive. Electron. J. Probab. **18**(60), 1–32 (2013)
3. E. Csaki, Y. Hu, Lengths and heights of random walk excursions. Discrete Math. Theor. Comput. Sci. AC, 45–52 (2003)
4. G. Giacomin, *Random Polymer Models* (Imperial College Press, World Scientific, 2007)
5. W.D. Kaigh, An invariance principle for random walk conditioned by a late return to zero. Ann. Probab. **4**, 115–121 (1976)
6. J.F. Le Gall, Random trees and applications. Probab. Surv. **2**, 245–311 (2005)
7. T. Liggett, An invariance principle for conditioned sums of independent random variables. J. Math. Mech. **18**, 559–570 (1968)
8. P. Révész, Local times and invariance. *Analytic Methods in Probab. Theory.* LNM 861 (Springer, New York, 1981), pp. 128–145
9. P. Révész, *Random Walk in Random and Non-random Environments*, 2nd edn. (World Scientific, Singapore, 2005)
10. D. Revuz, M. Yor, *Continuous Martingales and Brownian Motion*, 3rd edn. (Springer, New York, 1999)
11. R.P. Stanley, *Enumerative Combinatorics*, vol. 2 (Cambridge University Press, Cambridge, 1999)

A Short Proof of Stein's Universal Multiplier Theorem

Dario Trevisan

Abstract We give a short proof of Stein's universal multiplier theorem, purely by probabilistic methods, thus avoiding any use of harmonic analysis techniques (complex interpolation or transference methods).

1 Introduction

The celebrated Stein's universal multiplier theorem [10, Corollary IV.6.3] provides strong (L^p, L^p)-bounds for a general family of operators related to a Markovian semigroup $(T^t)_{t \geq 0}$, virtually without any assumption on the underlying measure space (X, m). More recent proofs of this classical result are based on analytic methods (see [1,2] and the monograph [3]), which also shows that the Markovianity assumption on the semigroup can be removed, keeping only the L^p-contractivity assumption. On the other hand, Stein's original proof relies on deep connections with martingale theory; not much later, P.A. Meyer began to investigate the problem purely by stochastic methods (see e.g. [7] and subsequent articles, and also [8] for an exposition of the transference approach).

In [10], the multiplier theorem is actually a corollary of L^p-bounds for suitable Littlewood-Paley g-functions, which follows from a clever complex interpolation between the L^2 case, which holds by spectral theory, and an L^p-inequality, obtained by martingale tools. From a probabilist's viewpoint, this interpolation argument could be a mountain to climb: Meyer literally wrote that *on ne "comprend" pas ce qui se passe* [7, end of Section 1].

In this note we prove the multiplier theorem (Theorem 1 below) relying only on martingale tools, namely Rota's construction and Burkhölder-Gundy inequalities, the main contribution being therefore that we avoid the use of complex interpolation. With hindsight, this result could be considered as an analogue of the short proof of the maximal theorem for Markovian semigroups, sketched in [10, below Theorem IV.4.9]: like in that case, powerful analytical tools can prove results for rather general semigroups but, in the Markovian setting, probability is enough and gives

D. Trevisan (✉)
Scuola Normale Superiore, Pisa, Italy
e-mail: dario.trevisan@sns.it

© Springer International Publishing Switzerland 2014
C. Donati-Martin et al. (eds.), *Séminaire de Probabilités XLVI*, Lecture Notes in Mathematics 2123, DOI 10.1007/978-3-319-11970-0_20

much simpler proofs. It is remarkable that, apparently, this shortcut went unnoticed, maybe because both Stein and Meyer were focusing mainly on Littlewood-Paley functions.

The proof allows also for an easy computation of the constants involved (in terms of p) and also for the norm of operators given by imaginary powers of the generator of the semigroup [10, Corollary IV.6.4]. Assuming these bounds only, one can then deduce boundedness of g-functions [6, Theorem 1.1]. Another application (Corollary 1) comes from the fact that some form of Burkhölder-Gundy inequalities still holds true for $p = 1$ (Davis' Theorem): we remark that one might also obtain analog results in a general, non-Markovian, setting by extrapolation on the L^p bounds (for a detailed account on Yano's extrapolation theory, see [4]).

After writing this note, we discovered that a similar argument already appeared in the last section of [9]: there, however, continuous-time stochastic calculus is widely used and it is not fully recognized that Rota's construction and Burkhölder-Gundy inequalities suffice, without any assumption on the underlying measure space.

2 Setting

We briefly recall the setting and notation of [10, Chapters III and IV]. Let (X, dx) be a σ-finite measure space and let $(T^t)_{t \geq 0}$ be a strongly continuous semigroup of operators defined on $L^2(X, dx)$ such that the following conditions hold:

1. $\|T^t f\|_p \leq \|f\|_p$ $(1 \leq p \leq \infty)$ (contraction);
2. T^t is self-adjoint on $L^2(X, \mu)$, for every $t \geq 0$ (symmetry);
3. $T^t f \geq 0$ if $f \geq 0$ (positivity);
4. $T^t 1 = 1$ (conservation of mass),

where $T^t 1$ is defined by $\sup_n T^t I_{A_n}$, taking a sequence $A_n \uparrow X$, with $I_{A_n} \in L^2$ (this is well defined in general because of positivity). The infinitesimal generator A of $(T^t)_{t \geq 0}$ in $L^2(X, dx)$ is given by

$$A = \lim_{t \downarrow 0} \frac{f - T^t f}{t},$$

for any $f \in L^2(X, \mu)$, whenever the limit exists in $L^2(X, dx)$ (that defines its domain $D(A)$).

Because of symmetry and contraction assumptions on T^t, the generator A is a self-adjoint, non-negative and densely defined operator. By spectral theory, there exists a unique resolution of the identity $(E(\lambda))_{\lambda \in \mathbb{R}}$ associated to A. In particular, the representation of $T^t = e^{-tA}$, $t \geq 0$, holds in the following sense:

$$\langle T^t f, g \rangle = \int_0^\infty e^{-t\lambda} d \langle E(\lambda) f, g \rangle, \tag{1}$$

where, for any $f, g \in L^2(X, dx)$, $\lambda \mapsto \langle E(\lambda) f, g \rangle$ is a bounded variation function on $\mathbb{R}$, with total variation not greater than $\|f\|_2 \|g\|_2$.

On the other side, positivity and conservation assumptions allows for a dynamical realization of the semigroup as the transition semigroup associated to a Markov process, with state space X and dx as invariant measure: this is the content of [10, Theorem IV.4.9] (due to G.C. Rota), that we describe here in a more explicit form (and actually a bit simplified, as we require only a finite product space). Note that, having no assumption on X, it is not clear whether T^ε is induced by some probability kernel; still, the proof proceeds as in the case of existence of Markov chains.

Given $\varepsilon > 0$ and $N \in \mathbb{N}$, let $\Omega = X^{N+1}$, endowed with the product σ-algebra. For $k \in \{0, \ldots, N\}$, let π_k be the projection on the k-th factor, let

$$\mathscr{F}_k = \sigma(\pi_k, \pi_{k+1}, \ldots, \pi_N)$$

which defines an reverse (i.e. decreasing) filtration and let $\hat{\mathscr{F}} = \sigma(\pi_0)$. Then, there exists a σ-finite measure $\mathbb{P} = \mathbb{P}_{\varepsilon, n}$ on Ω such that the law of π_0 w.r.t. $\mathbb{P}$ is dx and for $k \in \{0, \ldots, N\}$, $\mathbb{P}$ is σ-finite on $\mathscr{F}_k$ and for every $f \in L^1(X, dx)$, it holds

$$T^\varepsilon f(\pi_0) = \hat{\mathbb{E}}[\mathbb{E}_k[f(\pi_0)]] = \hat{\mathbb{E}}[f_k],$$

where $\hat{\mathbb{E}}$ denotes the conditional expectation operator w.r.t. $\hat{\mathscr{F}}$, $\mathbb{E}_k$ is the same, w.r.t. $\mathscr{F}_k$ and $f_k = \mathbb{E}_k[f \circ \pi_0]$ is a reverse martingale: of course, being the set of times finite, there is no problem in applying the usual theory of martingales. Moreover, it is not difficult to check that all the properties and theorems used here and in what follows, which are well known to hold in probability spaces, extend verbatim to the σ-finite case, the only exception being Corollary 1 below.

We recall now the special case of spectral multipliers problem, addressed by Stein. Let M be a bounded Borel function on $(0, \infty)$ and define, for $\lambda > 0$,

$$m(\lambda) = -\lambda \int_0^\infty M(t) e^{-t\lambda} dt, \tag{2}$$

(let also $m(0) = 0$), which is a so-called multiplier of Laplace transform type, when we use it to define, by means of spectral calculus, the operator

$$T_m f = \int_0^\infty m(\lambda) dE(\lambda) f, \tag{3}$$

for $f \in L^2(X, dx)$. Since

$$\|m\|_\infty = \sup_\lambda |m(\lambda)| \leq \sup_t |M(t)| = \|M\|_\infty, \tag{4}$$

it follows by the spectral theorem that T_m is well defined and maps continuously $L^2(X, dx)$ into itself, with operator norm $\|T_m\|_{2,2} \leq \|M\|_\infty$. The problem consists in proving that, for $p \in]1, \infty[$, T_m maps continuously $L^2 \cap L^p(X, dx)$ into itself.

3   Proof of the Multiplier Theorem

We are in a position to state and prove Stein's result.

Theorem 1 (Stein's Multiplier Theorem) *Let $p \in]1, \infty[$. Then, T_m is a bounded linear operator on $L^2 \cap L^p(X, dx)$, with*

$$\|T_m\|_{p,p} \leq C_p \|M\|_\infty,$$

where $C_p = O((p-1)^{-1})$ as $p \downarrow 1$.

We sketch heuristically the line of reasoning. By substituting (2), which gives m in terms of M, into (3) and exchanging integrals, we obtain the expression

$$T_m f = \int_0^\infty M(t) \left[\int_0^\infty -\lambda e^{-t\lambda} dE(\lambda) f \right] dt = \int_0^\infty M(t) \frac{d}{dt} T^t f \, dt, \qquad (5)$$

where we also used (1). Then, we formally simplify the increments dt and recall that, by Rota's construction, it holds $T^t f = \hat{\mathbb{E}}[f_t]$, where f_t is some reverse martingale:

$$T_m f = \int_0^\infty M(t) \, dT^t f = \hat{\mathbb{E}} \left[\int_0^\infty M(t) \, df_t \right].$$

To estimate the L^p norm, we use the fact that $\hat{\mathbb{E}}$ is a contraction and Burkhölder-Gundy inequalities, obtaining

$$\|T_m f\|_p \leq C_p \|M\|_\infty \|f\|_p.$$

To make this reasoning rigorous, we first consider the case when M is a step function and then we pass to the limit. To do this, we state and prove two elementary lemmas, the first being in fact a special case of (5) for step functions.

Lemma 1 *Given $N \in \mathbb{N}$, let $0 = t_0 \leq t_1 \leq \ldots \leq t_N < \infty$ and let*

$$M = \sum_{i=0}^{N-1} M_i I_{[t_i, t_{i+1}[}. \qquad (6)$$

Then, for every $f \in L^2(X, dx)$ it holds

$$T_m f = \sum_{i=0}^{N} M_i \left(T^{t_{i+1}} f - T^{t_i} f \right). \tag{7}$$

Proof Since $M \mapsto m \mapsto T_m$ is linear, it is enough to consider the case $M = I_{[0,t[}$ and prove that $T_m = T^t - Id$. Integrating by parts, we have $m(\lambda) = e^{-t\lambda} - 1$ and so we conclude by (1). □

Lemma 2 *Let $(M^n)_{n \geq 0}$ be a sequence of Borel functions, with $\|M_n\|_\infty$ uniformly bounded and converging $\mathcal{L}^1$-a.e. to some function M. Then, for every $f \in L^2(X, dx)$, it holds*

$$\lim_{n \to \infty} T_n f = T_m f \quad \text{in } L^2(X, dx),$$

where T_n denotes the operator defined by M^n in place of M.

Proof As above, by linearity, it is enough to consider the case $M = 0$ (i.e. $m = 0$ and $T_m = 0$). By dominated convergence, from (2) we obtain that, for every $\lambda \in [0, \infty[$, $|m_n(\lambda)|$ converges to zero. From (4) and the assumption on $(M^n)_{n \geq 0}$ this convergence is dominated by some constant, and this suffices to pass to the limit. Indeed, given $f \in L^2(X, dx)$, by spectral theorem, it holds

$$\|T_n f\|_2^2 = \langle T_n f, T_n f \rangle = \int_0^\infty |m_n(\lambda)|^2 \, d \langle E(\lambda) f, f \rangle.$$

As already remarked, $d \langle E(\lambda) f, f \rangle$ is a finite positive measure and so we conclude by dominated convergence. □

Proof of Theorem 1 We may assume that $|M| \leq 1$. First, let M be a step function of the form (6), where $\varepsilon = t_{i+1} - t_i$ constant for $i \in \{0, \ldots, N-1\}$. If we apply Rota's theorem, as described in the previous section, we obtain from Lemma 1 above that

$$T_m f \circ \pi_0 = \sum_{i=0}^{N-1} M_i \left(\hat{\mathbb{E}}[f_{i+1}] - \hat{\mathbb{E}}[f_i] \right) = \hat{\mathbb{E}} \left[\sum_{i=0}^{N-1} M_i (f_{i+1} - f_i) \right], \quad \mathbb{P}\text{-a.s. in } \Omega.$$

It holds therefore

$$\|T_m f\|_p = \left\| \hat{\mathbb{E}} \left[\sum_{i=0}^{N-1} M_i (f_{i+1} - f_i) \right] \right\|_p,$$

where the first norm is computed in $L^p(X, dx)$ and the other in $L^p(\Omega, \mathbb{P})$, because the law of π_0 is dx. Since conditional expectations are contractions, we have

$$\|T_m f\|_p \leq \left\| \sum_{i=0}^{N} M_i (f_{i+1} - f_i) \right\|_p.$$

We apply Burkhölder-Gundy inequalities for martingale transforms (e.g. [10, Theorem IV.4.2]) to the reverse martingale above:

$$\left\| \sum_{i=0}^{N} M_i (f_{i+1} - f_i) \right\|_p \leq c_p \|f\|_p,$$

where $c(p)$ is a constant, depending only on p: the claimed bound for $p \downarrow 1$ follows from constants-chasing in Marcinkiewicz interpolation. In the general case, we approximate a M with a sequence of step functions $(M^n)_{n \geq 1}$ such that $|M^n| \leq 1$ for every n and $M^n(s) \rightarrow M(s)$, $\mathscr{L}^1$-a.e. $s \in [0, \infty[$: this possibility is well-known, as it follows e.g. by density in $L^1([0, \infty[, \mathscr{L}^1)$ of step functions and a diagonal argument. For a fixed $f \in L^2 \cap L^p(X, dx)$, Lemma 1 entails that, up to a subsequence, $(T_n f)$ converge dx-a.e. to $T_m f$. By Fatou's lemma, it holds

$$\|T_m f\|_p \leq \liminf_{n \rightarrow \infty} \|T_n f\|_p \leq c_p \|f\|_p,$$

that gives the thesis. $\square$

Corollary 1 *If (X, dx) has finite measure $|X|$, it holds for every $f \in L^2(X, dx)$,*

$$\|T_m f\|_1 \leq c |X| \|M\|_\infty \|f\|_{L \log L},$$

where $c > 0$ is some universal constant.

Proof We may assume that $|X| = 1$ and $\|M\|_\infty \leq 1$. Arguing as above, we apply Davis' Theorem instead of Burkhölder-Gundy inequalities, e.g. [5, Theorem 2.1]:

$$\mathbb{E}\left[\left|\sum_{i=0}^{N} M_i (f_{i+1} - f_i)\right|\right] \leq c\mathbb{E}\left[\left(\sum_{i=0}^{N-1} |f_{i+1} - f_i|^2\right)^{1/2}\right] \leq c\mathbb{E}\left[\sup_{i=0,\ldots N} |f_i|\right].$$

Then, we use the well-known corollary of Doob's inequality, that allows to control the L^1 norm of a maximal function of martingale (closed by f) in terms of the $L \log L$ norm of f. $\square$

Acknowledgements The author is member of the GNAMPA group of the Istituto Nazionale di Alta Matematica (INdAM). He also thanks G.M. Dall'Ara for many discussions on the subject.

References

1. M.G. Cowling, *On Littlewood-Paley-Stein Theory*. Proceedings of the Seminar on Harmonic Analysis (Pisa, 1980), 1981
2. R.R. Coifman, R. Rochberg, G. Weiss, *Applications of Transference: The L^p Version of von Neumann's Inequality and the Littlewood-Paley-Stein Theory*. Linear Spaces and Approximation (Birkhäuser, Basel, 1978)
3. R.R. Coifman, G. Weiss, *Transference Methods in Analysis* (American Mathematical Society, Providence, 1976)
4. G.E. Karadzhov, M. Milman, Extrapolation theory: new results and applications. J. Approx. Theory **133**(1), 38–99 (2005)
5. E. Lenglart, D. Lépingle, M. Pratelli, *Présentation unifiée de certaines inégalités de la théorie des martingales*. Séminaire de Probabilités, XIV , Lecture Notes in Math., vol. 784 (Springer, Berlin, 1980)
6. S. Meda, On the Littlewood-Paley-Stein g-function. Trans. Am. Math. Soc. **347**(6), 2201–2212 (1995)
7. P.A. Meyer, *Démonstration probabiliste de certaines inégalités de Littlewood-Paley. I. Les inégalités classiques*. Séminaire de Probabilités, X, Lecture Notes in Math., vol. 511 (Springer, Berlin, 1976), pp. 125–141
8. P.-A. Meyer, *Sur la théorie de Littlewood-Paley-Stein (d'après Coifman-Rochberg-Weiss et Cowling)*. Séminaire de probabilités, XIX, Lecture Notes in Math., vol. 1123 (Springer, Berlin, 1985), pp. 113–129
9. I. Shigekawa, *The Meyer Inequality for the Ornstein-Uhlenbeck Operator in L^1 and Probabilistic Proof of Stein's L^p Multiplier Theorem*. Trends in Probability and Related Analysis (Taipei, 1996) (World Scientific, River Edge, 1997), pp. 273–288
10. E.M. Stein, *Topics in Harmonic Analysis Related to the Littlewood-Paley Theory*. Annals of Mathematics Studies, vol. 63 (Princeton University Press, Princeton, 1970)

On a Flow of Operators Associated to Virtual Permutations

Joseph Najnudel and Ashkan Nikeghbali

Abstract In (Comptes Rend Acad Sci Paris 316:773–778, 1993), Kerov, Olshanski and Vershik introduce the so-called virtual permutations, defined as families of permutations $(\sigma_N)_{N \geq 1}$, σ_N in the symmetric group of order N, such that the cycle structure of σ_N can be deduced from the structure of σ_{N+1} simply by removing the element $N + 1$. The virtual permutations, and in particular the probability measures on the corresponding space which are invariant by conjugation, have been studied in a more detailed way by Tsilevich in (J Math Sci 87(6):4072–4081, 1997) and (Theory Probab Appl 44(1):60–74, 1999). In the present article, we prove that for a large class of such invariant measures (containing in particular the Ewens measure of any parameter $\theta \geq 0$), it is possible to associate a flow $(T^\alpha)_{\alpha \in \mathbb{R}}$ of random operators on a suitable function space. Moreover, if $(\sigma_N)_{N \geq 1}$ is a random virtual permutation following a distribution in the class described above, the operator T^α can be interpreted as the limit, in a sense which has to be made precise, of the permutation $\sigma_N^{\alpha_N}$, where N goes to infinity and α_N is equivalent to αN. In relation with this interpretation, we prove that the eigenvalues of the infinitesimal generator of $(T^\alpha)_{\alpha \in \mathbb{R}}$ are equal to the limit of the rescaled eigenangles of the permutation matrix associated to σ_N.

Keywords Central measure • Flow of operators • Infinitesimal generator • Random operator • Virtual permutation

J. Najnudel (✉)
Institut de mathématiques de Toulouse, Université Paul Sabatier, 118, route de Narbonne, 31062 Toulouse Cedex 9, France
e-mail: joseph.najnudel@math.univ-toulouse.fr

A. Nikeghbali
Institut für Mathematik, Universität Zürich, Winterthurerstrasse 190, 8057-Zürich, Switzerland
e-mail: ashkan.nikeghbali@math.uzh.ch

481

1 Introduction

A large part of the research in random matrix theory comes from the problem of finding the possible limit distributions of the eigenvalues of a given ensemble of random matrices, when the dimension goes to infinity. Many different ensembles have been studied (see, for example, Mehta [9]), the most classical one is the so-called *Gaussian Unitary Ensemble* (GUE), where the corresponding random matrix is Hermitian, the diagonal entries are standard Gaussian variables and the real and the imaginary part of the entries above the diagonal are centered Gaussian variables of variance $1/2$, all these variables being independent. Another well-known ensemble is the *Circular Unitary Ensemble* (CUE), corresponding to a random matrix which follows the Haar measure on a finite-dimensional unitary group. A remarkable phenomenon which happens is the so-called *universality property*: there are some particular point processes which appear as the limit (after suitable scaling) of the distribution of the eigenvalues (or eigenangles) associated with a large class of random matrix ensembles, this limit being independent of the detail of the model which is considered. For example, after scaling, the small eigenvalues of the GUE and the small eigenangles of the CUE converge to the same process, called *determinantal process with sine kernel*, and appearing as the limit of a number of other models of random matrices. This process is a point process on the real line, such that informally, for $x_1, \ldots, x_n \in \mathbb{R}$, the probability to have a point in the neighborhood of x_j for all $j \in \{1, \ldots, n\}$ is proportional to the determinant of the matrix $(K(x_j, x_k))_{1 \leq j, k \leq n}$, where the kernel K is given by the formula:

$$K(x, y) = \frac{\sin(\pi(x - y))}{\pi(x - y)}.$$

Another point process which enjoys some universality properties is the *determinantal process with Airy kernel*, which is involved in the distribution of the largest eigenvalues of the GUE, and which is defined similarly as the determinantal process with sine kernel, except that the kernel K is now given by:

$$K(x, y) = \frac{\mathrm{Ai}(x) \, \mathrm{Ai}'(y) - \mathrm{Ai}(y) \, \mathrm{Ai}'(x)}{x - y},$$

where Ai denotes the Airy function. The phenomenon of universality is not well-explained in its full generality: a possible way to have a good understanding of the corresponding point processes is to express them as the set of eigenvalues of some universal infinite-dimensional random operators, and to prove that these operators are the limits, in a sense which needs to be made precise, of the classical random matrix ensembles. Such a construction has been done by Ramirez, Rider and Virág [14], where, as a particular case, the authors express a determinantal process with Airy kernel as the set of the eigenvalues of a random differential operator on a function space. Moreover, this infinite-dimensional operator is naturally interpreted as the limit of an ensemble of tridiagonal matrices, which have the same eigenvalue distribution as the GUE. However, this operator is not directly constructed from GUE (or another classical model such as CUE), and the reason for its universal properties are not obvious. On the other hand, an operator whose eigenvalues form a determinantal process with sine kernel, and which is related in a natural way to a classical random matrix model, has not been yet defined: a construction which seems to be very promising involves the interpretation of the sine kernel process as a function of a stochastic process called the Brownian carousel and constructed by Valkó and Virág [17].

Now, despite the fact that it seems to be particularly difficult to associate, in a natural way, an infinite-dimensional operator to the most classical matrix ensembles, we shall prove, in the present paper, that such a construction is possible and very explicit for a large class of ensembles of permutations matrices. These ensembles, and some of their generalizations, have already be studied by several authors (see, for example, Evans [3] and Wieand [18]), including the authors of the present paper (see [10]). An important advantage of permutations matrices is the fact that their eigenvalues can directly be expressed as a function of the size of the cycles of the corresponding permutations. Hence, it is equivalent to study these matrices or to deal with the corresponding cycle structure. Another advantage is the existence of a quite convenient way to define models of permutation matrices in all the different dimensions, on the same probability space, which gives a meaning to the notion of almost sure convergence when the dimension goes to infinity (a rather unusual situation in random matrix theory). This also gives the possibility to define an infinite-dimensional limit model, which will be explicitly constructed in this paper. Once the permutations and their matrices are identified, the main objects involved in our construction are the so-called *virtual permutations*, first introduced by Kerov, Olshanski and Vershik in [4], and further studied by Tsilevich in [15] and [16]. A virtual permutation can be defined as follows: it is a sequence $(\sigma_N)_{N \geq 1}$ of permutations, σ_N being of order N, and such that the cycle structure of σ_N is obtained from the cycle structure of σ_{N+1} by simply removing the element $N + 1$. Then, for all $\theta \geq 0$, it is possible to define a unique probability measure on the space of virtual permutations, such that for all $N \geq 0$, its image by the N-th coordinate is equal to the Ewens measure on the symmetric group of order N, with parameter θ. The Ewens measures on the space of virtual permutations are particular cases of the so-called *central measures*, studied in [16], and defined as the probability measures which are invariant by conjugation with any permutation of finite order (see below for the details). The central measures are completely characterized in [16], using the properties of exchangeable partitions, stated by Kingman in [5,6] and [7] (see also the course by Pitman [13]). The following remarkable statement holds: if the distribution of a random virtual permutation $(\sigma_N)_{N \geq 1}$ is a central measure, and if for all $N \geq 1$, $(l_k^{(N)})_{k \geq 1}$ denotes the decreasing sequence of cycle lengths (completed by zeros) of the permutation σ_N, then for all $k \geq 1$, the sequence $(l_k^{(N)}/N)_{N \geq 1}$ converges a.s. to a limit random variable λ_k. By using Fatou's lemma, one immediately deduces that a.s., the random sequence $(\lambda_k)_{k\geq 1}$ is nonnegative and decreasing, with:

$$\sum_{k\geq 1} \lambda_k \leq 1.$$

If the distribution of $(\sigma_N)_{N\geq 1}$ is the Ewens measure of parameter $\theta \geq 0$, then $(\lambda_k)_{k\geq 1}$ is a Poisson-Dirichlet process with parameter θ (if $\theta = 0$, then $\lambda_1 = 1$ and $\lambda_k = 0$ for $k \geq 1$). This convergence of the renormalized cycle lengths can be translated into a statement on random matrices. Indeed, to any virtual permutation following a central measure, one can associate a sequence $(M_N)_{N\geq 1}$ of random permutation matrices, M_N being of dimension N. If for $N \geq 1$, X_N denotes the point process of the eigenangles of M_N, multiplied by N and counted with multiplicity, then X_N converges a.s. to the limit process X_∞ defined as follows:

- X_∞ contains, for all $k \geq 1$ such that $\lambda_k > 0$, each non-zero multiple of $2\pi/\lambda_k$.
- The multiplicity of any non-zero point x of X_∞ is equal to the number of values of k such that $\lambda_k > 0$ and x is multiple of $2\pi/\lambda_k$.
- The multiplicity of zero is equal to the number of values of $k \geq 1$ such that $\lambda_k > 0$ if $\sum_{k\geq 1} \lambda_k = 1$, and to infinity if $\sum_{k\geq 1} \lambda_k < 1$.

This convergence has to be understood in the following way: for all functions f from $\mathbb{R}$ to $\mathbb{R}_+$, continuous with compact support, the sum of f at the points of X_N, counted with multiplicity, tends to the corresponding sum for X_∞ when N goes to infinity. In the case of the Ewens probability measure with parameter $\theta > 0$, a detailed proof of this result is given by the present authors in [10], in a more general context. Once the almost sure convergence of the rescaled eigenangles is established, one can naturally ask the following question: is it possible to express the limit point process as the spectrum of a random operator associated to the virtual permutation which is considered? The main goal of this article is to show that the answer is positive.

   More precisely, we prove that in a sense which can be made precise, for almost every virtual permutation following a central probability distribution, for all $\alpha \in \mathbb{R}$, and for all sequences $(\alpha_N)_{N\geq 1}$ such that α_N is equivalent to αN for N going to infinity, $\sigma_N^{\alpha_N}$ converges to an operator T^α, depending on $(\sigma_N)_{N\geq 0}$, on α but not on the choice of $(\alpha_N)_{N\geq 1}$. The flow of operators $(T^\alpha)_{\alpha\in\mathbb{R}}$ (which a.s. satisfies $T^{\alpha+\beta} = T^\alpha T^\beta$ for all $\alpha, \beta \in \mathbb{R}$) is defined on a function space which will be constructed later, and admits a.s. an infinitesimal generator U. Moreover, we prove that the spectrum of iU is exactly given by the limit point process X_∞ constructed above. This spectral interpretation of such a limit point process suggests that it is perhaps possible to construct similar objects for more classical matrix models, despite the discussion above about the difficulty of this problem. Indeed, in [12], Olshanski and Vershik characterize the measures on the space of infinite dimensional Hermitian matrices, which are invariant by conjugation with finite-dimensional unitary matrices. These central measures enjoy the following property: if a random infinite matrix follows one of them, then after suitable scaling, the point process of the extreme eigenvalues of its upper-left finite-dimensional submatrice converges a.s. to a limit point process when the dimension of this submatrice tends to infinity, similarly as for the case of virtual permutations, where X_N tends to X_∞. By the Cayley transform, the sequence of the upper-left submatrices of an infinite Hermitian matrix is mapped to a sequence $(M_N)_{N \geq 1}$ of unitary matrices, such that the matrix M_N can be deduces from M_{N+1} by a projective mapping, explicitly described by Neretin [11]. Among the central measures described in [12], there exists a unique measure for which all its projections on the finite-dimensional unitary groups are equal to Haar measure. In [1] Borodin and Olshanski study the Hermitian version of this measure, and a family of generalizations, depending on a complex parameter and called Hua-Pickrell measures. In particular, they prove that the corresponding point process is determinantal and compute explicitly its kernel: for the Haar measure, they obtain the image of a sine kernel process by the mapping $x \mapsto 1/x$. On the other hand, in a paper with Bourgade [2], by using the splitting of unitary matrices as a product of reflections, we define the natural generalization of virtual permutations to sequences of general unitary matrices of increasing dimensions. In the case where none of the matrices has an eigenvalue equal to 1, these sequences, called virtual isometries, coincide (up to a sign change) with the sequences of unitary matrices defined by Neretin, Olshanski and Vershik. In our paper, we give a purely probabilistic proof, under Haar measure, of the almost sure convergence of the rescaled eigenangles when the dimension goes to infinity, and we give an estimate of the rate of convergence. Our result is then used in a paper with Maples [8], where we prove that the coordinates of the eigenvectors also converge a.s., if the vectors are suitably normalized. From this property, we are able to construct a flow of random operators associated to a virtual isometry following Haar measure, in a similar way as we construct operators associated to virtual permutations in the present paper. In [8], we construct our operators on two different random vector spaces, whose link with the unitary matrices is not as well understood as in the present situation.

This article is organized as follows: in Sect. 2, we define the notion of a virtual permutation in a more general setting than it is usually done; in Sect. 3, we study the central measures in this new setting, generalizing the results by Kerov, Olshanski, Tsilevich and Vershik; in Sect. 4, we use the results of Sect. 3 in order to associate, to almost every virtual permutation under a central measure, a flow of transformations of a topological space obtained by completing the set on which the virtual permutations act; in Sect. 5, we interpret this flow as a flow of operators on a function space, and we deduce the construction of the random operator U.

2 Virtual Permutations of General Sets

The virtual permutations are usually defined as the sequences $(\sigma_N)_{N \geq 1}$ such that $\sigma_N \in \Sigma_N$ for all $N \geq 1$, where Σ_N is the symmetric group of order N, and the cycle structure of σ_N is obtained by removing $N + 1$ from the cycle structure of σ_{N+1}.

In this definition, the order of the integers is involved in an important way, which is not very satisfactory since we are essentially interested in the cycle structure of the permutation, and not particularly in the nature of the elements inside the cycles. Therefore, in this section, we present a notion of virtual permutation which can be applied to any set and not only to the set of positive integers. This generalization is possible because of following result:

Proposition 1 *Let $(\sigma_N)_{N \geq 1}$ be a virtual permutation (in the usual sense). For all finite subsets $I \subset \mathbb{N}^* := \{1, 2, \ldots\}$, and for all $N \geq 1$ larger than any of the elements in I, let $\sigma_I^{(N)}$ be the permutation of the elements of I obtained by removing the elements outside I from the cycle structure of σ_N. Then $\sigma_I^{(N)}$ depends only on I and not on the choice of N majorizing I, and one can write $\sigma_I^{(N)} =: \sigma_I$. Moreover, if J is a finite subset of $\mathbb{N}^*$ containing I, then σ_I can be obtained from the cycle structure of σ_J by removing all the elements of $J \setminus I$.*

Proof Let $N \leq N'$ be two integers majorizing I. By the classical definition of virtual permutations, σ_N is obtained by removing the elements strictly larger than N from the cycle structure of $\sigma_{N'}$. Hence, $\sigma_I^{(N)}$ can be obtained from the cycle structure of $\sigma_{N'}$ by removing the elements strictly larger than N, and then the elements smaller than or equal to N which are not in I. Since all the elements of I are smaller than or equal to N, it is equivalent to remove directly all the elements of $\{1, \ldots, N'\}$ outside I, which proves that $\sigma_I^{(N')} = \sigma_I^{(N)}$. Now, let J be a finite subset of $\mathbb{N}^*$ containing I, and N'' an integer which majorizes J. The permutation σ_I is obtained by removing the elements of $\{1, \ldots, N''\} \setminus I$ from the cycle structure of $\sigma_{N''}$. It is equivalent to say that σ_I is obtained from $\sigma_{N''}$ by removing the elements of $\{1, \ldots, N''\} \setminus J$, and then the elements of $J \setminus I$, which implies that σ_I can be obtained by removing the elements of $J \setminus I$ from the cycle structure of σ_J.

We can then define virtual permutations on general sets as follows:

Definition 1 A virtual permutation of a given set E is a family of permutations $(\sigma_I)_{I \in \mathscr{F}(E)}$, indexed by the set $\mathscr{F}(E)$ of the finite subsets of E, such that $\sigma_I \in \Sigma_I$, where Σ_I is the symmetric group of I, and such that for all $I, J \in \mathscr{F}(E)$, $I \subset J$, the permutation σ_I is obtained by removing the elements of $J \setminus I$ from the cycle structure of σ_J.

If E is a finite set, a virtual permutation is essentially a permutation, more precisely, one has the following proposition:

Proposition 2 *Let E be a finite set and σ a permutation of E. Then, one can define a virtual permutation $(\sigma_I)_{I \in \mathscr{F}(E)}$ of E as follows: σ_I is obtained from σ by removing the elements of $E \setminus I$ from its cycle structure. Moreover, this mapping is bijective, and the inverse mapping is obtained by associating the permutation σ_E to any virtual permutation $(\sigma_I)_{I \in \mathscr{F}(E)}$.*

This proposition is almost trivial, so we omit the proof.

Notation Because of this result, we will denote the set of virtual permutations of any set E by Σ_E. In the case $E = \mathbb{N}^*$, we can immediately deduce the following result from Proposition 1:

Proposition 3 *If $(\sigma_N)_{N \geq 1}$ is a virtual permutation in the classical sense, then with the notation of Proposition 1, $(\sigma_I)_{I \in \mathscr{F}(\mathbb{N}^*)}$ is a virtual permutation of $\mathbb{N}^*$ in the sense of Definition 1. Moreover, the mapping which associates $(\sigma_I)_{I \in \mathscr{F}(\mathbb{N}^*)}$ to $(\sigma_N)_{N \geq 1}$ is bijective: the existence of the inverse mapping is deduced from the fact that $\sigma_N = \sigma_{\{1,\ldots,N\}}$.*

An example of a virtual permutation defined on an uncountable set is given as follows: if E is a circle, and if for $I \in \mathscr{F}(E)$, σ_I is the permutation of I containing a unique cycle, obtained by counterclockwise enumeration of the points of I, then $(\sigma_I)_{I \in \mathscr{F}(E)}$ is a virtual permutation of E. In this example, it is natural to say that $(\sigma_I)_{I \in \mathscr{F}(E)}$ contains E itself as a unique cycle. More generally, one can define a cycle structure for any virtual permutation, by the following result:

Proposition 4 *Let $(\sigma_I)_{I \in \mathscr{F}(E)}$ be a virtual permutation of a set E, and let x, y be two elements of E. Then, one of the following two possibilities holds:*

- *For all $I \in \mathscr{F}(E)$ containing x and y, these two elements are in the same cycle of the permutation σ_I.*
- *For all $I \in \mathscr{F}(E)$ containing x and y, these two elements are in two different cycles of σ_I.*

The first case defines an equivalence relation on the set E.

Proof Let $I \in \mathscr{F}(E)$ containing x and y. By the definition of virtual permutations, it is easy to check that x and y are in the same cycle of σ_I if and only if $\sigma_{\{x,y\}}$ is equal to the transposition (x, y), this property being independent of the choice of I. Moreover, if the first item in Proposition 4 holds for x and y, and for y and z, then the permutation $\sigma_{\{x,y,z\}}$ contains a unique cycle, which implies that the first item holds also for x and z.

From now, the equivalence relation defined in Proposition 4 will be denoted by $\sim_{(\sigma_I)_{I \in \mathscr{F}(E)}}$, or simply by $\sim$ if no confusion is possible. The corresponding equivalence classes will be called the *cycles* of $(\sigma_I)_{I \in \mathscr{F}(E)}$. The cycle of an element $x \in E$ will be denoted $\mathscr{C}_{(\sigma_I)_{I \in \mathscr{F}(E)}}(x)$, or simply $\mathscr{C}(x)$. One immediately checks that this notion of cycle is consistent with the classical notion for permutations of finite order. Another notion which should be introduced is the notion of conjugation, involved in the definition of central measures given in Sect. 3.

Proposition 5 *Let $\Sigma_E^{(0)}$ be the group of the permutations of E which fix all but finitely many elements of E. Then, the group $\Sigma_E^{(0)}$ acts on Σ_E by conjugation, because of the following fact: for all $g \in \Sigma_E^{(0)}$, and for all $\sigma = (\sigma_I)_{I \in \mathscr{F}(E)}$, there exists a unique virtual permutation $g \sigma g^{-1} = (\sigma_I')_{I \in \mathscr{F}(E)}$ such that for all $I \in \mathscr{F}(E)$ containing all the points of E which are not fixed by g, $\sigma_I' = g_I \sigma_I g_I^{-1}$,* *where g_I denotes the restriction of g to I. Moreover, the cycles of $g \sigma g^{-1}$ are the images by g of the cycles of σ.*

Proof Let $\sigma = (\sigma_I)_{I \in \mathscr{F}(E)} \in \Sigma_E$, $g \in \Sigma_E^{(0)}$ and denote $E(g)$ the set of points which are not fixed by g. For all $I \in \mathscr{F}(E)$, let us define σ_I' as the permutation obtained from $g_{E(g) \cup I} \sigma_{E(g) \cup I} g_{E(g) \cup I}^{-1}$ by removing the elements of $E(g) \backslash I$ from its cycle structure. In other words, the cycle structure of σ_I' can be obtained from the cycle structure of $\sigma_{E(g) \cup I}$ by replacing all the elements by their image by g, and then by removing the elements of $E(g) \backslash I$. Now, if $J \in \mathscr{F}(E)$ and $I \subset J$, then the structure of σ_I' can also be obtained from the structure of $\sigma_{E(g) \cup J}$ by removing the elements of $J \backslash (E(g) \cup I)$, by replacing the remaining elements by their image by g, and then by removing the elements of $E(g) \backslash I$. Since all the elements of $J \backslash E(g) \cup I$ are fixed by g, the order of the two first operations is not important, and then σ_I' is obtained from $\sigma_{E(g) \cup J}$ by replacing the elements by their image by g, and then by removing the elements of $(J \cup E(g)) \backslash I$. This implies that σ_I' is obtained from σ_J' by removing the elements of $J \backslash I$ from its cycle structure, in other words, $(\sigma_I')_{I \in \mathscr{F}(E)}$ is a virtual permutation of E, which proves the existence of $g \sigma g^{-1}$. Its uniqueness is a direct consequence of the fact that all its components corresponding to a set containing $E(g)$ are determined by definition. Since, for I containing $E(g)$, the cycle structure of $(g \sigma g^{-1})_I$ is obtained from the structure of σ by replacing the elements by their image by g, it is easy to deduce that the cycles of the virtual permutation $g \sigma g^{-1}$ are the images by g of the cycles of σ, and that the conjugation is a group action of $\Sigma_E^{(0)}$ on Σ_E. □

We have so far given the general construction of virtual permutations, and some of their main properties: in Sect. 3, we introduce probability measures on the space of virtual permutations.

3 Central Measures on General Spaces of Virtual Permutations

In order to construct a probability measure on the space of virtual permutations of a set E, one can expect that it is sufficient to define its images by the coordinate mappings, if these images satisfy some compatibility properties. The following result proves that such a construction is possible if the set E is countable:

Proposition 6 *Let E be a countable set, let $S(E)$ be the product of all the symmetric groups Σ_I for $I \in \mathscr{F}(E)$, and let $\mathscr{S}(E)$ be the σ-algebra on $S(E)$, generated by all the coordinates mappings $(\sigma_I)_{I \in \mathscr{F}(E)} \mapsto \sigma_J$ for $J \in \mathscr{F}(E)$. For all $I, J \in \mathscr{F}(E)$ such that $I \subset J$, let $\pi_{J,I}$ be the mapping from Σ_J to Σ_I which removes the elements of $J \backslash I$ from the cycle structure. Let $(\mathbb{P}_I)_{I \in \mathscr{F}(E)}$ be a family of probability measures, $\mathbb{P}_I$ defined on the finite set Σ_I, which is compatible in the following sense: for all $I, J \in \mathscr{F}(E)$ such that $I \subset J$, the image of $\mathbb{P}_J$ by $\pi_{J,I}$*

is $\mathbb{P}_I$. Then, there exists a unique probability measure $\mathbb{P}$ on the measurable space $(S(E), \mathscr{S}(E))$ satisfying the following two conditions:

- $\mathbb{P}$ is supported by the set Σ_E of virtual permutations.
- For all $I \in \mathscr{F}(E)$, the image of $\mathbb{P}$ by the coordinate mapping indexed by I is equal to the measure $\mathbb{P}_I$.

Proof For every family $(I_k)_{1 \le k \le n}$ of elements of $\mathscr{F}(E)$, and for all $J \in \mathscr{F}(E)$ containing I_k for all k, let $\mathbb{P}_{I_1,\ldots,I_n;J}$ be the image of the probability measure $\mathbb{P}_J$ by the mapping

$$\sigma \mapsto (\pi_{J,I_1}(\sigma), \ldots, \pi_{J,I_n}(\sigma)),$$

from Σ_J to the product space $\Sigma_{I_1} \times \cdots \times \Sigma_{I_n}$. Then, the following two properties hold:

- Once the sets $(I_k)_{1 \le k \le n}$ are fixed, the measure

$$\mathbb{P}_{I_1,\ldots,I_n} := \mathbb{P}_{I_1,\ldots,I_n;J}$$

does not depend on the choice of J.
- For $n \ge 2$, the image of $\mathbb{P}_{I_1,\ldots,I_n}$ by a permutation $\sigma \in \Sigma_n$ of the coordinates is equal to $\mathbb{P}_{I_{\sigma(1)},\ldots,I_{\sigma(n)}}$, and the image by the application removing the last coordinate is $\mathbb{P}_{I_1,\ldots,I_{n-1}}$.

These properties can easily be proven by using the compatibility property of the family $(\mathbb{P}_I)_{I \in \mathscr{F}(E)}$. Now, let us observe that a probability measure $\mathbb{P}$ on $(S(E), \mathscr{S}(E))$ satisfies the conditions given in Proposition 6 if and only if for all $I_1, \ldots, I_n \in \mathscr{F}(E)$, the image of $\mathbb{P}$ by the family of coordinates indexed by $I_1, \ldots, I_n$ is equal to $\mathbb{P}_{I_1,\ldots,I_n}$ (note that the fact that E is countable is used in this step). The property of compatibility given in the second item above and the Caratheodory extension theorem then imply the existence and the uniqueness of $\mathbb{P}$.

Remark 1 The set Σ_E is the intersection of the sets $S_{J,K}(E) \in \mathscr{S}(E)$, indexed by the pairs $(J, K) \in \mathscr{F}(E) \times \mathscr{F}(E)$ such that $J \subset K$, and defined as follows: a family $(\sigma_I)_{I \in \mathscr{F}(E)}$ is in $S_{J,K}(E)$ if and only if $\sigma_J = \pi_{K,J}(\sigma_K)$. If E is countable, this intersection is countable, and then $\Sigma_E \in \mathscr{S}(E)$. However, if E is uncountable, Σ_E does not seem to be measurable, and we do not know how to construct the measure $\mathbb{P}$ in this case.

An immediate consequence of Proposition 6 is the following:

Corollary 1 *Let E be a countable set, and let $\mathscr{S}_E$ be the σ-algebra on Σ_E, generated by the coordinates mappings $(\sigma_I)_{I \in \mathscr{F}(E)} \mapsto \sigma_J$ for $J \in \mathscr{F}(E)$. Let $(\mathbb{P}_I)_{I \in \mathscr{F}(E)}$ be a family of probability measures, $\mathbb{P}_I$ defined on Σ_I, which is compatible in the sense of Proposition 6. Then, there exists a unique probability measure $\mathbb{P}_E$ on the measurable space $(\Sigma_E, \mathscr{S}_E)$ such that for all $I \in \mathscr{F}(E)$, the image of $\mathbb{P}_E$ by the coordinate indexed by I is equal to $\mathbb{P}_I$.*

An example of measure on Σ_E which can be constructed by using Proposition 6 and Corollary 1 is the Ewens measure, which, for $E = \mathbb{N}^*$, is studied in detail in [15]. The precise existence result in our more general setting is the following:

Proposition 7 *Let θ be in $\mathbb{R}_+$ and E a countable set. For all $I \in \mathscr{F}(E)$, let $\mathbb{P}_I^{(\theta)}$ be the Ewens measure of parameter θ, defined as follows: for all $\sigma \in \Sigma_I$,*

$$
\mathbb{P}_I^{(\theta)}(\sigma) = \frac{\theta^{n-1}}{(\theta + 1)\ldots(\theta + N - 1)},
$$

where n is the number of cycles of σ and N the cardinality of I. Then the family $(\mathbb{P}_I^{(\theta)})_{I \in \mathscr{F}(E)}$ is compatible in the sense of Proposition 6: the measure $\mathbb{P}_E^{(\theta)}$ on $(\Sigma_E, \mathscr{S}_E)$ which is deduced from Corollary 1 is called the Ewens measure of parameter θ.

The Ewens measures are particular cases of the so-called *central measures*. Indeed, for all countable sets E and for all $g \in \Sigma_E^{(0)}$, the conjugation by g defined in Proposition 5 is measurable with respect to the σ-algebra $\mathscr{S}_E$. Therefore it defines a group action on the set of probability measures on $(\Sigma_E, \mathscr{S}_E)$: by definition, a probability measure $\mathbb{P}$ is central if and only if it is invariant by this action. It is easy to check that this condition holds if and only if all the image measures of $\mathbb{P}$ by the coordinate mappings are invariant by conjugation (which is clearly the case for the Ewens measures). In the case of $E = \mathbb{N}^*$, Tsilevich (in [16]) has completely characterized the central measures, by using the properties of the partitions of countable sets, described by Kingman (see [5–7]). If the cardinality of a finite set I is denoted by $|I|$, then the result by Tsilevich can easily be translated in our framework as follows:

Proposition 8 *Let E be a countable set, $\mathbb{P}$ a central measure on $(\Sigma_E, \mathscr{S}_E)$, and σ a random virtual permutation following the distribution $\mathbb{P}$. Then, for all $x, y \in E$, the indicator of the event $\{x \sim_\sigma y\}$ is measurable with respect to the σ-algebra $\mathscr{S}_E$, and there exists a family of random variables $(\lambda(x))_{x \in E}$, unique up to almost sure equality, taking their values in the interval $[0, 1]$, and satisfying the following properties:*

- *For $I \in \mathscr{F}(E)$,*

$$
\frac{|I \cap \mathscr{C}_\sigma(x)|}{|I|} \xrightarrow[|I| \to \infty]{} \lambda(x),
$$

  in L^1, and then in all the spaces L^p for $p \in [1, \infty)$ since all the variables involved here are bounded by one.
- *For any strictly increasing sequence $(I_n)_{n \geq 1}$ of nonempty sets in $\mathscr{F}(E)$,*

$$
\frac{|I_n \cap \mathscr{C}_\sigma(x)|}{|I_n|} \xrightarrow[n \to \infty]{} \lambda(x)
$$

  a.s.

- *Almost surely, for all $x \in E$, $\lambda(x) = 0$ if and only if x is a fixed point of σ, i.e. a fixed point of σ_I for all $I \in \mathscr{F}(E)$ containing x.*
- *Almost surely, for all $x, y \in E$, $\lambda(x) = \lambda(y)$ if $x \sim_\sigma y$.*

For all $k \geq 1$, let us then define λ_k as the supremum of $\inf_{1 \leq j \leq k} \lambda(x_j)$ over sequences $(x_j)_{1 \leq j \leq k}$ of elements in E satisfying $x_j \not\sim_\sigma x_{j'}$ for $j \neq j'$, the supremum of the empty set being taken to be equal to zero. The sequence $(\lambda_k)_{k \geq 1}$ is a random variable with values in the space of the sequences of elements in $[0, 1]$, endowed with the σ-algebra generated by the coordinate mappings, $(\lambda_k)_{k \geq 1}$ is uniquely determined up to almost sure equality, and it lies a.s. in the (measurable) simplex Λ of non-increasing random sequences in $[0, 1]$, satisfying the inequality:

$$\sum_{k \geq 1} \lambda_k \leq 1.$$

Moreover, the central measure $\mathbb{P}$ is uniquely determined by the distribution of the sequence $(\lambda_k)_{k \geq 1}$, as follows:

- *For any probability measure ν on the space Λ endowed with the σ-algebra of the coordinate mappings, there exists a unique central measure $\mathbb{P}_\nu$ on $(\Sigma_E, \mathscr{S}_E)$ such that the corresponding random sequence $(\lambda_k)_{k \geq 1}$ has distribution ν.*
- *For all sequences $\lambda \in \Lambda$, let $\mathbb{P}_\lambda := \mathbb{P}_{\delta_\lambda}$, where δ_λ denotes the Dirac measure at λ. Then, for all $A \in \mathscr{S}(E)$, the mapping $\lambda \mapsto \mathbb{P}_\lambda(A)$ is measurable and for all probability measures ν, one has:*

$$\mathbb{P}_\nu(A) = \int_\Lambda \mathbb{P}_\lambda(A) d\nu(\lambda).$$

Remark 2 Intuitively, $\lambda(x)$ represents the asymptotic length of the cycle of x and $(\lambda_k)_{k \geq 1}$ is the non-increasing sequence of cycle lengths. For all $\theta \geq 0$, the Ewens measure of parameter θ is equal to $\mathbb{P}_{\nu^{(\theta)}}$, where $\nu^{(\theta)}$ is the Poisson-Dirichlet distribution of parameter θ (Dirac measure at the sequence $(1, 0, 0, \ldots)$ if $\theta = 0$).

Proof The measurability of the equivalence relation $\sim_\sigma$ is an immediate consequence of the fact that the event $x \sim_\sigma y$ depends only on the permutation $\sigma_{\{x,y\}}$. Therefore, $\pi_0 : \sigma \mapsto \sim_\sigma$ is a measurable mapping from $(\Sigma_E, \mathscr{S}_E)$ to the space of equivalence relations of E, endowed with the σ-algebra generated by the events of the form $\{x \sim y\}$ for $x, y \in E$. Since this measurable space is canonically identified with a measurable space $(\Pi_E, \mathscr{V}_E)$ such that Π_E is the space of partitions of the set E, the mapping π_0 can be identified to a measurable mapping π from $(\Sigma_E, \mathscr{S}_E)$ to $(\Pi_E, \mathscr{V}_E)$. This mapping induces a mapping from the probability measures on $(\Sigma_E, \mathscr{S}_E)$ to the probability measures on $(\Pi_E, \mathscr{V}_E)$, and Proposition 5 implies that a central measure is always mapped to the distribution of an exchangeable partition. Let us now prove that this correspondence is bijective, i.e. for any distribution $\mathbb{Q}$ on $(\Pi_E, \mathscr{V}_E)$ inducing an exchangeable partition, there exists a unique central measure $\mathbb{P}$ on $(\Sigma_E, \mathscr{S}_E)$ such that the image of $\mathbb{P}$ by π is equal to $\mathbb{Q}$. Indeed, this condition is satisfied if and only if for all $J \in \mathscr{F}(E)$, the law of the partition of J induced by the cycle structure of σ_J, where $(\sigma_I)_{I \in \mathscr{F}(E)}$ follows the distribution $\mathbb{P}$, is equal to the law of the restriction to J of a partition following the distribution $\mathbb{Q}$. Since a probability on Σ_J which is invariant by conjugation is completely determined by the corresponding distribution of the cycle lengths, the uniqueness of $\mathbb{P}$ follows. Now, for all $J \in \mathscr{F}(E)$, and all distributions $\mathbb{Q}_J$ on the partitions of J, let us define $\pi^{-1}(\mathbb{Q}_J)$ as the distribution of a random permutation $\sigma_J \in \Sigma_J$ satisfying the following conditions:

- The partition of J induced by the cycle structure of σ_J follows the distribution $\mathbb{Q}_J$.
- Conditionally on this partition, the law of σ_J is uniform.

The existence of a measure $\mathbb{P}$ satisfying the conditions above is then a consequence of the following property of compatibility, which can be easily checked: for all $I, J \in \mathscr{F}$ such that $I \subset J$, for all distributions $\mathbb{Q}_J$ on the partitions of J, the image of $\pi^{-1}(\mathbb{Q}_J)$ by the mapping from Σ_J to Σ_I which removes the elements of $J \setminus I$ from the cycle structure is equal to $\pi^{-1}(\mathbb{Q}_I)$, where $\mathbb{Q}_I$ is the image of $\mathbb{Q}_J$ by the restriction of the partitions to the set I. Now, the bijective correspondence induced by π implies that it is sufficient to show the equivalent of Proposition 8 for exchangeable partitions. Let $(x_n)_{n \geq 1}$ be an enumeration of the set E, and let us define the strictly increasing family of sets $(I_n^{(0)})_{n \geq 1}$, by:

$$I_n^{(0)} := \{x_1, x_2, \ldots, x_n\}.$$

If in Proposition 8 for exchangeable partitions, the convergence of $|I \cap \mathscr{C}(x)|/|I|$ to $\lambda(x)$ in L^1 when $|I|$ goes to infinity is removed, and if the almost sure convergence along any strictly increasing sequence of nonempty sets in $\mathscr{F}(E)$ is replaced by the convergence only along the sequence $(I_n^{(0)})_{n \geq 1}$, then the remaining of the proposition is a direct consequence of classical results by Kingman. Let us now prove the convergence in L^1, fixing $x \in E$. By dominated convergence,

$$\frac{|\mathscr{C}(x) \cap I_n^{(0)}|}{|I_n^{(0)}|} \xrightarrow[n \to \infty]{} \lambda(x), \tag{1}$$

in L^1 when n goes to infinity. Moreover, for all $J, K \in \mathscr{F}(E)$ such that $\{x\} \subset J \subset K$, the joint law of $|\mathscr{C}(x) \cap J|$ and $|\mathscr{C}(x) \cap K|$ depends only on $|J|$ and $|K|$, since the underlying probability measure on $(\Sigma_E, \mathscr{S}_E)$ is central. Now, for $|J|$ large enough, $I_{|J|}^{(0)}$, and a fortiori $I_{|K|}^{(0)}$, contain x. In this case, one has

$$\mathbb{E}\left[\left|\frac{|\mathscr{C}(x) \cap J|}{|J|} - \frac{|\mathscr{C}(x) \cap K|}{|K|}\right|\right] = \mathbb{E}\left[\left|\frac{|\mathscr{C}(x) \cap I_{|J|}^{(0)}|}{|I_{|J|}^{(0)}|} - \frac{|\mathscr{C}(x) \cap I_{|K|}^{(0)}|}{|I_{|K|}^{(0)}|}\right|\right]. \tag{2}$$

By the convergence (1), the right-hand side of 2 tends to zero when $|J|$ goes to infinity. In other words, there exists a sequence $(\varepsilon(n))_{n \geq 1}$, decreasing to zero at infinity, such that

$$\mathbb{E}\left[\left|\frac{|\mathscr{C}(x) \cap J|}{|J|} - \frac{|\mathscr{C}(x) \cap K|}{|K|}\right|\right] \leq \varepsilon(|J|),$$

for all $J, K \in \mathscr{F}(E)$ such that $\{x\} \subset J \subset K$. Now, for all $J, K \in \mathscr{F}(E)$ containing x, we obtain, by taking $J \cup K$ as an intermediate set and by using the triangle inequality:

$$\mathbb{E}\left[\left|\frac{|\mathscr{C}(x) \cap J|}{|J|} - \frac{|\mathscr{C}(x) \cap K|}{|K|}\right|\right] \leq \varepsilon(|J|) + \varepsilon(|K|). \tag{3}$$

Moreover, if $J \in \mathscr{F}(E)$ does not contain x, we have:

$$\frac{|\mathscr{C}(x) \cap (J \cup \{x\})|}{|J \cup \{x\}|} - \frac{1}{|J|} \leq \frac{|\mathscr{C}(x) \cap J|}{|J|} \leq \frac{|\mathscr{C}(x) \cap (J \cup \{x\})|}{|J \cup \{x\}|}. \tag{4}$$

Combining (3) and (4), we deduce, for any $J, K \in \mathscr{F}(E)$:

$$\mathbb{E}\left[\left|\frac{|\mathscr{C}(x) \cap J|}{|J|} - \frac{|\mathscr{C}(x) \cap K|}{|K|}\right|\right] \leq \varepsilon(|J|) + \varepsilon(|K|) + \frac{1}{|J|} + \frac{1}{|K|}.$$

In particular, by taking $K = I_n^{(0)}$, one has for $n \geq |J|$:

$$\mathbb{E}\left[\left|\frac{|\mathscr{C}(x) \cap J|}{|J|} - \frac{|\mathscr{C}(x) \cap I_n^{(0)}|}{|I_n^{(0)}|}\right|\right] \leq 2\varepsilon(|J|) + \frac{2}{|J|},$$

and, after letting $n \to \infty$:

$$\mathbb{E}\left[\left|\frac{|\mathscr{C}(x) \cap J|}{|J|} - \lambda(x)\right|\right] \leq 2\varepsilon(|J|) + \frac{2}{|J|},$$

which proves the convergence in L^1. Now, let $(I_n)_{n \geq 1}$ be a strictly increasing sequence of sets in $\mathscr{F}(E)$, containing x and such that $|I_n| = n$. Since the underlying measure on Σ_E is central, the law of the sequence $(|\mathscr{C}(x) \cap I_n|/n)_{n \geq 1}$ is independent of the choice of $(I_n)_{n \geq 1}$. Hence, $(|\mathscr{C}(x) \cap I_n|/n)_{n \geq 1}$ is a.s. a Cauchy sequence, since it is the case when we suppose that $I_n = I_n^{(0)}$ for n large enough. By the convergence in L^1 proven above, the limit of $(|\mathscr{C}(x) \cap I_n|/n)_{n \geq 1}$ is necessarily $\lambda(x)$ a.s. The almost sure convergence is then proven for any strictly increasing sequence $(I_n)_{n \geq 1}$ such that $x \in I_n$ and $|I_n| = n$ for all $n \geq 1$: the first condition can be removed simply by putting x into all the sets $(I_n)_{n \geq 1}$, which changes the quotient $|\mathscr{C}(x) \cap I_n|/|I_n|$ by at most $1/|I_n|$, the second one can be suppressed by inserting any strictly increasing sequence of nonempty sets in $\mathscr{F}(E)$ into an increasing sequence $(I_n)_{n\geq 1}$ such that $|I_n| = n$ for all $n \geq 1$.

As we have seen in its proof, the result of Proposition 8 gives some information about the asymptotic cycle lengths of a random virtual permutation following a central measure, but does not tell anything about the relative positions of the elements inside their cycle. It is a remarkable fact that these relative positions also admit an asymptotic limit when the size of the sets in $\mathscr{F}(E)$ goes to infinity:

Proposition 9 *Let E be a countable set, x, y two elements of E and σ a virtual permutation of E, which follows a central probability measure. Then, on the event $\{x \sim_\sigma y\}$:*

- *For all $I \in \mathscr{F}(E)$ containing x and y, there exists a unique integer $k_I(x, y) \in \{0, 1, \ldots, |I \cap \mathscr{C}_\sigma(x)| - 1\}$ such that $\sigma_I^k(x) = y$.*
- *The variable $k_I(x, y)/|I|$ converges in L^1, and then in all the spaces L^p for $p \in [1, \infty)$, to a random variable $\Delta(x, y) \in [0, 1]$ when $|I|$ goes to infinity.*
- *For any strictly increasing sequence $(I_n)_{n\geq 1}$ of sets in $\mathscr{F}(E)$ containing x and y, $k_{I_n}(x, y)/|I_n|$ converges a.s. to $\Delta(x, y)$ when n goes to infinity.*

Proof We may assume that $x \neq y$ since otherwise the result is obvious. If $x \sim y$, and if $I \in \mathscr{F}(E)$ contains x and y, then these two elements of E are in the same cycle of σ_I, which has length $|I \cap \mathscr{C}_\sigma(x)|$. This implies the existence and the uniqueness of $k_I(x, y)$. Now, let $(I_n)_{n\geq 1}$ be a strictly increasing sequence of sets in $\mathscr{F}(E)$ such that $I_1 = \{x, y\}$, and for all $n \geq 1$, $I_{n+1} = I_n \cup \{z_n\}$ with $z_n \notin I_n$. Then, for all $n \geq 1$, conditionally on σ_{I_n}, and on the event $x \sim y \sim z_n$, the position of z_n inside the cycle structure of $\sigma_{I_{n+1}}$ is uniform among all the possible positions in the cycle containing x and y, since the σ follows a central probability measure. Hence, again conditionally on σ_{I_n}, $x \sim y \sim z_n$, one has $k_{I_{n+1}}(x, y) = k_{I_n}(x, y) + 1$ with probability $k_{I_n}(x, y)/|\mathscr{C}(x) \cap I_n|$ (if z_n is inserted between x and y in their common cycle), and $k_{I_{n+1}}(x, y) = k_{I_n}(x, y)$ with probability $1 - (k_{I_n}(x, y)/|\mathscr{C}(x) \cap I_n|)$. One deduces that $(k_{I_n}(x, y)\mathbb{1}_{x\sim y}/|\mathscr{C}(x) \cap I_n|)_{n\geq 1}$ is a martingale with respect to the filtration generated by $(\sigma_{I_n})_{n\geq 1}$. Since it takes its values in $[0, 1]$, it converges a.s. to a limit random variable. Moreover, on the event $\{x \sim y\}$, the quotient $|\mathscr{C}(x) \cap I_n|/|I_n|$ converges a.s. to $\lambda(x)$, which is a.s. strictly positive, since x is not a fixed point of σ. Hence, again on the set $x \sim y$, $k_{I_n}(x, y)/|I_n|$ converges to a random variable $\Delta(x, y)$ a.s., and then in L^1 by dominated convergence. Now, let J, K be two finite subsets of E containing x and y. If J is included in K, then by the L^1 convergence of $k_{I_n}(x, y)/|I_n|$ at infinity, and by the centrality of the law of σ, one has:

$$\mathbb{E}[|(k_J(x, y)/|J|) - (k_K(x, y)/|K|)|\mathbb{1}_{x\sim y}]$$
$$= \mathbb{E}[|(k_{I_{|J|-1}}(x, y)/|J|) - (k_{I_{|K|-1}}(x, y)/|K|)|\mathbb{1}_{x\sim y}]$$
$$\leq \varepsilon(|J|),$$

where $\varepsilon(n)$ decreases to zero at infinity. If J is not supposed to be included in K, then by using $J \cup K$ as an intermediate step, one obtains:

$$\mathbb{E}[|(k_J(x,y)/|J|) - (k_K(x,y)/|K|)| \, \mathbb{1}_{x\sim y}] \leq \varepsilon(|J|) + \varepsilon(|K|).$$

In particular, for $J \in \mathscr{F}(E)$ and for all $n \geq |J|$:

$$\mathbb{E}[|(k_J(x,y)/|J|) - (k_{I_n}(x,y)/|I_n|)| \, \mathbb{1}_{x\sim y}] \leq 2\varepsilon(|J|),$$

which implies that

$$\mathbb{E}[|(k_J(x,y)/|J|) - \Delta(x,y)| \mathbb{1}_{x\sim y}] \leq 2\varepsilon(|J|).$$

Therefore, on the set $x \sim y$, $k_J(x,y)/|J|$ converges in L^1 to $\Delta(x,y)$. Now, let $(J_n)_{n\geq 1}$ be a strictly increasing sequence of sets in $\mathscr{F}(E)$, containing x and y. This sequence can be inserted in a sequence with satisfies the same assumptions as $(I_n)_{n\geq 1}$, which implies that on the event $\{x \sim y\}$, $k_{J_n}(x,y)/|J_n|$ converges a.s. to a random variable, necessarily equal to $\Delta(x,y)$ since $k_{J_n}(x,y)/|J_n|$ converges to $\Delta(x,y)$ in L^1.

For $x \sim y$, $\Delta(x,y)$ represents the asymptotic number of iterations of the virtual permutation σ which are needed to go from x to y, and $\Delta(x,y)/\lambda(x)$ is the asymptotic proportion of the cycle $\mathscr{C}(x)$, lying between x and y. The distribution of this proportion is deduced from the following result:

Proposition 10 *Let E be a countable set, and let σ be a virtual permutation of E, which follows a central probability measure. For $m, n_1, \ldots n_m \geq 0$, let $(x_j)_{1\leq j\leq m}$ and $(y_{j,k})_{1\leq j\leq m, 1\leq k\leq n_j}$ be distinct elements of E. Then, conditionally of the event:*

$$\{\forall j \in \{1,\ldots,m\}, \forall k \in \{1,\ldots,n_j\}, x_j \sim_\sigma y_{j,k}\} \cap \{\forall j_1 \neq j_2 \in \{1,\ldots,m\}, x_{j_1} \nsim x_{j_2}\},$$

the variables $(\Delta(x_j, y_{j,k})/\lambda(x_j))_{1\leq j\leq m, 1\leq k\leq n_j}$ are a.s. well-defined, independent and uniform on $[0,1]$, and they form a family which is independent of the variables $(\lambda(x))_{x\in E}$.

Remark 3 It is possible to take some indices j such that $n_j = 0$. In this case, there is no point of the form $y_{j,k}$, and then no variable of the form $\Delta(x_j, y_{j,k})/\lambda(x_j)$, involved in Proposition 10.

Proof Let $q \geq 1$, and let $(x_j)_{m+1\leq j\leq m+q}$ be a family of elements of E such that all the elements $(x_j)_{1\leq j\leq m+q}$ are distinct. Let us denote:

$$\mathscr{E} := \{\forall j \in \{1,\ldots,m\}, \forall k \in \{1,\ldots,n_j\}, x_j \sim_\sigma y_{j,k}\}$$
$$\cap \{\forall j_1 \neq j_2 \in \{1,\ldots,m\}, x_{j_1} \nsim x_{j_2}\}.$$

Moreover, for all $I \in \mathscr{F}(E)$ containing x_j for all $j \in \{1, \ldots, m + q\}$ and $y_{j,k}$ for all $j \in \{1, \ldots, m\}$, $k \in \{1, \ldots n_j\}$, and for all sequences $(p_j)_{1 \leq j \leq m+q}$ of strictly positive integers such that $p_j > n_j$ for all $j \in \{1, \ldots, m\}$, let us define the following event:

$$\mathscr{E}_{I,(p_j)_{1 \leq j \leq m}} := \mathscr{E} \cap \{\forall j \in \{1, \ldots m + q\}, |\mathscr{C}(x_j) \cap I| = p_j\}.$$

By the centrality of the law of σ, conditionally on $\mathscr{E}_{I,(p_j)_{1 \leq j \leq m+q}}$, the sequences $(k_I(x_j, y_{j,k}))_{1 \leq k \leq n_j}$, for $j \in \{1, \ldots, m\}$ such that $n_j > 0$, are independent, and the law of $(k_I(x_j, y_{j,k}))_{1 \leq k \leq n_j}$ is uniform among all the possible sequences of n_j distinct elements in $\{1, 2, \ldots, p_j - 1\}$. Now, for all $j \in \{1, \ldots, m\}$, let Φ_j be a continuous and bounded function from $\mathbb{R}^{n_j}$ to $\mathbb{R}$ (Φ_j is a real constant if $n_j = 0$). Since for $n_j > 0$, the uniform distribution on the sequences of n_j district elements in $\{1/p, 2/p, \ldots, (p-1)/p\}$ converges weakly to the uniform distribution on $[0, 1]^{n_j}$ when p goes to infinity, one has:

$$\mathbb{E}\left[\prod_{j=1}^{m} \Phi_j\left((k_I(x_j, y_{j,k})/p_j)_{1 \leq k \leq n_j}\right) \mid \mathscr{E}_{I,(p_j)_{1 \leq j \leq m+q}}\right]$$

$$= \prod_{j=1}^{m} \mathbb{E}\left[\Phi_j\left((k_I(x_j, y_{j,k})/p_j)_{1 \leq k \leq n_j}\right) \mid \mathscr{E}_{I,(p_j)_{1 \leq j \leq m+q}}\right]$$

$$= \left(\prod_{j=1}^{m} \int_{[0,1]^{n_j}} \Phi_j\right) + \alpha((\Phi_j, p_j)_{1 \leq j \leq m}),$$

where for fixed functions $(\Phi_j)_{1 \leq j \leq m}$, $|\alpha((\Phi_j, p_j)_{1 \leq j \leq m})|$ is uniformly bounded and tends to zero when the minimum of p_j for $n_j > 0$ goes to infinity. One deduces that for all continuous, bounded functions Ψ from $\mathbb{R}^m$ to $\mathbb{R}$:

$$\mathbb{E}\left[\Psi\left(\left(\frac{|\mathscr{C}(x_j) \cap I|}{|I|}\right)_{1 \leq j \leq m+q}\right) \prod_{j=1}^{m} \Phi_j\left((k_I(x_j, y_{j,k})/(|\mathscr{C}(x_j) \cap I|))_{1 \leq k \leq n_j}\right) \mid \mathscr{E}\right]$$

$$= \mathbb{E}\left[\Psi\left(\left(\frac{|\mathscr{C}(x_j) \cap I|}{|I|}\right)_{1 \leq j \leq m+q} \mid \mathscr{E}\right)\right]\left(\prod_{j=1}^{m} \int_{[0,1]^{n_j}} \Phi_j\right)$$

$$+ \mathbb{E}\left[\Psi\left(\left(\frac{|\mathscr{C}(x_j) \cap I|}{|I|}\right)_{1 \leq j \leq m+q}\right) \alpha((\Phi_j, |\mathscr{C}(x_j) \cap I|)_{1 \leq j \leq m}) \mid \mathscr{E}\right].$$

Now, for any strictly increasing sequence $(I_r)_{r\geq 1}$ of sets satisfying the same assumptions as I, and on the event $\mathscr{E}$:

- For all $j \in \{1, \ldots, m + q\}$, $|\mathscr{C}(x_j) \cap I_r|/|I_r|$ converges a.s. to $\lambda(x_j)$ when r goes to infinity.
- For all $j \in \{1, \ldots, m\}$ and $k \in \{1, \ldots, n_j\}$, $k_{I_r}(x_j, y_{j,k})/(|\mathscr{C}(x_j) \cap I_r|)$ tends a.s. to $\Delta(x_j, y_{j,k})/(\lambda(x_j))$, which is well-defined since $\lambda(x_j) > 0$ a.s. (by assumption, x_j is not a fixed point of σ if $n_j > 0$).
- For all $j \in \{1, \ldots, m\}$ such that $n_j > 0$, and then $\lambda(x_j) > 0$ a.s., $|\mathscr{C}(x_j) \cap I_r|$ tends a.s. to infinity, which implies that $\alpha((\Phi_j, |\mathscr{C}(x_j) \cap I|)_{1 \leq j \leq m})$ goes to zero.

By dominated convergence, one deduces:

$$\mathbb{E}\left[\Psi\left((\lambda(x_j))_{1\leq j\leq m+q}\right)\prod_{j=1}^{m}\Phi_j\left(\Delta(x_j, y_{j,k})/(\lambda(x_j)))_{1\leq k\leq n_j}\right)\Big|\mathscr{E}\right]$$

$$= \mathbb{E}\left[\Psi\left((\lambda(x_j))_{1\leq j\leq m+q}\right)\Big|\mathscr{E}\right]\left(\prod_{j=1}^{m}\int_{[0,1]^{n_j}}\Phi_j\right).$$

This proves Proposition 10 with $(\lambda(x))_{x\in E}$ replaced by $\lambda(x_j)_{j\in\{1,\ldots,m+q\}}$. By increasing q and by using monotone class theorem, we are done.

A particular case of Proposition 10 is the following:

Corollary 2 *Let E be a countable set, x and y two distinct elements of E, and σ a virtual permutation of E, which follows a central probability measure. Then, conditionally on the event $\{x \sim_\sigma y\}$, the variable $\Delta(x, y)/\lambda(x)$ is a.s. well-defined and uniform on the interval $[0, 1]$ (which implies that $\Delta(x, y) \in (0, \lambda(x))$ a.s.).*

Recall now that by definition, for $x \sim y$, $I \in \mathscr{F}(E)$ containing x and y, $k_I(x, y)$ is the unique integer between 0 and $|\mathscr{C}(x) \cap I| - 1$ such that $\sigma^{k_I(x,y)}(x) = y$. If the condition $0 \leq k_I(x, y) \leq |\mathscr{C}(x) \cap I| - 1$ is removed, then $k_I(x, y)$ is only defined up to a multiple of $|\mathscr{C}(x) \cap I|$. Hence, by taking $|I| \to \infty$, it is natural to introduce the class $\delta(x, y)$ of $\Delta(x, y)$ modulo $\lambda(x)$. This class satisfies the following properties:

Proposition 11 *Let E be a countable set, x, y, z three elements of E (not necessarily distinct), and σ a virtual permutation of E, which follows a central probability measure. Then, a.s. on the event $\{x \sim_\sigma y \sim_\sigma z\}$:*

- $\delta(x, x) = 0$;
- $\delta(x, y) = -\delta(y, x)$;
- $\delta(x, y) + \delta(y, z) = \delta(x, z)$.

Proof If $x = y = z$ and if x is a fixed point of σ, then Proposition 11 is trivial (with all the values of δ equal to zero). Hence, we can suppose that x is not a fixed point of σ. Let us now prove the third item, which implies immediately the first (by taking $x = y = z$) and then the second (by taking $x = z$). For any $I \in \mathscr{F}(E)$ containing

x, y and z, one has necessarily, on the event $\{x \sim y \sim z\}$:

$$\sigma_I^{k_I(x,y)+k_I(y,z)}(x) = \sigma_I^{k_I(x,z)}(x) = z,$$

which implies

$$k_I(x, y) + k_I(y, z) - k_I(x, z) \in \{0, |\mathscr{C}(x) \cap I|\}$$

since $k_I(x, y)$, $k_I(y, z)$ and $k_I(x, z)$ are in the set $\{0, 1, \ldots, |\mathscr{C}(x) \cap I| - 1\}$. Let $(I_n)_{n\geq1}$ be a strictly increasing sequence of sets in $\mathscr{F}(E)$ containing x, y and z: the sequence

$$\left(\frac{k_{I_n}(x, y) + k_{I_n}(y, z) - k_{I_n}(x, z)}{|\mathscr{C}(x) \cap I_n|} \right)_{n\geq1}$$

of elements in $\{0, 1\}$ converges a.s. to $(\Delta(x, y) + \Delta(y, z) - \Delta(x, z))/(\lambda(x))$ (recall that $\lambda(x) > 0$ since x is not a fixed point of σ). One deduces that $\Delta(x, y) + \Delta(y, z) - \Delta(x, z)$ is a.s. equal to 0 or $\lambda(x)$, which implies that $\delta(x, y) + \delta(y, z) = \delta(x, z)$.

The properties of δ suggest the following representation of the cycle structure of σ: for $x \in E$ which is not a fixed point of σ, one puts the elements of the cycle of x for $\sim_\sigma$ into a circle of perimeter $\lambda(x)$, in a way such that for two elements $y, z \in \mathscr{C}(x)$, $\delta(y, z)$ is the length, counted counterclockwise, of the circle arc between x and y. In order to make the representation precise, we shall fix the asymptotic cycle lengths of σ, by only considering the central measures of the form $\mathbb{P}_\lambda$, for a given sequence λ in the simplex Λ. When we make this assumption, we do not lose generality, since by Proposition 8, any central measure on Σ_E can be written as a mixture of the measures of the form $\mathbb{P}_\lambda$ (i.e. an integral with respect to a probability measure ν on Λ), and we obtain the following result:

Proposition 12 *Let $\lambda := (\lambda_k)_{k\geq1}$ be a sequence in the simplex Λ, let E be a countable set, and let $\sigma = (\sigma_I)_{I\in\mathscr{F}(E)}$ be a virtual permutation of E, following the central probability measure $\mathbb{P}_\lambda$ defined in Proposition 8. For $k \geq 1$ such that $\lambda_k > 0$, let C_k be a circle of perimeter λ_k, the circles being pairwise disjoint, and for all $x, y \in C_k$, let $y - x \in \mathbb{R}/\lambda_k\mathbb{Z}$ be the length of the arc of circle from x to y, counted counterclockwise and modulo λ_k. Let L be a segment of length $1 - \sum_{k\geq1}\lambda_k$ (the empty set if $\sum_{k\geq1}\lambda_k = 1$), disjoint of the circles C_k. Set*

$$C := (\cup_{k\geq0}C_k) \cup L.$$

Let μ the uniform probability measure on C (endowed with the σ-algebra generated by the Borel sets of L and C_k, $k \geq 1$), i.e. the unique measure such that $\mu(A)$ is equal to the length of A, for any set A equal to an arc of one of the circles C_k, or a segment included in L. Then, after enlarging, when necessary, the probability space *on which σ is defined, there exists a family $(X_x)_{x \in E}$ of random variables on C, independent and μ-distributed, and such that the following conditions hold a.s.:*

- *For all $x \in E$, $X_x \in L$ if and only if x is a fixed point of σ.*
- *For all x, y, distinct elements of E, $x \sim_\sigma y$ if and only if X_x and X_y are on the same circle C_k, and in this case, $\lambda(x) = \lambda(y) = \lambda_k$.*
- *For all x, y, distinct elements of E such that $x \sim_\sigma y$, $\delta(x, y)$ is equal to $X_y - X_x$ modulo $\lambda(x)$.*

Proof After enlarging, when necessary, the underlying probability space, it is possible to define, for all $\lambda \in (0, 1)$ such that there exist several consecutive indices k satisfying $\lambda_k = \lambda$, a uniform random permutation τ_λ of the set of these indices, for all $k \geq 1$ such that $\lambda_k > 0$, a uniform random variable U_k on C_k, and for all $n \geq 1$, a uniform variable V_n on L, in a way such that the virtual permutation σ, the variables $(U_k)_{k \geq 1}$, $(V_n)_{n \geq 1}$ and all the permutations of the form τ_λ are independent. Now, let $(x_n)_{n \geq 1}$ be an enumeration of E. By the results on partitions by Kingman (see [7]), it is a.s. possible to define a sequence $(k_n)_{n \geq 1}$ of integers by induction, as follows:

- If x_n is a fixed point of σ, then $k_n = 0$.
- If x_n is equivalent to x_m for some $m < n$, then $k_n = k_m$, independently of the choice of the index m.
- If x_n is not a fixed point of σ and is not equivalent to x_m for any $m < n$, then k_n is an integer such that $\lambda(x_n) = \lambda_{k_n}$, and if this condition does not determine k_n uniquely, then this integer is chosen in a way such that k_n is different from k_m for all $m < n$ and $\tau_{\lambda(x_n)}(k_n)$ is as small as possible.

The random permutations of the form τ_λ are used in order to guarantee the symmetry between all the circles in C which have the same perimeter. Moreover, one checks that for all $m, n \geq 1$ such that x_m and x_n are not fixed points of σ, $\lambda(x_n) = \lambda_{k_n} > 0$, and $x_m \sim x_n$ if and only if $k_m = k_n$. Let us now define the variables $(X_{x_n})_{n \geq 1}$ by induction, as follows:

- If $k_n = 0$, then $X_{x_n} = V_n$.
- If $k_n > 0$ and $k_n \neq k_m$ for all $m < n$, then $X_{x_n} = U_{k_n} \in C_{k_n}$.
- If $k_n > 0$, if $k_n = k_m$ for some $m < n$, and if m_0 denotes the smallest possible value of m satisfying this equality, then X_{x_n} is the unique point of $C_{k_n} = C_{k_{m_0}}$ such that $X_{x_n} - U_{k_{m_0}} = \delta(x_{m_0}, x_n)$, modulo $\lambda_{k_n} = \lambda_{k_{m_0}}$.

For all $n \geq 1$, $X_{x_n} \in L$ if and only if $k_n = 0$, otherwise, $X_{x_n} \in C_{k_n}$. Moreover, if $k_m = k_n > 0$ for some $m, n \geq 1$, and if m_0 is the smallest index such that $k_m = k_{m_0}$, then modulo λ_{k_m}, a.s.,

$$X_{x_n} - X_{x_m} = (X_{x_n} - U_{k_{m_0}}) - (X_{x_m} - U_{k_{m_0}}) = \delta(x_{m_0}, x_n) - \delta(x_{m_0}, x_m) = \delta(x_m, x_n).$$

One deduces that Proposition 12 is satisfied, provided that the variables $(X_{x_n})_{n \geq 1}$ are independent and uniform on C. In order to show this fact, let us first observe that by Proposition 10, the following holds: for all $n \geq 1$, conditionally on the restriction of the equivalence relation $\sim$ to the set $\{x_1, \ldots, x_n\}$, the variables $\Delta(x_{m_0}, x_m)/\lambda(x_{m_0})$, for $m_0 < m \le n$ and $x_{m_0} \sim x_m$, where x_{m_0} is the element of its equivalence class with the smallest index, are uniform on $[0, 1]$, independent, and form a family which is independent of $(\lambda(x_m))_{1 \le m \ge n}$. Now, once the restriction of $\sim$ to $\{x_1, \ldots, x_n\}$ is given, the sequence $(k_m)_{1 \le m \le n}$ is uniquely determined by the sequence $(\lambda(x_m))_{1 \le m \ge n}$, and the permutations of the form τ_λ (which form a family independent of σ), and then it is independent of the family of variables $\Delta(x_{m_0}, x_m)/\lambda(x_{m_0}) = \Delta(x_{m_0}, x_m)/\lambda_{k_{m_0}}$ stated above. Since the restriction of $\sim$ to $\{x_1, \ldots, x_n\}$ is a function of the sequence $(k_m)_{1 \le m \le n}$, one deduces that conditionally on this sequence, the variables $(X_{x_m})_{m \ge 1}$ are independent, X_{x_m} being uniform on C_{k_m} if $k_m > 0$, and uniform on L if $k_m = 0$. It is now sufficient to prove that the variables $(k_n)_{n \ge 1}$ are independent and that for all $k \ge 1$, $\mathbb{P}[k_n = k] = \lambda_k$. By symmetry, for $k^{(1)}, \ldots, k^{(p)} \ge 1$, and for any distinct integers $n_1, \ldots, n_p \ge 1$, the probability that $k_{n_j} = k^{(j)}$ for all $j \in \{1, \ldots, p\}$, does not depend on $n_1, \ldots, n_p$ (note that the permutations τ_λ play a crucial role for this step of the proof of Proposition 12). Hence, for $m > p$,

$$\mathbb{P}[\forall j \in \{1, \ldots, p\}, k_j = k^{(j)}] = \frac{(m-p)!}{m!} \mathbb{E}\left[\sum_{1 \le n_1 \ne \cdots \ne n_p \le m} \mathbb{1}_{\forall j \in \{1, \ldots, p\}, k_{n_j} = k^{(j)}}\right],$$

and then:

$$\mathbb{E}\left[\prod_{j=1}^{p} \frac{\left(|\{r \in \{1, \ldots m\}, k_r = k^{(j)}\}| - p\right)_+}{m}\right]$$

$$\le \mathbb{P}[\forall j \in \{1, \ldots, p\}, k_j = k^{(j)}]$$

$$\le \mathbb{E}\left[\prod_{j=1}^{p} \frac{|\{r \in \{1, \ldots m\}, k_r = k^{(j)}\}|}{m - p}\right].$$

By using Proposition 8 and dominated convergence for m going to infinity, one deduces:

$$\mathbb{P}[\forall j \in \{1, \ldots, p\}, k_j = k^{(j)}] = \prod_{j=1}^{p} \lambda_{k_j}.$$

Now, we observe that it is possible to recover the virtual permutation σ from the variables $(X_x)_{x \in E}$. More precisely, one has the following:

Proposition 13 *Let us consider the setting and the notation of Proposition 12. For all $I \in \mathscr{F}(E)$ and $x \in I$, the element $\sigma_I(x)$ can a.s. be obtained as follows:*

- *If $X_x \in L$, then $\sigma_I(x) = x$.*

- *If for* $k \geq 1$, $X_x \in C_k$, *then* $\sigma_I(x) = y$, *where* X_y *is the first point of the intersection of* C_k *and the set* $\{X_z, z \in I\}$, *encountered by moving counterclockwise on* C_k, *starting just after* X_x *(for example, if* X_x *is the unique element of* $C_k \cap \{X_z, z \in I\}$, *then* x *is a fixed point of* σ_I).

Proof If x is a fixed point of σ_I, then either $X_x \in L$, or the unique $y \in I$ such that X_y is on the same circle as X_x is x itself. If x is not a fixed point of σ_I, then $X_{\sigma_I(x)}$ is on the same circle C_k as X_x. Moreover, for any $y \in I$ different from x and $\sigma_I(x)$, but in the same cycle of σ_I, one has $0 < \Delta(x, \sigma_I(x)) < \Delta(x, y)$. Hence, there is no point of $C_k \cap \{X_z, z \in I\}$ on the open circle arc coming counterclockwise from X_x to $X_{\sigma_I(x)}$.

A consequence of Propositions 12 and 13 is the following:

Corollary 3 *Let* E *be a countable set,* λ *a sequence in the simplex* Λ, C *the set constructed in Proposition 12, and* $(X_x)_{x \in E}$ *a sequence of i.i.d. random variables, uniform on* C. *For all* $I \in \mathscr{F}(E)$, *it is a.s. possible to define a permutation* $\sigma_I \in \Sigma_I$, *by the construction given in Proposition 13. Moreover,* $(\sigma_I)_{I \in E}$ *is a.s. a virtual permutation, and its distribution is the measure* $\mathbb{P}_\lambda$ *defined in Proposition 8.*

Proof The possibility to define a.s. σ_I for all $I \in \mathscr{F}(E)$ comes from the fact that the points $(X_x)_{x \in E}$ are almost surely pairwise distinct. Moreover, since $(\sigma_I)_{I \in \mathscr{F}(E)}$ is a deterministic function of the sequence of points $(X_x)_{x \in E}$, its distribution is uniquely determined by the assumptions of Corollary 3. Since Propositions 12 and 13 give a particular setting on which $(\sigma_I)_{I \in \mathscr{F}(E)}$ is a virtual permutation following the distribution $\mathbb{P}_\lambda$, we are done.

Proposition 8 and Corollary 3 give immediately the following description of all the central measures on Σ_E:

Corollary 4 *Let* ν *be a probability measure on* Λ, λ *a random sequence following the distribution* ν, *and* C *the random set constructed from the sequence* λ *as in Proposition 12. Let* $(X_x)_{x \in E}$ *be a sequence of random points of* C, *independent and uniform conditionally on* λ. *For all* $I \in \mathscr{F}(E)$, *it is a.s. possible to define a permutation* $\sigma_I \in \Sigma_I$, *by the construction given in Proposition 13. Moreover,* $(\sigma_I)_{I \in E}$ *is a.s. a virtual permutation, and its distribution is the measure* $\mathbb{P}_\nu$ *defined in Proposition 8.*

This construction will be used in the next section, in order to study how σ acts on the completion of the space E with respect to a random metric related to δ.

4   A Flow of Transformations on a Completion of E

When one looks at the construction of the Ewens measures on the set of virtual permutations which is given at the end of Sect. 3, it is natural to consider the random metric given by the following proposition:

Proposition 14 *Let E be a countable set, and σ a virtual permutation of E, which follows a central distribution. Let d be a random function from E^2 to $\mathbb{R}_+$, a.s. defined as follows:*

- *If $x, y \in E$ and $x \not\sim_\sigma y$, then $d(x, y) = 1$.*
- *If $x, y \in E$ and $x \sim_\sigma y$, then $d(x, y) = \inf\{|a|, \delta(x, y) = a \,(\mathrm{mod}.\,\lambda(x))\}$.*

Then, d is a.s. a distance on E.

Proof For all $x \in E$, $\delta(x, x) = 0$ a.s., which implies that $d(x, x) = 0$. Conversely if $d(x, y) = 0$ for $x, y \in E$, then $x \sim y$ and $\delta(x, y) = 0$, which holds with strictly positive probability only for $x = y$. Finally if $x, y, z \in E$, then there are two cases:

- If x, y, z are not equivalent, then $d(x, y) = 1$ or $d(y, z) = 1$, and $d(x, z) \leq 1$, which implies that $d(x, y) + d(y, z) \geq d(x, z)$.
- If $x \sim y \sim z$, then the triangle inequality holds because $\delta(x, z) = \delta(x, y) + \delta(y, z)$ a.s.

Once the metric space (E, d) is constructed, it is natural to embed it in another space which is better known.

Proposition 15 *Let E be a countable set, let σ be a virtual permutation following a central measure, let $(\lambda_k)_{k \geq 1}$ be the non-increasing sequence of the asymptotic cycle lengths of σ, defined in Proposition 8, and let C be the random space defined from $(\lambda_k)_{k \geq 1}$ as in Proposition 12. Moreover, let us define the random metric D on C as follows:*

- *If x, y are in C_k for the same value of k, then*

$$D(x, y) = \inf\{|a|, x - y = a \,(\mathrm{mod}.\,\lambda_k)\}.$$

- *If $x = y \in L$, $D(x, y) = 0$.*
- *Otherwise, $D(x, y) = 1$.*

Then, if the probability space is sufficiently large to guarantee the existence of the random variables $(X_x)_{x \in E}$ described in Proposition 12, and if $K := \{X_x, x \in E\}$, then the mapping:

$$\phi : x \mapsto X_x$$

is a.s. a bijective isometry from (E, d) to (K, D).

Proof Since the variables $(X_x)_{x \in E}$ are a.s. pairwise disjoint, ϕ is a.s. bijective from E to K. By comparing the definitions of d and D and by using the fact that $\delta(x, y) = X_y - X_x$ for $x \sim y$, we easily see that ϕ is isometric.

The isometry defined in Proposition 15 gives an intuitive idea on how the completion of (E, d) looks like:

Proposition 16 *Let us take the assumptions and the notation of Proposition 15 and let us define the random metric space* $(\widehat{E}, \widehat{d})$ *as the completion of* (E, d). *Then the following properties hold a.s.:*

- *The mapping* ϕ *can be extended in a unique way to a bijective and isometric mapping* $\widehat{\phi}$ *from* $(\widehat{E}, \widehat{d})$ *to* (H, D), *where*

$$H := (K \cap L) \cup \bigcup_{k \geq 1} C_k.$$

- *There exists a unique extension* $\widehat{\sim}_\sigma$ *of the equivalence relation* $\sim_\sigma$ *to the set* $\widehat{E}$, *such that the set* $\{(x, y) \in \widehat{E}^2, x \widehat{\sim}_\sigma y\}$ *is closed in* $\widehat{E}^2$, *for the topology induced by the distance* $\widehat{d}$.
- *For all* x, y *distinct in* $\widehat{E}$, $x \widehat{\sim}_\sigma y$ *if and only if* $\widehat{\phi}(x)$ *and* $\widehat{\phi}(y)$ *are on a common circle* C_k, *for some* $k \geq 1$.
- *There exists a unique mapping* $\widehat{\lambda}$ *from* $\widehat{E}$ *to* $[0, 1]$, *extending* λ *in a continuous way.*
- *For all* $x \in \widehat{E}$, $k \geq 1$, $\widehat{\lambda}(x) = \lambda_k$ *if* $\widehat{\phi}(x) \in C_k$ *and* $\widehat{\lambda}(x) = 0$ *if* $\widehat{\phi}(x) \in L$.
- *There exists a unique mapping* $\widehat{\delta}$ *from the set* $\{(x, y) \in \widehat{E}^2, x \widehat{\sim}_\sigma y\}$ *to* H, *extending* δ *to a continuous mapping, for the topologies induced by the distances* $\widehat{d}$ *and* D;
- *For all* x, y *distinct in* $\widehat{E}$ *such that* $x \widehat{\sim}_\sigma y$, $\widehat{\delta}(x, y) = \widehat{\phi}(y) - \widehat{\phi}(x)$, *modulo* λ_k, *if* $\widehat{\phi}(x)$ *and* $\widehat{\phi}(y)$ *are on the circle* C_k.

Proof Since, conditionally on $(\lambda_k)_{k \geq 1}$ the variables $(X_x)_{x \in E}$ are independent and uniform on C, the space (K, D) is a.s. dense in (H, D), which implies the first item. The description of the third item proves the existence of $\widehat{\sim}_\sigma$: its uniqueness comes from the fact that $\sim_\sigma$ is defined on a dense subset of $\widehat{E}^2$. Similarly, the existence of $\widehat{\lambda}$ is deduced from the description given in the fifth item, and its uniqueness comes from the density of E in $\widehat{E}$. Finally, the uniqueness of $\widehat{\delta}$ comes from the density of the set $\{x, y \in E, x \sim_\sigma y\}$ in the set $\{(x, y) \in \widehat{E}^2, x \widehat{\sim}_\sigma y\}$, and its existence is due to the continuity, for all $k \geq 1$, of the mapping $(x, y) \mapsto \widehat{\phi}(y) - \widehat{\phi}(x)$ from $\{(x, y) \in \widehat{E}^2, \widehat{\phi}(x), \widehat{\phi}(y) \in C_k\}$ to $\mathbb{R}/\lambda_k \mathbb{Z}$.

Remark 4 If L is empty, which happens if and only if $\sum_{k \geq 1} \lambda_k = 1$ (for example under Ewens measure of any parameter), then H is equal to C. Otherwise, H is the union of the circles included in C, and a discrete countable set.

From now, the notations $\widehat{\phi}$, $\widehat{\lambda}$, $\widehat{\delta}$, $\widehat{\sim}_\sigma$, will be replaced by ϕ, λ, δ, $\sim_\sigma$ (or $\sim$), since this simplification is consistent with the previous notation. Note that the mapping ϕ is not determined by σ, since it depends to the choice of the variables $(X_x)_{x \in E}$ in Proposition 12, which is not unique in general (for example, if $\lambda_1 > 0$ a.s., and if $a \in \mathbb{R}$ is fixed, then one can replace each point $X_x \in C_1$ by the unique point $X'_x \in C_1$ such that $X'_x - X_x = a$, modulo λ_1). However, as stated in Proposition 16, the extensions of $\sim$, λ and δ are a.s. uniquely determined. We now have all the ingredients needed to construct the flow of transformations on $\widehat{E}$ indicated in the title of this section.

Proposition 17 *With the notation above and the assumptions of Proposition 15, there exists a.s. a unique family* $(S^\alpha)_{\alpha \in \mathbb{R}}$ *of bijective isometries of the set* $\widehat{E}$, *such that for all* $x \in \widehat{E}$, $S^\alpha(x) \sim x$ *and in the case where* x *is not a fixed point of* σ, $\delta(x, S^\alpha(x)) = \alpha$ *modulo* $\lambda(x)$. *This family is a.s. given as follows:*

- *If* x *is a fixed point of* σ, *then* $S^\alpha(x) = x$.
- *If* x *is not a fixed point of* σ, *then* $S^\alpha(x)$ *is the unique point of* $\widehat{E}$ *such that* $\phi(S^\alpha(x))$ *is on the same circle as* $\phi(x)$, *and* $\phi(S^\alpha(x)) - \phi(x) = \alpha$ *modulo* $\lambda(x)$.

Moreover, a.s., $S^{\alpha+\beta} = S^\alpha S^\beta$ *for all* $\alpha, \beta \in \mathbb{R}$.

Proof For all $x, y \in \widehat{E}$, $\alpha \in \mathbb{R}$, let us denote by $C(x, y, \alpha)$ the condition described as follows:

- If x is a fixed point of σ, then $y = x$.
- If x is not a fixed point of σ, then $x \sim y$ and $\phi(y) - \phi(x) = \alpha$, modulo $\lambda(x)$.

It is clear that the condition $C(x, y, \alpha)$ determines uniquely y once x and α are fixed, which proves that a.s., the mapping S^α from $\widehat{E}$ to $\widehat{E}$ is well-defined for all $\alpha \in \mathbb{E}$, and that its explicit description is given in Proposition 17. By using this description, it is immediate to deduce that for all $\alpha, \beta \in \mathbb{R}$, S^α is isometric and $S^{\alpha+\beta} = S^\alpha S^\beta$. In particular, $S^\alpha S^{-\alpha} = S^0$ is the identity mapping of $\widehat{E}$, and then S^α is bijective.

Now, as written in the introduction, the flow $(S^\alpha)_{\alpha \in \mathbb{R}}$ can be seen as the limit, for large sets $I \in \mathscr{F}(E)$, of a power of σ_I, with exponent approximately equal to $\alpha|I|$. The following statement gives a rigorous meaning of this idea.

Proposition 18 *Let us take the assumptions and the notation above, and let us define a sequence* $(\alpha_n)_{n \geq 1}$ *in* $\mathbb{R}$, *such that* α_n / n *tends to a limit* α *when* n *goes to infinity. Then, for all* $\varepsilon > 0$, *there exist* $C(\varepsilon), c(\varepsilon) > 0$ *such that for all* $I \in \mathscr{F}(E)$:

$$\mathbb{P}\left[\exists x \in I,\ d(\sigma_I^{\alpha_{|I|}}(x), S^\alpha(x)) \geq \varepsilon\right] \leq C(\varepsilon)e^{-c(\varepsilon)|I|}.$$

Proof Let us suppose $\alpha > 0$. Let $x \in E$, let I be a finite subset of E containing x, and let A be the set of $y \in I$ which are equivalent to x but different from x. Then, by Proposition 10, conditionally on A, and on the event where this set is not empty, the variables $\delta(x, y)/\lambda(x)$ for $y \in A$ are independent, uniform on $\mathbb{R}/\mathbb{Z}$ and they form of family which is independent of $\lambda(x)$. Moreover, if $(b_k)_{k \in \mathbb{Z}}$ denotes the increasing family of the reals in the class of $\delta(x, y)/\lambda(x)$ modulo 1, for some $y \in A \cup \{x\}$, with b_0 equal to zero, then $\delta(x, \sigma_I^{\alpha_{|I|}}(x))/\lambda(x)$ is the class of $b_{\alpha_{|I|}}$ modulo 1. One deduces that conditionally on $|A|$, this cardinality being different from zero,

$$\delta(S^\alpha(x), \sigma_I^{\alpha_{|I|}}(x)) = \lambda(x)b(|A|, \alpha_{|I|}) - \alpha,$$

where $(b(|A|, k))_{k \in \mathbb{Z}}$ is a 1-periodic random increasing family of reals, independent of $\lambda(x)$, and such that its elements in $[0, 1)$ are $b(|A|, 0) = 0$, and $|A|$ independent variables, uniform on $(0, 1)$. One deduces that for $|A| \geq 1$,

$$d(S^\alpha(x), \sigma_I^{\alpha_{|I|}}(x)) \leq \left| \lambda(x) b(|A|, \alpha_{|I|}) - \alpha \right| \wedge \lambda(x).$$

Moreover, one has obviously, in any case:

$$d(S^\alpha(x), \sigma_I^{\alpha_{|I|}}(x)) \leq \lambda(x).$$

One deduces, for all $\varepsilon > 0$, and $|I|$ large enough so that

$$\left| \frac{\alpha_{|I|}}{|I|} - \alpha \right| \geq \varepsilon/3,$$

$$\mathbb{P}\left[d(S^\alpha(x), \sigma_I^{\alpha_{|I|}}(x)) \geq \varepsilon \right] \leq \mathbb{P}\left[\lambda(x) \geq \varepsilon, \left| \frac{\lambda(x)\alpha_{|I|}}{|A|} - \frac{\alpha_{|I|}}{|I|} \right| \geq \varepsilon/3 \right]$$

$$+ \mathbb{P}\left[\lambda(x) \geq \varepsilon, |A| \geq 1, \left| \frac{\lambda(x)\alpha_{|I|}}{|A|} - \alpha \right| \leq 2\varepsilon/3, \lambda(x) \left| b(|A|, \alpha_{|I|}) - \frac{\alpha_{|I|}}{|A|} \right| \geq \varepsilon/3 \right].$$
(5)

In (5), the event involved in the second term of the sum is always supposed to occur if A is empty. Now,

$$\mathbb{P}\left[\lambda(x) \geq \varepsilon, \left| \frac{\lambda(x)\alpha_{|I|}}{|A|} - \frac{\alpha_{|I|}}{|I|} \right| \geq \varepsilon/3 \right] = \mathbb{P}\left[\lambda(x) \geq \varepsilon, ||A| - \lambda(x)|I|| \geq \frac{\varepsilon|A||I|}{3|\alpha_{|I|}|} \right]$$

$$\leq \mathbb{P}\left[\lambda(x) \geq \varepsilon, ||A| - \lambda(x)|I|| \geq \frac{\varepsilon^2|I|^2}{6|\alpha_{|I|}|} \right]$$

$$+ \mathbb{P}\left[\lambda(x) \geq \varepsilon, |A| \leq |I|\varepsilon/2 \right]$$

$$\leq 2\mathbb{P}\left[||A| - \lambda(x)|I|| \geq \frac{\varepsilon^2|I|}{6\alpha + 2\varepsilon + 1} \right],$$

if $|I|$ is large enough (depending only on the sequence $(\alpha_n)_{n \geq 1}$). Now, conditionally on $\lambda(x)$, $|A|$ is the sum of $|I| - 1$ independent Bernoulli random variables, with parameter $\lambda(x)$. Hence, there exist $c_1, c_2 > 0$, depending only on $(\alpha_n)_{n \geq 1}$ and ε, such that:

$$\mathbb{P}\left[\lambda(x) \geq \varepsilon, \left| \frac{\lambda(x)\alpha_{|I|}}{|A|} - \frac{\alpha_{|I|}}{|I|} \right| \geq \varepsilon/3 \right] \leq c_1 e^{-c_2|I|}.$$
(6)

In order to evaluate the last term of (5), let us observe that if the corresponding event holds, then for $|I|$ large enough, $\alpha_{|I|} > |I|\alpha/2$, which implies:

$$|A| \geq \frac{\lambda(x)\alpha_{|I|}}{\alpha + 2\varepsilon/3} \geq \frac{\varepsilon\alpha|I|}{2(\alpha + \varepsilon)}. \tag{7}$$

Moreover, one has:

$$b(|A|, \alpha_{|I|}) = \beta + [\alpha_{|I|}/(|A| + 1)],$$

where the brackets denote the integer part, and where, conditionally on $|A|$, β is a beta random variable of parameters $k := \alpha_{|I|} - (|A| + 1)[\alpha_{|I|}/(|A| + 1)]$ and $(|A| + 1) - k$. One deduces that, conditionally on $|A|$, the probability that $|b(|A|, \alpha_{|I|}) - \alpha_{|I|}/(|A| + 1)|$ is greater than or equal to $\varepsilon/6$ decreases exponentially with $|A|$, independently of $\alpha_{|I|}$. Moreover, for $|I|$ large enough, by (7):

$$\left| \frac{\alpha_{|I|}}{|A|} - \frac{\alpha_{|I|}}{|A| + 1} \right| \leq \frac{\alpha_{|I|}}{|A|^2} \leq \frac{5(\alpha + \varepsilon)^2}{\varepsilon^2 \alpha |I|} \leq \varepsilon/7.$$

One deduces that there exist $c_3, c_4 > 0$, depending only on $(\alpha_n)_{n \geq 1}$ and ε, such that:

$$\mathbb{P}\left[\lambda(x) \geq \varepsilon, \left| \frac{\lambda(x)\alpha_{|I|}}{|A|} - \alpha \right| \leq 2\varepsilon/3, \lambda(x) \left| b(|A|, \alpha_{|I|}) - \frac{\alpha_{|I|}}{|A|} \right| \geq \varepsilon/3 \right] \leq c_3 e^{-c_4|I|}, \tag{8}$$

By (5), (6), (8), there exist $c_5, c_6 > 0$ such that

$$\mathbb{P}\left[d(S^\alpha(x), \sigma_I^{\alpha_{|I|}}(x)) \geq \varepsilon \right] \leq c_5 e^{-c_6|I|}.$$

By adding these estimates for all $x \in I$, one deduces the statement of Proposition 18 for $\alpha > 0$. The proof is exactly similar for $\alpha < 0$. Now, let $(\alpha_n)_{n \geq 1}$ and $(\beta_n)_{n \geq 1}$ be two sequences such that α_n/n and β_n/n tend to 1. Then,

$$\sup_{x \in I} d(x, \sigma_I^{\alpha_{|I|} - \beta_{|I|}}(x))$$

$$\leq \sup_{x \in I} d(x, S^1(\sigma_I^{-\beta_{|I|}}(x))) + \sup_{x \in I} d(S^1(\sigma_I^{-\beta_{|I|}}(x)), \sigma_I^{\alpha_{|I|} - \beta_{|I|}}(x))$$

$$\leq \sup_{x \in I} d(S^{-1}(x), \sigma_I^{-\beta_{|I|}}(x)) + \sup_{x \in I} d(S^1(x), \sigma_I^{\alpha_{|I|}}(x)),$$

since S^1 is an isometry of $\widehat{E}$ and $\sigma_I^{-\beta_{|I|}}$ is a bijection of $|I|$. One deduces that Proposition 18 holds also for $\alpha = 0$.

Corollary 5 *With the assumptions of Proposition 18, if $|I|$ goes to infinity, then the supremum of $d(\sigma_I^{\alpha_{|I|}}(x), S^\alpha(x))$ for $x \in I$ converges to zero in probability, in L^p for all $p \in [1, \infty)$, and a.s. along any deterministic, strictly increasing sequence $(I_n)_{n \geq 1}$ of sets in $\mathscr{F}(E)$. In particular, if $(x_n)_{n \geq 1}$ is a random sequence of elements in E, such that $x_n \in I_n$ for all $n \geq 1$ and x_n converges a.s. to a random limit*

$x \in \widehat{E}$ when n goes to infinity (this situation holds if $x \in E$ and $x_n = x$ for n large enough), then $\sigma_{I_n}^{\alpha|I_n|}(x_n)$ converges a.s. to $S^\alpha(x)$ when n goes to infinity.

Proof The convergence in probability is directly implied by Proposition 18, and it implies convergence in L^p for all $p \in [1, \infty)$, since the distance is bounded by one. The almost sure convergence is proven by using the Borel-Cantelli lemma.

We have now constructed a flow of transformations of $\widehat{E}$ and we have related it in a rigorous way to the iterations of σ_I for large sets I. In Sect. 5, we interpret this flow as a flow of operators on a suitable random function space, and we construct its infinitesimal generator, as discussed in the introduction.

5 A Flow of Operators on a Random Function Space

Let us now define the random function space on which the flow of operators described below acts. We first take, as before, a random virtual permutation σ on a countable set E, which follows a central probability measure. By the results given in the previous sections, the following events hold a.s.:

- The variables $\lambda(x)$ are well-defined, strictly positive for all $x \in E$ which are not fixed points of σ, and $\lambda(x) = \lambda(y)$ for all $x, y \in E$ such that $x \sim_\sigma y$.
- With the definitions of Proposition 8, the non-increasing sequence $(\lambda_k)_{k \geq 1}$ of the cycle lengths is an element of the simplex Λ.
- The quantity $\delta(x, y)$ exists for all $x, y \in E$ such that $x \sim_\sigma y$, and $\delta(x, x) = 0$, $\delta(x, y) = -\delta(y, x)$, $\delta(x, y) + \delta(y, z) = \delta(x, z)$ for all x, y, z such that $x \sim_\sigma y \sim_\sigma z$.
- There exists a bijective isometry ϕ from E to a dense subset of

$$H := L_0 \cup \bigcup_{k \geq 1, \lambda_k > 0} C_k,$$

where C_k is a circle of perimeter λ_k, the set L_0 is empty if $\sum_{k \geq 1} \lambda_k = 1$ and countable if $\sum_{k \geq 1} \lambda_k < 1$, and the union is disjoint, for the distances d and D defined similarly as in Propositions 14 and 15.
- The mapping λ, the distance d and the equivalence relation $\sim_\sigma$ extend in a unique continuous way to the completed space $\widehat{E}$ of E, for the distance d.
- The mapping δ extends in a unique continuous way to the set $\{(x, y) \in \widehat{E}^2, x \sim_\sigma y\}$.
- The isometry ϕ extends in a unique way to a bijective isometry from $(\widehat{E}, d)$ to (H, D).
- For all distinct $x, y \in \widehat{E}$, $x \sim_\sigma y$ if and only if $\phi(x)$ and $\phi(y)$ are on the same circle included in H.
- For all distinct $x, y \in \widehat{E}$ such that $x \sim_\sigma y$, $\delta(x, y) = \phi(y) - \phi(x)$, modulo $\lambda(x)$.

- There exists a unique flow $(S^\alpha)_{\alpha \in \mathbb{R}}$ of isometric bijections of $\widehat{E}$ such that for all $\alpha \in \mathbb{R}$, $x \in \widehat{E}$, $\phi(S^\alpha(x)) - \phi(x) = \alpha$ modulo $\lambda(x)$ if x is not a fixed point of σ, and $S^\alpha(x) = x$ if x is a fixed point of σ.
- For all $\alpha, \beta \in \mathbb{R}$, $S^\alpha S^\beta = S^{\alpha+\beta}$.

From now, let us fix σ such that all the items above are satisfied: no randomness is involved in the construction of the operator U given below. We can interpret the flow $(S^\alpha)_{\alpha \in \mathbb{R}}$ as a flow of operators on a function space defined on E. The first step in the corresponding construction is the following result:

Proposition 19 *Let f be a function from E to $\mathbb{C}$. If f can be extended to a continuous function from $\widehat{E}$ to $\mathbb{C}$, then this extension is unique: in this case we say that f is continuous. For example, if f is uniformly continuous from E to $\mathbb{C}$, then it is continuous and its continuous extension to $\widehat{E}$ is also uniformly continuous.*

When a function from E to $\mathbb{C}$ is continuous, we can use its extension to $\widehat{E}$ in order to make the flow $(S^\alpha)_{\alpha \in \mathbb{R}}$ acting on it, as follows:

Proposition 20 *In the setting above, one can define a unique flow $(T^\alpha)_{\alpha \in \mathbb{R}}$ of linear operators on the space of continuous functions from E to $\mathbb{C}$, satisfying the following properties:*

- *For any continuous function f from E to $\mathbb{C}$, $T^\alpha(f)(x) = \hat{f}(S^\alpha(x))$ for all $\alpha \in \mathbb{R}$, where $\hat{f}$ is the continuous extension of f to $\widehat{E}$.*
- *For all $\alpha, \beta \in \mathbb{R}$, $T^{\alpha+\beta} = T^\alpha T^\beta$.*

Proof The first property is in fact a definition of T^α. It is straightforward to check that T^α is a linear operator on the space of continuous functions from E to $\mathbb{C}$, and to show the identity $T^{\alpha+\beta} = T^\alpha T^\beta$, so we omit the details.

Let us now define the space of continuously differentiable functions with respect to the flow of operators $(T^\alpha)_{\alpha \in \mathbb{R}}$:

Definition 2 In the setting above, let f be a continuous function from E to $\mathbb{C}$. We say that f is continuously differentiable, if and only if there exists a continuous function Uf from E to $\mathbb{C}$, necessarily unique, such that for all $x \in \widehat{E}$,

$$\frac{\widehat{T^\alpha(f)}(x) - \hat{f}(x)}{\alpha} \xrightarrow[\alpha \to 0]{} \widehat{Uf}(x),$$

where $\hat{f}$, $\widehat{T^\alpha(f)}$ and $\widehat{Uf}$ are the continuous extensions of f, $T^\alpha(f)$ and Uf to $\widehat{E}$.

The following result is immediate:

Proposition 21 *The mapping $f \mapsto Uf$ from the space of continuously differentiable functions to the space of continuous functions from E to $\mathbb{C}$, constructed in Definition 2, is a linear operator.*

Once the operator U is defined, it is natural to study its eigenfunctions and eigenvalues. The following result holds:

Proposition 22 *The eigenvalues of U are zero, and all the nonzero integer multiples of $2i\pi/\lambda_k$ for $\lambda_k > 0$. The corresponding eigenspaces are described as follows:*

- *The space corresponding to the eigenvalue zero consists of all the functions f from E to $\mathbb{C}$ such that $x \sim_\sigma y$ implies $f(x) = f(y)$.*
- *The space corresponding to the eigenvalue ai for $a \in \mathbb{R}^*$ consists of all the functions f such that $f(x) = 0$ if $\lambda(x) = 0$ or $\lambda(x)$ is not a multiple of $2\pi/a$, and such that for $\lambda(x)$ nonzero and divisible by $2\pi/a$, the restriction of f to the equivalence class of x for $\sim_\sigma$ is proportional to $y \mapsto e^{ai\delta(x,y)}$.*

Consequently, the dimension of the space corresponding to the eigenvalue zero is equal to the number of indices $k \geq 1$ such that $\lambda_k > 0$ if $\sum_{k\geq 1}\lambda_k = 1$, and to infinity if $\sum_{k\geq 1}\lambda_k < 1$. Moreover, for $a \in \mathbb{R}^*$, the dimension of the space corresponding to the eigenvalue ia is equal to the number of indices $k \geq 1$ such that λ_k is a nonzero multiple of $2\pi/a$, in particular it is finite.

Proof Let f be an eigenfunction of U for an eigenvalue $b \in \mathbb{C}$, and let $\hat{f}$ be its extension to $\widehat{E}$. One has $Uf = bf$, and then by continuity, $\widehat{Uf} = b\hat{f}$. For all $x \in E$, let g_x be the function from $\mathbb{R}$ to $\mathbb{C}$, given by:

$$g_x(\alpha) := \hat{f}(S^\alpha(x)).$$

For all $\alpha, \beta \in \mathbb{R}$,

$$g_x(\alpha + \beta) = \hat{f}(S^\alpha S^\beta(x)) = \widehat{T^\beta(f)}(S^\alpha(x)),$$

and then, for $\beta \neq 0$,

$$\frac{g_x(\alpha+\beta)-g_x(\alpha)}{\beta} = \frac{\widehat{T^\beta(f)}(S^\alpha(x))-\hat{f}(S^\alpha(x))}{\beta}$$
$$\xrightarrow[\beta\to 0]{} \widehat{Uf}(S^\alpha(x)) = b\,\hat{f}(S^\alpha(x)) = b\,g_x(\alpha).$$

Hence, g_x is continuously differentiable and satisfies the differential equation $g'_x = b\,g_x$, which implies that g_x is proportional to the function $\alpha \to e^{b\alpha}$. Since $S^\alpha(x) = x$ for $\alpha = \lambda(x)$, and for all $\alpha \in \mathbb{R}$ if $\lambda(x) = 0$, one has $\lambda(x) > 0$ and $e^{b\lambda(x)} = 1$, $b = 0$ and g_x constant, or g_x identically zero. Therefore, one of the three following possibilities holds for all $x \in E$:

- f is identically zero on the cycle of x.
- $b = 0$ and f is constant on the cycle of x.
- $\lambda(x) > 0$, b is multiple of $2i\pi/\lambda(x)$ and the restriction of f to the cycle of x is proportional to $y \mapsto e^{b\delta(x,y)}$.

Conversely, it is easy to check that any function which satisfies one of the three items above for all $x \in E$ is an eigenfunction of U for the eigenvalue b, which completes the proof of Proposition 22.

Remark 5 The function spaces, on which the operators $(T^\alpha)_{\alpha \in \mathbb{R}}$ and U are defined, are spaces of functions on E. It means that the completion of E is used only as an intermediate step of our construction. The notion of infinitesimal generator involved here is not the classical one, since our definition of U involves only pointwise convergence. However, one can easily check that, for example, in the case where Uf is uniformly continuous, we have the uniform convergence

$$\sup_{x \in E} \left| \frac{T^\alpha(f)(x) - f(x)}{\alpha} - Uf(x) \right| = \sup_{x \in \widehat{E}} \left| \frac{\widehat{T^\alpha(f)}(x) - \hat{f}(x)}{\alpha} - \widehat{Uf}(x) \right| \xrightarrow[\alpha \to 0]{} 0.$$

Another way to construct a flow of operators, similar to $(T^\alpha)_{\alpha \in \mathbb{R}}$, is to make the construction on the function space $L^2(\widehat{E}, \mu)$, where μ is the image by ϕ^{-1} of a distribution μ_H on H, such that the restriction of μ_H to C_k is λ_k times the uniform probability measure on C_k, and the restriction of μ_H on L_0 has an atom at each point of L_0 and total measure $1 - \sum_{k \geq 1} \lambda_k$ (when L_0 is empty, μ can be considered as the uniform measure on $\widehat{E}$). Then, the mappings $(S^\alpha)_{\alpha \in \mathbb{R}}$ induce a flow $(V^\alpha)_{\alpha \in \mathbb{R}}$ of unitary operators on the Hilbert space $L^2(\widehat{E}, \mu)$. Then, one can write $V^\alpha = e^{\alpha W}$, where iW is a self-adjoint operator (whose domain is a closed subspace of $L^2(\widehat{E}, \mu)$). The operator W is the infinitesimal generator of $(V^\alpha)_{\alpha \in \mathbb{R}}$: it has the same eigenvalues and eigenfunctions as U. We observe that in this construction, the completion of E cannot be removed at the end, since except in the trivial case where $\lambda_k = 0$, $L_0 = H$ and $\widehat{E} = E$, the set E has not full measure in $\widehat{E}$ (if L_0 is empty, we even have $\mu(E) = 0$).

As discussed before, the operator T^α can be viewed as a limit of $\sigma_I^{\alpha/|I|}$ for large $I \in \mathscr{F}(E)$ and $\alpha_{|I|}$ equivalent to $\alpha/|I|$. It is then natural to relate the permutation σ_I to the operator $T^{1/|I|}$, and then the operator $|I|(\sigma_I - \text{Id})$ to $|I|(T^{1/|I|} - \text{Id})$, where σ_I is identified with a permutation matrix. Now, the eigenvalues of $|I|(\sigma_I - \text{Id})$ are equal to $|I|(e^{i\kappa} - 1)$, where κ is an eigenangle of σ_I, and this quantity is expected to be close to $i\kappa|I|$, on the other hand, $|I|(T^{1/|I|} - \text{Id})$ is expected to be close to U. Hence, it is natural to compare the renormalized eigenangles of σ_I (i.e. multiplied by $i|I|$), and the eigenvalues of U computed in Proposition 22. The rigorous statement corresponding to this idea is the following:

Proposition 23 *Let σ be a random virtual permutation on a countable set E, following a central measure. Let X be the set of the eigenvalues of the random operator iU (which is a.s. well-defined), and for $I \in \mathscr{F}(E)$, let X_I be the set of the eigenangles of σ_I, multiplied by $|I|$. If $\gamma \in X$ (resp. $\gamma \in X_I$), let $m(\gamma)$ (resp. $m_I(\gamma)$) be the multiplicity of the corresponding eigenvalue (resp. rescaled* *eigenangle). Then X and X_I, $I \in \mathscr{F}(E)$, are included in $\mathbb{R}$, and for all continuous functions f from $\mathbb{R}$ to $\mathbb{R}_+$, with compact support, the following convergence holds:*

$$\sum_{\gamma \in X_I} m_I(\gamma) f(\gamma) \xrightarrow[|I| \to \infty]{} \sum_{\gamma \in X} m(\gamma) f(\gamma),$$

in probability, and a.s. along any fixed, strictly increasing sequence of sets in $\mathscr{F}(E)$.

Proof Since X and X_I have no point in the interval $(-2\pi, 2\pi)$ except zero, it is sufficient to prove the convergence stated in Proposition 23 for $f = \mathbb{1}_{\{0\}}$ and for f nonnegative, continuous, with compact support and such that $f(0) = 0$. Let $(I_n)_{n \geq 1}$ be an increasing sequence of subsets of E, such that $|I_n| = n$. Let us suppose that the underlying probability space is large enough to apply Proposition 12, and let us take the same notation. The multiplicity $m_{I_n}(0)$ is equal to the sum of the number of indices $k \geq 1$ such that there exists $x \in I_n$ with $X_x \in C_k$, and the number of elements $x \in I_n$ such that $X_x \in L$. By the fact that conditionally on $(\lambda_k)_{k \geq 1}$, the variables $(X_x)_{x \in E}$ are independent and uniform on C, a weak form of the law of large numbers implies that $m_{I_n}(0)$ increases a.s. to the number of indices $k \geq 1$ such that $\lambda_k > 0$ if $\sum_{k \geq 1} \lambda_k = 1$, and to infinity otherwise. In other words, $m_{I_n}(0)$ increases a.s. to $m(0)$, and then also in probability. Since the law of σ is central, the convergence in probability holds also for $|I|$ going to infinity and not only along the sequence $(I_n)_{n \geq 1}$, which proves the convergence in Proposition 23 for $f = \mathbb{1}_{\{0\}}$. Let us now suppose that f is nonnegative, continuous, with compact support and satisfies $f(0) = 0$. One has a.s., for all $n \geq 1$,

$$\sum_{\gamma \in X_{I_n}} m(\gamma) f(\gamma) = \sum_{m \in \mathbb{Z} \setminus \{0\}} \sum_{k \geq 1} f(2\pi mn / |I_n \cap \mathscr{C}_k|),$$

where $\mathscr{C}_k$ denotes the set of $x \in E$ such that $X_x \in C_k$, and with the convention:

$$f(2\pi mn / |I_n \cap \mathscr{C}_k|) := 0$$

for $|I_n \cap \mathscr{C}_k| = 0$. Since f has compact support, there exists $A > 0$ such that $f(t) = 0$ for $|t| \geq A$. Hence, the condition $f(2\pi mn / |I_n \cap \mathscr{C}_k|) > 0$ implies that $2\pi |m| < A$ and a fortiori $|m| \leq A$, on the other hand, it implies that $2\pi n / |I_n \cap \mathscr{C}_k| < A$, $|I_n \cap \mathscr{C}_k| / n \geq 1/A$, and in particular,

$$\frac{1}{n} \left| I_n \cap \left(\bigcup_{l \geq k} \mathscr{C}_l \right) \right| \geq 1/A. \tag{9}$$

Now, conditionally on $(\lambda_n)_{n \geq 1}$, the left hand side of (9) has the same law as the average of n i.i.d. Bernoulli random variables, with parameter $\sum_{l \geq k} \lambda_l$. Hence, by law of large numbers, if $k_0 \geq 1$ denotes the smallest integer such that $\sum_{l \geq k_0} \lambda_l \leq 1/2A$, one has a.s., for n large enough, $f(2\pi mn / |I_n \cap \mathscr{C}_k|) = 0$ if

$k \geq k_0$, and then

$$\sum_{\gamma \in X_{I_n}} m(\gamma)f(\gamma) = \sum_{m \in (\mathbb{Z}\setminus\{0\})\cap[-A,A]} \sum_{1 \leq k \leq k_0} f(2\pi mn/|I_n \cap \mathscr{C}_k|).$$

By the continuity of f and the fact that $|I_n \cap \mathscr{C}_k|/n$ tends to λ_k when n goes to infinity, one deduces that a.s.,

$$\sum_{\gamma \in X_{I_n}} m(\gamma)f(\gamma) \xrightarrow[n\to\infty]{} \sum_{m \in (\mathbb{Z}\setminus\{0\})\cap[-A,A]} \sum_{1 \leq k \leq k_0} f(2\pi m/\lambda_k) = \sum_{\gamma \in X} f(\gamma),$$

which gives the almost sure convergence stated in Proposition 23. By the centrality of the law of σ, one then deduces the convergence in probability.

References

1. A. Borodin, G. Olshanski, Infinite random matrices and ergodic measures. Commun. Math. Phys. **223**(1), 87–123 (2001)
2. P. Bourgade, J. Najnudel, A. Nikeghbali, A unitary extension of virtual permutations. Int. Math. Res. Not. **2013**(18), 4101–4134 (2012)
3. S.-N. Evans, Eigenvalues of random wreath products. Electr. J. Probab. **7**(9), 1–15 (2002)
4. S.-V. Kerov, G.-I. Olshanski, A.-M. Vershik, Harmonic analysis on the infinite symmetric group. Comptes Rend. Acad. Sci. Paris **316**, 773–778 (1993)
5. J.-F.-C. Kingman, Random discrete distribution. J. Roy. Stat. Soc. B **37**, 1–22 (1975)
6. J.-F.-C. Kingman, Random partitions in population genetics. Proc. Roy. Soc. Lond. (A) **361**, 1–20 (1978)
7. J.-F.-C. Kingman, The representation of partition structures. J. Lond. Math. Soc. (2) **18**, 374–380 (1978)
8. K. Maples, J. Najnudel, A. Nikeghbali, Limit operators for circular ensembles. Preprint (2013). arXiv:1304.3757
9. M.-L. Mehta, *Random Matrices*. Pure and Applied Mathematics Series (Elsevier Academic Press, Amsterdam, 2004)
10. J. Najnudel, A. Nikeghbali, The distribution of eigenvalues of randomized permutation matrices. Ann. de l'institut Fourier **63**(3), 773–838 (2013)
11. Y.-A. Neretin, Hua type integrals over unitary groups and over projective limits of unitary groups. Duke Math. J. **114**, 239–266 (2002)
12. G. Olshanski, A. Vershik, Ergodic unitarily invariant measures on the space of infinite Hermitian matrices. Am. Math. Soc. Trans. **175**, 137–175 (1996)
13. J. Pitman, *Combinatorial Stochastic Processes*. Lecture Notes in Math., vol. 1875 (Springer, Berlin, 2006)
14. J. Ramirez, B. Valkó, B. Virág, Beta ensembles, stochastic Airy spectrum, and a diffusion, http://arxiv.org/pdf/math/0607331 (2006)
15. N.-V. Tsilevich, Distribution of cycle lengths of infinite permutations. J. Math. Sci. **87**(6), 4072–4081 (1997)
16. N.-V. Tsilevich, Stationary random partitions of positive integers. Theory Probab. Appl. **44**(1), 60–74 (1999); Translation from Teor. Veroyatn. Primen. **44**(1), 55–73 (1999)
17. B. Valkó, B. Virág, Continuum limits of random matrices and the Brownian carousel. Invent. Math. **177**, 463–508 (2009)
18. K. Wieand, Eigenvalue distributions of random permutation matrices. Ann. Probab. **28**(4), 1563–1587 (2000)

LECTURE NOTES IN MATHEMATICS

 Springer

Edited by J.-M. Morel, B. Teissier; P.K. Maini

Editorial Policy (for Multi-Author Publications: Summer Schools / Intensive Courses)

1. Lecture Notes aim to report new developments in all areas of mathematics and their applications - quickly, informally and at a high level. Mathematical texts analysing new developments in modelling and numerical simulation are welcome. Manuscripts should be reasonably selfcontained and rounded off. Thus they may, and often will, present not only results of the author but also related work by other people. They should provide sufficient motivation, examples and applications. There should also be an introduction making the text comprehensible to a wider audience. This clearly distinguishes Lecture Notes from journal articles or technical reports which normally are very concise. Articles intended for a journal but too long to be accepted by most journals, usually do not have this "lecture notes" character.

2. In general SUMMER SCHOOLS and other similar INTENSIVE COURSES are held to present mathematical topics that are close to the frontiers of recent research to an audience at the beginning or intermediate graduate level, who may want to continue with this area of work, for a thesis or later. This makes demands on the didactic aspects of the presentation. Because the subjects of such schools are advanced, there often exists no textbook, and so ideally, the publication resulting from such a school could be a first approximation to such a textbook. Usually several authors are involved in the writing, so it is not always simple to obtain a unified approach to the presentation.

   For prospective publication in LNM, the resulting manuscript should not be just a collection of course notes, each of which has been developed by an individual author with little or no coordination with the others, and with little or no common concept. The subject matter should dictate the structure of the book, and the authorship of each part or chapter should take secondary importance. Of course the choice of authors is crucial to the quality of the material at the school and in the book, and the intention here is not to belittle their impact, but simply to say that the book should be planned to be written by these authors jointly, and not just assembled as a result of what these authors happen to submit.

   This represents considerable preparatory work (as it is imperative to ensure that the authors know these criteria before they invest work on a manuscript), and also considerable editing work afterwards, to get the book into final shape. Still it is the form that holds the most promise of a successful book that will be used by its intended audience, rather than yet another volume of proceedings for the library shelf.

3. Manuscripts should be submitted either online at www.editorialmanager.com/lnm/ to Springer's mathematics editorial, or to one of the series editors. Volume editors are expected to arrange for the refereeing, to the usual scientific standards, of the individual contributions. If the resulting reports can be forwarded to us (series editors or Springer) this is very helpful. If no reports are forwarded or if other questions remain unclear in respect of homogeneity etc, the series editors may wish to consult external referees for an overall evaluation of the volume. A final decision to publish can be made only on the basis of the complete manuscript; however a preliminary decision can be based on a pre-final or incomplete manuscript. The strict minimum amount of material that will be considered should include a detailed outline describing the planned contents of each chapter.

   Volume editors and authors should be aware that incomplete or insufficiently close to final manuscripts almost always result in longer evaluation times. They should also be aware that parallel submission of their manuscript to another publisher while under consideration for LNM will in general lead to immediate rejection.

4. Manuscripts should in general be submitted in English. Final manuscripts should contain at least 100 pages of mathematical text and should always include

   – a general table of contents;
   – an informative introduction, with adequate motivation and perhaps some historical remarks: it should be accessible to a reader not intimately familiar with the topic treated;
   – a global subject index: as a rule this is genuinely helpful for the reader.

   Lecture Notes volumes are, as a rule, printed digitally from the authors' files. We strongly recommend that all contributions in a volume be written in the same LaTeX version, preferably LaTeX2e. To ensure best results, authors are asked to use the LaTeX2e style files available from Springer's web-server at

   ftp://ftp.springer.de/pub/tex/latex/svmonot1/ (for monographs) and
   ftp://ftp.springer.de/pub/tex/latex/svmultt1/ (for summer schools/tutorials).

   Additional technical instructions, if necessary, are available on request from:
   lnm@springer.com.

5. Careful preparation of the manuscripts will help keep production time short besides ensuring satisfactory appearance of the finished book in print and online. After acceptance of the manuscript authors will be asked to prepare the final LaTeX source files and also the corresponding dvi-, pdf- or zipped ps-file. The LaTeX source files are essential for producing the full-text online version of the book. For the existing online volumes of LNM see:
   http://www.springerlink.com/openurl.asp?genre=journal&issn=0075-8434.

   The actual production of a Lecture Notes volume takes approximately 12 weeks.

6. Volume editors receive a total of 50 free copies of their volume to be shared with the authors, but no royalties. They and the authors are entitled to a discount of 33.3 % on the price of Springer books purchased for their personal use, if ordering directly from Springer.

7. Commitment to publish is made by letter of intent rather than by signing a formal contract. Springer-Verlag secures the copyright for each volume. Authors are free to reuse material contained in their LNM volumes in later publications: a brief written (or e-mail) request for formal permission is sufficient.

Addresses:
Professor J.-M. Morel, CMLA,
École Normale Supérieure de Cachan,
61 Avenue du Président Wilson, 94235 Cachan Cedex, France
E-mail: morel@cmla.ens-cachan.fr

Professor B. Teissier, Institut Mathématique de Jussieu,
UMR 7586 du CNRS, Équipe "Géométrie et Dynamique",
175 rue du Chevaleret,
75013 Paris, France
E-mail: teissier@math.jussieu.fr

For the "Mathematical Biosciences Subseries" of LNM:

Professor P. K. Maini, Center for Mathematical Biology,
Mathematical Institute, 24-29 St Giles,
Oxford OX1 3LP, UK
E-mail: maini@maths.ox.ac.uk

Springer, Mathematics Editorial I,
Tiergartenstr. 17,
69121 Heidelberg, Germany,
Tel.: +49 (6221) 4876-8259
Fax: +49 (6221) 4876-8259
E-mail: lnm@springer.com